T0327553

APPLIED TRIBOLOGY

Tribology Series

Editor: G Stachowiak

Khonsari and Booser	Applied Tribology: Bearing Design and Lubrication, 3rd Edition	June 2017
Bhushan	Introduction to Tribology, 2nd Edition	March 2013
Bhushan	Principles and Applications to Tribology, 2nd Edition	March 2013
Lugt	Grease Lubrication in Rolling Bearings	January 2013
Honary and Richter	Biobased Lubricants and Greases: Technology and Products	April 2011
Martin and Ohmae	Nanolubricants	April 2008
Khonsari and Booser	Applied Tribology: Bearing Design and Lubrication, 2nd Edition	April 2008
Stachowiak (ed)	Wear: Materials, Mechanisms and Practice	November 2005
Lansdown	Lubrication and Lubricant Selection: A Practical Guide, 3rd Edition	November 2003
Cartier	Handbook of Surface Treatment and Coatings	May 2003
Sherrington, Rowe and Wood (eds)	Total Tribology: Towards an Integrated Approach	December 2002
Kragelsky and	Tribology: Lubrication, Friction and Wear	April 2001
Stolarski and Tobe	Rolling Contacts	December 2000
Neale and Gee	Guide to Wear Problems and Testing for Industry	October 2000

APPLIED TRIBOLOGY

BEARING DESIGN AND LUBRICATION

Third Edition

Michael M. Khonsari
Dow Chemical Endowed Chair in Rotating Machinery
Department of Mechanical and Industrial Engineering
Louisiana State University, USA

E. Richard Booser
Engineering Consultant (Retired), USA

This edition first published 2017
© 2017 John Wiley & Sons Ltd

The right of Michael M. Khonsari and E. Richard Booser to be identified as authors of this work has been asserted in accordance with law.

Registered Offices
John Wiley & Sons, Inc., 111 River Street, Hoboken, NJ 07030, USA
John Wiley & Sons, Ltd., The Atrium, Southern Gate, Chichester, West Sussex, PO19 8SQ, UK

Editorial Office
The Atrium, Southern Gate, Chichester, West Sussex, PO19 8SQ, UK

For details of our global editorial offices, customer services, and more information about Wiley products visit us at www.wiley.com.

Wiley also publishes its books in a variety of electronic formats and by print-on-demand. Some content that appears in standard print versions of this book may not be available in other formats.

Library of Congress Cataloging-in-Publication Data

Names: Khonsari, Michael M., author. | Booser, E. Richard, author.
Title: Applied tribology: bearing design and lubrication / Michael M. Khonsari, E. Richard Booser.
Description: Third edition. | Hoboken, NJ: John Wiley & Sons Inc., 2017. |
   Includes bibliographical references and index.
Identifiers: LCCN 2016059982 (print) | LCCN 2017000217 (ebook) | ISBN 9781118637241 (cloth) |
   ISBN 9781118700259 (Adobe PDF) | ISBN 9781118700266 (ePub)
Subjects: LCSH: Tribology. | Bearings (Machinery)–Design and construction.
Classification: LCC TJ1075 .K46 2017 (print) | LCC TJ1075 (ebook) | DDC 621.8/22–dc23
LC record available at https://lccn.loc.gov/2016059982

Cover Design: Wiley
Cover Image: © prescott09/Gettyimages

Set in 10/12pt Times by Aptara Inc., New Delhi, India

Dedicated to:
Karen, Maxwell, Milton and Mason Khonsari and in memory
of Katherine Booser

Contents

Series Preface   ix

Preface (Third Edition)   xi

Preface (Second Edition)   xiii

About the Companion Website   xv

Part I   GENERAL CONSIDERATIONS

1   Tribology – Friction, Wear, and Lubrication   3

2   Lubricants and Lubrication   23

3   Surface Texture, Interaction of Surfaces and Wear   65

4   Bearing Materials   135

Part II   FLUID-FILM BEARINGS

5   Fundamentals of Viscous Flow   161

6   Reynolds Equation and Applications   189

7   Thrust Bearings   221

8   Journal Bearings   255

9   Squeeze-Film Bearings   329

10   Hydrostatic Bearings   373

11   Gas Bearings   395

12   Dry and Starved Bearings   433

Part III   ROLLING ELEMENT BEARINGS

13   Selecting Bearing Type and Size   463

14   Principles and Operating Limits   495

15   Friction and Elastohydrodynamic Lubrication   529

Part IV   SEALS AND MONITORING

16   **Seals Fundamentals**   573

17   **Condition Monitoring and Failure Analysis**   619

Appendix A Unit Conversion Factors   639

Appendix B Viscosity Conversions   641

Index   643

Series Preface

The first and the second edition of Applied Tribology Bearing Design and Lubrication were published in 2001 and 2008, respectively. Both books enjoy a five star ranking on the Amazon book list. The new edition follows the same style and structure as the two previous ones. The book starts with general description of the tribological concepts such as friction, wear, lubricants and lubrication, surface texture and bearing materials. The text then follows with detailed explanation of the lubrication regimes and the corresponding bearing designs. Two chapters that follow contain information on bearing selection and operating limits, very useful for practicing engineers. The book finishes with a chapter on seal fundamentals. In comparison to the second edition, this new version is much improved. The text has been updated, some of the chapters significantly extended, additional references included, and occasional errors removed. Throughout the text several new numerical examples, supporting the problems discussed, have also been added.

Substantial new sections have been included in several chapters. In particular, Chapter 3 has been thoroughly revised and enhanced by new segments on contact between surfaces, friction and wear phenomena, classification of wear, failure modes and wear maps. Chapter 6 has been supplemented by a discussion on optimizing a finite Rayleigh step bearing. In Chapter 7 a useful section on the maximum bearing temperature calculations, based on the thermohydrodynamic analysis, has been included. In Chapter 8 sections on tilted pad journal bearings, maximum bearing temperature and whirl instability have been added. Updated Chapter 9 now contains a discussion on the combined squeeze and rotational motion and vertical bearing configuration, commonly found in industrial applications. Chapter 15 has been updated by including discussions on mixed film lubrication, lubricant starvation, models of non-Newtonian behavior of lubricants, greases and solid lubricants. Chapter 16, dedicated to seal foundations, contains now a discussion on thermal effects in mechanical face seals and ways of reducing the surface temperature.

From the viewpoint of the reader, a great advantage of this book is that complex lubrication mechanisms and their practical applications in bearings are presented in an easily accessible form. Clear and straightforward style makes this book a highly attractive text for both undergraduate and postgraduate engineering students. The book is not only a good textbook but also a very useful reference. Thus it is highly recommended for undergraduate and postgraduate students studying tribology and machine design and as well as a solid reference for mechanical engineers. It is expected that this new edition, as its two predecessors, would be equally popular and continue to maintain its five star ranking on the engineering book list.

Gwidon Stachowiak
Tribology Laboratory
Department of Mechanical Engineering
Curtin University
Perth, Australia

Preface

Third Edition

The Second Edition of the Applied Tribology was published nearly a decade ago. Since then, the book has been adopted for courses in many universities as has served as the basis for many short courses and lectures in industries. With great appreciation to many positive comments received, the Third Edition has expanded chapters on micro-contact of surface asperities, friction and wear, thermal effect, and elastohydrodynamic and mixed lubrication. Nearly all chapters have been revised with expanded sections, updates with an emphasis on new research and development, and many additional references. As in the previous editions, we strive to maintain a balance between application and theory. Each chapter contains numerous example problems to illustrate how different concepts can be applied in practice. We hope that this book can serve the need for a single resource for a senior elective or a graduate-level course in tribology and a reference book for practicing engineers in the industry.

This edition would not have been possible without the inspiring research and assistance of numerous colleagues and former students associated with the Center for Rotating Machinery at Louisiana State University. In particular, we gratefully acknowledge the feedback by Drs. J. Jang, S. Akbarzadeh, X. Lu, J. Wang, W. Jun, Z. Luan, M. Mansouri, K. ElKholy, Z. Peng, Y. Qiu, M. Amiri, M. Naderi, A. Beheshti, M. Masjedi, J. Takabi, M. Fesanghary, A. Aghdam, A. Kahirdeh, and A., Mesgarnejad, A., N. Xiao, A. Rezasolatani, and C. Shen.

Michael M. Khonsari
E. Richard Booser

Preface

Second Edition

Tribology is a diverse field of science involving lubrication, friction, and wear. In addition to covering tribology as involved in bearings, the same basic principles are also demonstrated here for other machine elements such as piston rings, magnetic disk drives, viscous pumps, seals, hydraulic lifts, and wet clutches.

In this second edition of *Applied Tribology* all the chapters were updated to reflect recent developments in the field. In addition, this edition contains two new chapters: one on the fundamentals of seals and the other on monitoring machine behavior and lubricants, as well as bearing failure analysis. These topics are of considerable interest to the industry practitioners as well as students and should satisfy the needs of the tribology community at large.

Computer solutions for basic fluid film and energy relations from the first edition have been extended to seal performance. New developments in foil bearings are reviewed. For ball and roller bearings, conditions enabling infinite fatigue life are covered along with new ASME and bearing company life and friction factors.

Properties of both mineral and synthetic oils and greases are supplemented by an update on the greatly extended service life with new Group II and Group III severely hydrocracked mineral oils. Similar property and performance factors are given for full-film bearing alloys, for dry and partially lubricated bearings, and for fatigue-resistant materials for rolling element bearings. Gas properties and performance relations are also covered in a chapter on gas bearing applications for high-speed machines and for flying heads in computer read–write units operating in the nanotribology range.

Problems at the close of each chapter aid in adapting the book as a text for university and industrial courses. Many of these problems provide guidelines for solution of current design and application questions. Comprehensive lists of references have been brought up to date with 145 new entries for use in pursuing subjects to greater depth.

Both SI and traditional British inch-pound-second units are employed. Units in most common use are generally chosen for each section: international SI units for ball and roller bearing dimensions and for scientific and aerospace illustrations, traditional British units for oil-film bearings in industrial machinery. Many analyses are cast in dimensionless terms, enabling use of either system of units. Conversion factors are tabulated in two appendices, and guidelines for applying either system of units are given throughout the book.

The authors are indebted to past coworkers and students for their participation in developing topics and concepts presented in this book. Comprehensive background information has been assembled by the authors from their combined 80 years of laboratory, industrial, teaching, and consulting experience. Much of this background has been drawn from well over 250 technical publications, most of them in archival literature. Theses and dissertation projects at Ohio State University, the University of Pittsburgh, Southern Illinois University, Louisiana State University; and the Pennsylvania State University; numerous industrial projects at the Center for Rotating Machinery at Louisiana State University; four tribology handbooks organized and edited for the Society of Tribologists and Lubrication Engineers; and a 1957 book *Bearing Design and Application* coauthored with D. F. Wilcock.

This book is intended (1) for academic use in a one-semester senior engineering elective course, (2) for a graduate-level course in engineering tribology and (3) as a reference book for practicing engineers and machine designers. It is hoped that these uses will provide paths for effectively designing, applying, and lubricating bearings and other machine elements while taking advantage of concepts in tribology—the developing science of friction, wear, and lubrication.

Michael M. Khonsari
E. Richard Booser

About the Companion Website

Applied Tribology: Bearing Design and Lubrication, 3rd Edition is accompanied by a website:

www.wiley.com/go/Khonsari/Applied_Tribology_Bearing_Design_and_Lubrication_3rd_Edition

The website includes:

- Solutions Manual

Part I

General Considerations

1

Tribology – Friction, Wear, and Lubrication

Tribology is a relatively new term derived from the Greek word *tribos* for 'rubbing'. It is now universally applied to the emerging science of friction, wear, and lubrication involved at moving contacts. In its broad scope, it involves mechanical, chemical, and material technology. Usual tasks in tribology are reduction of friction and wear to conserve energy, enabling faster and more precise motions, increased productivity, and reduced maintenance.

Tribology was formally identified as an important and unified technical field in a report issued by a committee of the British Ministry of State for Education and Science chaired by Peter Jost (1966). The report concluded that huge savings would be possible in the United Kingdom by fully utilizing improved design and lubrication procedures. The unified approach to this field filled an existing void, and the American Society of Mechanical Engineers (ASME) then adopted the term for its Tribology Division in 1983 and the American Society of Lubrication Engineers revised its name to the Society of Tribologists and Lubrication Engineers in 1985.

Fundamental interest in tribology now exists for lubricant formulation, industrial operations, aerospace and transportation equipment, material shaping and machining, computers and electronic devices, power generation, and almost all phases of life where motion is encountered. This text focuses primarily on tribology of bearings and a number of related applications. Fundamentals are applied in such a way as to allow application of the principles of tribology to a variety of other machine elements.

1.1 History of Tribology

Many of the advances in tribology and bearing technology evolved over years, decades, or even centuries to meet the needs of new machinery. The Industrial Revolution, with its increase in rotational speeds beyond those of the windmill and cart axle, brought full hydrodynamic lubrication into normal use. Theory and technical understanding commonly followed actual machinery applications. In many cases, this understanding of technical details played a

Applied Tribology: Bearing Design and Lubrication, Third Edition. Michael M. Khonsari and E. Richard Booser.
© 2017 John Wiley & Sons Ltd. Published 2017 by John Wiley & Sons Ltd.
Companion Website: www.wiley.com/go/Khonsari/Applied_Tribology_Bearing_Design_and_Lubrication_3rd_Edition

vital role in continued improvements in bearing design, lubricants, and surface treatments for industrial machinery, aerospace units, transportation equipment, magnetic storage, and micro-electromechanical devices.

A few historical stepping stones related to the primary subjects covered in this textbook will now be reviewed. Many of these events are covered in more detail in the references presented at the end of this chapter. The interested reader is referred to Dowson (1998) for an excellent review of the history of tribology.

Friction

The notebooks of the famed engineer and artist Leonardo da Vinci revealed his postulation in 1508 of the concept of a characteristic coefficient of friction as the ratio of friction force to normal load. The French physicist Guillaume Amontons (1699) again established the significance of a coefficient of friction, which was independent of the apparent area of contact. The French physicist C.A. Coulomb (1785) further distinguished between static friction and kinetic friction, which was independent of velocity. Mechanisms for reduction of friction and wear with soft coatings and adherent molecular and lubricant surface layers were elucidated by Bowden and Tabor (1950).

Wear

This subject has proven to be quite complex, and generalizations are still elusive. Hundreds of empirical equations have been generated for specific materials and conditions. The most useful one appears to be that of Archard (1953), which enables a generalized dimensionless wear coefficient $k = VH/Wd$ to relate wear volume V to sliding distance d, normal load W, and indentation hardness H of the wearing material (see Chapter 3). This equation best describes the nature of the so-called adhesive wear. Other types of wear such as abrasive and corrosive are also described in Chapter 3.

Bearing Materials

For many centuries wood, stone, leather, iron, and copper were common bearing materials. Almost all engineering materials have now been employed in the continuing search for the best bearing material (see Chapter 4). In an early consideration of improved materials, Robert Hooke (1684) suggested steel shafts and bell-metal bushes as preferable to wood shod with iron for wheel bearings (Bhushan, 1999). High-lead and high-tin alloys patented in the United States in 1839 by Isaac Babbitt are unsurpassed for a wide range of industrial, automotive, and railway applications. An early German study of railway journal bearings by F.A. von Pauli (1849) established a composition similar to that of Babbitt (91% tin, 6% copper, and 3% zinc) as the best of 13 bearing metals (Cameron, 1966; 1976).

Suitably hard, fatigue-resistant rolling element bearing materials were achieved with modification of tool steel in Europe about 1900. This led to the development of AISI 52100 steel and its derivatives, which have since been used for all types of commercial and automotive rolling bearings.

Porous metal bearings were introduced in the 1920s. Plastics and composites involving polymers compounded with a wide variety of solid filler materials have found wide use, gaining their greatest impetus with the invention of nylon and polytetrafluoroethylene (PTFE) during World War II. The search continues, with ceramics among the materials being developed for high-temperature capability in aircraft engines and for high-speed rolling element bearings.

Lubricants

Tallow was used to lubricate chariot wheels before 1400 BC. Although vegetable oils and greases were used later, significant advances in the development of lubricants occurred only after the founding of the modern petroleum industry with the opening of the Drake well in Titusville, Pennsylvania, in 1859. Lubricant production reached $9500 \, m^3$/yr (2 500 000 gal/yr) in the following 20 years; worldwide production now exceeds 1000 times that volume at 1 million barrels/day. Petroleum lubricants still constitute well over 95% of total oil and grease production volume.

The past 20 years have seen a dramatic world-wide growth in hydrocracking petroleum oils to give Grade II and III lubricants with a three-fold life upgrade in many automotive and industrial applications (see Chapter 2).

Polymerized olefins were the first synthetic oils to be produced. This occurred in 1929 in an effort to improve on properties of petroleum oils. Interest in ester lubricants dates from 1937 in Germany, and their production and use expanded rapidly during and following World War II to meet military needs and for use in the newly developed jet aircraft engines. A broad range of other synthetic lubricants came into production during that same period for wide-temperature use, fire resistance, and other uses geared to a range of unique properties (see Chapter 2). Current production of synthetics approaches 100 million gal/yr, with nearly half being polyalphaolefin synthetic hydrocarbons.

Development of chemical additives to upgrade the properties and extend the lives of lubricating oils began about 1920. Commercial use has proceeded since about 1930 in step with increasing demands in automobiles, jet engines and other aerospace units, and high-speed and high-pressure hydraulic equipment (see Chapter 2).

Air, water, gasoline, solvents, refrigerant gases, air, and various fluids being processed in individual machines began to find use as 'lubricants' on a broadening scale in fluid-film bearings in the second half of the 1900s as improved designs and mating materials were developed on a customized basis.

Fluid-Film Bearings

The first studies of a shaft and bearing running under full hydrodynamic conditions were performed by F.A. von Pauli in 1849 and by G.A. Hirn in 1854 (see Cameron, 1966). In 1883 the celebrated Russian Nikilay Petroff concluded that friction in fluid-film bearings was a *hydrodynamic* phenomenon. His resulting power loss equation has continued to provide a foundation in this field (see Chapter 5).

Beauchamp Tower in 1883 found hydrodynamic lubrication experimentally and reported the generation of pressure in the oil film of a journal bearing. In considering Tower's findings,

Osborne Reynolds (both working under stimulus from the British Institution of Mechanical Engineers) in 1886 developed a mathematical expression for this pressure buildup that has become the foundation of hydrodynamic analysis of bearing performance (see especially Chapters 5 and 6). Thrust and journal bearing design guides are covered in Chapters 7 and 8, respectively.

Solution of the Reynolds equation was difficult, and Arnold Sommerfeld in 1904 developed a direct integration that enabled 'infinite length' analyses. Alastair Cameron and Mrs W.L. Wood in 1949 made an extremely useful solution to the Reynolds equation for finite-length journal bearings by a relaxation procedure carried out with a mechanical desktop calculator. Initiated in 1958 by Oscar Pinkus and by Albert Raimondi and John Boyd, digital computer analysis of journal bearing performance has come into widespread use for obtaining numerical solutions of the Reynolds equation. Recent advances have focused on solving the Reynolds equation with consideration of cavitation. Particularly fruitful has been evaluation of dynamics effects in bearings that take into account the orbit of the shaft, along with simultaneous solutions of the energy equation taking lubricant property variations into consideration.

Rolling Element Bearings

Several isolated examples have been recorded of application of rollers in 484–424 BC for transporting vessels on land and for launching military missiles. While historical sources increasingly mention the use of balls and rollers for bearing purposes following about AD 1500, widespread application of rolling contact bearings occurred during the twentieth century (see especially Allan, 1945).

1780   Perhaps the earliest ball bearing: a 610 mm bore bearing was developed for a rotating mill in Norwich, UK.
1868   General use of ball bearings in bicycles began.
1899   Timken began the manufacture of tapered roller bearings.
1902   Conrad obtained a British patent for the present design of the deep-groove ball bearing with its cage.
1904   Ball and roller bearings were used in electric automobiles.
1907   SKF founded.
1949   Grubin established the elastohydrodynamic principles involved in the lubrication of rolling contacts.

Recent research has been geared toward understanding the relationship between surface topography (Chapter 3), film thickness, and pressure distribution in rolling element bearings. Under highly loaded conditions, when surface asperities tend to interact, a mixed-film lubrication regime may have to be considered. Prediction of subsurface stress fields and their relationship to fatigue are also subjects of current research.

Nanotribology and Surface Effects

Atomic-scale studies have been expanding rapidly to develop new materials with improved tribological properties. With an atom being approximately 1/3 nanometer (nm) in diameter,

Table 1.1   Representative nanometer dimensions in tribology

	Size range (nm)
Atomic scale[a]	
C–C bond length	0.15
C–H bond length	0.11
Lubricant molecules	
Lube oil	1–5
Absorbed rust inhibitors[a]	3
Viscosity index improvers (VIs)	100–250
Lithium soap grease fibers, diameter × length[b]	0.2×2
Computer magnetic heads[a]	
Air film	0.15–0.3
Liquid film	1–2

[a] *Source*: Bhushan, 1997, 1999.
[b] *Source*: Klamann, 1984.

Krim in 1991 initiated naming this field 'nanotribology' to parallel the broad field of 'nanotechnology' (Krim, 1991; Cantor, 2004). The nanometer (10^{-12} m, or 10 ångstroms) then provides a convenient scale for atomic and molecular sizes such as those in Table 1.1.

An early definition of atomic surface compositions that produces satisfactory sliding characteristics was provided in General Motors rubbing tests of all available metallic elements on steel (Roach *et al.*, 1956). As detailed in Chapter 4, good scoring resistance was found only with elements that had atomic diameters at least 15% greater than iron and that were in the B subgroup of the periodic table, implying lack of tenacious metallic bonds at any localized junctions with steel.

Bushan (1997, 1999) updated the consideration of atomic structure and the related surface and lubricant details for avoiding wear under demanding operation with 0.15–0.3 nm thick air films in magnetic storage and micromechanical devices such as those involved in computer systems. Analysis of related air film bearings is provided in Chapter 11.

1.2 Tribology Principles

Several distinct regimes are commonly employed to describe the fundamental principles of tribology. These range from *dry sliding* to complete separation of two moving surfaces by *fluid-film lubrication*, with an intermediate range involving partial separation in *boundary* or *mixed* lubrication. When elastic surface deformation exerts a strong influence on fluid-film behavior, as in ball and roller bearings, *elastohydrodynamic lubrication (EHL)* introduces its distinctive characteristics.

Dry Sliding

In the absence of a fluid film, actual contact between two rubbing surfaces involves only sufficient high spots, or asperities, of the softer material so that their yield pressure balances the total load (Rabinowicz, 1995). Under load *W* in Figure 1.1, the real contact area is a relatively

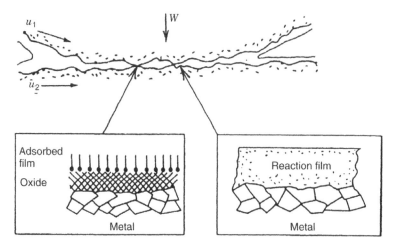

Figure 1.1   Surface films at asperity contacts between rubbing surfaces. *Source*: Harris, 1991. Reproduced with permission of John Wiley & Sons.

minute portion of the apparent total area and tends to increase proportionately with the load (see Chapter 4).

   A measurable force is required to slide a surface along on the contacting asperities (surface peaks) of another surface. This force must overcome the friction force associated with asperity yield strength. As suggested in Figure 1.1, friction can be significantly reduced by a thin, soft coating of a solid film lubricant such as graphite, molybdenum disulfide, or PTFE plastic, or by a sulfur- or phosphorus-rich layer formed by adsorption of additives from a lubricating oil. Wear volume can be related to the generation of wear fragments from a portion of these asperity contacts during sliding (see Chapters 2–4).

Fluid-Film Lubrication

In this regime, the moving surfaces are completely separated by a film of liquid or gaseous lubricant. A load-supporting pressure is commonly generated by one of the following types of action:

1. *Hydrodynamic lubrication (HL)* results from a film of separating fluid being drawn into a converging, wedge-shaped zone (as in Figure 1.2a) by the self-acting pumping action of a moving surface (Booser, 1995). Both the pressure and the frictional power loss in this film are a function of the lubricant's viscosity in combination with the geometry and shear rate imposed by bearing operating conditions (see Chapter 5). Hydrodynamic bearings are commonly in the form of either (a) a *sleeve bearing* (also called a *journal bearing*) surrounding its mating journal surface on the shaft of a machine as a radial bearing or (b) a *thrust bearing* to provide an oil film at the face of a shaft shoulder or collar for location and axial support of a rotor.
2. *The squeeze-film* action shown in Figure 1.2(b) is encountered in dynamically loaded bearings in reciprocating engines and under shock loads (see Chapter 9). Because time is required to squeeze the lubricant film out of a bearing, much higher loads can be carried

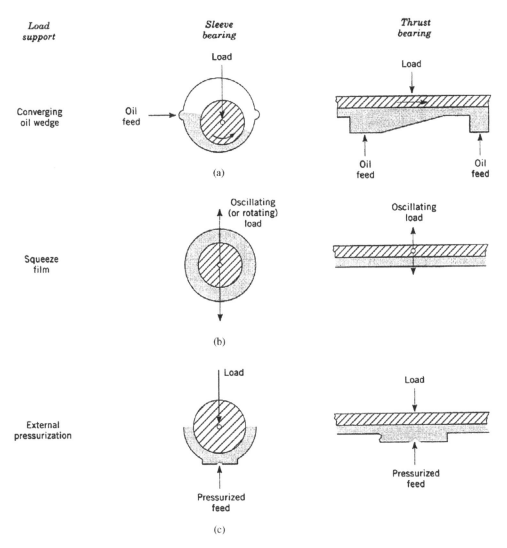

Figure 1.2 Principles of fluid-film action in bearings. *Source*: Booser, 1995. Reproduced with permission of John Wiley & Sons.

in sleeve bearings for automotive engines and metal rolling mills than with a steady, unidirectional load, as reflected in the typical values presented in Table 1.2. The much lower load capacity of bearings lubricated with low-viscosity fluids such as water and air is also indicated in Table 1.2.

3. *Externally pressurized* feed by an external pumping source may be used to generate pressure in the fluid before its introduction into the bearing film, as shown in Figure 1.2(c). Such a procedure is common in cases where limited hydrodynamic pumping action is available within the bearing itself, as during starting and slow-speed running with heavy machines or with low-viscosity fluids (see Chapter 10).

Table 1.2 Typical design loads for hydrodynamic bearings

Bearing type	Load on projected area MPa (psi)
Oil lubricated	
Steady load	
Electric motors	1.4 (200)
Turbines	2.1 (300)
Railroad car axles	2.4 (350)
Dynamic loads	
Automobile engine main bearings	24 (3 500)
Automobile connecting-rod bearings	34 (5 000)
Steel mill roll necks	35 (5 000)
Water lubricated	0.2 (30)
Air bearings	0.2 (30)

Source: Khonsari and Booser, 2004. Reproduced with permission of John Wiley & Sons.

Elastohydrodynamic Lubrication (EHL)

This is a form of hydrodynamic lubrication in which pressures are large enough to cause significant elastic deformation of the lubricated surfaces. As with HL, converging film thickness, sliding motion, and fluid viscosity play an essential role. Temperature effects, inadequate lubricant supply to the EHL film, and boundary lubrication play roles of varying importance, much as with fluid films in HL.

Significant differences between HL and EHL involve the added importance of material hardness, viscosity increase under high pressure, and degree of geometric conformation of the contacting surfaces. *Conformal surfaces* match snugly, such as the journal in a sleeve bearing with hydrodynamic lubrication shown in Figure 1.2, so that the load is carried on a relatively large area. With *nonconformal surfaces*, as with the two contacting rollers in Figure 1.3, the load must be carried on a small area: typically of the order of 1000-fold smaller than with a

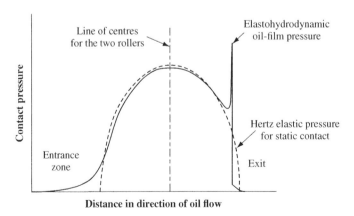

Figure 1.3 Pressure distribution between two rollers under load.

conformal conjunction. The following two distinct regimes exist in EHL:

1. *Hard EHL*. This involves materials of high elastic modulus in *nonconformal* contacts, commonly contacts involving primarily rolling action or combined rolling and sliding in ball and roller bearings, gear teeth, cams, and some friction drives. Since the surfaces do not conform well, the load is concentrated on small, elastically deformed areas, and hydrodynamic lubrication of these EHL contacts is commonly characterized by a very thin separating oil film that supports local stresses that tax the fatigue strength of the strongest steels.

   Oil-film pressures in these EHL contacts commonly range up to 0.5–3 GPa, 1000 times that in most hydrodynamic bearings, with a pattern which closely follows the Hertz elastic contact stress for a static contact, as illustrated in Figure 1.3. The overall oil-film thickness (often about 0.1–0.5 μm) is set primarily by oil viscosity, film shape, and velocity at the entry of the contact zone (see Chapter 15).

2. *Soft EHL*. Elastic deformation also plays an important role in film formation in rubber seals, tires, human joints, water-lubricated rubber bearings, and babbitt journal bearings under heavy loads and low speeds, as well as in similar applications of hydrodynamic bearings using soft bearing materials having low elastic modulus. Since the low contact pressures involved have a negligible effect on fluid viscosity in the conjunction, analytical relations are simpler for *soft EHL* in contrast to the *hard EHL* encountered with rolling element bearings. Hamrock (1994) and Khonsari and Hua (1997) give analytical tools for performance analysis within the various realms of EHL.

Boundary Lubrication

As the severity of operation increases, the speed N and viscosity μ eventually become incapable of generating sufficient oil-film pressure P to support the entire load. Asperities of the mating surfaces then penetrate with increasing contact area, plastic deformation, higher asperity temperatures, and finally surface tearing and seizure on a broad scale. The region of lubrication in Figure 1.4 shifts from full film with a friction coefficient (ratio of friction force to

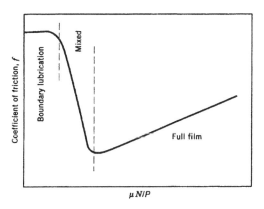

Figure 1.4 Stribeck dimensionless $\mu N/P$ curve relating lubrication regime and friction coefficient to absolute viscosity μ, rotational speed N, and unit load P. *Source*: Booser, 1995. Reproduced with permission of John Wiley & Sons.

Table 1.3   Characteristics of common classes of bearings

	Fluid film bearings	Dry bearings	Semilubricated	Rolling element bearings
Start-up friction coefficient	0.25	0.15	0.10	0.002
Running friction coefficient	0.001	0.10	0.05	0.001
Velocity limit	High	Low	Low	Medium
Load limit	High	Low	Low	High
Life limit	Unlimited	Wear	Wear	Fatigue
Lubrication requirements	High	None	Low/none	Low
High-temperature limit	Lubricant	Material	Lubricant	Lubricant
Low-temperature limit	Lubricant	None	None	Lubricant
Vacuum	Not applicable	Good	Lubricant	Lubricant
Damping capacity	High	Low	Low	Low
Noise	Low	Medium	Medium	High
Dirt/dust	Need seals	Good	Fair	Need seals
Radial space requirement	Small	Small	Small	Large
Cost	High	Low	Low	Medium

Source: Kennedy *et al.*, 1998. Reproduced with permission of Taylor & Francis.

normal force) of the order of 0.001 to *boundary* (or *mixed-film*) lubrication, where the friction coefficient rises to 0.03–0.1, and finally to complete loss of film support, where the friction coefficient may reach 0.2–0.4, typical for dry sliding.

When operating with an adsorbed surface layer or a chemical reaction coating in boundary lubrication, chemical additives in the oil and chemical, metallurgical, and mechanical factors involving the two rubbing surfaces determine the extent of wear and the degree of friction.

1.3   Principles for Selection of Bearing Types

Most bearings can be classified as either fluid-film bearings, dry or semilubricated sliding bearings, or rolling element bearings (Khonsari and Booser, 2004). Their relative advantages are listed in Table 1.3. A general preliminary guide to selection of each of these types for different load, speed, and size relations is provided in Figure 1.5.

Example 1.1   A radial load of 500 N (112 lb) is to be carried by a bearing on a 25 mm (0.98 in.) diameter shaft. Which bearing types could be considered for speeds of 100, 1000, 10 000, and 100 000 rpm?

Follow horizontally across Figure 1.5 for a 'Typical maximum load' of 500 N. Speed and load limits are then found to be adequate for the following bearing types:

Speed (rpm)	Possible bearing types
100	Rubbing, porous metal, rolling element, oil film
1 000	Porous metal, rolling element, oil film
10 000	Rolling element, oil film
100 000	Oil film

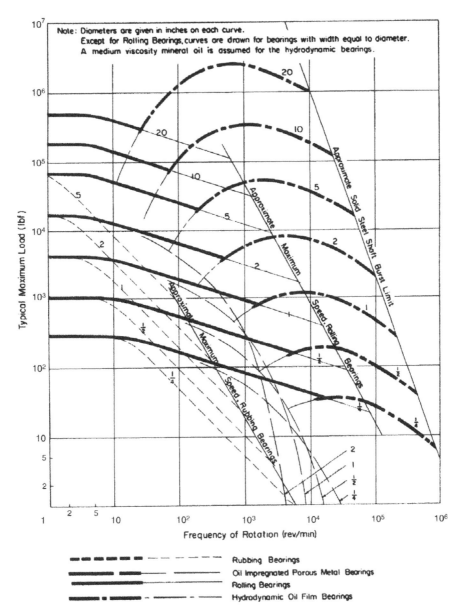

Figure 1.5   General guide for selecting the journal bearing type. Except for rolling element bearings, the length/diameter ratio = 1. Medium-viscosity mineral oil is assumed for hydrodynamic bearings. (From ESDU Data Item No. 65007, 1965.)

While the Engineering Sciences Data Unit (ESDU) also has a companion chart for thrust bearings, the same general type used for the journal bearing should be considered first.

Final selection of the most suitable basic type is then related to the following factors:

1. Mechanical requirements;
2. Environmental conditions; and
3. Economics.

Representative considerations follow for each of these factors, and later chapters give further guidelines for performance to be expected with individual bearing types. In many cases, preliminary pursuit of design details for several possibilities is helpful in making a final selection.

Mechanical Requirements

Unless the bearing fulfills all mechanical requirements imposed by the machine of which it is a part, other considerations such as environment and cost are of no importance. Brief comparisons of these items are included in Table 1.3.

Friction and Power Loss

Low starting friction, especially under a load, is a prime advantage of ball and roller bearings. While involving added complexity, externally pressurized oil lift pockets provide oil-film bearings with zero starting friction in a variety of large, heavy machines such as electric generators at hydroelectric dams and in utility power plants. For running machines, a coefficient of friction of the order of 0.001–0.002 is typical for both rolling-element and oil-film bearings. Start-up coefficients at breakaway of 0.15–0.25 are typical for oil-film bearings and for dry and semilubricated plastic and porous metal bearings. With dry and semilubricated surfaces, friction then drops by about half as motion gets underway.

The localized temperature increase generated by friction in rubbing asperities was analyzed by Blok (1937) to place a limit on the maximum speed and load which can be tolerated in bearing and gear contacts.

Speed

Each of the four classes of bearings presented in Table 1.3 has a practical speed limit. Common practice limits rolling bearings using oil lubrication to a (DN) value (mm bore × rpm) of 500 000 to 1 000 000, corresponding to a surface speed of 1600–3100 m/min (5000–10 000 ft/min). Limits with fluid-film bearings are much higher and less well defined, but surface speeds in turbine bearings range up to about 8700 m/min (30 000 ft/min). A typical dental drill uses air for journal bearing lubrication while spinning smoothly at speeds of 300 000 rpm. Much lower speed limits in the range of 100–500 m/min are imposed by surface heating effects with dry and semilubricated bearings, for which the appropriate operating zone is illustrated in Figure 1.6. (Note: Figure 1.6 does not show scales on its axes and is only meant to show the trend. See Chapter 12 for typical numerical values.)

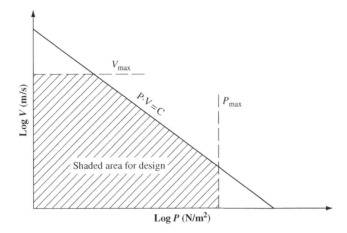

Figure 1.6 Operating zone for the design of dry and semilubricated bearings. *Source*: Fuller, 1984. Reproduced with permission of John Wiley & Sons.

Load

Rolling element bearings are generally more versatile in being able to carry their fatigue-rated load at all speeds from zero up and in all directions with normal lubrication. The load capacity of oil-film bearings, on the other hand, is very much a function of speed and oil viscosity, with their influence on oil-film formation. Dry and semilubricated porous metal and plastic bearings encounter a surface heating limit in their *PV* factor (contact pressure × surface velocity) which gives a much lower load limit with rising speed, but high loads are possible with appropriate material combinations at low speeds.

Hydrostatic bearings using externally pressurized oil have been used to support enormous structures such as observatory domes, telescopes, and large radio antennas where weight requirements range from 250 000 to over 1 000 000 pounds (see Chapter 10).

Momentary shock loads can be reasonably tolerated by both fluid film and rolling element bearings. Rotor unbalance loads and cyclic loads in internal combustion engines are well carried by oil-film bearings. Combined radial and thrust load capacity is a useful attribute of conventional deep-groove, single-row ball bearings.

Life

Rolling element bearings have a distinct fatigue life limit which results from the repeated contact stressing of the balls and rollers on their raceways, while fluid-film bearings in the usual rotating equipment provide essentially unlimited life. The same life pattern has been found with rolling element and oil-film bearings, however, in industrial electric motors, where maintenance needs and contamination commonly dictate life. Fatigue life is also a limiting factor for oil-film bearings under cyclic loading in internal combustion engines.

In dry and semilubricated bearings, life is estimated from a *wear coefficient*, which relates wear rate to unit loading and peripheral sliding distance. In the rapidly expanding use of this class of bearings, life is also related to temperature, contamination, and other environmental conditions.

Lubrication

In general, a rolling element bearing requires only enough lubricant to provide a film coating over the surface roughness of working surfaces. Less than one drop supplies this need for many small and medium-sized ball and roller bearings. Under conditions of heavy load and high speed, additional lubricant must be supplied, not for lubrication needs but to remove heat and maintain a reasonable limit on temperature. Many rolling element bearings depend only on an initial grease fill for years of operation when DN values are less than 300 000 (mm bore × rpm).

While no oil is usually fed to small sliding-type bearings which are to operate dry or semilubricated, oil-film bearings generally require relatively large quantities of oil to maintain the separating film between the bearing and its mating surface. This feed rate is proportional to the bearing length, width, clearance, and surface velocity and ranges up to $4 \, \text{m}^3/\text{min}$ (1000 gal/min) for the oil-film bearings in a steam turbine generator at an electric power station. Specially designed fluid-film bearings are unique in being able to operate with gas or ambient liquids such as water or gasoline as their hydrodynamic fluid.

Space Requirement

Dry and semilubricated bearings require a minimum of space. A porous metal, plastic, or plastic-lined bearing is commonly a bushing or thrust washer, such as that illustrated in Figure 1.7. These bearings have just enough wall thickness to provide the needed strength for insertion in a supporting housing. The bearing can even consist of no more than a formed or machined hole in a suitable plastic housing of an appliance or instrument.

In the case of ball and roller bearings, the outside diameter commonly ranges from about one and a half to three times that of the bore, and the axial dimension ranges from one-fifth to

Figure 1.7   Representative plastic composite bushings and thrust washers. *Source*: Courtesy of SKF.

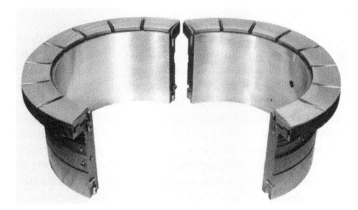

Figure 1.8   Oil-film bearing combining babbitted journal and thrust surfaces. *Source*: Courtesy of Kingsbury, Inc.

one-half of the shaft diameter. Oil-film journal bearings are more compact in their radial dimension, and range in axial length from about one-third to equal to the shaft diameter. Considerable additional volume is commonly required with oil-film bearings to accommodate seals plus feed and drain passages. Figure 1.8 illustrates the general proportions for a babbitt journal bearing combined with an integral thrust face with its oil-distributing grooves.

Environmental Conditions

Temperature range, moisture, dirt, and corrosive atmospheres are among the important environmental factors that require consideration. Within each bearing type, established design methodology is now available to meet most of the usual environmental demands.

Fluid-film bearings generally are most limited in their temperature range. Common soft metal bearing surfaces of tin and lead babbitt limit their upper temperature to the 125–150 °C (260–300 °F) range. Resistance to flow from the high viscosity of cold mineral oil in feed and drain passages, at the other extreme, limits the low temperature range of many oil-film bearings to −20 to +10 °C (−5 to 50 °F).

Ball and roller bearings can be adapted to meet wide-ranging environmental demands. For general applications, multipurpose greases and double-sealed and double-shielded enclosures provide water and contamination resistance, rust protection, and sufficiently long life to eliminate the need for routine maintenance. Synthetic oils and greases used with tool steel and ceramic bearing materials allow applications in an ever-broadening range of temperatures and other environmental conditions.

Economics

Overall economic considerations in selecting a bearing type include not only the first cost of the bearing itself, but also later costs in maintenance and eventually in replacement at the end of the useful life.

Cost

Where other factors do not dictate the choice, initial cost commonly is a dominant selection factor. For many mass-produced items such as small electric motors, household appliances, and automotive auxiliaries with low surface speeds and light loads, dry and semilubricated bushings and thrust washers have by far the lowest cost and are finding continually broadening use in bore sizes up to about 25 mm. Mass-produced oil-film bearings of smaller sizes also fall in this low-cost range for automobile engines, appliances, and fractional-horsepower motors. In general, they avoid the speed limits for porous metals and polymer compositions.

If these sliding element bearings appear marginal or impractical, ball and roller bearings should be considered as the next higher-priced alternative. While available from very small diameters in the 3–5 mm range, their use is common for 25–100 mm shaft diameters, and their use in larger diameters is growing. Chapters 13–15 provide application, lubrication, and performance guidelines for their use over a broad range of conditions.

More expensive oil-film bearings are common in larger sizes for high loads and high speeds, such as those used with auxiliary oil feed systems involving cooling and filtration of contaminants.

Maintenance and Life

Fluid-film bearings operating with parts separated by an oil film require little or no maintenance and their life is virtually unlimited, except under cyclic loading, where the bearing metal may suffer fatigue. In contrast, the repeated cyclic loading on the contacts in rolling element bearings results in limited life due to fatigue.

Maintenance of greased rolling bearings may at most require occasional purging or grease change. With oil lubrication of either sliding or rolling bearings, the supply system requires occasional attention for a change of oil, a new filter, or general cleaning.

With dry and semilubricated bearings, calculated bearing wear life is commonly selected to at least match the expected life of the overall unit. Required maintenance is usually limited to the replacement of worn bearings.

1.4 Modernization of Existing Applications

Reassessing the bearing class and bearing design is a common need for existing machines or products. This need may arise from problems with the present bearings, from extension of the product line, or in updating a machine design.

Industrial electric motors provide an example of modernization over an extended time. The earliest electric motors around 1900 used fluid-film bearings with an oil ring for a lubricant supply (see Chapter 12). With increasing availability of ball bearings by the time of World War I, their small size and reduced maintenance needs allowed their use in fractional horsepower motors and in motors for electric automobiles and small mine locomotives. In the 1920s, ball bearing use was extended to motor sizes up to about 200 horsepower (hp), but only at a premium price and with concerns over reliability. With lower bearing costs and the availability of long-life greases, ball bearings became standard for industrial motors up to 25–50 hp in the 1950s. Today, following subsequent redesigns, very few motors using oil-film bearings are available in sizes below about 1000 hp.

On the other hand, bearings continue to be of the oil-film type in stationary steam and gas turbines, particularly those in electric power service. The value of their essentially unlimited life far surpasses the higher cost of these bearings and their extended lubrication system compared with rolling element bearings. The trend toward lightweight and high-speed construction also depends to a large extent on the inherent damping and higher speed limit available with oil-film bearings.

For jet aircraft turbine engines, however, rolling element bearing use is firmly established. Dominant factors are a small lubricant supply, easy low-temperature starting, high thrust load capacity, and light weight. Moreover, ball bearings are better suited to these applications, where there may be an instantaneous interruption in the oil supply. This is an important consideration since aircraft experience a variety of operating conditions including takeoff, landing, and maneuvering, which can be conveniently handled by ball bearings.

With turbines, as with many other machines, selection of the bearing type will continue to be a question. Oil-film bearings may be considered for land-based derivatives of aircraft turbines, and ball and roller bearings will probably continue to be evaluated for small land-based turbines, compressors, and their accessory units, where oil-film bearings have been traditional.

Automotive chassis bearings have undergone a wide range of modifications to eliminate the need for relubrication. Automobile engine bearings will likely also undergo detailed evaluation of their type and materials with the advent of new electric propulsion and fuel cell components in the drive system. Such factors and innovations such as ceramic rolling elements, new plastic and composite sliding bearing materials, and new high-performance lubricants will call for continuing reevaluation of almost all bearing design and performance details.

1.5 A Look Ahead

Coming trends for bearing applications will likely be set by the needs for lower maintenance while operating at higher speeds and higher temperatures in more compact designs. The following are likely trends for the future of classification of bearings covered in this chapter (Khonsari and Booser, 2004).

Dry and Semilubricated Bearings

Plastics and their composites will continue to find broadening use for mild operating conditions as small bushings in household appliances, machine tools, instruments, business machines, automobile chassis and construction equipment. Many of these applications will involve direct integration of the bearings with housing and structural elements.

Ball and Roller Bearings

While continuing their dominance in jet engines and aerospace, rolling element bearing use will inctrease in small electric motors, automobile accessories, machine tools, construction and agriculture machines, and railroad equipment. Their very small lubrication needs will likely bring innovations in greases, self-contained lubricant impregnation in ball and roller bearing cages, and solid lubricant films. Further developments of ceramics, tool steels, and special lubricants will enable continually higher speeds and temperatures.

Fluid-Film Bearings

Conventional oil-lubricated bearings will continue to fill their traditional role in reciprocating automotive and diesel engines, electrical turbine generators, and other large machinery. Use of gases and low-viscosity liquids will likely escalate. Air-film bearings now used in flying heads on computer discs and in airliner cabin compressors will gradually expand in their use for aerospace, industrial, and instrument units. Water, gasoline, liquid ammonia, and a wide variety of liquid chemical and food process streams will likely be employed in bearings to avoid the complexity and contamination with conventional mineral oils and greases.

These are but a sampling of challenges coming in the twenty-first century: new bearing materials, new lubricants, advanced analysis and designs techniques, and improved surface profiles to match extremely thin fluid films.

Thermal effects will continue to be a major bottleneck for liquid-lubricated bearings (Khonsari, 1987a, 1987b; Pinkus, 1990; Jang and Khonsari, 2004; Yang, 2013). The adverse effect of heat on the lubricant viscosity and the concomitant drop in the lubricant film thickness call for innovative designs and heat management strategies in bearings and mechanical seals (Xiao and Khonsari, 2013). Progress will continue by taking advantage of bearing materials with superior thermal conductivity such as polycrystalline diamonds for use in severe operating conditions, surface texturing methods for improving performance of tribological components (Etsion, 2010; Kato, 2013; Talke, 2013), and the use of alternative lubricants that can withstand high temperatures such as granular lubrication (Heshmat, 2010; Khonsari, 2013).

Research progress will be in made in expanding the current state of the art at the interface of different lubrication regimes (hydrodynamic to mixed and to boundary) and with the recognition that progress can be made by a holistic view of the subject rather than compartmentalization. A Borromean rings structure in Figure 1.9 depicts how these regimes are interconnected and that the whole structure falls apart if one of the rings comes apart (Khonsari and Booser, 2010).

Recent advances in the unification of friction and wear (Bryant, 2010) with new developments in the science of degradation (Bryant *et al.*, 2008) with application to wear (Amiri and Khonsari, 2010) and Fatigue (Khonsari and Amiri, 2013) hold great promise not only for a

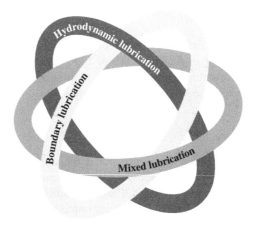

Figure 1.9   The Borromean rings structure.

better understanding of the key factors involved but also for exploring new horizons in a manner that could not have been envisioned in the past.

Problems

1.1 Select a bearing type for applications with a 2.0 in. diameter shaft, medium-viscosity mineral oil lubricant, and the following specifications:

   (a) 10 rpm, 500 lbf load;
   (b) 1000 rpm, 500 lbf load;
   (c) 10 000 rpm, 500 lbf load;
   (d) 1000 rpm, 10 000 lbf load.

1.2 List three general advantages and disadvantages to be expected in applying water-lubricated bearings rather than oil-film or rolling element bearings. Give five applications in which you conclude that water bearings would be advantageous.

1.3 Repeat Problem 1.2 for air bearings.

1.4 A 30 mm bore tapered roller bearing is being considered for use in supporting the impeller in a household food-disposal unit for kitchen sinks. High-shock loading is expected, as solids and possibly tableware enter the unit. Suggest two other possible types of bearings and compare their major attributes, as selected from items in Table 1.3. Which type of bearing is your primary recommendation?

1.5 A 5000 rpm gas turbine with a 6 in. shaft diameter at the bearing locations is to be used in driving a natural gas line pumping unit. List the major factors that might be expected to dictate the selection of rolling or fluid-film bearings. Should magnetic bearings be considered that would suspend and center the shaft while using an electronic control for the magnet elements?

1.6 State two types of bearings or bearing systems that you would consider for a conveyor carrying 10 lb castings in 20 min passage at a speed of 2 m/min through a 350 °C heat-treating oven. List five selection factors to be considered. Which factor appears to be most critical?

References

Allan, R.K. 1945. *Rolling Bearings*. Pitman & Sons, Ltd, London.

Amiri, M. and Khonsari, M.M. 2010. 'On the Thermodynamics of Friction and Wear—A Review,' *Entropy*, Vol. 12, pp. 1021–1049.

Archard, J.F. 1953. 'Contact and Rubbing of Flat Surfaces,' *Journal of Applied Physics*, Vol. 24, pp. 981–988.

Bhushan, B. 1997. 'Tribology of Magnetic Storage Systems,' *Handbook of Tribology and Lubrication Engineering*, Vol. 3. CRC Press, Boca Raton, Fla, pp. 325–374.

Bhushan, B. 1999. *Principles and Applications of Tribology*. John Wiley & Sons, New York.

Blok, H. 1937. 'Theoretical Study of Temperature Rise at Surfaces of Actual Contact under Oiliness Operating Conditions,' *Institution of Mechanical Engineers, Lubrication Discussion*, Vol. 2, pp. 222–235.

Booser, E.R. 1995. 'Lubrication and Lubricants,' *Encyclopedia of Chemical Technology*, 4th edn., Vol. 15. John Wiley & Sons, New York, pp. 463–517.

Bowden, F.P. and Tabor, D. 1950. *The Friction and Lubrication of Solids*. Clarendon Press, Oxford (republished in 1986).

Bryant, M.D., Khonsari, M.M., and Ling, F. 2008. 'On the Thermodynamic of Degradation,' *Proceedings of the Royal Society, Series A*, Vol. 464, pp. 2001–2014.

Bryant, M.D. 2010. 'Unification of Friction and Wear,' *Recent Development in Wear Prevention, Friction, and Lubrication*, G.K. Nikas, ed., Research Signpost, Kerala, India, pp. 159–196.

Cameron A. 1976. *Basic Lubrication Theory*, 2nd edn. Ellis Horwood, Ltd, Chichester, UK.

Cameron, A. 1966. *Principles of Lubrication*. Longman Green & Co., Ltd, London.

Cantor, N. 2004. 'Nanotribology: the Science of Thinking Small,' *Tribology and Lubrication Technology*, Society of Tribologists and Lubrication Engineers, Vol. 60(6), pp. 43–49.

Dowson, D. 1998. *History of Tribology*, 2nd edn. Institution of Mechanical Engineers, London.

Engineering Sciences Data Unit (ESDU). 1965. *General Guide to the Choice of Journal Bearing Type*, Item 65007. Institution of Mechanical Engineers, London.

Etsion, I. 2010. 'Laser Surface Texturing and Applications,' *Recent Development in Wear Prevention, Friction, and Lubrication*, G.K. Nikas, ed., Research Signpost, Kerala, India, pp. 137–158.

Fuller, D.D. 1984. *Theory and Practice of Lubrication for Engineers*, 2nd edn. John Wiley & Sons, New York.

Hamrock, B.J. 1994. *Fundamentals of Fluid Film Lubrication*. McGraw-Hill Book Co., New York.

Harris, T.A. 1991. *Rolling Bearing Analysis*, 3rd edn. John Wiley & Sons, New York.

Heshmat, H. 2010. *Tribology of Interface Layers*, CRC Press, Boca Raton, Fla.

Jang, J.Y. and Khonsari, M.M. 2004. 'Design of Bearings on the Basis of Thermohydrodynamic Analysis,' *Proceedings of the Institute of Mechanical Engineers, Part I. Journal of Engineering Tribology*, Vol. 218, pp. 355–363.

Jost, H.P. 1966. *Lubrication (Tribology): a Report on the Present Position and Industry's Needs*. Department of Education and Science, Her Majesty's Stationery Office, London.

Kato, K. 2013. 'Surface Texturing for Water Lubrication,' *Encyclopedia of Tribology*, Y.W. Chung and Q.J. Wang, eds, Springer Science, New York, pp. 3489–3493.

Kennedy, F.E., Booser, E.R., and Wilcock, D.F. 1998. 'Tribology, Lubrication, and Bearing Design,' *The CRC Handbook of Mechanical Engineering*, F. Kreith (ed.). CRC Press, Boca Raton, Fla, pp. 3.128–3.169.

Khonsari, M.M. 1987a. 'A Review of Thermal Effects in Hydrodynamic Bearings. Part I: Slider and Thrust Bearings,' *ASLE Transactions*, Vol. 30, pp. 19–25.

Khonsari, M.M. 1987b. 'A Review of Thermal Effects in Hydrodynamic Bearings. Part II: Journal Bearings,' *ASLE Transactions*, Vol. 30, pp. 26–33.

Khonsari, M.M. 2013. 'Granular Lubrication,' *Encyclopedia of Tribology*, Y.W. Chung and Q.J. Wang, eds, Springer Science, New York, pp. 1546–1549.

Khonsari, M.M. and Amiri, M. 2013. *Introduction to Thermodynamics of Mechanical Fatigue*, CRC Press, Boca Raton, Fla.

Khonsari, M.M. and Booser, E.R. 2004. 'An Engineering Guide for Bearing Selection,' *STLE Tribology and Lubrication Technology*, Vol. 60(2), pp. 26–32.

Khonsari, M.M. and Booser, E.R. 2010. 'On the Stribeck Curve,' *Recent Development in Wear Prevention, Friction, and Lubrication*, G.K. Nikas, ed., Research Signpost, Kerala, India, pp. 263–278.

Khonsari, M.M. and Hua, D.Y. 1997. 'Fundamentals of Elastohydrodynamic Lubrication,' *Tribology Data Handbook*, E.R. Booser (ed.). CRC Press, Boca Raton, Fla, pp. 611–637.

Klamann, D. 1984. *Lubricants and Related Products*. Verlag Chemie, Deerfield Beach, Fla.

Krim, I.J. 1991. 'Nanotribology of a Kr Monolayer: a Quartz-Crystal Microbalance Study of Atomic-Scale Friction,' *Physical Review Letters*, Vol. 66, pp. 181–184.

Pinkus, O. 1990. *Thermal Aspects of Fluid Film Tribology*. ASME Press, New York.

Rabinowicz, E. 1995. *Friction and Wear of Materials*, 2nd edn. John Wiley & Sons, New York.

Roach, A.E., Goodzeit, C.L., and Hunnicut, R.P. 1956. *Transactions ASME*, Vol. 78, p. 1659.

Talke, F. 2013. 'Surface Texture for Magnetic Recording,' *Encyclopedia of Tribology*, Y.W. Chung and Q.J. Wang, eds, Springer Science, New York, pp. 3485–3489.

Xiao, N. and Khonsari, M.M. 2013. 'A Review of Mechanical Seals Heat Transfer Augmentation Techniques,' *Journal of Patents on Mechanical Engineering*, Vol. 6, pp. 87–96.

Yang, 2013 Thermal Effect on EHL, *Encyclopedia of Tribology*, Y.W. Chung and Q.J. Wang, eds, Springer Science, New York, pp. 3586–3592.

2

Lubricants and Lubrication

As will be seen throughout this textbook, the most important property of a lubricant is usually its viscosity. While mineral oils are a common first choice, the basic decision about the lubricant type to be employed involves a number of other factors. These include the temperature range, speed, load, materials to be encountered, and general design considerations of the machine itself. Corresponding coverage in this chapter includes viscosity and related lubricant properties, synthetic lubricants, greases, solid lubricants, and lubricant supply methods.

2.1 Mineral Oils

Mineral oils produced by refining petroleum crude oil are the starting ingredients for over 95% of all lubricants now in production. These oils consist essentially of hydrocarbons ranging in molecular weight from approximately 250 (with a skeleton of about 18 carbon atoms) for low-viscosity lubricants up to 1000 for very high-viscosity lubricants. Typical molecular structures involved are shown in Figure 2.1 (Furby, 1973). For a given molecular size, paraffins have relatively low viscosity, low density, and higher freezing temperatures (pour points). Aromatics involve six-member unsaturated carbon rings, which introduce higher viscosity, more rapid change in viscosity with temperature, and dark-colored, insoluble oxidation products which are the source of sludge and varnish that accompany their oxidation in high-temperature service.

The initial step in producing mineral lubricating oil is distillation of the crude petroleum to remove lower-boiling gasoline, kerosene, and fuel oil while dividing lubricating oil fractions into several grades by their boiling point ranges. Subsequent refining steps remove straight chain paraffins (wax), undesirable aromatics, and sulfur and nitrogen contaminants. Colorless base oils of unusual purity and stability are now being produced by *catalytic hydrocracking* in an ongoing shift from solvent refining for Group I base stocks (see Table 2.1). For conversion to Group II cycloparaffins this hydrocracking step typically employs 500 psi hydrogen at 750 °F, and 1500 psi at 800–900 °F for obtaining Group III cycloparaffins and isoparaffins. Catalytic dewaxing follows this hydrocracking for Groups II and III to convert high-melting, straight-chain paraffin wax to low-pour-point branched chain paraffins (Booser, 2001; Kramer *et al.*, 2003; Khonsari and Booser, 2004a, b).

Applied Tribology: Bearing Design and Lubrication, Third Edition. Michael M. Khonsari and E. Richard Booser.
© 2017 John Wiley & Sons Ltd. Published 2017 by John Wiley & Sons Ltd.
Companion Website: www.wiley.com/go/Khonsari/Applied_Tribology_Bearing_Design_and_Lubrication_3rd_Edition

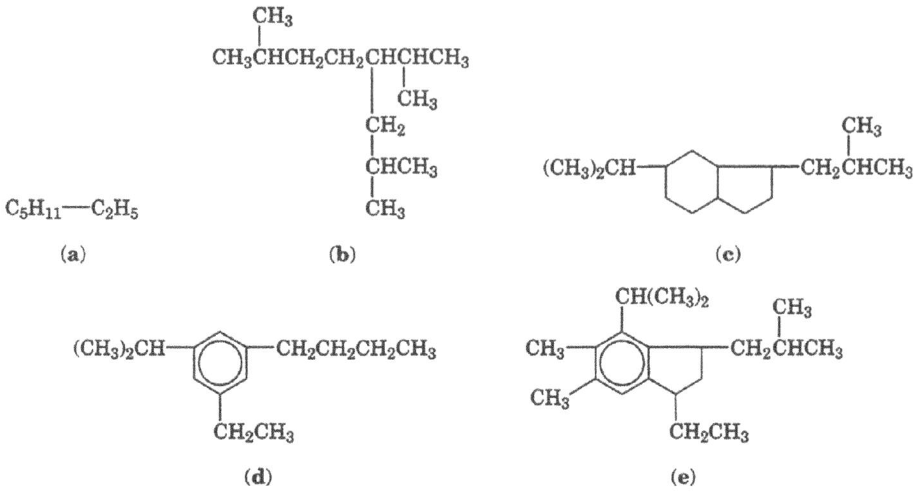

Figure 2.1 Typical structures in mineral lubricating oils: (a) *n*-paraffin, (b) isoparaffin, (c) cycloparaffin, (d) aromatic hydrocarbon, and (e) mixed aliphatic and aromatic ring. *Source*: Furby, 1973.

While eliminating sulfur, nitrogen, oxygen impurities and most aromatic hydrocarbons, Group II base stocks can be produced from less expensive crude oils to provide improved oxidation life in high-temperature service, fewer environmental questions, lower pour points, and higher viscosity index. Following their more severe hydrocracking, Group III base stocks involve nearly complete conversion to branched-chain paraffins with a very high viscosity index (VHVI) above 120. Table 2.2 indicates the increased service life available in representative applications of Group II and III base oils (Khonsari and Booser, 2004a).

About half of the base stocks in North America are now Group II and the trend away from Group I is also underway elsewhere. Group III oils, in more limited production, are now being used in 'synthetic' automotive oils where their low viscosities and limited solvency for additives have required tailored compounding. Many gear and compressor oils still employ Group I base stocks to provide the higher required viscosities.

Almost all current lubricants are fortified with chemical additives such as the following to meet service demands (Rizvi, 1997):

- *Oxidation inhibitors* – Improve oil life by limiting oxidation of the hydrocarbon molecules.
- *Rust inhibitors* – Preferentially absorb on iron and steel to prevent their corrosion by moisture.

Table 2.1 American Petroleum Institute base stock categories

Group	General description	Saturates (%)	Aromatics (%)	Sulfur (%)	Viscosity index
I	Solvent-refined mineral oil	<90 and/or	>10	>0.03	80–<120
II	Hydroprocessed	>90	<10	<0.03	80–<120
III	Hydrocracked (VHVI)	>90	<10	<0.03	120+
IV	PAO polyalphaolefins	(100)	(0)	(0)	
V	All others–other synthetics				

Source: API, 1993. Reproduced with permission of API Publication.

Table 2.2 Expected increase in service life with Group II/III base stocks

Application	Solvent-refined Group I	Hydrocracked Group II/III
Automobile engines	3000–4000 miles	7500–20 000 miles
Steam turbines	10 years	25 years
Gas turbines	5 years	10 years
Electric motors	2 years	5 years
Gear sets, EP oil	6 months	1 year

- *Antiwear and extreme pressure (EP) agents* – Commonly involve sulfur, phosphorus, and zinc compounds that form protective low shear strength films on metallic surfaces to limit friction and wear, particularly in concentrated contacts.
- *Friction modifiers* – Form adsorbed surface films to reduce or control friction in bearings.
- *Detergents and dispersants* – Reduce deposits of sludge in internal combustion engines.
- *Pour-point depressants* – Lower the temperature at which a mineral oil is immobilized by wax.
- *Foam inhibitors* – Often silicone polymers that enhance separation of air bubbles from oil.
- *Viscosity index improvers* – Long-chain polymers used in multigrade oils to reduce the effect of temperature on viscosity.

Table 2.3 lists the properties of some of the more common types of petroleum lubricants, along with the additive types typically incorporated. To meet the severe demands of internal combustion engines, the additive content can range up to 20–25% while incorporating most of the common types listed above. With the high-quality turbine oils used for long-life service in many industrial circulating systems, the additive content of rust and oxidation inhibitors is usually limited to 1–2%. The widely used industrial hydraulic oils are similar to turbine oils but usually contain an antiwear additive to minimize wear in high-pressure hydraulic pumps.

2.2 Synthetic Oils

Table 2.4 lists the physical properties of representative synthetic oils that are finding increasing use for extreme temperatures, fire resistance, and other applications requiring their unique physical and chemical characteristics. High cost is a common deterrent to their use where mineral oils give satisfactory performance. Environmental questions in handling and disposing of some types of synthetic oils have also been a problem.

Hydrocarbon poly(alpha-olefins) (PAOs) lead the synthetics in production volume. These are manufactured in a two-step polymerization of ethylene available from petroleum cracking. While they are quite similar chemically to the paraffinic oils available directly from refining petroleum, their tailored structure enables their use at both higher and lower temperatures, provides less sludge and varnish deposits under severe conditions, and gives improved operation under elastohydrodynamic conditions in gears. To improve their low solvency for additives, the combination of a PAO with about 15% of a synthetic ester fluid is common for automotive engine use. Production and use of PAO synthetic hydrocarbons and esters would probably

Table 2.3 Representative petroleum lubricating oils

Type	Viscosity (mm²/s(= cST))		Flash point (°C)	Pour point (°C)	Specific gravity at 15°C	Viscosity index	Common additives[a]	Uses
	At 40°C	At 100°C						
Automobile (SAE)								
10W	28	4.9	204	−28	0.878	106	R, O, D, VI, P, W, F, M	Automobile, truck, and marine reciprocating engines
20W	48	7.0	218	−24	0.884	103		
30	93	10.8	228	−20	0.890	100		
40	134	13.7	238	−16	0.895	97		
50	204	17.8	250	−10	0.901	94		
5W-30	65	10.4	210	−36	0.875	147		
10W-30	62	10.3	208	−36	0.880	155		
20W-40	138	15.3	246	−21	0.897	114		Railroad diesels
15W-40	108	15.0	218	−27	0.885	145		Diesels
Gear (SAE)								
80W-90	144	14.0	192	−22	0.900	93	EP, O, R, P, F	Automative and industrial gear units
85W-140	416	27.5	210	−14	0.907	91		
Automatic transmission	38	7.0	188	−40	0.867	140	R, O, W, F, VI, P, M	Automative hydraulic systems

Turbine								
Light	31	5.4	206	−10	0.863	107	R,O	Steam turbines, electric motors industrial systems
Medium	64	8.7	220	−6	0.876	105		
Heavy	79	9.9	230	−6	0.879	103		
Hydraulic fluids								
Light	30	5.3	206	−24	0.868	99	R, O, W	Machine tool hydraulic systems
Medium	43	6.5	210	−23	0.871	98		
Heavy	64	8.4	216	−22	0.875	97		
Extra low temperature	14	5.1	96	−62	0.859	370	R, O, W, VI, P	Aircraft hydraulic systems
Aviation								
Grade 65	98	11.2	218	−23	0.876	100	D, P, F	Reciprocating aircraft engines
Grade 80	139	14.7	232	−23	0.887	105		
Grade 100	216	19.6	244	−18	0.893	100		
Grade 120	304	23.2	244	−18	0.893	95		

[a] R, rust inhibitor; O, oxidation inhibitor; D, detergent-dispersant; VI, viscosity index improver; P, pour-point depressant; W, antiwear; EP, extreme pressure; F, antifoam; and M, friction modifier.

Source: Raimondi and Szeri, 1984. Reproduced with permission of Taylor & Francis.

Table 2.4 Properties of representative synthetic oils

Type	Viscosity (cSt) at 100 °C	40 °C	−54 °C	Pour point (°C)	Flash point (°C)	Typical uses
Synthetic hydrocarbons						
Mobil 1, 5W-30[a]	11	58	–	−54	221	Auto engines
SHC 824[a]	6.0	32	–	−54	249	Gas turbines
SHC 629[a]	19	141	–	−54	238	Gears
Organic esters						
MIL-L-7808	3.2	13	12,700	−62	232	Jet engines
MIL-L-23699	5.0	24	65,000	−56	260	Jet engines
Synesstic 68[b]	7.5	65	–	−34	266	Air compressors, hydraulics
Polyglycols						
LB-300-X[c]	11	60	–	−40	254	Rubber seals
50-HB-2000[c]	70	398	–	−32	226	Water solubility
Phosphates						
Fyrquel 150[d]	4.3	29	–	−24	236	Fire-resistant fluids for die casting, air compressors and hydraulic systems
Fyrquel 220[d]	5.0	44	–	−18	236	
Silicones						
SF-96 (50)[e]	16	37	460	−54	316	Hydraulic and damping fluids
SF-95 (1000)[e]	270	650	7,000	−48	316	Hydraulic and damping fluids
F-50[e]	16	49	2,500	−74	288	Aircraft and missiles
Fluorochemicals						
Halocarbon 27[f]	3.7	30	–	−18	None	Oxygen compressors, liquid oxygen systems
Krytox 103[g]	5.2	30	–	−45	None	

[a] ExxonMobil.
[b] ExxonMobil.
[c] Union Carbide Chemicals Co.
[d] Akzo Chemicals.
[e] General Electric Co.
[f] Halocarbon Products Corporation.
[g] DuPont Co.
Source: Booser, 1995.

expand dramatically if automotive manufacturers used them for initial fill of new cars to achieve longer oil-change intervals.

Organic esters have a distinctive market for aircraft and military use, where their wide temperature range has led to their general use in jet engines and related aircraft accessories. Table 2.4 indicates typical uses for other common types of synthetics. Vegetable oils are also finding specialty use as biodegradable lubricants. Common applications are as hydraulic fluids in farm machinery and in two-cycle engines where the oil is lost to soil or water (Lawate *et al.*, 1997).

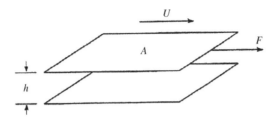

Figure 2.2   Illustration for the definition of viscosity: μ = shear stress (F/A)/shear rate (U/h).

2.3   Viscosity

The viscosity of a fluid is a measure of its resistance to motion. It represents the fluid stiffness or internal friction (see Figure 2.2). With two plane surfaces moving over each other separated by a fluid, the force F required to maintain constant velocity U is proportional to the fluid viscosity μ, area A, and the reciprocal of separating distance h.

$$F = \frac{\mu UA}{h}; \quad \mu = \frac{(F/A)}{(U/h)} = \frac{\text{shear stress } \tau}{\text{shear rate } \dot{\gamma}}. \tag{2.1}$$

Using SI units with force F (newtons, N), fluid film thickness h (m), velocity U (m/s), and area A (m^2), units of viscosity become N s/m^2(= pa s). Multiplication of this pa s viscosity value by 1.450 E-04 gives viscosity in British engineering units of lb(force) s/in.^{2} (= reyns).

This *absolute viscosity* divided by density is called the *kinematic viscosity* measured by most laboratory instruments in which the density of the fluid drives the oil through a calibrated capillary. Kinematic viscosity is usually provided in lubricant data sheets in units of centistokes (cSt = mm^2/s in SI units). The stokes is the metric viscosity in poise (dynes force. s/cm^2) divided by the lubricant density (g/cm^3). Table 2.5 provides common conversion factors,

Table 2.5   Viscosity conversion factors

Multiply	By	To obtain
Dynamic (or absolute) viscosity (units = force · time/area)		
Poise (P) (dyne s/cm^2)	1.000E-1	pa s (SI preferred unit), also Ns/m^2
Centipoise (cP) (mPa s)	1.000E-3	pa s
cP	1.450E-7	Reyn (lb$_f$ s/in.^{2})
Reyns (lb$_f$ s/in.^{2})	6.895E + 3	pa s
Kinematic viscosity: absolute viscosity/density (units = area/time)		
Stokes (St) (cm^2/s)	1.000E-04	m^2/s (SI preferred unit)
Centistokes (cSt) (mm^2/s)	1.000E-06	m^2/s
in.^{2}/s	6.452E-04	m^2/s
Conversion between kinematic and absolute viscosity		
cSt	density (g/cm^3)	cP (mPa s)
cSt	1.450E $-$ 07 · density	Reyns (lb$_f$ s/in.^{2})
m^2/s	1.000E + 03 · density	pa s

particularly to international SI units. Appendices A and B in this textbook provide conversions for some of the many other units used for viscosity.

Centistoke viscosity values now commonly used to characterize oil viscosity can be obtained approximately from the once common Saybolt Universal seconds values (SUS) through the following relationship:

$$\text{Kinematic viscosity (cSt)} = 0.22(\text{SUS}) - 180/(\text{SUS}). \tag{2.2}$$

Viscosity Classifications

The general International Organization for Standardization (ISO) viscosity classification for industrial oils from the American Society for Testing and Materials (ASTM) D2422 is given in Table 2.6 in comparison with the Society of Automotive Engineers (SAE) viscosity grades for automotive oils and those of the American Gear Manufacturers Association (AGMA) for gear lubricants (Booser, 1997b). The ISO classification is based solely on viscosity ranges at 40 °C.

The SAE grade equivalents are approximate since their basis is primarily their 100 °C viscosity (with an added low-temperature requirement at temperatures down to −35 °C for winter W grades), as shown in Table 2.7. With synthetic automotive oils, or with mineral oils using polymer additives to obtain lower viscosity at winter temperatures, two grade requirements

Table 2.6 Equivalent ISO industrial, SAE automotive, and AGMA gear oil grades

ISO VG grade	Viscosity (cSt) (at 40°) Minimum	Viscosity (cSt) (at 40°) Maximum	SAE crankcase oil grades	SAE aircraft oil grades	SAE gear lube grades	AGMA gear lube grades Regular	AGMA gear lube grades EP
2	1.98	2.42	–	–	–	–	–
3	2.88	3.52	–	–	–	–	–
5	4.14	5.06	–	–	–	–	–
7	6.12	7.48	–	–	–	–	–
10	9.00	11.0	–	–	–	–	–
15	13.5	16.5	–	–	–	–	–
22	19.8	24.2	5W	–	–	–	–
32	28.8	35.2	10W	–	–	–	–
46	41.4	50.6	15W	–	75W	1	
68	61.2	74.8	20W	–	–	2	2 EP
100	90.0	110	30	65	80W–90	3	3 EP
150	135	165	40	80	–	4	4 EP
220	198	242	50	100	90	5	5 EP
320	288	352	60	120	–	6	6 EP
460	414	506	–	–	85W–140	7 comp	7 EP
680	612	748	–	–	–	8 comp	8 EP
1000	900	1100	–	–	–	8A comp	8A EP
1500	1350	1650	–	–	250	–	–

Comparisons are nominal since SAE grades are not specified at 40° viscosity; VI of lubes could change some of the comparisons.
Source: Kennedy *et al.*, 1998.

Table 2.7   Viscosity grades for automotive engine oils

SAE viscosity grade	Low-temperature viscosities		High-temperature viscosities (cSt at 100°)	
	Cranking[a] (cP) max at temp °C	Pumping[b] (cP) max with no yield stress at temp °C	Min	Max
0W	3 250 at −30	30 000 at −35	3.8	−
5W	3 500 at −25	30 000 at −30	3.8	−
10W	3 500 at −20	30 000 at −25	4.1	−
15W	3 500 at −15	30 000 at −20	5.6	−
20W	4 500 at −10	30 000 at −15	5.6	−
25W	6 000 at −5	30 000 at −10	9.3	−
20	−	−	5.6	<9.3
30	−	−	9.3	<12.5
40	−	−	12.5	<16.3
50	−	−	16.3	<21.9
60	−	−	21.9	<26.1

[a] ASTM D5293.
[b] ASTM D455.
Source: Courtesy of SAE J300.

can be met with a single oil. An SAE 5W-30 oil, for instance, will fall in the SAE 30 viscosity range at 100 °C while also meeting the 5W low-temperature viscosity requirements given in Table 2.7.

For many types of machinery, ISO 32 grade industrial oil or an equivalent SAE 10 automotive grade is given first consideration (see Table 2.3). Oils of this viscosity have been widely used in automobiles, turbines, compressors, electric motors and generators, various instruments, and household equipment. For high loads, low speeds, and high temperatures, a shift is commonly made to higher viscosities such as SAE 40 for diesel engines and ISO 68, 100, 150, and occasionally higher grades for industrial machinery. Lower viscosities may be used, on the other hand, for low-temperature and high-speed operation.

Viscosity–Temperature Relations

Oil viscosity decreases rapidly with increasing temperature in the general pattern shown in Figure 2.3. A straight line drawn through viscosity values for an oil at any two temperatures (typically 40 °C and 100 °C) enables estimation of viscosity at any other temperature down to just above the pour point where separating wax distorts the flow. Such a straight line relates kinematic viscosity v in cSt (mm^2/s) to absolute temperature T (°C + 273) by the Walther equation:

$$\log\log(v + 0.7) = A - B\log T. \qquad (2.3)$$

where A and B are constants for any given oil.

This relation holds well for most mineral and synthetic oils, at high pressures, and at a variety of given shear rates for polymer solutions in multigrade automotive oils (Song *et al.*, 1991). With this versatility, Equation (2.3) is a common choice for establishing viscosity at

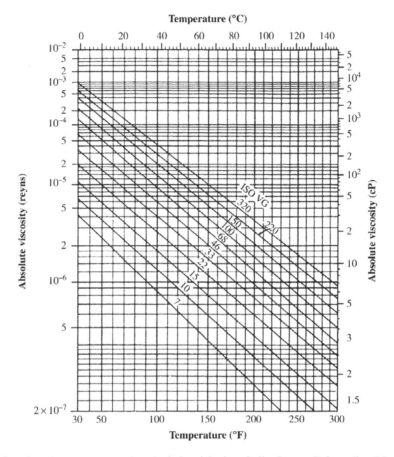

Figure 2.3 Viscosity–temperature chart for industrial mineral oils. *Source*: Raimondi and Szeri, 1984. Reproduced with permission of Taylor & Francis.

various temperatures in computer programs for analyses of bearing oil films, in calculating losses in gears and bearings, for sizing oil feed lines, and in analyzing hydraulic systems.

The following simplified version, often involving much less manipulation than Equation (2.3) in formal mathematical use, can be applied over limited temperature ranges:

$$\log \mu = A' - \frac{B'}{T} \quad \text{or} \quad \mu = a'e^{b'/T}. \tag{2.4}$$

This linear relation of logarithm of viscosity to reciprocal of absolute temperature is analogous to temperature relations for vapor pressure and chemical reaction rates. While absolute viscosity $\mu(\text{Pa} \cdot \text{s})$ is usually employed, the same general form of relation holds also for kinematic viscosity in cSt (mm^2/s).

Much improved representation of viscosity over an extended temperature range is available through the use of a modified temperature term in Equation (2.4) to give the following Vogel equation (Jang and Khonsari, 2013a):

$$\mu = \mu_{\text{ref}}e^{-\beta(T-T_{\text{ref}})}. \tag{2.5}$$

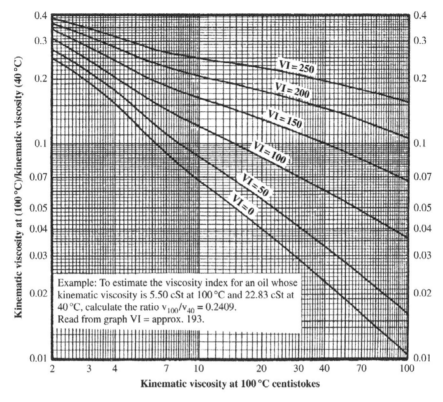

Figure 2.4 VI from kinematic viscosities at 40 °C and 100 °C. *Source*: ExxonMobil Corp., 1992. Reproduced with permission of Exxon Corp.

where μ_{ref} represents the viscosity at reference temperature T_{ref} (often at the film inlet) and β is a viscosity–temperature coefficient. The value of β for a given lubricant can be estimated from the viscosities at two temperatures in the range of interest from the following relation:

$$\beta = \frac{\ln(\mu_2/\mu_1)}{T_1 - T_2}. \tag{2.5a}$$

As an initial approximation, β can be obtained from the 40 °C and 100 °C viscosity values commonly supplied in lubricant data sheets. For extended temperature ranges and for use in computer codes, ASTM has established Equation (2.3) as a useful general representation of viscosity–temperature characteristics.

Viscosity index (VI), although empirical, is commonly used to indicate the relative decrease in viscosity with increasing temperature. As indicated in Figure 2.4 (Exxon Corp., 1992), the viscosity of an oil with a higher VI drops less in passing from 40 °C to 100 °C. A VI of 100 reflects the least temperature-sensitive lubricant which was normally available in the past as a conventional Pennsylvania petroleum oil, and a VI of 0 represents the most temperature-sensitive Gulf Coast oil. The following equation gives the viscosity index of any oil in question with a VI below 100:

$$VI = \frac{L - U}{L - H} \tag{2.6}$$

where, for the same 100 °C kinematic viscosity measured on the oil in question, L is the 40 °C viscosity of the 0 VI oil, H is the 40 °C viscosity of the 100 VI oil, and U is the 40 °C viscosity of the unknown VI oil. Detailed tables are available for obtaining the VI of any oil from its 40 °C and 100 °C viscosities (ASTM D2270).

Since this method gave confusing results and even some double values when VI exceeded 100, the following empirical fit now replaces Equation (2.6) in the ASTM procedure for VI in this high range:

$$VI = \frac{(\text{antilog } N) - 1}{0.00715} + 100 \tag{2.7}$$

where $N = (\log H - \log U) / \log Y$ and Y = viscosity in centistokes at 100 °C for the fluid of interest.

Mineral oils formulated with base stocks having a VI above about 90 are generally desirable since they are composed primarily of paraffinic-type hydrocarbons, which can be expected to give longer life, more freedom from sludge and varnish, and superior performance when compounded with proper additives for a given application. Much higher VI values ranging up to 150 and higher are available with additives, with special refining of petroleum oils, and with synthetic lubricants.

Viscosity–temperature coefficient (VTC), given by the following relation, is useful for some synthetic oils and multigrade mineral oils not well defined by the usual VI system (Wilcock and Booser, 1957):

$$VTC = \frac{(V_{40} - V_{100})}{V_{40}} = 1 - \frac{V_{100}}{V_{40}}. \tag{2.8}$$

VTC gives a measure of the fractional change in kinematic viscosity v in going from 40 °C to 100 °C, varying from 0 for an oil unaffected by temperature to unity at the other extreme. VTC is nearly constant at about 0.60 for essentially the full range of viscosity for dimethyl silicone fluids and ranges up to 0.78 for phenyl methyl silicones, despite a wide variation in VI. VTC also is helpful in considering the addition of VI-improving polymers to mineral oils: despite a significant increase in VI with the addition of polymer, the VTC value commonly remains relatively constant at the VTC of the base stock.

Viscosity–Pressure Relations

The relations of viscosity to both temperature and pressure outlined here are generally satisfactory for analysis of the usual hydrodynamic oil films and behavior in conventional hydraulic and lubricating oil systems. On the other hand, much uncertainty still exists as to the state of oil under the rapid transit times and much higher contact pressures ranging up to 1400–3500 MPa (20 000–500 000 psi) in elastohydrodynamic films in rolling element bearings and gear contacts.

The dramatic increase in oil viscosity shown in Figure 2.5 is analogous to the effect of lowering temperature. Like lowered temperature, sufficiently raised pressure will eventually cause an actual phase change with solidification of an oil. The following exponential relationship

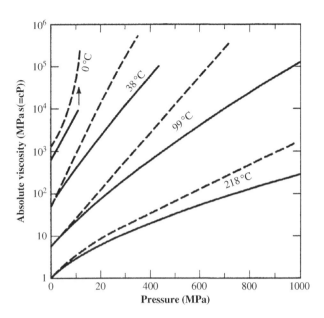

Figure 2.5 Increase in viscosity with increasing pressure for typical mineral oils: (—) paraffinic, (- - -) naphthenic, (↑) solid. (ASME, 1952). *Source*: Booser, 1995. Reproduced with permission of John Wiley & Sons.

often attributed to Barus applies for many oils at pressures up to about 70 MPa (10 000 psi) such as those encountered in conventional fluid-film bearings and in hydraulic systems:

$$\mu = \mu_0 e^{\alpha P} \quad \text{or} \quad \ln\left(\frac{\mu}{\mu_0}\right) = \alpha P \qquad (2.9)$$

where μ_0 is the centipoise viscosity (MPa. s) at atmospheric (essentially zero) pressure, μ is the viscosity at pressure P at the same temperature, and α is the *pressure–viscosity coefficient*. Values follow the general trend for viscosity level and are about 40% lower at 100 °C than at 40 °C.

The isothermal relation of centipoise viscosity μ (mPa. s) to pressure over a broad range of gauge pressures P up to moderately high values is represented more precisely by the following Roelands equation (Hamrock, 1994; Fein, 1997):

$$\log_{10}\mu + 1.200 = (\log_{10}\mu_0 + 1.200) \cdot \left(1 + \frac{P}{C}\right)^Z \qquad (2.10)$$

where μ_0 is the viscosity at atmospheric pressure and the viscosity pressure index Z is a constant which is characteristic of the lubricant and is relatively independent of temperature. Z is commonly taken as 0.6 for conventional mineral oils and falls in the range of 0.4–0.8 for many synthetics, as shown by the representative values in Table 2.8 (Fein, 1997). The value of constant C depends on the pressure units employed (196.1 MPa or 28 440 psi), as required to make P/C dimensionless.

Table 2.8 Representative values of viscosity–pressure index Z

Material	Z
Mineral oil, typical value	0.60
Synthetic oils	
PAO synthetic hydrocarbon	0.45
Polyglycol	0.37–0.40
Diester	0.47
Polyol ester	0.48
Polybutenes	0.91
Polymethylsiloxane	0.49
Castor oil	0.43
Rapeseed oil	0.42
Glycerol	0.18
Water	0.10
Water-glycol fluid	0.14

Source: Fein, 1997.

Values of Z for most mineral oils, synthetic hydrocarbons, diesters, and polyesters, both with and without the usual additives, can be estimated more closely from their 40 °C and 100 °C centipoise viscosities by use of the following relations (Fein, 1997):

$$Z = [7.81(H_{40} - H_{100})]^{1.5}(F_{40}) \tag{2.11}$$

where

$$H_{40} = \log(\log(\mu_{40}) + 1.200)$$
$$H_{100} = \log(\log(\mu_{100}) + 1.200)$$
$$F_{40} = (0.885 - 0.864\, H_{40})$$

EHL Pressure–Viscosity Coefficients

Viscosity–pressure relations for oil films involved in elastohydrodynamic (EHL) contacts of gear teeth and in ball and roller bearings commonly use a Barus-type α_{EHL} that is equivalent to the following (Hamrock, 1994; Fein, 1997):

$$\alpha_{EHL} = \left\{ \int_0^\infty \left[\frac{\mu_0 dP}{\mu} \right] \right\}^{-1} \tag{2.12}$$

The value of the α_{EHL} pressure–viscosity coefficient can be estimated rather simply by entering Figure 2.6 with the pressure–viscosity index Z from Table 2.8, along with the viscosity μ_0 at atmospheric pressure and at the temperature involved. Figure 2.6 summarizes Roelands' (1966) calculations of the relations involved in Equation (2.12) and is commonly adequate for estimating α_{EHL} since the accuracy of most EHL equations is rather limited by uncertainty about lubricant film temperature and other factors (Fein, 1997).

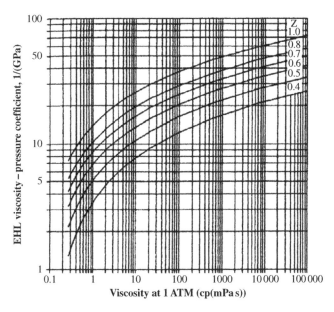

Figure 2.6   EHL pressure–viscosity coefficient. *Source*: Fein, 1997. Reproduced with permission of Taylor & Francis.

An alternative approach for determining this EHL pressure–viscosity coefficient is to use the following empirical equation involving viscosity μ_0 (pa s) at atmospheric (zero) pressure and Roelands' pressure–viscosity index Z from Equation (2.11) (Hamrock, 1994):

$$\alpha_{EHL} = Z[5.1 \times 10^{-9}(\ln \mu_0 + 9.67)] \qquad (2.13)$$

Relations are provided in Chapter 15 for applying EHL pressure–viscosity coefficients in calculating oil-film thickness in EHL contacts.

Interestingly, oil viscosity may drop slightly below its value at atmospheric pressure when an oil is exposed to high pressures while in equilibrium with nitrogen and some other gases. Thinning effect of the dissolved gas tends to balance the increase in viscosity usually to be expected with the higher pressure (Klaus and Tewksbury, 1984).

2.4 Free Volume Viscosity Model

The preceding sections have illustrated how important it is to include the effect of pressure and temperature on lubricant viscosity. This is particularly crucial in EHL applications where the effect of pressure on viscosity can become enormous. The viscosity-pressure relationships presented thus far work well up to moderate pressures beyond which their validity becomes questionable. As discussed by Bair (2007), accurate quantitative EHL analyses require further consideration.

Let us begin from the equation of state, which describes a relationship between the lubricant density (or volume) to pressure and temperature. The most commonly used equation of state in EHL application originates from the work of Dowson and Higginson (1966) which under isothermal conditions reads:

$$\frac{V}{V_0} = \frac{1 + \dfrac{K_0' - 1}{2K_0}p}{1 + \dfrac{K_0' + 1}{2K_0}p} \tag{2.14}$$

where K_0 is the bulk modulus at ambient pressure with $P = 0$ and K_0' denotes the rate of change of K with P at $P = 0$. V represents the volume and V_0 is the volume at $P = 0$.

As pointed out by Bair (2007), assuming that the isothermal condition holds, with $K_0 = 1.67$ GPA and $K_0' = 6.67$, this equation appropriately predicts the lubricant viscosity for moderate pressures up to 350 MPa.

The so-called free-volume viscosity model developed by Doolittle in 1951 provides a much more realistic assessment of viscosity variation for use in quantitative predictions in EHL applications (Bair, 2007).

$$\mu = \mu_0 \exp\left[B \frac{V_\infty}{V_0} \left(\frac{1}{\dfrac{V}{V_0} - \dfrac{V_\infty}{V_0}} - \frac{1}{1 - \dfrac{V_\infty}{V_0}} \right) \right] \tag{2.15}$$

where B and V_∞/V_0 are constants and V/V_0 can be determined by using the following equation of state attributed to Tait:

$$\frac{V}{V_0} = 1 - \frac{1}{1 + K_0'} \ln\left[1 + \frac{P}{K_0}(1 + K_0') \right] \tag{2.16}$$

Constants K_0 and K_0' can be experimentally determined. For polydimethylsilonxane, for example, the constants are $K_0 = 0.85$ GPa and $K_0' = 10.19$ (Sachdev et al., 1998).

The Tait equation can be modified to take into account the variation of the lubricant's volume, V, with both pressure and temperature:

$$\frac{V}{V_R} = \frac{\rho_R}{\rho} = [1 + \alpha_V(T - T_R)] \left\{ 1 - \frac{1}{1 + K_0'} \ln\left(1 + \frac{P}{K_{00} \exp(-\beta_K T)}(1 + K_0') \right) \right\} \tag{2.17}$$

where the subscript R denotes the reference state at $P = 0$ and $T = T_R$, K_{00}, α_V and β_K are parameters that should be specified for a given lubricant along with K_0 and K_0'. Bair (2007) suggests the following values for numerical EHL simulations. $K_0' = 11$; $\alpha_V = 8 \times 10^{-4}$ K^{-1}; $K_{00} = 9$ GPa; $\beta_K = 6.5 \times 10^{-3}$ K^{-1}. This equation is capable of accurately predicting the density at a given pressure and temperature. Finally, once the volume ratio, V/V_R, is known, viscosity can be predicted using the free-volume equation developed by Doolittle (1951)

$$\frac{\mu}{\mu_R} = \exp\left[BR_0 \left(\frac{1 + \varepsilon(T - T_R)}{\dfrac{V}{V_R} - R_0(1 + \varepsilon(T - T_R))} - \frac{1}{1 - R_0} \right) \right] \tag{2.18}$$

where μ is the lubricant viscosity and parameters B, R_0 and ε are parameters unique for a given lubricant.

Additional information is given by Bair (2007). Applications of the free-volume viscosity model in EHL can be found in series of papers by Jang *et al.* (2008), Kumar and Khonsari (2008, 2009), Krupka *et al.* (2010), and Masjedi and Khonsari (2014). These considerations are particularly important in prediction of traction coefficient in EHL contacts.

2.5 Density and Compressibility

The density of a lubricant is necessary in performance calculations since the kinematic viscosity measured in laboratory instruments (usually in centistokes) must be multiplied by density to obtain the absolute viscosity for oil-film characterization. In general, densities of petroleum mineral oils range from about 850 to $900 \, \text{kg/m}^3$ at $15.6 \, °\text{C}$ (0.85–$0.90 \, \text{g/cm}^3$ or 0.85–0.90 specific gravity, which is the oil density divided by the density of water of about $1.0 \, \text{g/cm}^3$). Multiplying values in g/cm^3 by 0.03613 gives a characteristic density of about $0.0307 \, \text{lb/in.}^3$ for many mineral oils.

The density ρ of an oil at some temperature T ($°\text{C}$) other than the $15.6 \, °\text{C}$ ($60 \, °\text{F}$) given in Table 2.3 can be calculated by use of the typical coefficient of expansion for mineral oils of 0.00063 per $°\text{C}$ (0.00035 per $°\text{F}$) in the following relation:

$$\rho_t = \rho_{15.6}[1 - 0.00063(T - 15.6)] \tag{2.19}$$

Increasing pressure affects oil density in a pattern analogous to that of decreasing temperature. As pressure increases, the oil molecules become more and more closely compacted until they eventually reach their solidification pressure. For mineral oils, the following traditional relation reflects the diminishing compressibility as pressure is increased from atmospheric until the oil solidifies after it has lost about 30% of its volume (Hamrock, 1994):

$$\frac{\rho}{\rho_0} = 1 + \frac{0.6P}{1 + 1.7P} \tag{2.20}$$

where

$$\rho_0 = \text{density when } P = 0$$
$$P = \text{gage pressure, GPa}$$

Lubricant compressibility under increasing pressure is often given as an isothermal secant bulk modulus K, which simply is a measure of resistance to uniform compression. If an initial volume V subjected to a differential pressure dP experiences a differential volume change of dV, then its bulk modulus is defined as $K = -dP/(dV/V) = -V\frac{dP}{dV}$. An average value of bulk modulus over a pressure range at a particular temperature (Chen, 1997):

$$K = \frac{V_0(P - P_0)}{V_0 - V_P} \tag{2.21}$$

where

K = isothermal secant bulk modulus at pressure P (GPa) and a given temperature
V_P = specific volume at pressure P, cm^3/g; also equal to the reciprocal of oil density in g/cm^3
P_0 = atmospheric pressure, 0.0001 GPa
V_0 = specific volume at atmospheric pressure, cm^3/g

For mineral oils and pure hydrocarbons at a given temperature, isothermal secant bulk modulus B_m at pressure P is related to oil viscosity and temperature as follows (Song et al., 1991):

$$K_m = K_{0m} + A \cdot P \tag{2.22}$$

where

K_{0m} = isothermal secant bulk moduls of minerals oil at atmospheric pressure (GPa)
$\log_{10}(K_{0m}) = 0.3766[\log_{10}(v)]^{0.3307} - 0.2766$
v = kinematic viscosity at atmospheric pressure and at a given temperature (cSt)
P = pressure (GPa)

For synthetic oils, the deviation factor S is added to give the isothermal secant bulk modulus K_s:

$$K_s = K_m + S \tag{2.23}$$

where S is given as follows for various classes of synthetic oils:

Fluid class	Deviation factor S (GPa)
Methyl silicone	−0.755
Phenyl silicone	−0.160
Perfluoropolyether	−0.823
Polybutene	−0.268
Polyalphaolefin (PAO)	−0.091
Ester	+0.092
Pentaaerythritol ester	+0.219
Phosphate ester	+0.301
Polyphenyl ether	+0.709

Once the bulk modulus is calculated as B_m or B_s from Equation (2.22) or (2.23), the specific volume or density can be obtained from Equation (2.21).

If the pressure exceeds the solidification pressure, commonly in the range of 0.8 to 1.5 GPa, the rate of decrease in relative volume with increasing pressure, $-d\mathrm{V}_r/dP$, becomes relatively constant, with values ranging from about 0.04 to 0.055 Gpa^{-1} for different lubricants (Hamrock, 1994). Relative volume Z is taken as final specific volume (cm^3/g) divided by the initial specific volume as pressure is raised from an initial to a final value.

2.6 Thermal Properties

Specific heat of petroleum oils increases gradually with increasing temperature and can be approximated by the relation

$$C_p = 1800[1 + 0.002(T)] \quad \mathrm{J/kg\,°C} \tag{2.24}$$

Thermal conductivity of petroleum oils diminishes slightly with increasing temperature, as given in the following approximation:

$$k = 0.1312[1 - (6.3 \times 10^{-4})(T)] \quad \text{W/m}^\circ\text{C} \tag{2.25}$$

where T is $^\circ$C.

Thermal properties of synthetic oils with a major portion of hydrocarbon structure, such as synthetic hydrocarbons and esters, are reasonably similar to those given above for mineral oils. Detailed information on other types is available from suppliers (see also *Tribology Data Handbook*, CRC Press, 1997).

Example 2.1   Consider ISO 32 light mineral turbine oil with the properties listed in Table 2.3. This oil is to be used at a contact pressure of 1.4 GPa (200 000 psi) in an EHL contact at 65 $^\circ$C in a ball bearing. Determine (a) the viscosity of the oil at 65 $^\circ$C at atmospheric pressure and (b) the viscosity and EHL pressure–viscosity coefficient at 65 $^\circ$C and 1.4 GPa pressure.

(a) Atmospheric pressure: applying Equation (2.3) twice, once for the 40 $^\circ$C viscosity of 31 cSt, then for the 100 $^\circ$C viscosity of 5.4 cSt:

$$\log\log(31 + 0.7) = A - B\log(40 + 273.1)$$
$$\log\log(5.4 + 0.7) = A - B\log(100 + 273.1).$$

Solving these simultaneous equations:

$$A = 9.39795; \quad B = 3.69500.$$

Reapplying Equation (2.3) for 65 $^\circ$C:

$$\log\log(v_{65} + 0.7) = 9.39795 - 3.69500\log(65 + 273.1).$$

The resulting kinematic viscosity is

$$v_{65} = 12.794 \, \text{cSt(mm}^2\text{/s)}.$$

Referring to Equation (2.19) (see Section 2.5), we determine the density at 65 $^\circ$C:

$$\rho_{65} = \rho_{15.6}[1 - 0.00063(T - 15.6)]$$
$$\rho_{65} = 0.863[1 - 0.00063(65 - 15.6)] = 0.836 \, \text{g/cc}.$$

For dynamic viscosity at atmospheric pressure:

$$\mu_{65} = 12.794 \times 0.836 = 10.7 \, \text{cP(mPa s)} = 10.7 \times 10^{-3} \, \text{pa s}.$$

(b) EHL contact pressure of 1.4 GPa: adjusting 15.6 $^\circ$C density 0.863 by Equation (2.19):

$$\rho_{40} = 0.863[1 - 0.00063(40 - 15.6)] = 0.850 \, \text{g/cc}$$
$$\rho_{100} = 0.863[1 - 0.00063(100 - 15.6)] = 0.817 \, \text{g/cc}.$$

The resulting dynamic viscosities at atmospheric pressure are

$$\mu_{40} = 31\,\text{cSt} \times 0.850 = 26.34\,\text{cP (mPa s)}$$
$$\mu_{100} = 5.4\,\text{cSt} \times 0.817 = 4.4\,\text{cP (mPa s)}.$$

Calculating Roelands' Z in Equation (2.11) by first calculating H_{40} and H_{100}:

$$H_{40} = \log(\log 26.4 + 1.200) = 0.4186$$
$$H_{100} = \log(\log 4.4 + 1.200) = 0.2656$$
$$F_{40} = (0.885 - 0.864 \times 0.4186) = 0.523.$$

Inserting these values in Equation (2.11) for Z:

$$Z = [7.81(H_{40} - H_{100})]^{1.5}(F_{40})$$
$$= [7.81(0.4186 - 0.2656)]^{1.5}(0.523) = 0.683. \tag{2.26}$$

Finally, at 65 °C this gives the EHL pressure–viscosity coefficient from Equation (2.13) with atmospheric pressure viscosity $\mu_0 = 10.7 \times 10^{-3}$ pa s:

$$\alpha_{\text{EHL}} = Z[5.1 \times 10^{-9}(\ln \mu_0 + 9.67)]$$
$$= 0.683\{5.1 \times 10^{-9}[\ln(10.7 \times 10^{-3}) + 9.67]\}$$
$$= 1.78 \times 10^{-8}\,\text{Pa}^{-1} = 17.8\,\text{GPa}^{-1}. \tag{2.27}$$

The value of the EHL pressure–viscosity coefficient can also be estimated as $18\,\text{GPa}^{-1}$ from Figure 2.6 for this case with $\mu_0 = 10.7$ cP and $Z = 0.68$.

2.7 Non-Newtonian Lubricants

Pure mineral oils and synthetic oils of similar molecular size exhibit viscosities which are independent of shear rate, a characteristic which is termed *Newtonian* and tends to hold up to shear rates of at least $10^6\,\text{s}^{-1}$ (Klamann, 1984). The following two lubricant types, however, involve deviations from Newtonian behavior in the patterns illustrated in Figure 2.7, and these are classified as *non-Newtonian* (Jang and Khonsari, 2013b).

1. Oils Thickened with a Polymer Viscosity Index Improver

Several percent of a polymer additive in the 10 000–100 000 molecular weight range can be compounded with a low-viscosity mineral oil base stock to produce a multigrade automotive oil such as SAE 10W-30. These multigrade oils take advantage of the thickening effect of the polymer in a low-viscosity base stock (see Figure 2.8). Added polymers maintain the superior viscosity–temperature characteristic of the low-viscosity base stock while providing an approximately constant percentage viscosity increase throughout much of the temperature spectrum. The viscosity increase for a specific polymer and polymer concentration is given as *specific viscosity* $\mu_{\text{sp}} = (\mu - \mu_0)/\mu_0$, where μ_0 and μ are, respectively, viscosities of the base oil and the polymer containing oil at a given temperature. Values of specific viscosity increase with increasing addition of polymer, and a *reduced viscosity* μ_{sp}/c is sometimes employed as being more independent of the volume fraction of polymer, c.

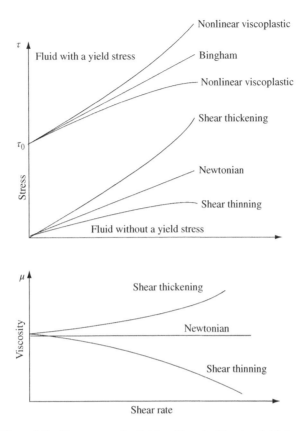

Figure 2.7   Flow curves for fluids with and without a yield stress.

Typical non-Newtonian behavior of polymer-containing multigrade oil is illustrated in Figure 2.9 (Klaus and Tewksbury, 1984). Constant viscosity (Newtonian character) is maintained until the shear rate rises to a critical value in the 5000–$10\,000\,s^{-1}$ range. With a further increase in shear rate, viscosity then falls rather linearly to a stable viscosity or 'second Newtonian' region, which begins in the 10^4–$10^6\,s^{-1}$ range. This $10^6\,s^{-1}$ shear rate is reached, for instance, in a 160 mm (6.3 in.) diameter turbine bearing rotating at 60 cycles per second with an oil-film thickness of 0.05 mm (0.002 in.). In an automotive engine, oils of 3–15 cP viscosity are subjected to a $5 \times 10^5\,s^{-1}$ shear rate. On cold starting with oil viscosities of 3000–50,000 cP, shear rates are of the order of 10^3–$10^4\,s^{-1}$. Progress on EHL analysis with consideration or second Newtonian is reported by Bair and Khonsari (1996).

At high rates of shear, the long polymer chains become oriented in flow streamlines and a *temporary* and reversible viscosity loss of up to 75% is encountered in the 'second Newtonian' region illustrated in Figure 2.9. Continuous mechanical stress on these polymer additive molecules by turbulence, cavitation, and high shear rates can lead to their breakage and permanent size reduction. This, in turn, results in *permanent* loss of the order of 10–20% for the $100\,°C$ viscosity of the additive-treated oil. Fluids exhibiting a drop in viscosity with increasing shear rate are referred to as *shear thinning* or *pseudoplastic*. The term *apparent viscosity* is used to quantify their resistance to motion at a specified shear rate.

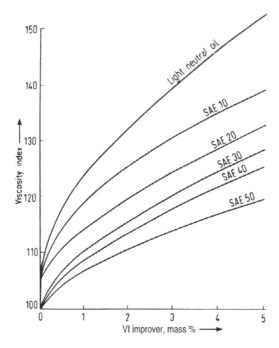

Figure 2.8 Effect of VI improvers as a function of base oil viscosity. *Source*: Klamann, 1984. Reproduced with permission of John Wiley & Sons.

Paranjpe (1992) suggests the following relation for expressing the general nature of this thinning tendency as apparent viscosity μ of multigrade oil drops from its first to its second Newtonian region with increasing shear rate $\dot{\gamma}$:

$$\mu = \frac{\mu_0(\sigma + \mu_\infty \dot{\gamma})}{\sigma + \mu_0 \dot{\gamma}} \tag{2.28}$$

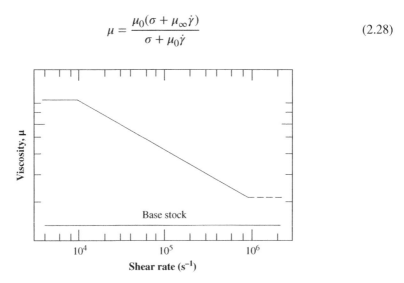

Figure 2.9 Loss of apparent viscosity of polymer-thickened oil with increasing shear rate. *Source*: Klaus and Tewksbury, 1984. Reproduced with permission of Taylor & Francis.

where μ_0 is the viscosity at low shear rates, μ_∞ is the viscosity at high shear rates in the second Newtonian region, and σ is a curve-fitting variable related to the shear stability of the lubricant. A comprehensive EHL analysis of shear thinning lubricants with comparison to experimental measurements is provided by Jang *et al.* (2007).

2. Lubricating Greases

These commonly consist of 80–90% mineral oil gelled with 10–20% of an organometallic soap or a nonmelting powder to create a *Bingham* solid. The typical non-Newtonian behavior of greases is indicated in Figure 2.7. The semisolid nature of grease provides some shear strength, and flow commences when the applied shear stress is greater than a critical value τ_{cr}. Once motion is initiated, however, apparent viscosity of the grease drops until it approaches the viscosity of the base oil for shear rates above about $10\,000$–$20\,000\,\mathrm{s^{-1}}$, where grease flow becomes essentially Newtonian (Klamann, 1984; see also Section 2.7).

Viscoelastic Effect

Even more complex non-Newtonian behavior is encountered in EHL conjunctions in rolling element bearings and gears where the lubricant encounters rapid and extreme pressure variations (Hamrock, 1994). The proposed rheological model in Figure 2.10 suggests that with increasing shear rate, the shear stress reaches a plateau at τ_L known as the limiting shear stress. An inverse exponential approach is then made to a limiting shear stress τ in the following pattern, which reflects behavior of the lubricant as a plastic solid at high shear rates (Bair and Winer, 1990):

$$\tau = \tau_L(1 - e^{-\mu_0\dot\gamma/\tau_L}) \tag{2.29}$$

This rheological equation, which is based on experimental data, suggests that the lubricating oil becomes solid for $\mu_0\dot\gamma/\tau_L > 5$ and at relatively low shear rates of 10–$100\,\mathrm{s^{-1}}$ in EHL

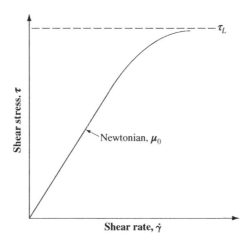

Figure 2.10   Limiting shear stress τ_L for oil in EHL contact.

contacts. The value of the limiting stress increases at lower temperatures; at higher pressures it rises somewhat linearly with increasing pressure P in the pattern $\tau_L = \tau_0 + kP$ (Bhushan, 1999). Higher limiting stress at higher pressure probably accompanies increasing density of the solidified oil, as discussed in the next section.

This is an area which appears to call for further study. Oils generally solidify under high pressure, and the solidified oil then has a characteristic shear strength. Attributing solidification to an increasing shear rate might imply that an increasing shear rate accelerates the solidification process under high pressure. Initial wax-like crystalline micelles initially separating from the liquid oil, for instance, may be oriented by a high shear rate to form an overall solid structure more quickly.

2.8 Oil Life

Lubricating oil supplied to the bearings in many machines deteriorates primarily due to oxidation. This involves an aging process in which the oil reacts chemically with atmospheric oxygen to increase oil acidity and viscosity, darken its color, and create varnish-like surface deposits (Booser, 1997c; Khonsari and Booser, 2003, 2004a, b, 2013a). Oxidation-inhibiting additives are employed in most lubricating oils to control this oxidation by breaking down hydroperoxides formed in the initial oxidation step, by breaking the chain reaction involved in oxidation, and by deactivating the catalytic effect of metal surfaces contacted by the oil.

Useful oil life continues through an initial induction period during which the oxidation inhibitor is slowly consumed by oxidation, evaporation, or other physical and chemical effects. While subsequent additions of new oil or an oxidation inhibitor will delay oxidation, useful oil life ends when the oxidation induction period finally expires and oxidation reactions accelerate. The remaining useful lubricating oil life can be evaluated by various laboratory techniques including (1) Fourier transform infrared spectroscopy (FTIR) to measure the remaining oxidation inhibitor concentration by light absorbed in the 2–50 µm wavelength range, (2) the ASTM D2272 rotating bomb oxidation test (RBOT) for turbine oils, and (3) rapid routines involving electrochemical methods, microscale oxidation tests, differential thermal analysis, and high-pressure differential scanning calorimetry (Kauffman, 1994).

Standardized laboratory tests for evaluating the oxidation life of new oils include ASTM D943, D2272, D2893, and D4742; Federal Test Method Standard 791, Method 5308.6; the ALCOR deposition test; International Harvester BT10: British IP280 and IP331B; and the US Army Mobility Equipment R and D Method (Smith, 1986). Based on such laboratory test results, oxidation life L (h) typically drops by a factor of 2 for a 10 °C rise in the 100–150 °C range. This is equivalent to factor $k_2 = 4750$ in the following Arrhenius equation for chemical reaction rates as a function of absolute temperature T (T °C + 273):

$$\log L = k_1 + \frac{k_2}{T} = k_1 + \frac{4750}{T} \tag{2.30}$$

where $1/k_1$ is the characteristic rate coefficient for oxidation of a specific oil, and k_2 reflects the kinetic energy needed in collision of two molecules to activate the oxidation reaction. The log of the oxidation life L then decreases (oxidation rate increases) as the Kelvin absolute temperature (T + 273) rises to increase the probability of molecular collisions with sufficient energy for an oxidation reaction (Khonsari and Booser, 2003).

Table 2.9 Representative oxidation life of mineral oils under ideal conditions

Oil type	k_1 in Equation (2.30)	Maximum temperature for indicated life (°C)			
		1 h	100 h	1 000 h	10 000 h
Uninhibited	−10.64	173	103	75	51
EP gear lubricant	−10.31	187	113	84	59
Hydraulic	−8.76	269	168	131	99
Turbine	−8.45	289	182	142	106
Heavily refined, hydrocracked	−8.05	317	200	157	121

Source: Booser, 1997c. Reproduced with permission of Taylor & Francis.

Table 2.9 gives typical values for k_1 for industrial mineral oil lubricants when subjected to agitation by air under laboratory conditions without the presence of catalysts or contamination. Oil types include (1) uninhibited oils such as those used in a once-through system, (2) EP gear oils, (3) conventional industrial hydraulic oils, (4) premium turbine-type oils used for long life in turbines, compressors, and electric motors, and (5) the longest-life Group II and Group III oils available following refining by severe extraction or hydrocracking. The temperatures calculated from Equation (2.30) for several oxidation life periods are also given in Table 2.9. Since similar products in each category vary widely, individual suppliers should be contacted for experience with their products.

Representative lives of inhibited synthetic oils of various types are compared with that of mineral oil in Figure 2.11 when in contact with steel in an air environment (Beerbower, 1982). While synthetics demonstrate longer life than mineral oils in this comparison, the added influence of hydrolysis will lead to shorter life for some phosphates, silicates, and esters even when exposed simply to atmospheric humidity. Where their unique temperature range or fire resistance is needed, special additives, desiccated air, and filtration to remove hydrolysis products may be necessary.

Lubrication application factors may reduce these life values. The combined effect of water plus iron and copper surfaces, for instance, reduces oxidation life by a factor of about 6 in the ASTM D943 Turbine Oil Stability Test. A reduction factor of about 3 seems typical in electric motors, hydraulic systems, steam turbines, and compressors, while a factor of 10 is involved in some land-based heavy-duty gas turbines to cover the added degradation from hot spots, oxidation inhibitor evaporation, and other degradation effects. For demanding applications with an oxidation life one-tenth that of uncatalyzed oxidation, limiting temperatures for a specified life in Table 2.9 should correspondingly be reduced 33 °C when using a life factor of 2 per 10 °C.

With several temperature zones in a machine or a lubrication system, the temperature in segment n can be considered to cause deterioration at its rate $1/L_n$ in its oil volume C_n. Summing the effect of all the segments gives the following overall deterioration rate for all of the oil in the system:

$$\frac{C}{L} = \frac{C_n}{L_n} + \frac{C_{n+1}}{L_{n+1}} + \frac{C_{n+2}}{L_{n+2}} + \cdots \tag{2.31}$$

Example 2.2 Compare the deterioration rate for 80 liters (20 gal) of turbine oil in a bearing assembly at 138 °C (280 °F) with that for 8000 liters in the main reservoir at 71 °C.

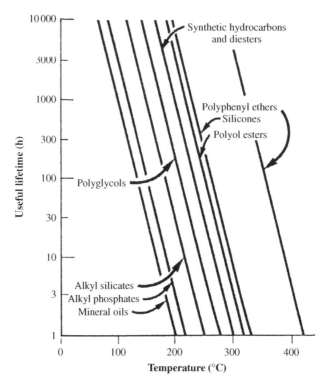

Figure 2.11 Life expectancy of inhibited lubricating oils in air. *Source*: Beerbower, 1982. Reproduced
with permission of STLE.

With $k_1 = -8.45$ from Table 2.9, giving oxidation life $L = 1280\,h$ from Equation (2.30), its
deterioration factor C/L becomes $80/1280 = 0.063$; by contrast, 8000 liters in the main reservoir at 71 °C (160 °F) with $L = 228\,000\,h$ give a C/L factor of only 0.035. Thus, nearly twice
as much loss of oil oxidation life (0.065/0.035) would be expected in the bearing assembly as
in the main reservoir. Evaporation of a low molecular weight oxidation inhibitor such as diter-
tiarybutyl paracresol (DBPC) would further accelerate deterioration of oil in the hot bearing
assembly.

2.9 Greases

Grease is lubricating oil thickened with a gelling agent such as a metallic soap or a nonmelting
powder. For design simplicity, decreased sealing requirements, and low maintenance, greases
are given first consideration for lubricating ball and roller bearings in electric motors, aircraft
accessories, household appliances, machine tools, automotive wheel bearings, instruments,
and railroad and construction equipment. Greases are also a common choice for slow-speed
sliding applications and small gear units. In general, the thickener maintains the grease in
contact with the machine elements so that the oil component can form a fluid lubricating film
(Booser and Khonsari, 2007, 2013b).

Table 2.10   Temperature range for synthetic greases

Grease type	Maximum temperature for 1000 h life (°C)	Lowest temperature for 1000 g/cm torque in 204 bearing (°C)
Petroleum	145	−28
Diester	125	−56
Polyester	160	−46
Synthetic hydrocarbon	145	−40
Conventional silicone	170	−35
Special silicone	230	−73
Perfluoroalkyl polyether	260	−35
Polyphenyl ether	280	10

Source: Booser, 1995.

Oils in Greases

Petroleum mineral oils are used in 98–99% of the grease produced, and these oils are commonly in the SAE 20–30 viscosity range, with 100–130 cSt viscosity at 40 °C. Oils in this viscosity range provide low volatility for long life at elevated temperatures together with sufficiently low torque for use down to subzero temperatures.

Higher-viscosity oils up to the 450–650 cSt range at 40 °C are employed for some high-temperature greases and for compounding with extreme pressure additives in greases for high-contact stresses at relatively low speeds. Less viscous oils ranging down to about 25 cSt at 40 °C are employed in special greases for low temperatures and for low torque in some high-speed equipment. Maximum tolerable oil viscosity in a grease at winter temperatures is about 100 000 cSt for starting medium-sized equipment using ball bearings.

Synthetic oils are employed in about 1–2% of current greases, where their higher cost is justified by unusual temperature conditions or other demands that cannot be met with mineral oil greases. Severe operating and temperature requirements have led to use of a broad range of synthetic greases for military applications. While synthetic hydrocarbons (PAOs) and diesters appear to enjoy the greatest volume production, almost all of the synthetic oils can be used. Comparisons of their typical temperature limits are given in Table 2.10 (Booser, 1995).

Thickeners

Gelling agents used include fatty acid soaps of lithium, calcium, aluminum, and sodium in concentrations of 6–20 wt%. Lithium soaps predominate, with use in about 75% of grease production. Fatty acids employed are usually oleic, palmitic, stearic, and other carboxylic acids obtained from tallow, hydrogenated fish oil, and castor oil. The relatively low upper temperature limit of 65–80 °C with traditional simple-soap calcium and aluminum greases can be raised to the 120–125 °C range with new complex soaps. Calcium complex soaps, for instance, are prepared by reacting both a high molecular weight fatty acid such as stearic and a low molecular weight acetic acid with calcium hydroxide dispersed in mineral oil.

Fine inorganic powders, such as bentonite clay, can be simply mixed with mineral oils to provide inexpensive, nonmelting greases with excellent water resistance for long life in automotive, industrial, and agricultural equipment at temperatures up to about 140 °C. Polyurea

Table 2.11 Consistency classification of the National Lubricating Grease Institute (NLGI)

Grade number	Worked penetration (0.1mm)	Approximate yield strength (p_a)	Approximate self-supporting height (cm) (inches)	Description
000	445–475			Very fluid
00	400–430	90		Fluid
0	355–385	130	1.30 (0.5)	Semifluid
1	310–340	180	1.80 (0.7)	Very soft
2	265–295	300	3.00 (1.2)	Soft
3	220–250	560	5.60 (2.2)	Semifirm
4	175–205	1 300	13.0 (5.2)	Firm
5	130–160	3 800	38.0 (15.0)	Very firm
6	85–115			Hard

nonmelting organic powders are used in premium petroleum greases for applications ranging up to 150–170 °C. PTFE organic powder and various nonmelting powders are used with stable synthetic perfluoroalkyl polyether oils for service up to the 260–270 °C range. Polyphenyl ether greases are used from 10 to 315 °C in high vacuum diffusion pumps and for their radiation resistance.

Mechanical Properties

Greases vary in consistency with increasing thickener content from slurries of lubricating oils, which are fluid at room temperature, to hard brick types that are cut with a knife.

The most common measurement of grease consistency is the depth of penetration of a standard cone into the grease in tenths of a millimeter in 5 s at 25 °C after the grease is worked for 60 strokes with a perforated disk plunger (ASTM D217). Table 2.11 gives the consistency classification of the National Lubricating Grease Institute (NLGI) based on this worked penetration. The approximate yield stress τ_0 to be expected for each NLGI penetration is also listed. A yield value of 98 p_a indicates that a grease layer 1 cm high (and with a density of $1.0 \, g/cm^3$) will just withstand the force of gravity and not slump under its own weight. Yield stress usually decreases and slumping tendency increases with increasing temperature.

For analytical use, greases are often considered as Bingham solids (see Figure 2.7), with a yield stress τ_0 being required to initiate flow and a linear increase in shear stress τ with shear rate $\dot{\gamma}$ above the yield point according to the following relation (Klamann, 1984):

$$\tau = \tau_0 + a\dot{\gamma} \tag{2.32}$$

where a represents 'plastic' viscosity. For a given grease type, the significance of the Bingham equation usually decreases as the NLGI consistency grade drops from 6 to 000 in Table 2.11. Greases of the 000 class are fluid and so similar to the base oil that the yield stress τ_0 term disappears and Newtonian viscous flow behavior predominates.

The following relation, referred to as the *Sisko equation*, has been used for shear rates in the range from 10^{-2} to $10^4 \, s^{-1}$:

$$\tau = \tau_0 \, a\dot{\gamma} + b\dot{\gamma}^n \tag{2.33}$$

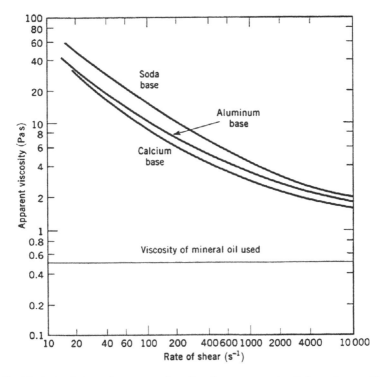

Figure 2.12   Typical relation of apparent viscosity of grease to rate of shear. *Source*: Boner, 1954. Reproduced with permission of Reinhold Publishing Corp.

where b and n relate the pseudoplastic behavior involved. When $n = 1$, the Bingham relation holds as increasing shear rate gives a linear increase in shear stress, just as in Equation (2.32). When $n < 1$, shear thinning is encountered as shear stress falls below the linear Bingham relation in Figure 2.7 for a rising shear rate. When $n > 1$, shear thickening occurs, as characterized by the power law relationship in the last term of Equation (2.33).

Unfortunately, great caution is needed in applying any formal relations to grease behavior. Yield stress and the other factors depend heavily on undefined patterns in temperature, the nature and state of the grease thickener, and the history of involvement of the grease in mechanical operating conditions.

General flow properties of greases are illustrated in Figure 2.12. Apparent viscosity drops rapidly as the shear rate increases to about $1000\,\text{s}^{-1}$. For very high shear rates such as those experienced in most rolling element bearings, grease thins further to give essentially Newtonian behavior with an apparent viscosity of the order of one and a half to three times that of the base oil in the grease.

NLGI Grade 2 greases are the most widely used. Their 'stiffness' is adequate to avoid the slumping and churning action in most ball bearings for shaft diameters up to about 75 mm, and yet they are soft and oily enough to provide lubrication needs of most bearings. Softer greases (down to Grade 000) are used for better feeding characteristics in some multiple-row roller bearings and gear mechanisms. Stiffer Grade 3 greases are used in many prepacked ball bearings where a Grade 2 might be broken down by its close contact with the churning action of the rotating balls and their separator. Blocks of hard brick greases are inserted directly in the sleeve-bearing box of paper mills.

Additives in greases usually include oxidation and rust inhibitors. Antiwear and extreme pressure additives, such as those of sulfur and phosphorus types, are included for heavy loads at low speeds where elastohydrodynamic action is insufficient to provide a full separating film between mating surfaces (see Booser and Khonsari (2013b) as well as Chapter 15 for other details relating to grease life and performance factors).

2.10 Solid Lubricants

Solid lubricants provide thin films of a solid between two moving surfaces to reduce friction and wear, usually for high temperatures, aerospace, vacuum, nuclear radiation, and other environments not tolerated by conventional oils and greases. Of the wide range of solid lubricants, the more common types listed in Table 2.12 include inorganic compounds and organic polymers, both commonly used in a bonded coating on a metal substrate, plus chemical conversion coatings and metal films (Lipp, 1976; Gresham, 1997).

Most important of the inorganic compounds listed in Table 2.12 are layer-lattice solids in which atomic bonding on an individual surface is by strong covalent or ionic forces and that between layers is by relatively weak van der Waals forces. Graphite is widely used as a dry powder or dispersed in water, mineral oil, or solvents for use in metalworking operations, wire drawing, high-temperature conveyors, and a variety of industrial applications. Since adsorbed water normally plays a decisive role in graphite lubrication, high friction is encountered in vacuum and may occur in air with water desorption above 100 °C. Its lubricating ability can be restored, however, by adding organic materials, cadmium oxide, or molybdenum disulfide to the graphite, or by impregnating the graphite structure with an organic material. Oxidation usually sets an upper limit of about 550 °C for graphite use (Lancaster, 1983).

Molybdenum disulfide has increasingly displaced graphite based on its independence of a need for water vapor, superior load capacity, and more consistent properties (Lancaster, 1983). It oxidizes above 400 °C to the trioxide, which may be abrasive.

Above 550 °C wear can often be minimized by the oxide layer formed on nickel-base and cobalt-base superalloys. Fluorides such as calcium fluoride, barium fluoride, lithium fluoride, and magnesium fluoride can be used as ceramic-bonded coatings with fusion bonding, or as components of plasma-sprayed composite coatings (Sliney, 1992).

Soft inorganic powders without a layer-lattice structure are also used as solid lubricants: lime as a carrier in wire drawing, talc and bentonite as grease fillers in cable pulling, zinc oxide in greases for high loads, and milk of magnesia for bolts. Toxicity has reduced the once common use of basic white lead and lead carbonate.

Various polymers provide self-lubricating properties when used as thin films, as bearing materials, and as binders for lamellar solids (Lancaster, 1983; Blanchet, 1997; Gresham, 1997). Of these polymers, PTFE is outstanding for use from −200 to 250 °C in providing a friction coefficient in the 0.03–0.1 range. Coatings of 25 μm and greater in thickness are typically applied in powder or dispersion form over an etched metal surface or over a chemical conversion coating such as a phosphate on ferrous metals and zinc; an anodized coating on aluminum, magnesium, or zinc; a passivated surface on corrosion-resistant steels; black oxide on copper and iron; or chromate on various metals (Gresham, 1997).

Table 2.12 Common solid lubricants

Material	Acceptable usage temperature (°C)				Average friction coefficient, f		Remarks
	Minimum		Maximum				
	In air	In N_2 or vacuum	In air	In N_2 or vacuum	In air	In N_2 or vacuum	
Molybdenum disulfide, MoS_2	−240	−240	370	820	0.10–0.25	0.05–0.10	Low f, carries high load, good overall lubricant, can promote metal corrosion
Graphite	−240	–	540	Unstable in vacuum	0.10–0.30	0.02–0.45	Low f and high load capacity in air, high f and wear in vacuum, conducts electricity
PTFE	−70	−70	290	290	0.02–0.15	0.02–0.15	Lowest f of solid lubricants, load capacity moderate and decreases at elevated temperature
Calcium fluoride–barium fluoride eutectic, Ca_2–BaF_2	430	430	820	820	0.10–0.25 above 540 °C 0.25–0.40 below 540 °C	Same as in air	Can be used at higher temperature than other solid lubricants, high f below 540 °C

Source: Kennedy et al., 1998. Reproduced with permission of Taylor & Francis.

Table 2.13 Properties of soft metals

Metal	Mohs' hardness	Melting point (°C)
Gallium		30
Indium	1	155
Thallium	1.2	304
Lead	1.5	328
Tin	1.8	232
Gold	2.5	1063
Silver	2.5–3	961

Source: Peterson *et al.*, 1960. Reproduced with permission of STLE.

Films of the soft metals listed in Table 2.13 are often used as solid lubricants for their low shear strength, easy bonding to metal substrates, and high thermal conductivity. Electroplated silver and copper are used as bolt lubricants, as are pastes incorporating powders of nickel, silver, copper, and lead. Tin, zinc, copper, and silver coatings are used as lubricants in metalworking (Schey, 1983). Silver films are used in sliding and rolling contacts in vacuum and at high temperatures since silver is unique in forming no alloys with steel and is soft at high temperatures. Gallium has been effective as a coating in vacuum.

2.11 Lubricant Supply Methods

Lubrication supplies can generally be grouped into three classifications: (1) self-contained units for supplying lubrication to individual bearings; (2) circulating systems dedicated to a single piece of equipment such as a motor, turbine, or compressor; and (3) centralized systems in manufacturing plants, paper mills, and metalworking. Upper speed limits for common feed arrangements are indicated in Figure 2.13 (Wilcock and Booser, 1987).

Self-Contained Units

Simple devices for individual bearings usually use simple mechanical recirculation or bath immersion. Cooling must also be provided if frictional power losses are high.

Capillary action within a wick, in a grease, or in a porous metal bearing is common for circulating oil in small devices. Oil lifted from a small reservoir is used to feed bearings in small machines, timer motors, and electrical instruments. The height h to which oil will rise between closely spaced plates or in capillary tubes is given by

$$h = \frac{2\sigma \cos\theta}{r\rho g} \tag{2.34}$$

where a typical value for surface tension σ of a mineral oil is 30 dynes/cm (mN/m), oil density ρ is commonly about 0.85 g/cm^3, and acceleration due to gravity is 981 cm/s. Because oils wet most surfaces readily, the cosine of contact angle θ is usually unity. For a capillary radius

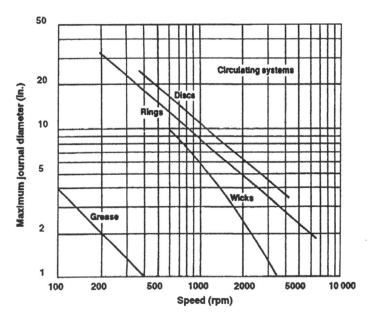

Figure 2.13   Size–speed limits for journal bearing lubrication methods. *Source*: Wilcock and Booser, 1987. Reproduced with permission of John Wiley & Sons.

r of 0.013 cm (0.005 in.) or for 0.013 cm spacing between two parallel plates, oil can then be expected to rise:

$$h = \frac{2(30)(1)}{0.013(0.85)(981)} = 5.5 \, \text{cm}$$

Capillary action in wick lubrication is used in millions of fractional horsepower motors. While commonly used for lifting oil up to about 2 in., raising oil up to 5 in. or more is common for railway journal bearings. Application considerations and oil delivery calculations for wick-oiled bearings are presented in Chapter 12.

Oil rings, disks, chains, and rotating machine elements can be brought into use for sizable oil delivery rates in many medium-sized and large machines at low to medium speeds. Motion of these delivery surfaces lifts oil by a combination of inertia and surface tension effects. Chapter 12 gives procedures for estimating journal bearing performance under the usually starved conditions encountered as these devices commonly deliver somewhat less oil than needed for a full hydrodynamic bearing oil supply.

Circulating Oil Systems

For greater oil demands in larger and high-speed equipment, a circulating system is brought into use with an oil reservoir, piping, a cooler, and filters. Typical characteristics of systems for use with industrial equipment are given in Table 2.14 (Kennedy *et al.*, 1998). Typical arrangement of components is indicated in Figure 2.14, with oil from a reservoir being pumped from the reservoir through a filter, cooler, and piping to the lubricated machine components before draining back to the reservoir.

Table 2.14 Typical circulating oil systems

Application	Duty	Oil viscosity at 40 °C (cSt)	Oil feed (gpm)	Pump type	Reservoir dwell time (min)	Filter type	Filter rating (µm)
Electrical machinery	Bearings	32–68	2	Gear	5	Strainer	50
General	Bearings	68	10	Gear	8	Dual cartridge	100
Paper mill dryer section	Bearings, gears	150–220	20	Gear	40	Dual cartridge	120
Steel mill	Bearings	150–460	170	Gear	30	Dual cartridge	150
	Gears	68–680	240	Gear	20	Dual cartridge	
Heavy duty gas turbines	Bearings, controls	32	600	Gear	5	Pleated paper	5
Steam turbine-generators	Bearings	32	1000	Centrifugal	5	Bypass 15%/h	10

Source: Kennedy *et al.*, 1998. Reproduced with permission of Taylor & Francis.

The system lubricant volume to be held in the reservoir represents the total feed to all bearings, gears, controls, and other machine elements multiplied by the oil dwell time. For the 1000 gpm oil feed to bearings in a turbine generator, as given in Table 2.14, for instance, a 5000 gal oil supply is needed for a 5 min dwell to allow for separation of entrained air and contaminants. An added 1 min of flow rate or equivalent added reservoir height is usually added to accommodate thermal expansion, foam, and air venting. Longer dwell time is required for separation of water and more extensive contaminants in some paper and steel mill systems. With restricted space and weight in aircraft, marine, and automobile engines, the reservoir may be simply the machine sump with less than a 1 min oil supply.

Maximum tolerable viscosity above which oil stiffness blocks adequate oil flow, lubrication, and hydraulic functions in a machine is given in Table 2.15. For example, the size of piping and other components for an industrial turbine are typically designed to accommodate oil viscosities up to about 175 cSt (~ 145 centipoise) at low temperatures. From the viscosity–temperature plots in Figure 2.3, this corresponds to a 45 °F low-temperature limit when using an industrial ISO VG-32 grade circulating oil.

The pipe diameter must be large enough to avoid an undue pressure drop in feed lines, to avoid backup in drain lines at the minimum oil temperature, and to avoid cavitation in pump suction lines. Feed lines usually operate at a velocity of 5–10 ft/s, drain lines 0.5–1.0 ft/s, and pump suction lines 3–5 ft/s (Khonsari and Booser, 2006).

Example 2.3 Establish the lubrication system requirements for a machine to be fed 200 gpm of ISO VG-32 light turbine oil at 115 °F and 30 psig to its bearings located 22 ft above the reservoir.

For 5 min dwell time, the lubricating oil charge becomes 5 × 200 gal. With additional free space for 1 min of flow (200 gal) above the static oil level, total reservoir size becomes 1200 gal.

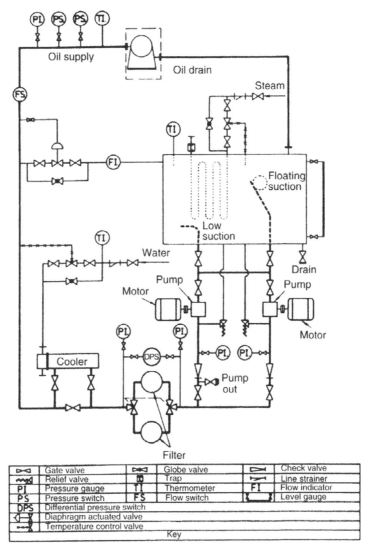

Figure 2.14   Typical circulating oil system arrangement. *Source*: Twidale and Williams, 1984. Reproduced with permission of Taylor & Francis.

Selecting a 3 in. inside diameter feed pipe ($D = 3$) from Figure 2.15 gives 9.1 ft/s (109 in./s) flow velocity V. With kinematic viscosity v of VG-32 oil at 115 °F being 26 cSt ($26/645.2 = 0.0403$ in.^{2}/s), the feed line flow is characterized by the following dimensionless Reynolds number Re:

$$Re = DV/v = \frac{3 \times 109}{0.0403} = 8100 \qquad (2.35)$$

Table 2.15 Representative viscosity limits for low-temperature operation

Type of service	Low-temperature viscosity limit (cSt)
Heavy-duty turbine	175
Ring oiling, motors and line shafts	1 000
Hydraulic pumps	1 300
Industrial electric motors	2 000
Small gear pumps	2 000
Automobile engines	3 000
Jet aircraft engines	13 000
Industrial gearboxes	50 000

Source: Khonsari and Booser, 2007. Reproduced with permission of Machinery Lubrication.

Since the Reynolds number is above 2500, flow is turbulent and the psi pressure drop P_ℓ in the feed line is given by

$$P_\ell = \frac{0.069\rho Q^2 L}{Re^{0.2}D^5}$$

$$= \frac{0.069(0.037)(200)^2(74)}{(8100^{0.2})(3^5)} = 5.14\,\text{psi} \tag{2.36}$$

where ρ is oil density (lb/in.^{3}), Q is oil flow in gpm, L is equivalent line length (ft), and D is inside pipe diameter (in.). Pipe length is 40 ft. Equivalent length of three elbows and a tee fitting from Table 2.16 is $3(25) + 60 = 135$ pipe diameters $= 34$ ft. This gives the equivalent line length L of $40 + 34 = 74$ ft.

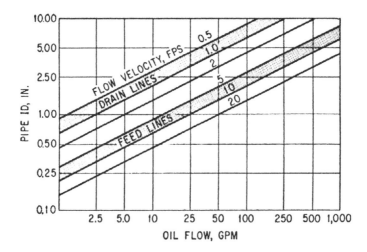

Figure 2.15 Typical oil line sizes for circulating oil systems. *Source*: Booser and Smeaton, 1968. Reproduced with permission of McGraw-Hill.

Table 2.16   Equivalent length of pipe fittings

Fitting/valve	Equivalent length in pipe diameters
45° Elbow	15
90° Elbow	25
Tee	60
Gate valve, open	6
Globe valve, open	300

Source: Twidale and Williams, 1984. Reproduced with permission of Taylor & Francis.

If the Reynolds number had been below 2500, the flow pressure drop P_l (psi) would have been given by

$$P_\ell = \frac{Q_{\rho v}}{132D^4} \tag{2.37}$$

Output pressure P (psi) required from the pump becomes

$$P = (P_2 - P_0) + \rho \Delta h + P_f \tag{2.38}$$

with the delivery pressure P_2 of 30 psi (commonly in the 15–40 psi range), atmospheric pressure in the reservoir for $P_0 = 0$, delivery height h of 22 ft (264 in.) above the reservoir oil level, and flow friction of 4.3 psi in the line +10 psi cooler pressure drop +10 psi filter drop for a total $P_f = 24.3$ psi:

$$P = 30 + 0.0307(264) + 24.3 = 62.4 \, \text{psi}$$

Pump driving horsepower E_p needed to deliver $Q = 200$ gpm becomes

$$E_P = \frac{QP}{1714\eta_p} = \frac{200 \times 62.4}{1714 \times 0.7} = 10.4 \, \text{hp} \tag{2.39}$$

where pumping efficiency η_p is taken as 0.7 and is commonly in the 0.6–0.8 range.

Feed piping to bearings is often sized to provide oil velocities of 5–10 ft/s to avoid undue pressure drop. An orifice normally drops the 25–30 psi line pressure to about 10–15 psi before oil is fed to individual bearings. Drain piping is sized for 0.5–1 ft/s flow velocity. Minimum slope s for running with a drain line half full is given by the following with viscosity v (cSt) taken at the minimum operating temperature:

$$s = \frac{0.00125Qv}{D^4} \tag{2.40}$$

A minimum slope of 1 in 40 toward the reservoir is common for VG-32 oil. A greater slope of 1 in 10 would be appropriate for more viscous oils in steel and paper mills. Running with the drain line half full gives space for escape of foam and air entrained and pulled along by the draining oil.

A filter using a medium such as pleated paper is commonly located just downstream of the feed pump to remove particulates from the warm oil leaving the reservoir. Often two units are used in parallel to allow servicing of one while continuing to use the other. Typical filter ratings are given in Table 2.14 in microns (1 μm = 1 × 10⁻⁶ m = 0.00004 in.). A cooler commonly follows the filter to lower the oil temperature by 20–40 °F (10–20 °C) before passing the oil in the 115–130 °F (45–55 °C) range to bearings and other machine elements.

Centralized Lubrication Systems

Limitations in using self-contained lubrication or a circulating system for individual locations have led to use of centralized distribution systems for production line equipment, steel and paper mills, and construction and mining machines (Simpson, 1997). A central reservoir combined with a pumping or pressurized air system distributes metered oil flow, oil mist, or timed pulses of oil or grease. Distribution is commonly through a single pressurized line. Flow-resistance fittings apportion lubricant to each delivery point. Dual feed lines are used for long distances, as in a transfer machine, with double-acting metering pistons being actuated by pressurized feed from one of the two lines. A reversing valve vents the first line back to the reservoir and supplies feed pressure to the second line. Then lubricant is delivered with the pistons moving in the opposite direction.

Oil mist orifice metering systems are used for supplying limited oil feed rates for demanding applications in petrochemical plants and in steel and paper mills (Khonsari and Booser, 2005; Rieber, 1997). 'Dry' mist involving oil particles up to about 6–7 μm in size can be piped for distances up to 300 ft. At delivery points the dry mist is converted to wet mist by agglomerating oil particles with classifier fittings to form larger drops, or wet sprays, which then run down adjacent surfaces. The air stream must be kept in the laminar regime below about 24 ft/s since turbulence will cause oil particles to impact the pipe wall and be removed from the air stream before reaching their delivery points. The interested reader is referred to Bloch (1998) for a detailed description of oil mist lubrication systems.

Problems

2.1 A low-viscosity mineral oil has viscosities of 22 cSt at 40 °C and 4.3 cSt at 100 °C. Addition of 3% of a polymer VI improver is expected to raise the viscosity by 40% at both temperatures. Calculate the viscosity in centistokes, reyns, and pascal seconds for both the original and additive-treated oils at −18 °C (0 °F) and 120 °C (248 °F). Estimate the VI of both.

2.2 Fifty gallons of VG-32 mineral turbine oil has been used at 80 psi air pressure and 160 °F in a compressor. Calculate the volume of air released from the oil when the unit is vented to atmospheric pressure. Assume 10% air solubility by volume.

2.3 The maximum viscosity oil that can be fed through the nozzles to a roller bearing in a jet engine was found to be 1000 cSt. What is the lowest temperature that can be tolerated with a VG-32 mineral turbine oil, a VG-32 synthetic hydrocarbon oil, or a Mil-L-7808 diester?

2.4 What is the percent change in viscosity per °C for a VG-32 mineral turbine oil at 10 °C (50 °F)? At 120 °C (248 °F)? What are the comparable values for an 85W-140 gear oil?

2.5 Calculate the 100 °C viscosity for Mil-L-7808 diester jet engine oil at 5000 psi pressure, 150 000 psi. Repeat for Mobil 1, 5W-30 synthetic engine oil. See Tables 14.1 and 14.2 for oil properties.

2.6 The lowest temperature at which an outdoor compressor can be started with an SAE 10W-30 oil is found to be 15 °C. Calculate the size of electrical heaters required to bring the 30 gal of oil from a winter temperature of 0 °C up to 15 °C in 15 min, assuming that half of the heat applied is lost to the surroundings.

References

American Petroleum Institute. 1993. *API Publication 1509*.

ASME. 1952. *Viscosity and Density of Over 40 Lubricating Fluids of Known Composition at Pressures to 150,000 psi and Temperatures to 425 °F*. ASME Special Publication. ASME, New York.

Bair, S. 2007. *High Pressure Rheology for Quantitative Elastohydrodynamics*, Tribology and Interface Engineering Series No. 54, Elsevier, Amsterdam.

Bair, S. and Khonsari, M.M. 1996. 'An EHD Inlet Zone Analysis Incorporating the Second Newtonian,' *ASME Journal of Tribology*, Vol. 118, pp. 341–343.

Bair, S. and Winer, W.O. 1990. 'The High Shear Stress Rheology of Liquid Lubricants at Pressures of 2 to 200 MPa,' *Journal of Tribology Transactions*, Vol. 112, pp. 246–252.

Beerbower, A. 1982. 'Environmental Capabilities of Liquid Lubricants,' *STLE Special Publication SP-15*. Society of Tribologists and Lubrication Engineers, Park Ridge, Ill., pp. 58–69.

Bhushan, B. 1999. *Principles and Applications of Tribology*. John Wiley & Sons, New York.

Blanchet, T.A. 1997. 'Friction, Wear and PV Limits of Polymers and Their Composites,' *Tribology Data Handbook*, E.R. Booser (ed.). CRC Press, Boca Raton, Fla, pp. 547–562.

Bloch, H.P. 1998. *Oil Mist Lubrication: Practical Applications*. Fairmont Press, Lilburn, Ga.

Boner, C.J. 1954. *Manufacture and Application of Lubricating Greases*. Reinhold Publishing Corp., New York.

Booser, E.R. 1995. 'Lubrication and Lubricants,' *Encyclopedia of Chemical Technology*, 4th edn., Vol. 15. John Wiley & Sons, New York, pp. 463–517.

Booser, E.R. 1997a. 'Solid Lubricants,' *Tribology Data Handbook*. CRC Press, Boca Raton, Fla, pp. 156–158.

Booser, E.R. 1997b, 'Comparison of Viscosity Classifications,' *Tribology Data Handbook*. CRC Press, Boca Raton, Fla, pp. 180–181.

Booser, E.R. 1997c. 'Life of Oils and Greases,' *Tribology Data Handbook*. CRC Press, Boca Raton, Fla, pp. 1018–1028.

Booser, E.R. 2001. 'The Push toward Lifetime Lubes,' *Machine Design*, 17 May, pp. 77–84.

Booser, E.R. and Khonsari, M.M. 2007. 'Systematically Selecting the Best Grease for Equipment Reliability,' *Machinery Lubrication*, Jan.–Feb., pp. 19–25.

Booser, E.R. and Khonsari, M.M. 2013a. 'Oil Life,' *Encyclopedia of Tribology*, Y.W. Chung and Q.J. Wang (eds.), Springer Science, New York, pp. 2475–24478.

Booser, E.R. and Khonsari, M.M. 2013b. 'Grease Life,' *Encyclopedia of Tribology*, Y.W. Chung and Q.J. Wang (eds.), Springer Science, New York, pp. 1555–1561.

Booser, E.R. and Smeaton, D.A. 1968. 'Circulating-Oil-System Design,' *Standard Handbook of Lubrication Engineering*. McGraw-Hill Book Co., New York, Section 25, pp. 23–44.

Booser, E.R. and Wilcock, D.F. 1987. 'Simplified Journal Bearing Design,' *Machine Design*, Vol. 59(9), pp. 101–107.

Chen, C-I. 1997. 'Typical Lubricating Oil Properties,' *Tribology Data Handbook*. CRC Press, Boca Raton, Fla, pp. 3–33.

Doolittle, A.K. 1951. 'Studies in Newtonian Flow II: The Dependence of the Viscosity of Liquids on Free-Space,' *Journal of Applied Physics*, Vol. 22, p. 1471.

Dowson, D. and Higginson, G.R. 1966. *Elastohydrodynamic Lubrication*, Pergamon Press, Oxford.

Exxon Corp. 1992. *Tables of Useful Information*, Lubetext DG 400. Exxon Corp., Houston, Tex.

Fein, R.S. 1997. 'High Pressure Viscosity and EHL Pressure–Viscosity Coefficients,' *Tribology Data Handbook*. CRC Press, Boca Raton, Fla, pp. 638–644.

Furby, N.W. 1973. 'Mineral Oils,' *Interdisciplinary Approach to Liquid Lubricant Technology*, P.M. Ku (ed.). NASA SP-318, NTIS N74-12219-12230, pp. 57–100.

Gresham, R.M. 1997. 'Bonded Solid Film Lubricants,' *Tribology Data Handbook*. CRC Press, Boca Raton, Fla, pp. 600–607.

Hamrock, B.J. 1994. *Fundamentals of Fluid Film Lubrication*. McGraw-Hill Book Co., New York.

Jang, J.Y., Khonsari, M.M., and Bair, S. 2007. 'On the Elast hydrodynamic Analysis of Shear-Thinning Fluids,' *Proceedings of Royal Society*, Series A, Vol. 463, pp. 3271–3290.

Jang, J.Y. and Khonsari, M.M. 2013a. 'Lubrication with Newtonian Fluids,' *Encyclopedia of Tribology*, Y.W. Chung and Q.J. Wang eds, Springer Science, New York, pp. 2142–2146.

Jang, J.Y. and Khonsari, M.M. 2103b. 'Lubrication with Non-Newtonian Fluids,' *Encyclopedia of Tribology*, Y.W. Chung and Q.J. Wang, eds, Springer Science, New York, pp. 2146–2151.

Jang, J.Y., Khonsari, M.M., and Bair, S. 2008. 'Correction Factor Formula to Predict the Central and Minimum Film Thickness,' *ASME Journal of Tribology*, Vol. 130, pp. 024501:1–4.

Kauffman, R.E. 1994. 'Rapid Determination of Remaining Useful Lubricant Life,' *Handbook of Lubrication and Tribology*, Vol. 3, E.R. Booser (ed.). CRC Press, Boca Raton, Fla, pp. 89–100.

Kennedy, F.E., Booser, E.R., and Wilcock, D.F. 2005. 'Tribology, Lubrication, and Bearing Design,' *The CRC Handbook of Mechanical Engineering*. 2nd edition, CRC Press, Boca Raton, Fla, Section 3, pp. 129–170.

Khonsari, M.M. and Booser, E.R. 2003. 'Predicting Lube Life–Heat and Contaminants are the Biggest Enemies of Bearing Grease and Oil,' *Machinery Lubrication*, Sept.–Oct. issue, pp. 50–56.

Khonsari, M.M. and Booser, E.R. 2004a. 'New Lubes Last Longer,' *Machinery Lubrication Magazine*, May–June, pp. 52–60.

Khonsari, M.M. and Booser, E.R. 2004b. 'Matching Lube Oil Systems to Machinery Requirements,' *Machinery Lubrication*, Nov.–Dec. issue, pp. 56–62.

Khonsari, M.M. and Booser, E.R. 2005. 'Guidelines for Oil Mist Lubrication,' *Machinery Lubrication*, Sept.–Oct. issue, pp. 58–63.

Khonsari, M.M. and Booser, E.R. 2006. 'Matching Oil Filtration to Machine Requirements,' *Machinery Lubrication*, Nov.–Dec. issue, pp. 17–22.

Khonsari, M.M. and Booser, E.R. 2007. 'Low Temperature and Viscosity Limits,' *Machinery Lubrication*, March–April issue, pp. 36–44.

Klamann, D. 1984. *Lubricants and Related Products*. Verlag Chemie, Deerfield Beach, Fla.

Klaus, E.E. and Tewksbury, E.J. 1984. 'Liquid Lubricants,' *Handbook of Lubrication*, Vol. II, E.R. Booser (ed.). CRC Press, Boca Raton, Fla, pp. 229–254.

Kramer, D.C., Lok, B.K., Krug, R.R., and Rosenbaum, J.M. 2003. 'The Advent of Modern Hydroprocessing–the Evolution of Base Oil Technology–Part 2,' *Machinery Lubrication Magazine*, May–June issue.

Krupka, I., Kumar, P. Bair, S. and Khonsari, M.M., and Hartl, M. 2010. 'The Effect of Load (Pressure) for Quantitative EHL Film Thickness,' *Tribology Letters*, Vol. 37, pp. 613–622.

Kumar, P. and Khonsari, M.M. 2008. 'EHL Circular Contact Film Thickness Correction Factor for Shear-Thinning Fluids,' *ASME Journal of Tribology*, Vol. 130, pp. 041506: 1–7.

Kumar, P. and Khonsari, M.M. 2009. 'On The Role Of Lubricant Rheology and Piezo-viscous Properties in Line and Point Contact EHL,' *Tribology International*, Vol. 42, pp. 1522–1530.

Lancaster, J.K. 1983. 'Solid Lubricants,' *Handbook of Lubrication*, Vol. 2. CRC Press, Boca Raton, Fla, pp. 269–290.

Lawate, S.S., Kai, K., and Huang, C. 1997. 'Vegetable Oils: Structure and Performance,' *Tribology Data Handbook*. CRC Press, Boca Raton, Fla, pp. 103–116.

Lipp, I.C. 1976. *Lubrication Engineering*, Vol. 32, pp. 574–584.

Masjedi, M. and Khonsari, 2014 'Theoretical and Experimental Investigation of Traction Coefficient in Line-Contact EHL of Rough Surfaces,' *Tribology International*, Vol. 70, pp. 179–189.

Paranjpe, R. 1992. 'Analysis of Non-Newtonian Effects in Dynamically Loaded Finite Journal Bearing Including Mass Conserving Cavitation,' *Journal of Tribology Transactions*, Vol. 114, pp. 736–744.

Peterson, M.B., Murray, S.F., and Florek, J.J. 1960. *STLE Transactions*, Vol. 3, pp. 225–234.

Raimoindi, A.A. and Szeri, A.Z. 1984. 'Journal and Thrust Bearings,' *Handbook of Lubrication*, Vol. 2. CRC Press, Boca Raton, Fla, pp. 413–462.

Rieber, S.C. 1997. 'Oil Mist Lubrication,' *Tribology Data Handbook*. CRC Press, Boca Raton, Fla, pp. 396–403.

Rizvi, Q.A. 1997. 'Additives: Chemistry and Testing,' *Tribology Data Handbook*. CRC Press, Boca Raton, Fla, pp. 117–137.

Roelands, C.J.A. 1966. 'Correlational Aspects of the Viscosity–Temperature–Pressure Relationship of Lubricating Oils'. Doctor's thesis, Technical High School of Delft, Druk, VRB, Groningen, the Netherlands.

Sachdev, V.K., Ugur, Y. and Jain, R.K. 1998. 'Equation of State of Poly(dimethylsiloxane) Melts,' *Journal of Polymer Science B*, Vol. 36, pp. 841–850.

Schey, J.A. 1983. *Tribology in Metalworking: Friction, Lubrication and Wear*. ASM International, Metals Park, Ohio.

Simpson, J.H., III. 1997. 'Centralized Lubrication of Industrial Machines,' *Tribology Data Handbook*. CRC Press, Boca Raton, Fla, pp. 385–395.

Sliney, H.F. 1992. 'Solid Lubricants,' *ASM Handbook*, Vol. 18. ASM International, Metals Park, Ohio, pp. 113–122.

Smith, A.N. 1986. *Turbine Lubricant Oxidation: Testing, Experience, and Prediction*. ASTM STP 916. ASTM, Philadelphia, Pa, pp. 1–26.

Song, H.S., Klaus, E.E., and Duda, J.L. 1991. 'Prediction of Bulk Moduli for Mineral Oil Based Lubricants, Polymer Solutions and Several Classes of Synthetic Fluids,' *Journal of Tribology*, Vol. 113, p. 675.

Twidale, A.J. and Williams, D.C.J. 1984. 'Circulating Oil Systems,' *Handbook of Lubrication*, Vol. 2. CRC Press, Boca Raton, Fla. pp. 395–409.

Wilcock, D.F. and Booser, E.R. 1957. *Bearing Design and Application*. McGraw-Hill, New York.

3

Surface Texture, Interaction of Surfaces and Wear

All surfaces are rough to some degree. Texture of surfaces involves variations from large-scale features down to those of microscopic size at crystal faces and dislocation steps. While interactions on a molecular scale are of vital importance in selecting bearing materials and in lubricant chemistry, covered in Chapters 2 and 4, this chapter covers surface texture primarily in relation to mechanical aspects of friction, lubrication, and wear in bearing and seal applications.

Surface roughness is a significant factor in planning and evaluating the needs of a new bearing or seal design; with dry and semilubricated bearings; in boundary lubrication; during starting, stopping, and slow-speed running of machinery; with elastohydrodynamic films; and with magnetic recording, gas bearings, and other situations involving very small lubricant film thicknesses. Although many fluid-film bearings covered in later chapters operate with thick fluid films where surface finish plays only a minor role, surface finish is still a common design parameter related to the minimum film thickness.

The first part of this chapter covers the pertinent parameters that are commonly used to characterize surface texture. It then outlines measuring techniques for surface texture. The second part of the chapter considers what happens when two surfaces come into intimate contact and introduces the basic features of contact mechanics. Finally, in the third part of the chapter, different types of wear commonly encountered in machinery are presented.

3.1 Geometric Characterization of Surfaces

Surfaces produced by machining and other shaping processes are generally complex. With surface roughness being the primary consideration, the following basic terms, illustrated in Figure 3.1, are commonly employed in characterizing surface texture:

- *Roughness*: Fine (short-wavelength) irregularities such as those produced by machining processes.

Applied Tribology: Bearing Design and Lubrication, Third Edition. Michael M. Khonsari and E. Richard Booser.
© 2017 John Wiley & Sons Ltd. Published 2017 by John Wiley & Sons Ltd.
Companion Website: www.wiley.com/go/Khonsari/Applied_Tribology_Bearing_Design_and_Lubrication_3rd_Edition

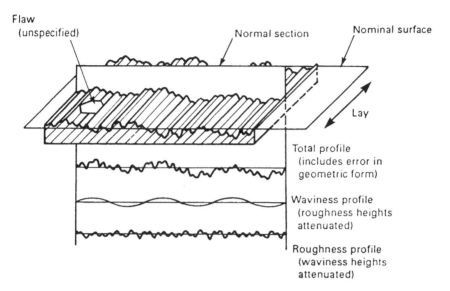

Figure 3.1 Basic characteristics of surface structure. *Source*: ASME B46.1-1995. Reproduced with permission of ASME.

- *Waviness*: Widely spaced (long-wavelength) surface irregularities resulting from manufacturing process problems such as cutting tool runout, deflections, uneven wear in tools, and workpiece misalignment. Distinction between waviness and longer-wavelength form error are not always made in practice.
- *Lay*: Direction of the predominant surface pattern determined by the direction of the machining process used. Turning, milling, drilling, grinding, and other cutting tool machining processes usually produce a surface that has lay: striations or peaks and valleys in the direction in which the cutting tool was drawn across the surface. Other processes such as casting, peening, and grit blasting which produce no characteristic direction are said to have a nondirectional, particulate, or protuberant lay.
- *Flaws*: Unintentional, unexpected, and unwanted interruptions in the topography such as scratches, gouges, pits, and cracks.

In applying surface profile measurements to finishing processes such as turning, grinding, polishing, and sand blasting, two types of statistical parameters are employed: a single surface parameter such as average roughness R_a or root mean square (rms) roughness R_q, and statistical functions such as power spectral density.

The wide dimensional range of roughness for conventional finishing extends from surfaces produced by rough planing and turning to fine lapping in which features may have an amplitude of a few tenths of a micrometer. With energy beam machining, the scale drops to atomic dimensions with peak-to-valley heights ranging to less than a few thousandths of a micrometer. Since height features in a profile are generally small compared with the horizontal spacing of crests and asperities, compressed graphs are used which exaggerate the vertical profile. As an example, Figure 3.2 compares a short length of the profile of a ground surface for which vertical magnifications are both 1/10 and 50 times the horizontal.

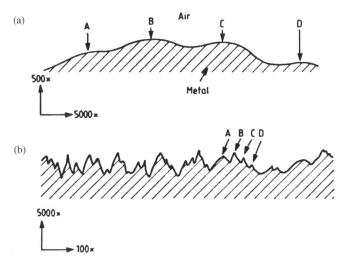

Figure 3.2   Profile distortion caused by making a usable chart length. *Source*: Whitehouse, 1994. Reproduced with permission of IOP Publication.

3.2 Surface Parameters

The following are the standard parameters of many that are used to describe surface roughness. Many firms have developed their own measures to match perceived requirements for specific applications. Fortunately, new computerized systems enable ready evaluation of complex parameters used in a variety of surface studies and industrial specifications.

Amplitude Parameters

Average Roughness (R_a)

The most common designation of roughness is the traditional roughness average R_a, illustrated in Figure 3.3. This simply gives the average deviation of the individual point measurements of heights and depths from the arithmetic mean elevation of the profile. Table 3.1 gives typical ranges of surface roughness R_a produced by common production methods. Roughness average was formerly known as *arithmetic average* (AA) and *centerline average* (CLA). Evaluation length L is often set to be five times the longest wavelength of profile fluctuations that may be measured.

Root-Mean-Square Roughness (R_q)

Closely related is the root-mean-square (rms) roughness R_q, which takes the square root of the sum of the squares of the individual deviations from the mean line within the sampling length (Fig. 3.3). By involving the square of height and depth terms, rms values are more sensitive to occasional highs and lows in a profile, which makes R_q a valuable supplement to R_a for fine finishes of honed surfaces and for optical components. For a simple sine wave, $R_q/R_a = 1.11$. This ratio becomes 1.25 for the Gaussian distribution of surface heights, as found with many

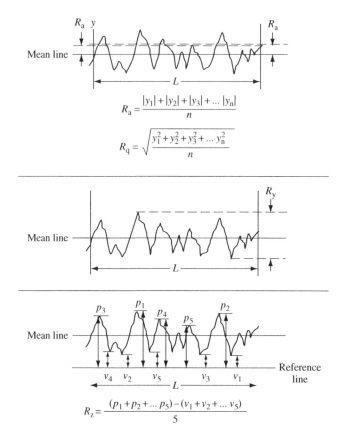

$$R_a = \frac{|y_1| + |y_2| + |y_3| + \dots |y_n|}{n}$$

$$R_q = \sqrt{\frac{y_1^2 + y_2^2 + y_3^2 + \dots y_n^2}{n}}$$

$$R_z = \frac{(p_1 + p_2 + \dots p_5) - (v_1 + v_2 + \dots v_5)}{5}$$

Figure 3.3 Commonly used surface amplitude characterizations. *Source*: Tabenkin, 1984. Reproduced with permission of John Wiley & Sons.

finishes, and 1.45 for honing (Dunaevsky *et al.*, 1997). In considering the mechanics of surface contacts and in statistical analyses, the symbol σ is also used for the rms surface roughness. If the reference line for establishing R_q is the same as the mean line of a surface profile, as expected, $\sigma = R_q$.

Peak-to-Valley Height

Two measures of extremes in the roughness profile are the 10-point height R_z given in Figure 3.3 and maximum peak-to-valley height R_y. The 10-point R_z is the average distance between the five highest peaks and the five deepest valleys within the sampling length, measured from a reference line parallel to the mean line but not crossing the roughness profile. This is the most repeatable of the extreme roughness values and is recognized by the ISO along with R_y. Maximum peak-to-valley height R_y is the maximum within the sampling length and is commonly about three times R_a. This most sensitive indicator of high peaks and deep scratches is used where extreme height of a roughness profile is important, for example precision tolerancing or wear of a soft surface by a hard one.

Table 3.1 Typical surface roughness values

	Arithmetic roughness average, R_a	
	μm	μin.
Production method		
Sand casting, flame cutting, hot rolling	12.5–25	500–1000
Sawing	3.2–25	125–1000
Planing, shaping, forging	1.6–12.5	63–500
Drilling, milling	1.6–6.3	63–250
Investment and cold mold casting	1.6–3.2	63–125
Laser, electrochemical	0.8–6.3	32–250
Broaching, reaming, cold rolling, drawing, extruding	0.8–3.2	32–125
Die casting	0.8–1.6	32–63
Boring, turning	0.4–3.2	16–125
Electrolytic grinding, burnishing	0.2–0.5	8–20
Grinding, honing[a]	0.1–1.2	4–50
Polishing, electropolishing, lapping[a]	0.08–0.6	3–25
Superfinishing[a]	0.025–0.2	1–8
Components		
Gears	0.25–10	10–400
Hydrodynamic oil-film bearings		
Journal bearings, thrust bearings	0.2–1.6	8–64
Journals, thrust runners	0.1–0.8	4–32
Rolling bearings		
Raceways	0.05–0.3	2–12
Rolling elements	0.025–0.12	1–5

[a]Less frequently finished in the range down to 0.012 μm 0.5 μin.
Source: Hamrock, 1994; ASME, 1995.

Spacing and Shape Parameters

Mean Peak Spacing (S_m)

As a wavelength parameter used to characterize the spacing of peaks and valleys, the mean peak spacing S_m is defined for a profile as the average spacing between two successive negative crossings of the mean line (Figure 3.4). These spacings often characterize the surfacing process, such as shot size in abrasive blasting, grit size of a grinding wheel, or feed of a cutting tool.

Skewness

Shape profile can be described by its asymmetry, or *skewness*, and the sharpness of its peaks, or *kurtosis*. Skewness is reflected in the shape profiles in Figure 3.5. All have the same roughness average R_a and wavelength S_m, but the shapes involved would be expected to provide very different performance in bearing and seal surfaces. The profile of (b), for instance, would represent a good bearing surface with some wedge action to help generate an oil film over the considerable load supporting area. Profile (a), on the other hand, would provide very little

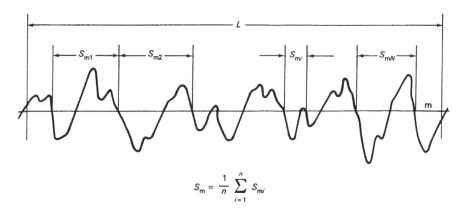

$$S_m = \frac{1}{n} \sum_{i=1}^{n} S_{mi}$$

Figure 3.4 Peak spacing parameter S_m for defining the wavelength. *Source*: Song and Vorburger, 1992. Reproduced with permission of ASM International.

bearing support area on its asperity peaks, and their sharp profile would tend to promote wear. Skewness parameter R_{sk}, the most important shape parameter in measuring symmetry, is defined as

$$R_{sk} = \frac{1}{nR_q^3} \sum_{i=1}^{n} y_i^3 \tag{3.1}$$

where $y(x)$ is the surface profile, as sampled by the set of n points y_i. According to this definition, profile (a) in Figure 3.5 has positive skewness, (b) is negative, and (c) and (d) are symmetrical around the mean line to give zero skewness. Negative skewness with $R_{sk} < -1.5$ often indicates a good bearing surface.

Peak Sharpness (Kurtosis)

The sharpness of the profile peaks, illustrated in Figure 3.6, is measured as the kurtosis parameter R_{ku}, defined as follows:

$$R_{ku} = \frac{1}{nR_q^4} \sum_{i=1}^{n} y_i^4 \tag{3.2}$$

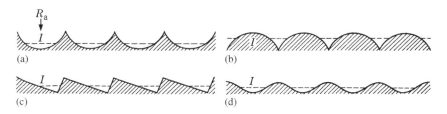

Figure 3.5 Profiles with the same roughness average R_a and wavelength but varying skewness. *Source*: Song and Vorburger, 1992. Reproduced with permission of ASM International.

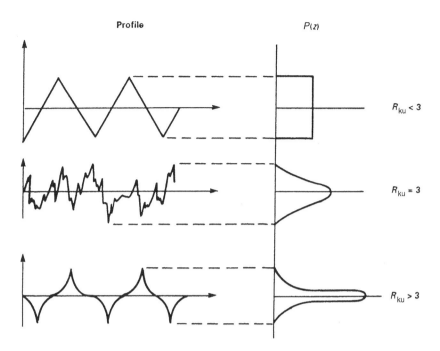

Figure 3.6 Profiles with different kurtosis values. *Source*: ASME B46.1-1995. Reproduced with permission of ASME.

The Gaussian (normal) distribution of surface profile peaks has a skewness value of zero and a kurtosis of 3.0. Kurtosis less than 3.0 indicates a broad, flat distribution curve, while a greater value corresponds to a more sharply peaked distribution. The following values of skewness and kurtosis are typical (ASME B46.1):

$$\text{Grinding} \quad R_{sk} = 0; \quad R_{ku} = 3$$
$$\text{Honing} \quad R_{sk} < 0; \quad R_{ku} > 3$$
$$\text{Turning, electro-discharge machining} \quad R_{sk} > 0; \quad R_{ku} < 3$$

Skewness and kurtosis values have been found to have special value in evaluating critical surfaces such as the honed internal surface of an engine cylinder wall. By imposing slow rotation in conjunction with axial oscillatory motion, the stone on the honing head floats in the cylinder bore under light load to give a distinctive surface texture (DeGarmo *et al.*, 1984).

Hybrid Parameters

Combining height and spacing information for complete definition of surface texture usually requires differentiation of the waveform. While digital methods have improved the ability to define derivatives of the surface profile, evaluations are often difficult.

The most common hybrid parameter is *average absolute slope*, as given by

$$\Delta_a = \left(\frac{1}{L}\right) \int_0^L \left|\frac{dy}{dx}\right| dx \tag{3.3}$$

where dy/dx is the absolute value of the instantaneous slope of the profile over the evaluation length L. As with profile heights, *root mean square slope* values which define curvatures of asperities are given by

$$\Delta_q = \left[\left(\frac{1}{L} \right) \int_0^L \left(\frac{dy}{dx} \right)^2 dx \right]^{1/2} \tag{3.4}$$

Ranges of values of surface parameters normally encountered with fluid-film lubrication are the following (Hamrock, 1994):

> Density of asperities, $10^2 - 10^6$ peaks/mm^2
> Asperity spacing, 1–75 μm
> Asperity slopes, 0° to 25° but mainly 5° to 10°
> Radii of peaks, mostly 10–30 μm

3.3 Measurement of Surface Texture

Surface profile is commonly measured in a line across the lay by a contacting stylus, which traverses the surface. Other methods use a noncontacting probe or optical interferometric fringes to make measurements either along a profile line or over an area (raster area methods) in point-by-point measurement of peaks and valleys. A summary of representative surface roughness measuring techniques is given in Table 3.2, and literature references are given in ASME B46.1-1995 and Bhushan (1999).

Contacting Methods

Principles involved in a stylus profilometer are illustrated in Figure 3.7. As the stylus is pulled steadily over a surface, its vertical rise and fall as it passes over surface peaks (asperities) and valleys is converted into an electrical analog signal. Modern profilometers pass this signal through a filter to separate long-wavelength (waviness) components from short-wavelength roughness. The signal is then amplified for further processing through a chart recorder and/or digitizing for further processing via a computer with appropriate software. While stylus instruments are normally applied in traversing a single line, a three-dimensional map of surface texture can be obtained with special fixturing to add a second axis of motion for repeated traces along closely spaced parallel traces over a surface.

Skid versus Skidless Instruments

Stylus instruments are usually of two types: skid or skidless. The least expensive systems use a rounded skid, which rides along the surface, very close to the stylus, to trace a reference line. The vertical position of the stylus is then measured with respect to the skid. Since surface waviness is filtered out to some extent by the skid and the skid averages over a range of peaks, the skid serves as a high-pass mechanical filter. Skidless profilometers use a precision reference surface built into the instrument itself. An undistorted profile is gathered, with the stylus itself being the only limiting factor. Flexure pivot mounting of the stylus in some commercial designs

Table 3.2 Representative surface measuring techniques

| Type of measurement | Resolution (μm^a) | | Dimensions | Comments |
	Lateral	Vertical		
Noncontacting methods				
Optical methods				
Optical microscopy	0.25–0.35	0.18–0.35	3D	Depends on optical system quality
Taper sectioning	0.25–0.35	0.18–0.35	2D	Destructive technique
Optical interferometry	0.5	0.001	2D	
Optical reflectance		0.002–0.003	1D, 3D	Very smooth surfaces, <100 nm
Electron microscopy				
Scanning electron microscopy (SEM)	0.01		3D	Requires vacuum
Transmission electron microscopy (TEM)	0.005	0.0025	3D	Requires vacuum
Atomic force microscopy	Atomic resolution		3D	No vacuum required
Contact methods				
Stylus profilometry	1.3–2.5	0.005–0.25	1D, 2D, 3D	Stylus size is ultimate limit on resolution
Replicas (negative impressions)	Surface details duplicated down to 2			Used when surface cannot be examined directly Between two conducting surfaces
Capacitance	1.5		1D	Between two conducting surfaces
Air gauging	0.1		1D	Flow between two surfaces

[a] $1\ \mu m = 1 \times 10^{-6}\ m = 39.4\ \mu in.$ $1\ \mu in. = 1 \times 10^{-6}\ in.$ $1\ nm = 1 \times 10^{-9}\ m.$
Source: Dunaevsky *et al.*, 1997. Reproduced with permission of Taylor & Francis.

has eliminated the need for any sliding reference surfaces, but the trace length is limited by the degree to which the flexure may be bent (Song and Vorburger, 1992).

Cutoff Lengths

Stylus instruments use electrical filters to set the spacing width, or *cutoff*, over which to respond to irregularities (ASME B46.1-1995). The common short-wavelength cutoff selected for roughness sampling lengths must be large enough to provide adequate information for evaluating roughness, yet small enough to avoid distorting roughness values with waviness evaluations. International standard cutoff values have been 0.08, 0.25, and 0.8 mm, with 0.8 mm to be used if no other value is specified. As the current American national standard, ASME B46.1-1995 gives cutoff lengths of 0.08, 0.25, 0.80, 2.5, and 8.0 mm.

Following preliminary evaluation of surface roughness, this cutoff length is usually set at about 10 times mean peak spacing S_m for periodic profiles, about eight times the roughness average R_a for nonperiodic profiles. Total evaluation length L (see Figure 3.3) is correspondingly set at five times this roughness cutoff sampling length. A separate long-wavelength cutoff will attenuate widely spaced undulations of the waviness profile, and waviness evaluation

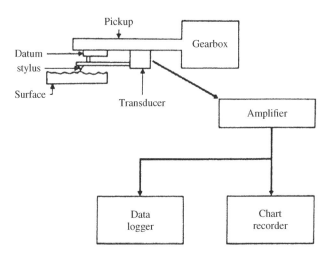

Figure 3.7   Principles involved in stylus profilometry. *Source*: Hutchings, 1992. Reproduced with permission of Taylor & Francis.

length should be several times the long-wavelength cutoff to achieve better statistics. In some cases, as for surfaces in actual contact with a mating surface, the largest convenient cutoff will be used to evaluate the overall contact area. For irregularities of small width, such as a surface subject to fatigue cracks, a short cutoff is used.

Stylus Sensitivity

The normal diamond stylus used is basically either conical, with a nominal tip radius of 2, 5, or 10 μm (with a corresponding maximum static measuring force of 0.0007, 0.004, or 0.016 N), or a truncated pyramid with a rectangular flat tip 2–4 μm on a side (ASME B46.1-1995). Since this tip size is relatively large compared with typical surface roughnesses, special chisel edges with tip radii as small as 0.1 μm are used to examine very fine surface details. For many industrial applications, the tip radius ranges up to 5 or 10 μm (200–400 μin.).

A major distorting effect is encountered in the difference between the horizontal and vertical magnifications of a stylus trace, as illustrated in Figure 3.2. Since the vertical extent of surface irregularities is much less than their horizontal extent, magnification in the vertical direction is generally 100–100 000, and in the horizontal direction 10–5000. The ratio of vertical to horizontal magnification will depend on the roughness of the surface but typically ranges from 10 to 5000.

As a stylus traverses a surface, it generates a profile which is smoother than the measured surface, with broadened peaks and narrowed valleys, because of the finite tip dimension. Lateral resolution is also a factor with stylus instruments. At their low end, wavelength measurements are limited by stylus width and sampling frequency. For long wavelengths, the distance traversed by the stylus and high-pass electrical filtering are limits. For industrial instruments, spatial wavelength bandwidth commonly extends approximately over 10–800 μm (400 to 0.3×10^4 μin.). Larger and smaller wavelengths are detected with reduced or no sensitivity (Song and Vorburger, 1992). Further error is encountered with very compliant and delicate surfaces by distortion or even damage to the surface by the diamond surface despite the light load on the stylus.

Ultimate resolution is provided by the atomic force microscope, which is similar to a contacting stylus instrument using a probe tip chemically etched, often to a radius of less than 100 nm. Electromechanical force between the sample surface and the probe tip produces a cantilever deflection, which is sensed by a laser beam or one of several other techniques to nanometer resolution. The surface roughness of tapes and discs is measured with a lateral resolution in the 1 μm range with a noncontacting optical profiler and down to 2 nm with an atomic force microscope (Bhushan, 1994).

Noncontacting Methods

Noncontacting optical methods avoid distortion and scratching of the surface, which might otherwise occur with a diamond stylus. Optical microscopy, light section microscopy, and optical interferometry enable examination and three-dimensional mapping of both very fine surface features and surfaces which are deflected or scratched by a stylus. When used in combination with an array of photodiodes in an image sensor, the surface contours can be digitized to generate a surface profile along a line on the surface, just like a stylus profilometer. Alternatively, a two-dimensional array of photodiodes could be used to record variation in height over a surface area. Electron microscopic methods offer better resolution and depth of field than optical methods because of the extremely short wavelength of the electron beams. Optical and electron microscopic devices are invaluable in demanding studies such as those of wear mechanisms and of fine surface features for tapes and discs in magnetic recording.

Pneumatic techniques using air-gauging devices provide simple, quick, portable assessment of surface roughness on the shop floor. These methods use a flat, finely finished measurement head which contains an orifice for feeding air. When the head is placed on a surface, the air flow rate gives a measure of the surface finish.

3.4 Measurement of Surface Flatness

The most accurate means for measuring surface flatness is based on the light bands interference readings emanating from a monochromatic light source, i.e. a light with a single wavelength, λ. To use this method, the surface must be capable of reflecting light, often requiring lapping for hard materials and polishing for soft materials. The technique the description of which follows is routinely used for measuring the surface flatness of mechanical seals covered in Chapter 16.

The instrument used for measuring flatness typically utilizes monochromatic light source from a tube filled with helium augmented with an appropriate interference filter. One full wavelength of yellow helium light source is 5876 Å (23.2 μin.) The increment of measure for one fringe is one half of the full wavelength, i.e. 11.6 μin. To measure the surface flatness, the part is placed on the instrument stage under the light source, as shown schematically in Figure 3.8 (Cliffone, 1996). An optical flat is then positioned on the surface to be checked. Where the optical flat touches the surface (point A), the light reflects 180° out of phase so that the incoming light and the reflected light waves 'interfere' and a *dark band* appears. Interference also occurs at all locations where the gap between the optical flat and surface is one or multiple of a whole wavelength (Point B). If the gap accommodates only half of the wavelength, then the

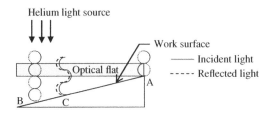

Figure 3.8 Exaggerated illustration of interference with a helium light source and an optical flat. *Source*: Courtesy of John Crane Inc.

reflected light is in phase with the incoming light and a *light band* appears (Point C). If helium light source is used, then typically one observes a series of light and dark bands with 11.6 μin. thickness measured from center to center of each set of dark bands. These bands can be used to measure deviation of the work surface from flat.

To interpret the degree of flatness, consider Figure 3.9. If the bands for a rectangular work piece are perfectly straight and parallel to one another, then the surface is said to be flat within 1 μ in (illustration (a)). In most cases, however, the light bands are curved ((*b*) and (*c*)). If the curvature is pointed away from the line of contact, then the part is convex; if the curvature

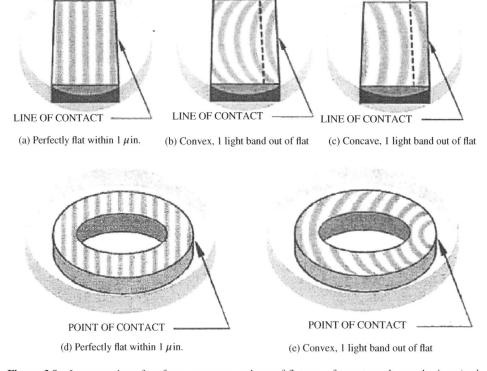

Figure 3.9 Interpretation of surfaces curvature and out of flatness of a rectangular work piece (*a*, *b*, and *c*) and a circular, seal-like work piece (*d* and *e*). *Source*: Cliffone 1996. Courtesy of John Crane Inc.

is pointed toward the line of contact, the part is concave. The lack of flatness is measured by counting the number of full bands crossed by an imaginary tangent line. If one full band is intersected, then the part is said to be one helium light band (11.6 μin.) out of flat. Similarly, if two full bands are intersected, then the part is two helium light bands (23.2 μin.) out of flat. Additional guidelines are available from seal manufacturers (see, for example, Ciffone, 1996).

For a circular, seal-like work piece, a similar procedure can be applied, except that the imaginary tangent line is considered with respect to a point of contact as illustrated in Figure 3.9 (d) and (e). Typically, most face seals are flat to within 2–3 helium light bands with a surface finish of 4–5 μin.

3.5 Statistical Descriptions

Various statistical measures have been employed to describe the overall geometric nature of surfaces. Complexity of statistical parameters ranges from high-order probability functions to match the diversity of surface features down to a minimum number of parameters for practical use.

A generalized description of the distribution of surface heights is provided by the *amplitude density function* (ADF). Figure 3.10 shows a typical height distribution, often similar to a normal probability curve. The value of ADF, $P(y)$, at any height y is proportional to the probability of finding a point on the surface at height y above the mean line. The quantity $P(y)dy$ is then the fraction of the surface profile which lies at heights between y and $y + dy$ above the mean line. The rms roughness R_q is the standard deviation of ADF (Hutchings, 1992).

A function related to the ADF with appeal for tribology studies is the *bearing ratio*, also known as the *air/metal ratio* or *material ratio*, as represented in an *Abbott-Firestone curve*. This is also called a *bearing area curve* (BAC), which represents a plane, smooth surface being lowered to the profile of the surface under investigation. When the plane first touches the surface at a point, the bearing ratio (contact length/total profile length) is zero. As the line representing the plane surface moves further down, its contact length with the surface profile increases and the corresponding bearing ratio increases. Finally, when the line reaches the deepest valley of the surface profile, the bearing ratio rises to 100%. Figure 3.10 shows a plot of bearing ratio. Differentiation of the bearing ratio curve gives the amplitude density function. Some surface analyzers provide a video display of height distributions and BAC.

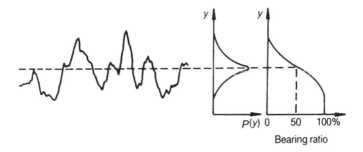

Figure 3.10 Amplitude density function (ADF) and bearing area curve (BAC) for a profile trace. *Source*: Hutchings, 1992. Reproduced with permission of Taylor & Francis.

3.6 Surface Texture Symbols

The standard symbols shown in Figure 3.11 have been adopted for use on drawings and specifications for controlling roughness, waviness, and lay (ANSI 14.36-1978, Rev. 1993). While metric units are regarded as standard, nonmetric units may be used which are consistent with other units used on drawings or documents. The basic check symbol shown in Figure 3.11(a) is modified as shown for specific requirements. As indicated in Figure 3.11(e), a horizontal extension is added at the top right corner of the check mark when surface values other than roughness average are specified.

Many finishing processes generate a directional pattern on the surface, or lay. The following are standard lay symbols:

Symbol	Meaning
=	Lay approximately parallel to the line representing the surface to which the symbol is applied
⊥	Lay approximately perpendicular to the line representing the surface to which the symbol is applied
X	Lay angular in both directions to the line representing the surface to which the symbol is applied
M	Lay multidirectional
C	Approximately circular relative to the center of the surface to which the symbol is applied
R	Approximately radial relative to the center of the surface to which the symbol is applied
P	No lay, e.g. pitted, protuberant, particulate, or porous

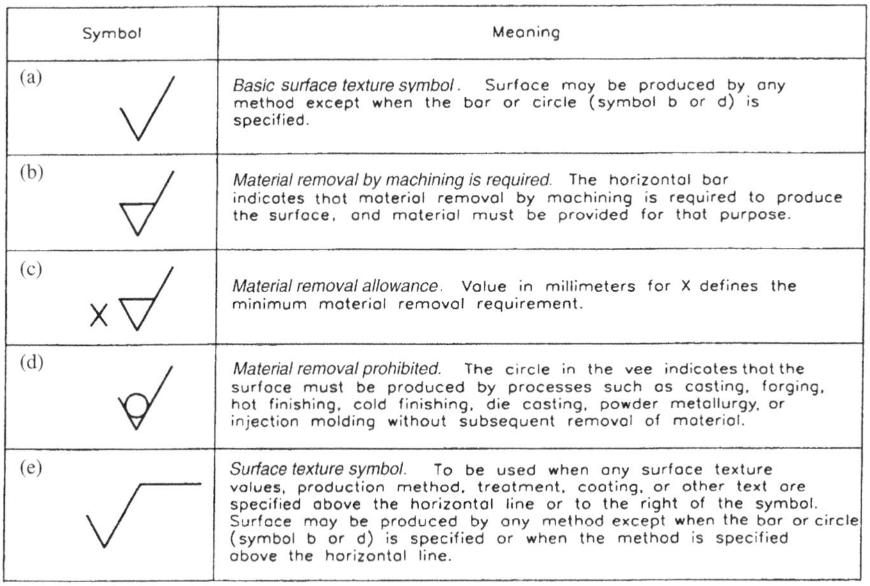

	Symbol	Meaning
(a)		*Basic surface texture symbol*. Surface may be produced by any method except when the bar or circle (symbol b or d) is specified.
(b)		*Material removal by machining is required*. The horizontal bar indicates that material removal by machining is required to produce the surface, and material must be provided for that purpose.
(c)		*Material removal allowance*. Value in millimeters for X defines the minimum material removal requirement.
(d)		*Material removal prohibited*. The circle in the vee indicates that the surface must be produced by processes such as casting, forging, hot finishing, cold finishing, die casting, powder metallurgy, or injection molding without subsequent removal of material.
(e)		*Surface texture symbol*. To be used when any surface texture values, production method, treatment, coating, or other text are specified above the horizontal line or to the right of the symbol. Surface may be produced by any method except when the bar or circle (symbol b or d) is specified or when the method is specified above the horizontal line.

Figure 3.11   Surface texture symbols. *Source*: ASME Y14.36M-1996. Reproduced with permission of ASME.

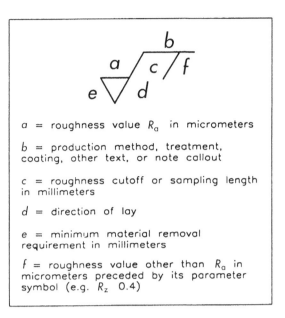

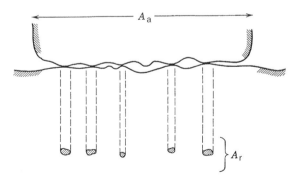

a = roughness value R_a in micrometers

b = production method, treatment, coating, other text, or note callout

c = roughness cutoff or sampling length in millimeters

d = direction of lay

e = minimum material removal requirement in millimeters

f = roughness value other than R_a in micrometers preceded by its parameter symbol (e.g. R_z 0.4)

Figure 3.12   Application of surface texture values and lay direction. *Source*: ASME Y14.36M-1996. Reproduced with permission of ASME.

Peak heights of a profile are independent of the orientation of the stylus track, wavelength of features, and filter cutoff; various other details, however, depend on the angle between the profile track and the lay. Standard practice calls for measuring roughness at right angles to the lay, and this is assumed in specifications. Application of symbol values and lay direction are illustrated in Figure 3.12.

3.7 Contact Between Surfaces

When two parallel surfaces are brought together with a light load, initial contact occurs only at a very few microscopic peak areas or asperities, as suggested in Figure 3.13. As the normal load

A_a

A_r

Figure 3.13   Schematic illustration of an interface showing apparent and real areas of contact. *Source*: Rabinowicz, 1995. Reproduced with permission of John Wiley & Sons.

increases, these initially contacting asperities are compressed to bring more and more areas into contact. In portions of these real areas of contact with low stress and involving plastics and elastomers with their low elastic modulus, deformation is elastic; but with most practical cases involving contact of metal surfaces, plastic deformation of asperities is likely.

Several approaches have been employed to predict rough surface contact characteristics (see, e.g. McCool (1986); Liu *et al.* (1999); Adams (2000) and Pugliese *et al.* (2008) for detailed reviews). Broadly, they can be classified into two categories: deterministic and statistical. Depending on the application and the desired results, each of these two approaches has its own advantages and disadvantages. The deterministic approach involves simulation of the real surface profile and hence provides a more detailed description of the pressure in the contact area, the deformation field and the sub-surface stress distribution. However, it requires direct measurement of the surface profile and extensive numerical calculations. In contrast, the statistical approach, although relatively approximate, is more convenient to formulate as it only requires the specification of a few surface parameters and thus lends itself to the generalization of predictions for different geometries, surface properties, and loading conditions.

The modeling methodology in statistical approach consists of two steps. The first step is to develop an appropriate relationship for the contact of a single asperity and nominally flat surface, and the second step is to extend the solution for that individual asperity to an ensemble of asperities in order to determine the contact behavior of rough surfaces. This is achieved through employing the statistical distribution of asperities and specification of surface parameters. In what follows, these two steps and the associated developed approaches are described.

Micro-Contact Considerations: Deformation of Single Asperity

As noted in Section 3.2, surface profiles involve asperity slopes generally in the range below 10°. In considering a single asperity, it is convenient to model it as a smooth protuberance of spherical shape (Greenwood and Williamson, 1966). Elastic contact between a sphere of radius β and a plane (Figure 3.14) under normal load W, results in a circular contact patch of radius a according to Hertz elastic contact analysis

$$a = \left(\frac{3W\beta}{4E}\right)^{1/3} \tag{3.5}$$

where β is the radius of the spherical asperity. Equivalent elastic modulus E for the contact depends on Young's moduli E_1 and E_2, and Poisson's ratios v_1 and v_2 for the sphere and plane,

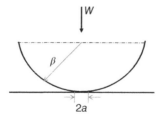

Figure 3.14   Contact of spherical asperity with a flat surface.

as follows[1]:

$$\frac{1}{E} = \frac{1 - v_1^2}{E_1} + \frac{1 - v_2^2}{E_2} \tag{3.6}$$

Contact area A between the sphere and a flat surface becomes

$$A = \pi a^2 = 2.6 \left(\frac{W\beta}{E}\right)^{2/3} \tag{3.7}$$

For purely elastic deformation, the contact area is then proportional to $W^{2/3}$. Mean normal stress on the spherical asperity, W/A, varies as $W^{1/3}$.

According to Hertzian solution, normal approach (interference) between the sphere and the plate is

$$\omega = \left[\frac{(9/16)W^2}{(E^2\beta)}\right]^{1/3} \tag{3.8}$$

and solving for W in terms of the normal approach yields

$$W = \left(\frac{4}{3}\right) E\beta^{1/2}\omega^{3/2} \tag{3.9}$$

The contact area can be also written is terms of the interference

$$A = \pi\beta\omega \tag{3.10}$$

As load on the contact increases, a critical point is reached where the elastic limit is exceeded within the asperity of the softer material. The real area of contact A for the asperity then grows so that its mean pressure P_m will balance the total normal load W according to the following relation (see also Section 4.1). The area of contact then becomes

$$A = \frac{W}{P_m} \tag{3.11}$$

At fully plastic condition the mean pressure is, in fact, the hardness (mean yield pressure) of the softer material symbolized as H, hereafter. If hardness remains constant, the area of contact will then be directly proportional to the load. Since hardness is often of the order of 1000 times the design load on a bearing surface, this contact area on surface asperities is only a small fraction of the total nominal area involved.

Great progress has been made toward an understanding the complex nature of the contact of rough surfaces and the subject remains an active area of research. While the details are beyond the scope of this book, it is nevertheless useful to briefly describe some of the pertinent developments in this exciting field.

---

[1] While Equation (3.6) is used here in analyses of asperity temperature rise and distortions in sliding surface contacts, the left-hand side of an otherwise identical relation is almost universally taken as $2/E'$ in defining the equivalent modulus for determining raceway contacts (Equation 14.10) and related EHL elastohydrodynamic relations in ball and roller bearings (see Chapter 14).

Twenty years after the pioneering work of Greenwood and Williamson (1966), Chang, Etsion and Bogy (1987) proposed a model (often referred to as CEB model) that extended the GW to take into account the effect of elasto-fully plastic deformation of asperities. The treatment of the intermediate regime of deformation was introduced by Zhao, Maietta, and Chang (2000) who provided an elastic-elasto/plastic-fully plastic model that bridges the elastic and plastic behavior of the asperity. This model which uses a mathematical function to bridge the elastic and fully plastic regime is referred to as the ZMC model. More recent refinement is due to the work of Kogut and Etsion (2002, 2003) and Jackson and Green (2005, 2006) and involves the use of the finite element analysis (FEA) to obtain very detailed features of asperity deformation. The models of Kogut-Etsion and Jackson-Green for the deformation of a single asperity were described in the previous section. What follows is the extension of these models to the contact of rough surfaces using statistical method of Greenwood and Williamson.

The following empirical expressions are provided by Kogut and Etsion (2002), the so-called KE model, in piece-wise form for the contact load and contact area

$$P(\omega) = P_c \left(\frac{\omega}{\omega_c} \right)^{\frac{3}{2}} \qquad 0 < \frac{\omega}{\omega_c} \le 1,$$

$$P(\omega) = 1.03 P_c \left(\frac{\omega}{\omega_c} \right)^{1.425} \qquad 1 < \frac{\omega}{\omega_c} \le 6$$

$$P(\omega) = 1.4 P_c \left(\frac{\omega}{\omega_c} \right)^{1.263} \qquad 6 < \frac{\omega}{\omega_c} \le 110 \qquad (3.12)$$

$$P(\omega) = \frac{3}{K} P_c \left(\frac{\omega}{\omega_c} \right) \qquad 110 < \frac{\omega}{\omega_c}$$

and for contact area-interference:

$$A(\omega) = A_c \left(\frac{\omega}{\omega_c} \right) \qquad 0 < \frac{\omega}{\omega_c} \le 1$$

$$A(\omega) = 0.93 A_c \left(\frac{\omega}{\omega_c} \right)^{1.136} \qquad 1 < \frac{\omega}{\omega_c} \le 6$$

$$A(\omega) = 0.94 A_c \left(\frac{\omega}{\omega_c} \right)^{1.146} \qquad 6 < \frac{\omega}{\omega_c} \le 110 \qquad (3.13)$$

$$A(\omega) = 2 A_c \left(\frac{\omega}{\omega_c} \right) \qquad 110 < \frac{\omega}{\omega_c}$$

where

$$\omega_c = \left(\frac{\pi K H}{2E} \right)^2 \beta, \quad P_c = \frac{4}{3} E \beta^{1/2} \omega_c^{3/2}, \quad A_c = \pi \beta \omega_c \qquad (3.14)$$

The parameter ω_c is the critical interference at the inception of the plastic deformation and P_c and A_c represent the associated contact load and contact area (see Eqs. 3.9 and 3.10). The parameter K is the proportionality factor between the maximum contact pressure and the hardness at the onset of plastic deformation ($P_{yield} = KH$). Tabor (1951) suggested the constant value of 0.6 for K, while other researchers showed that it depends on the Poisson's ratio. For example, Chang et al. (1988) proposed:

$$K = 0.454 + 0.41\upsilon \qquad (3.15)$$

Kogut and Etsion consider hardness to be a fixed material parameter. The hardness for most metals is typically taken to be equal to 2.8 times the yield strength, S_y, for untreated surfaces based on the Tabor model.

Jackson and Green (2005) took the effects of material properties and geometry into account during the deformation. In contrast to the KE model, which considers the hardness as a fixed intrinsic material property ($H = 2.8S_y$), Jackson and Green (JG) showed that the hardness is not constant during the deformation and hence they considered hardness variation during the asperity deformation. Also, they showed that the assumption of the elastic deformation for the asperity contact is valid not only within the critical interference limit but also up to 1.9 times the critical interference. Based on the finite element analysis, they provided the following empirical expressions:

$$P(\omega) = P'_c \left(\frac{\omega}{\omega'_c} \right)^{\frac{3}{2}} \qquad 0 < \frac{\omega}{\omega'_c} \leq 1.9$$

$$P(\omega) = P'_c \Gamma_{JG} \qquad 1.9 < \frac{\omega}{\omega'_c} \qquad (3.16)$$

and Γ_{JG} is the elasto-plastic term given by:

$$\Gamma_{JG} = \left(\frac{\omega}{\omega'_c} \right)^{3/2} e^{-\frac{1}{4}\left(\frac{\omega}{\omega'_c} \right)^{5/12}} + \frac{2.84 \times 4}{C} \left[1 - e^{-0.82\left(\sqrt{\frac{\sigma}{\beta}} \sqrt{\omega} \left(\frac{1}{1.9}\left(\frac{\omega}{\omega'_c} \right) \right)^{D/2} \right)^{-0.7}} \right] \left(\frac{\omega}{\omega'_c} \right)$$

$$\times \left(1 - e^{-\frac{1}{25}\left(\frac{\omega}{\omega'_c} \right)^{5/9}} \right)$$

and for contact area-interference:

$$A = \pi \beta \omega'_c \qquad 0 < \frac{\omega}{\omega'_c} \leq 1.9$$

$$A = A_c \times \left(\frac{\omega}{\omega'_c} \right) \left(\frac{\omega}{1.9\omega'_c} \right)^D \qquad 1.9 < \frac{\omega}{\omega'_c} \qquad (3.17)$$

where

$$C = 1.295.e^{0.736\nu}, \quad D = 0.14.e^{23\frac{S_y}{E}}$$

$$\omega'_c = \left(\frac{\pi C S_y}{2E}\right)^2 \beta, \quad P'_c = \frac{4}{3}E'\beta^{1/2}\omega'^{3/2}_c, \quad A_c = \pi\beta\omega'_c \tag{3.18}$$

Note that the parameter ω'_c is the critical interference at the inception of the plastic deformation based on the Jackson and Green model and P_c and A_c represent the associated contact load and contact area.

Contact of a Rough Flat Surface and a Smooth Flat Surface (Greenwood and Williamson-Based Models)

The best-known treatment of rough surfaces is due to the work of Greenwood and Williamson in 1966, who modeled the contact problem of spherical asperities pressed against a nominally flat rough surface by extending the load-interference and contact area-interference solutions of an individual hemispherical asperity to a population of asperities. Referring to Figure 3.15, let us assume that the rough surface and the flat surface are separated by a distant d. This separation represents the distance from the mean of the asperity (summit) heights to the flat surface. Also shown is the mean of the surface heights and the separation distance between the surfaces, h. As shown in Figure 3.15, y_s denotes the distance between the mean line of the summit heights and that of the surface heights $(d = h - y_s)$.

Let η represent the asperity density in a nominal contact area A_n, then the total number of asperities is:

$$N = \eta A_n \tag{3.19}$$

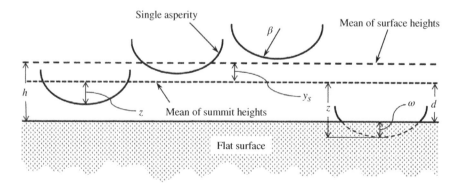

Figure 3.15 Contact of a rough surface with asperities with a smooth flat surface.

The asperities whose heights, z, are greater than d will come into intimate contact with the flat surface and create an interference ω, where $\omega = z - d$. The probability of an asperity of height z making contact can be described by the asperity height distribution function $\phi(z)$:

$$prob(z > d) = \int_d^\infty \phi(z)dz \qquad (3.20)$$

Therefore, the total number of asperities in contact is given by

$$N_c = N \int_d^\infty \phi(z)dz \qquad (3.21)$$

Similarly, let $W_i(\omega)$ and $A_i(\omega)$ represent, respectively, the contact load and the contact area for each individual asperity, then the expected contact load, W and the expected total area of contact (real area of contact, A_r) can be determined from the following expressions

$$W = N \int_d^\infty W_i(\omega)\phi(z)dz \qquad (3.22)$$

$$A_r = N \int_d^\infty A_i(\omega)\phi(z)dz \qquad (3.23)$$

Note that these equations are general. So far, no assumptions have been made in terms of the deformation regime of The asperities (elastic, plastic, or the elasto-plastic) or their distribution.

The Greenwood-Williamson Model (GW)

The simplest case is considered in the classical article by Greenwood and Williamson (1966). They assume that for an individual asperity the contact is purely elastic and they make use of the Hertzian contact equations described earlier for an asperity contacting a nominally flat surface (Eqs. 3.9 and 3.10). Substituting Eq. 3.9 into Eq. 3.22 and Eq. 3.10 into Eq. 23 yields a very well-known model of Greenwood and Williamson to describe the contact of a rough flat surface and a smooth flat surface, often referred to as the GW model

$$W = N \int_d^\infty \left(\frac{4}{3} E \beta^{1/2} \omega^{3/2} \right) \phi(z)dz = \frac{4}{3} E \beta^{1/2} N \int_d^\infty (z-d)^{3/2} \phi(z)dz \qquad (3.24)$$

$$A_r = N \int_d^\infty \pi \beta \omega \phi(z)dz = N\pi\beta \int_d^\infty (z-d)\, \phi(z)dz \qquad (3.25)$$

Now, typically, the distribution function $\phi(z)$ is assumed to obey the Gaussian distribution expressed as

$$\phi(z) = \frac{1}{\sigma_s \sqrt{2\pi}} e^{-\frac{z^2}{2\sigma_s^2}} \qquad (3.26)$$

where σ_s is the standard deviation of asperity (summits) height distribution. It will be shown later that the standard deviation of the asperity height distribution (σ_s) and the standard deviation of surface heights distribution or surface roughness (σ) are related for Gaussian surfaces. It is also useful to present a dimensionless form of the above equations

$$\frac{W}{EA_n} = \frac{4}{3}E\eta\beta^{1/2}\sigma^{3/2}\int_{d^*}^{\infty}(\omega^*)^{3/2}\phi^*(z^*)dz^* \tag{3.27}$$

$$\frac{A_r}{A_n} = \pi\eta\beta\sigma\int_{d^*}^{\infty}(z^*-d^*)\,\phi^*(z^*)dz^* \tag{3.28}$$

$$\phi^*(z^*) = \frac{1}{\sqrt{2\pi}}\left(\frac{\sigma}{\sigma_s}\right)e^{-\left(\frac{\sigma}{\sigma_s}\right)^2\frac{z^{*2}}{2}} \tag{3.29}$$

where

$$\omega^* = z^* - d^*, \omega^* = \omega/\sigma, z^* = z/\sigma, d^* = d/\sigma.$$

The Kogut-Etsion Model (KE)

Given the relationships for a single asperity contact, Kogut and Etsion extended the approach to the contact of rough surfaces using the statistical method of Greenwood and Williamson. Substituting Eqs. 3.12 and 3.13 into Eqs. 3.27 and 3.28, respectively, results in the KE model written in dimensionless form as follows (Kogut and Etsion, 2003):

$$\frac{W}{EA_n} = \eta\beta\sigma\left(\frac{2\omega_c^*\pi K\Omega}{3}\right)\left\{\int_{d^*}^{d^*+\omega_c^*}I^{\frac{3}{2}} + 1.03\int_{d^*+\omega_c^*}^{d^*+6\omega_c^*}I^{1.425}\right.$$

$$\left.+ 1.4\int_{d^*+6\omega_c^*}^{d^*+110\omega_c^*}I^{1.263} + \frac{3}{K}\int_{d^*+110\omega_c^*}^{\infty}I^1\right\} \tag{3.30}$$

$$\frac{A_r}{A_n} = \pi\,\eta\beta\sigma\,\omega_c^*\left\{\int_{d^*}^{d^*+\omega_c^*}I^1 + 0.93\int_{d^*+\omega_c^*}^{d^*+6\omega_c^*}I^{1.136}\right.$$

$$\left.+ 0.94\int_{d^*+6\omega_c^*}^{d^*+110\omega_c^*}I^{1.146} + 2\int_{d^*+110\omega_c^*}^{\infty}I^1\right\} \tag{3.31}$$

where

$$I^\alpha = \left(\frac{\omega^*}{\omega_c^*}\right)^\alpha\varphi^*(z)dz^*; \Omega = \frac{H}{E} \tag{3.31a}$$

The Jackson-Green Model (JG)

Similar to the KE model, the JG model can be obtained by substituting the Jackson and Green relationships for a single asperity contact into the load-separation and the real area-separation relationships (Eqs. 3.10 and 3.11). The dimensionless JG model is given as follows

$$\frac{W}{EA_n} = \eta\beta\sigma\left(\frac{2\omega_c'^{*}\pi CS_y}{3E}\right)\left\{\int_{d^*}^{d^*+1.9\omega_c'^{*}} J^{3/2} + \int_{d^*+1.9\omega_c'^{*}}^{\infty} \Gamma_{JG}\varphi^*(z^*)dz^*\right\} \qquad (3.32)$$

$$\frac{A_r}{A_n} = \pi\eta\beta\sigma\omega_c'^{*}\left\{\int_{d^*}^{\frac{h-y_s}{\sigma}+1.9\omega_c'^{*}} J^1 + (1.9)^{-D}\int_{d^*+1.9\omega_c'^{*}}^{\infty} J^{D+1}\right\} \qquad (3.33)$$

where

$$J^\alpha = \left(\frac{\omega^*}{\omega_c^*}\right)^\alpha \varphi^*(z)dz^*$$

These equations (GW, KE and JG models) are extensively used in tribology to characterize the behavior of surface roughness. An application of these equations in the evaluation of thermoelastic instability is given by Jang and Khonsari (1999, 2003, 2004). They are also useful for the prediction of the machine element performance that operates in a mixed lubrication regime (see the Stribeck curve, Figure 1.4). Examples include a heavily-loaded roller and gears operating under slow rolling speeds in which the hydrodynamic pressure developed in the fluid is not sufficient to separate the surfaces and, as a result, the surface asperities come into intimate contact. In the mixed lubrication regime, the imposed load on the system is shared: a part of the load is carried by the fluid and part by the tip of the asperities. The above equations, when used in conjunction with the hydrodynamic equations (Chapter 6), are useful in characterizing the friction behavior of the Stribeck curve. The interested reader may refer to Lu *et al.* (2005, 2005), Akbarzadeh and Khonsari (2008, 2009, 2010a, 2010b), Khonsari and Booser (2010), Beheshti and Khonsari (2012, 2013, 2014), and Masjedi and Khonsari (2012, 2014a, 2014b, 2015a, 2015b, 2015c, 2015d, 2016).

Contact of Two Rough Surfaces

All relationships described in the preceding sections are derived assuming that one of the surfaces is ideally smooth and the other one is rough. It would be useful to discuss how these relationships can be applied to the practical situations where both surfaces are rough. Greenwood and Tripp (1967) show that the contact of two rough surfaces can be treated as a contact of one rough surface with equivalent (effective) parameters and an ideal smooth surface. Based on the relationships provided by McCool (1987), it is possible to extend the formulas to treat the contact of two rough surfaces.

The surface parameters σ, β and η, can be expressed in terms of three quantities, m_0, m_2 and m_4 called the 'spectral moments' evaluated based on the surface profile $\lambda(x)$. The appropriate relationships are:

$$m_0 = AVG[\lambda^2(x)] = \sigma^2 \qquad (3.34)$$

$$m_2 = AVG\left[\left(\frac{d\lambda(x)}{dx}\right)^2\right] \qquad (3.35)$$

$$m_4 = AVG\left[\left(\frac{d^2\lambda(x)}{dx^2}\right)^2\right] \qquad (3.36)$$

Note that the square root of spectral moment of order zero, m_0, is simply the rms of the surface heights (surface roughness). Now, assuming that the surface profile $\lambda(x)$ can be described by a Gaussian distribution—as is the case in most engineering surfaces—then one could derive the following relationships between the asperity radius β and density η and the spectral moments (Bush et al., 1976; Longuethiggins, 1957):

$$\beta = 0.375\sqrt{\frac{\pi}{m_4}} \tag{3.37}$$

$$\eta = \frac{m_4}{m_2 6\pi\sqrt{3}} \tag{3.38}$$

The equivalent spectral moments of two rough surfaces can then be written as follows (McCool, 1987)

$$(m_0)_{eq} = (m_0)_1 + (m_0)_2 \tag{3.39}$$

$$(m_2)_{eq} = (m_2)_1 + (m_2)_2 \tag{3.40}$$

$$(m_4)_{eq} = (m_4)_1 + (m_4)_2 \tag{3.41}$$

The equivalent roughness, asperity radius and asperity density can be written as follows (Beheshti and Khonsari, 2012)

$$\sigma_{eq} = \sqrt{\sigma_1^2 + \sigma_2^2} \tag{3.42}$$

$$\frac{1}{\beta_{eq}} = \sqrt{\frac{1}{\beta_1^2} + \frac{1}{\beta_2^2}} \tag{3.43}$$

$$\frac{1}{\eta_{eq}} = \frac{1}{\eta_1}\left(\frac{\beta_{eq}}{\beta_1}\right)^2 + \frac{1}{\eta_2}\left(\frac{\beta_{eq}}{\beta_2}\right)^2 \tag{3.44}$$

Relationship Between Surface Features and GW Parameters

It is useful to relate the parameters of interest in the formulations of Greenwood-Williamson to the surface features parameters obtained by a profilometer. Using the surface spectral moments, a method described by Bush et al. (1976) and McCool (1987) can be used for this purpose. For the contact of isotropic surfaces with Gaussian distribution of surface heights, the following relationships exist between σ, σ_s and y_s

$$\frac{\sigma}{\sigma_s} = \frac{\eta\beta\sigma}{\sqrt{(\eta\beta\sigma)^2 - 3.71693 \times 10^{-4}}} \tag{3.45}$$

$$\frac{y_s}{\sigma} = \frac{0.045944}{\eta\beta\sigma} \tag{3.46}$$

The Asperity Plasticity Index

The severity of plastic flow at asperities can be predicted by a *plasticity index*, originally introduced by Greenwood and Williamson (1966). The plasticity index can be conveniently expressed in terms of the critical interference and the standard deviation of asperity heights (σ_s) as:

$$\psi = \sqrt{\frac{\sigma_s}{\omega_c}} \tag{3.47}$$

Using Eq. 3.45 and the definition of the critical interference (Eq. 3.14) based on the KE model, the plasticity index can be written as follows:

$$\psi = \frac{2E'}{\pi KH}\sqrt{\frac{\sigma}{\beta}}\left(1 - \frac{3.71693 \times 10^{-4}}{(\eta\beta\sigma)^2}\right)^{1/4} \tag{3.48}$$

This equation combines the material and topographic properties of the surfaces in contact. Plastic flow can indeed occur at micro contacts with the plasticity index of 1 and higher and to account for this, consideration should be given to plastic and elasto-plastic effects associated with asperities. It is worth mentioning that plasticity index can be alternatively calculated using the critical interference based on the JG model.

Example 3.1   A flat Al_2O_3 -TiC read–write head surface contacts a thin-film magnetic computer hard disc with characteristics given by Bhushan (1999). Using the values tabulated below, calculate the plasticity index ψ from Equation (3.48) (Assume that $\eta\beta\sigma = 0.05$). Generalized equations follow this property table for equivalent contact elastic modulus E, asperity height σ, and tip radius β.

	Magnetic hard disk	Head slider
Hardness, H (GPa)	6	23
Modulus, E (GPa)	113	450
Poison ratio, v	0.3	0.22
Asperity peak heights, σ	$7\,nm = 7 \times 10^{-6}\,mm$	$2\,nm = 2 \times 10^{-6}\,mm$
Inverse radius of asperity peaks, $1/\beta$	$4.9\,mm^{-1}$	$0.53\,mm^{-1}$

Calculations:

$$1/E = (1 - v_1{}^2)/E_1 + (1 - v_2{}^2)/E_2 \qquad (1 - 0.3^2)/113 + (1 - 0.22^2)/450 = 0.0102\,GPa^{-1}$$

$$\sigma = (\sigma_1{}^2 + \sigma_2{}^2)^{1/2} \qquad (7^2 + 2^2)^{1/2} = 7.28\,nm = 7.28 \times 10^{-6}\,mm$$

$$1/\beta - 1/\beta_1 + 1/\beta_2 \qquad 4.9 + 0.53 - 5.43\,mm^{-1}$$

$$\psi = \frac{2(0.0102)^{-1}}{\pi \times 0.6 \times 6}\sqrt{\frac{7.28 \times 10^{-6}}{5.43^{-1}}}\left(1 - \frac{3.71693 \times 10^{-4}}{(0.05)^2}\right)^{1/4} = 0.1$$

As can be judged from this value of 0.10 for ψ and from Figure 3.16, these asperity contacts are in the elastic range. Plastic flow of asperities for ψ values less than about 0.6–1.0 would result only from high nominal pressure on very smooth surfaces.

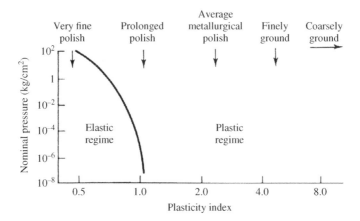

Figure 3.16 Dependence of asperity deformation on the plasticity index for aluminum surfaces. *Source*: Hutchings, 1992. Reproduced with permission of Taylor & Francis.

Values of ψ in the plastic range from about 3 up to 100 are typical for most metal finishing processes (see Figure 3.16). The load supported on each asperity is then proportional to its contact area, and the total area of real contact is proportional to the normal load and independent of the detailed form and distribution of asperity heights. Steel ball and roller bearings with their smooth surfaces and high hardness are an important exception in giving primarily elastic contacts with ψ values around 1. With ceramics and polymers, values of E/H about one-tenth those for metals commonly lead to elastic contact.

Three main categories have been suggested in which the real contact area is larger (Rabinowicz, 1995). In the first category, surfaces are smooth, as with highly polished contacting surfaces in ball bearings which involve elastic contact.

The second category of increased contact area involves shear force, as well as normal force, acting on the junction. In a typical situation, the final area of contact before sliding commences under the influence of a lateral force might be about three times the value that existed before the shear force was first applied. In the third category of factors leading to increased contact area, surface properties and surface energy may sometimes play a significant role (Rabinowicz, 1995).

Contact of Curved Surfaces

All the models mentioned above are applicable to the contact of two flat surfaces. The macro-level contact of two smooth curved surfaces can be described by the Hertzian theory. Table 3.3 summarizes the analytical Hertzian contact expressions for maximum pressure and contact dimensions in line contact (e.g. rectangular conjunction between two cylindrical rollers or a roller and flat surface), the spherical point (e.g. circular conjunction between a ball and flat surface or two balls of identical radius) and the ellipsoidal point (e.g. elliptical conjunction between barrel-shaped bodies). The parameter E is the effective (equivalent) modulus of the two bodies (i.e. Equation 3.6) repeated below for convenience:

$$\frac{1}{E} = \frac{1 - v_1^2}{E_1} + \frac{1 - v_2^2}{E_2}$$

Table 3.3 Hertzian contact expressions for line, circular and elliptical point-contact configurations

Configuration	Maximum pressure	Contact dimension	Description
Line contact (Rectangular conjunction)	$p_H = \dfrac{2W}{\pi Lc}$	$a = \sqrt{\dfrac{4WR_l}{\pi LE}}$	L: contact length a: contact half-width R_l: cylinder effective radius
Spherical Point contact (Circular conjunction)	$p_H = \dfrac{3W}{2\pi a^2}$	$a = \sqrt[3]{\dfrac{3R_sW}{4E}}$	a: contact radius R_s: sphere effective radius
Ellipsoidal point contact (Elliptical conjunction)[1]	$p_H = \dfrac{3W}{2\pi ab}$	$b = \sqrt[3]{\dfrac{3R_xDW\tilde{S}}{\pi E\kappa(D+1)}}$ $a = \kappa b \quad a \geq b$	$D = \dfrac{R_y}{R_x} \quad R_y \geq R_x$ $\kappa = D^{2/\pi}$ $\tilde{S}(\kappa) = \displaystyle\int_0^{\pi/2} \sqrt{1 - \left(1 - \dfrac{1}{\kappa^2}\right)\sin^2\tau}\, d\tau$ R_x: effective ellipsoid minimum radius of curvature R_y: effective ellipsoid maximum radius of curvature a: contact half-length b: contact half-width $\tilde{S}$: elliptical integral of the first kind

[1]*Source*: Hamrock and Brewe, 1983. Reproduced with permission of ASME.

where v_1 and v_2 are the Poisson's ratios for each body. The parameter R denotes the effective (equivalent) radius defined as

$$\frac{1}{R} = \frac{1}{R_1} \pm \frac{1}{R_2}$$

where $+$ is for the outer contact and $-$ is for inner contact. Note that for a ball of radius R_1 on a flat surface, for example, the equivalent radius becomes $R = R_1$, since $R_2 \to \infty$.

In Table 3.3, the surfaces are assumed to be smooth. When surfaces are rough, the contact properties deviate from those predicted by the Hertzian approach. Pioneered by Greenwood and Tripp (Greenwood and Tripp, 1967), multiple studies have investigated the effect of roughness on the contact of curved bodies for line-contact (Gelinck and Schipper, 2000; Beheshti and Khonsari, 2012), circular point-contact (Greenwood and Tripp, 1967; Kagami *et al.*, 1983; Bahrami *et al.*, 2005; and Li *et al.*, 2010) and the general case of elliptical-point contact (Beheshti and Khonsari, 2014).

To fully appreciate the importance of roughness on curved bodies contact, Kagami *et al.* (1983) conducted an experimental study where a smooth steel sphere was pressed under a series of loads against a rough steel plate as well as a rough copper plate. See Figure 3.17. The rough plates were coated with a lamp black film after they had been coated with an evaporated carbon film in a vacuum. This allowed direct observation and measurement of the indentations using a differential transformer. Their results are shown in Figures 3.18 and 3.19 for both copper and steel plates along with analytical predictions of Beheshti and Khonsari (2014) using

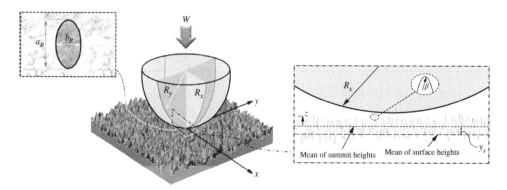

Figure 3.17 Illustration of a smooth sphere in contact with a rough surface. *Source*: Beheshti and Khonsari, 2014. Reproduced with permission of ASME.

the KE model and the JG model, where both yield nearly the same results. The notable feature here is that the prediction of Hertzian equation when applied to rough surfaces substantially deviates from the results of measurements, whereas with the proper consideration of roughness characteristics, one can accurately determine the characteristics of the surfaces.

Before leaving the subject, it is revealing to examine how the real area of contact is influenced by the load and surface roughness. According to Equation 3.7, the Hertzian analysis simply states that $A \propto W^{2/3}$. For fairly smooth surfaces with $\Psi = 1$, the GW model predicts $A \propto W^{0.89}$ while the KE simulations reveal that $A \propto W^{0.95}$. For rough surfaces with $\Psi = 9.2$, the GW prediction changes to $A \propto W^{0.95}$ and KE model predicts that $A \propto W^{0.99}$, i.e. the real area changes with load in nearly linear fashion (Beheshti and Khonsari, 2014).

The above discussion represents a brief presentation of statistical micro-contact treatment of rough surfaces in tribology, which is currently an active area of research. To apply these

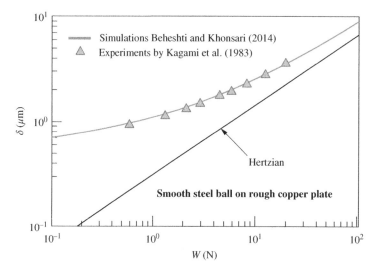

Figure 3.18 Contact of a smooth steel ball on a rough copper plate.

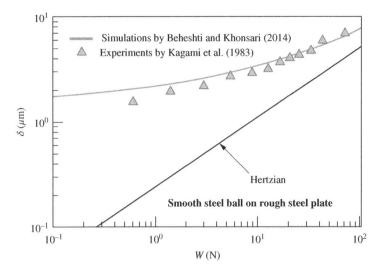

Figure 3.19   Contact of a smooth steel ball on a rough steel plate.

findings to machine elements, one often needs to predict the pertinent contact parameters such as the maximum pressure, contact dimensions, compliance, real area of contact, and pressure distribution for different types of curved surfaces. Recently, based on simultaneous solution of micro-contact models (e.g. KE and JG) and surface bulk deformation, Beheshti and Khonsari have provided convenient formulas for the prediction of these parameters for line-contact configuration (Beheshti and Khonsari, 2012) and circular and elliptical-point contact configuration (Beheshti and Khonsari, 2014). In these papers, the interested reader will find the formulation details, application examples, and additional references.

3.8 Temperature Rise in Sliding Surfaces

Most sliding contacts involve only a small fraction of the total area, commonly limited to junctions between local surface asperities. Since these local contacts are subject to concentrated frictional heating, their temperature may rise to a 'flash temperature,' much higher than their neighboring surface areas. Calculation of the flash temperature – while only an approximation – is helpful in estimating the onset of possible melting at asperities, changes in material properties, and in assessing boundary lubrication effects by lubricants, especially by antiwear and extreme-pressure additives.

Direct measurement of the flash temperature rise has proven to be difficult. Different techniques involve using embedded thermocouples, taking advantage of the thermocouple effect between dissimilar electrical conductors, or relying on the color change in liquid crystals and special coatings. Infrared observations of hot spots through a sliding glass surface have also aided in establishing relations such as those which follow.

Figure 3.20 provides a circular junction model to illustrate typical flash temperature factors. Total frictional heat Q (watts) generated is given by fWV with coefficient of friction f experienced in sliding at velocity V (m/s) under load W (N). This total frictional heat is then

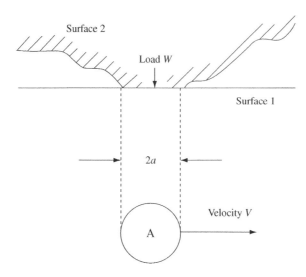

Figure 3.20 Moving circular frictional heat source. *Source*: Hutchings, 1992. Reproduced with permission of Taylor & Francis.

conducted in components Q_1 and Q_2 from the heated contact zone at its elevated average temperature T into the two sliding materials which remain at their bulk ambient temperature $T_0(°C)$. By analogy with electrical current flow (Bowden and Tabor, 1950), thermal conductance $4ak_1$ gives contact temperature rise $(T - T_0)$ for body 1 and $4ak_2$ gives the same $(T - T_0)$ at body 2. Then

$$Q = Q_1 + Q_2 = 4ak_1(T - T_0) + 4ak_2(T - T_0) \tag{3.49}$$

Hence

$$(T - T_0) = \frac{fWV}{4a(k_1 + k_2)} \tag{3.50}$$

where k_1 and k_2 are the thermal conductivities (of elements 1 and 2), and a is the radius of the circular contact.

This simplified relation applies only to low sliding speeds. At higher surface speeds this relation becomes more complex with a drop in contact temperature T as more and more heat flows into the cool, fresh surface coming into the contact zone. The following relation covers the entire speed range for each of the surfaces (Tian and Kennedy, 1994):

$$(T - T_0) = \frac{Q}{4.56ak(0.66 + Pe)^{0.5}} \tag{3.51}$$

where Pe is the Peclet number defined as $Pe = Va/2\kappa$. For a stationary heat source $Pe = 0$ and for low speeds $Pe = 0.1-5$, for intermediate speeds $Pe = 5-100$, and $Pe > 100$ for high to very high speeds. The interested reader may also refer to Mansouri and Khonsari (2005) and

Jun and Khonsari (2007a, 2007b) for temperature rise in systems undergoing unidirectional as well as oscillatory sliding motions. Oscillatory sliding motions, particularly those with small amplitude, are important in analyzing problems involving fretting (Aghdam and Khonsari, 2013; Beheshti et al., 2013).

Example 3.2 Following an example by Dunaevsky et al. (1997), a contact of radius $r = 0.01$ m by stationary carbon steel surface 2 in Figure 3.20 carries $W = 3000$ N load against carbon steel surface 1 moving at $V = 0.5$ m/s with coefficient of friction $f = 0.1$. The carbon steel has thermal conductivity $k_1 = k_2 = 54$ W/m°C. and thermal diffusivity $\kappa = 1.474 \times 10^{-5}$ m^2/s.

Heat flow into stationary surface 2 by Equation (3.49) is

$$Q_2 = 4ak_1(T - T_0) = 4 \times 0.01 \times 54 \times (T - T_0) = 2.16(T - T_0).$$

For moving surface 1,

$$Pe = Va/2k = 0.5(0.01)/(2 \times 1.474 \times 10^{-5}) = 170$$

For this high-speed case with $Pe > 100$, Equation (3.23) gives

$$Q_1 = 4.56 \times 0.01 \times 54 \times (0.66 + 170)^{0.5} = 32.2(T - T_0)$$

Taking into account frictional heat flow into the two surfaces:

$$Q = Q_1 + Q_2 = fWV$$
$$32.2(T - T_0) + 2.16(T - T_0) = 0.1 \times 3000 \times 0.5 = 150$$
$$(T - T_0) = 150/(32.1 + 2.16) = 4.4\,°C$$

If all frictional heat generated had gone into stationary surface 1, its rise would have been

$$(T - T_0) = fWV/4ak_1 = 150/(4 \times 0.01 \times 54) = 69\,°C$$

The much lower 4.4 °C net rise demonstrates the cooling action by the fresh surface as it moves through the contact zone.

For the special case involving the same material (Cowan and Winer, 1992) for surfaces 1 and 2 in Figure 3.20, contact radius a (in meters) is determined by the plastic asperity deformation:

$$a = [W/(\pi \cdot H)]^{1/2} \tag{3.52}$$

since $A = W/H$, where H is the material hardness (N/m^2). For metals the hardness is roughly three times the yield strength. If the deformation is elastic:

$$a \approx 0.91(WR/E_v)^{1/3} \tag{3.53}$$

where R is the non-deformed radius of curvature of body 2; $E_v = E/(1 - v^2)$ in units of N/m^2; E is the Young's modulus; and v is the Poisson's ratio.

3.9 Lubrication Regime Relation to Surface Roughness

Surface roughness plays a distinctive role in setting the performance of fluid-film bearings. When mating surfaces are so smooth that no contact exists between their asperities, an ideal operation occurs, with no wear and minimum friction. As a full separating film is lost under conditions of high surface roughness, low speed, high load, low viscosity, or limited feed of fluid, an increasing portion of the load is carried by the direct contact of asperities and by any surface films formed by the lubricant.

The following dimensionless film thickness parameter Λ is commonly used to describe this relative magnitude of fluid-film thickness in comparison with composite surface roughness σ:

$$\Lambda = \frac{h_{\min}}{\sigma} = \frac{h_{\min}}{\sqrt{\sigma_a^2 + \sigma_b^2}} \tag{3.54}$$

where

σ_a = rms surface finish of surface a
σ_b = rms surface finish of surface b

Arithmetic roughness average values should be multiplied by 1.25 to obtain rms values for the surfaces with Gaussian distribution of asperity heights.

While initially adopted primarily to characterize the elastohydrodynamic lubrication of ball and roller bearings, Λ values are equally useful for lubricant films in hydrodynamic journal and thrust bearings. The following range of values generally apply to these oil-film bearings:

Full-film lubrication $\qquad\qquad\qquad\qquad\qquad\qquad$ $\Lambda > 3$
Partial- or mixed-film and boundary lubrication $\quad$ $\Lambda < 3$

Full-film hydrodynamic separation of surfaces in a wide variety of bearings using a range of oils and low-viscosity fluids, such as water, requires a minimum Λ of about 3. Lower values generally lead to increasing friction and wear as some asperities come into direct contact. While separation of two surfaces might be hypothesized when the nominal film thickness equals surface roughness $\Lambda = 1.0$, the maximum asperity height commonly ranges up to three times the rms value. Further elastic, alignment, and thermal distortions with industrial journal and thrust bearings may raise the minimum full-film Λ to the range of about 10 (Elwell and Booser, 1972).

These Λ values are not well defined in the operation of conventional machinery bearings. With industrial electric motors, for instance, an initial $R_a = 32\,\mu\text{in}$. journal finish is commonly used in combination with a babbitt with $R_a = 64\,\mu\text{in}$. After a few start-ups and running for a few hours at 900 rpm, the babbitt is commonly polished to a much finer finish in the load zone and the journal is gradually burnished to a 16 μin. or better R_a finish. Using a factor of 1.25 to convert the arithmetic average roughness to rms and with a nominal 0.0008 in. minimum oil-film thickness, initial $\Lambda = 0.0008/\{1.25\,[(0.000032)^2 + (0.000064)^2]^{1/2}\} = 9$. With the surface roughness of both surfaces dropping to about 16 μin. Λ then increases to a value of 28 in service.

In ball and roller bearing contacts, composite surface roughness σ for use in Equation (3.54) can be taken as 0.12 μm for aerospace bearings, 0.25 μm for off-the-shelf bearings, and 0.65 μm for very large industrial bearings (Zaretsky, 1992). Fatigue life generally reaches a maximum value when Λ exceeds about 3.0 as a full elastohydrodynamic film is formed. Above $\Lambda = 6$, there has even been some indication of infinite fatigue life (see Chapter 13). Wear and surface deformation of rolling surfaces are to be expected, along with a reduction in fatigue life, as Λ drops below about 1.0; and extreme-pressure and antiwear additives and higher oil viscosity are commonly brought into use to minimize wear under the boundary lubrication conditions. For values between 1.0 and 3.0, minor surface glazing and distress are usually encountered. The following generalizations reflect general experience with rolling element bearings:

Full elastohydrodynamic lubrication $\Lambda > 3$
Partial (mixed-film) lubrication $1 < \Lambda < 3$
Boundary lubrication $\Lambda < 1$

In a detailed evaluation of gas turbine main shaft bearings operating under partial elastohydrodynamic lubrication, damage and fatigue were found to begin at the surface for low film thickness values ($\Lambda < 1$). For high values of $\Lambda (\Lambda > 5)$, fatigue was found to begin below the surface, presumably at a depth for high shear stress in the steel (Averbach and Bamberger, 1991).

Designation of regimes with less than a full elastohydrodynamic film can involve a range of definitions, as shown in Figure 3.21. With industrial oil film bearings, elastic and elastohydrodynamic support diminishes for Λ values that drop below about 3. This gives rising friction, which finally peaks at essentially the level for unlubricated metal-to-metal sliding.

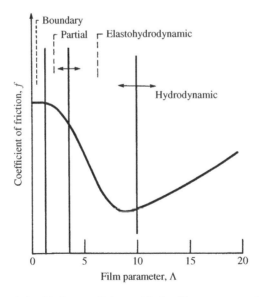

Figure 3.21   Variation of the friction coefficient with the film parameter. *Source*: Hamrock, 1994. Reproduced with permission of McGraw-Hill.

3.10 Friction

When two surfaces rub in dry sliding, the real area of contact A involves only sufficient asperities of the softer material so that their mean yield pressure P will balance the total load W as previously expressed in Equation 3.11 in Section 3.7:

$$A = W/P \qquad (3.11)$$

Referring to Figure 3.22, if the junction shear strength is greater than that of the soft material, shear will take place along the dashed lines to produce a wear fragment.

During dry sliding, the force F required to displace an asperity of the softer material with shear strength s, as in Figure 3.22, becomes:

$$F = As \qquad (3.55)$$

Although this shearing of asperity contact may account for 90% or more of total friction force, other factors do contribute. Lifting force may be needed to raise asperities over the roughness of the mating surface. Scratching by dirt and wear particles may introduce plowing. Internal material damping and surface films may also play a role.

The coefficient of friction f is defined by the ratio of friction force F to the applied load W. Combining Equations 3.11 and 3.55 yields:

$$f = \frac{F}{W} = \frac{s}{P} \qquad (3.56)$$

Shear strength s and yield pressure P are usually both related to the same material properties such as lattice structure and bond strength. For example, in metals, typically $s \approx 0.2P$ and thus $f \approx 0.2$. In general, f often falls within a rather narrow range of about 0.20–0.35 for a wide variety of materials, as indicated by typical data in Table 3.4 for bearing materials varying several hundredfold in yield strength sliding on steel.

The relatively narrow variation of friction coefficient can be conveniently explained by Equation (3.56). Take, for example, the case of displacing an asperity on two contacting hard metals. Here, P would be large and since a relatively high shear would be required to displace an asperity, s would be large. Similarly, two relatively soft metals (small P) in contact require a proportionally lower s value and the coefficient of friction remains relatively constant.

Friction coefficients can rise to 10 or higher with the welding of mating materials such as two very clean copper surfaces in contact. Unlubricated ceramics sliding on themselves commonly have friction coefficients in the 0.3–0.9 range (Jahanmir and Fischer, 1994).

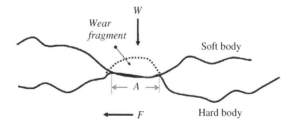

Figure 3.22   Asperity contact between two rubbing surfaces and formation of a wear fragment.

Table 3.4   Typical friction coefficients on steel

Material	f
Carbon-graphite	0.19
Lead babbitt	0.24
Bronze	0.30
Aluminum alloy	0.33
Polyethylene	0.33

The following are typical (not exclusive) magnitudes of other sliding friction coefficients (Blau, 2012):

Boundary lubricated surfaces, ~0.07–0.12
Oxidized metal surfaces, ~0.15–0.4
Dry metal surfaces, ~0.4–0.9
Aluminum alloys, and very clean metals in a vacuum, >1.0

Static friction coefficients typically range up to and above 50% higher than sliding values as a result of adhesion and related interactions between the surfaces. Tables of friction coefficients are available in the literature for a wide variety of material combinations (Bowden and Tabor, 1954; Fuller, 1963; Ludema, 1984; Ruff, 1997). The difficulty in directly using these friction coefficient values stems from the fact that their error in applications commonly ranges up to 20% and more (Rabinowicz, 1995).

Friction coefficients down to about 0.05 can be obtained with low-shear-strength surface layers such as those obtained with PTFE sliding on steel; solid lubricant surface treatments; soft metal coatings of tin, lead, indium, etc., or additive layers provided by some lubricants (Figure 3.23). Provided the thin surface layer does not break down, or is not worn away, shear strength s in Figure 3.23 will be low, while contact area A is held small as load is borne with little deformation of the hard substrate. Referring again to Equation (3.56), in this case, P is controlled by the substrate and remains high while s is reduced. Thus, the friction coefficient f is lowered. This in part explains of the advantages of using a babbitted bushing in bearings, particularly during the start/stop operations.

The usual coefficient of friction encountered with bearing materials can also be dramatically lowered by minimizing the sliding contact or by avoiding it altogether. With ball and roller bearings, for instance, the coefficients of friction values are usually in the 0.001–0.002 range (see Table 15.1). In hydrodynamic bearings, the value of 0.001 is common when operating

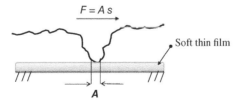

Figure 3.23   Achieving low friction force $F = As$ by depositing a very thin soft film with low shear strength s on a hard surface to give small contact area A.

with a full oil film (see Chapters 7 and 8). With hydrostatic lifts to provide a separating oil film to start a large turbine generator, starting friction drops essentially to zero (see Chapter 10).

3.11 Wear

Wear occurs to some degree whenever surfaces are in sliding contact. While four main types of wear are covered here: adhesive, abrasive, surface fatigue, corrosive, each one acts in such a way as to involve other forms (Archard, 1980; Rabinowicz, 1995; Bhushan, 1999; Bruce, 2012). A few additional wear types are briefly introduced at the end of this section, and Chapter 17 gives examples of how to monitor and troubleshoot wear problems.

Adhesive Wear

This is the most common type of wear as schematically illustrated in Figure 3.24. With the formation and rupture of asperity junctions during sliding, there can be (a) shearing along the interface, or (b) shearing within one of the asperities – usually that of the weaker material. Strong 'cold welds' are formed at some asperity junctions that upon shearing action must break to generate wear. This occurs if the junction strength becomes much greater than the bulk strength of one of the contacting materials, which typically occurs in the weaker part of the junction, i.e. the softer material along the exaggerated path line shown in Figure 3.24. These fragments may then be transferred back to their original surface, or else form loose wear particles. Material transfer is typically from the soft body to the harder body, but the reverse is also possible. The fact that wear fragments made of harder material are often found in tribopair interaction is explained by hypothesizing that there exists local weak spots or regions within the harder material that tend to break off and form small, but hard wear fragments (Figure 3.25). This takes place if these regions coincide with local regions of relatively high strength in the softer material at a junction. This *statistical theory* of wear contends that the junctions and asperities at the interface experience mean stress values superimposed by some 'statistical fluctuations' of these stress values (Rabinowicz, 1995).

Prediction of Adhesive Wear

The initial run-in operation of a bearing commonly involves higher than normal wear related to the machined surface finish, misalignment, and contaminant particles. Steady wear then tends

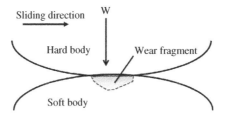

Figure 3.24 Formation of a wear fragment from the softer body along the junction between a hard and soft body being sheared.

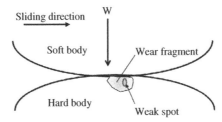

Figure 3.25 Hypothesis of formation of a small but hard wear fragment from the hard body in contact with a soft body.

to reach a steady rate proportional to the total normal load W, as described by the so-called Archard equation:

$$\frac{V}{s} = k\frac{W}{H} \tag{3.57}$$

Equation 3.57 implies that wear volume V of the softer material is (1) proportional to total load W on the two sliding surfaces, (2) proportional to sliding distance s, and (3) inversely proportional to hardness H of the softer of the two contacting surfaces.

Typical values of the dimensionless *wear coefficients k* in are given in Table 3.5 and graphically shown in Figure 3.26 (Rabinowicz, 1980, 1995). Clearly, the greatest wear occurs when identical materials slide against each other without lubrication. The corresponding relationships with the coefficient of friction based on large number of experiments on different types of metals are shown in Figure 3.27 (Rabinowicz, 1995).

This ranking reflects Figure 3.28 that gives relative metallurgical compatibility of pairs of metallic elements according to the degree of their mutual solubility. When gold and silver are melted together, for instance, their mutual solubility gives ready adhesion and wear when sliding together as solids. Conversely, a non-soluble, metallurgically incompatible pair, like silver and iron, gives ideal sliding contact with no adherence of their asperities – and low wear coefficients. Thus, in terms of good sliding compatibility and low wear, the darker the circles, the better the performance.

Unfortunately wear coefficient k may vary by several orders of magnitude for seemingly similar conditions. The most common cause of erratic variation is the effect of local heating of contacting asperities to some 'flash' temperature (Equation 3.51) to alter their physical hardness and structural make-up, even cause melting (Ludema, 1984).

Table 3.5 Wear coefficient depending on the state of lubrication

	Identical	Compatible	Partially compatible Partially incompatible	Incompatible
Unlubricated	1500×10^{-6}	500×10^{-6}	1500×10^{-6}	15×10^{-6}
Poor lubricant	300×10^{-6}	100×10^{-6}	20×10^{-6}	3×10^{-6}
Good lubricant	30×10^{-6}	10×10^{-6}	2×10^{-6}	0.3×10^{-6}
Excellent lubricant	1×10^{-6}	0.3×10^{-6}	0.1×10^{-6}	0.03×10^{-6}

Source: Rabinowicz, 1980. Reproduced with permission of John Wiley & Sons.

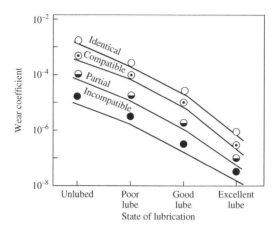

Figure 3.26 Typical wear coefficients for adhesive and abrasive wear. *Source*: Rabinowicz, 1980. Reproduced with permission of John Wiley & Sons.

Since a clear definition of surface hardness is often difficult, Equation 3.57 is commonly modified by eliminating the uncertain hardness term H to include it with k as a dimensional specific wear rate (mm^3/N-m) for use with various metals, ceramics, and self-lubricating materials (see Chapter 12).

Appropriate values of k can easily be measured using a laboratory tribometer (Figure 3.29). Often a pin-on-disk configuration is used where the pin or a flat head pin installed in an upper specimen holder contacts a disk made of the mating pair. The load is specified using the vertical downward motion of the z carriage to create the desired stress and the disk is set in motion

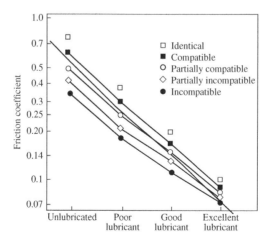

Figure 3.27 Typical friction coefficient for metal on metal. *Source*: Rabinowicz, 1995. Reproduced with permission of John Wiley & Sons.

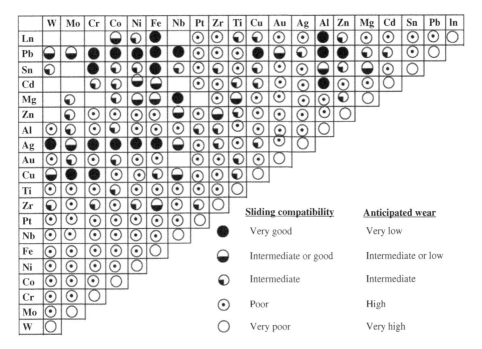

Figure 3.28 Compatibility pairs, based on mutual solubility of elements. *Source*: Adapted from Rabinowicz, 1983.

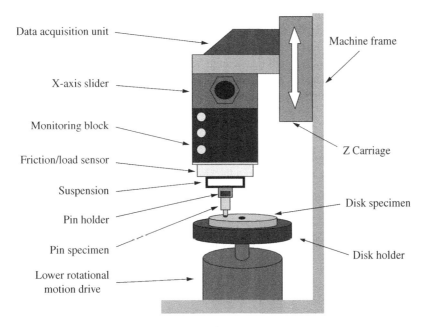

Figure 3.29 Schematic of a commercial tribometer. *Source*: Beheshti and Khonsari, 2010. Reproduced with permission of John Wiley & Sons.

at a constant speed and allowed to run for a specified length of time. The sensors continuously measure friction and results are recorded and stored in a computer via a data acquisition unit. Estimate of wear as a function of time is also made possible in some tribometers equipped with a sensor such as a Linear Variable Differential Transformer (LVDT) sensor – an electrome-chanical transducer that converts the rectilinear motion in the z-direction into an electrical signal to measure minute movements due to wear. But one can simply use an accurate digital scale to weigh the specimen before and after the test. The use of a surface profilometer also facilitates the identification and estimation of the worn volume. With this information, the Archard's equation can be used to evaluate k. Such a tribometer can be used for unlubricated friction and wear tests as well as lubricated contact by mounting a lubricant reservoir on the top of the disk holder, into which the disk can be submerged.

   Note that in a pin-on-disk configuration, the pin is always in contact with the disk and wears continuously, whereas only a certain portion of the disk comes into contact with the pin in a cyclic fashion. An alternative to the pin-on-disk configuration is to replace the pin with a disk, which convert the apparatus to a disk-on-disk tribometer. Such a configuration is more representative of a mechanical seal, where a rotating ring is attached to the shaft so that it rotates as it rubs against a stationary ring affixed in the gland. See Chapter 16.

Derivation and Interpretation of Archard's Adhesive Wear Equation

To gain further insight into the nature of adhesive wear, it is useful to return to Archard's equation once more and examine its physical interpretation. Archard's equation is an empirical relationship based on examination of experimental data for unlubricated friction pair. Archard considered the asperity interaction of two surfaces pressed together by a normal load W and subjected to sliding motion. Asperities are assumed to be spherical in shape with an average radius of a. We will consider the interaction between one asperity that could generate a wear particle when pressed against and slid across a flat surface. Note that not all asperities will produce a wear particle and this will be accounted for later.

   Now, if we assume that under the maximum load of dW, the material plastically yields and that there is a definite probability of producing a wear particle, then

$$dW = \pi a^2 H \tag{3.58}$$

where H represents the hardness.

   This contact results in a worn volume in the shape of a hemisphere of the magnitude

$$dV = \frac{2}{3}\pi a^3 \tag{3.59}$$

The required distance traveled by the asperity to produce this wear volume is simply

$$ds = 2a \tag{3.60}$$

Therefore,

$$\frac{dV}{ds} = \frac{1}{3}\pi a^2 = \frac{1}{3}\frac{dW}{H} \tag{3.61}$$

Of course, only a small fraction, k_1, of the asperity comes into contact to produce wear particles. The total worn volume V resulting from a sliding distant s can be thus obtained summing up all the asperities and taking into the account the fraction k_1:

$$\frac{V}{s} = \sum \frac{dV}{ds} = \frac{1}{3}k_1\frac{W}{H}$$

(3.62)

Therefore, the wear rate (V/s) is directly proportional to the load and inversely proportional to the hardness of the softer material. If each asperity interaction were to result in a wear fragment, then $k_1 = 1$, which, of course, is not the case.

Letting $k = \frac{1}{3}k_1$, we arrive at Equation (3.57) repeated below.

$$V = k\frac{W}{H}s$$

(3.63)

Dividing the above equation by the *apparent area* of contact yields:

$$\frac{V}{A} = k\frac{W/A}{H}s$$

(3.64)

The left-hand side is simply the *wear depth, d*, and the W/A on the right-hand side is the *nominal (mean) pressure*. Therefore, the wear depth can be determined from the following relationship for design purposes where one is interested in guarding against failure during its useful life.

$$d = k\frac{P_m}{H}s$$

(3.65)

Note that interestingly the coefficient of friction does not directly appear in Archard's wear law. To explain this, Table 3.6 shows the results of a series of unlubricated wear tests reported by Archard (1980) with wide range of materials where the load 3.9 N (0.4 kgf) at a sliding speed of 1.8 m/s. It shows that the lowest coefficient of friction was 0.18 for PTFE against tool steel and the maximum friction coefficient is 0.62 for the case of unlubricated mild steel

Table 3.6 Friction coefficient, wear rate and wear coefficient for unlubricated materials based on tests conducted with 3.9N load at 1.8 m/s

Wearing surface	Counter surface	Friction coefficient	Hardness (kg/mm^2)	Wear coefficient	Wear rate (10^{-6} mm^3 per m sliding)
Mild Steel	Mild Steel	0.62	186	7×10^{-3}	1.57×10^4
60/40 Lead Brass	Tool Steel	0.24	95	6×10^{-4}	2.4×10^3
PTFE	Tool Steel	0.18	5	2.4×10^{-5}	2.0×10^3
Stellite	Tool Steel	0.60	690	5.5×10^{-5}	32
Ferritic Stainless Steel	Tool Steel	0.53	250	1.7×10^{-5}	27
Polyethylene	Tool Steel	0.53	17	1.3×10^{-5}	3
Tungsten Carbide	Tungsten Carbide	0.35	1300	1×10^{-6}	0.3

Source: Archard, 1980. Reproduced with permission of ASME.

rubbing against mild steel. Archard reasons that this difference is insignificant compared to the corresponding wear coefficient that varies by several orders of magnitude.

It should be emphasized again that the hardness H in Archard's equation corresponds to that of the softer (wearing) material. Based on large number of experiments with wide scatter, Rabinowicz (1980, 1995) suggests that the wear coefficient of the hard material, k_H, can be taken to be $k_H \approx \frac{k_s}{3}$, where k_s is the wear coefficient for the softer material.

Example 3.3 Verify the value of volumetric wear rate (mm^3 per m sliding distance) for PTFE rubbing against tool steel shown in Table 3.6.

The wear rate can easily be calculated using the following equation.

$$\frac{V}{s} = k\frac{W}{H}$$

$$\frac{V}{s} = 2.4 \times 10^{-5}\frac{0.4}{5}$$

$$= 1.92 \times 10^{-6} \ mm^3/mm$$
$$= 1.92 \times 10^{-3} \ mm^3/m$$
$$\sim 2.0 \times 10^3 \quad (\times 10^{-6} \ mm^3/m)$$

Example 3.4 A 15-mm-long unlubricated copper-lead bushing of 25 Brinell hardness ($25 \ kg/mm^2$) and diameter of 30.4 mm carries a 30-mm diameter steel shaft (150 Brinell hardness, $150 \ kg/mm^2$) at 50 rpm under a load of 10 N (1.02 kgf).

a. Determine the wear volume after 100 hours of operation.
b. Assuming that the contact between the shaft and the bushing can be categorized as Hertzian, estimate the radial wear depth. The modulus of elasticity and Poisson's ratio of Steel are 200 GPa and 0.28, respectively and the modulus and Poisson's ratio of copper are 117 GPa and 0.33, respectively.
c. Estimate the wear of shaft.

a. The sliding distance corresponding to 100 hours of operation is:

$$s = \pi D(\text{revs}) = \pi(30)(50 \times 60 \times 100) = 2.83 \times 10^7 \ mm \ (2.83 \times 10^4 \ m)$$

Taking as partially compatible (Figure 3.22), the wear coefficient from Figure 3.20 is estimated to be $k = 1.5 \times 10^{-5}$.

Applying Equation 3.63 for wear volume V (mm^3) of the softer bronze:

$$V = ksW/H = (1.5 \times 10^{-5})(2.83 \times 107)(1.02)/25 = 17.3 \ mm^3$$

b. Assuming that Hertzian analysis applies, we now proceed to determine the mean pressure between the shaft and the bushing. Referring to Figure 3.30, first, the equivalent radius should be calculated.

$$\frac{1}{R} = \frac{1}{R_b} - \frac{1}{R_s} = \frac{1}{15 \times 10^{-3}} - \frac{1}{15.2 \times 10^{-3}} = 8.77 \ m$$
$$R = 1.140 \ m$$

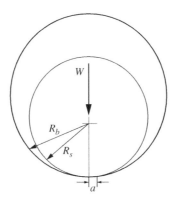

Figure 3.30   Schematic of the contact between the shaft and the bushing.

The equivalent modulus of elasticity is:

$$\frac{1}{E} = \frac{1 - v_b}{E_b} + \frac{1 - v_s}{E_s}$$

$$E = 81.8 \text{ GPa}$$

The load per unit length is: $P_L = \dfrac{W}{L} = \dfrac{10}{15 \times 10^{-3}} = 666.7 \text{ N/m}$

Referring now to Table 3.3, the semi-width of the Hertzian contact can be predicted using the following expression.

$$a = 2 \left(\frac{P_L R}{\pi E} \right)^{1/2} = 2 \left(\frac{666.7 \times 1.140}{\pi \times 81.8 \times 10^9} \right)^{1/2} = 1.09 \times 10^{-4} \text{ m}$$

The mean pressure is $P_m = \dfrac{P_L}{2a} = \dfrac{666.7}{2 \times 1.09 \times 10^{-4}} = 3.06 \times 10^6 \text{ N/m}^2$

Therefore, the wear depth is estimated using Equation (3.65):

$$d = k \frac{P_m}{H} s = 1.5 \times 10^{-5} \times \frac{3.06 \times 10^6}{25 \times 10^6 \times 9.8} \times 2.83 \times 10^4 = 5.3 \times 10^{-3} \text{ m} \quad (5.3 \text{ mm})$$

Note that this is only an approximate estimate since the validity of Hertzian assumption for this configuration must be examined. This analysis also tacitly assumes that wear is confined to the Hertzian contact area and that it does not change with time.

c. Shaft Wear. To assess the wear of a shaft three or more times harder than its bushing, Rabinowicz (1995) suggests surface wear volume will be less by three times the hardness ratio. With this assumption, the shaft wear volume becomes a factor of $3(150/25) = 18$ times less, or $17.3/18 = 0.96 \text{ mm}^3$. With this taking place over the shaft area of $\pi DL = (\pi \times 30)(15) = 1414 \text{ mm}^2$, shaft wear depth would be only 0.0007 mm. For a shaft less than three times as hard as the softer bushing, the harder material wear volume would be less by a factor of the hardness difference squared.

The complex nature of wear and its prediction have long been of interest and will likely remain a subject of further research for years to come. It is appropriate to briefly discuss some of the more recent interpretations and progress made toward understanding of wear characteristics. Of particular interest here is the prediction and interpretation of wear coefficient by the theory of fatigue analysis, the delamination wear theory, and via thermodynamic laws.

Physical Meaning of the Wear Coefficient in Adhesive Wear

Further insights into the meaning of the adhesive wear coefficient can be gained by performing a dimensional analysis, often known as the Buckingham Pi theorem. Heresy (1966) presents the application of dimensional analysis in tribology where he shows the derivation of dimensionless parameter $Z = \mu U/P$, known as Hersey number, where μ, U, and P represent the lubricant viscosity, sliding speed, and load per unit length, respectively. Hersey number is useful for characterizing different lubrication regimes as prescribed by the Stribeck curve (Figure 1.4). Relevant to the subject of adhesive wear, Keikichi and Kunikazu (1960) performed a detailed dimensional analysis of the steady-state unlubricated, adhesive wear and showed that Archard's law can, in fact, be deduced from dimensional analysis. Shaw (1977) provides an adhesive wear analysis with special interest in relating the parameters to what one measures using a tribometer with a single pair of metals. He begins his analysis by postulating that the wear volume, V, is a function of sliding distance, s, applied load, W, the hardness of the softer material, H and the sliding velocity, U, as follows.

$$V = \psi(W, H, s, U) \tag{3.66}$$

where ψ represents some functions of the quantities within the parentheses. After dimensional analysis, he obtains:

$$VH/Ws = K = const. \tag{3.67}$$

Equation (3.67) is, of course, similar to Archard's wear law with K representing the non-dimensional wear coefficient. Shaw goes on to offer the following interesting interpretation of wear coefficient K. The non-dimensional wear coefficient is proportional to the ratio of wear volume to the volume that is plastically deformed during the sliding distance s. Hardness, H, in Eq. (3.67) can be replaced with W/A_H, where A_H represents the area of the contact of the harder material. This area is proportional to the plastically deformed area of softer material, A_S. Therefore, one can rewrite Equation (3.67) in the following form.

$$K = \frac{VH}{Ws} = \frac{VW/A_H}{Ws} = \frac{V}{sA_H} \approx \frac{V}{sA_S} = \frac{V}{V_p} \tag{3.68}$$

where V_p represent the plastically deformed wear volume.

According to Equation (3.68), the wear coefficient is directly proportional to the ratio of the worn volume to that of the volume of the plastically deformed softer material, $K \approx V/V_p$. Archard (1980) reasons that this is why, depending on the friction pair, the wear coefficient can be as low as 10^{-6} or 10^{-7}.

During the running-in phase, when the asperities undergo polishing, the behavior of wear is time-dependent and significantly more involved. A recent study on the associated dimensional analysis is reported by Mortazavi and Khonsari (2016).

The Fatigue Theory of Adhesive Wear

If the junction does not break in the softer material and instead occurs at the original interface, then the interacting asperities are effectively subjected to cyclic stress that eventually leads to fatigue crack formation. While fatigue is known to be the classic failure mode of roller contact in ball and roller bearings, there are many studies that contend that sliding wear and fatigue are intimately related. According to Kragelskii (1965) and Rozeanu (1963), loose wear fragment are generated in successive steps as the surface asperities are stressed and eventually break off. Kimura (1981) also considers fatigue to be responsible for wear simply due to the repetitive nature of practical sliding system.

Experimental evidence for fatigue during sliding wear is provided by Heilmann and Rigney (1981). They conducted a series of block-on-ring wear tests with Cu-Ni alloys test blocks sliding against AISI 440 C stainless steel rings with 58-62 Rockwell C hardness subjected to 20.45 kg (45 lb) normal load. Their experiments showed that the coefficient of friction repeatedly drops and recovers to its previous level after a few minutes of sudden decrease (Figure 3.31). Their experiments also showed that both the block and the ring plastically deform. Wear particles generated had mixed composition, and that some of the particles had transferred to the surface of the specimen. Further observations using Scanning Electron Microscope (SEM) of the wear track on the Cu-Ni block revealed clear evidence of formation of striations similar to those commonly associated with fatigue failure (Figure 3.32). As a classic feature representative of fatigue crack growth, striation-like signatures are produced by relative motion between fracture surfaces during cyclic loading.

More recently, using fatigue theory of adhesive wear, Arnell *et al.* (1991) provide a more plausible explanation for the observation that a harder material can be worn by a softer one,

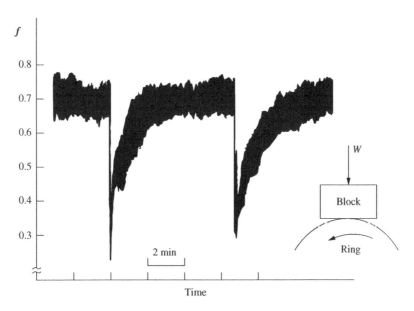

Figure 3.31 Friction coefficient measurement via of block on ring test apparatus. *Source*: Heilmann and Rigney, 1981. Reproduced with permission of Springer.

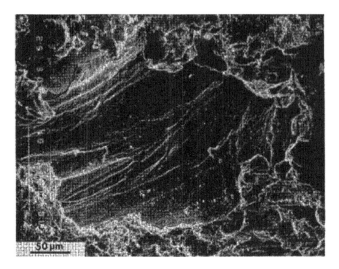

Figure 3.32 An SEM picture showing evidence of the occurrence of striations as the result of sliding wear similar to that commonly observed in fatigue failure. *Source*: Heilmann and Rigney, 1981. Reproduced with permission of Springer.

a phenomenon which cannot be adequately explained by adhesion theory alone. Clearly, the complex process of adhesive wear requires consideration of both adhesion and fatigue mechanisms.

Interpretation of Wear Coefficient by Fatigue Analysis

The fatigue theory of adhesive wear postulates that the wear coefficient can be interpreted as the inverse of $\mathcal{N}$ 'number of events' required for the formation and detachment of a wear particle from the surface asperities of the softer body. Kragelskii (1965) proposes that these events are associated with the number of the loading cycle. He defines the wear coefficient to be 1/3 of probability ($\mathcal{P}$) that an asperity forms a wear particle:

$$k = \frac{\mathcal{P}}{3} = \frac{1}{3\mathcal{N}} \tag{3.69}$$

Based on this relationship, adhesive wear can be interpreted as the detachment of wear particles from the surface asperities as a result of fatigue fracture. Given a sufficient number of cycles, this can occur not only from the softer surface but also from that of the harder one. As described before, this phenomenon is difficult to explain using the adhesive theory without considering the interior of the weak regions (Figure 3.33).

Equation (3.69) directly relates the coefficient of sliding wear to the fatigue properties, and if, for a specific material and the loading condition, the number of cycles to fatigue failure $\mathcal{N}$ is known, then one can estimate the wear coefficient. $\mathcal{N}$ can typically range from 10^3 to 10^6 and its precise determination is a subject of considerable interest in many branches of applied mechanics. See, for example, an extensive review given by Fatemi and Yang (1998).

Recently, Beheshti and Khonsari (2010) conducted a study that uses the modern continuum damage mechanic approach to estimate the strength of an asperity and wear coefficient

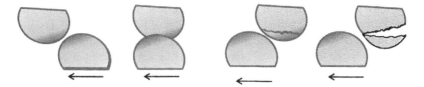

Figure 3.33   Progression of asperity fatigue from the early undamaged stage (left) to eventual formation of wear fragment (right) after $\mathcal{N}$ number of cycles.

analytically based on the bulk properties of the material. The results of their predictions were compared with the measured wear coefficient using a pin-on-disk apparatus. The stationary pin (the wearing, i.e. the softer material) was 8 mm in diameter, made of AL 6061-T6, and the rotating disk was 100 mm in diameter, made of SS 304. The range of sliding velocity tested was 30 mm/s to 120 mm/s, which corresponds to 10–40 rpm. The normal load ranged from 3–30 N. The speed and load were held constant during each test, which lasted from 2–5 hours which corresponds to a sliding distance of 250–2500 m. The amount of wear and consequently the wear coefficient were obtained by measuring the weight of the pin before and after the test by means of precise digital scale with the accuracy of 0.0001 gr. The results of the experiment and predictions are shown in Figure 3.34. Reasonable agreement between these predictions and the measured value shows encouraging results. Additional predictions of the theory with different materials and operating conditions together with comparison with published experimental work are also given in Beheshti and Khonsari (2010).

The Delamination Theory of Wear

While the adhesion theory is very useful due to its simplified nature, it is less satisfactory in providing answers to some critical questions. For example, the assertion that the asperity-to-asperity interaction is the only mechanism for particle generation implies that wear rate at

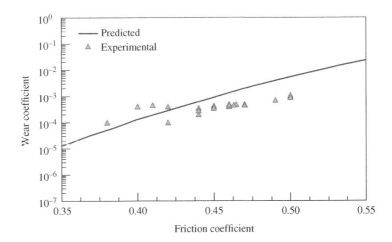

Figure 3.34   Predicted wear coefficient as a function of the measured coefficient of friction. *Source*: Beheshti, 2010. Reproduced with permission of John Wiley & Sons.

steady state must necessarily depend on surface roughness. This is contrary to experimental observation. Further, the explanation that the formation of hard wear fragments are due to a weak region in the hard surface is perhaps over-simplified and not entirely satisfactory. Nevertheless, it points to the importance of consideration of microstructural effects.

In pursuit of answers to some of these intriguing questions, Suh (1973) developed a so-called delamination theory of wear with a comprehensive overview published in 1977 (Suh, 1977) with further development and implications (Suh *et al.*, 1977; Jahanmir, 1977). According to this theory, the surface asperities of the softer body are repeatedly deformed by the passage of the harder body generating wear fragments as its surface polishes and becomes smooth in the process. Concurrently, the subsurface experiences plastic deformation due to the movement of dislocations. Now, the dislocations very near the surface tend to 'pull out' of the surface and break through the oxide layer on the surface, whereas the substrate dislocations at a certain depth below the surface tend to be stable and cause strain hardening instead. Thus, a flow stress gradient is developed that increases from the surface inward, reaching a maximum value after which it again decreases. This stress gradient in the soft body accommodates penetration of the harder surface asperities and permits a plowing action while the substrate is further deformed. The deformation results in the nucleation of voids where the stress created by deformation exceeds the compressive stress. This is confined to the substrate since very near the surface the tri-axial state of compressive stress is at its maximum and does not permit voids to form. Cracks are then nucleated at these voids and tend to propagate and coalesce upon repeated stress in the direction than runs parallel to the surface. Eventually after elongating to a critical length, the crack moves upward toward the surface and sheet-like wear fragments (Figure 3.35) are dislodged from the surface, hence the name delamination. These laminated sheets extend

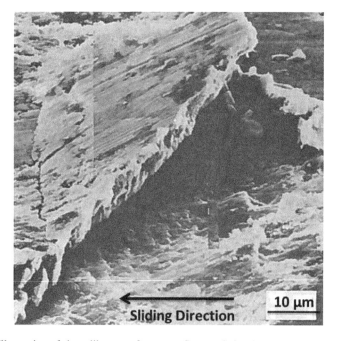

Figure 3.35   Illustration of sheet-like wear fragment. *Source*: Suh, 1977. Reproduced with permission of John Wiley & Sons.

over a distance, one to two orders of magnitude longer than the diameter of the asperity contact (Suh *et al.*, 1974).

Interpretation of the Wear Coefficient by the Delamination Theory

As described earlier, the delamination theory of wear postulates that cracks nucleated below the surface tend to propagate parallel to the surface and coalesce upon linking with other cracks to reach a critical length where the crack breaks through the surface and forms a sheet-like wear fragment.

In general, crack propagation for *each cycle* depends on the coefficient of friction at the asperity contact with the surface, the depth below the surface where the crack is located, the crack length, the applied stress, and material properties.

To quantify the prediction of this theory, Suh and Sin (1983) derived the following relationship for wear rate due to an existing crack of length C located at the depth d_c below the surface. Referring to Figure 3.36, the wear rate is:

$$\frac{V}{s} = \frac{L_c^2.d_c.(\Delta C_L + \Delta C_R)}{\lambda.l_c} \tag{3.70}$$

where ΔC_L and ΔC_R represent the *average* crack-length propagation rates at the end of the crack (left and right) due to the passage of an asperity traveling from left to right. Parameters λ and l_c represent the asperity contact spacing and crack spacing.

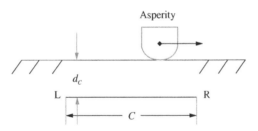

Figure 3.36 Illustration of subsurface crack below a surface acted upon an asperity moving from left (L) to right (R) at the surface.

Suh and Sin (1983) performed an analysis using the finite element method (FEM) to examine the plastic formation zone around a crack subjected to a moving asperity load at eight specific locations along the surface. Input to the program was the friction coefficient as well as material properties such as the modulus of elasticity, Poisson's ratio, and the shear yield stress. This analysis allowed estimation of the average values of ΔC_L and ΔC_R. Suh and Sin also provided the following example to illustrate how Equation (3.70) can be used.

Example 3.5 Consider a 20-μm crack located at the depth of $d_c = 5$ μm below the surface. Let the asperity contact length be $L = 2$ mm. The contact spacing and contact length are assumed to be $\lambda = l_c = 100$ μm, and $\Delta C_L = \Delta C_R = 0.0035$ μm.

Determine the wear rate and wear coefficient assuming that $H = 100$ kg/mm^2 and $W = 1$ kg.

Using Equation (3.70), we have

$$\frac{V}{S} = \frac{L_c^2 . d_c . (\Delta C_L + \Delta C_R)}{\lambda . l_c}$$

$$= \frac{(2 \times 10^{-3})^2 (5 \times 10^{-6})(0.007 \times 10^{-6})}{(100 \times 10^{-6})(100 \times 10^{-6})}$$

$$= 1.4 \times 10^{-11} \text{ m}^3/\text{m} \quad (1.4 \times 10^{-2} \text{ mm}^3 \text{ per m sliding})$$

The wear coefficient is simply

$$k = \frac{2H}{W}\left(\frac{V}{s}\right) = \frac{3 \times 100}{1}(1.4 \times 10^{-11}) = 4.2 \times 10^{-3}$$

Abrasive Wear

There are different types of abrasive wear. In the simplest case the source of abrasive wear may be the transition of sliding wear at it progresses in time. This can occur even in the absence of an external source. Wear fragments generated during sliding in a friction pair give birth to the third-body effect – a term coined by Godet in the 1970s to describe the particulate medium (i.e. the third body) formed at the interface of two bodies in contact (Godet, 1984; Berthier, 1996). In the case of dry (unlubricated) contacts, the third body can accommodate shearing action and thus protect the surfaces.

Abrasive wear can also come about when hard and abrasive particles enter the system, causing abrasion of both the soft and hard components of the friction pair simultaneously, albeit at a different rate. The result is the formation of grooves on the surfaces as a hard particle plows through and leaves a wear track with ridges on the surfaces with often troublesome outcomes. This type of wear is widespread in the field, particularly in applications such as in drilling and mining as well as earth-moving machinery where abrasive contaminants are likely to be present. Unfortunately, excessive replacement costs and associated maintenance requirements make it impractical to keep the system entirely clean by filtering out all the particles. As depicted in Figure 3.37, when an abrasive particle enters a journal bearing, for example, it is quickly forced to travel with the shaft in the direction of rotation as the particle simultaneously

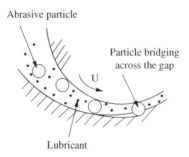

Figure 3.37   Illustration of an abrasive contaminant entering the clearance space of a journal bearing. The particle scores the shaft as it plows through the protective soft layer where it partially embeds itself.

begins to penetrate into and plow through the soft metal on the bushing inner surface (Khonsari *et al.*, 1999; Pascovici and Khonsari, 2001). In troublesome cases, the particle locks itself into a partially embedded position and bridges the gap between the shaft and the bushing. Large local temperatures at particle-shaft interface are known to cause bearing failure (Khonsari and Booser, 2006).

The third type of abrasive encountered is the deliberate introduction of abrasive powder for the purpose of polishing and lapping often required in finishing processes such as in the preparation of mechanical face seals. In these instances, the wear rate could be quite high. Of course, polishing without the use of abrasive is also possible. *Broaching*, for example, is a machining process to remove material using a toothed tool and a *burnishing* process up to certain level can help to polish a surface by removing the asperities while improving the surface hardness at the same time. It has been suggested that burnishing wear can be characterized by developing a model of plastic waves in front of hard asperities with small slopes.

Abrasive Wear Rate and Abrasive Wear Coefficient

A very simple model can be derived to gain some insight into the abrasive wear process. As in the case of adhesive wear, we consider a single asperity. Here, the worst case scenario would be a conically shaped tip angle of 2θ that has plowed through a distance s in the softer material as shown in Figure 3.38.

Referring to Figure 3.38, note that only half of the area pushed by the conical indentor is active. That is:

$$dA = \frac{\pi}{2} h^2 \tan^2 \theta$$

with the associated load of

$$dW = dA.H = \frac{\pi}{2} h^2 \tan^2 \theta$$

The grooved volume formed over a distance ds is: $dV = h^2 ds \tan \theta$

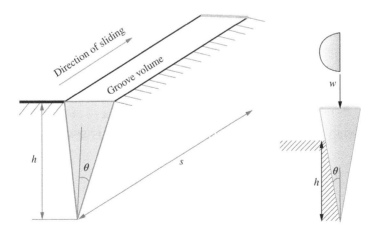

Figure 3.38   A simple model of a conical shape abrasive particle: perspective view and details.

Therefore, the wear rate becomes

$$\frac{dV}{ds} = \frac{2}{\pi} \cot \theta \frac{dW}{H}$$

which sums up to

$$\frac{V}{s} = K_a \frac{2 \cot \theta}{\pi} \frac{W}{H}$$

Finally, letting $k_a = K_a \dfrac{2 \cot \theta}{\pi}$, we arrive at the following abrasive wear rate equation

$$\frac{V}{s} = k_a \frac{W}{H} \tag{3.71}$$

Thus, the abrasive wear equation has an identical form to that of Archard's equation for adhesive wear. This, of course, should not be surprising because they both should be governed by the same dimensional analysis as presented in the previous section.

Equation (3.71) is of limited utility since it over-simplifies a very complex process. In practice, only a portion of the grooved volume is detached as wear fragment, while some material piles up at the edges.

Typical abrasive wear coefficients are given in the upper two curves of Figure 3.39 (Kennedy *et al.*, 2005). Wear coefficients in the two-body case range from 6×10^{-2} to 3×10^{-3}, while they are about an order of magnitude smaller for the three-body case, namely 3×10^{-3} to 3×10^{-4}. This difference suggests that the abrasive grains in the three-body case spend about 90% of their time in rolling action and only 10% in sliding and abrading the surfaces (Rabinowicz, 1995). This is borne out in the low coefficient of friction during three-body abrasion with $f = 0.25$, versus $f = 0.60$ for two-body abrasion.

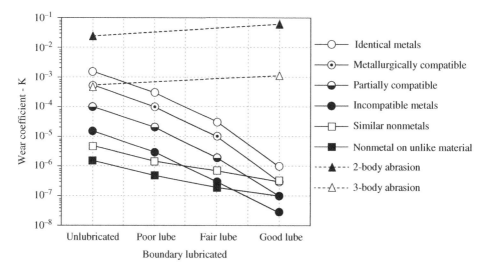

Figure 3.39 Coefficient of friction for sliding and abrasive wear. *Source*: Kennedy, 2005. Reproduced with permission of Taylor & Francis.

Table 3.7   Abrasion-resistant materials

Material	Hardness (kg/mm^2)
Bearing steel	700–950
Tool steel	700–1000
Chromium electroplated	900
Carburized steel	900
Nitrided steel	900–1250
Tungsten carbide, cobalt binder	1400–1800

Source: Rabinowicz, 1995. Reproduced with permission of John Wiley & Sons.

The most common contaminant in nature is silica with a hardness of 800 kg/mm^2. A number of potentially hard enough materials to resist its abrasion are listed in Table 3.7. Eliminating abrasive particles by filtering both a circulating lubricant and any intake air is also helpful, especially as augmented by having a soft bearing material of 20–30 kg/mm^2 from Table 4.3 to collect and bury any abrasive particles entering the system.

A useful alternative in many cases is a highly elastic bearing material to take up wear and contaminant particles elastically rather than by plastic deformation. Water-lubricated rubber bearings for ship propeller shafts, for instance, have proved much more effective with significantly less wear in sandy water than bronze or lignum vitae bearings.

Corrosive Wear

When the environment reacts chemically with surfaces, their sliding contact can then produce corrosive wear. The initial stage involves a corrosive attack on a surface identical to ordinary corrosion. The second stage then consists of wearing away this corroded surface. If the corrosion layer is harder and more brittle than the original surface, the total thickness of the corrosion layer may flake off completely at one time to give a relatively rapid corrosion rate, as in Figure 3.40. With common journal bearing materials on which a sulfide layer forms from oil additives, only part of the soft corrosion layer is removed to give a far smaller corrosion rate as in Figure 3.41.

Corrosive wear is a common concern in oil-film bearings (see Section 4.1). Most susceptible to corrosion are bearing materials containing lead, cadmium, zinc and copper – with silver distinctively subject to sulfur attack. Oxidized lubricating oil, and especially with water contamination, is usually the most likely corrosive media.

With ferrous metals in gears, cams, and ball and roller bearings, water and atmospheric moisture are the primary culprits.

Surface Fatigue, Brittle Fracture, Impact, Erosion

Various other factors may also contribute to wear. Life-limiting wear, commonly from dirt or lack of sufficient lubrication with aged grease, has been a major life factor with many ball and roller bearings – as has been *surface fatigue* from the repeated stressing in rolling contacts (see Chapters 14 and 15).

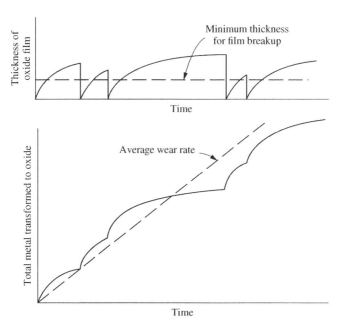

Figure 3.40 Schematic view of high rate of wear with periodic complete removal of a brittle oxide corrosion layer. *Source*: Rabinowicz, 1995. Reproduced with permission of John Wiley & Sons.

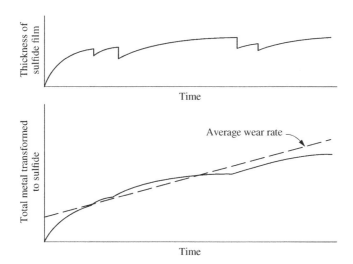

Figure 3.41 Steps for slower wear rate with only partial removal of an adherent oil additive sulfide film without exposing the base metal surface. *Source*: Rabinowicz, 1995. Reproduced with permission of John Wiley & Sons.

Fracture wear may occur in brittle materials such as glass and sintered ceramic. A characteristic series of initial cracks in the wear track tends to be followed by large wear particles during subsequent surface breakup.

Surface wear also results from repeated *impact* of engineering components such as cams and gears. The cause has been attributed to a wear mechanism such as in sliding, to surface fatigue, or surface fracture wear. When impact energies are very large, rapid material removal occurs by surface fracture wear; when impact energies are low, wear is by surface fatigue (Rabinowicz, 1995).

Thermodynamics of Wear

With its different features and attributes, wear is an extremely complex phenomenon to characterize for predictive purposes. As noted by many authors—and in particular by Meng and Ludema in a comprehensive study of more than 100 influential parameters in friction—there exists a scarcity of predictive models to account for the multifarious processes involved in friction and wear (Meng and Ludema, 1995). Accordingly, the identification of the unifying principles that govern these processes is, indeed, an important applied scientific endeavor and an active area of current research (Bryant, 2010; Aghdam and Khonsari, 2014, 2016).

In *essence*, friction and wear are examples of dissipative processes in which the system's *free energy*, Ψ, responsible for doing useful work, decays with time as degradation progresses. Put in other words, if we let Ψ_i represent the initial free energy of a pristine material, then after completion of the dissipative process, its free energy decreases to Ψ_f such that $\Psi_f < \Psi_i$. This decay in the free energy continues until the system attains a minimum at the equilibrium state in accordance with the principle of minimum free energy (Figure 3.42). Of course, the free energy and the internal energy, U, are related via temperature, T, and system's entropy, S, by $\Psi = U - TS$. Thus, the system's path to the minimum free-energy is always accompanied by increasing entropy until it reaches its peak value at the equilibrium state (Amiri *et al.*, 2012).

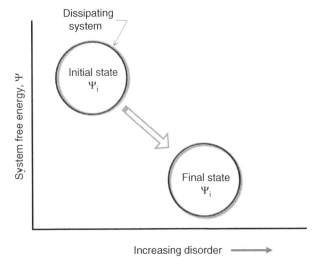

Figure 3.42   The evolution of degradation in a system.

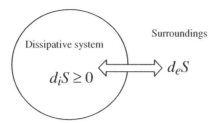

Figure 3.43 Entropy of a system consists of entropy generation and entropy flow.

The increase in entropy is a consequence of increasing disorder in the system with time. Hence, entropy is often referred to as the 'arrow of time' (Prigogine, 1997).[2]

The concept of the characterization of dissipative systems via entropy provides a new path forward in treating material degradation behavior in systems involving friction, wear, and fatigue. These dissipative processes involve a variety of complex and physically diverse interfacial phenomena that often occur in an inextricably intertwined fashion. Notwithstanding the multiplicity of underlying dissipative processes in friction, they share one unique feature: they all produce entropy. Therefore, thermodynamic entropy production is believed to be a propitious measure for a systematic study of degradation. To this end, methodologies that are based on irreversible thermodynamics have recently emerged as a unifying tool for the analysis of degradation in a wide range of engineering applications (Aghdam and Khonsari, 2011).

While the full description of the methodology is beyond the scope of this book, it is nonetheless worthwhile to briefly review how dissipative processes evolve within the context of irreversible thermodynamics (Amiri *et al.*, 2010; Khonsari and Amiri, 2013; Amiri and Modarres, 2014).

The total entropy of a system dS is composed of two terms: entropy generation within the system, d_iS, and entropy flow (flux), d_eS, as a result of exchange of heat or matter between the system and its surroundings (Figure 3.43):

$$dS = d_iS + d_eS \tag{3.72}$$

According to the second law of thermodynamics, $d_iS \geq 0$, but d_eS can be negative, or positive for an open system or zero in an isolated system that is not permitted to exchange with the surroundings.

For a system subjected to irreversible process (e.g. friction), entropy is generated via the following equation:

$$\frac{d_iS}{dt} = \sum XJ \tag{3.73}$$

where X and J are referred to as the thermodynamic forces and flows, respectively. If we assume that the tribosystem only exchanges heat with its surroundings at the rate $\dot{Q}$, then the system's

---

[2] Ilya Prigogine was awarded the Nobel Prize for his contribution to dissipative structures, complex systems, and thermodynamic irreversibility in 1977.

entropy flow is:

$$\frac{d_e S}{dt} = \frac{d\dot{Q}}{T} \tag{3.74}$$

where T is the temperature.

Experimentally, the concept of entropy flow provides a convenient means for measuring the entropy generation since at steady state, the rate of entropy generation must be identical to the entropy flow:

$$d_i S = -d_e S \geq 0 \tag{3.75}$$

Now, the interfacial interactions between surfaces are influenced by the operating conditions (e.g. sliding speed, load, interfacial temperature), the material properties (e.g. hardness) and coefficient of friction. All these parameters directly affect the entropy flow, and their influence can be assessed by proper measurement. Doelling $et\ al.$ (2000) applied this concept to a boundary lubricated system consisting of a copper rider (D) on a steel slider (F) instrumented with three thermocouples as shown in Figure 3.44. Doelling $et\ al.$ (2000) measured the entropy flow $d_e S/dt$ by calculating $S_n = \sum^n \frac{\Delta Q^{(n)}}{T^{(n)}}$, where $\Delta Q^{(n)}$ is the heat input to the slider during the nth time interval (n indexes time), and $T^{(n)}$ is the corresponding average absolute surface temperature of the stationary slider rubbing a rotating cylinder. The slider and the rider were made of AISI 1020 steel and C110 copper, respectively. Also measured was wear via surface recession. The results of their measured normalized wear were plotted as a function of normalized entropy and shown in Figure 3.44. These correspond to $W = 9.1$ kg load and a speed of 3.3 m/s.

Starting from Archard's wear law equation, Doelling $et\ al.$ (2000) show that

$$V = k \frac{TS}{fH}$$

where T is temperature, S is the measured entropy flow and f is the friction coefficient. It can also be shown that the Archard's wear coefficient is

$$k = B \frac{\mu H}{T}$$

where B is the slope of the measured results from the wear-entropy flow (Figure 3.44).

Using these results, Doelling $et\ al.$ (2000) calculated an average wear coefficient for this system and obtained $k = 1.01 \times 10^{-4}$. Interestingly, for some metals under poor lubrication, Rabinowicz (1980) gives $k = 1.0 \times 10^{-4}$, which is remarkably close. It is important to note that in these experiments, the wear coefficient was estimated by measuring wear, temperatures and forces, whereas Rabinowicz's results are obtained by measuring wear, forces and distance.

To further explore the nature of this finding, Bryant $et\ al.$ (2008) put forward a general theorem that can be used to analyze the degradation of a system via entropic characterization. They showed that the slope of measured wear-entropy represents a so-called degradation coefficient which is directly related to the Archard's wear coefficient. Tribometer tests instrumented with LDVT to measure wear and thermocouples to measure temperature in dry sliding contact with brass on steel and bronze on steel also resulted in a remarkably close estimation of Archard's wear coefficient (Amiri $et\ al.$, 2010). These results are indicative of the usefulness of the approach. The interested reader may refer to Quraishi $et\ al.$ (2005), Dai (2008), Bryant

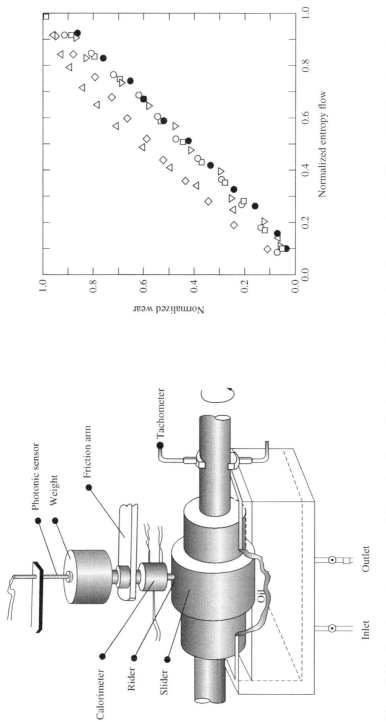

Figure 3.44 Experimental apparatus and the results of normalized wear plotted as a function of entropy flow (Doelling *et al.*, 2000). The experimental symbols correspond to measurements at different surface temperatures ranging from approximately 50–65 °C. $W = 9.1$ kg and $U = 3.3$ m/s.

(2010), and Amiri and Khonsari (2010), Aghdam and Khonsari (2011, 2013, 2014a, 2014b, 2016), Abdel-Aal (2013) for further discussion of this exciting area of research.

Classification of Wear, Failure, and Wear Maps

Wear strongly depends on the operating conditions and its behavior varies depending on which wear regimes it operates in. The most influential parameters are the load and speed. Load affects the level of stress and speed controls the interfacial temperature.

Broadly, wear can be categorized into mild, moderate, and severe. *Mild wear coefficient* is typically in the range of $k \approx 10^{-8}$ to 10^{-6}. *Moderate wear* occurs when the wear coefficient increases by two orders of magnitude, i.e. $k \approx 10^{-6}$ to 10^{-4}; and the coefficient of wear in the case of *severe wear* is expected to be at least two orders of magnitude higher than moderate wear, in the range of $k \approx 10^{-4}$ to 10^{-2}.

Increasing the load/speed can cause failure in the form of scuffing or seizure, the definition of which follows.

Scuffing is a form of localized surface damage wherein sliding surfaces exhibit welding. Scuffed surfaces can easily be detected with the naked eye even though surfaces do not show evidence of localized melting. Yet there is significant wear associated with scuffed surfaces.

Galling is a term used by some authors to identify severe form of scuffing and associated failure. However, the term galling is not descriptive of the form of failure. Thus, its use is discouraged (Peterson and Winer, 1980).

Seizure is a failure wherein the relative motion between sliding surfaces is halted due to intimate contact between surfaces and locking of the bodies together often with catastrophic consequences.

Lim and Ashbey (1987) developed an extensive series of maps that are useful for distinguishing different mechanisms of wear in the unlubricated sliding of a tribopair made of steel. Figure 3.45 shows one of the maps developed based on pin-on-disk tests. The contour lines represent the lines that have constant normalized wear rates. The normalized (dimensionless) velocity is defined by multiplying the velocity by the contact radius divided by the thermal diffusivity (ratio of thermal conductivity/(density × specific heat). Normalized pressure is simply the contact pressure divided by the hardness.

The thick boundary lines mark the transition between different mechanisms thought to govern the character of wear. At the lower left corner, at very low speeds and pressure, wear is ultra-mild with minor effect. The boundary of this mechanism is given by line AB. Line DE marks the transition to severe wear and seizure occurs at pressures higher than those shown by line CD. To the left of DE, wear process is considered to be governed by mechanical action so that the wear rate depends primarily on the contact pressure and to a much lesser extent on the velocity. To the right of DE, the effect of the sliding velocity becomes more pronounced and other effects such as thermal and chemical oxidation will dominate. More recent development of wear maps with discussion on thermal effects is reported by Lim (1998).

Before leaving the subject, it would be useful to describe some general characteristics of mild and severe wear. Typically, mild wear occurs at light loads and transitions to severe wear occur when load is increased. Mild wear is a slow and gradual process involving removal of the tip of the asperity peaks. These are primarily confined to the surface with little to no damage to the subsurface. In contrast, in severe wear, plastic deformation extends to well below the surface, and results in the formation of 'patches' of visible surface damage. Over time,

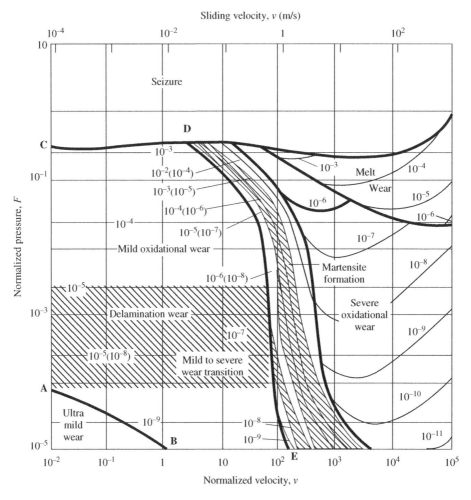

Figure 3.45 Wear map showing different wear regimes and transitions. *Source*: Lim and Ashbey, 1987. Reproduced with permission of Elsevier.

these patches grow, become rougher, and cover the entire contact area. Another interesting characteristic of mild wear is the oxidation that forms an insulating skin on the surface. Archard and Hirst (1956) report an appreciate contact resistance of up to 50 Ω and contend that wear is confined to this surface skin with wear particles that are on the order of 100 Å (1 Å = 10^{-10} m). In severe wear, particle fragments are much larger (0.1 mm) and considerably greater in number. Here the contact resistance is very small.

Dry Bearing Wear Life

For polymers, other dry bearing materials, and porous metal bearings, Equation 3.57 can conveniently be rearranged for life t in terms of a PV limit for an acceptable bearing wear depth d:

$$t = d(H/k)[1/PV]$$

with unit load (at mean pressure) P while sliding at velocity V for the softer bearing material with its hardness H and wear coefficient k. Typical PV limits for various non-metallic materials are given later in Table 12.1, with porous metals included in Table 12.3.

Lubricated Wear

Degree of wear encountered in lubricated bearing elements can broadly be categorized in terms of film thickness parameter Λ (see Section 3.9) where

$$\Lambda = \frac{\text{lubricant minimum film thickness, } h}{\text{composite surface roughness, } \sigma}$$

Expected film thickness h may be calculated using the information provided in later chapters for various bearing types and operating conditions. Effective surface roughness σ can normally be taken as the root-mean-square value or 1.25 times the arithmetic average peak-to-valley value (see Section 3.2), assuming that the surface heights obey the Gaussian distribution. As reviewed in Section 3.9, full-film separation of bearing surfaces ideally requires a minimum Λ of about 3.0 to avoid asperity contacts. Further elastic, thermal and alignment distortions raise this value to about 10 for most rolling-element and oil-film bearings. For Λ values below 10, where some surface contact may be expected, Figure 3.26 will give some idea as to the magnitude of wear rate to be expected.

Analytical prediction of lubricated wear is also a fruitful area of investigation (Beheshti and Khonsari, 2013). Clearly in hydrodynamic lubrication regime, wear is nil since by definition the surfaces are fully separated by a layer of lubricant. Therefore, lubricated wear can occur in boundary-lubricated and less severely in mixed lubrication regimes, where asperity contact is likely. Here, of course, the asperity contacts are less severe than dry contact since the tips of the asperities have trapped oil molecules that help reduce the friction coefficient. A simple analytical treatment (Stolarski, 1990) is to modify the Archard's wear coefficient for dry contact k by multiplying it by a so-called fractional film defect coefficient ψ ($k_l = k \times \psi$). ψ simply represents the ratio of the direct metal-on-metal contact area to the real area of contact. It also can be interpreted as the probability that an asperity comes into contact directly with another asperity while it passes over the mating surface in a region that is not occupied by the 'adsorbed lubricant molecules' (Stolarski, 1990). With this modification, k_l can be interpreted as the Archard wear coefficient for lubricated surfaces when asperity contact exists. In lubricated wear and especially mixed-lubrication regime, only part of the load is carried by the asperities and the other part is sustained by the fluid pressure. The load supported by the fluid does not contribute to wear and, thus, the total load needs to be replaced by the asperity load in the Archard wear relationship. For more details of the calculation of wear in a mixed-lubrication regime, the reader is referred to Wu and Cheng (1993), Akbarzadeh and Khonsari (2009, 2011) and Beheshti and Khonsari (2013), Masjedi and Khonsari (2012, 2014a, 2014b, 2015a, 2015b, 2015c, 2016).

General Progression of Wear

Figure 3.46 (Jones, 1970) shows the progression of the wear rate as a function of cycles of operational use. Initially, the surfaces tend to polish as the surfaces run in and the asperities are removed. This occurs concomitantly as the real area of contact increases substantially

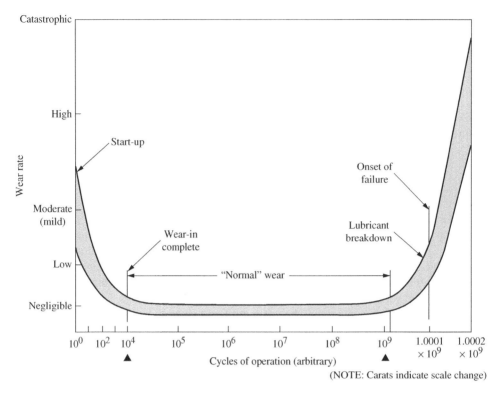

Figure 3.46 Wear rate during the life cycle of a bearing from the start through the running-in period to normal wear and eventual failure. Note that oscillatory motion can aggravate the process. *Source*: Jones, 1970. Reproduced with permission of NASA.

from the new bearing as the asperities plastically deform and their summits are truncated so that more asperities begin to participate in supporting the imposed load. While fairly rapid compared to the normal bearing life, the running-in mode represents quite a substantial wear rate (Akbarzadeh and Khonsari, 2013; Mehdizadeh *et al.*, 2015; Mortazavi and Khonsari, 2016). Once broken in, the process of normal wear begins with constant wear rate and continues over the normal life of the bearing. In the absence of contamination or misalignment issues during installation, a properly designed bearing with an adequate supply of lubricant experiences negligible wear, and can in theory function indefinitely without wear, except those that occur during stop/start. Practically, however, contamination, heat, lubricant breakdown, surface fatigue and external elements from the surroundings including perturbations in the operating conditions begin to deteriorate the bearing performance and cause wear to begin to set in. If unattended, wear rate then begins to increase until the bearing fails.

Effect of Load and Speed in Bearings

Operating conditions can significantly affect wear, not just the magnitude but also the regime. If, for example, the load in a bearing is gradually increased, the regime can change from being initially hydrodynamic with negligible wear to mixed lubrication where wear begins to set in. Further increase in load can easily move the operation into boundary lubrication with an

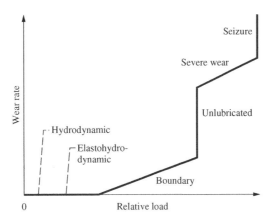

Figure 3.47   Expected trend in wear rate in different regimes. *Source*: Hamrock, 2004. Reproduced with permission of McGraw-Hill.

associated increase in the friction coefficient and wear. Increasing the load still further will increase the wear rate from initially mild to severe wear and eventually surfaces lock up, motion is halted, and the bearing catastrophically seizes (Figure 3.47).

Note that bearing seizure can occur very rapidly even at moderate loads during the start-up period if the contact is devoid of lubricant (Burton, 1965; Bishop and Ettles, 1982; Dufrane and Kannel, 1989; Kim and Khonsari, 1989; Hazlett and Khonsari, 1992a, 1992b). This has been reported in journal bearings that have been out of service for a relatively long period of time (e.g. in the compressor of air conditioning units). The problem is more severe in bearings with misaligned shafts (Khonsari and Kim, 1989), and may also be expected if the lubricant flow is temporarily interrupted (Krithivasan and Khonsari, 2003). Other examples include grease-lubricated pin-bushing systems used in excavators and earth-moving machinery that operate under heavy load in an oscillatory fashion (Jun and Khonsari, 2009).

The underlying problem in most cases is due to thermomechanical interaction and the rapid thermal expansion of the surfaces. Thus, it should be no surprise that, depending on the material properties of surfaces and the operating clearance, bearing seizure can occur even in the case of unloaded lubricated journal bearings. These troublesome problems are seen to occur in vertical pumps and mixers (Jang *et al.*, 1998a, 1998b). Wang (1997) provides a review of seizure in conformal bearings.

Seizure has also been reported in ball and rolling element bearings due to the change in the preload (Tu, 1995; Wang *et al.*, 1996). Chun (1988) finds increasing the cooling rate of the shaft decreases the thermal preload and stabilizes the thermal balance of the system so that seizure is avoided. A recent review of the state of the art covering both conformal and nonconformal bearings seizure is given by Takabi and Khonsari (2015).

Means of Wear Reduction

The following simple remedies may be used to reduce wear (Kennedy *et al.*, 2005):

- Use dissimilar materials for sliding pairs.
- Use hardened surfaces or tough materials.
- Add wear-resistant coatings.

- Reduce normal loads.
- Reduce surface temperatures (particularly for polymers).
- Reduce surface roughness of hard surfaces contacting soft surfaces.
- Reduce size of and remove abrasive particles.
- Grease and solid lubricants can be effective when liquid lubricants cannot be used.

Problems

3.1 For a simple sine wave surface profile, show that $R_q/R_a = 1.11$.

3.2 Calculate R_a, R_q, R_{sk}, and R_{ku} for the following symmetrical surface profiles: sinusoidal, square steps, and triangular. Each of the surfaces has a peak-to-valley height of 5.0 μm and a wavelength of 25 μm.

3.3 Sketch the amplitude density function (ADF) and bearing area curve (BAC) for each of the three cases in problem 3.2.

3.4 If the upper half of the simple sine wave surface in problem 3.2 is reduced to a flat plateau by honing, repeat the calculation of the same four surface texture parameters.

3.5 A 20 mm diameter pin slides at a velocity of 1m/s under 200N load on a flat carbon steel disc. Calculate the interface temperature rise for pins of carbon steel, copper, and nylon using the following characteristics:

Pin material	Friction coefficient, f	Thermal conductivity, k(W/m°C)	Diffusivity, κ(m^2/s)
Carbon steel	0.50	54	01.47×10^{-5}
Copper	0.36	386	11.23×10^{-5}
Nylon	0.30	0.25	00.013×10^{-5}

3.6 A flat carbon-graphite pump seal is to run against either a cast iron or Al_2O_3 ceramic seal face. Surface finish produces asperity slopes of 7° with $(\sigma/R) = 0.122$. What is the plasticity index for each combination based on the following:

	Seal carbon-graphite	Seal face Cast iron	Seal face Al_2O_3 ceramic
Elastic modulus, E (GPa)	24	120	370
Hardness, H (GPa)	0.90	0.235	15
Poisson's ratio, v	0.17	0.25	0.26

What general changes could be made to achieve elastic interface contact with a plasticity index of less than 1.0?

References

Abbott, E.J. and Firestone, F.A. 1933. 'Specifying Surface Quality A Method Based On Accurate Measurement and Comparison,' *Mechanical Engineering*, Vol. 55, p. 569.

Abdel-Aal, H.A. 2013. 'Thermodynamic modeling of Wear,' in *Encyclopedia of Tribology*. Q. Wang and Y.H. Chung, Eds. Springer, New York, NY, pp. 3622–3636.

Aghdam A.B., Beheshti. A., and Khonsari, M.M. 2012. 'On the Fretting Crack Nucleation With Provision For Size Effect,' *Tribology International*, Vol. 47, pp. 32–43.

Aghdam, A.B., Beheshti. A., and Khonsari, M.M. 2014a. 'Prediction of Crack Nucleation in Rough Line-Contact Fretting via Continuum Damage Mechanics Approach,' *Tribology Letters*, Vol. 53, pp. 631–643.

Aghdam, A.B. and Khonsari, M.M. 2011. 'On the Correlation Between Wear and Entropy in Dry Sliding Contact,' *Wear*, Vol. 279, pp. 761–790,

Aghdam, A.B. and Khonsari, M.M. 2013. 'Prediction of Wear In Reciprocating Dry Sliding Via Dissipated Energy and Temperature Rise,' *Tribology Letters*, Vol. 50, pp. 365–378.

Aghdam, A.B. and Khonsari, M.M. 2014b. 'Prediction of Wear in Grease-Lubricated Oscillatory Journal Bearings Via Energy-Based Approach,' *Wear*, Vol. 318. 188–201.

Aghdam, A.B. and Khonsari, M.M. 2016. 'Application of a Thermodynamically Based Wear Estimation Methodology,' *ASME Journal of Tribology*, Vol. 138, 041601:1–8.

Akbarzadeh, S. and Khonsari, M.M. 2008. 'Thermoelastohydrodynamic Analysis of Spur Gears with Consideration of Surface Roughness,' *Tribology Letters*, Vol. 32, pp. 129–141.

Akbarzadeh, S. and Khonsari, M.M. 2009. 'Prediction of Steady State Adhesive Wear in Spur Gears Using the EHL Load Sharing Concept,' *ASME Journal of Tribology*, Vol. 131, 024503–2:1–5.

Akbarzadeh, S. and Khonsari, M.M. 2010a. 'Effect of Surface Pattern on Stribeck Curve,' *Tribology Letters*, Vol. 37, pp. 477–486.

Akbarzadeh, S. and Khonsari, M.M. 2010b. 'On the Prediction of Running-in Behavior in Mixed-Lubrication Line Contact,' *ASME Journal of Tribology*, Vol. 132 / 032102:1–11.

Akbarzadeh, S. and Khonsari, M.M. 2011. 'Experimental and Theoretical Investigation of Running-in,' *Tribology International*, Vol. 44 (2), pp. 92–100.

Akbarzadeh, S. and Khonsari, M.M. 2013. 'On the Optimization of Running-In Operating Conditions in Applications Involving EHL Line Contact,' *Wear*, Vol. 303, pp. 130–137.

American Society of Mechanical Engineers (ASME). 1995. *Surface Texture (Surface Roughness, Waviness, and Lay)*. ANSI/ASME B46.1. ASME, New York.

American Society of Mechanical Engineers (ASME). 1996. *Surface Texture Symbols*. ASME Y14.36M-1996, ASME, New York.

Amiri, M. and Khonsari, M.M. 2010. 'On the Thermodynamics of Friction and Wear–A Review,' *Entropy*, Vol. 12, pp. 1021–1049.

Amiri, M. and Khonsari, M.M. 2012. 'On the Role of Entropy Generation in Processes Involving Fatigue,' *Entropy*, Vol. 14, pp. 24–31.

Amiri, M., Khonsari, M.M., and Brahmeshwarkar, S. 2010. 'On the Relationship Between Wear and Thermal Response in Sliding Systems,' *Tribology Letters*, Vol. 38, pp. 147–154.

Amiri, M., Khonsari, M.M., and Brakhmeshwarkar, S. 2012. 'An Application of Dimensional Analysis to Entropy-Wear Relationship,' *ASME Journal of Tribology*, Vol. 134, 011604:1–5.

Archard, J.F. 1980. 'Wear Theory and Mechanisms,' in *Wear Control Handbook*. M.B. Peterson and W.O. Winer, Eds. ASME, New York.

Archard, J.F. and Hirst, W. 1956. 'Wear of Metals under Unlubricated Conditions,' *Proceedings of Royal Society, A*, Vol. 236, pp. 397–410.

Arnell, R.D., Davies, P.B., Halling, J., and Whomes, T.L. 1991. *Tribology-Principles and Design Applications*, Macmillan, Pittsburgh, PA, pp. 11–17.

Averbach, B. and Bamberger, E. 1991. 'Analysis of Bearing Incident in Aircraft Gas Turbine Main Shaft Bearings,' *STLE Tribology Transactions*, Vol. 34, p. 241.

Beheshti, A., Aghdam, A., and Khonsari, M.M. 2013. 'Deterministic Surface Tractions in Rough Contact under Stick-Slip Condition: Application to Fretting Fatigue Crack Initiation,' *International Journal of Fatigue*, Vol. 56, pp. 75–85.

Beheshti, A. and Khonsari, M.M. 2010. 'A Thermodynamic Approach For Prediction of Wear Coefficient Under Unlubricated Sliding Condition,' *Tribology Letters*, Vol. 38 (3), pp. 347–354.

Beheshti, A. and Khonsari, M.M. 2011. On the Prediction of Fatigue Crack Initiation in Rolling/Sliding Contacts with Provision for Loading Sequence Effect,' *Tribology International*, Vol. 44, pp. 1620–1628.

Beheshti, A. and Khonsari, M.M. 2012. 'Asperity Micro-Contact Models As Applied To The Deformation of Rough Line Contact,' *Tribology International*, Vol. 52, pp. 61–74.

Beheshti, A. and Khonsari, M.M. 2013. 'An Engineering Approach for the Prediction of Wear In Mixed Lubricated Contacts,' *Wear*, Vol. 308 (1), pp. 121–131.

Beheshti, A. and Khonsari, M.M. 2014. 'On the Contact of Curved Rough Surfaces: Contact Behavior and Predictive Formulas,' *ASME Journal of Applied Mechanics*, Vol. 81 (11), pp. 111004

Berthier, Y. 1996. 'Maurice Godet's Third Body.' Proc. of the 22nd Leeds-Lyon Symposium on Tribology, *Tribology Series*, Vol. 31, pp. 21–30.

Bhushan, B. 1994. 'Tribology of Magnetic Storage Systems,' *Handbook of Lubrication and Tribology*, Vol. III. CRC Press, Boca Raton, Fla, pp. 325–374.

Bhushan, B. 1999. *Principles and Applications of Tribology*. John Wiley & Sons, New York.

Bishop J.L, and Ettles C.M.M. 1982, 'The Seizure of Journal Bearings by Thermoelastic Mechanisms,' *Wear*, Vol. 79, pp. 37–52.

Blanchet, T.A. 1997. 'Friction, Wear, and PV Limits of Polymers and Their Composites,' *Tribology Data Handbook*. CRC Press, Boca Raton, FL. pp. 547–562.

Blau, P.J. 2012. 'Friction,' *Handbook of Lubrication and Tribology*, Vol. II, 2nd Ed. CRC Press, Boca Raton, FL. Chapter 5.

Bowden, F.P. and Tabor, D. 1950. *The Friction and Lubrication of Solids*. Oxford University Press, London.

Bowden, F.P. and Tabor, D. 1954, *The Friction and Lubrication of Solids*, Clarendon Press, Oxford, Part I.

Bruce, R.W. 2012. *Handbook of Lubrication and Tribology*, Vol. II, 2nd Ed. Chapters 7–13, 42, CRC Press, Boca Raton, Fla.

Bryant, M.D. 2010. 'Unification of Friction and Wear,' Recent Developments in Wear Preventino, *Friction and Lubrication*, (G.K. Nikas, Editor). Research Signpost Publication, Kerala, India.

Burton, R.A. 1965. 'Thermal Aspects of Bearing Seizure,' *Wear*, Vol. 8, pp. 157–172.

Burton, R.A. and Staph, H.E. 1967. 'Thermally Activated Seizure of Angular Contact Bearings,' *ASLE Transactions*.Vol. 10, pp. 408–415.

Bush, A.W., Gibson, R.D., and Keogh, G.P. 1976. 'Limit of Elastic Deformation in Contact of Rough Surfaces,' *Mechanical Research Communications*, Vol. 3, pp. 169–174.

Chang, W.R., Etsion, I., and Bogy, D.B. 1987. 'An Elastic-Plastic Model for the Contact of Rough Surfaces,' *ASME Journal of Tribology*, Vol. 109, pp. 257–263.

Chun, J. 1988. 'Experimental Investigation of Water Cooling Inside the Spindle Bearing System.' *Ji Chuang/Machine Tools*. Vol. 12, pp. 15–19.

Ciffone, J.G. 1996. 'Lapping. Polishing, and Flatness,' John Crane Mechanical Maintenance Training Center, Arlington Heights, Ill.

Cowan, R. and Winer, W.O. 1992. 'Frictional Heating Calculations,' *ASM Handbook*, Vol. 18, pp. 39–44.

Dai, Z. 2008. 'An Irreversible Thermodynamics Theory for Friction and Wear,' Proceedings of CIST 2008&ITS-IFToMM2008 Beijing, China, pp. 576–578.

DeGarmo, E.P., Black, J.T., and Kosher, R.A. 1984. *Materials and Processes in Manufacturing*. Macmillan Publishing Co. New York.

Doelling, K.K., Ling, F.F., Bryant, M.D., and Heilman, B.P. 2000. 'An Experimental Study of the Correlation Between Wear and entropy Flow in Machinery Components,' *Journal of Applied Physics*, Vol. 88, pp. 2999–3003.

Dufrane, K.F. and Kannel, J.W. 1989. 'Thermally Induced Seizures of Journal Bearings.' *ASME Journal of Tribology*. Vol. 111, pp. 288–292.

Dunaevsky, V.V., Jeng, Y.-R., and Rudzitis, J.A. 1997. 'Surface Texture,' *Tribology Data Handbook*. CRC Press, Boca Raton, Fla, pp. 415–434.

Elwell, R.C. and Booser, E.R. 1972. 'Low-Speed Limit of Lubrication,' *Machine Design*, June 15, pp. 129–133.

Fatemi, A., Yang, L. 1998. 'Cumulative Fatigue Damage and Life Prediction Theories: A Survey Of The State of the Art for Homogeneous Materials,' *International Journal of Fatigue*, Vol. 20. pp. 9–34.

Fuller, D.D. 1984. *Theory and Practice of Lubrication for Engineers*, 2nd Ed. Wiley, New York.

Godet, M. 1984. 'The Third Body Approach: A Mechanical View of Wear,' *Wear*, Vol. 100, pp. 437–452.

Greenwood, J.A. and Tripp, J.H. 1967. 'Elastic Contact of Rough Spheres,' *ASME Journal of Applied Mechanics*, Vol. 34, pp. 153–159.

Greenwood, J.A. and Williamson, J.B.P. 1966. 'Contact of Nominally Flat Surfaces,' *Proceedings of the Royal Society*, London, Series A, Vol. 295, pp. 300–319.

Hamrock, B.J. and Brewe, D. 1983. 'Simplified Solution for Stresses and Deformations,' *ASME Journal of Lubrication Technology*, Vol. 105, pp. 171–177.

Hamrock, B.J., Schmid, S.R., and Jacobson, B.O. 2004. *Fundamentals of Fluid Film Lubrication*, 2nd edition, McGraw-Hill, New York.

Hamrock, B.J. 1994. *Fundamentals of Fluid Film Lubrication*. McGraw-Hill, New York.

Hazlett, T.L. and Khonsari, M.M. 1992a. 'Thermoelastic Behavior Of Journal Bearings Undergoing Seizur,' *Tribology International*. Vol. 25, pp. 183–187.

Hazlett, T.L. and Khonsari, M.M. 1992b. 'Finite-Element Model of Journal bearings Undergoing Rapid Thermally Induced Seizure,' *Tribology International*, Vol. 25, pp. 177–182.

Heilmann, P. and Rigney, D.D. 1981. 'Experimental Evidence for Fatigue during Sliding Wear,' *Metallurgical Transactions A*, Vol. 12, pp. 686–688.

Hersey, M.D. 1966. *Theory and Research in Lubrication: Foundations for Future Developments*. John Wiley and Sons, New York, NY.

Hutchings, I.M. 1992. *Tribology: Friction and Wear of Engineering Materials*. CRC Press, Boca Raton, Fla.

Jackson, R.L. and Green, I. 2005. 'A Finite Element Study of Elasto-Plastic Hemispherical Contact Against a Rigid Flat,' *ASME Journal of Tribology*, Vol. 127, pp. 343–354.

Jackson, R.L. and Green, I. 2006. 'A Statistical Model of Elasto-Plastic Asperity Contact Between Rough Surfaces,' *Tribology International*, Vol. 39, pp. 906–914.

Jahanmir, S. 1977. 'A Fundamental Study on the Delamination Theory of Wear,' PhD dissertation, Department of Mechanical Engineering, Massachusetts Institute of Technology.

Jahanmir, S. and Fischer, T.E. 1994. 'Friction and Wear of Ceramics,' *Handbook of Lubrication and Tribology*, Vol. III. CRC Press, Boca Raton, FL. pp. 103–120.

Jahanmir, S., Suh, N.P., and Abrahamson, E.P. 1974. 'Microscopic Observations of the Wear Sheets Formation by Delamination,' *Wear*, Vol. 28, pp. 235–249.

Jang, J.Y. and Khonsari, M.M. 1999. 'Thermoelastic Instability Including Surface Roughness Effects,' *ASME Journal of Tribology*, Vol. 121, pp. 648–654.

Jang, J.Y. and Khonsari, M.M. 2003. 'A Generalized Thermoelastic Instability Analysis,' *Proceedings of Royal Society*, Series A, Vol. 459, pp. 309–329.

Jang, J.Y. and Khonsari, M.M. 2004. 'Thermoelastic Instability of Two-Conductors Friction System Including Surface Roughness,' *ASME Journal of Applied Mechanics*, Vol. 71, pp. 57–68.

Jang, J.Y., Khonsari, M.M., and Pascovici, M.D. 1998a. 'Thermohydrodynamic Seizure: Experimental and Theoretical Analysis,' *ASME Journal of Tribology*, Vol. 120, pp. 8–15.

Jang J.Y., Khonsari, M.M., and Pascovici, M.D. 1998b. 'Modeling Aspects of a Rate-Controlled Seizure in an Unloaded Journal Bearing,' *Tribology Transactions*. Vol. 41, pp. 481–488.

Jones, W. 1971. 'Lubrication, Friction and Wear,' *NASA SP* 8063.

Jun, W. and Khonsari, M.M. 2007a. 'Analytical Formulation of Temperature Profile by Duhamel's Theorem for Bodies Subjected to Oscillatory Heat Source,' *ASME Journal of Heat Transfer*, Vol. 129, pp. 236–240.

Jun, W. and Khonsari, M.M. 2007b. 'Transient Temperature Involving Oscillatory Heat Source,' *ASME Journal of Tribology*, Vol. 129, pp. 517–527.

Jun, W. and Khonsari, M.M. 2009. 'Thermomechanical Coupling in Oscillatory Systems with Application to Journal Bearing Seizure,' *ASME Journal of Tribology*, Vol. 131, 021601, pp. 1–14.

Keikichi, E. and Kunikazu, H. 1960. 'An Application of Dimensional Analysis to the Characteristics of Wear under Dry Condition,' *Bulletin of JSME*, *3*, pp. 6–12.

Kennedy, F.E., Booser, E.R., and Wilcock, D.F. 2005. 'Tribology, Lubrication and Bearing Design.' *Handbook of Mechanical Engineering*, 2nd edn. CRC Press, Boca Raton, Fla, pp. 4.124–4.170.

Khonsari, M.M. and Booser, E.R. 2006. 'Effect of Contamination on the Performance of Hydrodynamic Bearings,' *IMechE, Journal of Engineering Tribology*, Vol. 220, Part J, pp. 419–428.

Khonsari, M.M. and Booser, E.R. 2010. 'On the Stribeck Curve,' *Recent Developments in Wear Prevention, Friction and Lubrication*, G. Nikas, Editor, Old City Publishing, Philadelphia, PA, pp. 263–278.

Khonsari, M.M. and Kim, H.J. 1989. 'On Thermally Induced Seizure in Journal Bearings,' *Journal of Tribology*, Vol. 111, pp. 661–667.

Khonsari, M.M., Pascovici, M., and Kuchinschi, B. 1999. 'On the Scuffing Failure of Hydrodynamic Bearings in the Presence of an Abrasive Contaminant,' *ASME Journal of Tribology*, Vol. 121, pp. 90–96.

Kimura, Y. 1981. 'The role of fatigue in sliding wear,' *Fundamentals of Friction and Wear of Materials* Rigney, D.A. (ed.), pp. 187–219. ASM Materials Science Seminar, Metal Parks, OH.

Kogut, L. and Etsion, I. 2002. 'Elastic-Plastic Contact Analysis Of A Sphere And A Rigid Flat,' *ASME Journal of Applied Mechanics*, Vol. 69, pp. 657–662.

Kogut, L. and Etsion, I. 2003. 'A Finite Element Based Elastic-Plastic Model For The Contact Of Rough Surfaces,' *Tribology Transactions*, Vol. 46, pp. 383–390.

Komvopoules, K. 2012. ' Adhesive Wear,' *Handbook of Lubrication and Tribology*, Vol. II, 2nd Ed. CRC Press, Boca Raton, Fla, Chapter 7.

Kragelskii, I. Vol. 1965. *Friction and Wear*. Butterworths, London.

Krithivasan, R. and Khonsari, M.M. 2003. 'Thermally Induced Seizure in Journal Bearings During Startup and transient flow disturbance,' *ASME Journal of Tribology*, Vol. 125, pp. 833–841.

Lim, S.C. 1998. 'Recent Developments in Wear Mechanism,' *Tribology International*, Vol. 31, pp. 87–97.

Lim, S.C. and Ashby, M.F. 1987. 'Wear-Mechanism Maps,' *Acta Metall.* Vol. 35, pp. 1–24.

Lin, L.P. and Lin, J.F. 2005. 'An Elastoplastic Microasperity Contact Model For Metallic Materials,' *ASME Journal of Tribology*, Vol. 127, pp. 666–672.

Longuethiggins, M.S. 1957. 'The Statistical Analysis of a Random, Moving Surface,' *Philosophical Trancactions of the Royal Society - A*, Vol. 249, pp. 321–387.

Lu, X. and Khonsari, M.M. 2005. 'On the Lift-off Speed in Journal Bearings,' *Tribology Letters*, Vol. 20, pp. 299–305.

Ludema, K.C. 1984. 'Friction,' *Handbook of Lubrication*, Vol. 2, E.R. Booser, Editor, CRC Press, Boca Raton, Fla, pp. 31–48.

Ludema, K.C. 1996. *Friction, Wear, Lubrication*. CRC Press, Boca Raton, FL.

Mansouri, M. and Khonsari, M.M. 2005. 'Surface Temperature in Oscillating Sliding Interfaces,' *ASME Journal of Tribology*, Vol. 127, pp. 1–9.

Masjedi, M. and Khonsari, M.M. 2012. 'Film Thickness and Asperity Load Formulas for Line-Contact EHL with Provision for Surface Roughness,' *ASME Journal of Tribology*, Vol. 134, 011503:1–10.

Masjedi, M. and Khonsari, M.M. 2014a. 'Theoretical and Experimental Investigation of Traction Coefficient in Line-Contact EHL of Rough Surfaces,' *Tribology International*, Vol. 70, pp. 179–189.

Masjedi, M. and Khonsari, M.M. 2014b. 'Mixed Elastohydrodynamic Lubrication Line-contact Formulas with Different Surface Patterns, Proceedings of the Institution of Mechanical Engineers,' *Part J: Journal of Engineering Tribology*, Vol. 228, pp. 849–859.

Masjedi, M. and Khonsari, M.M. 2015a. 'On the Prediction of Steady-State Wear Rate in Spur Gears,' *Wear*, V. *342-343*, pp. 234–243.

Masjedi, M. and Khonsari, M.M. 2015b. 'An Engineering Approach for Rapid Evaluation of Traction Coefficient and Wear in Mixed EHL,' *Tribology International*, Vol. 92, pp. 184–190.

Masjedi, M. and Khonsari, M.M. 2015c. 'On the Effect of Surface Roughness in Point-Contact EHL: Formulas for Film Thickness and Asperity Load,' *Tribology International*, Vol. 82, Part A, pp. 228–244.

Masjedi, M. and Khonsari, M.M. 2015d. 'A Study on the Effect of Starvation in Mixed Elastohydrodynamic Lubrication,' *Tribology International*, Vol. 85, pp. 26–36.

Masjedi, M. and Khonsari, M.M. 2016. 'Mixed lubrication of Soft contacts: An Engineering Look,' *IMechE Journal of Engineering Tribology*, Part J, Vol. 208–210.

McCool, J.I. 1987. 'Relating Profile Instrument Measurements to the Functional Performance of Rough Surfaces,' *ASME Journal of Tribology*, Vol. 109, pp. 264–270.

Mehdizadeh, M., Akbarzadeh, S. Kayghobad Shams, K., and Khonsari, M.M. 2015. 'Experimental Investigation on the Effect of Operating Conditions on the Running-in Behavior of Lubricated Elliptical Contacts,' *Tribology Letters*, Vol. 59, pp. 1–6.

Mortazavi, V. and Khonsari, M.M. 2016. 'On the Predication of Transient Wear,' *ASME Journal of Tribology*, Vol. 138, 041604:1–8.

Pascovici, M.D. and Khonsari, M.M. 2001. 'Scuffing Failure of Hydrodynamic Bearings Due to an Abrasive Contaminant Partially Penetrated in the Bearing Over-Layer,' *ASME Journal of Tribology*, Vol. 123, pp. 430–433.

Pascovici, M.D., Khonsari, M.M., and Jang, J.Y. 1995. 'On the Modeling of a Thermomechanical Seizure,' *ASME Journal of Tribology*, Vol. 117, pp. 744–747.

Quraishi, S.M., Khonsari, M.M., and Baek, D.K. 2005. 'A Thermodynamic Approach for Predicting Fretting Fatigue Life,' *Tribology Letters*, Vol. 19, No. 3, pp. 169–175.

Rabinowicz, E. 1995. *Friction and Wear of Materials*. John Wiley & Sons, New York.

Rozeanu, L. 1963. Fatigue Wear as a Rate Process,' *Wear*, Vol. 6, pp. 337–340.

Ruff, A.W. 1997. 'Typical Friction and Wear Data,' *Tribology Data Handbook*, CRC Press, Boca Raton, FL. pp. 435–444.

Shaw, M.C. 1977, 'Dimensional Analysis for Wear Systems,' *Wear*, Vol. 43, pp. 263–266.

Song, J.F. and Vorburger, T.V. 1992. 'Surface Texture, Friction, Lubrication, and Wear Technology,' *ASM Handbook*, Vol. 18. ASM International, Materials Park, Ohio, pp. 334–345.

Suh, N.P. 1973. 'The Delamination Theory of Wear,' *Wear*, Vol. 25, pp. 111–124.

Suh, N.P., Jahanmir, S., Abrahamson, E.P. II, and Turner, A.P.L. 1974. 'Further Investigation of the Delamination Theory of Wear.' *Journal of Lubrication Technology, Trans. ASME*, Vol. 94, pp. 631–637.

Tabenkin, A.N. 1984. 'The Growing Importance of Surface Finish Specs,' *Machine Design*, 20 Sept. pp. 99–102.

Tabor, D. 1951. *The Hardness of Metals*, Clarendon Press, Oxford.

Takabi, J. and Khonsari, M.M. 2015. 'On the Thermally Induced Seizure in Bearings: A Review,' to appear in *Tribology International*, Vol. 91, pp. 118–130.

Tian, X. and Kennedy, F.E. 1994. 'Maximum and Average Flash Temperatures in Sliding Contacts,' *ASME Journal of Tribology*, Vol. 116, pp. 167–174.

Tu, J.F. 1995. 'Thermoelastic Instability Monitoring for Preventing Spindle Bearing Seizure. *Tribology Transactions*. Vol. 38, pp. 11–18.

Wang, H., Conry, T.F., and Cusano, C. 1996. 'Effects of Cone/Axle Rubbing Due to Roller Bearing Seizure on the Thermomechanical Behavior of a Railroad Axle,' *ASME Journal of Tribology*, Vol. 118, pp. 311–319.

Wang, Q.A. 1997. 'Seizure Failure of Journal-Bearing Conformal Contacts,' *Wear*, Vol. 210, pp. 8–16.

Whitehouse, D.J. 1994. *Handbook of Metrology*. Institute of Physics Publication, Bristol, UK.

Wu, S.F. and Cheng, H.S. 1993. 'Sliding Wear Calculation in Spur Gears,' *ASME Journal of Tribology*, Vol. 115, pp. 493–500.

Zaretsky, E.V. 1992. *Life Factors for Rolling Bearings*. STLE, Park Ridge, Ill.

Zhao, Y.W. Maietta, D.M., and Chang, L. 2000. 'An Asperity Microcontact Model Incorporating The Transition from Elastic Deformation To Fully Plastic Flow,' *ASME Journal of Tribology*, Vol. 122, pp. 86–93.

4

Bearing Materials

Selection of materials for either fluid-film or rolling element bearings depends on matching material properties with the need for low friction, low wear rate, and long life. While almost all engineering materials have been used at some time in a search for optimum bearing materials, final selection is commonly based on judgment as to the most essential material properties, type of application, and cost.

A review is provided in this chapter of the common factors involved in selecting bearing materials. This is followed by consideration of specific bearing materials in the following four categories: (1) oil-film bearings such as those used in a wide variety of machinery; (2) porous metal, plastic, and other nonmetallic materials used in dry and semilubricated bearings under relatively mild operating conditions in household appliances, instruments, office machines, toys, and many less demanding applications; (3) high-temperature materials for aeronautical, industrial, nuclear and rock drilling applications; and (4) high-hardness, high-fatigue-strength materials for ball and roller bearings.

Recent literature references are provided throughout the chapter on specific materials, and more detailed information can be obtained from bearing material suppliers and from a number of comprehensive surveys (Petersen and Winer, 1980; Blau, 1992; Booser, 1992; Hamrock, 1994; Kingsbury, 1997). Further details are also provided in later chapters, where material considerations are integrated with performance analyses for specific applications such as gas bearings.

4.1 Distinctive Material Selection Factors

Even bearings that operate with a full oil film can be expected to contact their shaft during initial break-in, while starting and stopping, or during interruptions in lubricant supply. During this direct material contact, the bearing material must avoid to locally weld to the shaft with scoring or galling under the high stress, high strain and elevated temperature at localized areas. Strong forces developed between asperities in the real area of contact can result in removal of small particles from the softer surface. These particles can then reattach to either surface, mix with material from other surfaces as they tend to embed themselves fully or partially penetrate inward into the soft protective layer, or expelled as loose wear debris. Therefore, for

Applied Tribology: Bearing Design and Lubrication, Third Edition. Michael M. Khonsari and E. Richard Booser.
© 2017 John Wiley & Sons Ltd. Published 2017 by John Wiley & Sons Ltd.
Companion Website: www.wiley.com/go/Khonsari/Applied_Tribology_Bearing_Design_and_Lubrication_3rd_Edition

Table 4.1 Scoring resistance of metallic elements sliding against low carbon steel

Good	Fair	Poor	Very poor	
Germanium	Carbon	Magnesium	Beryllium	Molybdenum
Silver	Copper	Aluminum	Silicon	Rhodium
Cadmium	Selenium	Copper	Calcium	Palladium
Indium	Cadmium	Zinc	Titanium	Cerium
Tin	Tellurium	Barium	Chromium	Tantalum
Antimony		Tungsten	Iron	Iridium
Thallium			Cobalt	Platinum
Lead			Nickel	Gold
Bismuth			Zirconium	Thorium
			Columbium	Uranium

design purposes, consideration should be given to the compatibility of the friction pair, the embedability and conformability as well as the fatigue strength of the bearing materials.

Compatibility

Compatibility testing of metallic elements rubbing against common low carbon AISI 1045 steel is classified in Table 4.1 in terms of scoring resistance (Roach *et al.*, 1956). Good scoring resistance is provided only by elemental metals which (1) had atomic diameters at least 15% greater than that of iron for a minimum tendency to form atomic junctions and for limited mutual solubility and (2) were in the B subgroups of the periodic table, which implies covalent atomic bonds rather than the more tenacious metallic bonds. Cadmium and copper are seen to bridge two ratings. In the much more complex bearing alloys, element content gives a guideline to their performance. Adding more of a good element like lead in a copper-base bearing alloy, for instance, can be expected to improve scoring resistance, while an increase in poorly performing zinc would be detrimental (DeHart, 1984).

Various plastics and other nonmetallics also provide excellent compatibility as bearing materials. Their application is generally limited to low surface speeds where their low thermal conductivity does not lead to overheating and damage to the shaft.

Embedability and Conformability

The hardness and modulus of elasticity of oil-film and boundary-lubricated bearing materials should be as low as possible while providing sufficient strength to carry the applied load. In initial operation, this provides the maximum possible surface softness and elasticity to aid in compensating for misalignment and other geometric errors. In addition, the soft bearing material is better adapted to absorb foreign dirt particles, machining chips, and grinding debris, and thus minimize scoring and wear of the bearing and shaft. Generally, journal hardness should be at least two to three times the hardness of the bearing material, as listed in Table 4.2 (Rabinowicz, 1995). The soft tin and lead babbitts are unsurpassed for both embedability and conformability.

Table 4.2   Representative properties of sliding bearing materials

	Hardness	Density (g/cc)	Tensile strength (MPa)	Modulus of elasticity (GPa)	Thermal conductivity (W/(m K))	Coefficient of expansion $(10^6/\,°C)$	Poisson ratio at 20 °C
Metals							
Lead babbitt	21 B	10.1	69	29	24	25	–
Tin babbitt	25 B	7.4	79	52	55	23	–
Copper lead	25 B	9.0	55	52	290	20	–
Silver	25 B	10.5	160	76	410	20	–
Aluminum alloy	45 B	2.9	150	71	210	24	0.33
Lead bronze	60 B	8.9	230	97	47	18	0.33
Tin bronze	70 B	8.8	310	110	50	18	0.33
Zinc alloy	95 B	5.1	320	79	125	26	0.27
Steel	150 B	7.8	520	210	50	12	0.30
Cast iron	180 B	7.2	240	160	52	10	0.26
Porous metals							
Bronze	40 B	6.4	120	11	29	16	0.22
Iron	50 B	6.1	170	97	28	10	0.22
Aluminum	55 RH	2.3	100	7	137	22	–
Plastics							
Acetal	94 RM	1.42	69	2.8	0.22	80	–
Nylon	79 RM	1.14	79	2.8	0.24	80	0.40
PTFE	60 SD	2.17	21	0.4	0.24	130	–
Phenolic	100 RM	1.36	69	6.9	0.28	28	–
Polyester	78 SD	1.45	59	2.3	0.19	95	–
Polyethylene, high density	69 SD	0.95	–	0.9	0.5	126	0.35
Polyimide	52 RE	1.43	73	3.2	0.43	50	–
Ceramics							
Alumina	1470 V	3.8	220	370	35	7.1	0.22
Silicon carbide	2460 V	3.3	500	430	85	4.5	0.14
Silicon nitride	1700 V	3.2	520	310	30	3.5	0.28
Tungsten carbide	91 RA	14.2	900	560	70	6	–
Zirconium oxide	1200 V	5.5	690	150	1.7	8	0.3
Other nonmetallics							
Carbon graphite	75 SS	1.7	14	14	9	3	–
Wood	–	0.68	8	12	0.19	5	–
Rubber	65 SA	1.2	10	0.04	0.16	77	0.50

Notes: B = Brinell, kg/mm^2; R = Rockwell, SD = Shore durometer, SS = Shore scleroscope, V = Vickers, kg/mm^2.
Conversion factors: psi = MPa/6.895 × 10^{-6}; kg/mm^2 = MPa/10. As a broad approximation for metals, hardness in kg/mm^2 = 3× tensile yield strength in kg/mm^2.
Source: Ruff, 1977; Hamrock, 1994; misc.

Table 4.3 Fatigue strength of typical
oil-film bearing alloys

Material	Load capacity (MPa)
Thick babbitt	5–10
Thin babbitt	
0.25–0.50 mm	14
0.10 mm	17
Copper lead	38
Aluminum alloys	55
Silver	80
Bronze	100

Source: DeHart, 1984; Kingsbury, 1997.

Strength

While a bearing alloy must have sufficient strength to avoid extrusion or fatigue under load, too high strength introduces brittleness, poor embedding of foreign particles, and inability to conform to misalignment. Material strength in itself is often less important for high load capacity than design details, lubrication, and application experience.

Fatigue strength is particularly important for reciprocating loads encountered in connecting rod and main bearings of reciprocating internal combustion engines. The need for high fatigue strength often brings into use aluminum, copper – lead, bronze, or silver bearings for automotive or diesel engines. A thin babbitt overlay is commonly employed for improved compatibility (Kingsbury, 1997). Table 4.3 gives a rating of relative fatigue strength provided by various bearing materials (DeHart, 1984).

High fatigue strength is especially critical for ball and roller bearing materials. Repetitive high-contact stresses at ball and roller contacts demand materials with high hardness and the greatest available fatigue strength.

Corrosion Resistance

Bearing alloys containing lead, copper, zinc, and cadmium are susceptible to varying degrees to corrosion by organic acids and peroxides formed in lubricating oils that have become oxidized in service. This difficulty can be minimized by selecting oxidation-inhibited oils, avoiding high bearing operating temperatures, and changing the oil regularly. Water contamination and sulfur- and chlorine-containing oil additives may also contribute to bearing corrosion, as may an attack on silver bearings in some diesel engines by sulfur in oil additives.

Particular attention must be given to selection of bearing materials that may be exposed to attack by corrosive chemical fluids, high-temperature gases, liquid metals, and water.

Thermal Properties

High thermal conductivity minimizes temperature increases, particularly during initial run-in of bearings and during high-speed and high-load operation. Plastic bearings usually are avoided for high-speed equipment, where their poor heat transfer may lead to charring of the plastic and overheating of the journal.

Matched thermal expansion coefficients are important to maintain both suitable clearance of a bearing with its shaft and the desired fit in its housing. Suitable accommodations are required when applying a bearing material with a unique coefficient of expansion. With water lubrication, for instance, a carbon-graphite bearing liner, with its low expansion coefficient, is commonly shrunk in a steel shell to minimize loss of clearance with a steel shaft at elevated temperature. Designs using aluminum and bronze bearings or aluminum housings should take into account the difference in their thermal expansion from that of steel.

General drop in bearing material properties should also be considered for elevated temperatures (Khonsari and Booser, 2004). The yield strength of lead and tin babbitts, for instance, drops by about half as their temperature rises to the 135 °C (275 °F) range.

4.2 Oil-Film Bearing Materials

The operating demands on bearings running with full oil-film lubrication have led to the use primarily of lead- and tin-based white-metal babbitts and various copper and aluminum alloys. Despite their shortcomings, steel and cast iron structural parts are often used as bearing materials for low cost and design simplicity. Silver and zinc also find limited use. A brief discussion in this chapter is followed by a more general review in Chapter 12 of plastics and porous metal for small bearings and bushings for light-duty and intermittent service.

Table 4.2 gives representative property values for a variety of materials used in oil-film and other sliding bearing applications. Since values for relatively minor processing and compositional differences may vary greatly, individual suppliers should be contacted for details on specific materials.

Babbitts

Lead and tin white-metal alloys patented by Isaac Babbitt in 1839 are commonly the first choice for bearing materials in offering superior compatibility with steel shafts, their ability (due to their softness) to embed foreign particles, and their unique ability to adapt to misalignment by mild wiping on initial run-in as enabled by their low melting points. Table 4.4 covers representative physical properties of typical babbitt compositions (ASTM, 1990; SAE, 1991).

Tin babbitt alloys commonly contain about 3–8% copper and 5–8% antimony. Within a soft solid – solution matrix of antimony in tin, small, hard Cu_6Sn_5 copper – tin intermetallic particles are dispersed (Dean and Evans, 1976). Increasing copper increases the proportion of Cu_6Sn_5 needles or stars in the microstructure. An increase in antimony above 7.5% results in antimony – tin cubes. Hardness and tensile strength increase with greater copper and antimony content, while ductility decreases. Low antimony (3–7%) and low copper content (2–4%) provide maximum resistance to fatigue cracking. Since these low-alloy compositions are relatively soft and weak, a compromise is often made with fatigue resistance and compressive strength.

Despite their higher cost, tin babbitts are often used in preference to lead babbitts for their excellent corrosion resistance, easy bonding, and less tendency toward segregation. SAE 12 (ASTM Grade 2) is widely used in industrial and automotive bearings. SAE 11 and ASTM Grade 3 also find extensive industrial use.

Table 4.4 Babbitt alloys (white metals)

	Tin base			Lead base		
Designation						
ASTM B23	11	2	3	13	7	15
SAE	–	12	–	13	14	15
Nominal composition (%)						
Tin	87	89	84	6	10	1
Lead				84	75	84
Antimony	7	7.5	8	10	15	15
Copper	6	3.5	8			
Arsenic						1
Specific gravity	7.4	7.39	7.45	10.5	9.7	10.1
Melting point (°C)	240	241	240	240	240	247
Complete liquefaction (°C)	400	354	422	256	268	353
Brinell hardness	26	24	27	19	22.5	20
Ultimate tensile strength (MPa)	90	77	69	69	72	71
Compressive yield strength (MPa)	45	42.1	45.5	22.8	24.5	24.8
Approximate % compressive yield strength retained at:						
100 °C	49	52	52	47	51	58
150 °C	23	24	24	25	29	32
200 °C	5	7	7	7	11	11

Note: Properties are for chill cast alloys.
Source: Kingsbury, 1997. Reproduced with permission of Taylor & Francis.

Lead babbitts commonly contain 9–16% antimony and up to 12% tin to provide hard cuboid crystals of antimony – tin (SbSn) in a eutectic matrix of the three metals (Dean and Evans, 1976). To minimize segregation during casting, up to 0.5% copper is usually added. SAE 15, containing 1% arsenic for finer grain structure, has been used in automotive bearings for its resistance to fatigue and better high-temperature properties. This alloy also gives excellent service in large hydroelectric generator thrust bearings.

Sufficient tin is added in SAE grades 13–16 for reasonable corrosion resistance. With 10% tin, SAE grade 14 has frequently been used in railroad, industrial, and automotive bearings. Softer SAE 13 is used in similar applications. Corrosion problems can usually be avoided by using oxidation-inhibited lubricating oils and regular relubrication to avoid buildup of acidic oxidation products. Corrosion becomes most likely in the presence of oxidized oil contaminated with water.

Babbitt application methods vary. For larger bearings in electric motors, turbines, compressors, pumps, and other industrial equipment, babbitt is usually centrifugally cast in a steel or bronze shell and finished to 1.5–10 mm thickness. Sound bonding to the shell requires care in a series of steps: cleaning the shell, rinsing, fluxing, tinning, babbitt casting, and finally rapid quenching. For smaller bearings and bushings, such as used in automotive engines and small industrial equipment, a bimetal strip is first produced by casting the babbitt on a continuous steel strip or by electroplating lead babbitt of 10% tin and 3% copper as specified by SAE 19 and SAE 190 (DeHart, 1984). After formation of oil-distributing grooves and broaching of oil feed holes, the strip is cut to size and the individual segments are rolled into finished bearings.

For high fatigue strength in automotive bearings, a very thin layer of babbitt enables much of the stress to be accommodated in the stronger backing material. Relative fatigue resistance was found to be as follows with decreasing thickness of tin babbitt (ASM, 1961):

Babbitt thickness (mm)	Relative fatigue resistance
1.00	1
0.50	1
0.25	1.5
0.13	3.2
0.08	4.6

Three-layer strip bearings are frequently used for connecting rod, camshaft, and main bearings for heavy-duty service in reciprocating engines (Kingsbury, 1992, 1997). These consist of a low carbon steel backing; an intermediate layer about 0.3–0.8 mm thick of copper – lead, leaded bronze, aluminum, or electroplated silver; and a thin babbitt layer 0.025–0.50 mm thick. The thin babbitt layer functions primarily during run-in, after which the higher-strength intermediate layer carries the load.

Copper Alloys

Characteristics of typical copper-based bearing materials are given in Table 4.5. Alloys at the top of the list, with their higher lead content, are commonly given first consideration because of their better compatibility and resistance to scoring. When higher strength, hardness, and wear

Table 4.5 Copper-base bearing alloys

Material	Designation[a] UNS	Designation[a] SAE	Composition (%) Cu	Composition (%) Sn	Composition (%) Pb	Composition (%) Zn	Brinell hardness	Ultimate tensile strength (MPa)	Initial melting point (°C)
Copper–lead		485	48		51		45	150	1050
Copper–lead		48	70		30		50	165	955
High lead tin bronze	C94300		70	5	25		48	185	900
Semiplastic bronze	C93800		78	6	16		55	205	855
Leaded red brass	C83600		85	5	5	5	60	255	855
Phosphor bronze	C93700	792	80	10	10		60	240	762
Bearing bronze	C93200	660	83	7	7	3	65	240	855
Navy G	C90300		88	8		4	70	310	854
Gunmetal	C90500		88	10		2	75	310	854
Leaded gunmetal	C92700		88	10	2		77	290	850
Aluminum bronze	C95400		85 (4 Fe, 11 Al)				195	620	

[a] UNS, Unified Numbering System; SAE, Society of Automotive Engineers.
Source: Kingsbury, 1997. Reproduced with permission of Taylor & Francis.

resistance are required, alloys toward the bottom of the list, with increasing tin and aluminum content, are selected.

With binary copper – lead, a continuous copper phase provides the primary load support, while pockets of 20–50% lead supply a soft, compatible bearing surface film. Tin content of 3–4% is commonly incorporated with the lead to minimize corrosion. Leaded bronzes containing about 5–25% lead and up to 10% tin find use in a wide range of applications for higher hardness and better fatigue strength. The traditional 10% tin–10% lead phosphor bronze has been replaced in many applications with C93200 containing 3% zinc for easier casting into continuous rods and tubing and for easy forming into final bearing shapes.

The higher-hardness tin bronzes listed in Table 4.5, such as gunmetal alloys, require reliable lubrication, good alignment, and 300–400 minimum Brinell shaft hardness. With their inherent strength, they are commonly not cast on a steel backing. Aluminum bronzes provide excellent strength, as well as shock and wear resistance in bushings and bearing plates for machine tools, aircraft landing gear, and other rather special applications. Age-hardened beryllium copper containing 2% beryllium also provides high strength and has been used in airframe bearings for loads of up to $315 \, MN/m^2$ (50 000 psi) (Glaeser, 1983).

While copper alloy bearings give excellent service in mill equipment, farm machinery, machine tools, pumps, and motors, their utility is limited at high surface speeds and with marginal lubrication by their tendency to form a copper transfer film on a mating steel shaft. At surface speeds above about 8–15 m/s, selective plucking may occur with softer copper material from hotter load zones in the bearing surface welding in lumps onto the cooler, stronger copper transfer film on a steel journal. Adequate lubrication, and especially a thin overlay of a good-compatibility material from Table 4.1 in 'bimetal' and 'trimetal' bearings, however, provide the advantages of high strength and high temperature capability of copper bearing materials in many demanding applications (Kingsbury, 1992, 1997).

Aluminum

Alloys such as those listed in Table 4.6 offer high fatigue strength, excellent corrosion resistance, high thermal conductivity, and low cost (DeHart, 1984; Kingsbury, 1992). Alloying with tin is common for improved compatibility, along with 1.5–11% silicon and up to 2% copper.

Table 4.6 Aluminum-base bearing alloys

Material, SAE designation	Alloying elements (%)							Type bearing
	Si	Cu	Sn	Ni	Pb	Cd	Mg	
770		1	6.5	1				Solid, cast
780	1.5	1	6.5	1				Bi- or trimetal, rolled
781	4	0.1				1	0.1	Bi- or trimetal, wrought
782		1		1		3		Bi- or trimetal, wrought
783		1	20					Bimetal
784	11	1						Intermediate trimetal layer
787	4	0.5	1		6			Bimetal
788	3	1	12		2			Bimetal, wrought

Source: Kingsbury, 1997. Reproduced with permission of Taylor & Francis.

Although finding only minor use in general industrial applications because of their limited compatibility, aluminum bearings are widely used in automotive and diesel engines, reciprocating compressors, and aircraft. Good surface finish, minimum shaft hardness of 85 Rockwell B, and good lubrication are required.

Venerable 6.5% tin – aluminum SAE 770 and SAE 780 are widely used for solid wall and bimetal aluminum bearings in medium- and heavy-duty diesel engines. With silicon and cadmium added for improved compatibility, SAE 781 and 782 are applied as 0.5–3.0 mm thick intermediate layers on steel backing and with a thin electroplated babbitt overlay. In Europe 11% silicon alloys are used for highly loaded diesels.

Special aluminum alloys with much higher tin (SAE 783 and 786) and lead content (SAE 787) have been developed with specialized processing techniques to eliminate the need for an overlay. The performance of high-tin alloys led to their dominant use in European automotive bearings. Medium-tin SAE 788 is employed as a surface layer in bimetal bearings for connecting rod, main, and camshaft bearings, as well as for thrust washers and pump and transmission bushings (Kingsbury, 1997). While high-lead alloys have found widespread and successful use in automotive bearings in the United States, high cost has limited their continued use.

Cast Iron and Steel

Flake graphite in gray cast iron allows bearings machined directly in structural parts to carry loads up to about 1.0 MPa (145 psi) at surface speeds up to about 0.8 m/s for such applications as pivots and lightly loaded transmissions. With good alignment, copious oil feed, and hardened and ground journals, loads range up to 4.5 MPa (650 psi) for main bearings in cast iron refrigeration compressors and up to 5.5 MPa (800 psi) for connecting rods (Dean and Evans, 1976). A phosphate-etched surface is commonly applied as an aid for initial run-in.

Guide surfaces and journal bearings can also be inexpensively machined in structural steel parts for loads up to 1.4 MPa (200 psi) at speeds up to 0.8 m/s. Oils incorporating extreme pressure and antiwear additives assist in avoiding wear and scoring problems.

Silver

Bearings consisting of about 0.3 mm of silver electrodeposited on a steel backing and with a 0.025–0.10 mm lead babbitt overlay have given excellent performance in diesels, superchargers, and connecting rod and main bearings of reciprocating aircraft engines (DeHart, 1984). Unique self-healing and excellent compatibility in rubbing on steel make thin electroplated silver useful as a bearing material under severe sliding conditions in a variety of special, prototype, and experimental machines.

Zinc

Zinc alloys find use for better wear life and lower-cost replacements for SAE 660 and other bronzes. Compositions containing about 10–30% aluminum for improved properties are formed as cast tubes for use as solid alloys and bushings (Kingsbury, 1997). Oscillating and rotating applications are made at speeds up to 7 m/s (1400 ft/min) at temperatures up to 90–125 °C.

4.3 Dry and Semilubricated Bearing Materials

A extensive variety of polymers and their composites, porous metal bearings, carbon-graphite, wood, and rubber are widely used in dry sliding; with process fluids such as air, water and solvents; or under conditions of sparse lubrication. They commonly allow design simplification, freedom from regular maintenance, and reduced sensitivity to contamination. While these materials are usually applied at low speeds with minimal lubrication, performance normally improves the closer the approach to full-film lubrication. Material properties are covered in this section; performance and design details follow in Chapter 12.

Plastics

While almost all commercial plastics find use both dry and lubricated for sliding contact at low speeds and light loads, the most commonly used thermoplastics are nylon, acetal resins, PTFE, and those such as polyethylene and polypropylene produced in large volume by polymerizing gases from petroleum cracking. Thermosetting resins typically used for bearing applications are phenolics, polyesters, and polyimides.

Injection-molded acetal and nylon provide inexpensive small bearings for myriad lightly loaded applications in household appliances, office machines, small industrial devices, toys, and instruments. The polyamide nylons require little or no lubrication, involve low friction, and provide quiet operation. Troublesome cold flow of nylon under load is minimized by incorporating fillers such as inorganic powders or glass fibers, or by applying the nylon as a thin layer on steel backing. Acetal and ultrahigh molecular weight polyethylene are often used to mold inexpensive housings, gears, and other machine elements of which bearings are only a portion in appliance, automotive, and industrial applications.

PTFE is uniquely useful as a bearing material in providing a coefficient of friction ranging from 0.04 to 0.10 at temperatures ranging from cryogenic to 250 °C. When used alone, PTFE has several limitations: a maximum sliding speed of about 0.3 m/s; maximum loading of 0.04 MN/m^2, beyond which cold flow is to be expected; and inability to accept supplementary oil lubrication. Conventional petroleum and synthetic oils do not wet PTFE well, and any oil present increases the wear rate by interfering with the usual back-and-forth exchange of PTFE wear fragments between the bearing and its normal transfer layer on a steel surface. These limitations are avoided by the use of PTFE powder in thermoplastic composites, where a 15% addition of PTFE will commonly reduce the coefficient of friction of the neat thermoplastic by 50% and its wear rate by a factor of 10 or more (Blanchet, 1997). Addition of inorganic powders, metal powders, graphite, and glass fibers will, in turn, reduce the wear rate of PTFE by orders of magnitude while still providing the low coefficient of friction exhibited by the PTFE polymer itself. For low speeds and semistatic use, PTFE fabric enables loads up to 400 MN/m^2 (60 000 psi) in automotive ball and socket joints, bridge bearings, and aircraft controls. Interweaving a secondary fiber such as polyester, glass, or cotton with the PTFE enables bonding to a steel backing for support.

Inexpensive phenolic resins are commonly used as composites with cotton fibers, cellulose, or graphite. In small applications such as appliances, business machines, and instruments, bearings are often formed simply as holes in phenolic or polyester structural elements. High processing costs generally restrict their use in larger bearings in industrial and marine applications. Below a 50 mm bore size, injection-molded thermoplastics are more common. Polyimide molding compounds incorporating graphite and other fillers are used in ball bearing retainers, bearing seals, aircraft bushings, and piston rings at temperatures up to 260 °C. Polysulfones and poly(phenylene sulfide) also find specialized application at high temperatures.

Carbon-Graphite

A range of properties are obtained by high-pressure molding of mixtures of graphite powder, petroleum coke, lamp black, and coal tar pitch followed by curing at temperatures up to 1440 °C. The resulting porosity is impregnated with phenolic or epoxy resins, copper, babbitt, bronze, glass, or silver to give a wide range of strength, hardness, and wear properties. Common uses include pump bearings for water, gasoline, and solvents; conveyor and furnace bearings at temperatures up to 400 °C; and in foods, drugs, and machinery where oil and grease contamination is to be avoided.

Common shaft surfaces are hardened tool steel, chrome plate, high-strength bronze, and carbide and ceramic overlays with a 0.25 μm or better finish. Tests in dry operation at speeds from 0.05 to 47 m/s (10–9200 fpm) indicate that a coefficient of friction of 0.16–0.20 and a specific wear rate (volume wear per unit force, per unit of sliding distance, (see Equation 3.64) of 14×10^{-16} m/N are typical for well-applied grades (Booser and Wilcock, 1978).

Rubber

Synthetic rubber is commonly used in water-lubricated journal bearings as a series of axial segments separated by longitudinal grooves for supplying the water. The rubber staves, in turn, are enclosed in a rigid bronze cylindrical shell. Resilience of rubber and its ability both to embed dirt particles and to resist abrasion bring rubber bearings into general use for marine propeller shafts, water turbines and pumps, and conveyors for slurries of gravel and ore. Non-corrodable shaft surfaces or sleeves are commonly composed of bronze, Monel, chrome plate, or stainless steel.

Wood

Traditional lignum vitae and oil-impregnated maple and oak have largely been replaced as bearing materials by plastics, porous metal, and rubber. Typical applications are with water and other low-viscosity fluids at relatively low speeds and temperatures below 70 °C in food and chemical processing machinery, ship propeller shafts, conveyors, and hydraulic pumps and turbines.

4.4 Air Bearing Materials

Unique compositions of materials are being employed in the growing use of air as the bearing operating fluid (see also Chapter 11). Major applications have been in foil bearings for high-speed aircraft accessories, in air lift bearings for nearly frictionless moving of heavy loads, and for hard disk drives in computers.

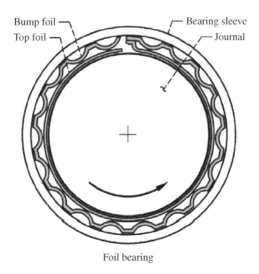

Figure 4.1   Compliant bump foil supports active top bearing surfaces of foil bearing. *Source*: NASA, 2000.

Foil Air Bearings

These are finding wide use in oil-free aerospace turbomachinery such as turbochargers and cabin air cooling and pressurizing units. With construction of a compliant foil support structure, such as shown in Figure 4.1, load capacity W ranges up to a pound for each inch of bearing diameter and square inch of projected area for every 1000-shaft rpm in the following relation (NASA, 2000):

$$W = K_W(L \cdot D)(D \cdot \Omega) \tag{4.1}$$

where

W = maximum steady-state load that can be supported (lb)
K_W = bearing load capacity coefficient (lb/(in^3 krpm)). K_W has increased to 1.0 or higher in recent applications from 0.3–0.5 in earlier foil bearings. On this basis, loadings now are generally on the order of 30 psi on the projected bearing area
L = bearing axial length (in.)
D = shaft diameter (in.)
Ω = shaft speed in thousand rpm (krpm)

The load capacity thus becomes proportional to the product of projected bearing area $(D \cdot L)$ multiplied by surface velocity $(D \cdot \Omega)$.

Example 4.1   A 2 in. diameter air-lubricated foil bearing is to carry 180 lb radial load at 20 000 rpm. What is the minimum axial bearing length required?

Rearranging Equation (4.1) and letting $K_W = 1$:

$$L = \frac{W}{K_W D(D.\Omega)} = \frac{180}{1.0 \times 2(2 \times 20)} = 2.25 \text{ in.}$$

It is worthwhile noting that this 180 lb load is about 10 times greater than a conventional rigid air bearing of this size can handle.

Running at high speeds with very low friction loss, most foil bearings have used a thin PTFE polymer coating on the supporting foil to minimize wear during starting and stopping when the shaft surface velocity is insufficient to generate a full separating air film. With suitable flexible steel backing, a porous bronze layer can provide a suitable backing for the solid lubricant surface layer (DellaCorte, 2004). Nickel-chromium alloy INCX-750 has been used for its strength and elastic properties for the foils. These foil units have commonly been mated with INCX shafts, or with lower cost stainless steels like A286 (Laskowski and DellaCorte, 1996).

While PTFE is limited to an upper temperature of about 400 °F, plasma spray coatings raised this upper temperature limit to 1200 °F. This has been further raised to 1650 °F by self-lubricating NASA powder metallurgy duplex composites consisting of a hard nickel – chrome/chrome oxide phase with a soft noble metal and fluoride phases. A corresponding demand on the foil and shaft materials leads to use of high-temperature materials of the nature of those listed in Tables 4.7 and 4.8. These high-temperature foil bearings range up to about 150 mm bore size and have found interesting use by NASA and Mohawk Innovative Technology for small developmental turbofan units for a twin engine jet (Fields, 2004). Significant research efforts are underway to better understand the thermal characteristics of foil bearings and limiting speeds. The interested reader is referred to Peng and Khonsari (2004a, 2004b, 2006).

Table 4.7 Representative hard metals and superalloys for high-temperature bearings

Material	Approximate temperature limit (°C)
Nitrided steels	500
Tool steels	500
Hastalloy C[a]	750
Stellite 6[b]	750
Rene 41[c]	850
Tribaloy T-400[d]	850

[a] 57% Ni, 17% Mo, 16% Cr, 5% Fe + Mn.
[b] 67% Co, 28% Cr, 4% W, 1% C.
[c] 55% Ni, 19% Cr, 10% Co, 10% Mo, 3% Ti + Al, Fe, Si, Mn, C, B.
[d] 62% Co, 28% Mo, 8% Cr, 2% Si. *Source*: Lancaster, 1973. Reproduced with permission of John Wiley & Sons.

Table 4.8 Wear rates of high-temperature materials
sliding on themselves at 500 °C

Material	Specific wear rate ($mm^3/N\,m$)
Ceramics[a]	10^{-3}–10^{-5}
Nickel-base alloys	10^{-3}–10^{-5}
Tool steels	10^{-4}–10^{-5}
Cobalt-base alloys	10^{-5}–10^{-6}
Cermets[b]	10^{-5}–10^{-7}

[a] Al_2O_3, Zr_2O_3, SiC.
[b] WC-Co, TiC-Ni-Mo, Cr_3C_2-NiCr, Al_2O_3-CrMo.
Source: Lancaster, 1973. Reproduced with permission of
John Wiley & Sons.

Air Lifts

Externally pressurized air bearings find use for pallets and other lifts for moving heavy loads of materials and machinery. All-terrain vehicles and bulky semistationary loads can be raised up to 3 in. These air bearings depend on the same principles as given in Chapter 10 for externally pressurized bearings employing oil. Material selection is often customized to the application. Unique use is being made of porous structures of graphite, ceramics and sintered metals to uniformly distribute the air flow. This distribution pattern avoids misalignment of the bearing unit with air escaping uselessly to the high clearance side as discussed in Chapter 10.

Computer Hard Disk Drives

Magnetic recording in a hard disk system employs motion between a small read–write head and the thin-film magnetic medium on a relatively high-speed disk. In continually evolving technology that is enabling massive information storage and transfer in ever-smaller units, air bearings have involved lower-and-lower air film flying heights that are now down to as little as 3 nm (about 10 average atoms) between the head and its hard disk (*Wikipedia Encyclopedia –* Anon., 2006).

Typical hard disks (ranging from 3.5 in. diameter for desktop computers down to 1.8 in. in subnotebooks) are usually aluminum or glass coated on both sides with a thin complex magnetic film that typically incorporates a cobalt alloy applied by sputtering (Figure 4.2). For wear protection during starting, stopping, and incidental surface contacts, this magnetic film is covered by a diamond-like carbon layer topped with a fluoroether synthetic oil (see Chapter 2) for lubrication. To minimize wear at low speeds during starts and stops, the head may be lifted either onto a plastic ramp near the outer edge of the disk, or moved to a smooth landing zone near the inner edge of the disk where smooth laser-generated bumps of nanometer scale greatly improve friction and wear performance.

Read–write heads are usually fabricated on Al_2O_3/TiC ceramic wafers developed for their high hardness, wear resistance, and generally excellent tribological properties. After being precision lapped, the thin magnetic operating film is deposited in complex steps. Representative slider length is 2 mm, width 1.6 mm with two 0.25 mm guide rails, and a 50% length taper on its running surface to generate the supporting air film (see Chapter 11). This head slider

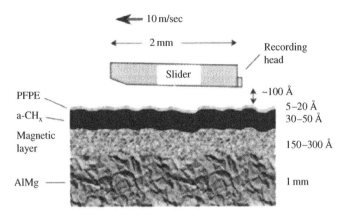

Figure 4.2   Cross-section of a hard disk surface. *Source*: Carnes, 2004. Reproduced with permission of John Wiley & Sons.

is mounted on a suspension (see Figure 4.3) enabling it to record and retrieve data from the surface of the hard disk as it rotates at constant speed in the 5000–15,000 rpm range.

4.5 High-Temperature Materials

When temperature limits for lubricating oils (150–250 °C) and the usual solid lubricants (350–400 °C) are exceeded, quite different bearing materials must be selected to operate either dry in slow-speed sliding or with poor lubricating fluids such as gas, liquid metals, or pressurized water. Prototype testing is a common step in material selection for operation at these high temperatures in gas turbines, diesel engines, rocket engines, and nuclear plant equipment.

High-temperature strength and reasonable friction and wear characteristics lead to alloys of cobalt, nickel, chromium, and molybdenum such as those listed in Table 4.7 for use up to 500–850 °C (Lancaster, 1973). Among these are stellites with chrome carbides dispersed in a cobalt matrix, and cobalt or nickel base tribaloys, which develop hardness by an intermetallic laves phase rather than massive carbides (Glaeser, 1997). Table 4.8 gives a general range of

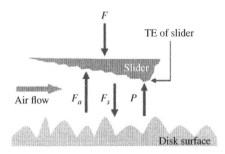

Figure 4.3   Forces at head – disk interface with TE as trailing edge: F_a = air bearing force, F_s = adhesion force, P = contact force, F = preload. *Source*: Carnes, 2004. Reproduced with permission of John Wiley & Sons.

wear rates experienced with these high-temperature materials when sliding against themselves in pin-on-disk tests. Above the following approximate transition temperatures, a smooth oxide surface forms in air with sufficient rapidity to minimize metal-to-metal contact and provide a low wear rate (Peterson *et al.*, 1960):

Alloy	Temperature (°C)
Steel	185
Cobalt	350
Molybdenum	460
Titanium	575
Chromium	630
Nickel	630

Ceramic materials used for bearings and seals are listed in Table 4.9. Of these, aluminum oxide (Al_2O_3) and silicon carbide (SiC) are well established. Silicon nitride (Si_3N_4) is an effective ball material for use with tool-steel raceways. Partially stabilized zirconia (ZrO_2 + Y_2O_3 or MgO) is of interest for low heat rejection diesel engines (Murray, 1997).

These ceramics offer high compressive strength and fatigue resistance, corrosion resistance, low density, and retention of mechanical properties up to temperatures well above 1000 °C. Except for some lightly loaded applications at low speeds, ceramics need lubrication to prevent high friction and excessive wear. High wear rates of 1×10^{-4}–1×10^{-6} mm^2/N m for unlubricated ceramic couples can be reduced with lubrication to 1×10^{-7}–1×10^{-10} or less, as needed for most practical devices. Candidate lubricants include oils, greases, water and water vapor, aqueous solutions, low shear strength solids, reactive vapors, and friction polymers. At temperatures above 250–300 °C, the list of candidates drops to vapor phase lubrication, solid lubricant films, and oxide films formed by reaction with the environment. Typical unlubricated coefficients of friction range from 0.5 to 1.0 (Murray, 1997).

Table 4.9 Ceramics used for bearing and seal applications

Property	Hot pressed alumina Al_2O_3	Cold pressed alumina AD999	Silicon carbide α-SiC	Silicon nitride Si_3N_4	Partially stabilized zirconia PSZ MS
Density (g/cm^3)	3.9–3.98	3.96	3.10	3.21	5.5–6.05
Hardness (kg/mm^3)	2050$^+$	1500	2800	1500	1050
Modulus of elasticity (GPa)	380–400	390	410	325	150–220
Tensile strength (MPa)	260	220	–	524	689
Flexural strength (MPa)	260	276	–	524	689
Compressive strength (MPa)	2760	2070	3900	3000	1850
Fracture toughness (K_{IC}, MPa m$^{1/2}$)	3–3.2	3	4.6	6–7	10–12
Thermal shock resistance (TSR)	4.10	3–4	10	20	0.6
Maximum use temperature in air (°C)	2000	1760	1650	1300	900–1400

Source: Murray, 1997. Reproduced with permission of Taylor & Francis. See also Table 4.2.

Cracking under thermal shock is always a concern with ceramic components that must undergo a rapid rise in surface temperature. Frictional surface heating during dry sliding can involve rapid rise and fall of temperatures, especially at localized contact spots, that will produce spalling and accelerated wear. Low thermal conductivity and high elastic modulus make a material more prone to this thermal shock and rupture as reflected in its thermal shock resistance (TSR):

$$\text{TSR} = \frac{\sigma k}{E\alpha} \times 10^{-3} \qquad (4.2)$$

where

σ = tensile fracture stress (MPa)
k = thermal conductivity (W/m °C)
E = elastic modulus (MPa)
α = thermal expansion coefficient (°C^{-1})

The lower the TSR, the higher the sensitivity to thermal shock fractures. With TSR values ranging down to 0.6, sensitivity to thermal shock is always a concern with partially stabilized zirconia. While silicon nitride offers a superior value among ceramics of approximately 20, this value compares with the much greater thermal shock resistance of tool steels with TSR values ranging to 57 (Glaeser, 1997). The critical sliding velocity above which thermoelastic instability (TEI) is encountered may approximately be proportional to (TSR)^2 (Glaeser, 1992). Thermoelastic instability is associated with high-speed sliding contact where pressure perturbations appear in the contact. As a result of local thermoelastic expansion and high contact pressures, an unstable thermal loop is set up that often leads to the formation of macroscopic hot spots, observable with the naked eye. This phenomenon is commonly observed in mechanical seals (Burton, 2000; Peng et al., 2003; Jang and Khonsari, 2003, 2013), brakes (Lee and Barber, 1993), and clutches (Jang and Khonsari, 2002) and many other types of application involving the contact between conductor-insulator or conductor-conductor surfaces (Jang and Khonsari, 2004a, 2004b, 2006).

Nickel, cobalt, molybdenum, or chromium are often incorporated with ceramic to form a composite having increased toughness, ductility, and shock resistance. Plasma and other thermal spray processes deposit wear-resistant ceramic coatings as powders of aluminum oxide (Al_2O_3), chromium oxide (Cr_2O_3), titanium nitride (TiN), tungsten carbide (WC), and titanium oxide (TiO_2) on metal substrates with or without cobalt, nickel, or chromium incorporated to improve mechanical properties.

Example 4.2 Determine the thermal shock resistance for a tool steel with the following properties: tensile strength $\sigma = 2000$ MPa, thermal conductivity $k = 51.9$ W/m °C, elastic modulus $E = 200\,000$ MPa, and thermal expansion coefficient $\alpha = 12.1 \times 10^{-6}$ m/m °C.
  Using these values in Equation (4.2):

$$\text{TSR} = \frac{2000 \times 51.9}{200\,000 \times 12.1 \times 10^{-6}} \times 10^{-3} = 43$$

4.6 Rolling Bearing Materials

Steels for ball and roller bearings are commonly selected to provide Rockwell C hardness in the 58–63 range at load-carrying contacts. Below this level, fatigue strength drops rapidly for contact stresses commonly in the 700–2800 MPa (100 000–400 000 psi) range; troublesome brittleness is encountered at higher hardness levels.

Representative steels employed for ball and roller bearings are listed in Table 4.10. AISI 52100 through-hardened steel is the primary choice for industrial ball bearings and many spherical and cylindrical roller bearings. With its higher chromium content, 440C is commonly employed for corrosion resistance. Carburized steels (e.g. AISI 8620) are used for tapered roller bearings and other heavily loaded types that benefit from the tougher core material and from the case compressive residual stress developed during carburizing. Both types contain 0.6–1.1% carbon at their highly stressed contact surfaces to give high hardness. For standard bearing steels, an elastic modulus of 207 GPa (30 × 10^6 psi) and 0.3 Poisson's ratio are usually used for calculating contact stresses and EHD film thickness.

With their lower carbon content, case-hardened steels are used for the rollers and races of many roller bearings in automotive and railroad equipment for better resistance both to shock load and to cracking with heavy interference fits when mounted on shafts. Many automotive front wheel drive bearings now use case-hardening alloys such as 1570. Large bearings for excavators and cranes employ heat-treatable steel with an increased alloy content and 0.4–0.5% carbon. Their raceway surfaces are frequently gas carburized or induction hardened (Eschmann *et al.*, 1985).

The operating temperature limit for AISI 52100 steel bearings commonly ranges from 125 to 175 °C. As operating temperatures exceed this range and approach the final stabilizing temperature for the steel, the small amount of retained austenite phase transforms into less dense martensite, with an increase in bearing dimensions and with some loss of fatigue strength. Case-carburized bearings are usually limited to an upper operating temperature of 150–175 °C.

Table 4.10   Representative materials for rolling element bearings

Material	Composition (%)							Maximum operating temperature	
	C	Cr	Mn	Si	Mo	Ni	V	°C	°F
Steels									
52100 through hardened	1.0	1.5	0.35	0.25				150	300
1570 case carburized	0.70		0.95					150	300
8620 case carburized	0.20	0.5	0.80	0.22	0.20	0.55	1.0	150	300
440C stainless	1.00	17.0	0.40	0.30	0.50			175	350
M50 tool steel	0.85	4.1			4.25		1.0	315	600
M50-NiL	0.13	4.2	0.25	0.20	4.25	3.40	1.2	315	600
Ceramics									
Si$_3$N$_4$ ceramic								650	1200
Si$_3$N$_4$-M50 hybrid	Ceramic balls, steel rings							425	800

Source: Moyer, 1997. Reproduced with permission of Taylor & Francis.

Table 4.11   Cage materials for rolling element bearings

Operating conditions	Typical material	Manufacturing method	Remarks
Good lubrication			
Low speed or noncritical	Low-carbon steel	Riveted strips	Standard material
Low speed, corrosive environment	AISI 430 stainless	Riveted strips	Standard for 440C bearings
Medium speed and temperature	Iron silicon bronze	Machined	Jet engine bearings
High speed	Phenolic-fabric laminates	Machined	
High temperature	S Monel, 17-4PH, or 430 stainless	Machined riveted strips	High strength, low speed
Marginal lubrication	Silver plate	Plate	Temperature to 150–180 °C
No lubrication	PTFE/MoS_2/glass fiber composite	Hot press, machine	Low loads

Source: Moyer, 1997. Reproduced with permission of Taylor & Francis.

M50 tool steel and M50NiL are commonly used for temperatures above 175 °C under the severe conditions encountered in aircraft jet engines.

Ceramics, especially silicon nitride (Si_3N_4), are being applied in an increasing range of severe conditions. Their unique properties include high-temperature capability to 1000 °C and above, inertness to corrosion and hostile environments, low coefficient of expansion, and low density, which greatly reduces troublesome self-loading by centrifugal force on balls and rollers at speeds of 50 000 rpm and above in machine tools and aerospace equipment. Hybrid bearings using ceramic balls with M50 tool steel rings enable operation up to 300 °C while avoiding the difficulties in manufacturing ceramic rings.

Typical cage materials used for spacing the rolling elements are listed in Table 4.11. Simple low-carbon steel strip formed and riveted to provide ball pockets is widely used for the low stress levels on cages in most ball bearings. AISI 430 stainless strip is used in the same fashion for 440C stainless steel bearings for corrosion resistance. Fabric-base phenolic laminates and other nonmetallics provide better rubbing characteristics and easier lubrication at high speeds. For high temperatures, a shift is made to high-temperature alloys, which are often silver-plated for high speeds or where there may be marginal or interrupted lubrication, as in some jet engine and aerospace equipment. In vacuum, silver plate and composites of PTFE, molybdenum disulfide (MoS_2), and glass fibers enable operation at light loads.

Polycrystalline Diamond (PCD)

While their use is widespread in down hole drilling applications for oil and gas, polycrystalline diamonds (PCD) have recently emerged as promising bearing materials that can withstand harsh operating conditions (high load, high speed, and high temperatures), particularly when the operating environment contains abrasive contaminants (Figure 4.4). PCD is formed by

Figure 4.4 A PCD thrust-like diamond bearing with circular pads on the runner and stator. *Source*: Courtesy of US Synthetic.

Table 4.12 Comparison of PCD properties with several materials

	Thermal Conductivity (W/m K)	Tensile Strength (MPa)	Compressive Strength (GPa)	Modulus of Elasticity (GPa)	Hardness (GPA, Koop)
Polycrystalline diamond	543	1300–1600	6.9–7.6	841	49.8
Tungsten carbide	70	334	2.68	669–696	1.8
Silicon nitride	30	520	–	296	1.8
Silicon carbide	85	500	2.5	434	2.4
Steel (4140)	42.6	415	–	205	0.2

Source: Lingwall *et al.*, 2013.

subjecting diamond grit to high temperature and pressure (about 1500 °C, and pressure and 6 GPa). PCD exhibits remarkable thermal and mechanical properties as shown in Table 4.12 (Lingwall *et al.*, 2013). Comparison with 4140 steel and silicon nitride shows more than a ten-fold increase in thermal conductivity, which offers superior thermal performance.

Problems

4.1 Friction loss of 0.2 hp is encountered in a 2 in. long, 2 in. diameter sleeve bearing. If half of the heat loss is conducted out through a 0.030 in. thick liner of bearing material, calculate the temperature drop across the liner if the bearing material is tin babbitt, aluminum, or nylon.

4.2   What general type of sleeve bearing material would you recommend for the following applications: blower in an electric hair dryer; 10 000 kW gas turbine-generator for standby electric power; rolling mill to be lubricated with a water emulsion; space station electric generator to be lubricated with liquid ammonia; propeller shaft bearings for an ocean-going fishing boat.

4.3   A sleeve bearing having a 1/8 in. thick layer of lead babbitt failed due to fatigue in a reciprocating gas engine after 1100 h of operation at 110 °C. Suggest three possible alternatives to give 50 000 h of life, along with the advantages and disadvantages of each.

4.4   Porous bronze sleeve bearings are to be used for a rotating flow meter in an air line. Maximum radial force on the two-bearing impeller will be 50 lb. Rotation will be at speeds up to 1000 rpm on its 1/2 in. shaft. How long should each bearing be? Assume $PV = 50\,000$ psi ft/ min.

4.5   A 1350 lb radial belt load is applied to a 3 in. diameter journal bearing on a line shaft. Using a breakaway coefficient of friction of 0.35 for tin bronze, calculate the starting torque. Suggest means for reducing this starting torque.

4.6   A compressor drive shaft at rest exerts a maximum contact compressive stress of 1200 psi on a sleeve bearing. What is the maximum transient bearing temperature that can be tolerated on shutdown of the unit with tin babbitt, lead babbitt, or SAE 660 bronze (assuming a linear drop in bronze strength to zero at its initial melting point and a compressive yield strength 0.5 times its ultimate tensile strength)?

4.7   What is the elastic deflection of a 20 mm thick layer of each of the following bearing materials under 600 psi compressive contact stress: nylon, lead babbitt, aluminum, rubber, tungsten carbide?

4.8   Give three methods for controlling the corrosion of babbitt sleeve bearings in industrial electric motors.

4.9   Suggest three alternative constructions for connecting rod bearings in a reciprocating engine, with the advantages and disadvantages of each.

References

Anonymous. 2006. 'Hard Disk,' Internet *Wikipedia Encyclopedia*.

ASM. 1961. *Metals Handbook*, Vol. 1. ASM International, Materials Park, Ohio, pp. 843–863.

ASTM. 1990. *ASTM Standards*, Vol. 02.04. ASTM, Philadelphia, pp. 9–11.

Blanchet, T.A. 1997. 'Friction, Wear, and PV Limits of Polymers and Their Composites,' *Tribology Data Handbook*, E.R. Booser (ed.). CRC Press, Boca Raton, Fla, pp. 547–562.

Blau, P.J. (ed.). 1992. *Friction, Lubrication, and Wear Technology, ASM Handbook*, Vol. 18. ASM International, Materials Park, Ohio.

Booser, E.R. 1992. 'Bearing Materials,' *Encyclopedia of Chemical Technology*, 4th edn, Vol. 4. John Wiley & Sons, New York, pp. 1–21.

Booser, E.R. and Wilcock, D.F. 1978. 'Assessment of Carbon-Graphite Wear Properties,' MEP-69. Mechanical Engineering Publications, Ltd. Worthington, UK, pp. 169–201.

Burton, R. 2000. *Heat, Bearings, and Lubrication*, Springer-Verlag New York, Inc.

Carnes, K. 2004. 'Hard-Driving Lubrication,' *Tribology and Lubrication Technology*, Vol. 60 (11), pp. 11–38.

Dean, R.R. and Evans, C.J. 1976. *Tribology*, Vol. 9, pp. 101–108.

DeHart, A.O. 1984. 'Bearing Materials,' *Handbook of Lubrication*, Vol. 2. CRC Press, Boca Raton, Fla, pp. 463–476.

DellaCorte, C. 2004. 'A Systems Approach to the Solid Lubrication of Foil Air Bearings for Oil-Free Turbomachinery,' *ASME Journal of Tribology*, Vol. 126, pp. 201–207.

Dunaevsky, V.V. *et al.* 1997. *Tribology Data Handbook*, E.R. Booser (ed.). CRC Press, Boca Raton, Fla, Chapters 40, 42, 43, 44.

Eschmann, L., Hasbargen, L., and Weigand, K. 1985. *Ball and Roller Bearings*. John Wiley & Sons, New York.

Fields, S. 2004. 'Foil Air Bearing Technology Takes Flight,' *Tribology and Lubrication Technology*, Vol. 60 (4), pp. 29–33.

Glaeser, W.A. 1983. 'Wear Properties of Heavy-Loaded Copper-Base Alloys,' *Journal of Metals*, Vol. 35, pp. 50–55.

Glaeser, W.A. 1992. 'Friction and Wear of Ceramics,' *ASM Handbook*, Vol. 18, pp. 812–815.

Glaeser, W.A. 1997. 'Wear-Resistant Hard Materials,' *Tribology Data Handbook*. CRC Press, Boca Raton, Fla, pp. 540–546.

Hamrock, B.J. 1994. *Fundamentals of Fluid Film Lubrication*. McGraw-Hill Book Co., New York.

Jang, J.Y. and Khonsari, M.M. 2002. 'On the Formation of Hot Spots in Wet Clutch Systems,' *ASME Journal of Tribology*, Vol. 124, pp. 336–345.

Jang, J.Y. and Khonsari, M.M. 2003. 'A Generalized Thermoelastic Instability Analysis,' *Proceedings of the Royal Society of London*, Series A, Vol. 459, pp. 309–329.

Jang, J.Y. and Khonsari, M.M. 2004a. 'A Note on the Rate of Growth of Thermoelastic Instability,' *ASME Journal of Tribology*, Vol. 126, pp. 50–55.

Jang, J.Y. and Khonsari, M.M. 2004b. 'Thermoelastic Instability of Two-Conductors Friction System Including Surface Roughness,' *ASME Journal of Applied Mechanics*, Vol. 71, No. 1, pp. 57–68.

Jang, J.Y. and Khonsari, M.M. 2013. 'Thermoelastic Instability in Mechanical Systems with Provision for Surface Roughness,' *Encyclopedia of Elasticity*, Springer Science, New York.

Kennedy, F.E., Booser, E.R., and Wilcock, D.F. 1998. 'Tribology, Lubrication, and Bearing Design,' *Handbook of Mechanical Engineering*. CRC Press, Boca Raton, Fla, pp. 3.128–3.169.

Khonsari, M.M. and Booser, E.R. 2004. 'Bearings–Keeping their Cool,' *Machine Design*, 4 May, pp. 93–95.

Kingsbury, G.R. 1992. 'Friction and Wear of Sliding Bearing Materials,' *ASM Handbook – Friction, Lubrication, and Wear Technology*, Vol. 18. ASM International, Materials Park, Ohio, pp. 741–757.

Kingsbury, G.R. 1997. 'Oil Film Bearing Materials,' *Tribology Data Handbook*. CRC Press, Boca Raton, Fla, pp. 503–525.

Lancaster, J.K. 1973. 'Dry Bearings: A Survey of Materials and Factors Affecting their Performance,' *Tribology*, Vol. 6, pp. 220–251.

Laskowski, J.A. and DellaCorte, C. 1996. 'Friction and Wear Characteristics of Candidate Foil Bearing Materials from 25 C to 800 C,' NASA TM-107082.

Lee, K. and Barber, J.R., 1993. 'Frictionally Excited Thermoelastic Instability in Automotive Disk Brakes,' *ASME Journal of Tribology*, Vol. 115, pp. 607–614.

Lingwall, B.A., Lu, X., Cooley, C.H., Sexton, T.N. and Khonsari, M.M. 2013. 'On the Performance of Diamond Bearings Designed for Boundary, Mixed-mode and Hydrodynamic Lubrication Regimes,' paper presented at 2013 World Tribology Congress, Torino, Italy, Sept. 8–13.

Moyer, C.A. 1997. 'Rolling Element Bearing Materials,' *Tribology Data Handbook*. CRC Press, Boca Raton, Fla, pp. 495–502.

Murray, S.F. 1997. 'Properties of Advanced Ceramics,' *Tribology Data Handbook*. CRC Press, Boca Raton, Fla, pp. 563–572.

NASA. 2000. 'Load Capacity Estimation of Foil Air Journal Bearings for Oil-Free Turbomachinery Applications,' NASA/TM-2000-209782.

Peng, Z-C, Khonsari, M.M. and Pascovici, M.D. 2003. 'On the Thermoelastic Instability of a Thin-film Lubricated Sliding Contact: A Closed-form Solution,' *Journal of Engineering Tribology*, Proceedings of Institution of Mechanical Engineers, Part J, Vol. 217, pp. 197–204.

Peng, Z-C and Khonsari, M.M. 2004a. 'Hydrodynamic Analysis of Compliant Foil Bearing with Compressible Air Flow,' *ASME Journal of Tribology*, Vol. 126, pp. 542–546.

Peng, Z-C and Khonsari, M.M. 2004b. 'On the Limiting Load Carrying Capacity of Foil Bearings,' *ASME Journal of Tribology*, Vol. 126, pp. 117–118.

Peng, Z.C. and Khonsari, M.M. 2006. 'A Thermohydrodynamic Analysis of Foil Journal Bearings,' *ASME Journal of Tribology*, Vol. 128, pp. 534–541.

Peterson, M.B., Lee, R.L., and Florek, J.J. 1960. *ASLE Transactions*, Vol. 3(1), p. 101.

Peterson, M.B. and Winer, W.O. (eds). 1980. *Wear Control Handbook*. ASME, New York.

Rabinowicz, E. 1984. 'Wear Coefficients,' *Handbook of Lubrication*, Vol. 2. CRC Press, Boca Raton, Fla, pp. 201–208.

Rabinowicz, E. 1995. *Friction and Wear of Materials*, 2nd edn. John Wiley & Sons, New York.

Roach, A.E., Goodzeit, C.L., and Hunnicutt, R.P. 1956. *Transactions of the ASM*, Vol. 78, p. 1696.

Ruff, A.W. 1997. 'Typical Properties of Sliding Contact Materials,' *Tribology Data Handbook*, E.R. Booser (ed.). CRC Press, Boca Raton, Fla, p. 494.

SAE 1991. *SAE Handbook*, Vol. 1. Warrendale, Pa, Sections 10.35–10.37.

Part II
Fluid-Film Bearings

5

Fundamentals of Viscous Flow

This chapter concentrates on the fundamentals of viscous flow. While several pertinent works on fluid mechanics are referenced at the close of this chapter, our objective here is to use this knowledge to develop a theoretical foundation for analyzing hydrodynamic lubrication problems.

For this purpose, we will review the equations that govern the conservation of mass, momentum, and energy. Applications of these conservation laws are illustrated through examples. Also introduced in this chapter are nondimensionalization techniques for generalizing results in a variety of complex fluid mechanics problems. We conclude with order-of-magnitude analyses for determining the relative importance of terms in equations in both this and later chapters.

5.1 General Conservation Laws

The three basic conservation laws are those governing conservation of mass, momentum, and energy. Formal derivations of these equations are available in most textbooks of fluid mechanics (see e.g. Schlichting, 1975; White, 1991) and therefore are not repeated here.

The governing equations in vector notation are as follows:

Conservation of mass (continuity equation):

$$\frac{\partial \rho}{\partial t} + \nabla \cdot (\rho V) = 0 \tag{5.1}$$

Conservation of momentum (Navier-Stokes equations):

$$\underbrace{\rho \frac{DV}{Dt}}_{\text{inertia forces}} = \underbrace{F_{\text{B}}}_{\text{body forces}} - \underbrace{\nabla P}_{\text{pressure forces}} + \underbrace{\nabla \cdot \tau_{ij}}_{\text{viscous forces}} \tag{5.2}$$

Applied Tribology: Bearing Design and Lubrication, Third Edition. Michael M. Khonsari and E. Richard Booser.
© 2017 John Wiley & Sons Ltd. Published 2017 by John Wiley & Sons Ltd.
Companion Website: www.wiley.com/go/Khonsari/Applied_Tribology_Bearing_Design_and_Lubrication_3rd_Edition

Conservation of energy (energy equation):

$$\rho\underbrace{\frac{De}{Dt}}_{\text{convective terms}} = \underbrace{\nabla \cdot (k\nabla T)}_{\text{conductive terms}} + \underbrace{\frac{DP}{Dt}}_{\substack{\text{work done} \\ \text{by compression}}} + \underbrace{\Phi}_{\substack{\text{viscous} \\ \text{dissipation}}} \tag{5.3}$$

where V = velocity vector, ρ = density, t = time, F_B = body forces, τ = shear stress, P = pressure, T = temperature, e = enthalpy $= i + P/\rho$, i = internal energy, Φ = viscous dissipation function $= \tau_{ij}(\partial u_i/\partial x_j)$ and k = thermal conductivity.

Operator D/Dt = total derivative, material derivative, or particle derivative, defined as follows:

$$\frac{D}{Dt} = \underbrace{\frac{\partial}{\partial t}}_{\text{local derivative}} + \underbrace{u\frac{\partial}{\partial x} + v\frac{\partial}{\partial y} + w\frac{\partial}{\partial z}}_{\text{convective derivatives}} \tag{5.4}$$

Brief descriptions of individual terms in these equations follow.

5.2 Conservation of Mass

Equation (5.1) is simply the statement of conservation of mass: the *net* rate of a mass flow into and out of a differential control volume must be zero. The second term is simply divergence of ρV. Making use of appropriate operators, these equations can be written in Cartesian (rectangular), cylindrical, or spherical coordinate systems. In most lubrication problems treated in this book, the spherical coordinate system is of limited use; we shall focus our attention on the Cartesian and cylindrical coordinate systems.

Cartesian Coordinates

In the Cartesian coordinate system, $V \equiv V(x, y, z)$, the continuity equation takes the following form (Figure 5.1):

$$\frac{\partial \rho}{\partial t} + \frac{\partial(\rho u)}{\partial x} + \frac{\partial(\rho v)}{\partial y} + \frac{\partial(\rho w)}{\partial z} = 0 \tag{5.5}$$

where u, v, and w are the components of the velocity in the x, y, and z directions, respectively.

Cylindrical Coordinates

In the cylindrical coordinate system, $V \equiv V(r, \theta, z)$, and the continuity equation is

$$\frac{\partial \rho}{\partial t} + \frac{1}{r}\frac{\partial}{\partial r}(r\rho v) + \frac{1}{r}\frac{\partial}{\partial \theta}(\rho w) + \frac{\partial}{\partial z}(\rho u) = 0 \tag{5.6}$$

where u, v, and w are the components of the velocity in the axial z, radial r, and circumferential θ directions, respectively (Figure 5.2).

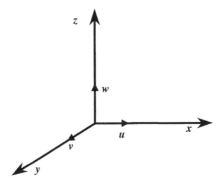

Figure 5.1 Cartesian coordinate system.

If the flow is incompressible, then there is no change in the density and $\partial \rho / \partial t = 0$ regardless of whether the flow is steady or unsteady. Then Equation (5.1) reduces to

$$\nabla \cdot V = 0 \qquad (5.7)$$

and density does not appear in the continuity equation.

5.3 Conservation of Momentum

The Navier-Stokes equations for conservation of momentum simply follow Newton's second law, $F = ma$, which on a per unit volume basis is: $(\rho DV/Dt) = F$. The terms on the left-hand side of Equation (5.2) are inertia terms involving acceleration. When inertia terms are present, Equation (5.2) is a nonlinear partial differential equation. Body forces are those that apply to the entire body of the fluid. The most common types of body forces are gravitational, buoyancy, and electromagnetic. The remaining two terms are classified as surface forces, i.e. pressure and viscous forces applied to a differential element of the fluid. Note that τ_{ij} is the stress field, a tensor with three normal stress components τ_{ii} and six shear stress components $\tau_{ij}(i \neq j)$.

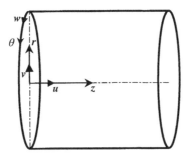

Figure 5.2 Cylindrical coordinate system.

Cartesian Coordinates (x, y, z)

In Cartesian coordinates, conservation of momentum takes on the following form in three equations corresponding to the x, y, and z directions, respectively:

x-Momentum:

$$\rho\left(\frac{\partial u}{\partial t} + u\frac{\partial u}{\partial x} + v\frac{\partial u}{\partial y} + w\frac{\partial u}{\partial z}\right) = \left(\frac{\partial}{\partial x}\tau_{xx} + \frac{\partial}{\partial y}\tau_{yx} + \frac{\partial}{\partial z}\tau_{zx}\right) - \frac{\partial P}{\partial x} + F_{Bx} \quad (5.8a)$$

y-Momentum:

$$\rho\left(\frac{\partial v}{\partial t} + u\frac{\partial v}{\partial x} + v\frac{\partial v}{\partial y} + w\frac{\partial v}{\partial z}\right) = \left(\frac{\partial}{\partial x}\tau_{xy} + \frac{\partial}{\partial y}\tau_{yy} + \frac{\partial}{\partial z}\tau_{zy}\right) - \frac{\partial P}{\partial y} + F_{By} \quad (5.8b)$$

z-Momentum:

$$\rho\left(\frac{\partial w}{\partial t} + u\frac{\partial w}{\partial x} + v\frac{\partial w}{\partial y} + w\frac{\partial w}{\partial z}\right) = \left(\frac{\partial}{\partial x}\tau_{xz} + \frac{\partial}{\partial y}\tau_{yz} + \frac{\partial}{\partial z}\tau_{zz}\right) - \frac{\partial P}{\partial z} + F_{Bz} \quad (5.8c)$$

Cylindrical Coordinates (r, θ, z)

There are three equations corresponding to the r, θ, and z directions, respectively:

r-Momentum:

$$\rho\left(\frac{\partial v}{\partial t} + v\frac{\partial v}{\partial r} + \frac{w}{r}\frac{\partial v}{\partial \theta} - \frac{w^2}{r} + u\frac{\partial v}{\partial z}\right) = \left(\frac{1}{r}\frac{\partial}{\partial r}(r\tau_{rr}) + \frac{1}{r}\frac{\partial}{\partial \theta}\tau_{\theta r} + \frac{\partial}{\partial z}\tau_{zr} - \frac{\tau_{\theta\theta}}{r}\right) - \frac{\partial P}{\partial r} + F_{Br}$$

$$(5.9a)$$

θ-Momentum:

$$\rho\left(\frac{\partial w}{\partial t} + v\frac{\partial w}{\partial r} + \frac{w}{r}\frac{\partial w}{\partial \theta} + \frac{vw}{r} + u\frac{\partial w}{\partial z}\right) = \left(\frac{1}{r^2}\frac{\partial}{\partial r}(r^2\tau_{r\theta}) + \frac{1}{r}\frac{\partial}{\partial \theta}\tau_{\theta\theta} + \frac{\partial}{\partial z}\tau_{z\theta}\right) - \frac{1}{r}\frac{\partial P}{\partial \theta} + F_{B\theta}$$

$$(5.9b)$$

z-Momentum:

$$\rho\left(\frac{\partial u}{\partial t} + v\frac{\partial u}{\partial r} + \frac{w}{r}\frac{\partial u}{\partial \theta} + u\frac{\partial u}{\partial z}\right) = \left(\frac{1}{r}\frac{\partial}{\partial r}(r\tau_{rz}) + \frac{1}{r}\frac{\partial}{\partial \theta}\tau_{\theta z} + \frac{\partial}{\partial z}\tau_{zz}\right) - \frac{\partial P}{\partial z} + F_{Bz} \quad (5.9c)$$

Once the constitutive equation for a given fluid is given, the above equations can be used to determine the components of the velocity in the appropriate directions.

Newtonian Fluids

Many fluids, including most mineral and synthetic lubricants other than those containing polymer additives for improving VI, are classified as linearly viscous or Newtonian. The relationship between the shear stress and strain rate is

$$\tau = \mu\dot{\gamma} \tag{5.10}$$

where μ is the fluid viscosity and

$$\dot{\gamma} = [\nabla \cup + (\nabla \cup)^{\mathsf{T}}]$$

where τ = shear stress tensor, $\dot{\gamma}$ = strain rate tensor, $\nabla \cup$ = deformation rate tensor, and $(\nabla \cup)^{\mathsf{T}}$ = transpose of $\nabla \cup$.

General relationships for the deformation rate and shear stress tensors depend on the coordinate system used. For Cartesian and cylindrical coordinates, they are as follows.

Cartesian Coordinates

$$\nabla \cup = \begin{bmatrix} \dfrac{\partial u}{\partial x} & \dfrac{\partial v}{\partial x} & \dfrac{\partial w}{\partial x} \\[2mm] \dfrac{\partial u}{\partial y} & \dfrac{\partial v}{\partial y} & \dfrac{\partial w}{\partial y} \\[2mm] \dfrac{\partial u}{\partial z} & \dfrac{\partial v}{\partial z} & \dfrac{\partial w}{\partial z} \end{bmatrix}; \qquad (\nabla \cup)^{\mathsf{T}} = \begin{bmatrix} \dfrac{\partial u}{\partial x} & \dfrac{\partial u}{\partial y} & \dfrac{\partial u}{\partial z} \\[2mm] \dfrac{\partial v}{\partial x} & \dfrac{\partial v}{\partial y} & \dfrac{\partial v}{\partial z} \\[2mm] \dfrac{\partial w}{\partial x} & \dfrac{\partial w}{\partial y} & \dfrac{\partial w}{\partial z} \end{bmatrix}; \qquad \tau = \begin{bmatrix} \tau_{xx} & \tau_{xy} & \tau_{xz} \\ \tau_{yx} & \tau_{yy} & \tau_{yz} \\ \tau_{zx} & \tau_{zy} & \tau_{zz} \end{bmatrix}$$

$$\tag{5.11}$$

Normal stress components:

$$\tau_{xx} = 2\mu\frac{\partial u}{\partial x}; \qquad \tau_{yy} = 2\mu\frac{\partial v}{\partial y}; \qquad \tau_{zz} = 2\mu\frac{\partial w}{\partial z} \tag{5.12}$$

Shear stress components:

$$\tau_{xy} = \mu\left(\frac{\partial u}{\partial y} + \frac{\partial v}{\partial x}\right) = \tau_{yx}; \qquad \tau_{xz} = \mu\left(\frac{\partial u}{\partial z} + \frac{\partial w}{\partial x}\right) = \tau_{zx}; \qquad \tau_{yz} = \mu\left(\frac{\partial w}{\partial y} + \frac{\partial v}{\partial z}\right) = \tau_{zy}$$

$$\tag{5.13}$$

Cylindrical Coordinates

$$\nabla U = \begin{bmatrix} \dfrac{\partial v}{\partial r} & \dfrac{\partial w}{\partial r} & \dfrac{\partial u}{\partial r} \\[2ex] \dfrac{1}{r}\dfrac{\partial v}{\partial \theta} - \dfrac{w}{r} & \dfrac{1}{r}\dfrac{\partial w}{\partial \theta} + \dfrac{v}{r} & \dfrac{1}{r}\dfrac{\partial u}{\partial \theta} \\[2ex] \dfrac{\partial v}{\partial z} & \dfrac{\partial w}{\partial z} & \dfrac{\partial u}{\partial z} \end{bmatrix}; \qquad (\nabla U)^{\mathsf{T}} = \begin{bmatrix} \dfrac{\partial v}{\partial r} & \dfrac{1}{r}\dfrac{\partial v}{\partial \theta} - \dfrac{w}{r} & \dfrac{\partial v}{\partial z} \\[2ex] \dfrac{\partial w}{\partial r} & \dfrac{1}{r}\dfrac{\partial w}{\partial \theta} + \dfrac{v}{r} & \dfrac{\partial w}{\partial z} \\[2ex] \dfrac{\partial u}{\partial r} & \dfrac{1}{r}\dfrac{\partial u}{\partial \theta} & \dfrac{\partial u}{\partial z} \end{bmatrix};$$

$$\tau = \begin{bmatrix} \tau_{rr} & \tau_{r\theta} & \tau_{rz} \\ \tau_{\theta r} & \tau_{\theta\theta} & \tau_{\theta z} \\ \tau_{zr} & \tau_{z\theta} & \tau_{zz} \end{bmatrix} \tag{5.14}$$

Normal stress components:

$$\tau_{rr} = 2\mu\frac{\partial v}{\partial r}; \qquad \tau_{\theta\theta} = 2\mu\left(\frac{1}{r}\frac{\partial w}{\partial \theta} + \frac{v}{r}\right); \qquad \tau_{zz} = 2\mu\frac{\partial u}{\partial z} \tag{5.15}$$

Shear stress components:

$$\tau_{r\theta} = \mu\left[r\frac{\partial}{\partial r}\left(\frac{w}{r}\right) + \frac{1}{r}\frac{\partial v}{\partial \theta}\right] = \tau_{\theta r}; \qquad \tau_{rz} = \mu\left(\frac{\partial u}{\partial r} + \frac{\partial v}{\partial z}\right) = \tau_{zr};$$

$$\tau_{\theta z} = \mu\left(\frac{1}{r}\frac{\partial u}{\partial \theta} + \frac{\partial w}{\partial z}\right) = \tau_{z\theta} \tag{5.16}$$

Using the above relationships, the conservation of momentum equations for an incompressible Newtonian fluid can be simplified to give the following.

Cartesian Coordinates

x-Momentum:

$$\rho\left(\frac{\partial u}{\partial t} + u\frac{\partial u}{\partial x} + v\frac{\partial u}{\partial y} + w\frac{\partial u}{\partial z}\right) = \mu\left(\frac{\partial^2 u}{\partial x^2} + \frac{\partial^2 u}{\partial y^2} + \frac{\partial^2 u}{\partial z^2}\right) - \frac{\partial P}{\partial x} + F_{Bx} \tag{5.17a}$$

y-Momentum:

$$\rho\left(\frac{\partial v}{\partial t} + u\frac{\partial v}{\partial x} + v\frac{\partial v}{\partial y} + w\frac{\partial v}{\partial z}\right) = \mu\left(\frac{\partial^2 v}{\partial x^2} + \frac{\partial^2 v}{\partial y^2} + \frac{\partial^2 v}{\partial z^2}\right) - \frac{\partial P}{\partial y} + F_{By} \tag{5.17b}$$

z-Momentum:

$$\rho\left(\frac{\partial w}{\partial t} + u\frac{\partial w}{\partial x} + v\frac{\partial w}{\partial y} + w\frac{\partial w}{\partial z}\right) = \mu\left(\frac{\partial^2 w}{\partial x^2} + \frac{\partial^2 w}{\partial y^2} + \frac{\partial^2 w}{\partial z^2}\right) - \frac{\partial P}{\partial z} + F_{Bz} \tag{5.17c}$$

Cylindrical Coordinates

$$V \equiv V(r, \theta, z)$$

r-Momentum:

$$\rho \left(\frac{\partial v}{\partial t} + v \frac{\partial v}{\partial r} + \frac{w}{r} \frac{\partial v}{\partial \theta} - \frac{w^2}{r} + u \frac{\partial v}{\partial z} \right)$$

$$= \mu \left[\frac{\partial}{\partial r} \left(\frac{1}{r} \frac{\partial}{\partial r} (rv) \right) + \frac{1}{r^2} \frac{\partial^2 v}{\partial \theta^2} + \frac{\partial^2 v}{\partial z^2} - \frac{2}{r^2} \frac{\partial w}{\partial \theta} \right] - \frac{\partial P}{\partial r} + F_{Br}$$

(5.18a)

θ-Momentum:

$$\rho \left(\frac{\partial w}{\partial t} + v \frac{\partial w}{\partial r} + \frac{w}{r} \frac{\partial w}{\partial \theta} + \frac{vw}{r} + u \frac{\partial w}{\partial \theta} \right)$$

$$= \mu \left[\frac{\partial}{\partial r} \left(\frac{1}{r} \frac{\partial}{\partial r} (rw) \right) + \frac{1}{r^2} \frac{\partial^2 w}{\partial \theta^2} + \frac{\partial^2 w}{\partial z^2} + \frac{2}{r^2} \frac{\partial v}{\partial \theta} \right] - \frac{1}{r} \frac{\partial P}{\partial \theta} + F_{B\theta}$$

(5.18b)

z-Momentum:

$$\rho \left(\frac{\partial u}{\partial t} + v \frac{\partial u}{\partial r} + \frac{w}{r} \frac{\partial u}{\partial \theta} + u \frac{\partial u}{\partial z} \right) = \mu \left[\frac{1}{r} \frac{\partial}{\partial r} \left(r \frac{\partial u}{\partial r} \right) + \frac{1}{r^2} \frac{\partial^2 u}{\partial \theta^2} + \frac{\partial^2 u}{\partial z^2} \right] - \frac{\partial P}{\partial z} + F_{Bz} \quad (5.18c)$$

5.4 Conservation of Energy

The conservation of energy (Equation 5.3) is a statement of the first law of thermodynamics for a differential element of fluid. Simply, it states that the total energy of a system (internal energy plus kinetic and potential energy) is equal to the energy added to and the work done by the system.

The left-hand terms of Equation (5.3) are often referred to as the *convective terms*, for they account for motion of the fluid relative to the boundaries in which it flows. The first term on the right-hand side of Equation (5.3) presents the rate of energy transfer by means of conduction in a differential element. The second term expresses the rate of work done by the differential fluid volume in expansion or compression. For incompressible fluids, this term is negligible.

The last term plays an important role in most lubrication problems. It represents the rate of viscous energy dissipated into heat because of internal friction within the differential fluid element. This so-called viscous dissipation produces an irreversible heat flow that contributes to raising the temperature of the fluid.

The energy equations for incompressible Newtonian fluids written in the Cartesian and cylindrical coordinate systems are as follows.

Cartesian Coordinates

$$\rho C_p \left(\frac{\partial T}{\partial t} + u \frac{\partial T}{\partial x} + v \frac{\partial T}{\partial y} + w \frac{\partial T}{\partial z} \right) = \frac{\partial}{\partial x} \left(k \frac{\partial T}{\partial x} \right) + \frac{\partial}{\partial y} \left(k \frac{\partial T}{\partial y} \right) + \frac{\partial}{\partial z} \left(k \frac{\partial T}{\partial z} \right) + \Phi \quad (5.19)$$

where

$$\Phi = \tau_{ij}\frac{\partial u_i}{\partial x_j} = \tau_{xx}\left(\frac{\partial u}{\partial x}\right) + \tau_{yy}\left(\frac{\partial v}{\partial y}\right) + \tau_{zz}\left(\frac{\partial w}{\partial z}\right) + \tau_{xy}\left(\frac{\partial u}{\partial y} + \frac{\partial v}{\partial x}\right)$$

$$+ \tau_{yz}\left(\frac{\partial v}{\partial z} + \frac{\partial w}{\partial y}\right) + \tau_{zx}\left(\frac{\partial w}{\partial x} + \frac{\partial u}{\partial z}\right)$$

$$= 2\mu\left(\frac{\partial u}{\partial x}\right)^2 + 2\mu\left(\frac{\partial v}{\partial y}\right)^2 + 2\mu\left(\frac{\partial w}{\partial z}\right)^2 + \mu\left(\frac{\partial u}{\partial y} + \frac{\partial v}{\partial x}\right)^2$$

$$+ \mu\left(\frac{\partial v}{\partial z} + \frac{\partial w}{\partial y}\right)^2 + \mu\left(\frac{\partial w}{\partial x} + \frac{\partial u}{\partial z}\right)^2$$

Cylindrical Coordinates

$$\rho C_p\left(\frac{\partial T}{\partial t} + v\frac{\partial T}{\partial r} + \frac{\omega}{r}\frac{\partial T}{\partial \theta} + u\frac{\partial T}{\partial z}\right) = \frac{1}{r}\frac{\partial}{\partial r}\left(kr\frac{\partial T}{\partial r}\right) + \frac{1}{r^2}\frac{\partial}{\partial \theta}\left(k\frac{\partial T}{\partial \theta}\right) + \frac{\partial}{\partial z}\left(k\frac{\partial T}{\partial z}\right) + \Phi$$

$$(5.20)$$

where

$$\Phi = \tau_{rr}\left(\frac{\partial v}{\partial r}\right) + \tau_{\theta\theta}\left(\frac{1}{r}\frac{\partial w}{\partial \theta} + \frac{v}{r}\right) + \tau_{zz}\left(\frac{\partial u}{\partial z}\right) + \tau_{r\theta}\left[r\frac{\partial}{\partial r}\left(\frac{w}{r}\right) + \frac{1}{r}\frac{\partial v}{\partial \theta}\right]$$

$$+ \tau_{\theta z}\left(\frac{1}{r}\frac{\partial u}{\partial \theta} + \frac{\partial w}{\partial z}\right) + \tau_{rz}\left(\frac{\partial u}{\partial r} + \frac{\partial v}{\partial z}\right)$$

$$= 2\mu\left(\frac{\partial v}{\partial r}\right)^2 + 2\mu\left(\frac{1}{r}\frac{\partial w}{\partial \theta} + \frac{v}{r}\right)^2 + 2\mu\left(\frac{\partial u}{\partial z}\right)^2 + \mu\left[r\frac{\partial}{\partial r}\left(\frac{w}{r}\right) + \frac{1}{r}\frac{\partial v}{\partial \theta}\right]^2$$

$$+ \mu\left(\frac{1}{r}\frac{\partial u}{\partial \theta} + \frac{\partial w}{\partial z}\right)^2 + \mu\left(\frac{\partial u}{\partial r} + \frac{\partial v}{\partial z}\right)^2$$

In order to determine the temperature distribution in a fluid undergoing deformation, Equation (5.19) or (5.20) must be solved. To solve for the temperatures which are a function of the velocity components, the appropriate momentum equations must first be solved to determine the flow velocity components.

Note that the terms in the viscous dissipation often simplify significantly, as we shall see in Example 5.1 since, depending on the problem, many of the gradients of the velocity components are nil. In the absence of motion, when the velocity components are all nil, all the convective velocity terms and viscous dissipation drop out and the energy equation reduces to the governing equation for heat conduction, as expected.

In many lubrication problems, viscosity variation with temperature is significant. This implies that momentum equations are 'coupled' with the energy equation and therefore must be solved simultaneously to obtain a meaningful solution to the problems of temperature and velocity components. These problems are often solved iteratively by an appropriate numerical scheme (Khonsari and Beaman, 1986; Khonsari *et al.*, 1996a; Fillon and Khonsari, 1996;

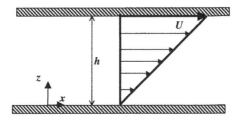

Figure 5.3 Couette velocity profile.

Keogh *et al.*, 1997; Jang and Khonsari, 1997, 2004). Within the context of lubrication theory associated with bearings, these problems are often referred to as the *thermohydrodynamic analyses* (Khonsari, 1987a, 1987b).

Example 5.1 (Couette flow) Consider the flow between two infinitely wide flat plates with a narrow separation gap of h, as shown in Figure 5.3. The bottom plate is stationary, and the top moves with a constant velocity U. Assume that the flow is steady state, laminar, Newtonian, incompressible, and has constant properties. Determine the following:

(a) Fluid velocity profile across the gap;
(b) Fluid temperature distribution, assuming that the top and bottom plates are maintained at temperatures T_U and T_L, respectively.

Fluid velocity profile To determine the fluid velocity profile, we must use the Navier-Stokes equations. The Cartesian coordinate system fits the problem. Since the plates are infinitely wide in the y direction, $\partial/\partial y = 0$ and the problem is two-dimensional, and $V \equiv V(x, z)$. Let us first examine the components of the shear stress from Equation (5.13):

$$\tau_{xy} = \tau_{yx} = \mu \left(\frac{\partial u}{\partial y} + \frac{\partial v}{\partial x} \right) = 0$$

$$\tau_{yz} = \tau_{zy} = \mu \left(\frac{\partial w}{\partial y} + \frac{\partial v}{\partial z} \right) = 0$$

$$\tau_{xz} = \tau_{zx} = \mu \left(\frac{\partial u}{\partial z} + \frac{\partial w}{\partial x} \right) = \mu \frac{\partial u}{\partial z}$$

Therefore, τ_{zx} is the only shear stress component which exists in this problem, as expected.

This problem deals with the steady state, i.e. $\partial/\partial t = 0$, and there is no body force in the x direction. Therefore, $F_{Bx} = 0$ and Equation (5.17a) reduces to

x-Momentum:

$$\rho \left(u \frac{\partial u}{\partial x} + w \frac{\partial u}{\partial z} \right) = \mu \left(\frac{\partial^2 u}{\partial x^2} + \frac{\partial^2 u}{\partial z^2} \right) - \frac{\partial P}{\partial x}$$

Assuming that the flow is fully developed, $\partial u/\partial x = 0$. This implies that velocity does not vary along the x direction and that u is only a function of z, that is, $u = u(z)$. Since the surfaces are parallel, w, the component of velocity in the z direction, is zero. Therefore, the momentum equation reduces further to

$$0 = \mu \frac{\partial^2 u}{\partial z^2} - \frac{\partial P}{\partial x}$$

In a Couette-type flow, pressure is constant, so $\partial P/\partial x = 0$. Therefore, the momentum equation reduces to an ordinary differential equation:

$$\frac{d^2 u}{dz^2} = 0$$

The appropriate boundary conditions are:

$$u = 0 \quad \text{at } z = 0$$
$$u = U \quad \text{at } z = h$$

where the no-slip condition at the bounding walls is assumed.

Integrating the above equation twice and evaluating the constants of integration yields

$$u = \frac{U}{h} z$$

i.e. a linear velocity profile, as shown in Figure 5.3. This is a shear-induced flow, and there is no pressure gradient.

Film temperature distribution   To obtain the temperature distribution, we begin by writing the energy equation in the two-dimensional Cartesian coordinate system. Equation (5.19) reduces to

$$\rho C_p \left(u \frac{\partial T}{\partial x} + w \frac{\partial T}{\partial z} \right) = k \left(\frac{\partial^2 T}{\partial x^2} + \frac{\partial^2 T}{\partial z^2} \right) + \Phi$$

where the viscous dissipation term is

$$\Phi = 2\mu \left(\frac{\partial u}{\partial x} \right)^2 + 2\mu \left(\frac{\partial w}{\partial z} \right)^2 + \mu \left(\frac{\partial w}{\partial x} + \frac{\partial u}{\partial z} \right)^2$$

Since $w = \partial u/\partial x = 0$, the viscous dissipation term is further reduced to $\Phi = \mu(\partial u/\partial z)^2$, where the velocity gradient across the gap is determined from the velocity profile obtained previously.

From the boundary conditions specified, temperature does not vary along the direction of motion, that is $\partial T/\partial x = 0$. Therefore

$$0 = k \frac{\partial^2 T}{\partial z^2} + \mu \left(\frac{\partial u}{\partial z} \right)^2$$

using the velocity profile determined in part (a), $\partial u/\partial z = U/h$ and the energy equation becomes a simple ordinary differential equation:

$$k \frac{d^2 T}{dz^2} = -\mu \frac{U^2}{h^2}$$

Integrating twice gives the following solution:

$$T = -\frac{\mu}{k}\left(\frac{U}{h}\right)^2 \frac{z^2}{2} - \frac{C_1}{k}z + C_2$$

Boundary conditions are

$$T = T_L \text{ at } z = 0; \qquad T = T_U \text{ at } z = h$$

From these the integration constants C_1 and C_2 can be determined. The result for the temperature distribution is

$$\frac{T - T_L}{T_U - T_L} = \left(\frac{z}{h}\right) + \frac{1}{2}\frac{\mu U^2}{k\left(T_U - T_L\right)}\left(\frac{z}{h}\right)\left(1 - \frac{z}{h}\right)$$

The left-hand side is dimensionless. To consider the dimensionless form of the right-hand side, let

$$\bar{z} = \frac{z}{h} \quad \text{and} \quad Br = \frac{\mu U^2}{k\left(T_U - T_L\right)} \equiv \text{Brinkman number}[1]$$

The above equation now reads:

$$\frac{T - T_L}{T_U - T_L} = \bar{z} + \frac{1}{2}Br\bar{z}\left(1 - \bar{z}\right)$$

The Brinkman number, Br, is an important parameter in assessing the extent of the role of viscous dissipation in raising the temperature relative to $(T_U - T_L)$. The higher the Brinkman number, the larger the temperature rise due to viscous dissipation.

In most lubrication problems, the viscous dissipation term is an important parameter since the film gap is very small and velocity gradients are large. It is instructive to plot the temperature profile across the gap for a range of Br and examine the physical implications of the temperature profile in terms of heat transfer (Figure 5.4).

When $Br = 0$, temperature distribution is linear across the gap and heat is transferred from the hotter plate (say, the sliding surface) to the cooler one. That is, the sliding surface becomes cooler. When $Br > 2$, the direction of heat flow changes. Figure 5.4 shows that the fluid could actually heat up the sliding surface (examine $dT/dz\,|_{z=h}$).

The location and magnitude of the maximum fluid temperature can easily be determined by letting $dT/dz = 0$ and solving for z. The result is

$$T_{max} = T_L + (T_U - T_L)\left(\frac{0.5}{Br} + 0.125Br + 0.5\right) \quad \text{at } \bar{z} = 0.5 + \frac{1}{Br}$$

For the example as shown in Figure 5.4, when $Br = 4$, $T_{max} = 1.125$ at $\bar{z} = 0.625$. If both plates are maintained at the same temperature, i.e. $T_L = T_U = T_0$, the temperature profile

---

[1]This is one form of the Brinkman number and it is simply a product of two dimensionless numbers known as the *Prandtl number*, *Pr*, and the *Eckert number*, *Ec* : $Br = Pr \cdot Ec$.

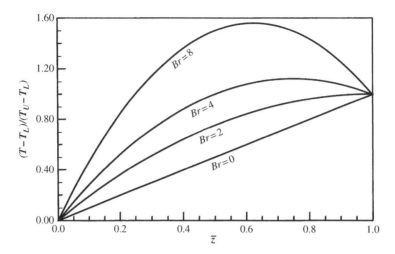

Figure 5.4   Temperature profile across the gap.

across the gap will be symmetric, with maximum temperature $T_{max} = T_0 + 0.125\,\mu U^2/k$ at the center. Taking $U = 6.28\,\text{m/s}$, $\mu = 0.25\,\text{Pa.\,s}$ (engine oil at 40 °C), and $k = 0.144\,\text{W/m\,K}$, this simple relationship predicts that $\Delta T_{max} = T_{max} - T_0 = 0.125\,\mu U^2/k = 8.6\,°\text{C}$.

This problem illustrates the use of the energy equation in conjunction with the momentum equation. The analytical results obtained in this simple problem are often used as a first approximation for characterizing the temperature in a journal bearing. Theoretically, this procedure gives a viable solution if the radius of curvature of the system is neglected and it is assumed that the shaft rotates concentrically within the bearing so that the Couette flow assumption holds (see Petroff's derivation in Section 5.5).

It is important to review some important features of this problem and assumptions made in order to arrive at these simplified exact, analytical solutions. Temperature boundary conditions were simply assumed to be constant and known. If, for example, the inner surface of the stationary plate, representing the bushing, were assumed to be perfectly insulated, then the adiabatic boundary condition would apply at $z = 0: \ dT/d\bar{z} = 0$. The resulting temperature profile across the gap is

$$T = T_U + \mu\frac{U^2}{k}\left(1 - \frac{z^2}{h^2}\right)$$

which predicts that the maximum temperature is $T_{max} = T_U + 0.5\,\mu U^2/k$ at $z = 0$. Using the same typical numbers for an engine oil at 40 °C, sheared at 6.28 m/s, this equation predicts that $\Delta T_{max} = T_{max} - T_U = 0.5\,\mu U^2/k = 34\,°\text{C}$.

In a journal bearing, for example, because of the loading imposed, the shaft position will be eccentric, and as a result, there will be a circumferential pressure gradient. Therefore, the Couette flow will be superimposed with a pattern of pressure-induced (Poiseuille) flow. Furthermore, the temperature around the circumference will not be uniform.

Another assumption is that fluid properties remain constant, independent of temperature. This assumption requires careful examination for practical applications. In particular, the lubricant's viscosity can change drastically with temperature. Property variations as a function of

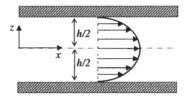

Figure 5.5   Poiseuille flow between two infinitely long parallel plates.

temperature are indeed important in the great majority of full-film lubrication problems and add to the complexity of the resulting equations and their solution. In many cases, one must resort to numerical schemes since the governing equations do not easily lend themselves to closed-form analytical solutions.

Example 5.2   (Poiseuille flow)   Determine the fully developed velocity profile for a laminar, incompressible fluid flowing between two infinitely wide parallel plates.

Referring to Figure 5.5, since both plates are stationary, the flow is driven by an existing pressure gradient between the inlet and outlet. This gradient is generated by the supply pressure through an external means, such as a pump at the inlet.

The x-momentum equation reduces to

$$\frac{d^2 u}{dz^2} = \frac{1}{\mu} \frac{dP}{dx}$$

The boundary conditions are

$$u = 0 \quad at \; z = \pm \frac{h}{2}$$

Solving for the velocity profile:

$$u = \frac{1}{2\mu} \frac{dP}{dx} \left(z^2 - \frac{h^2}{4} \right)$$

The mean velocity is obtained by integrating the above expression across the gap and dividing by the result by the gap width:

$$u_{\text{mean}} = \frac{2}{2h} \int_{-h/2}^{h/2} u \, dz = +\frac{h^2}{6\mu} \frac{dP}{dx}$$

Subject to a given set of appropriate boundary conditions, the temperature profile across the gap can be determined using a procedure similar to that in Example 5.1.

5.5   Petroff's Formula

The following formula was originally developed by Petroff, a Russian scientist, in 1883. It gives a simple relationship for evaluating torque and power loss in a journal bearing. While the derivation assumes a concentric shaft, i.e. zero eccentricity, the relationships derived are useful for evaluating the performance of many lightly and moderately loaded journal bearings.

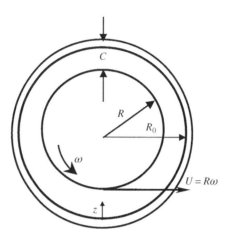

Figure 5.6   Petroff's concentric bearing.

Consider two concentric cylinders of length L, with the outer cylinder stationary and the inner one rotating at a given rotation speed of ω (rad/s). The clearance containing a fluid with viscosity μ is constant around the gap; therefore, $h = C = R_0 - R$ (Figure 5.6). Petroff's development assumes that the clearance is small and the curvature negligible. The problem is therefore posed in the Cartesian coordinate system.

Total friction force can be evaluated by integrating the shear stress over the area:

$$F = \int_A \tau \, dA \tag{5.21}$$

Shear stress is simply

$$\tau = \mu \frac{\partial u}{\partial z} = \mu \frac{U}{h} \tag{5.22}$$

where U is the linear speed. Assuming that μ, U, and h are constants,

$$F = \mu \frac{U}{h} \int_A dA = \mu \frac{U}{h} A \tag{5.23}$$

The appropriate area is simply $A = 2\pi RL$; therefore

$$F = \mu \frac{2\pi RLU}{h} \tag{5.24}$$

Since $h = C$ and $U = R\omega$,

$$F = 2\mu \frac{\pi R^2 \omega L}{C} \tag{5.25}$$

The torque is

$$\mathbf{T} = F \cdot R = \frac{2\pi \mu R^3 \omega L}{C} \tag{5.26}$$

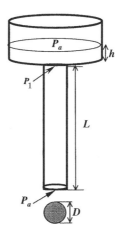

Figure 5.7   Capillary viscometer.

The power loss is

$$E_P = F \cdot U = \frac{2\pi\mu R^3 \omega^2 L}{C} \tag{5.27}$$

One can then estimate the friction coefficient using the following definition, although theoretically a concentric journal bearing does not offer any load-carrying capacity:

$$f = \frac{F}{W} \tag{5.28}$$

where W represents the load. The above equations provide a useful estimate for the coefficient of friction and power loss in many journal bearings operating with no more than moderate eccentricity in the range up to 0.5–0.7.

5.6   Viscometers

A variety of procedures are available for measuring the viscosity of a liquid. We will focus on Newtonian fluids.

Capillary Tube Viscometer

This device consists simply of a small fluid reservoir of height h connected to which is a capillary tube in the vertical position whose length and diameter are known (Figure 5.7). Volumetric flow rate in the capillary is given by the following general relation for a circular flow passage:

$$q = \frac{\pi R^4}{8\mu} \frac{\Delta P}{L} \tag{5.29}$$

where the pressure drop is

$$\Delta P = P_1 - \rho g - (P_a - \rho g) = P_1 - P_a \tag{5.30}$$

But $P_1 = p_a + \rho g h$. Therefore

$$\Delta P = \rho g h \tag{5.31}$$

The flow rate equation becomes

$$q = \frac{\pi R^4}{8\mu} \frac{\rho g h}{L} \tag{5.32}$$

Solving for viscosity yields

$$\mu = \frac{\pi R^4 h}{8L} \frac{\rho g}{q} \tag{5.33}$$

Rotational Viscometers

Several types are available. The most common ones are parallel disk, cylindrical, and cone-and-plate, viscometers. A brief description of each follows.

Parallel Disk Viscometer

The fluid is placed between two parallel disks of equal size. The top disk is kept stationary, and the bottom one is set in rotation. It can be shown that for a purely tangential flow under laminar conditions, the torque is simply

$$\mathbf{T} = \frac{\pi \mu \omega R^4}{2b} \tag{5.34}$$

where h is the separation gap. By measuring the resisting torque exerted on the top plate, the viscosity can be evaluated.

Cylindrical Viscometer

This rotational viscometer consists of two concentric cylinders with a small radial clearance containing the fluid whose viscosity is to be measured. A schematic of the device is shown in Figure 5.8. The inner cylinder is stationary, and the outer one rotates at speed ω (rad/s). The resisting torque of the stationary cylinder is measured by means of a torsional spring to determine the fluid viscosity.

The total torque measured is a result of the shearing forces in the direction of motion, $\mathbf{T}_\theta$, as well as the torque between the bottom of the two cylinders, $\mathbf{T}_b$:

$$\mathbf{T} = \mathbf{T}_\theta + \mathbf{T}_b$$

It can be shown (see Problem 5.6) that

$$\mathbf{T}_\theta = 4\pi\mu L\omega \frac{R_1^2 R_2^2}{R_2^2 - R_1^2} \tag{5.35}$$

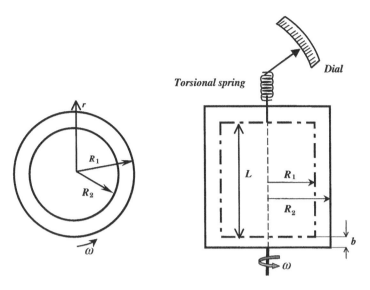

Figure 5.8   Cylindrical viscometer.

and

$$\mathbf{T}_b = \mu\omega\pi\frac{R_1^4}{2b} \tag{5.36}$$

$\mathbf{T}_b$ is normally much smaller than $\mathbf{T}_\theta$. Once the torque is measured, Equation (5.35) can be used to calculate viscosity.

While thermal effects may become significant and affect the measurement of viscosity, temperature rise can be *estimated* by solving the energy equation. If the temperature rise is significant, one may need to reformulate the energy equation. Furthermore, one can make provision for viscosity variation with temperature by using an appropriate relationship such as $\mu = \mu_0 e^{-\beta(T-T_0)}$. Note that the energy equation becomes nonlinear, making closed-form analytical solutions more difficult.

The momentum and energy equations for the fluid contained between two concentric cylinders of radii R_1 and R_2 maintained at temperatures T_1 and T_2 can be solved to obtain the temperature distribution across the film gap. An analytical solution is available assuming iso-viscous, laminar flow with negligible end effects (i.e. no side leakage). The result is as follows (Weltmann and Kuhns, 1952):

$$T = T_1 + (T_2 - T_1)\frac{\ln(R_1/r)}{\ln(R_1/R_2)} + \frac{A^2}{4\mu k}\left[\left(\frac{1}{R_2^2} - \frac{1}{r^2}\right) - \left(\frac{1}{R_2^2} - \frac{1}{R_1^2}\right)\frac{\ln(r/R_2)}{\ln(R_1/R_2)}\right] \tag{5.37}$$

Here A is a parameter related to torque, as measured by a rotational viscometer. It is given by

$$A = \frac{\mathbf{T}}{2\pi L} \tag{5.38}$$

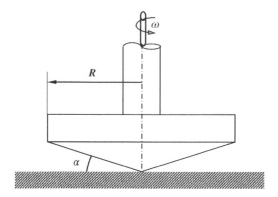

Figure 5.9   Cone-on-plate viscometer.

If the temperatures of both surfaces are known, the above equation can be integrated across the gap to obtain a reasonable mean temperature from which an effective viscosity can be determined.

Cone-and-Plate Viscometer

This viscometer consists of a flat plate and a cone (Figure 5.9). A small sample of a fluid is placed on the plate, and the cone is set to rotate at a known angular velocity. By measuring the torque required to turn the cone, the viscosity can be determined.

The appropriate derivation of the velocity profile and the shear stresses require consideration of the spherical coordinates system (Bird *et al.*, 1960). A simplified analysis shows that for a small angle α (typically one-half of one degree), the torque is:

$$\mathbf{T} = \frac{2}{3\alpha}\pi\mu\omega R^3 \tag{5.39}$$

Falling Body Viscometer

In this viscometer, a small steel ball or modified cylinder is dropped in a tube containing the fluid of interest (Figure 5.10). Initially the ball will experience linear acceleration and then will reach its terminal velocity. At this point, the forces acting on the sphere are in equilibrium and the ball descends at a constant speed V.

Using the Stokes formula for the drag on a sphere in an unbounded domain, it can be shown that

$$\mu = \frac{D_B^2 g(\rho_B - \rho)}{18V} \tag{5.40}$$

where D_B and ρ_B are the diameter and density of the ball, respectively, and ρ is the lubricant density. By timing the fall of the sphere between two graduation marks on the tube, the terminal velocity V and then the viscosity can be computed. Note that by placing the tube in a bath of water with controlled temperature, one can determine the fluid viscosity at various temperatures.

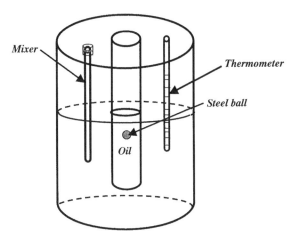

Figure 5.10 Falling sphere viscometer.

The Stokes formula assumes that the sphere is in an unbounded domain. Therefore, a correction factor is needed to account for the presence of the tube wall. Let $\propto = D_B/D_T$ denote the ratio of the ball to the tube diameter. The following factor, known as the *Faxen term*, is applied to correct the terminal velocity in Equation (5.40):

$$V_C = F_c V \tag{5.41}$$

where $F_c = 1 - 2.104\alpha + 2.09\alpha^3 - 0.9\alpha^5$.

5.7 Nondimensionalization of Flow Equations

In engineering analysis, one can gain significant insight into a problem by casting the governing equations in a nondimensional form. When this is done, a set of key dimensionless parameters often emerges which governs the behavior of the problem. As an example, consider the flow of a fluid in the geometry shown in Figure 5.11.

Consider the x-momentum equation:

$$\rho\left(\frac{\partial u}{\partial t} + u\frac{\partial u}{\partial x} + v\frac{\partial u}{\partial y} + w\frac{\partial u}{\partial z}\right) = F_{B_x} - \frac{\partial P}{\partial x} + \mu\left(\frac{\partial^2 u}{\partial x^2} + \frac{\partial^2 u}{\partial y^2} + \frac{\partial^2 u}{\partial z^2}\right) \tag{5.42}$$

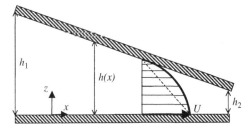

Figure 5.11 Slider bearing configuration illustrating combined Couette and Poiseuille components.

Clearly, $F_{B_x} = 0$. If the surfaces were parallel, the pressure gradient would have been nil. However, because of the wedge $\partial P/\partial x \neq 0$. We shall proceed by using appropriate characteristic parameters to normalize the above equation. Using U and L as our reference velocity and characteristic length, respectively, we obtain

$$\bar{v} = \frac{v}{U}; \qquad \bar{w} = \frac{w}{U}; \qquad \bar{u} = \frac{u}{U}; \qquad \bar{x} = \frac{x}{L}; \qquad \bar{y} = \frac{y}{L}; \qquad \bar{z} = \frac{z}{L} \tag{5.43}$$

where the overbar signifies nondimensional quantities. The logical choice for normalizing time is $\bar{t} = (U/L)t$. However, we can proceed by letting $\bar{t} = t/t_{ref}$ and let the problem guide us to evaluate a proper choice for t_{ref}. Similarly, we let $\bar{P} = P/P_{ref}$. The dimensionless form of the x-momentum equation is

$$\rho \left(\frac{U}{t_{ref}} \frac{\partial \bar{u}}{\partial \bar{t}} + \frac{U^2}{L} \bar{u} \frac{\partial \bar{u}}{\partial \bar{x}} + \frac{U^2}{L} \bar{v} \frac{\partial \bar{u}}{\partial \bar{y}} + \frac{U^2}{L} \bar{w} \frac{\partial \bar{u}}{\partial \bar{z}} \right) = -\frac{P_{ref}}{L} \frac{\partial \bar{P}}{\partial \bar{x}} + \mu \frac{U}{L^2} \left(\frac{\partial^2 \bar{u}}{\partial \bar{x}^2} + \frac{\partial^2 \bar{u}}{\partial \bar{y}^2} + \frac{\partial^2 \bar{u}}{\partial \bar{z}^2} \right)$$

$$\tag{5.44}$$

Multiplying through by $L/\rho U^2$ yields

$$\left(\frac{1}{t_{ref}} \frac{L}{U} \frac{\partial \bar{u}}{\partial \bar{t}} + \bar{u} \frac{\partial \bar{u}}{\partial \bar{x}} + \bar{v} \frac{\partial \bar{u}}{\partial \bar{y}} + \bar{w} \frac{\partial \bar{u}}{\partial \bar{z}} \right) = -\frac{P_{ref}}{\rho U^2} \frac{\partial \bar{P}}{\partial \bar{x}} + \frac{\mu}{\rho UL} \left(\frac{\partial^2 \bar{u}}{\partial \bar{x}^2} + \frac{\partial^2 \bar{u}}{\partial \bar{y}^2} + \frac{\partial^2 \bar{u}}{\partial \bar{z}^2} \right) \tag{5.45}$$

Note that $(1/t_{ref})(L/U)$ is a dimensionless quantity suggesting that $t_{ref} = (L/U)$. Also, $P_{ref} = \rho U^2$. Furthermore, the coefficient appearing in the front of the viscous forces is simply the inverse of the Reynolds number. The above equation takes on the following form:

$$\left(\frac{\partial \bar{u}}{\partial \bar{t}} + \bar{u} \frac{\partial \bar{u}}{\partial \bar{x}} + \bar{v} \frac{\partial \bar{u}}{\partial \bar{y}} + \bar{w} \frac{\partial \bar{u}}{\partial \bar{z}} \right) = -\frac{\partial \bar{P}}{\partial \bar{x}} + \frac{1}{Re} \left(\frac{\partial^2 \bar{u}}{\partial \bar{x}^2} + \frac{\partial^2 \bar{u}}{\partial \bar{y}^2} + \frac{\partial^2 \bar{u}}{\partial \bar{z}^2} \right). \tag{5.46}$$

Physically, the Reynolds number represents the ratio of inertial forces to viscous forces. When Re is large, the contribution of the viscous term becomes less significant and the inertia forces are dominant. Similarly, when Re is very small (creeping motion), the inertia terms may be assumed negligible compared to the viscous term. In many lubrication problems treated in this book, the creeping motion prevails.

Note that, for clarity and simplicity in presentation, in this example we have used L as the characteristic length throughout. For internal flows, the characteristic length is normally taken to be the width of the channel. However, in thin-film lubrication problems one must carefully choose a length scale that appropriately characterizes the film thickness and the bearing length.

5.8 Nondimensionalization of the Energy Equation

As with the flow relations discussed in the previous section, use of a nondimensional form also provides insight into temperature and energy relations. For this purpose, we will restrict our attention to the infinitely wide slider-bearing configuration shown in Figure 5.11.

In the Cartesian coordinate system, the steady-state energy equation with uniform thermal conductivity is

$$\rho C_p \left(u \frac{\partial T}{\partial x} + w \frac{\partial T}{\partial z} \right) = k \left(\frac{\partial^2 T}{\partial x^2} + \frac{\partial^2 T}{\partial z^2} \right) + \mu \left(\frac{\partial u}{\partial z} \right)^2 \tag{5.47}$$

Let

$$\bar{x} = \frac{x}{L}; \qquad \bar{z} = \frac{z}{h_2}; \qquad \bar{u} = \frac{u}{U}; \qquad \bar{w} = \frac{w}{U}; \qquad \bar{T} = \frac{T - T_i}{T_{\text{ref}} - T_i}; \qquad \bar{\mu} = \frac{\mu}{\mu_i} \tag{5.48}$$

Using the dimensionless parameters defined in Equation (5.48) the energy equation (5.47) becomes

$$\rho C_p (T_{\text{ref}} - T_i) \frac{U}{L} \left(\bar{u} \frac{\partial \bar{T}}{\partial \bar{x}} + \bar{w} \frac{L}{h_2} \frac{\partial \bar{T}}{\partial \bar{z}} \right)$$

$$= \left(\frac{k}{L^2}(T_{\text{ref}} - T_i) \frac{\partial^2 \bar{T}}{\partial \bar{x}^2} + \frac{k}{h_2^2}(T_{\text{ref}} - T_i) \frac{\partial^2 \bar{T}}{\partial \bar{z}^2} \right) + \frac{\mu_i U^2}{h_2^2} \bar{\mu} \left(\frac{\partial \bar{u}}{\partial \bar{z}} \right)^2 \tag{5.49}$$

Multiplying Equation (5.49) by $(T_{\text{ref}} - T_i)^{-1} h_2^2 / k$ and simplifying the result yields

$$\left(\frac{h_2}{L} \right)^2 Re_L \cdot Pr \left(\bar{u} \frac{\partial \bar{T}}{\partial \bar{x}} + \frac{L}{h_2} \bar{w} \frac{\partial \bar{T}}{\partial \bar{z}} \right) = \left(\frac{h_2}{L} \right)^2 \frac{\partial^2 \bar{T}}{\partial \bar{x}^2} + \frac{\partial^2 \bar{T}}{\partial \bar{z}^2} + Br \bar{\mu} \left(\frac{\partial \bar{u}}{\partial \bar{z}} \right)^2 \tag{5.50}$$

where $Re_L = \rho U L / \mu$ is the Reynolds number based on the slider length, $Pr = C_p \mu / k$ is the Prandtl number, and $Br = \mu_i U^2 / k(T_{\text{ref}} - T_i)$ is the Brinkman number.

The product of $Re \cdot Pr$ is often referred to as the *Peclet number*, defined as $Pe = \rho C_p U L / k$. The Peclet number is a measure of convective terms divided by conduction terms. Making use of this definition, Equation (5.50) becomes

$$\left(\frac{h_2}{L} \right)^2 Pe \left(\bar{u} \frac{\partial \bar{T}}{\partial \bar{x}} + \frac{L}{h_2} \bar{w} \frac{\partial \bar{T}}{\partial \bar{z}} \right) = \left(\frac{h_2}{L} \right)^2 \frac{\partial^2 \bar{T}}{\partial \bar{x}^2} + \frac{\partial^2 \bar{T}}{\partial \bar{z}^2} + Br \bar{\mu} \left(\frac{\partial \bar{u}}{\partial \bar{z}} \right)^2 \tag{5.51}$$

Equation (5.51) reveals several interesting facts. The first term on the right-hand side, the conduction term in the direction of motion, is scaled by a very small number of the order 10^{-6}. Hence, it is negligibly small. The same multiplier also appears in front of the Peclet number on the left-hand side. Therefore, it is important to examine the product of the two terms to determine the significance of the convective terms. For this purpose, let us use the typical numbers for specifying a slider bearing:

$$L = 0.02 \, \text{m}; \qquad \frac{h}{L} = 10^{-3}; \qquad U = 6.28 \, \text{m/s}; \qquad \rho = 876 \, \text{kg/m}^3;$$

$$\mu = 0.25 \, \text{pa s (engine oil at } 40\,^\circ C); \qquad C_P = 1964 \, \text{J/kg K}; \qquad k = 0.144 \, \text{W/m K}$$

Using these numbers, $Pe = \rho C_p U L / k \approx 1.5 \times 10^6$. Therefore, $Pe(h_2/L)^2 \approx 1.5$ which is indeed significant. Taking $\Delta T = T_{\text{ref}} - T_i = 20$ and computing the Brinkman number gives

$Br = 3.4$, which is of the same order as $Pe(h_2/L)^2$. Therefore, both the convective terms and the viscous dissipation terms must be retained in the energy equation. These terms can be neglected if $Pe(h_2/L)^2 \ll 1$ or $Br \ll 1$.

It is also interesting to note that the component of the velocity across the gap is scaled by a relatively large multiplier (L/h_2) and the entire transverse convective term is not negligible. Therefore, the final energy equation is

$$\left(\frac{h_2}{L}\right)^2 Pe \left(\bar{u}\frac{\partial \bar{T}}{\partial x} + \frac{L}{h_2}\bar{w}\frac{\partial \bar{T}}{\partial z}\right) = \frac{\partial^2 \bar{T}}{\partial \bar{z}^2} + Br\bar{\mu}\left(\frac{\partial \bar{u}}{\partial \bar{z}}\right)^2 \qquad (5.52)$$

5.9 Order-of-Magnitude Analysis

Another powerful analytical tool is order-of-magnitude analysis. Proper application of this technique enables one to determine the relative importance of different terms in an equation. Therefore, one can decide which terms are likely to be negligibly small.

Consider flow in a slider bearing with inlet and outlet film thicknesses of h and h_2, respectively. Bearing length is L and the top plate is stationary, while the bottom plate undergoes a sliding motion at speed U and fluid is dragged into the bearing clearance space (Figure 5.11). We shall assume that the fluid is an incompressible Newtonian liquid with constant viscosity μ.

Let us examine the x-momentum equation:

$$\rho \left(\frac{\partial u}{\partial t} + u\frac{\partial u}{\partial x} + v\frac{\partial u}{\partial y} + w\frac{\partial u}{\partial z}\right) = F_{B_x} - \frac{\partial P}{\partial x} + \mu \left(\frac{\partial^2 u}{\partial x^2} + \frac{\partial^2 u}{\partial y^2} + \frac{\partial^2 u}{\partial z^2}\right) \qquad (5.53)$$

Assuming that the problem is steady state $\partial u/\partial t = 0$ and since there is no body force in the x direction, $F_{B_x} = 0$.

Comparison of Inertia Terms and Viscous Terms

Let us examine the inertia terms individually. If the two plates were perfectly parallel, $w = 0$ (see Example 5.1). However, in this problem there exists a component w. Yet because of the small inclination angle, $w \ll u$. Therefore it is reasonable to assume that $u(\partial u/\partial x) > w(\partial u/\partial z)$.

Next, we compare $u(\partial u/\partial x)$ with $v(\partial u/\partial y)$. Let us assume that the bearing width is of the same order as its length:

$$u \sim U; \qquad x \sim L; \qquad y \sim L \qquad (5.54)$$

where $\sim$ should be read as 'order of'. Therefore, $\partial u/\partial x \sim U/L$ and $\partial u/\partial y \sim U/L$. Now v represents the leakage flow velocity in the y direction. Physically, if the bearing length is of the same order as its width, then the leakage flow velocity would be a fraction of that in the direction of motion. Therefore, $u(\partial u/\partial x) > v(\partial u/\partial y)$; hence the inertial term $u(\partial u/\partial x)$ can be considered to be the dominant term for the purpose of comparison with viscous forces.

The viscous forces contain these terms:

$$\mu(\partial^2 u/\partial x^2), \ \mu(\partial^2 u/\partial y^2), \quad \text{and} \quad \mu(\partial^2 u/\partial z^2).$$

Let us compare $\mu(\partial^2 u/\partial x^2)$ and $\mu(\partial^2 u/\partial z^2)$ by evaluating their ratio:

$$\frac{\mu(\partial^2 u/\partial x^2)}{\mu(\partial^2 u/\partial z^2)} = \frac{(\partial/\partial x)(\partial u/\partial x)}{(\partial/\partial z)(\partial u/\partial z)} \sim \frac{(1/L)\cdot(U/L)}{(1/h)\cdot(U/h)} = (h/L)^2 \ll 1 \tag{5.55}$$

Note that typically in hydrodynamic lubrication $h/L = O\left(10^{-3}\right)$. Thus $\mu(\partial^2 u/\partial x^2) \ll \mu(\partial^2 u/\partial z^2)$. This is an important finding which is generally true in most lubrication problems: the viscous forces with gradient across the film gap are the dominant terms.

Let us now compare viscous forces and inertia forces:

$$\frac{\rho u(\partial u/\partial x)}{\mu(\partial^2 u/\partial z^2)} \sim \frac{\rho U \cdot (U/L)}{\mu(U/h^2)} = \frac{\rho U}{\mu}\left(\frac{h^2}{L}\right) = \frac{\rho UL}{\mu}\left(\frac{h}{L}\right)^2 = Re\left(\frac{h}{L}\right)^2 = Re_{\mathrm{m}} \tag{5.56}$$

where Re_{m} is a modified Reynolds number. Again, $(h/L)^2$ – a quantity of the order 10^{-6} – shows up as a multiplier. Clearly, unless $Re = (\rho UL/\mu)$ becomes very large, the viscous forces tend to dominate the inertia forces. It is instructive to examine the magnitude of this term for a typical bearing. Consider the following specifications:

$$L = 0.02\,\mathrm{m}; \qquad \frac{h}{L} = 10^{-3}; \qquad U = 6.28\,\mathrm{m/s}; \qquad \rho = 876\,\mathrm{kg/m^3}; \qquad \mu = 0.25\,\mathrm{pa\,s} \tag{5.57}$$

$$\frac{\rho UL}{\mu}\left(\frac{h}{L}\right)^2 = \frac{876(\mathrm{kg/m^3})\cdot 6.28(\mathrm{m/s})\cdot 0.02\,\mathrm{m}}{0.25(\mathrm{N/m^2})\,\mathrm{s}}(10^{-3})^2 \cong 0.0004$$

Therefore, the inertia forces are negligible compared to the viscous forces. While this is typically the case for the majority of hydrodynamic bearings, turbulence is encountered when Re exceeds the 1000–2000 range, with inertia effects coming into predominance.

Contribution of Gravity

In this example, the component of the body forces (i.e. gravity) did not enter the problem. We can, however, assess the role of gravity in a typical lubrication problem.

$$F_{B_y} = \text{gravity per unit volume} = \rho g$$

Comparing gravity force ρg to inertia forces $\rho u(\partial u/\partial x)$:

$$\frac{\rho u(\partial u/\partial x)}{\rho g} \sim \frac{U \cdot (U/L)}{g} = \frac{U^2}{Lg} = Fr \tag{5.58}$$

where Fr is a dimensionless quantity sometimes referred to as the *Froude number*. With the above bearing specification we have

$$Fr = \frac{(6.28)^2}{(0.02)(9.8)} \cong 201$$

This implies that body forces due to gravity are much smaller than inertia forces and therefore are negligible compared to viscous forces. In general, one can neglect the gravity forces in lubrication problems.

Contribution of the Pressure Term

Next, we will compare inertia forces and pressure. Let $P \sim P_{ref}$:

$$\frac{\partial P / \partial x}{\rho u (\partial u / \partial x)} \sim \frac{P_{ref}/L}{\rho (U^2/L)} = \frac{P_{ref}}{\rho U^2} = Eu \qquad (5.59)$$

where Eu is a dimensionless parameter known as the *Euler number*. The average pressure based on the projected load in a hydrodynamic bearing is on the order of 5 MPa. Therefore

$$Eu = \frac{5 \times 10^6 \text{ N/m}^2}{(876)(6.28)^2} \cong 145$$

This implies that the contribution of pressure is much greater than the inertia effect and must be considered in conjunction with viscous forces.

Comparison of Pressure and Viscous Forces

Finally, it is instructive to compare viscous forces with pressure:

$$\frac{\mu (\partial^2 u / \partial z^2)}{\partial P / \partial x} \sim \frac{\mu (U/h_{ref}^2)}{P_{ref}/L} = \frac{\mu U L}{P_{ref} h_{ref}^2} = \Lambda_B \qquad (5.60)$$

where Λ_B is sometimes referred to as the *bearing number*. Using the same typical numerical values, the bearing number is

$$\Lambda_B = \frac{0.25(6.28)(0.02)}{5 \times 10^6 (0.02 \times 10^{-3})^2} \approx 17$$

Hence pressure and viscous forces are both important in a typical lubrication problem.

Similar conclusions are obtainable by a proper nondimensionalization procedure (see e.g. Hamrock, 1994).

The above analysis indicates that the dominant terms in the x-component of the Navier-Stokes equation for a typical lubrication problem are

$$\frac{\partial P}{\partial x} = \mu \frac{\partial^2 u}{\partial z^2} \qquad (5.61)$$

The above equation is the starting point for solving many lubrication problems. It is sometimes referred to as the *Navier-Stokes equation subject to a thin-film approximation*. As it stands, the viscosity is assumed to remain constant and the fluid is Newtonian. We can generalize the above equation to read:

$$\frac{\partial P}{\partial x} = \frac{\partial \tau_{xz}}{\partial z} \qquad (5.62)$$

For a Newtonian fluid, Equation (5.62) simply reduces to the following:

$$\frac{\partial P}{\partial x} = \frac{\partial}{\partial z}\left(\mu \frac{\partial u}{\partial z}\right) \qquad (5.63)$$

Example 5.3 (Combined Couette and Poiseuille flow)   Consider the flow of an incompressible Newtonian fluid in an infinitely wide slider bearing where there is no side leakage (Figure 5.11). The bottom plate slides at a constant velocity U, and the stationary top plate is tilted so that the film thickness h across the gap changes along the slider's length. Determine the velocity profile and volumetric flow rate.

Assuming that viscosity remains uniform across the gap, i.e. $\mu \neq \mu(z)$, Equation (5.61) can be used to determine the velocity profile. Integrating Equation (5.62) twice yields

$$u = \frac{1}{2\mu}\frac{dP}{dx}z^2 + C_1 z + C_2$$

where the constants of integration C_1 and C_2 are evaluated using the following boundary conditions:

$$u = U \text{ at } z = 0 \quad \text{and} \quad u = 0 \text{ at } z = h(x)$$

The result is

$$u = \underbrace{U\left(1 - \frac{z}{h}\right)}_{\text{Couette}} - \underbrace{\frac{h^2}{2\mu}\frac{z}{h}\left(1 - \frac{z}{h}\right)\frac{dP}{dx}}_{\text{Poiseuille}}$$

Here the velocity is composed of two distinct terms: the Couette component and the Poiseuille component.

The Couette component is induced by the shearing action alone. If the gap width h were constant, it would predict a linear velocity profile identical to that obtained in Example 5.1, except that here the bottom plate is the sliding member. The Poiseuille component is the pressure-driven flow. A hydrodynamic bearing is self-acting, meaning that pressure is developed automatically because the top and bottom plates form a physical wedge with $dh/dx \neq 0$. This is in contrast to Example 5.2, where the pressure gradient responsible for driving the flow between two stationary parallel plates had to be generated via the pressure difference between the inlet at supply pressure P_s and outlet and ambient pressure p_a.

The volumetric flow rate can be determined by integrating the flow velocity across the gap:

$$q = \int_0^{h(x)} u\,dz$$

Substituting for u and performing the integration gives

$$q = \frac{Uh}{2} - \frac{h^3}{12\mu}\frac{dP}{dx}$$

This flow rate remains constant at any x location, as required by the law of conservation of mass. That is, $dq/dx = 0$. Therefore

$$\frac{d}{dx}\left(\frac{Uh}{2} - \frac{h^3}{12\mu}\frac{dP}{dx}\right) = 0$$

which simplifies to the following equation:

$$\frac{d}{dx}\left(\frac{h^3}{\mu}\frac{dP}{dx}\right) = 6U\frac{dh}{dx}$$

With a given film gap profile, this equation can be solved to determine the pressure distribution in the bearing. The above equation forms the basis of the Reynolds equation described in Chapter 6.

Problems

5.1 Show that the general equations of motion written in terms of τ as given in Equations (5.8) and (5.9) reduce to Equations (5.17) and (5.18) for Newtonian fluids. It is sufficient to show the x-momentum equation in the Cartesian coordinate system and the z-momentum equation in the cylindrical coordinate system.

5.2 Consider the flow of a laminar incompressible Newtonian fluid flowing between infinitely wide parallel plates separated by distance h and tilted at angle α. Assuming that the flow is fully developed, determine the following:
(a) An expression for the fluid velocity profile;
(b) Mean flow velocity;
(c) Volumetric flow rate.
Plot the velocity profile and the shear stress profile across the gap.

5.3 Determine the fluid temperature distribution in Problem 5.2, assuming the following:
(a) The bottom and top plates are maintained at uniform temperatures T_L and T_U, respectively.
(b) The bottom plate is maintained at temperature T_L, but the top plate is insulated.
(c) Determine the location and magnitude of the maximum temperature in parts (a) and (b).
Estimate the maximum temperature in part (b) with the following specifications: SAE 30 oil, channel length of $L = 4$ in., channel gap of $h = 0.001$ in., and $T_L = 80\,°F$ for tilt angles of $\pm = 0$, 45, and 90°. Evaluate the properties at 80 °F. The inlet pressure is 40 psi.

5.4 Two infinitely wide parallel plates separated by distance h contain a Newtonian fluid. The top and bottom plates are set in motion at the speeds $U_T = U_1$ and $U_B = -2U_1$.
Determine the following:
(a) Velocity profile;
(b) Volumetric flow rate;
(c) Shear stress profile across the gap.

5.5 Consider the purely tangential flow of a Newtonian fluid in a parallel plate viscometer. Show that for a very small separation gap when $b = R$, the torque exerted on the top disk by setting the bottom disk in motion can be evaluated from

$$T = \frac{\pi\mu\omega R^4}{2b}$$

A sample of oil is placed in the parallel disk viscometer of $R = 5$ cm with $b/R = 0.09$. The torque measured is 0.016 N m when $\omega = 20$ rad/s. Determine the viscosity of the fluid.

5.6 Consider two concentric cylinders with radii R_1 and R_2 rotating at angular velocities ω_1 and ω_2. Assume that only the circumferential component of the velocity exists.

(a) Using the Navier-Stokes equations in the cylindrical coordinates, derive the equation for the velocity profile $w(r)$.

(b) Evaluate the shear stresses at the inner wall *and* the outer wall.

(c) Suppose that the inner cylinder is stationary and the outer one rotates, and that a torsional spring is connected to the inner cylinder to measure the torque resulting from the shearing force in the fluid resisting motion. Show that the torque can be determined from the following equation:

$$T_\theta = 4\pi \mu L \omega_2 \frac{R_1^2 R_2^2}{R_2^2 - R_1^2} \tag{1}$$

where L is the length of the inner cylinder.

(d) Show that Equation (1) reduces to Petroff's equation for $c = 1$, where $c = R_2 - R_1$. *Hint*:

$$R_2^2 - R_1^2 = 2R_2(R_2 - R_1) - (R_2 - R_1)^2$$

(e) A sample of oil is placed in a rotational viscometer as described above. $R_1 = 7.62$ cm, $R_2 = 7.67$ cm, $l = 20.32$ cm, and $b = 0.32$ cm. When the outer cylinder rotates at 30 rpm, the dial indicator connected to the spring indicates that a torque of 400 cm N is required. Compute the value of viscosity using Equation (1). Show your answers in three different units: $N s/m^2$, reyns (lbf s/in.^{2}), and Cp (centipoise). What is the percentage of error if Petroff's equation is used instead of Equation (1)?

(f) A correction must be made to Equation (1) since the dial measures an additional torque due to the small gap, b, between the bottom of the inner cylinder and that of the outer cylinder. Show that the magnitude of this additional torque is

$$T_b = \mu \omega_2 \pi \frac{R_1^4}{2b}$$

so that the total torque is

$$T_b = \mu \omega_2 \pi \frac{R_1^4}{2b}$$

Using the values given in part (e), determine an improved value for the viscosity of the fluid. What is the percentage of error if Equation (1) is not corrected?

5.7 Determine the fluid temperature distribution in the rotational viscometer in Problem 5.6. Assume that the rotating cylinder is insulated and that the stationary cylinder remains at the ambient temperature, $T_a = 20\,°C$. Determine the maximum temperature rise in the fluid with the specifications in Problem 5.6.

5.8 Using the Petroff formula, calculate the bearing power loss (watts) for a moderately loaded 100 mm diameter journal running at 5000 rpm in a sleeve bearing. Axial length of the bearing is 75 mm, and diametral clearance is 0.20 mm. Bearing temperature is 60 °C and the bearing is lubricated with a light ISO VG 32 industrial mineral oil.

References

Bird, R.B., Stewart, W.E., and Lightfoot, E.N. 1960. *Transport Phenomena*. John Wiley & Sons, New York.

Fillon, M. and Khonsari, M.M. 1996. 'Thermohydrodynamic Design Charts for Tilting-Pad Journal Bearings,' *ASME Journal of Tribology*, Vol. 118, pp. 232–238.

Hamrock, B. 1994. *Fundamentals of Fluid Film Lubrication*. McGraw-Hill Book. Co., New York.

Jang, J. and Khonsari, M.M. 1997. 'Thermohydrodynamic Design Charts for Slider Bearings,' *ASME Journal of Tribology*, Vol. 119, pp 733–740.

Jang, J.Y. and Khonsari, M.M. 2004. 'Design of Bearings Based on Thermohydrodynamic Analysis,' *Journal of Engineering Tribology*, Proceedings of Institution of Mechanical Engineers, Part J, Vol. 218, pp. 355–363.

Keogh, P., Gomiciaga, R., and Khonsari, M.M. 1997. 'CFD Based Design Techniques for Thermal Prediction in a Generic Two-axial Groove Hydrodynamic Journal Bearing,' *ASME Journal of Tribology*, Vol. 119, pp. 428–436.

Khonsari, M.M. 1987a. 'A Review of Thermal Effects in Hydrodynamic Bearings. Part I: Slider and Thrust Bearings,' *ASLE Transactions*, Vol. 30(1), pp. 19–25.

Khonsari, M.M. 1987b. 'A Review of Thermal Effects in Hydrodynamic Bearings. Part II: Journal Bearings,' *ASLE Transactions*, Vol. 30(1), pp. 26–33.

Khonsari, M.M. and Beaman, J. 1986. 'Thermohydrodynamic Analysis of Laminar Incompressible Journal Bearings,' *ASLE Transactions*, Vol. 29(2), pp. 141–150.

Khonsari, M.M., Jang, J., and Fillon, M. 1996. 'On the Generalization of Thermohydrodynamic Analysis for Journal Bearings,' *ASME Journal of Tribology*, Vol. 118, pp. 571–579.

Schlichting, H. 1975. *Boundary Layer Theory*, 7th edn. McGraw-Hill Book Co., New York.

Weltmann, R.N. and Kuhns, P.W. 1952. 'Effect of Shear Temperature on Viscosity in a Rotational Viscometer Measurement,' *Journal of Colloid Science*, Vol. 7, pp. 218–226.

White, F. 1991. *Viscous Fluid Flow*, 2nd edn. McGraw-Hill Book Co., New York.

6

Reynolds Equation and Applications

We begin this chapter by presenting the derivation of the classical Reynolds equation based on the work of Osborne Reynolds (Reynolds, 1886). It reflects tremendous insight into fluid behavior in bearing lubricant films, which then gave birth to the science of hydrodynamic lubrication. Solution of the Reynolds equation enables one to determine the pressure distribution in a bearing with an arbitrary film shape. Once the pressure profile is evaluated, all other bearing performance parameters such as load-carrying capacity, friction force, flow rates, etc., can easily be determined.

6.1 Assumptions and Derivations

The following five simplifying assumptions made in this derivation are valid for the majority of applications that we encounter. They are generally referred to as the *lubrication assumptions for laminar, Newtonian, inertialess, and thin-film flows*. The order-of-magnitude analysis discussed in Chapter 5 should be referred to in order to justify the validity of these assumptions.

1. The fluid is assumed to be Newtonian, with direct proportionality between shear stress and shearing velocity.
2. Inertia and body force terms are assumed to be negligible compared to the viscous terms, i.e.

$$\rho \frac{DU}{Dt} = 0 \quad \text{and} \quad F_B = 0$$

3. Variation of pressure across the film is assumed to be negligibly small, i.e. $\partial P/\partial z = 0$, so that $P = P(x, y)$ only. This can be justified by a proper analysis (Dowson, 1962).
4. Flow is laminar. In the presence of turbulence, one must use a modified form of the Reynolds equation.
5. Curvature effects are negligible. This implies that the thickness of the lubricant film is much smaller than the length or width of the bearing, so that the physical domain of the flow can be unwrapped. This allows use of the Cartesian coordinate system.

Applied Tribology: Bearing Design and Lubrication, Third Edition. Michael M. Khonsari and E. Richard Booser.
© 2017 John Wiley & Sons Ltd. Published 2017 by John Wiley & Sons Ltd.
Companion Website: www.wiley.com/go/Khonsari/Applied_Tribology_Bearing_Design_and_Lubrication_3rd_Edition

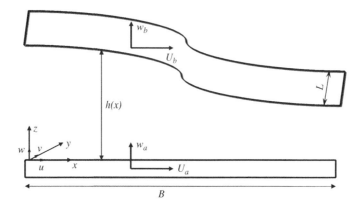

Figure 6.1   Coordinate system.

We shall now derive the Reynolds equation, which is valid for three-dimensional flows. However, note that pressure is two-dimensional because of assumption 3. In the discussion that follows, we shall use the equations for conservation of mass and momentum (Navier-Stokes).

Navier-Stokes Equations

Consider a three-dimensional bearing configuration with an arbitrary film gap $h = h(x, y)$, as shown in Figure 6.1. The bottom and top surfaces are assumed to undergo a sliding motion at U_a and U_b. Also, a normal squeeze motion may exist in the z direction, caused by either or both of the surfaces. Therefore, in addition to sliding in the x direction, the surfaces are free to approach or move away from each other. The squeeze components of the bottom and top surfaces are denoted by w_a and w_b, respectively.

Referring to Equation (5.63), and subject to the assumptions listed above, we can write the x- and y-momentum equations as follows:

x-Momentum equation:

$$\frac{\partial P}{\partial x} = \frac{\partial}{\partial z}\left(\mu\frac{\partial u}{\partial z}\right) \tag{6.1}$$

y-Momentum equation:

$$\frac{\partial P}{\partial y} = \frac{\partial}{\partial z}\left(\mu\frac{\partial v}{\partial z}\right) \tag{6.2}$$

Assuming that $\mu \neq \mu(z)$, we integrate the above equations to determine the velocity field. Note that this assumption needs to be carefully reexamined for accurate consideration of thermal effects in hydrodynamic lubrication analyses (Dowson, 1962).

First, integration of the above partial differential equations (6.1) and (6.2) gives

$$\frac{\partial u}{\partial z} = \frac{z}{\mu}\frac{\partial P}{\partial x} + \frac{A}{\mu}$$

$$\frac{\partial v}{\partial z} = \frac{z}{\mu}\frac{\partial P}{\partial y} + \frac{C}{\mu}$$

where A and C are constants of integration. Integrating again yields

$$u = \frac{z^2}{2\mu}\frac{\partial P}{\partial x} + A\frac{z}{\mu} + B \tag{6.3}$$

$$v = \frac{z^2}{2\mu}\frac{\partial P}{\partial y} + C\frac{z}{\mu} + D \tag{6.4}$$

where B and D are additional constants of integration to be determined using the appropriate boundary conditions.

Boundary Conditions

$$\text{At } z = 0, \begin{cases} u = U_a \\ v = 0 \\ w = w_a \end{cases} \quad \text{and} \quad \text{at } z = h, \begin{cases} u = U_b \\ v = 0 \\ w = w_b \end{cases} \tag{6.5}$$

where we assume that the surfaces do not undergo a sliding motion in the y (i.e. axial) direction. After the constants of integration are evaluated, the following velocity distributions emerge:

$$u = \underbrace{\frac{1}{2\mu}\frac{\partial P}{\partial x}(z^2 - zh)}_{\substack{\text{Poiseuille flow term} \\ \text{(pressure gradient)}}} + \underbrace{\left(1 - \frac{z}{h}\right)U_a + \frac{z}{h}U_b}_{\substack{\text{Couette flow (constant pressure)} \\ \text{(shear team)}}} \tag{6.6}$$

and

$$v = \frac{1}{2\mu}\frac{\partial P}{\partial y}(z^2 - zh) \tag{6.7}$$

According to Equation (6.6), the velocity profile in the direction of motion is composed of both the Poiseuille term, due to the pressure gradient in the x direction, and the Couette term, which is due purely to the motion of the surfaces, with a zero pressure gradient. Equation (6.7) represents the flow velocity in the y direction. In other words, it represents the velocity profile in fluid leaking out the sides. In the absence of sliding motion in that direction, this velocity profile contains only a Poiseuille-type term and no shear term. Actual volumetric flow rates in the direction of sliding and leakage flow rates can be determined by integrating Equations (6.6) and (6.7), respectively, across the gap.

Conservation of Mass

The general equation for conservation of mass for a compressible flow is

$$\frac{\partial \rho}{\partial t} + \frac{\partial(\rho u)}{\partial x} + \frac{\partial(\rho v)}{\partial y} + \frac{\partial(\rho w)}{\partial z} = 0 \tag{6.8}$$

Integrating the continuity equation across the film thickness yields

$$\int_0^h \frac{\partial \rho}{\partial t} + \int_0^h \frac{\partial(\rho u)}{\partial x}\,dz + \int_0^h \frac{\partial(\rho v)}{\partial y}\,dz + \int_0^h \frac{\partial(\rho w)}{\partial z}\,dz = 0 \tag{6.9}$$

Let us examine each of the above terms.

$$\int_0^h \frac{\partial \rho}{\partial t} \, dz = h \frac{\partial \rho}{\partial t} \tag{6.10}$$

$$\int_0^h \frac{\partial(\rho w)}{\partial z} \, dz = \rho w|_0^h = \rho(w_h - w_0) = \rho(w_b - w_a) \tag{6.11}$$

$$\int_0^h \frac{\partial(\rho u)}{\partial x} \, dz = \frac{\partial}{\partial x} \int_0^h (\rho u) dz - (\rho u)|_{z=h} \frac{\partial h_*}{\partial x} \tag{6.12}$$

Assuming that $\rho \neq \rho(z)$, the right-hand side of Equation (6.12) can be rewritten as

$$\text{RHS} = \frac{\partial}{\partial x} \left(\rho \underbrace{\int_0^h u dz}_{q_x} \right) - U_b \rho \frac{\partial h}{\partial x} \tag{6.13}$$

Since the volumetric flow rate

$$q_x = -\frac{h^3}{12\mu} \frac{\partial P}{\partial x} + \frac{(U_a + U_b)}{2} h,$$

Equation (6.13) becomes

$$\begin{aligned}
\text{RHS} &= \frac{\partial}{\partial x} \left[-\frac{\rho h^3}{12\mu} \frac{\partial P}{\partial x} + \frac{\rho(U_a + U_b)h}{2} \right] - U_b \rho \frac{\partial h}{\partial x} \\
&= \frac{\partial}{\partial x} \left(-\frac{\rho h^3}{12\mu} \frac{\partial P}{\partial x} \right) + \frac{1}{2} \frac{\partial}{\partial x} [\rho(U_a + U_b)h] - U_b \rho \frac{\partial h}{\partial x}
\end{aligned} \tag{6.14}$$

The third term in Equation (6.9) is simply

$$\int_0^h \frac{\partial(\rho v)}{\partial y} \, dz = \frac{\partial}{\partial y} \left(\rho \underbrace{\int_0^h v \, dz}_{q_y} \right) - v|_{z=h} \rho \frac{\partial h}{\partial y} \tag{6.15}$$

---

*Recall Leibnitz's integration rule:

$$\int_0^h \frac{\partial}{\partial x} [f(x, y, z)] dz = \frac{\partial}{\partial x} \int_0^h [f(x, y, z)] dz - f(x, y, h) \frac{\partial h}{\partial x}$$

Substituting

$$q_y = -\frac{h^3}{12\mu}\frac{\partial P}{\partial y}$$

Since there is only Poiseuille contribution in the leakage direction, and noting that $v|_{z=h} = 0$, Equation (6.15) reduces to

$$\int_0^h \frac{\partial(\rho v)}{\partial y}\,dz = \frac{\partial}{\partial y}\left(-\frac{\rho h^3}{12\mu}\frac{\partial P}{\partial y}\right) \tag{6.16}$$

Substituting Equations (6.10)–(6.16) into Equation (6.9) yields

$$\frac{\partial}{\partial x}\left(-\frac{\rho h^3}{12\mu}\frac{\partial P}{\partial x}\right) + \frac{1}{2}\frac{\partial}{\partial x}[\rho(U_a + U_b)h] - U_b\rho\frac{\partial h}{\partial x}$$

$$+ \frac{\partial}{\partial y}\left(-\frac{\rho h^3}{12\mu}\frac{\partial P}{\partial y}\right) + \rho(w_b - w_a) + h\frac{\partial\rho}{\partial t} = 0 \tag{6.17}$$

Combining terms, we get

$$\underbrace{\frac{\partial}{\partial x}\left(\frac{\rho h^3}{12\mu}\frac{\partial P}{\partial x}\right) + \frac{\partial}{\partial y}\left(\frac{\rho h^3}{12\mu}\frac{\partial P}{\partial y}\right)}_{\substack{\text{Poiseuille terms: net flow rate} \\ \text{due to pressure gradients}}} = \underbrace{\frac{1}{2}\frac{\partial}{\partial x}[\rho(U_a + U_b)h]}_{\substack{\text{Couette term: net flow rate} \\ \text{due to shear}}}$$

$$-\underbrace{\rho U_b\frac{\partial h}{\partial x}}_{\substack{\text{geometric} \\ \text{squeeze}}} + \underbrace{\rho(w_b - w_a)}_{\substack{\text{normal} \\ \text{squeeze}}} + \underbrace{h\frac{\partial\rho}{\partial t}}_{\text{local expansion}} \tag{6.18}$$

In order to gain physical insight, we expand the Couette term on the right-hand side of Equation (6.18) as follows:

$$\underbrace{\frac{1}{2}\frac{\partial}{\partial x}[\rho(U_a + U_b)h]}_{\substack{\text{Couette term: net rate} \\ \text{due to shear}}} = \underbrace{\frac{1}{2}\rho h\frac{\partial}{\partial x}(U_a + U_b)}_{\text{physical stretch}} + \underbrace{\frac{1}{2}\rho(U_a + U_b)\frac{\partial h}{\partial x}}_{\substack{\text{physical} \\ \text{wedge}}} + \underbrace{\frac{1}{2}(U_a + U_b)h\frac{\partial\rho}{\partial x}}_{\substack{\text{density} \\ \text{wedge}}}$$

$$\tag{6.19}$$

A brief physical interpretation of each term follows.

Physical stretch, $\partial(U_a + U_b)/\partial x$, takes into account the variation in tangential velocities. For steady loads, this means a rubber-like stretching of the bearing material. Assuming that bearing surfaces are inelastic (i.e. made of metal), $\partial(U_a + U_b)/\partial x = 0$. For most practical bearings with surfaces made of relatively rigid materials, this term is nil.

Physical wedge, $\partial h/\partial x$, is the most important term responsible for pressure generation in hydrodynamic bearings. Basically, to generate a positive load-carrying capacity, the surfaces

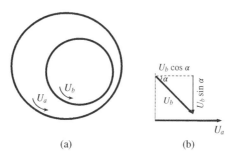

(a)   (b)

Figure 6.2   Geometric squeeze action.

should possess a convergent gap, $\partial h/\partial x < 0$. For $\partial h/\partial x > 0$ and divergence with an incompressible liquid, the lubricant film cavitates and pressure drops below ambient to approach a limiting value of zero absolute.

Density wedge, $\partial \rho/\partial x$, similar to the physical wedge, can generate positive pressure if $\partial \rho/\partial x < 0$. However, this term does not generally have a substantial influence on the pressure generation capacity and is negligible compared to the physical wedge. If $\partial h/\partial x \approx 0$ and in the absence of normal squeeze – as in a flat-plate thrust bearing – the density wedge may be responsible for some load-carrying capacity when the density change in the direction of motion becomes significant as a result of the viscous dissipation effect raising the temperature of the fluid film.

Geometric squeeze, $U_b(\partial h/\partial x)$, is an important term which accounts for the squeeze-like action brought about by the geometrical configuration. Even if no physical squeeze action is present (i.e. $\partial h/\partial t = 0$), there may be a component of velocity in the z direction. To illustrate the concept, consider the motion of two cylinders rotating at different speeds U_a and U_b. See Figure 6.2(a). This situation is modeled in Figure 6.2(b).

The angle of approach α is very small, so that the small angle approximation applies: $\cos \alpha \approx 1$ and $\sin \alpha \approx \tan \alpha$. The normal (squeeze-type) velocity $= U_b \sin \alpha \approx U_b \tan \alpha = U_b (\partial h/\partial x)$. Therefore, $U_b (\partial h/\partial x)$ is a squeeze-like term that arises because of the geometry. In other words, an observer sitting on the bottom plate will see the top plate approaching him at $V_b = U_b (\partial h/\partial x)$, where V_b is the geometric squeeze velocity due to x-direction velocity U_b. Note that the tangential velocity is $U_b \cos \alpha \approx U_b$.

Normal squeeze, $\rho(w_b - w_a)$, is brought about by the relative motion of the surfaces normal to the direction of motion so that the fluid is physically squeezed. A substantial load-carrying capacity could occur due to the normal squeeze alone. We shall discuss this particular type of load-carrying capacity under the heading of squeeze bearings in Chapter 9.

Local expansion, $\partial \rho/\partial t$, provides a measure of the change in volume of a given mass of fluid due to expansion as a result of the heat supply to a bearing, for instance, or with changing pressure in the case of a gas. The contribution of this term to the pressure-generating capacity in the majority of lubrication problems is negligibly small.

Note that in the derivations presented in this chapter, it is assumed that surfaces do not move axially. Hamrock (1994) gives a general derivation including axial motion and presents a thorough physical description of the terms in the Reynolds equation. The interested reader may refer to Wilcock and Booser (1957) and Pinkus and Sternlicht (1961) for further discussion.

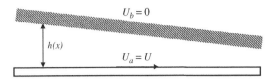

Figure 6.3   Slider bearing.

General Reynolds Equation

Neglecting the stretch terms, density wedge, and local expansion terms yields the following simplified Reynolds equation, which is sufficient for most practical applications:

$$\frac{\partial}{\partial x}\left(\frac{\rho h^3}{12\mu}\frac{\partial P}{\partial x}\right) + \frac{\partial}{\partial y}\left(\frac{\rho h^3}{12\mu}\frac{\partial P}{\partial y}\right) = \frac{1}{2}\rho(U_a - U_b)\frac{\partial h}{\partial x} + \rho(w_b - w_a) \qquad (6.20)$$

The following examples illustrate important features of this equation in a variety of geometries. We will concentrate on the right-hand-side of Equation (6.20).

Example 6.1   Steady-state operation of a slider bearing (Figure 6.3) with the bottom plate sliding at U and the top plate stationary is

$$\begin{cases} U_a = U \\ U_b = 0 \end{cases} \text{ and } \begin{cases} w_a = 0 \\ w_b = 0 \end{cases}$$

Therefore

$$\text{RHS} = \frac{1}{2}\rho U\frac{\partial h}{\partial x}$$

Example 6.2   Steady-state operation of a journal bearing (Figure 6.4) with the bushing stationary and the shaft rotating is

$$\begin{cases} U_a = 0 \\ U_b = U \end{cases} \text{ and } \begin{cases} w_a = 0 \\ w_b = U\frac{\partial h}{\partial x} \end{cases}$$

Therefore

$$\text{RHS} = \frac{1}{2}\rho U\frac{\partial h}{\partial x}$$

The general form of the Reynolds equation for situations where film thickness is invariant with time (i.e. steady state, without normal squeeze motion) is

$$\frac{\partial}{\partial x}\left(\frac{\rho h^3}{12\mu}\frac{\partial P}{\partial x}\right) + \frac{\partial}{\partial y}\left(\frac{\rho h^3}{12\mu}\frac{\partial P}{\partial y}\right) = \frac{1}{2}\left(U_1 + U_2\right)\frac{\partial(\rho h)}{\partial x} \qquad (6.21)$$

where U_1 and U_2 are the actual surface velocities, *not* the components.

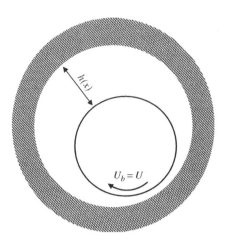

Figure 6.4   Journal bearing.

If the film thickness varies with time, i.e. if the surfaces actually move in the z direction – as in squeeze film bearings – the general Reynolds equation becomes

$$\frac{\partial}{\partial x}\left(\frac{\rho h^3}{12\mu}\frac{\partial P}{\partial x}\right) + \frac{\partial}{\partial y}\left(\frac{\rho h^3}{12\mu}\frac{\partial P}{\partial y}\right) = \frac{1}{2}\left(U_1 + U_2\right)\frac{\partial(\rho h)}{\partial x} + \rho\left(w_2 - w_1\right) \qquad (6.22)$$

where w_1 and w_2 are the velocities of the surfaces in the z direction.

In the majority of applications, only one surface moves and one is stationary:

$$\begin{aligned} U_1 &= 0; & U_2 &= U \neq 0 \\ w_1 &= 0; & w_2 &= w \neq 0 \end{aligned}$$

Therefore, the right-hand-side of the Reynolds equation becomes

$$\frac{1}{2}U\frac{\partial(\rho h)}{\partial x} + \underbrace{\rho w}_{\displaystyle \rho\frac{\partial h}{\partial t}} \; ; \qquad w = \partial h/\partial t \text{ (squeeze velocity)}$$

Standard Reynolds Equation

The most commonly encountered form of the Reynolds equation with impermeable walls is

$$\frac{\partial}{\partial x}\left(\frac{\rho h^3}{12\mu}\frac{\partial P}{\partial x}\right) + \frac{\partial}{\partial y}\left(\frac{\rho h^3}{12\mu}\frac{\partial P}{\partial y}\right) = \frac{1}{2}U\frac{\partial(\rho h)}{\partial x} + \rho\frac{\partial h}{\partial t} \qquad (6.23a)$$

For an incompressible fluid the above equation reduces to

$$\frac{\partial}{\partial x}\left(\frac{h^3}{12\mu}\frac{\partial P}{\partial x}\right) + \frac{\partial}{\partial y}\left(\frac{h^3}{12\mu}\frac{\partial P}{\partial y}\right) = \frac{1}{2}U\frac{\partial h}{\partial x} + \frac{\partial h}{\partial t} \qquad (6.23b)$$

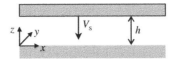

Figure 6.5 Hydrodynamic squeeze action.

The solution to this equation gives the hydrodynamic pressure distribution. It accounts for both the physical wedge and pressure generation capability due to the normal approach of surfaces, i.e. squeeze action.

Steady Film

The second term on the right-hand side of Equation (6.23) vanishes for a steady film. The source term then becomes the physical wedge and, as explained previously to generate positive pressure, must have $\partial h/\partial x < 0$.

Squeeze Film

In the absence of a physical wedge, it is possible to generate a load-carrying capacity by squeeze action alone. This requires that $\partial h/\partial x < 0$. Let V_s denote the squeeze velocity, as shown in Figure 6.5. The Reynolds equation governing this situation is

$$\frac{\partial}{\partial x}\left(\frac{\rho h^3}{12\mu}\frac{\partial P}{\partial x}\right) + \frac{\partial}{\partial y}\left(\frac{\rho h^3}{12\mu}\frac{\partial P}{\partial y}\right) = -\rho V_s \tag{6.24}$$

Cylindrical Coordinates

In some situations it is necessary to formulate a problem in polar-cylindrical coordinates, where the Reynolds equation takes the following form:

$$\frac{\partial}{\partial r}\left(\frac{\rho r h^3}{12\mu}\frac{\partial P}{\partial r}\right) + \frac{1}{r}\frac{\partial}{\partial \theta}\left(\frac{\rho h^3}{12\mu}\frac{\partial P}{\partial \theta}\right) = \frac{1}{2}U\frac{\partial(\rho h)}{\partial \theta} - r\rho V_s \tag{6.25}$$

6.2 Turbulent Flows

While neglected in the original derivation by Reynolds, turbulence has a remarkable influence on the performance of a hydrodynamic bearing. Experimental evidence for the influence of turbulence in bearings in the field and in various laboratories has been documented in the literature. Relatively large bearings are susceptible to turbulence when they operate under high rotational speeds (large ω), when using low-viscosity lubricants (such as water or cryogenic fluids), or when operating with large clearance so as to produce Reynolds numbers (Re) in the range 1000–2000 and higher.

A complete derivation of the Reynolds equation with provisions for turbulence is quite lengthy and will not be presented here. The following equation, based on the linearized turbulent theory, is generally accepted for incompressible fluids (Ng and Pan, 1965):

$$\frac{\partial}{\partial x}\left(\frac{G_x h^3}{\mu}\frac{\partial P}{\partial x}\right) + \frac{\partial}{\partial y}\left(\frac{G_y h^3}{\mu}\frac{\partial P}{\partial y}\right) = \frac{1}{2}U\frac{\partial h}{\partial x} + \frac{\partial h}{\partial t} \tag{6.26}$$

where G_x and G_y are the so-called turbulent coefficients, defined as follows:

$$\begin{aligned}
\frac{1}{G_x} &= 12 + 0.0136(Re)^{0.90} \\
\frac{1}{G_y} &= 12 + 0.0043(Re)^{0.96}
\end{aligned} \tag{6.27}$$

where $Re \equiv \rho Uh/\mu$ and $U = R\omega$. Note that the turbulent coefficients only influence the pressure gradient terms and that Equation (6.26) applies to both laminar regimes (with $G_x = G_y = 1/12$) and turbulent regimes.

Turbulence in general tends to enhance the load-carrying capacity but also increases the power loss. Numerical solution of Equation (6.27) predicts an increase in the pressure and therefore in the load. To examine the effect of turbulence on power loss, one needs to revisit the shear stress on the sliding component. Recall that the velocity profile consists of both Poiseuille and Couette terms as follows:

$$\begin{aligned}
u &= \frac{1}{2\mu}\frac{\partial P}{\partial x}(z^2 - zh) + \frac{z}{h}U \\
\tau_{xz} &= \mu\frac{\partial u}{\partial z} = \frac{\partial P}{\partial x}\left(z - \frac{h}{2}\right) + \mu\frac{U}{h}
\end{aligned} \tag{6.28}$$

To determine the power loss, friction force must be evaluated at the sliding surface. Denoting $\tau_s \equiv \tau_{xz}|_{z=h}$ and $\tau_c \equiv$ Couette shear stress $= \mu\,(U/h)$, we can write the following expression for shear stress on the surface of the sliding member:

$$\tau_s = \frac{h}{2}\frac{\partial P}{\partial x} + \tau_C \tag{6.29}$$

The above expression must be modified for the turbulent regime (Sneck and Vohr, 1984). Since the increase in pressure is taken into account by solving Equation (6.26), it is only needed to apply a correction factor to the Couette shear stress. According to the linearized theory, the following coefficient is introduced for this purpose:

$$\tau_s = \frac{h}{2}\frac{\partial P}{\partial x} + G_f \tau_C \tag{6.30}$$

where $G_f = 1 + 0.0012\,(Re)^{0.94}$.

Power loss is simply the product of linear velocity and the friction force (the integrated effect of shear stress on the sliding member over the appropriate area) as follows:

$$E_p = U\int_A \tau_S \, dA \tag{6.31}$$

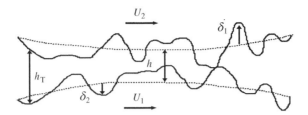

Figure 6.6   A friction pair consisting of two rough surfaces in relative motion.

6.3 Surface Roughness

The importance of roughness in predicting bearing performance has gained considerable attention in the tribology literature. Recent designs in a variety of tribological components call for smaller operating film thicknesses where surface topography may play an important role.

Consider two 'real' surfaces with a nominal film gap h in sliding motion, as shown in Figure 6.6. Local film thickness h_T for this friction pair is defined to be of the form

$$h_T = h + \delta_1 + \delta_2 \tag{6.32}$$

where h is the nominal film thickness (compliance), defined as the distance between the mean levels of the two surfaces. δ_1 and δ_2 are the random roughness amplitudes of the two surfaces measured from their mean levels. The combined roughness $\delta = \delta_1 + \delta_2$ has a variance $\sigma = \sigma_1{}^2 + \sigma_1{}^2$.

Assuming that the fluid is incompressible, the following form of the Reynolds equation governs the pressure field:

$$\frac{\partial}{\partial x}\left(\frac{h_T^3}{12\mu}\frac{\partial P}{\partial x}\right) + \frac{\partial}{\partial y}\left(\frac{h_T^3}{12\mu}\frac{\partial P}{\partial y}\right) = \frac{U_1 + U_2}{2}\frac{\partial h_T}{\partial x} + \frac{\partial h_T}{\partial t} \tag{6.33}$$

To account for the effect of roughness on the pressure, Patir and Cheng (1978) derived the following equation for average pressure:

$$\frac{\partial}{\partial x}\left(\phi_x\frac{h^3}{12\mu}\frac{\partial P_m}{\partial x}\right) + \frac{\partial}{\partial y}\left(\phi_y\frac{h^3}{12\mu}\frac{\partial P_m}{\partial y}\right) = \frac{U_1 + U_2}{2}\frac{\partial \overline{h}_T}{\partial x} + \frac{U_1 - U_2}{2}\sigma\frac{\partial \overline{\phi}_s}{\partial x} + \frac{\partial \overline{h}_T}{\partial t} \tag{6.34}$$

where P_m is the mean pressure and $\overline{h}_T$ is the average gap. The so-called pressure flow factors ϕ_x and ϕ_y are the ratio of average flow in a bearing with rough surfaces to that of a bearing with smooth surfaces. The shear flow factor ϕ_s represents the additional flow transport due to sliding in a rough bearing. These factors were obtained through flow simulation of a rough surface having a Gaussian distribution for surface height. The pressure flow factors ϕ_x and ϕ_y approach 1 (i.e. smooth) as h/σ approaches ∞. The shear flow factor ϕ_S tends toward 0 for a large value of h/σ. These flow factors are also functions of the statistical properties in terms of the roughness heights and the directional properties of the asperities.

To study surfaces with directional properties, the surface characteristic γ defined by Kubo and Peklenik (1967) and Peklenik (1967–68) can be used. The parameter γ can be viewed as the

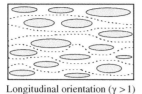
Longitudinal orientation ($\gamma > 1$)

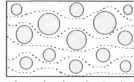
Isotropic orientation ($\gamma = 1$)

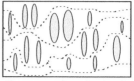
Transverse orientation ($\gamma < 1$)

Figure 6.7 Illustration of three types of roughness orientations.

length-to-width ratio of a representative asperity. In Figure 6.7 three sets of asperity patterns are identified: purely transverse, isotropic, and longitudinal roughness patterns corresponding to $\gamma = 0, 1$, and ∞, respectively. For an isotropic roughness structure ($\gamma = 1$), there is no preferred orientation; hence, the pressure flow factors ϕ_x and ϕ_y are identical. The flow factors computed by Patir and Cheng are listed in Table 6.1.

The average gap $\bar{h}_T$ is defined as

$$\bar{h}_T = \int_{-h}^{\infty} (h + \delta) f(\delta) \mathrm{d}\delta \tag{6.35}$$

where $f(\delta)$ is the probability density function of δ. Experimental results show that for many engineering surfaces, the height distribution is Gaussian to a very good approximation (cf. Greenwood and Williamson, 1966). Using the Gaussian density function with zero mean, the average gap becomes

$$\bar{h}_T = \frac{1}{2} h \left[1 + \mathrm{erf}\left(\frac{h}{\sqrt{2}\sigma}\right) \right] + \frac{\sigma}{\sqrt{2\pi}} e^{-h^2/2\sigma^2} \tag{6.36}$$

Table 6.1 Surface roughness flow factors

ϕ_x				ϕ_s				
$\gamma \leq 1$	$\phi_x = 1 - Ce^{-rH}$			$H \leq 5$	$\Phi_s = A_1 H^{\alpha_1} e^{-\alpha_2 H + \alpha_3 H^2}$			
$\gamma > 1$	$\phi_x = 1 + CH^{-r}$			$H > 5$	$\Phi_s = A_2 e^{-0.25H}$			
	$\phi_y(H,\gamma) = \phi_x(H, 1/\gamma)$				$\phi_s = \left(\frac{\sigma_1}{\sigma}\right)^2 \Phi_s\left(H, \gamma_1\right) - \left(\frac{\sigma_2}{\sigma}\right)^2 \Phi_s(H, \gamma_2)$			
	where $H = h/\sigma$							
γ	C	R	Range	γ	A_1	α_1	α_2	α_3	A_2
1/9	1.48	0.42	$H > 1$	1/9	2.046	1.12	0.78	0.03	1.856
1/6	1.38	0.42	$H > 1$	1/6	1.962	1.08	0.77	0.03	1.754
1/3	1.18	0.42	$H > 0.75$	1/3	1.858	1.01	0.76	0.03	1.561
1	0.90	0.56	$H > 0.5$	1	1.899	0.98	0.92	0.05	1.126
3	0.225	1.5	$H > 0.5$	3	1.560	0.85	1.13	0.08	0.556
6	0.520	1.5	$H > 0.5$	6	1.290	0.62	1.09	0.08	0.388
9	0.870	1.5	$H > 0.5$	9	1.011	0.54	1.07	0.08	0.295

Similarly, the average Reynolds equation in the cylindrical coordinate system with provision for surface roughness is

$$\frac{\partial}{\partial r}\left(r\phi_r\frac{h^3}{12\mu}\frac{\partial P_m}{\partial r}\right) + \frac{1}{r^2}\frac{\partial}{\partial\theta}\left(\phi_\theta\frac{h^3}{12\mu}\frac{\partial P_m}{\partial\theta}\right) = \frac{\omega_1+\omega_2}{2}\frac{\partial\overline{h}_T}{\partial\theta} + \frac{\omega_1-\omega_2}{2}\sigma\frac{\partial\phi_s}{\partial r} + \frac{\partial\overline{h}_T}{\partial t}$$

(6.37)

6.4 Nondimensionalization

It is instructive to consider a dimensionless form of the Reynolds equation. Without lack of generality, we will consider the geometry of a slider bearing shown in Figure 6.8 and restrict our attention to laminar flow, without roughness considerations. The appropriate Reynolds equation for incompressible fluid, equation 6.23b, is repeated below:

$$\frac{\partial}{\partial x}\left(\frac{h^3}{12\mu}\frac{\partial P}{\partial x}\right) + \frac{\partial}{\partial y}\left(\frac{h^3}{12\mu}\frac{\partial P}{\partial y}\right) = \frac{1}{2}U\frac{\partial h}{\partial x} + \frac{\partial h}{\partial t}$$

(6.23b)

Let

$$\overline{x} = \frac{x}{B}; \quad \overline{y} = \frac{y}{L}; \quad \overline{h} = \frac{h}{h_2}; \quad \overline{t} = \frac{t}{t_{ref}}; \quad \overline{P} = \frac{P}{P_{ref}}$$

(6.38)

where t_{ref} and P_{ref} are yet to be determined.

Using the dimensionless group defined in Equation (6.38), the Reynolds equation takes on the following form:

$$\frac{P_{ref}h_2^2}{12\mu B^2}\frac{\partial}{\partial\overline{x}}\left(\overline{h}^3\frac{\partial\overline{P}}{\partial\overline{x}}\right) + \frac{P_{ref}h_2^2}{12\mu L^2}\frac{\partial}{\partial\overline{y}}\left(\overline{h}^3\frac{\partial\overline{P}}{\partial\overline{y}}\right) = \frac{U}{2B}\frac{\partial\overline{h}}{\partial\overline{x}} + \frac{1}{t_{ref}}\frac{\partial\overline{h}}{\partial\overline{t}}$$

(6.39)

Multiplying through by $12\mu B^2/P_{ref}h_2^2$ yields

$$\frac{\partial}{\partial\overline{x}}\left(\overline{h}^3\frac{\partial\overline{P}}{\partial\overline{x}}\right) + \left(\frac{B}{L}\right)^2\frac{\partial}{\partial\overline{y}}\left(\overline{h}^3\frac{\partial\overline{P}}{\partial\overline{y}}\right) = \frac{6\mu BU}{P_{ref}h_2^2}\frac{\partial\overline{h}}{\partial\overline{x}} + \frac{12\mu B^2}{P_{ref}h_2^2 t_{ref}}\frac{\partial\overline{h}}{\partial\overline{t}}$$

(6.40)

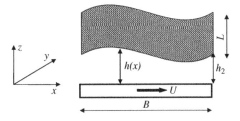

Figure 6.8   Finite slider bearing geometry.

Letting $P_{ref} = \mu B U / h_2^2$ and $t_{ref} = B/U$ further simplifies the above equation to

$$\frac{\partial}{\partial \overline{x}} \left(\overline{h}^3 \frac{\partial \overline{P}}{\partial \overline{x}} \right) + \left(\frac{B}{L} \right)^2 \frac{\partial}{\partial \overline{y}} \left(\overline{h}^3 \frac{\partial \overline{P}}{\partial \overline{y}} \right) = 6 \frac{\partial \overline{h}}{\partial \overline{x}} + 12 \frac{\partial \overline{h}}{\partial \overline{t}} \tag{6.41}$$

This equation is a second-order partial differential equation which does not lend itself to a closed-form, analytical solution. Therefore, one generally resorts to a numerical finite difference scheme to obtain the pressure field for the lubricant film (Vohr, 1984). The most common method is to replace the partial derivatives in Equation (6.41) by an appropriate finite difference operator, thus sampling the pressure field at discrete points in a rectangular grid representing the unwrapped film of a journal bearing. The pressure field is then determined by one of three methods: relaxation, direct matrix inversion, or columnwise influence coefficients. A brief description of the relaxation solution technique is given in Section 6.8.

6.5 Performance Parameters

Bearing performance parameters include the load-carrying capacity, center of pressure, friction force and friction coefficient, flow rates (Figure 6.9), and power loss. Once the pressure distribution is known, one can easily evaluate all of these parameters. Some of the appropriate relationships are summarized below:

- Load-carrying capacity:

$$W = \int_A P dA = \int_{y=0}^{L} \int_{x=0}^{B} P dx \, dy \tag{6.42}$$

where A represents the appropriate surface area on which pressure acts.
- Center of pressure:

$$x_{cp} = \frac{\int_{y=0}^{L} \int_{x=0}^{B} Px dx \, dy}{W} \tag{6.43}$$

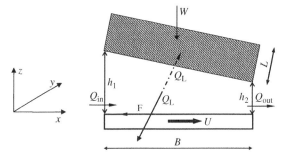

Figure 6.9   Illustration of the flow rates in a finite slider bearing.

- Friction force:

$$F = \int_A \tau_{zx} \, dA = \int_{y=0}^{L} \int_{x=0}^{B} \tau_{zx} \, dx \, dy \tag{6.44}$$

where τ_{zx} represents the appropriate shear stress evaluated on the moving surface.
- Friction coefficient:

$$f = \frac{F}{W} \tag{6.45}$$

- Inlet and outlet flow rates:

$$Q_{in} = \int_{y=0}^{L} \int_{z=0}^{h_1} u|_{x=0} \, dz \, dy \tag{6.46}$$

$$Q_{out} = \int_{y=0}^{L} \int_{z=0}^{h_2} u|_{x=L} \, dz \, dy \tag{6.47}$$

where u is the flow velocity in the direction of sliding. It is a function of the pressure gradient and the fluid viscosity.
- Side leakage (axial) flow rate:

$$Q_L = 2 \int_{x=0}^{B} \int_{z=0}^{h} v|_{y=0} \, dz \, dx = 2 \int_{x=0}^{B} \int_{z=0}^{h} v|_{y=B} \, dz \, dx \tag{6.48}$$

where v denotes the flow velocity in the axial direction. Note that the leakage flow rate with symmetry can be determined at either side of the bearing ($y = 0$ or $y = L$) and that the total flow rate would be twice the flow from one side. Length B of the active film area may end short of the total extent of the bearing segment. In a journal bearing, for instance, the clearance space begins to diverge following the minimum film thickness. As discussed in depth in Chapter 8, after a short distance beyond the minimum film thickness, the film pressure in the clearance drops below ambient and no further leakage occurs. In this case, the outflow can be taken as inflow Q_{in} minus the side flow for the active film length.
- Power loss:

$$E_P = FU \tag{6.49}$$

6.6 Limiting Cases and Closed-Form Solutions

Two distinct limiting solutions emerge from the dimensionless form of the Reynolds equation (6.41):

1. *Infinitely long approximation (ILA)*. Referring to Figure 6.8, if $L \gg B$, then $(B/L)^2 = 1$ and the second term on the left-hand side of Equation (6.41) can be neglected in comparison to the first term. Physically, this approximation implies that the term associated with the leakage is negligible. For this reason, this limiting case is often referred to as the *infinitely*

Table 6.2 Infinitely long approximation (ILA) for a slider bearing

Performance parameters	ILA solution
Film thickness formula	$h(x) = -\dfrac{h_1 - h_2}{B}x + h_1$
Film thickness ratio	$\lambda = \dfrac{h_1}{h_2}$
Characteristic film thickness	$H = \dfrac{2h_1 h_2}{h_1 + h_2}$
Pressure distribution	$P = P_0 + 6\mu U \dfrac{B}{h_1^2 - h_2^2}\dfrac{(h_1 - h)(h - h_2)}{h^2}$
Load-carrying capacity	$W = \dfrac{6\mu U B^2}{(\lambda - 1)^2 h_2^2}\left[\ln \lambda - \dfrac{2(\lambda - 1)}{\lambda + 1}\right]$
Friction force	$F = \dfrac{\mu U B}{(\lambda - 1)h_2}\left[4\ln \lambda - \dfrac{6(\lambda - 1)}{\lambda + 1}\right]$
Center of pressure	$x_{cp} = \dfrac{B}{2}\left[\dfrac{2\lambda}{\lambda - 1} - \dfrac{\lambda^2 - 1 - 2\lambda \ln \lambda}{(\lambda^2 - 1)\ln \lambda - 2(\lambda - 1)^2}\right]$
Flow rate	$Q_x = \dfrac{UH}{2}; \; Q_y = 0$

long approximation (ILA). Table 6.2 summarizes the performance parameters for the ILA case (Schlichting, 1979; Sneck and Vohr, 1984). The corresponding solutions for journal bearings are presented in Chapter 8.

2. *Infinitely short approximation (ISA)*. If $L \ll B$, then $(B/L)^2 \gg 1$ and the first term on the left-hand side of Equation (6.41) can be neglected in comparison to the second term. Physically, this approximation implies that the term associated with the flow in the direction of sliding motion is negligible compared to the leakage term. Table 6.3 summarizes slider bearing performance parameters for the ISA case. Similar ISA solutions for journal bearings are given in Chapter 8.

Table 6.3 Infinitely short approximation (ISA) for a slider bearing

Performance parameters	ISA solution
Film thickness formula	$h(x) = -\dfrac{h_1 - h_2}{B}x + h_1$
Film thickness ratio	$\lambda = \dfrac{h_1}{h_2}$
Pressure distribution	$P = \dfrac{3\mu U}{h^3}\dfrac{dh}{dx}(y^2 - by)$
Load-carrying capacity	$W = \dfrac{L^3}{4}\mu U\left(\dfrac{1}{h_2^2} - \dfrac{1}{h_1^2}\right)$
Friction force	$F = \dfrac{L\mu U B}{h_1 - h_2}\ln \lambda$
Flow rate	$Q_y = \dfrac{LU}{2}(h_1 - h_2)$

Both ILA and ISA approximations result in closed-form, analytical solutions under special circumstances. These approximations provide very useful and, indeed, realistic solutions for many bearings. A bearing is said to be finite if the length-to-width ratio B/L is of the order 1 and no ISA or ILA solutions would be sufficient. Nonetheless, one can use these approximations to check numerical solutions that must be formulated for finite bearings.

A question often arises: what is the appropriate B/L ratio for ILA and ISA? Generally, if $(B/L) \leq 0.5$, the ILA assumption will offer a sufficiently accurate result. Similarly, if $(B/L) \geq 2$, the ISA may be appropriate. Figure 6.10a shows a typical pressure distribution in a slider bearing of finite dimension. It was obtained by numerical solution with $B/L = 1$ (see Section 6.8). The corresponding solution for an infinitely wide bearing is shown in Figure 6.10b. Note that in the case of a finite bearing under isoviscous conditions, the axial

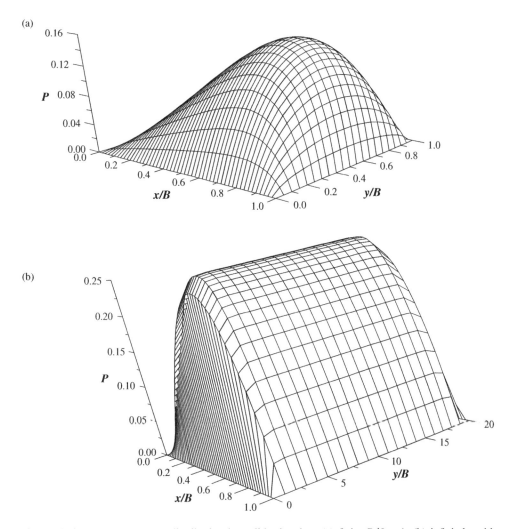

Figure 6.10   Typical pressure distribution in a slider bearing: (a) finite $B/L = 1$; (b) infinitely wide simulation using $L/B = 20$.

pressure is parabolic. Therefore, fluid leaks at the sides of a finite bearing, whereas in the wide bearing case, the axial pressure is flat. These figures demonstrate that for a given λ, the ILA predicts a much greater load-carrying capacity than is actually possible. It therefore follows that for the ILA, the results must be carefully examined before they are used in practice.

A Simplified Form of Reynolds Equation for Steady Film

If the ILA holds and the film is steady, the Reynolds equation reduces to the following simple ordinary differential equation:

$$\frac{d}{dx}\left(\frac{h^3}{12\mu}\frac{dP}{dx}\right) = \frac{U}{2}\frac{dh}{dx} \tag{6.50}$$

Integrating once, we obtain

$$\frac{h^3}{12\mu}\frac{dP}{dx} = \frac{Uh}{2} + C \tag{6.51}$$

The hydrodynamic pressure must reach a peak and then drop. Therefore, at some unknown distance from the inlet, where the film thickness is h_m, the pressure gradient $dP/dx = 0$. Using this information as a boundary condition to evaluate the integration constant, we get $C = -Uh_m/2$. Therefore, we arrive at the following convenient form of the Reynolds equation:

$$\frac{dP}{dx} = 6\mu U\frac{h - h_m}{h^3} \tag{6.52}$$

For a specified film thickness profile h, this equation can be easily integrated to get the pressure distribution. This simplified form is sometimes very convenient for analytical solutions.

6.7 Application: Rayleigh Step Bearing

Lord Rayleigh performed an intriguing analysis to determine the film shape profile for an optimum load-carrying capacity, assuming that the ILA holds (Rayleigh, 1918). Using the calculus of variations approach, he showed that a composite film thickness profile in the form of a step, as shown in Figure 6.11, can be selected to yield the optimum load-carrying capacity. The Rayleigh step bearing finds many industrial applications from high-speed turbomachinery to micro-electro-mechanical systems (MEMS) and devices (Ogata, 2005; Rahmani et al., 2009).

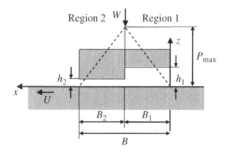

Figure 6.11 Rayleigh step bearing.

It is easy to manufacture and hence this shape is frequently used in constructing thrust bearings (Vedmar, 1993), which is covered in Chapter 7.

Optimization of Load-Carrying Capacity

The Rayleigh step bearing finds industrial applications from high-speed turbomachinery to micro-electro-mechanical systems (MEMS) and devices (Ogata, 2005; Rahmani *et al.*, 2009). It is easy to manufacture and hence this shape is frequently used in constructing thrust bearings (Vedmar, 1993), which is covered in Chapter 7.

The solution requires dividing the fluid-film domain into two regions. The governing equation for each region is simply the Reynolds equation. The key to solving problems with composite film thickness is the proper specification of the boundary conditions, particularly at the common joint where the step is formed.

The governing equation, assuming that ILA holds, is

$$\frac{d}{dx}\left(\frac{h^3}{\mu}\frac{dP}{dx}\right) = 6U\frac{dh}{dx} \tag{6.53}$$

where

$$h = \begin{cases} h_1 & 0 \le x \le B_1 \\ h_2 & B_1 \le x \le B \end{cases}$$

In Region 1, the film thickness is constant $h = h_1$; therefore, the right-hand side of the Reynolds equation vanishes. The governing equation and the boundary conditions are

$$\frac{d}{dx}\left(\frac{h_1^3}{\mu}\frac{dP_1}{dx}\right) = 0 \tag{6.54}$$

$$\begin{cases} P_1 = 0 & \text{at } x = 0 \\ P_1 = P_C & \text{at } x = B_1 \end{cases} \tag{6.55}$$

where P_c is the 'common pressure' where the two regions meet at B_1. Integrating Equation (6.54) twice and evaluating one of the constants of integration yields

$$P_1 = C_1 x \tag{6.56}$$

In Region 2, the governing equation is

$$\frac{d}{dx}\left(\frac{h_2^3}{\mu}\frac{dP_2}{dx}\right) = 0 \tag{6.57}$$

Boundary conditions are:

$$\begin{cases} P_2 = 0 & \text{at } x = B \\ P_2 = P_C & \text{at } x = B_1 \end{cases} \tag{6.58}$$

Integrating Equation (6.58) twice and evaluating one of the constants of integration yields

$$P_2 = C_3(x - B) \tag{6.59}$$

C_1 and C_3 are two unknowns to be determined. Two additional equations are needed for this purpose. They are:

$$\text{At } x = B_1 \quad \begin{cases} P_1 = P_2 = P_c \\ q_1 = q_2 \end{cases} \tag{6.60}$$

Using the first condition in Equation (6.58), we have

$$C_3 = -\frac{B_1}{B_2} C_1 \tag{6.61}$$

The second condition in Equation (6.58) is simply

$$-\frac{h_1^3}{12\mu}\left(\frac{dP_1}{dx}\right) + \frac{Uh_1}{2} = -\frac{h_2^3}{12\mu}\left(\frac{dP_2}{dx}\right) + \frac{Uh_2}{2} \tag{6.62}$$

Inserting the pressure gradients using Equations (6.56) and (6.59) and solving for C_1 yields

$$C_1 = \frac{6\mu U(h_1 - h_2)B_2}{h_1^3 B_2 + h_2^3 B_1} \tag{6.63}$$

The final expressions for the pressure profiles in regions 1 and 2 are:

Region 1:

$$P_1(x) = \frac{6\mu U(h_1 - h_2)B_2}{h_1^3 B_2 + h_2^3 B_1} x \tag{6.64}$$

Region 2:

$$P_2(x) = \frac{6\mu U(h_1 - h_2)B_1}{h_1^3 B_2 + h_2^3 B_1}(B - x) \tag{6.65}$$

Note that when

$$\begin{cases} h_1 = h_2 & P_1(x) = P_2(x) = 0 \\ B_1 = 0 \text{ or } B_2 = 0 & P_1(x) = P_2(x) = 0 \end{cases} \tag{6.66}$$

since, in each case, the geometry reduces to two parallel plates which theoretically cannot generate any load-carrying capacity. The pressure distribution in each region is a linear function of length, with the maximum pressure occurring where the two plates meet, i.e. $P_{max} = P_c$. The load per unit width, W/L, is computed by integrating the pressure in region and summing the results:

$$\frac{W}{L} = \frac{1}{L}(W_1 + W_2) = \int_0^{B_1} P_1 dx + \int_{B_1}^{B_2} P_2 dx \tag{6.67}$$

$$= \frac{3\mu UBB_2(B - B_2)(\lambda - 1)}{(B_2\lambda^3 + B - B_2)h_2^2} \tag{6.68}$$

where $\lambda = h_1/h_2$.

Optimization of Load-Carrying Capacity

In this section, we seek to determine the maximum load-carry capacity (W_{max}) of a Rayleigh step bearing. This is a simple optimization problem which can be analytically derived by satisfying the following conditions (Archibald, 1950):

$$\frac{dW}{dB_2} = 0 \text{ and } \frac{dW}{d\lambda} = 0 \tag{6.69}$$

To satisfy the first condition in Equation (6.69), differentiating Equation (6.68) with respect to B_2 and setting the result equal to zero yields

$$\lambda = \left(\frac{B - B_2}{B_2} \right)^{2/3}$$

Or in terms of the length B_2, we have

$$B_2 = B(\lambda^{3/2} + 2)^{-1} \tag{6.70}$$

Similarly, satisfying the second condition in Equation (6.69) yields

$$B_2 = B(2\lambda^3 - 3\lambda^2 + 1)^{-1} \tag{6.71}$$

Dividing Equation (6.71) by Equation (6.70) to eliminate B and B_2 yields a single relationship in terms of λ. The solution is:

$$\lambda = 1 \pm \sqrt{3/2}$$

Since $\lambda > 1$, the only admissible solution is $\lambda = 1 + \sqrt{3/2} = 1.866$
   With this value of λ, $B_1 = 2.549B_2$.
   Substituting these values into Equation (6.88) yields the following expression for W_{max}:

$$W_{max} = \frac{0.206\mu UB^2 L}{h_2^2} \tag{6.72}$$

This surprisingly simple relationship has a general form that is applicable for a variety of other configurations. In general, the maximum load-carrying capacity has the following form:

$$W_{max} = C_L \frac{\mu UB^2 L}{h_2^2} \tag{6.73}$$

where C_L is the *load coefficient* (Fesanghary and Khonsari, 2012).

Rayleigh Step Bearing with Finite Dimensions

The above solution for an infinitely wide (1-D) step bearing can be extended to a finite two-dimensional case. However, the problem is more complex and requires a partial differential equation to be solved. Archibald (1950) provides an analytical treatment of this problem.

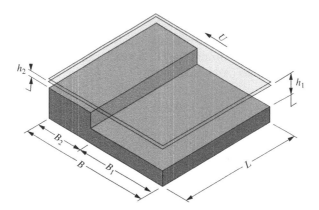

Figure 6.12   Finite Rayleigh step. *Source*: Fesanghary and Khonsari, 2012. Reproduced with permission of Elsevier.

Referring to Figure 6.12, the result for the load-carrying capacity is:

$$W = \frac{48\mu U L^3 (\lambda - 1)}{\pi^4 h_2^2} \sum_{n=1,3,5\ldots}^{\infty} \frac{1}{n^4} \frac{\tanh \frac{n\pi B_1}{2B} + \tanh \frac{n\pi B_2}{2B}}{\coth \frac{n\pi B_1}{B} + \lambda^3 \coth \frac{n\pi B_2}{B}} \tag{6.74}$$

To determine the optimum load-carrying capacity, unlike the 1-D case, it is not an easy task to find an analytical solution. For this purpose, Fesanghary and Khonsari (2012) numerically presented the summation term in Equation (6.74) with a sufficiently large number of terms ($n = 200$). Then, similar to the 1-D case, the derivatives with respect to B_1 and λ were evaluated and a relationship for the maximum load-carrying capacity was obtained in the identical form of Equation (6.73). Here, the load-coefficient C_L, λ and ξ ($= B_2/B_1$) are a function of the bearing aspect ratio, B/L only. The results are shown graphically in Figure 6.13. See

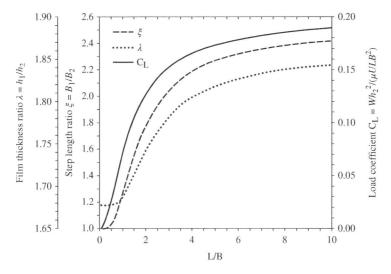

Figure 6.13   Parameters for finite step bearing with optimized load-carrying capacity. *Source*: Fesanghary and Khonsari, 2012. Reproduced with permission of Elsevier.

Fesanghary and Khonsari (2012) for the curve-fit expression for these results. Note that as the aspect ratio is increased, the optimum length and film-height ratios approach the values obtained for the 1-D case, as expected.

6.8 Numerical Method

To illustrate a simple numerical solution using the finite difference method, recall the dimensionless form of the Reynolds equation (6.41). Neglecting the transient term, we have

$$\frac{\partial}{\partial \overline{x}}\left(\overline{h}^3 \frac{\partial \overline{P}}{\partial \overline{x}}\right) + \left(\frac{B}{L}\right)^2 \frac{\partial}{\partial \overline{y}}\left(\overline{h}^3 \frac{\partial \overline{P}}{\partial \overline{y}}\right) = 6\frac{\partial \overline{h}}{\partial \overline{x}} \tag{6.75}$$

Equation (6.75) is an elliptic partial differential equation similar to the Poisson heat conduction equation. To further simplify the problem, assume that $h = h(x)$ only and drop the dimensionless bar notation for convenience, bearing in mind that all terms in Equation (6.76) are dimensionless. Therefore, the above equation can be written as follows:

$$\frac{\partial^2 P}{\partial x^2} + \Lambda^2 \frac{\partial^2 P}{\partial y^2} + \frac{3h'}{h}\frac{\partial P}{\partial x} = 6h' \tag{6.76}$$

where the aspect ratio $\Lambda = B/L$ and $h' = \partial h/\partial x$.

Referring to Figure 6.14, the appropriate finite difference operators for the partial derivatives are:

$$\frac{\partial^2 P}{\partial x^2} = \frac{P_{i-1,j} - 2P_{i,j} + P_{i+1,j}}{(\Delta x)^2}$$

$$\frac{\partial^2 P}{\partial y^2} = \frac{P_{i,j-1} - 2P_{i,j} + P_{i,j+1}}{(\Delta y)^2} \tag{6.77}$$

$$\frac{\partial P}{\partial x} = \frac{P_{i+1,j} - P_{i-1,j}}{2\Delta x}$$

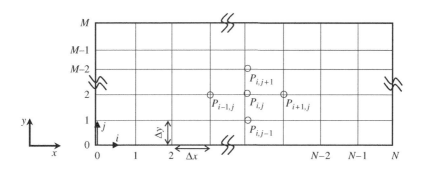

Figure 6.14 Finite difference mesh.

Substituting the operators (6.77) into equation (6.76), grouping terms, and solving for $P_{i,j}$, we obtain

$$P_{i,j} = a_i P_{i+1,j} + b_i \left(P_{i,j-1} + P_{i,j+1} \right) + c_i P_{i-1,j} - d_i \tag{6.78}$$

where

$$a_i = \frac{1 + 1.5\Delta x(h_i'/h_i)}{G_2}; \qquad b_i = \frac{G_1}{G_2}; \qquad c_i = \frac{1 - 1.5\Delta x(h_i'/h_i)}{G_2}; \qquad d_i = 6\frac{(\Delta x)^2}{G_2}\frac{h_i'}{h_i^3}$$

$$G_1 = \Lambda^2 \left(\frac{\Delta x}{\Delta y} \right)^2; \qquad G_2 = 2 + 2G_1$$

Equation (6.78) forms the basis for computing the pressure at each mesh point (i, j). Essentially, the partial differential equation is now replaced with a series of algebraic equations that can be easily solved using a matrix solver.

A very efficient iterative solution approach is the so-called successive over-relaxation (SOR) method. A brief description of the method follows. The above equation is rewritten in the following standard SOR form:

$$P_{i,j} = P_{i,j} + \omega \underbrace{\left[a_i P_{i+1,j} + b_i \left(P_{i,j-1} + P_{i,j+1} \right) + c_i P_{i-1,j} - d_i - P_{i,j} \right]}_{\text{residual terms}} \tag{6.79}$$

where ω is the so-called over-relaxation factor. The objective is to relax the residual terms to zero by successive iteration. The over-relaxation factor simply accelerates the convergence process (Figure 6.15). This factor always lies between $1 \le \omega < 2$ (note that $\omega \ne 2$). When $\omega = 1$, the iteration scheme simply becomes the same as the Gauss-Seidel method, which has a much slower convergence rate. By trial and error, one can determine an optimum value of the relaxation factor ω_{opt} for the fastest convergence. The optimum relaxation factor is a function of the mesh size. Normally, $\omega \cong 1.7$ is a good starting point for determining ω_{opt}. Figure 6.15 shows a typical plot of the iteration numbers needed to converge to the solution as a function of ω.

The procedure is to simply begin the computations by assuming an initial pressure distribution (zero everywhere is often a convenient choice) for the interior nodes (i.e. nodes $i = 1$, $N - 1; j = 1, M - 1$) and setting the boundary conditions for the pressure at the inlet, the outlet, and the sides (i.e. $P_{0,j}, P_{N,j}; P_{i,0}, P_{i,M}$). Starting at the entrance edge of the oil film and marching stepwise from node to node toward the trailing edge, Equation (6.79) predicts an updated pressure for each interior node. Any negative pressures calculated in a diverging clearance zone are arbitrarily set equal to zero (or ambient) pressure. The procedure can be repeated until the difference between the computed pressure values in two successive iterations falls below a specified tolerance value. When dealing with the dimensionless Reynolds equation, a typical tolerance value is $\varepsilon = 10^{-3}$, or $\varepsilon = 10^{-4}$ if more accuracy is desired.

Once this pressure field is established for the lubricant film, other performance parameters follow from the pressure distribution. Appropriate expressions for performance

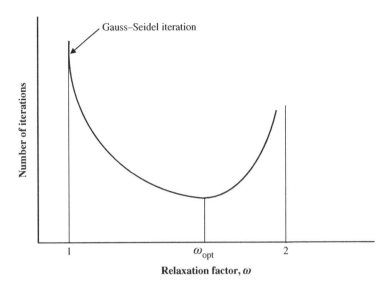

Figure 6.15 Convergence rate of SOR scheme.

parameters are given in Section 6.5. A numerical integration routine such as the Simpson rule is often used for evaluating the integrals in Equations (6.42)–(6.48). Power loss ($E_p = F \cdot U$) is easily calculated once the friction force F is known.

After the node pressures are generated, the temperature rise in the fluid can also be predicted by solving the energy equation with corresponding use of an appropriate lubricant viscosity at each grid point. The energy equation and the Reynolds equation are coupled through the viscosity–temperature relationship. Therefore, these equations must be solved simultaneously by a second iterative procedure. For the first iteration, all node temperatures (and the corresponding fluid properties) throughout the film are set equal to the inlet temperature corresponding to the first pressure iteration. Application of the energy equation then gives a revised set of node temperatures (and their viscosities) for use in a new iteration for pressure distribution. This cycle is repeated until steady pressure and temperature fields are reached, which then represent the simultaneous solution of the Reynolds equation and the energy relations equation for the bearing film.

Computer codes for numerical solution of the Reynolds equation are available from bearing suppliers and other sources for a variety of fluid-film bearings. These provide calculations of film thickness (load capacity), film temperature, oil flow, power loss, and dynamic response coefficients. Computer programs for predicting the temperature distribution in the lubricant film and the bounding solids are much more involved. Known as *thermohydrodynamic analyses*, such routines require careful assessment of conduction heat transfer to the boundaries and convection losses to the surroundings. General discussions of numerical procedures for thermohydrodynamic analyses and design guides are available from Jang and Khonsari (1997, 2004) for slider bearings and from Khonsari and Beaman (1986) and Khonsari *et al.* (1996) and Keogh *et al.* (1997) for journal bearings. Tilting-pad thrust bearings are covered by Fillon and Khonsari (1996).

Problems

6.1 Show that the steady-state Reynolds equation can be derived by simply using $q_x + q_y = 0$, where q_x and q_y are the flow rates in the direction of motion and the axial direction, respectively.

6.2 A journal bearing is designed so that both the shaft and the bushing can rotate at specified rotational speeds of U_b and U_a, respectively.
   (a) Show that if $U_a = -U_b$ (synchronous, counterrotating), theoretically the bearing does not generate load support.
   (b) Show that if $U_a = U_b$, the bearing generates twice as much load as it does when $U_a = 0$ (stationary bushing).

6.3 A slider bearing is designed so that both surfaces are capable of sliding.
   (a) Show that if $U_a = -U_b$, the bearing can theoretically generate twice the load of conventional slider bearings.
   (b) Show that if $U_a = U_b$, the bearing is devoid of generating load support.

6.4 A viscous pump is in the shape of a Rayleigh step bearing (Figure 6.16). Before the slider begins to move, there is contact between part of the step and the sliding surface, i.e. $h_2 = 0$. Assuming that the bearing is infinitely wide:
   (a) Show that the minimum sliding speed necessary for the bearing to take off is

$$U = \frac{W}{3\mu} \left(\frac{h_1}{B_1} \right)^2$$

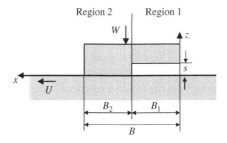

(a) Before takeoff

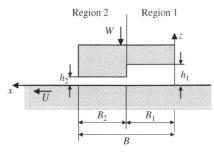

(b) After takeoff

Figure 6.16   Problem 6.4.

(b) Determine a numerical value for the critical speed, given that load $W =$ 20 N/mm, $B_1 = 0.050$ m, $h_1 = 0.050$ mm, $B_2 = 0.02$ m, and the oil is SAE 30 at $T = 20\,°C$.

(c) Determine the steady-state film thickness, h_2, after the bearing floats.

(d) Determine an expression for the coefficient of friction and a numerical value in terms of the specifications in part (b).

6.5 An infinitely wide step bearing is modified to include an axial groove whose depth is much greater than the film thickness (Figure 6.17). The inlet and outlet film thicknesses are h_1 and h_2, respectively. The sliding velocity of the slider is U, and the fluid viscosity is μ. Assuming that pressure remains constant in the groove:

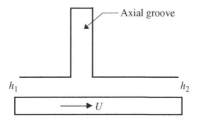

Figure 6.17 Problem 6.5.

(a) Sketch the pressure distribution for this bearing.

(b) Determine an expression for the load-carrying capacity in terms of the maximum pressure.

(c) Determine the critical speed needed for this bearing to take off from an initial position when $h_2 = 0$, assuming that the load per unit width of this bearing is $W = 15$ N/mm, $B_1 = B_2 = B_3 = 15$ mm, and $h_1 = 0.05$ mm.

6.6 Consider an infinitely wide taper-step slider bearing shown in Figure 6.18.

(a) Write down all the necessary boundary conditions for this problem.

(b) Starting from the Reynolds equation, determine an expression for the pressure distribution for this bearing.

(c) Plot the pressure profile. (A simple sketch is sufficient. If you wish to use a plotting package, then use some typical numbers for viscosity and dimensions.)

(d) Show that when $h_1(x) = h_3$ (constant film thickness) and $P_i = 0$, the expression for the pressure distribution reduces to that for the Rayleigh step bearing.

(e) Determine an expression for the velocity profile and volumetric flow rate.

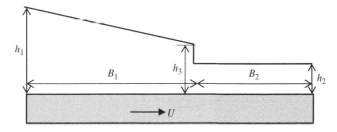

Figure 6.18 Problem 6.6.

6.7   Show that the pressure distribution of a short bearing (assuming that ISA holds) is given by the following expression:

$$P = \frac{3\mu U}{h^3} \frac{dh}{dx} \left(y^2 - \frac{B^2}{4} \right)$$

where y is measured from the bearing center line.

6.8   As a consultant, you have been asked to determine the relationship between the film thickness to load-carrying capacity of a short bearing. Show that theoretically the load-carrying capacity in a short bearing is not dependent upon the shape of the film thickness.

6.9   Verify the ISA solutions for load-carrying capacity, flow rate, and friction force given in Table 6.3. Extend the derivations to include an expression for the power loss and the temperature rise.

6.10   Two infinitely long cylinders in rolling motion are shown in Figure 6.19. The film thickness h is described as $h = h_0 + (x^2/2R)$, where R is the equivalent radius of curvature, defined as

$$\frac{1}{R} = \frac{1}{R_1} + \frac{1}{R_2}$$

Write down the appropriate governing equation and the boundary conditions for determining the pressure distribution in the fluid. Solve for the pressure and load-carrying capacity.

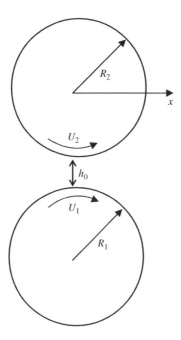

Figure 6.19   Problem 6.10.

6.11 In deriving the Reynolds equation, it was assumed that the viscosity does not vary across the film thickness. Research shows that this assumption may be unacceptable when thermal effects are to be considered. Show that when $\mu = \mu\,(x,\ y,\ z)$, the Reynolds equation for a slider bearing takes the following form:

$$\frac{\partial}{\partial x}\left(\rho F_1 \frac{\partial P}{\partial x}\right) + \frac{\partial}{\partial y}\left(\rho F_1 \frac{\partial P}{\partial y}\right) = \frac{\partial}{\partial x}\left(\rho U_a \frac{F_2}{F_3}\right) + \frac{\partial}{\partial t}(\rho h)$$

where

$$F_1 = \int_0^h \frac{z}{\mu}\left(z - \frac{F_2}{F_3}\right) dz; \qquad F_2 = \int_0^h \frac{z}{\mu}\,dz; \qquad F_3 = \int_0^h \frac{1}{\mu}\,dz$$

6.12 Suppose that in a particular application the surfaces are porous. Show that the appropriate form of the Reynolds equation is

$$\frac{\partial}{\partial x}\left(\frac{\rho h^3}{12\mu}\frac{\partial P}{\partial x}\right) + \frac{\partial}{\partial y}\left(\frac{\rho h^3}{12\mu}\frac{\partial P}{\partial y}\right) = \frac{1}{2}U\frac{\partial(\rho h)}{\partial x} + \rho\left(\omega|_h - \omega|_0\right) + \rho\frac{\partial h}{\partial t}$$

where $(\omega|_h - \omega|_0)$ represents the net flow through the porous boundaries.

6.13 An interesting application of the hydrodynamic lubrication equation is in analyzing the film thickness in foil bearings. These bearings are capable of operating at high speeds and high temperatures because of their resistance to thermally induced seizure (Gross *et al.*, 1980; Fuller, 1984). Application in tape drives for recording devices is also widespread. In the accompanying figure, the foil wrapped around a shaft of radius R is under tension T and travels with velocity U. As shown in Figure 6.20, the film thickness involves three distinct regions: entrance, central, and exit zones. Assuming perfectly flexible foil with no axial deflection, the equilibrium equation reduces to

$$P - P_o = \frac{T}{R}\left(1 - R\frac{d^2 h}{dx^2}\right)$$

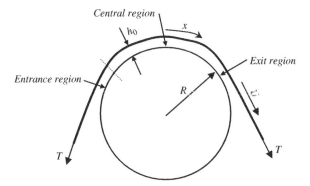

Figure 6.20   Problem 6.13 – nomenclature for a foil bearing.

where x represents the coordinate distance in the direction of motion. Assuming that the ILA solution applies, show that the Reynolds equation and the equilibrium equations combine to form the following relationship for film thickness:

$$\frac{d}{dx}\left(h^3\frac{d^3h}{dx^3}\right) = -\frac{6\mu U}{T}\frac{dh}{dx}$$

Show that the above equation reduces to the following nonlinear differential equation:

$$\frac{d^3\bar{h}}{d\bar{\zeta}^3} = \frac{1-\bar{h}}{\bar{h}^3}$$

where $\bar{h} = h/h_0$ (h_0 is the so-called central film thickness) and

$$\bar{\zeta} = \frac{1}{h_0}\left(\frac{6\mu U}{T}\right)^{1/3}x$$

To determine an analytical solution, linearize the above equation, taking advantage of the fact that film thickness *variation* is very small, i.e., $\Delta\bar{h} \ll 1$ (constant gap). Let $\bar{h} = 1 + \varepsilon$, where ε is a very small number (corresponding to the variation rom the constant gap, h_0). Show that the above equation reduces to

$$\frac{d^3\varepsilon}{d\bar{\zeta}^3} + \varepsilon = 0$$

which has the solution

$$\varepsilon = C_1e^{-\bar{\zeta}} + C_2e^{\bar{\zeta}/2}\cos\left(\frac{\sqrt{3}}{2}\bar{\zeta}\right) + C_3e^{\bar{\zeta}/2}\sin\left(\frac{\sqrt{3}}{2}\bar{\zeta}\right)$$

References

Archibald, F. 1950. 'A Simple Hydrodynamic Thrust Bearing,' *Transactions of ASME*, Vol. 72, pp. 393–400.

Dowson, D. 1962. 'A Generalized Reynolds Equation for Fluid-Film Lubrication,' *International Journal of Mechanical Engineering Sciences*, Vol. 4, pp. 159–170.

Fesanghary, M. and Khonsari, M.M. 2012. 'Topological and Shape Optimization of Thrust Bearings for Enhanced Load-Carrying Capacity,' *Tribology International*, Vol. 53, pp. 12–22.

Fillon, M. and Khonsari, M.M. 1996. 'Thermohydrodynamic Design Charts for Tilting-Pad Journal Bearings,' *ASME Journal of Tribology*, Vol. 118, pp. 232–238.

Fuller, D.D. 1984. *Theory and Practice of Lubrication for Engineers*. John Wiley & Sons, New York.

Greenwood, J.A. and Williamson, J.B.P. 1966. 'Contact of Nominally Flat Surfaces,' *Proceedings of the Royal Society*, London, Series A, Vol. 295, pp. 300–319.

Gross, W.A., Matsch, L.A., Castelli, V., Eshel, A., Vohr, J.H., and Wildmann, M. 1980. *Fluid Film Lubrication*. John Wiley & Sons, New York.

Hamrock, B.J. 1994. *Fundamentals of Fluid Film Lubrication*. McGraw-Hill Book Co., New York.

Jang, J. and Khonsari, M.M. 1997. 'Thermohydrodynamic Design Charts for Slider Bearings,' *ASME Journal of Tribology*, Vol. 119, pp. 859–868.

Jang, J.Y. and Khonsari, M.M. 2004. 'Design of Bearings Based on Thermohydrodynamic Analysis,' *Journal of Engineering Tribology, Proceedings of Institution of Mechanical Engineers, Part J*, Vol. 218, pp. 355–363.

Keogh, P., Gomiciaga, R., and Khonsari, M.M. 1997. 'CFD Based Design Techniques for Thermal Prediction in a Generic Two-Axial Groove Hydrodynamic Journal Bearing,' *ASME Journal of Tribology*, Vol. 119, pp. 428–436.

Khonsari, M.M. and Beaman, J. 1986. 'Thermohydrodynamic Analysis of Laminar Incompressible Journal Bearings,' *ASLE Transactions*, Vol. 29, No. 2, pp. 141–150.

Khonsari, M.M., Jang, J. and Fillon, M. 1996. 'On the Generalization of Thermohydrodynamic Analysis for Journal Bearings,' *ASME Journal of Tribology*, Vol. 118, pp. 571–579.

Kubo, M. and Peklenik, J. 1967. 'An Analysis of Microgeometrical Isotropy for Random Surface Structures,' presented at the 17th CIRP General Assembly, Ann Arbor, Mich.

Ng, C.W. and Pan, C.H.T. 1965. 'A Linearized Turbulent Lubrication Theory,' *Journal of Basic Engineering, Transactions ASME*, Vol. 87, p. 675.

Ogata, H. 2005. 'Thermohydrodynamic Lubrication Analysis Method of Step Bearings,' *IHI Engineering Review*, Vol. 38, pp. 6–10.

Patir, N. and Cheng, H.S. 1978. 'An Average Flow Model for Determining Effects of Three-Dimensional Roughness on Partial Hydrodynamic Lubrication,' *ASME Journal of Lubrication Technology*, Vol. 100, pp. 12–17.

Peklenik, J. 1967–68. 'New Developments in Surface Characterization and Measurement by Means of Random Process Analysis,' *Proceedings of the Institute of Mechanical Engineers*, Vol. 182, Part 3K, pp. 108–126.

Pinkus, O. and Sternlicht, B. 1961. *Theory of Hydrodynamic Lubrication*. McGraw-Hill Book Co., New York.

Rahmani, R., Shirvani, A. and Shirvani, H. 2009. 'Analytical Analysis and Optimisation of the Rayleigh Step Slider Bearing,' *Tribology International*, Vol. 42, pp. 666–674.

Rayleigh, L. 1918. 'Notes on the Theory of Lubrication,' *Philosophical Magazine*, Vol. 35, No. 1, pp. 1–12.

Reynolds, O. 1886. 'On the Theory of Lubrication and its Application of Mr. Beauchamp Tower's Experiments,' *Philosophical Transactions of the Royal Society*, Vol. 177, part 1, pp. 157–234.

Schlichting, H. 1979. *Boundary-Layer Theory*, 7th edn., McGraw-Hill Book Co., New York.

Sneck, H.J. and Vohr, J.H. 1984. 'Hydrodynamic Lubrication,' *Handbook of Lubrication*, Vol. II. CRC Press, Boca Raton, Fla, pp. 69–91.

Vedmar, L. 1993. *On the Optimum Design of Sectorial Step Thrust Bearings*, Machine Elements Division, Lund Technical University, Lund.

Vohr, J.H. 1984. 'Numerical Methods in Hydrodynamic Lubrication,' *Handbook of Lubrication*, Vol. II. CRC Press, Boca Raton, Fla, pp. 93–104.

Wilcock, D.F. and Booser, E.R. 1957. *Bearing Design and Application*. McGraw-Hill Book Co., New York.

7

Thrust Bearings

The hydrodynamic principles developed in the previous chapter are employed in a variety of thrust bearing designs to provide axial load support or simply axial location for a rotor. While the principles involved are the same, fluid-film designs range from coin-sized flat washers to sophisticated assemblies several feet in diameter. In each case, a fluid-film pressure is generated to balance an applied thrust load with sufficient separation of the stationary and rotating surfaces to avoid wear, to provide low friction, and to avoid excessive temperature rise.

The operating principles involved can be illustrated by considering a flat surface sliding over a tapered land, as shown in Figure 7.1. Motion of the flat surface, or thrust runner, draws fluid such as lubricating oil into a wedge-shaped zone over the tapered land. Pumping into the zone of reducing downstream clearance by shearing action of the runner then pressurizes the oil. The change in separation h adjusts the inflow and outflow to provide a balance of the integrated pressure over the bearing area with the applied load. A group of these tapered land thrust pads are arranged with oil distribution grooves in an annular configuration to form a complete thrust bearing, as in Figure 7.2.

7.1 Thrust Bearing Types

Six common types of thrust bearings are shown in Table 7.1. The first four are hydrodynamic, with their oil-film pressure being generated by internal pumping action. The hydrostatic bearing uses an external oil pump to supply the pressurized supporting film. Since parallel planes do not directly provide pumping action, flat land bearings depend on the thermal expansion of the oil, thermal warping and irregularities of bearing surfaces, and possibly other effects (see Section 7.1). Characteristic application practices of these various types follow.

Typical unit loads and size ranges for each type are listed in Table 7.1. For the typical babbitt surface used in most machinery thrust bearings, normal upper design load limits are commonly set in the 250 500 psi range to allow for uncertainties and load transients. Failure loads in test are usually in the 1200–1500 psi range. At operating speeds, these failure loads will commonly involve a peak film pressure of two to three times this nominal loading. The bearing babbitt will then flow when local oil-film pressure reaches the babbitt yield strength as temperature rises to the 275–300 °F range. At very low speeds, babbitt wear becomes a limiting factor when

Applied Tribology: Bearing Design and Lubrication, Third Edition. Michael M. Khonsari and E. Richard Booser.
© 2017 John Wiley & Sons Ltd. Published 2017 by John Wiley & Sons Ltd.
Companion Website: www.wiley.com/go/Khonsari/Applied_Tribology_Bearing_Design_and_Lubrication_3rd_Edition

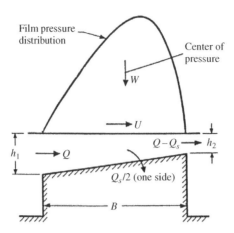

Figure 7.1   Fixed-pad slider bearing. *Source*: Raimondi and Szeri, 1984. Reproduced with permission of Taylor & Francis.

overall oil film thickness drops to less than 3–10 times the nominal surface roughness of the mating runner.

Tapered-land bearings provide reliable, compact designs for a large variety of midsize to large, high-speed machines such as turbines, compressors, and pumps. A flat land extending for about 10–20% of the circumference at the trailing edge of each segment is commonly included for higher load capacity and to minimize wear during starting, stopping, and at low speeds. Because the operation of tapered-land bearings is sensitive to load, speed, and lubricant viscosity, this bearing type is commonly designed to match the rated operating conditions of specific machines.

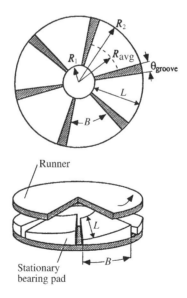

Figure 7.2   Configuration of a fixed-pad thrust bearing. *Source*: Khonsari and Jang, 1997. Reproduced with permission of Taylor & Francis.

Table 7.1 Common thrust bearings and their application range

	Type	OD (in.)	Unit load (psi)
Taper		2–35	150–400
Tilt pad		4–120	250–700
Spring support		50–120	350–700
Step		0.5–10	100–300
Flat		0.5–20	20–100
Hydrostatic		3–50	500–3000

Pivoted-pad thrust bearings are being used more frequently in turbines, compressors, pumps, and marine drives in much the same general size and load range as the tapered-land design. Each of the 3–10 or more pads in the bearing is commonly supported on a central pivot, which leaves it free to adjust to form a nearly optimum oil wedge for supporting high loads even with widely varying speeds and lubricants and both directions of rotation. Use of leveling links behind the pivots is common to equalize the load carried by individual pads and to accommodate some misalignment. Off-the-shelf units are available for shaft diameters ranging from about 5 to 30 cm (2–12 in.); custom designs are adaptable to a wide range of machinery for outside diameters ranging up to 4 m (170 in.).

Recent modifications for increasing the load capacity, lowering the temperature, and lowering the power loss include offsetting the pivots to 60–65% beyond the leading edge of the pad, replacing steel with copper for backing of the babbitt bearing material to give lower peak surface temperature, and using nonflooded lubrication to minimize parasitic power loss at high surface speeds.

Springs or other flexible supports for the thrust segments are used in some of the largest thrust bearings carrying millions of pounds of thrust. This flexible mounting avoids the high contact stresses imposed by loading individual pivots. Rubber backing for each thrust pad can be used to provide this flexible support in smaller bearings and where axial space is at a premium.

Step bearings provide a simple design for smaller bearings. With the use of a coined or etched step, they are well suited to mass production for small bearings and thrust washers for use with low-viscosity fluids such as water, gasoline, and solvents. Minimum film thickness is so small in these cases that it becomes impractical to machine features such as tapers or pivots. Step height must be small, with the same order of thickness as the minimum film thickness for optimum load capacity, yet large enough to allow for some wear. Erosion of the step by contaminants is sometimes a problem.

Flat-land bearings, with their simplicity and low cost, are the first choice for light loads such as those involved in simple positioning of a rotor in electric motors, appliances, crankshafts, and other machinery. They will carry only 10–20% of the load of the other thrust bearing types because flat parallel surfaces do not directly provide the pumping action needed to build oil-film pressure in a converging wedge-shaped clearance. Instead, a flat-land bearing relies on thermal expansion of the oil and warping of the bearing material by heating from the passing oil film. Minor misalignment and localized surface irregularities may also help to generate oil-film pressure. Introduction of radial distributing grooves in an otherwise flat thrust face for improved oil feed and cooling enables loading to the high end of the range indicated in Table 7.1.

Hydrostatic thrust bearings employ an external pump to provide bearing film pressure when sufficient load support cannot be generated by simple internal hydrodynamic pumping action. This type is considered primarily for low-viscosity fluids, very low speeds, very high loads, starting under high load, or limited space.

A very compact thrust bearing can involve, for instance, feed of high-pressure oil into a single pocket at the end of a rotor. Larger bearings usually use three or more pressurized pockets. Hydraulic flow resistors in the supply line to each pocket, or equal flow to each pocket from ganged gear pumps, then provides the asymmetric pocket pressures needed to support an off-center load. Unit loading on the bearing is usually limited to about 0.5 times (0.75 with fixed flow systems) the fluid feed pressure of up to about 5000 psi from the external pump. Analytical details for hydrostatic bearings are covered in Chapter 10.

7.2 Design Factors

The inside diameter of a thrust bearing is selected to be sufficiently larger than the adjacent shaft to allow for assembly, to accommodate a blend radius between the shaft and thrust runner surface, and to provide for any oil flow to be supplied to the thrust bearing at its inside diameter. This clearance typically ranges from about $\frac{1}{8}$ in. for a 2 in. shaft to $\frac{1}{2}$ in. for a 10 in. shaft.

Referring to Figure 7.2, the bearing outside radius R_2 is then selected to provide a bearing area sufficient to support a total thrust load W (N or lb) with an average loading $P(N/m^2$ or psi):

$$R_2 = \left(R_1^2 + \frac{W}{\pi k_g P_{avg}} \right)^{0.5} \tag{7.1}$$

where k_g is the fraction of area between R_1 and R_2 not used for oil feed passages, typically 0.80–0.85. For initial considerations, the value of P might be taken as the mean of the high and low values from Table 7.1. Final design details would be adjusted from the results of performance analyses such as those which follow.

The number of oil feed passages in Figure 7.2 is commonly set so that the individual bearing sectors (pads) are 'square,' with circumferential breadth B at their mean diameter approximately equal to their radial length L. While experience has demonstrated that square pads usually provide an optimum oil film, there are deviations. With very large bearing diameters, for instance, square sectors may become so long in their circumferential dimension B that the oil film overheats before reaching the next oil feed groove. On the other hand, keeping $B = L$ in sectors of small radial length L may introduce an impractical number of radial oil grooves. Even numbers of sectors simplify machining of oil feed grooves and other features, although odd numbers of pads are readily accommodated in pivoted-pad thrust bearings.

The width of radial oil feed grooves, or of gaps between individual pads, is commonly about 15–20% of circumferential breadth B. Small grooves are often machined about as deep as they are wide. In large bearings, groove depth is selected to keep radial oil flow velocity in the groove below about 1.5–3 m/s (5–10 ft/s). Oil feed may be either through an inlet hole in the bottom of each radial groove or from an opening at the thrust bearing inside diameter to feed oil radially out through the groove. The outer end of each oil feed groove is commonly dammed to maintain a full oil supply. Modest flow of 10–20% of total oil feed should then exit through a bleed notch in this dam to minimize accumulation of air and dirt in the groove. With tilting-pad thrust bearings, these oil feed grooves are replaced by similar circumferential gaps between each pivoted (or spring-supported) pad.

7.3 Performance Analysis

Behavior of the oil film on each sector of a thrust bearing follows the same hydrodynamic principles as in journal bearings. The Reynolds equation, as developed in Cartesian coordinates for journal bearings in Chapter 6, Section 6.1, takes the following form for cylindrical coordinates:

$$\frac{\partial}{\partial r}\left(\frac{\rho r h^3}{12\,\mu}\frac{\partial P}{\partial r}\right) + \frac{1}{r}\frac{\partial}{\partial \theta}\left(\frac{\rho h^3}{12\mu}\frac{\partial P}{\partial \theta}\right) = U\frac{\partial\,(\rho h)}{\partial \theta} - 12 r \rho V_s \qquad (7.2)$$

As with journal bearings, the desired method of analysis is numerical analysis using finite difference techniques for a uniform grid, with spacing of dr radially and $r d\theta$ in the direction of motion circumferentially. For fixed geometry sectors such as tapered lands, a minimum film thickness is first selected. This enables establishment of the clearance h at the center of each grid and its derivatives in the radial and circumferential directions. After the pressure distribution is obtained by a numerical procedure, as described in Chapter 6, total thrust load, power loss, and flow characteristics follow. To obtain these operating characteristics for a given applied load, this analytical routine is repeated while adjusting the assumed minimum film thickness in a stepwise procedure until the calculated thrust load matches the given applied load.

While Equation (7.2) implies constant viscosity (constant temperature), finite difference computer analyses commonly apply the following energy relation for each grid segment to obtain the temperature distribution. This is given from Equation (5.19) in Cartesian coordinates to correspond to the solutions which follow.

$$\rho C_p\left(u\frac{\partial T}{\partial x} + v\frac{\partial T}{\partial y} + w\frac{\partial T}{\partial z}\right) = k\left(\frac{\partial^2 T}{\partial x^2} + \frac{\partial^2 T}{\partial y^2} + \frac{\partial^2 T}{\partial z^2}\right) + \mu\left(\frac{\partial u}{\partial z}\right)^2 \qquad (7.3)$$

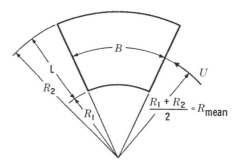

Figure 7.3   Sector for a fixed-land (or pivoted-pad) thrust bearing. *Source*: Raimondi and Szeri, 1984. Reproduced with permission of Taylor & Francis.

The usual procedure initially assumes a uniform temperature (and viscosity) over all grid points to obtain from the Reynolds equation a preliminary map of local grid pressures. Equation (7.3) then gives a first approximation of grid temperatures. Using a new distribution of viscosities at these temperatures then allows solution of the Reynolds equation with an updated pressure field. This iterative procedure typically converges within a few cycles.

Constant viscosity approximations are provided in the following sections while relating a rectangular bearing sector with dimensions $B \times L$ to the circular configuration of Figure 7.3. This rectangular representation allows more ready evaluation of a range of variables and gives results that are quite similar to those from more accurate polar analysis using the variable viscosity in Equation (7.2). After preliminary performance characteristics are obtained by the following procedures for general design layout and performance analysis, more exact evaluations should be made for critical bearings on a case-by-case basis, with a comprehensive computer code for cylindrical coordinates incorporating such factors as the Reynolds equation, energy relations for variable film tempeature, and viscosity. Significant thermal and mechanical deflections may require consideration of thermomechanical expansion. Further complications include consideration of hot oil carryover between pads and parasitic churning losses in the oil.

7.4 Tapered-Land Thrust Bearings

Constant-viscosity Reynolds equation solutions for tapered-surface rectangular bearing segments are provided as dimensionless performance coefficients in Table 7.2. These solutions are used to relate the performance of a rectangular bearing segment to the circular configuration of Figure 7.3 (circumferential breadth B, radial length L). The table covers a wide range of aspect ratios L/B from infinitely long radially down to 0.25 for a circumferentially broad sector, along with a range of inlet-to-outlet film thickness ratios h_1/h_2 of 1.2–10. These aspect and film thickness ratios suffice to describe a wide range of pad geometries, whereas a circumferential sector would also require the subtended angle for its description. This rectangular approximation and the assumption of constant viscosity permit analyses of a wide range of geometries and operating variables from the relatively compact tabulation of performance coefficients in Table 7.2.

Table 7.2 Performance coefficients for tapered-land thrust bearings[a]

L/B	0.25	0.5	0.75	1.0	1.5	2.0	∞
			$h_1/h_2 = 1.2$				
K_h	0.064	0.115	0.153	0.180	0.209	0.225	0.266
K_f	0.912	0.913	0.914	0.915	0.916	0.917	0.919
K_q	0.593	0.586	0.579	0.574	0.567	0.562	0.549
K_s	0.087	0.074	0.061	0.050	0.034	0.026	0.000
			$h_1/h_2 = 1.5$				
K_h	0.084	0.151	0.200	0.234	0.275	0.296	0.351
K_f	0.813	0.817	0.821	0.825	0.830	0.833	0.842
K_q	0.733	0.714	0.696	0.680	0.659	0.647	0.610
K_s	0.215	0.182	0.150	0.124	0.087	0.065	0.000
			$h_1/h_2 = 2$				
K_h	0.096	0.170	0.223	0.261	0.305	0.328	0.387
K_f	0.698	0.708	0.718	0.727	0.739	0.747	0.768
K_q	0.964	0.924	0.884	0.849	0.798	0.772	0.690
K_s	0.430	0.362	0.298	0.245	0.172	0.129	0.000
			$h_1/h_2 = 3$				
K_h	0.100	0.173	0.225	0.261	0.304	0.326	0.384
K_f	0.559	0.579	0.600	0.617	0.641	0.655	0.696
K_q	1.426	1.335	1.236	1.148	1.024	0.951	0.738
K_s	0.873	0.744	0.596	0.510	0.359	0.270	0.000
			$h_1/h_2 = 4$				
K_h	0.098	0.165	0.212	0.244	0.282	0.302	0.352
K_f	0.476	0.503	0.529	0.551	0.581	0.598	0.647
K_q	1.888	1.745	1.586	1.444	1.242	1.122	0.779
K_s	1.316	1.128	0.938	0.774	0.544	0.408	0.000
			$h_1/h_2 = 6$				
K_h	0.091	0.148	0.186	0.211	0.241	0.256	0.294
K_f	0.379	0.412	0.448	0.469	0.502	0.521	0.574
K_q	2.811	2.560	2.273	2.013	1.646	1.431	0.818
K_s	2.208	1.905	1.587	1.306	0.916	0.686	0.000
			$h_1/h_2 = 10$				
K_h	0.079	0.121	0.148	0.165	0.185	0.195	0.221
K_f	0.283	0.321	0.353	0.377	0.408	0.426	0.474
K_q	4.657	4.182	3.624	3.118	2.412	2.001	0.834
K_s	4.013	3.499	2.907	2.390	1.670	1.250	0.000

[a] For use in Equations (7.5)–(7.8).
Source: Khonsari and Jang, 1997. Reproduced with permission of John Wiley & Sons.

Common proportions of tapered-land segments are commonly 'square', with $L/B = 1$, and a typical taper is given by the following for oil-film bearings (Wilcock and Booser, 1957):

$$(h_1 - h_2) \approx 0.003B^{0.5} \tag{7.4a}$$

where the length B is in inches. Alternatively, when B is specified in meters, the above formula reads

$$(h_1 - h_2) \approx 0.00048B^{0.5} \qquad (7.4b)$$

Employing the nomenclature used in Figures 7.1 and 7.2, the following relations give minimum film thickness h_2, frictional power loss E_P per sector, flow into sector Q_1, total leakage flow from the two sides of segment Q_s, and the overall temperature rise ΔT for the total oil feed Q (per sector) as it is heated by frictional energy E_P in passing to the drain.

$$\text{Minimum film:} \qquad h_2 = K_h \left(\frac{\mu U B}{P_{avg}} \right)^{0.5} \qquad (7.5)$$

$$\text{Power loss:} \qquad E_P = \frac{K_f \mu U^2 B L}{h_2} \qquad (7.6)$$

$$\text{Inlet flow rate:} \qquad Q_1 = K_q U L h_2 \qquad (7.7)$$

$$\text{Side leakage flow rate:} \qquad Q_s = K_s U L h_2 \qquad (7.8)$$

$$\text{Temperature rise:} \qquad \Delta T = \frac{E_P}{Q_1 \rho C_P} = \left[\frac{K_f}{K_q K_h^2 \rho C_P} \right] P_{avg} \qquad (7.9)$$

where

B = circumferential breath of sector at mean radius (m) (in.)
C_P = oil specific heat (J/(kg °C) (in. lb/(lb °F)); typically 1979 J/kg °C or 4400 in. lb/°F lb (= 0.47 Btu/°F lb)
K_h, K_f, K_q, K_s = dimensionless coefficients
L = radial length of sector, (m) (in.)
P_{avg} = unit load on projected area of sector, W/BL (N/m^2) (lb/in.^{2})
U = surface velocity at mean diameter of sector (m/s) (in./s)
W = load on a sector (N) (lb)
h_1, h_2 = leading edge and trailing edge film thicknesses (m) (in.)
ρ = oil density (N/m) (lb/in^3.); typically 850 kg/m^3 (0.0307 lb/in.^{3})
μ = oil viscosity at operating temperature (N. s/m^2) (lb. s/in.^{2}, also termed *reyns*)

K_h relates the minimum film thickness at the trailing edge to viscosity, surface velocity, and load. K_f reflects the ratio of total power loss to the viscous shear loss for film thickness h_2. Flow coefficients K_q and K_s relate calculated inlet and side flows to the runner surface velocity and to the discharge film thickness at the trailing edge of the sector.

While film thickness and power loss relations are generally adequately defined with the rectangular representation, oil flow relations are more approximate. Rather than relatively equal leakage flow over both the inner and outer diameters, as would be indicated for a rectangle, most side leakage flow for a pie-shaped sector (Figure 7.3) will be at the outside diameter with its longer periphery. Generally, the oil feed rate should be set to maintain a reasonably cool operation, with no more than a 40 °F rise in oil temperature from feed to drain. This feed flow will primarily involve the sum of two factors: (1) Q_s side leakage resulting from hydrodynamic action over each sector (with a tapered land geometry selected to provide 70% or more of the total flow) and (2) additional flow to be discharged from an orifice at the outboard end of each radial oil distributing groove, where a slight pressure of 5–10 psi is normally maintained to provide full flow at the leading edge of the adjacent sector.

Temperature Rise

Overall energy balance given by Equation (7.9) is usually adequate for establishing the typical bearing temperature to be expected and the corresponding appropriate oil viscosity for performance calculations. If the temperature rise pattern is required to evaluate limiting operating load and speed conditions, the temperature profile along the flow line where both pressure and temperature are at their maximum is given by the following energy balance (Booser *et al.*, 1970):

$$\frac{\mu U^2}{h} + \frac{h^3}{12\mu}\left(\frac{dP}{dx}\right)^2 = \rho C_p \frac{dT}{dx}\left(\frac{Uh}{2} - \frac{h^3}{12\mu}\frac{dP}{dx}\right) \tag{7.10}$$

Transverse pressure and temperature gradients vanish in this representation, which equates mechanical energy input on the left side of the equation to oil temperature rise on the right side.

While a closed-form solution of Equation (7.10) can be obtained for the temperature rise pattern along the tapered land by an approximate relation of viscosity to temperature or film thickness, a numerical solution can be made in a finite difference pattern. Maximum temperature rise in the oil film typically will be about twice the general temperature rise given by Equation (7.9).

Design Procedure

1. Determine the outside radius R_2 using Equation (7.1). Make allowance for radial clearance along the shaft (normally $\frac{3}{8}$ in. suffices), uncertainty about the total expected thrust load (see the unit load in Table 7.1), and the area reserved for oil feed passages (typically, $k_g = 0.85$, employing 15% of the area for oil feed).
2. Determine the mean radius $R = (R_1 + R_2)/2$, $L = R_2 - R_1$ and surface speed $U = 2\pi R\omega$.
3. Select the number of pads (start by considering a square pad, $B = L$).
4. Assume a minimum film thickness compatible with surface roughness.
5. Assume a temperature rise (typically $\Delta T = 30\,^\circ$F), and determine the corresponding oil viscosity at $T = T_i + \Delta T$ (use the temperature viscosity chart [Figure 2.3] in Chapter 2).
6. Estimate K_h from Table 7.2.
7. Recompute h_2 using Equation (7.5) and check its appropriateness.
8. Determine K_h, K_f, and K_s from Table 7.2.
9. Determine power loss E_p using Equation (7.6).
10. Determine flow rates using Equation (7.8).
11. Estimate the temperature rise using Equation (7.9).
12. Check the temperature rise with the initially assumed value. Iterate.

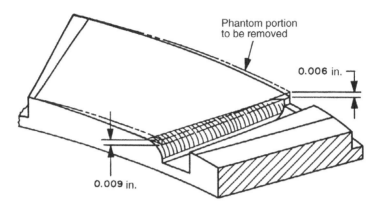

Figure 7.4   Thrust bearing segment with 10% flat and a compound taper. *Source*: Missana *et al.*, 1971. Reproduced with permission of Taylor & Francis.

Total oil feed to the bearing should satisfy two requirements: (1) match the hydrodynamic side leakage from each sector and (2) maintain cool operation, usually with no more than a 30–40 °F rise in oil temperature from feed to drain. If additional cooling oil is required, the volume should be set by the size of bleed orifices in dams at the outside end of each radial oil-distributing groove. Planning an excess feed of 10–20% should normally be arranged with appropriate orificing to maintain a pressure of 5–10 psi in each oil feed groove.

Several modifications of tapered land bearings are common. A flat plateau at the trailing edge, as in Figure 7.4, provides some thrust capacity for starting under a light load. Such a flat also increases film thickness (as in K_h) by about 13% for 10–20% flat, as indicated in Table 7.3, which gives performance coefficients for various percentages of flat. Use of an initial 10% flat is common to provide some tolerance for wear. Load capacity then actually increases as wear extends the flat area into the 20–30% range (Missana *et al.*, 1971).

A double taper, as shown in Figure 7.4, is often employed to give about a 50% greater ratio of h_1/h_2 at the inside radius of a land than at its outer periphery. Load capacity is adequately represented by analysis for the average taper, but needed oil feed is reduced by the smaller taper

Table 7.3   Performance coefficients with a flat incorporated at trailing edge of tapered land[a]

% Flat	Film thickness, K_h	Power loss, K_f	Fluid feed, K_q
0	0.260	0.727	0.849
10	0.283	0.767	0.827
20	0.292	0.801	0.805
30	0.294	0.834	0.782
40	0.289	0.866	0.755
50	0.277	0.893	0.725

[a] For $L/B = 1$, $h_2/h_1 = 2.0$. Dimensionless coefficients for Equations (7.5)–(7.7).
Source: Neale, 1970. Reproduced with permission of Sage Publications.

at the outside diameter (Wilcock and Booser, 1957). For high surface speeds in the largest bearing sizes where turbulence normally generates more than adequate oil-film thickness, a relatively large one-dimensional uniform taper induces high inlet oil flow for cooling. Replacing the usual steel backing plate with more thermally conductive copper will also commonly lower the maximum temperature of the babbitt surface layer about 5 °C. Analytical methods have been developed to evaluate the importance of high thermal conductivity in lowering the maximum temperature (Khonsari and Wang, 1991).

Analyses of sector thrust bearings using the performance coefficients for a fixed taper in Table 7.2 fortunately also hold *approximately* for most other bearing shapes covered in the following sections, with the same ratio of inlet to outlet film thickness (Neale, 1970; Fuller, 1984; Kennedy *et al.*, 1998).

Example 7.1 Design a thrust bearing for a steam turbine with a thrust load of 25 000 lb and speed of 3600 rpm. The rotor has a 5 in. diameter journal. ISO 32 viscosity oil is to be fed at $T_i = 120$ °F. Allowing a $\frac{3}{8}$ in. radial clearance along the shaft sets the thrust bearing bore at 5.75 in. ($R_1 = 2.875$ in.). Using a unit load of 300 psi from Table 7.1, which allows for some uncertainty in the expected total thrust load, outside radius of the bearing can be obtained from Equation (7.1).

Determine all the performance parameters, including flow rates, power loss, and temperature rise.

Using Equation (7.1) for the outside radius R_2:

$$R_2 = \left(2.875^2 + \frac{25\,000}{\pi \times 0.85 \times 300}\right)^{0.5} = 6.3\,\text{in.}$$

where it is assumed that 15% of the area is reserved for oil feed passages ($k_g = 0.85$). Radial length L for each sector then is $6.3 - 2.875 = 3.425$ in. At the mean radius $R_m = (2.875 + 6.3)/2 = 4.59$ in., the total circumference for all sectors is $2\pi R_m k_g = 24.5$ in. The number of square sectors with $B = L$ becomes $24.5/3.425 = 7.2$. Using seven lands, the circumferential length B for each pad is $24.5/7 = 3.5$ in. (six or eight lands might also be considered for simpler machining). The taper of each land ($h_2 - h_1$) from Equation (7.4) becomes $0.003 \times (3.5)^{0.5} = 0.006$ in. Surface velocity at the mean radius $U = 2\pi(4.59)(3600/60) = 1730\,\text{in/s}$.

Analysis requires the use of Equations (7.5)–(7.9), with a preliminary assumption of a bearing temperature (oil viscosity) and minimum film thickness. This trial film thickness is then adjusted in subsequent iterations until the calculated film thickness and bearing temperature match the assumed values.

Assuming the minimum film thickness $h_2 = 0.002$ in., $h_1/h_2 = (0.006 + 0.002)/0.002 = 4$. $L/B = 3.525/3.5 \approx 1$. Using $K_h = 0.244$ for these ratios from Table 7.2, the film thickness h_2 is first checked by Equation (7.5). Assuming a bearing temperature rise of 30 °F above the feed temperature sets the oil temperature at 150 °F, with a corresponding viscosity of 1.5×10^{-6} lb. s/in^2. from the temperature–viscosity chart (Figure 2.3) in Chapter 2.

$$h_2 = 0.244 \left[\frac{(1.5 \times 10^{-6})(1730)(3.5)}{300}\right]^{0.5} = 0.00134\,\text{in.}$$

Using a revised ratio of $h_1/h_2 = (0.006 + 0.00134)/0.00134 = 5.5$, interpolation for this film thickness ratio and $L/B = 1$ in Table 7.2 gives the following performance coefficients:

$$K_h = 0.219, \quad K_f = 0.490, \quad K_q = 1.871, \quad K_s = 1.173$$

Power loss E_P and oil side leakage Q_s are given as follows from Equations (7.5), (7.6), and (7.8):

$$E_P = 0.490 \frac{(1.5 \times 10^{-6})(1730)^2(3.5)(3.525)}{0.00134} = 20\,254 \text{ lb. in/s.}$$
$$Q_s = 1.173(1730 \times 3.525 \times 0.00134) = 9.59 \text{ in}^3/\text{s.}$$
$$Q_1 = K_q ULh_2 = 1.871 \times 1730 \times 5.525 \times 0.00126 = 15.285 \text{ in}^3/\text{s.}$$

The temperature rise given by Equation (7.9) assumes that the total frictional power loss E_P goes into heating the total oil flow exiting from each sector. Using the oil density of $0.0307/\text{in.}^3$ and a specific heat of $4400 \text{ in. lb/lbm} \,^\circ\text{F}$,

$$\Delta T = \frac{E_P}{(Q_1 \rho C_P)} = \frac{20\,254}{(15.285 \times 0.0307 \times 4400)} = 9.8\,^\circ\text{F}$$

Since this temperature rise does not match the $30\,^\circ\text{F}$ initially assumed, repeating the calculation routine with the new minimum film thickness of 0.00134 in., with a new assumed intermediate temperature rise of $25\,^\circ\text{F}$, and with viscosity at the new bearing temperature of $120 + 25 = 145\,^\circ\text{F}$ being 1.65×10^{-6} lb s/in.^{2} gives the following:

$$h_2 = 0.219 \left[\frac{(1.65 \times 10^{-6})(1730)(3.5)}{300} \right]^{0.5} = 0.00126 \text{ in.}$$

which gives a revised $h_1/h_2 = (0.006 + 0.00126)/0.00126 = 5.8$. For this ratio and $L/B = 1$, Table 7.2 provides by interpolation:

$$K_h = 0.214, \quad K_f = 0.477, \quad K_q = 1.94, \quad K_s = 1.253$$

Repeating the use of Equations (7.6), (7.8), and (7.9) with these revised values gives power loss per sector $E_P = 23\,065$ lb in./s $(23\,065/6600 = 3.5 \text{ hp})$, and $Q_1 = 14.9 \text{ in}^3/\text{s}$ $(6.64 \times 60/252 = 1.6 \text{ gpm})$. This gives a temperature rise of $11.9\,^\circ\text{F}$ which is only slightly higher than the last iteration. One more iteration can be done to bring the temperature rise results closer together. It should be remembered that the oil leakage and power loss should be multiplied by the number of sectors, seven, to obtain the results for the complete bearing assembly.

7.5 Pivoted-Pad Thrust Bearings

Performance of a pivoted pad is calculated on the same basis as for a fixed, tapered land with the same slope. As with fixed-pad bearings, the bearing is divided into a set of pie-shaped segments, with the circumferential breadth B of each sector commonly set approximately equal

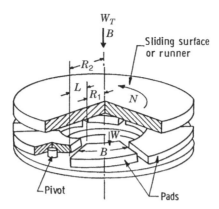

Figure 7.5 Pivoted-pad thrust bearing. *Source*: Raimondi and Szeri, 1984. Reproduced with permission of Taylor & Francis.

to its radial length L. Many off-the-shelf bearings consist of six pads, with their outer radius R_2 equal to twice their inner radius R_1.

For a pivoted-pad thrust bearing such as that illustrated in Figure 7.5, pivot location involves equilibrium of moments acting about the pivot point. Two forces participate in this moment balance: the resultant from summing film pressures over the bearing area and the reaction of oil-film forces normal to the pad surface. Once equilibrium is established during operation, a change in speed or load will cause a shift in pad inclination and film thickness to reestablish the moment balance about the pad pivot. Pivot location is usually set 55–58% radially out from R_1 to provide no radial tilt (Raimondi and Szeri, 1984). Only a modest change in pad temperature was observed when the pivot was moved 4% radially outward from this position for zero radial tilt (Gardner, 1975).

When varying the pivot location in the circumferential direction, Figure 7.6 indicates the variation in the h_1/h_2 ratio to be expected with an assumption of constant viscosity and no deflection of a flat pad bearing surface (Hamrock, 1994). When combined with other performance coefficients such as those in Table 7.1, this leads to an optimum pivot location 58% circumferentially beyond the inlet edge of the pad for maximum load capacity (and film thickness) and a bit farther along for minimum power loss. Unfortunately, the utility of analyses assuming constant viscosity and rigid, flat pads is somewhat limited. While these assumptions, for instance, indicate no load capacity for a central (50%) pivot location, the majority of thrust bearings that use conventional lubricating oils for industrial equipment employ a central pivot and still provide nearly optimum load capacities for operation in the load range indicated in Table 7.1. The commonly used central pivot also offers design and manufacturing simplicity and enables operation in both directions of rotation.

To accommodate the highest possible loads, location of the pivot in the 70–75% range circumferentially was found to give the greatest minimum oil-film thickness and lowest bearing metal temperature (Gardner, 1988). While reasonable performance was obtained even at an 80% location, oil-film pressure then rose rapidly toward the trailing edge of the pad. A secondary set of leveling plates is commonly used to equalize the load on individual thrust pads (Figure 7.7).

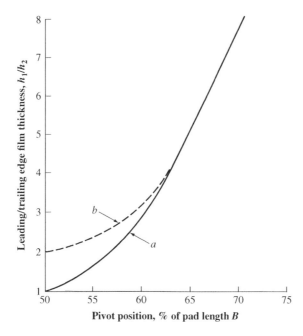

Figure 7.6   Variation in film thickness with circumferential pivot location. Solid line (a) is based on the rigid flat analysis and the dashed line (b) is based on experience indicating thermal and pressure-induced deflection with normal oil lubrication.

For heavy loads at high surface speeds, replacement of the usual steel pads with more thermally conductive copper helps to lower the peak temperature of the babbitt surface layer, typically by some 4–10 °C. Other steps helpful in reducing bearing temperature include (1) introducing the cool feed oil directly to the leading edge of each pad while (2) draining the overall housing so that it is free of warmer bulk oil, which then also minimizes parasitic oil churning losses at high surface speeds (see Section 7.9).

To minimize analytical deficiencies, pivoted-pad thrust bearing computer codes for demanding applications should incorporate the following: (1) energy relations for obtaining the oil-film

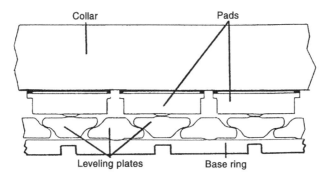

Figure 7.7   Leveling plates for equalizing the load on pivoted pads. *Source*: Wilcock and Booser, 1957. Reproduced with permission of McGraw-Hill.

temperature and viscosity distribution over the pad area, (2) bending of the pad by pressures in the oil film, (3) pad temperatures leading to thermal bending from the temperature difference between the hotter oil film and cooler oil at the back of the pad (Vohr, 1981), (4) hot oil carry-over from one pad to the next, (5) possible oil-film turbulence, and (6) parasitic power losses (Booser and Missana, 1990). Pad deflections with many larger bearings (giving an umbrella shape with its peak above the pivot) are similar to, or greater in magnitude than, the minimum thickness of the oil film.

With low-viscosity fluids such as water, liquid metals, and gases, film thickness is usually quite small and the lack of thermal and pressure-induced distortions renders a central pivot impractical. In such applications, the pad is commonly produced with a small spherical or cylindrical crown. A central pivot combined with a spherical crown height of 0.5–2 times the minimum film thickness will result in a load capacity nearly equivalent to that of a flat pad with its optimum pivot location 58% along the pad from its leading edge (Raimondi, 1960; Raimondi and Szeri, 1984).

For an estimate of performance in the following example by an approximation which has matched many test and field observations, a centrally pivoted pad with oil lubrication is considered equivalent to a rigid flat pad with the ratio of inlet to outlet film thickness $h_1/h_2 = 2.0$.

Example 7.2   Design a pivoted-pad thrust bearing for the steam turbine in Example 7.1. Applying the same requirements given in Example 7.1 for a tapered land bearing, select a design for a 25 000 lb thrust load at 3600 rpm to be fed with ISO 32 oil at 120 °F.

This leads again to seven nearly square pads ($L/B = 3.425/3.5 \approx 1$) in a pivoted-pad bearing. Taking $h_1/h_2 = 2$, which represents general operating experience for centrally pivoted pads with oil lubrication, Table 7.2 gives the following performance coefficients:

$$K_h = 0.261, \qquad K_f = 0.727, \qquad K_q = 0.849, \qquad K_s = 0.245$$

The temperature rise given by Equation (7.9) assumes that the total frictional power loss E_p goes into heating the total oil inlet flow Q_1 passing over each pad:

$$\Delta T = \frac{K_f}{(K_q K_h^2 \rho C_p)} P$$

$$= \left\{ \frac{0.727}{[(0.849)(0.261)^2(0.0307)(4400)]} \right\} \times 300 = 28\,°F$$

With viscosity of VG 32 oil at $T = 120 + 28 = 148\,°F$ being 1.6×10^{-6} reyns, minimum film thickness h_2 from Equation (7.5), power loss E_p from Equation (7.6), and oil feed Q_1 from Equation (7.7) become

$$h_2 = 0.261 \left[\frac{(1.6 \times 10^{-6})(1730)(3.5)}{300} \right]^{0.5} = 0.0015\,\text{in.}$$

$$E_p = \frac{0.727(1.6 \times 10^{-6})(1730)^2(3.5)(3.525)}{0.0013} = 28\,634\,\text{lb in./s} \quad (28\,634/6600 = 4.3\,\text{hp})$$

$$Q_1 = 0.849(1730)(3.525)(0.0013) = 7.77\,\text{in.}^3/\text{s} \quad (7.77 \times 60/252 = 1.85\,\text{gpm})$$

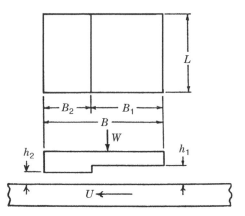

Figure 7.8   Simple step thrust bearing. *Source*: Wilcock and Booser, 1957. Reproduced with permission of McGraw-Hill.

7.6 Step Thrust Bearings

A step in the thrust bearing surface is fully as efficient as a tapered land in providing hydro-dynamic action for generating oil-film pressure. The step design of Figure 7.8 has found its greatest use in small bearings where coining, etching, or stamping of a simple thrust washer or machine surface can form the optimum small step height. This design has failed to find general use in large machinery, however, because of difficulty with dirt accumulation at the step and wear effects leading to diminishing step effectiveness during operation.

With zero gauge pressure on all exterior boundaries, i.e. atmospheric pressure, a solution of the full Reynolds equation is possible when one assumes that pressure on the common boundary is given by a Fourier sine series (Hamrock, 1994; see also Section 6.7). For maximum load capacity, the step location varies somewhat with both pad length-to-width ratio and ratio of step height to minimum film thickness. The optimum step is generally located at 50–70% of the distance from the inlet edge of the pad. Archibald (1950), applying hyperbolic functions to obtain solutions which included side leakage, found the following performance characteristics for a square pad ($L/B = 1$) with its optimum step location at 55% of pad length B from the leading edge, and step height $s = 0.7$ times the trailing edge film thickness h_2:

$$h_2 = 0.27 \left(\frac{\mu U B}{P} \right)^{0.5} \tag{7.11}$$

$$E_p = 0.84 \left(\frac{\mu U^2 B}{h_2} \right) \tag{7.12}$$

$$Q = 0.77 \, U L h_2 \tag{7.13}$$

These relations correspond to dimensionless coefficients $K_h = 0.27$, $K_f = 0.84$, and $K_q = 0.77$ in Equations (7.5)–(7.7); these numerical values are also quite similar to those in Table 7.2 for a tapered-land pad with the same $L/B = 1$ and $h_1/h_2 = 1.7$. This similarity suggests that Table 7.2 can be used to estimate step pad performance for various geometries by applying the same L/B and inlet-to-outlet film thickness ratio as for a tapered-land thrust pad.

A step thrust bearing, as shown in Figure 7.8, has considerable side leakage over inlet section B_1. If shrouds are introduced along the sides of this section, load capacity can be increased up to about 70%, with a corresponding increase of 30% in minimum film thickness and its coefficient K_h (Hamrock, 1994).

Example 7.3   Calculate the performance of a simple step thrust bearing for a gasoline pump which is to carry a thrust load of 25 lb at 10 000 rpm. The inner bearing diameter is $\frac{5}{8}$ in., and the outside diameter is 1 in., to provide a pad radial length $L = \frac{3}{16}$ in.

Circumference at the pitch line is $\pi\, 13/16 = 2.553$ in., which will accommodate 10 sectors with $B = \frac{3}{16}$ in. at the pitch line, with feed grooves approximately $\frac{1}{16}$ in. wide. Unit load $P = W/BL = 25(10 \times 3/16 \times 3/16) = 71$ psi. Viscosity of gasoline at the 140 °F operating temperature is 0.5 cP (7.25×10^{-8} lb s/in.^{2}). Pitch line velocity $U = \pi \times 13/16 \times 10\,000/60 = 425$ in/s. Distance from the leading edge of each sector to the step at the pitch line is $0.55 \times 3/16 = 0.103$ in.

Film thickness from Equation (7.11) is

$$h_2 = 0.27[(7.25 \times 10^{-8})(425)(3/16)/71]^{0.5} = 0.000077\,\text{in.}$$

This small film thickness and the marginal lubrication provided by gasoline would call for a custom selection of a bearing material such as carbon-graphite running against a hardened thrust face having a surface finish one-tenth or less than the minimum film thickness (8 μin. or better). If space permits, the bearing surface area should be doubled.

Power loss from Equation (7.12) is

$$E_P = 0.84[(7.25 \times 10^{-8})(425)^2(3/16)(3/16)/0.000077] = 5.02\,\text{in. lb/s}$$

Fluid flow into the leading edge at each pad from Equation (7.13) is

$$Q_1 = 0.77(3/16)(0.000077)(425) = 0.00477\,\text{in.}^3/\text{s}$$

Since power loss and flow are calculated per pad, they would be multiplied by 10 for the 10 sectors of the bearing. The temperature rise from Equation (7.9), assuming that the density of gasoline is 0.028 lb/in.^{3} and its specific heat as 4700 lb in./lb °F:

$$\Delta T = \frac{E_P}{(Q_1 \rho C_P)} = \frac{5.02}{(0.0047 \times 0.028 \times 4700)} = 8\,°\text{F}$$

7.7 Spring-Mounted Thrust Bearings

Individual pads for this type of bearing are geometrically similar to those for tapered-land or pivoted-pad types, but the pads are flexibly supported on a nest of numerous precompressed springs. These spring-mounted bearings are normally applied at speeds of 50–700 rpm at loads of 400–500 psi in large vertical machines such as hydroelectric generators at power dams (Wilcock and Booser, 1957). Use of either tin or lead babbitt is common, with ISO VG 68 oil as the lubricant.

Very accurate construction is required for the largest bearing sizes to accommodate thin oil films during starting and stopping and to provide appropriate oil-film thickness during normal running. Fine surface finish of the runner by honing to 3 μin. is common practice. Thrust runner stiffness and machining should provide axial runout at the outside diameter of less than about 0.001 in.

Complete analysis of hydrodynamic behavior involves complex computer programming. Solution of the Reynolds equation, energy relations for oil-film temperature and viscosity, and thermal and elastic deflections are obtained in much the same manner as with pivoted-pad bearings. An additional step is needed to match thermal and mechanical deflections induced by oil-film pressure and temperature distribution with compression of the supporting springs (Vohr, 1981). Successful performance has been obtained with springs selected for 0.018 in. compression per 100 psi of oil-film pressure. Springs are then precompressed to an equivalent bearing loading of about 250 psi to reduce the total deflection involved. For severe loading conditions, spring stiffness has been varied through the nest of springs to aid in oil-film formation. Spring stiffness at the leading edge can be reduced somewhat, for instance to induce a thicker inlet oil film. Hydrostatic oil lift pockets are commonly provided in each bearing pad to ease breakaway on starting.

Despite the complexity of a comprehensive analysis, a useful approximation of the overall performance of a spring-supported bearing is commonly provided by simply assuming that the inlet film thickness to each pad is twice the trailing edge film thickness, with $h_1/h_2 = 2.0$.

While individual pads are normally made 'square' ($L/B = 1$) for smaller size ranges, the largest-diameter bearings are elongated, with their circumferential breadth B shorter than their radial length L. The shortened path in the tangential direction of motion avoids overheating of the oil film and the babbitt bearing surface; this temperature rise can be approximated from Equation (7.10). Water cooling of pads is sometimes used with either pivoted-pad or spring support for the largest hydroelectric generators.

For small to medium-sized machinery, rubber or other elastic backing is useful for enhanced performance where a simple flat-land thrust bearing would otherwise be used.

7.8 Flat-Land Thrust Bearings

This type of thrust bearing is used in a wide variety of applications for mild conditions at light loads, as in axial positioning of a rotor or to accommodate occasional reverse thrust loads. Relatively small, ungrooved flat-land bearings are commonly used for loads ranging up to 20 to 35 psi. This can be extended to the 100 psi range with radial grooves to distribute oil feed over four, six, eight, or so sectors sized in the manner described in Section 7.2.

Under some low-speed and high-load conditions, a flat-plate thrust bearing operates in a region of boundary lubrication without a fully developed oil film. Under such conditions, some metal-to-metal contact occurs, and suitable oil additives and bearing material must be selected to minimize wear and heat generation.

Unfortunately, no simple analytical procedure has yet been developed for predicting performance, since operation involves the motion between two parallel surfaces without a distinct hydrodynamic wedge action (Booser and Khonsari, 2010; Yang and Kaneta, 2013). Test observations of power loss (or friction coefficient) can be used in Equation (7.6) to estimate film thickness and then fluid flow to a sector. An approximate estimation of performance can also be obtained from extrapolation of the performance coefficients in Table 7.2 for the ratio of leading edge to trailing edge film thickness $h_1/h_2 = 1.2$ to give the following:

$$K_h = 0.4 \text{ to } 0.8 \times \text{value for } h_1/h_2 = 1.2; \qquad K_f \approx 1.0; \qquad K_q \approx 0.5 \qquad (7.14)$$

From experimental observations, K_h for $h_1/h_2 = 1.2$ should be used with a multiplier of about 0.4 for common applications of small flat-land bearings at low speeds and low loads. A higher multiplier of about 0.8 appears appropriate for longer circumferential path B, higher speeds, and higher load conditions, where greater film heating creates more of a thermal wedge by thermal expansion of both the oil and bearing material toward the trailing edge of each sector. Mechanical distortions, surface irregularities, and misalignment between the thrust runner and bearing surface may all contribute to the load capacity. Use of a higher-viscosity grade of oil and creation of slight waviness in an otherwise flat thrust surface can contribute significantly to load capacity.

Example 7.4   Calculate the performance to be expected at 160 °F with ISO 32 oil (1.3×10^{-6} lb s/in.^{2} viscosity) with a flat-land thrust bearing having an inside diameter of 1.75 and an outside diameter of 3.00 in. carrying a 150 lb thrust load in an 1800 rpm electric motor.

Radial length of each sector $L = (3.00 - 1.75)/2 = 0.625$ in. Available circumference at the mean diameter, with 15% used for radial oil grooves, becomes $0.85\pi \times 2.375 = 6.34$. For square lands ($L/B = 1.0$), circumferential breadth B of each land becomes $6.34/0.625 = 10.1$. Since 10 lands unduly break up the thrust area, six radial oil grooves would be more appropriate, giving $B = 6.34/6 = 1.06$ in. with $L/B = 0.625/1.06 = 0.59$.

Following the relation of Table 7.2 with a film thickness multiplier of 0.6, $K_h = 0.6(0.129) = 0.077$, $K_f = 1.0$, and $K_q = 0.5$. Equations (7.5)–(7.9) then provide the following estimates of performance for each of the six sectors:

$$\text{Film thickness: } h_2 = K_h \left(\frac{\mu U B}{P} \right)^{0.5} = 0.077 \left[\frac{(1.3 \times 10^{-6})(224)(1.06)}{38} \right]^{0.5} = 0.00022 \text{ in.}$$

$$\text{Power loss: } E_P = \frac{K_f \, \mu U^2 \, BL}{h} = \frac{1.0(1.3 \times 10^{-6})(224)^2(1.06)(0.625)}{0.00022} = 196 \text{ lb in./s}$$

$$\text{Oil feed: } Q_1 = K_q U L h = 0.5(224)(0.625)(0.00022) = 0.0135 \text{ in.}^3/s$$

$$\text{Temperature rise: } \Delta T = \frac{E_P}{(Q_1 \, \rho C_P)} = \frac{196}{(0.0135 \times 0.0307 \times 4400)} = 107 \, °F$$

This high temperature rise should be reduced by steps such as increasing the number of lands (smaller B) and doubling the oil feed rate for added cooling.

Table 7.4 compares the performance coefficients of each of the bearing types covered. In view of the differences in geometry, the similarity of these coefficients is surprising. It might well be concluded that selection of the thrust bearing type to be used should usually be based on the cost, complexity, size, and other operating requirements rather than primarily on the calculated performance factors covered here.

It is interesting to note that early analytical solutions for finite bearings by Archibald showed that, compared to an inclined profile, a straight step-film profile could not offer a significant improvement in terms of load-carrying capacity (Archibald, 1950). Instead of a constant width,

Table 7.4 Comparison of dimensionless performance coefficients for various thrust bearing types[a]

Thrust bearing type	Film thickness, K_h	Power loss, K_f	Oil feed, K_q
Tapered land, 20% flat[b]	0.29	0.80	0.81
Pivoted pad			
Central pivot[c]	0.26	0.73	0.85
65% pivot[d]	0.23	0.51	1.66
Spring supported[c]	0.26	0.73	0.85
Optimum step	0.27	0.84	0.77
Flat land	~0.07	~1.0	~0.5

[a] Coefficients for square sectors with $L/B = 1$ as applied in Equations (7.5)–(7.7).
[b] For $h_1/h_2 = 2.0$.
[c] Generalized for $h_1/h_2 = 2.0$ from test and field observations.
[d] For $h_1/h_2 = 5$ from Figure 7.6.

Archibald suggested the load-carrying capacity would improve if the step width were designed to gradually decrease in the sliding direction. Kettleborough (1954, 1955) later provided experimental evidence for this assertion. Based on his experimental tests, he suggested that trapezoidal pocket geometry could result in considerable improvements over the straight step sliders. Many years later, Rhode and McAllister (1973) numerically solved for the optimum film profile in rectilinear sliders using a novel mathematical optimization algorithm. Their findings showed that, similar to Kettleborough's experimental results, the optimum film profile is a step-like profile with a curved interface. See additional discussions and more recent development for the optimization of step slider bearing and design procedure by Fesenghary and Khonsari (2012).

7.9 Maximum Bearing Temperature Based on Thermohydrodynamic Analysis

The iterative method for determining the bearing temperature rise refines the initial guess of the effective viscosity based on representing the entire bearing temperature field with a single effective or equilibrium temperature. Methods for estimating this equilibrium bearing temperature based on the temperature rise are not unique and vary in different references. However, most effective viscosity methods use a global energy balance with the common assumption that the majority of heat dissipated in the fluid is carried away by the fluid and thus neglect the conduction of heat into the solids and convection into the surroundings. This is merely a simplifying assumption that circumvents the analytical difficulties in the treatment of the energy equation in the differential form and its coupling to the Reynolds equation (Khonsari, 1987a, 1987b). The most troubling aspect of the effective viscosity method is that it does not provide any information regarding the bearing maximum temperature. The lubricant temperature rises along the length of the slider in the direction of motion and it typically reaches a maximum value in the outlet section in the vicinity of the stationary pad. Figure 7.9 shows a picture of a failed pivoted-pad thrust-bearing pad that clearly shows the operating temperature has exceeded its maximum limit near the outlet. This limit is governed by the plastic flow of the protective layer on the surface of the stationary surface. For example, under pressure, a babbitt

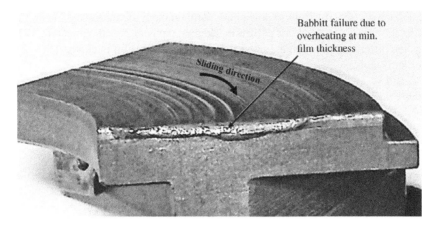

Babbitt failure due to
overheating at min.
film thickness

Sliding direction

Figure 7.9 A failed pivoted pad thrust bearing due to overheating.

plastically flows at about 150 °C. Thus, the maximum temperature, T_{max}, is an important parameter since if it exceeds a certain limit, bearing failure becomes imminent.

A complete model for a hydrodynamic bearing that accounts for the temperature-dependent viscosity with proper heat transfer characteristics is generally referred to as the thermohydrodynamic (THD) model (Dowson and Hudson, 1964a, 1964b) Prediction of a bearing's performance based on a THD analysis generally requires the simultaneous solution of the Reynolds equation with the energy equation in the fluid and the solids (Ezzat and Rhode, 1973). It also correctly incorporates the variation of viscosity along and across the film gap at every computational point in the lubricant. This requires deriving the Reynolds equation once more to take the variation of viscosity across the film into account, using a so-called generalized Reynolds equation (Dowson, 1962). Numerous reports show that a properly formulated THD theory yields remarkably close results compared to those measured experimentally for different bearing types and configurations (Boncompain *et al.*, 1986; Khonsari and Beaman, 1986; Khonsari *et al.*, 1996; Keogh *et al.*, 1997). Nonetheless, THD analyses are indeed computationally intensive and require a significant amount of time and effort to develop.

In this section, we provide a simple procedure that allows one to rapidly estimate the bearing maximum temperature as well as a more realistic effective temperature that can be used for the design of thrust bearings (Jang and Khonsari, 2004).

For this purpose, we introduce the following dimensionless temperature rise parameters: (Jang and Khonsari, 1997; Fillon and Khonsari, 1996)

$$\kappa_1 = \frac{\alpha\,\mu_i\,\beta\,U}{k\,B}\left(\frac{B}{S_h}\right)^2, \ \kappa_2 = U\sqrt{\frac{\mu_i\,\beta}{k}} \tag{7.15}$$

The first temperature-rise parameter κ_1 is associated with the viscous dissipation and it incorporates all the oil properties, velocity and the bearing geometry. The second temperature-rise parameter κ_2 incorporates all the oil properties and the linear velocity calculated at the mean radius of the pad.

In these equations, α represents the thermal diffusivity of the lubricant (m²/s). Thermal diffusivity is simply: $\alpha = k/\rho C_p$, k is the lubricant's thermal conductivity (W/m.K). μ_i is the inlet viscosity (Pa.s), β represent the temperature-viscosity coefficient (1/K), which is based on the fact that the viscosity drops exponentially with temperature increases. This follows the viscosity relation $\mu = \mu_i e^{-\beta(T-T_i)}$ described in Chapter 2, where T_i is the inlet temperature. s_h represents the difference between the maximum and minimum film thickness ($s_h = h_1 - h_2$), i.e., meter in SI units.

Once κ_1 and κ_2 are evaluated, then using Table 7.5, one can easily obtain a realistic value of an effective bearing temperature, T_{mean}, as well as an estimate of T_{max}. Note that the equations in Table 7.1 depend on the clearance ratio a_s defined as: ($a_s = h_1/h_2$). Further, the expressions have developed assuming that the bearing aspect ratio $\Lambda_s = B/L = 1$. Temperature expressions are in dimensionless form as indicated by an 'overbar' on the parameter. These can easily be converted to dimensional form by noting that $\bar{T} = \beta(T - T_i)$.

Once the effective (mean) temperature is obtained from Table 7.5, one can use the curve-fit expressions given in Table 7.6 to quickly predict the bearing performance parameters. The procedure is as follows (Jang and Khonsari, 2004):

1. Compute the bearing characteristic number defined as $K_f = \mu_{eff} LU/m^2 W$ at the inlet temperature, where W is the load per pad and m is the slope of the slider bearing, $m = (h_1 - h_2)/B = S_h/B$
2. Compute the clearance (film thickness) ratio a_s from the expression $a_s = a_s(K_f, \Lambda_s)$ from Table 7.6.
3. Compute the temperature-rise parameters using the bearing specifications.
4. Compute the maximum and mean temperatures from expressions in Table 7.5.
5. Compute the effective viscosity at the determined mean temperature.
6. Compute the bearing characteristic number again with the effective viscosity.
7. Compute the clearance ratio again.
8. Check the region of applicability in Table 7.5 once again with the new bearing characteristic number.
9. Compute the bearing performance parameters from Table 7.6 as a function of Λ_s and a.

Example 7.5 Consider the following slider bearing specifications (Raimondi *et al.*, 1968; Jang and Khonsari, 2004) $\Lambda_s = 1$, $m = 0.001$, $L = 0.076$ m, $\alpha = 0.9355 \times 10^{-7}$ m^2/s, $k = 0.13$ W/mK, $\beta = 0.0337/K$, $W = 16000$ N, $U = 30.48$ m/s, $T_s = 43.3\,^\circ$C, $\mu_i = 0.035$ Pa $\times$ s and $P_s = 0$.
Determine the maximum bearing temperature and a mean temperature

Step 1. Compute the bearing characteristic number K_f taking the viscosity at the inlet temperature

$$K_f = \frac{\mu_{eff} LU}{m^2 W} = \frac{0.035 \times 0.0762 \times 30.48}{(0.001)^2 \times 16000} = 5.081$$

Step 2. Compute the clearance ratio and the shoulder height using Table 7.6.

$$a_s = \frac{1}{0.28594 - \dfrac{0.09977}{\Lambda} + 0.11465 \ln\left(K_f\right)} = 2.685$$

$$S_h = m.B = (0.001)(0.076) = 7.62 \times 10^{-5} \text{ m}$$

Since $\Lambda_s = 1$, $B = L = 0.076$ m, and $\frac{S_h}{B} = m = 0.001$

Table 7.5 Maximum and mean temperature of slider bearings at $\Lambda = 1$

Valid over the range of $0.001 \leq \kappa_1 \leq 2$ $0.01 \leq \kappa_2 \leq 5$		Maximum temperature	Mean temperature
Film Thickness Ratio	Equation Constant	$\overline{T}_{max} = \left[a + \dfrac{b}{\kappa_1} + \dfrac{c}{\kappa_2^{1.5}} \right]^{-1}$	$\overline{T}_{mean} = \left[a + \dfrac{b}{\kappa_1} + \dfrac{c}{\kappa_2^{2}} \right]^{-1}$
$1.25 \leq a_s < 1.75$	a	0.24314	1.03892
Region I	b	0.69569	4.63987
	c	1.94883	3.03201
$1.75 \leq a_s < 2.25$	a	0.22144	0.84089
Region II	b	0.25883	1.90308
	c	1.44159	2.95531
$2.25 \leq a_s < 2.75$	a	0.21092	0.76222
Region III	b	0.15082	1.21812
	c	1.26859	3.07632
$2.75 \leq a_s < 3.25$	a	0.20586	0.71834
Region IV	b	0.10500	0.91341
	c	1.16447	3.20380

Table 7.6 Expressions for determining bearing performance parameters

	Bearing characteristic number $K_f = \dfrac{\mu_{eff} L U_L}{m^2 W}$	$a_s = \dfrac{1}{0.28594 - \dfrac{0.09977}{\Lambda} + 0.11465 \ln(K_f)}$ or $K_f = \exp\left[-1.68660 + \dfrac{2.09519}{\sqrt{\Lambda}} + \dfrac{8.74965}{a_s^2} \right]$
Slider Bearing	Friction coefficient f	$f = m \exp\left[-1.16655 + \dfrac{2.06060}{\sqrt{\Lambda}} + \dfrac{5.68202}{a^2} \right]$
Valid over the range of	Maximum pressure $\overline{P}_{max} = \dfrac{s_h^2(P_{max})}{\mu_{eff} BU}$	$\overline{P}_{max} = \exp\left[3.26497 - \dfrac{0.71501}{\Lambda} - \dfrac{8.80027}{a_s} \right]$
$0.5 \leq \Lambda \leq 1.5$ $1.3 \leq a \leq 3.2$	Side leakage flow rate $\overline{Q}_s = \dfrac{Q_s}{mBLU}$	$\overline{Q}_s = \dfrac{1}{2.09386 + 1.96421\Lambda^{1.5} - 0.02187a_s^{1.5}}$
	Inlet flow rate $\overline{Q}_i = \dfrac{Q_1}{mBLU}$	$\overline{Q}_i = 0.38858 - 0.26290\sqrt{\Lambda} + \dfrac{0.50023}{\ln(a_s)}$

Step 3. Calculate the temperature rise parameters.

$$\kappa_1 = \frac{\alpha \mu_i \beta U}{k B} \left(\frac{B}{s_h}\right)^2$$

$$\frac{(0.9355 \times 10^{-7})(0.035)(0.0337)(30.48)}{(0.13)(0.0762)} \left(\frac{1}{0.001}\right)^2 = 0.34$$

$$\kappa_2 = U \sqrt{\frac{\mu_i \beta}{k}}$$

$$= 30.48 \sqrt{\frac{(0.035)(0.0337)}{0.13}} = 2.903$$

According to Table 7.5, the film thickness ratio a_s falls in the Region III. Therefore,
Step 4. Compute the dimensionless maximum and mean temperatures using Table 7.5

$$\overline{T}_{max} = \left[0.21092 + \frac{0.15082}{\kappa_1} + \frac{1.26859}{\kappa_2^{1.5}}\right]^{-1} = 1.0977$$

$$\overline{T}_{mean} = \left[0.76222 + \frac{1.21812}{\kappa_1} + \frac{3.07632}{\kappa_2^2}\right]^{-1} = 0.2123$$

Step 5. Compute the effective viscosity at the determined mean temperature.

$$\mu_{eff} = \mu_i e^{-\overline{T}} = (0.035)\, e^{-(0.2123)} = 0.0283$$

Step 6. Compute the bearing characteristic number again at the effective viscosity

$$K_f = \frac{\mu_{eff} L U_L}{m^2 W} = \frac{(0.0283)(0.0762)(30.48)}{(0.001)^2 (16000)} = 4.108$$

Step 7. Compute the clearance ratio again

$$a = \frac{1}{0.28594 - 0.09977/\sqrt{\Lambda_S} + 0.11465 \ln (K_f)} = 2.872$$

Now the clearance ratio lies in Region IV in Table 7.5.
Step 8. Determine the maximum and mean temperatures in Region IV.

$$\overline{T}_{max} = \left[0.20586 + \frac{0.105}{\kappa_1} + \frac{1.16447}{\kappa_2^{1.5}}\right]^{-1} = 1.332$$

$$\overline{T}_{mean} = \left[0.71834 + \frac{0.91341}{\kappa_1} + \frac{3.2038}{\kappa_2^2}\right]^{-1} = 0.264$$

These non-dimensional temperatures can be converted to dimensional form by multiplying by the inlet temperature. Therefore,

$$T_{max} = \frac{\overline{T}_{max}}{\beta} + T_i$$

$$= \frac{1.332}{0.0414} + 36.8 = 82.8 \,^{\circ}\text{C}$$

$$T_{mean} = \frac{\overline{T}_{mean}}{\beta} + T_i$$

$$= \frac{0.264}{0.0414} + 36.8 = 51 \,^{\circ}\text{C}$$

Now the next iteration would involve determining a new effective viscosity at $T_{mean} = 51\,^{\circ}\text{C}$ is

$$\mu_{eff} = \mu_i \, e^{-\beta(T_{mean}-T_i)} = 0.027 \, \text{Pa.s}$$

Repeating the calculations. The updated baring characterization parameter and the clearance ratio are: $K_f = 3.902$ and $a_s = 2.903$.

Step 9. Determine the bearing performance parameters using the expressions in Table 7.6.

$$f = m \exp\left[-1.16655 + \frac{2.06060}{\sqrt{\Lambda}} + \frac{5.68202}{a_s^2}\right]$$

$$= 0.001 \exp\left[-1.16655 + \frac{2.06060}{1} + \frac{5.68202}{2.923^2}\right] = 0.0048$$

Maximum pressure is predicted to be:

$$\overline{P}_{max} = \exp\left[3.26497 - \frac{0.71501}{\Lambda} - \frac{8.80027}{a_s}\right]$$

$$P_{max} = \mu_{eff} B \frac{U}{S_h^2} \overline{P}_{max} = 6.78 \, \text{MPa}$$

The side leakage and inlet flows are:

$$\overline{Q}_s = \frac{1}{2.09386 + 1.96421\Lambda^{1.5} - 0.02187a_s^{1.5}} = 0.253$$

$$Q_s = mBU\overline{Q}_s = 0.001 \times 0.0762 \times 30.46 \times 0.253 = 44.8 \times 10^{-6} \, \text{m}^3/\text{s}$$

$$\overline{Q}_i = 0.38858 - 0.26290\sqrt{\Lambda} + \frac{0.50023}{\ln(a_s)} = 0.592$$

$$Q_i = mBU\overline{Q}_i = 0.001 \times 0.0762 \times 30.46 \times 0.592 = 104.8 \times 10^{-6} \, \text{m}^3/\text{s}$$

These predictions are remarkably close to numerical THD simulations reported by Jang and Khonsari (2004).

7.10 Parasitic Power Losses

Test results have indicated that 25–50% of total fluid-film bearing losses in turbomachinery may occur simply in feeding oil to the bearings (Booser and Missana, 1990). While commonly much higher in thrust bearings, with their larger diameter surfaces, these parasitic losses for both journal and thrust bearings can be assigned to the following two categories:

1. Through-Flow Loss

These losses result from acceleration of the oil as it enters feed ports, in entering and leaving feed grooves, and finally in discharging from the bearing. This power loss E_P (in. lb/s) for flow volume Q (in.^{3}) can be related to the mean linear surface velocity V (in./s), oil density ρ (typically 0.0307 lb/in.^{3}), and gravitational acceleration g (386 in./s) as follows:

$$E_P = kQ \left(\frac{\rho V^2}{2g} \right) \tag{7.16}$$

Multiplying factor k in tests varied from 1.6 to 1.9 for a number of tapered-land and pivoted pad thrust bearings and from 1.5 to 3.2 for journal bearings. Pivoted-pad bearings are at the upper end of the range.

Through-flow losses for a pivoted-pad thrust bearing can be assigned to the following sources in Figure 7.10:

1. One velocity head loss ($\rho V^2/2g$) calculated at the shaft surface velocity as the oil feed enters under the base ring, makes an abrupt turn, and is brought to the bulk oil velocity between the outside diameter of the shaft and the inside diameter of the base ring.
2. One velocity head loss calculated at the mean thrust runner diameter as the oil passes into the channels between the tilting pads and then into the bearing oil film.
3. Approximately 25% of discharge oil is assumed to be splashed onto the thrust runner to be reaccelerated at the outside diameter of the runner.

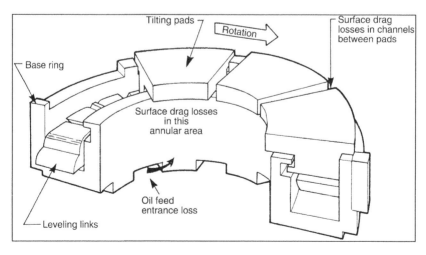

Figure 7.10   Parasitic losses in a pivoted-pad thrust bearing. *Source*: Booser and Missana, 1990. Reproduced with permission of Taylor & Francis.

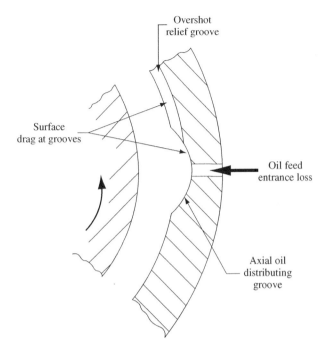

Figure 7.11   Parasitic losses in journal bearings. *Source*: Booser and Missana, 1990. Reproduced with permission of Taylor & Francis.

With a plain cylindrical axial-groove journal bearing as illustrated in Figure 7.11, a single parasitic loss element of 1.5 velocity heads (calculated at the shaft surface velocity) is encountered as feed oil is accelerated as it enters a turbulent vortex in the feed groove and then is further accelerated as it enters the bearing oil film itself. In pivoted-pad journal bearings, vortex action and churning in the square-sided cavity between pads offer more resistance to entrance feed oil.

2. Surface Drag Losses

Power loss from fluid friction drag on wetted shaft and thrust collar areas $A(\text{in}^2)$ follows this almost universal turbulent relation for hydraulic drag in channels:

$$\frac{E_p}{A} = f\left(\frac{\rho V^2}{2g}\right) V \qquad (7.17)$$

where the friction factor f is related as follows to the dimensionless Reynolds number ($Re = hV\rho/\mu g$):

$$f = \frac{0.016}{Re^{0.25}} \qquad (7.18)$$

with h being the clearance between the rotating surface and its adjacent stationary wall.

While journal bearings are covered in Chapter 8, it is worthwhile mentioning that for circular annular spaces around a journal – and for overshot internal grooves such as those used in the upper half of some journal bearings – the drag coefficient f should be modified to a 30–40% higher value according to the following relations provided by Vohr (1968):

$$f = m\left[(R_2 - R_1)R_2/R_1^2\right]^{0.25} Re^{-0.5} \tag{7.19}$$

where C is the radial clearance, $m = 0.46$ for $80(R_1/C)^{0.5} < Re < 10^4$, and $m = 0.073$ for $10^4 < Re < 10^5$.

Example 7.6 Calculate the surface drag loss on the rim of a 15 in. diameter, 3 in. thick thrust runner rotating at $N = 167$ rev/s (10 000 rpm) in a 15.5 in. diameter casing filled with oil having a viscosity of 2×10^{-6} reyns. What would be the corresponding through-flow power loss for 100 gpm (385 in^3/s) entering this clearance gap with loss of 1.0 velocity head?

Applying Equation (7.18) with surface velocity $V = \pi DN = \pi \times 15 \times 167 = 7854$ in./s, oil density $\rho = 0.0307$ lb/in^3, gap $h = (15.5 - 15)/2 = 0.25$ in., *surface drag loss* becomes

$$Re = hV\rho/\mu g = 0.25 \times 7854 \times 0.0307/(2 \times 10^{-6})(386) = 7854$$
$$f = 0.016/Re^{0.25} = 0.016/(7854)^{0.25} = 9.572 \times 10^{-4}$$
$$E_P/A = f(\rho V^2/2g)V = 9.572 \times 10^{-4}[0.0307 \times 7854^2/(2 \times 386)]7854 = 18\,440 \text{ in. lb/s/ in}^2$$
$$E_P = 18\,440A = 18\,440(\pi \times 15 \times 3) = 2.607 \times 10^6 \text{ in. lb/s} \quad (2.607 \times 10^6/6600 = 395 \text{ hp})$$

One velocity head gives the following for *through-flow loss:*

$$E_P = Q(\rho V^2/2g) = 385[0.0307 \times 7854^2/(2 \times 386)]$$
$$= 9.444 \times 10^5 \text{ in. lb/s} \quad (9.444 \times 10^5/6600 = 143 \text{ hp})$$

These losses would be minimized by appropriate draining of the housing to avoid fully flooded operation.

Possible Methods for Reducing Parasitic Losses

Reduction of Feed Oil

The simplest approach for reducing parasitic losses is to reduce oil feed. If the oil feed rate is reduced to the extent that internal grooves are not completely filled with oil, then the parasitic drag loss in the groove also drops frequently to a negligible value. As illustrated in Figure 7.12, a steam turbine with 350 in.^{2} tapered-land thrust bearing operating at 3600 rpm experienced a sharp drop in power loss with decreasing oil flow. As the design oil feed was gradually reduced, at 166 gpm the total power loss in the bearing dropped from 800 to 473 hp, i.e. 40% reduction. At this point, the back pressure dropped in the feed grooves with simultaneous elimination of through-flow loss for the entering oil (Booser and Missana, 1990).

Leading Edge Feed Grooves

There are pivoted-pad bearing designs available that use sprays or channels to feed oil directly into the leading edge of each thrust pad in a bearing cavity otherwise empty of oil. These

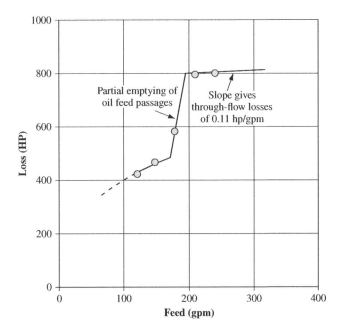

Figure 7.12 Decrease in power loss in a 27 in. OD. tapered land thrust bearing partially filled. *Source*: Booser and Missana, 1990. Reproduced with permission of Taylor & Francis.

arrangements practically avoid surface drag losses which would otherwise be encountered by rotating surfaces immersed in oil as well as portions of through-flow losses. Such designs are offered primarily for the surface speeds above that encountered in journal bearings with diameters about 10–12 in. operating at 3600 rpm. For lower surface velocities, parasitic power losses become so low as not to warrant the extra cost and complexity with specially directed feed (Khonsari and Booser, 2006). Figure 7.13 shows a pivoted-pad thrust bearing employing offset pivots and an inlet oil feed groove at the leading edge of each pad. Also included are oil lift pockets for low starting friction.

In high-speed bearings, leading edge oil feed grooves provide significant benefits. Oil flow rates can be as much as 60% lower with power loss down to 45%. While a variety of lubrication flow arrangements, design, and operating factors exert an influence, the resulting bearing temperature may be reduced by 16–35 °F.

As a surprising factor, reversing the rotation to place the oil feed groove at the trailing edge of each pad still provides satisfactory operation of a pivoted pad journal bearing at moderate speeds. This suggests that hot oil is still carried over from one pad to the next, and that absence of bulk oil from the bearing cavity is likely to be the major benefit to be realized from directed lubrication.

7.11 Turbulence

High surface speeds, low fluid viscosities, and large film thicknesses may introduce such high Reynolds numbers, $hV\rho/\mu g$, that the bearing fluid film will depart from its usual laminar

Figure 7.13 Pivoted pad thrust bearing with leading edge oil feed groove and oil lift pockets for low-friction starting. *Source*: Courtesy of Kingsbury, Inc.

viscous flow pattern to produce disordered turbulence. Transition to turbulence is to be expected in a thrust bearing oil film when the local Reynolds number in the film reaches the 1000–1200 range (Sneck and Vohr, 1984). A lower transition value of 580–800 has been found when an average Reynolds number for a bearing pad is considered. If there is doubt, both laminar and turbulent performance should be calculated. The one giving the higher calculated power loss will be the flow regime to be expected (Missana *et al.*, 1971).

With turbulence, both fluid-film thickness (load capacity) and power loss are increased as the film achieves a significantly higher apparent viscosity. While the thickness of laminar boundary layers and other details of turbulence are better handled in a comprehensive computer code, the following approximation was found in a test for a turbulent friction factor f (Missana *et al.*, 1971):

$$f = \frac{0.026}{Re^{0.25}} \tag{7.20}$$

This 0.026 coefficient is higher than the 0.016 for simple hydraulic drag on an exposed surface in Equation (7.18). However, since Equation (7.20) includes unevaluated parasitic friction losses, use of Equation (7.18) (together with Equation (7.17)) is recommended with addition of a separate evaluation of parasitic losses.

If power loss calculated with Equation (7.17) is higher than the calculated laminar value, say by 30%, the actual fluid viscosity at the bearing operating temperature can be multiplied by 1.30 to obtain an 'effective' viscosity for use in Table 7.2 to obtain a rough approximation of film thickness and flow values. A more formal calculation routine for oil film turbulence is provided by the linearized turbulent theory outlined in Chapter 6 (Section 6.2).

Turbulence introduces unique relations of thrust bearing performance to various operating conditions. Power loss and temperature rise in an operating bearing, for instance, commonly become relatively independent not only of total applied load, but also of fluid viscosity. While the power loss rises significantly with a turbulent oil film, increased oil-film thickness results in only a minor increase in the maximum oil-film temperature.

Problems

7.1 Performance of a fixed-pad slider bearing is to be evaluated when fed at 120 °F with an ISO VG 32 oil (see Chapter 2 for the viscosity–temperature plot for typical VG 32 oil). Specifications for the bearing are as follows:

$$W = 3000\,\text{lbf}; \qquad U = 960\,\text{in./s}; \qquad B = 4\,\text{in.}; \qquad L = 3\,\text{in.};$$
$$h_1 - h_2 = 0.004\,\text{in. taper}$$

Determine the following:
(a) inlet-to-outlet film thickness ratio;
(b) minimum film thickness;
(c) power loss;
(d) oil flow rates;
(e) temperature rise and bearing operating temperature.

7.2 Repeat the analysis of Problem 7.1 with tapers of 0.000 and 0.010 in. for comparison with the 0.004 in. taper in the problem.

7.3 Calculate the performance of a six-shoe pivoted-pad thrust bearing with (a) central pivots and (b) the pivots displaced 65% from the leading edges of the shoes.

7.4 For the tapered-land bearing of Problem 7.1 with an operating oil viscosity of 2×10^{-6} reyns (lbf s/in.^{2}) and a film thickness ratio $h_1/h_2 = 2.0$, calculate the bearing load capacity, power loss, oil flow, and temperature rise for a minimum film thickness $h_2 = 0.001$ in. using (a) the ILA from Table 6.2, (b) the ISA from Table 6.3, and (c) the performance coefficients of Table 7.2.

7.5 For the ILA in Problem 7.4(a), assume that the solution gives the circumferential pressure profile at the radial center line. Assume that this is the maximum pressure profile and that the pressure drops radially in a sinusoidal pattern to zero at the inner and outer edges of each land. Repeat the performance calculations called for in Problem 7.3 and compare this modified ILA load capacity with that calculated in Problem 7.4(a) using the performance coefficients of Table 7.2.

7.6 Apply the energy relation of Equation (7.10) to calculate the maximum temperature to be reached at the trailing edge of the pad in Problem 7.1. Use an iterative finite difference procedure along the center line of the pad with an initial temperature of 120 °F for VG 32 grade oil, oil density of 0.0307 lb/in.^{3}, and oil specific heat of 4400 in. lb/°F lb. (See Chapter 2 for the viscosity–temperature relation for VG 32 oil.) Use Equation (6.52) for the film pressure.

7.7 You are to recommend a design for a spring-supported thrust bearing to carry 4 million lb of thrust in a 90 rpm hydroelectric generator. The inside diameter is to be 48 in. Maximum unit loading on the pads is limited to 500 lb/in.^{2}, which will set the outside diameter. Viscosity of the VG 68 oil fed at operating conditions is 3.5×10^{-6} reyns. Using an assumed film thickness ratio $h_1/h_2 = 2.0$, calculate minimum film thickness, power loss, oil-film flow, and temperature rise with 6 pads and then 12 pads.

References

Archibald, F.R. 1950. 'A Simple Hydrodynamic Thrust Bearing,' *Transactions of the ASME*, Vol. 72, pp. 393–400.

Boncompain, R., Fillon, M., and Frene, J. 1986. 'Analysis of Thermal Effects in Hydrodynamic Bearings,' *ASME Journal of Tribology*, Vol. 108, pp. 219–224.

Booser, E.R. and Khonsari, M.M. 2010. 'Flat-Land Thrust Bearing Paradox,' *Machine Design*, Vol. 82, pp. 52–59.

Booser, E.R. and Missana, A. 1990. 'Parasitic Power Losses in Turbine Bearings,' *Tribology Transactions*, Vol. 33, pp. 157–162.

Booser, E.R., Linkinhoker, C.L., and Ryan, F.D. 1970. 'Maximum Temperature for Hydrodynamic Bearings under Steady Load,' *Lubrication Engineering*, Vol. 26, pp. 226–235.

Dowson, D. 1962. 'A Generalized Reynolds Equation for Fluid Film Lubrication,' *International Journal of Mechanical Science*, Vol. 4, pp. 159–170.

Dowson, D. and Hudson, J. 1964a. 'Thermohydrodynamic Analysis of Infinite Slider Bearing: Part I, The Plane-Inclined Slider-Bearing,' *Proceedings of the Institution of Mechanical Engineers*, Vol. 34, pp. 34–44.

Dowson, D. and Hudson, J. 1964b. 'Thermo-hydrodynamic Analysis of the Infinite Slider-Bearing: Part II, the Parallel-Surface Slider-bearing,' *Proceedings of the Institution of Mechanical Engineers*, Vol. 34, pp. 45–51.

Ezzat, H.A. and Rohde, S.M. 1973. 'A Study of the Thermohydrodynamic Performance of Finite Slider Bearings,' *ASME Journal of Lubrication Technology*, Vol. 95, pp. 298–307.

Fesanghary, M. and Khonsari, M.M. 2012. 'Topological and Shape Optimization of Thrust Bearings for Enhanced Load-Carrying Capacity,' *Tribology International*, Vol. 53, pp. 12–22.

Fillon, M. and Khonsari, M.M. 1996. 'Thermohydrodynamic Design Charts for Tilting-Pad Journal Bearings,' *ASME Journal of Tribology*, Vol. 118, pp. 232–238.

Fuller, D.D. 1984. *Theory and Practice of Lubrication for Engineers*, 2nd edn. John Wiley & Sons, New York.

Gardner, W.W. 1975. 'Performance Tests on Six-Inch Tilting Pad Thrust Bearings,' *ASME Journal of Lubrication Technology*, Vol. 18, pp. 430–438.

Gardner, W.W. 1988. 'Tilting Pad Thrust Bearing Tests – Influence of Pivot Location,' *ASME Journal of Tribology*, Vol. 110, pp. 609–613.

Hamrock, B.J. 1994. *Fundamentals of Fluid Film Lubrication*. McGraw-Hill Book Co., New York.

Jang, J.Y. and Khonsari, M.M. 1997. 'Thermohydrodynamic Design Charts for Slider Bearings,' *ASME Journal of Tribology*, Vol. 119, pp. 733–740.

Kennedy, F.E., Booser, E.R., and Wilcock, D.F. 1998. 'Tribology, Lubrication, and Bearing Design,' *Mechanical Engineering Handbook*. CRC Press, Boca Raton, Fla, Section 3.10.

Keogh, P., Gomiciaga, R., and Khonsari, M.M. 1997. 'CFD Based Design Techniques for Thermal Prediction in a Generic Two-axial Groove Hydrodynamic Journal Bearing,' *ASME Journal of Tribology*, Vol. 119, pp. 428–436.

Kettleborough, C. 1955. 'An Electrolytic Tank Investigation Into Stepped Thrust Bearings,' *Proceedings of the Institution of Mechanical Engineers*, Vol. 169, pp. 679–688.

Kettleborough, C.T. 1954. 'The Stepped Thrust Bearing-A Solution by Relaxation Methods,' *Transactions of ASME*, Vol. 76, p. 19.

Khonsari, M.M. and Beaman, J.J. 1986. 'Thermohydrodynamic Analysis of Laminar Incompressible Journal Bearings,' *ASLE Transactions*, Vol. 29, pp. 141–150.

Khonsari, M.M. and Booser, E.R. 2006. 'Parasitic Power Losses in Hydrodynamic Bearings,' *Machinery Lubrication*, March–April, pp. 36–39.

Khonsari, M.M., Jang, J.Y., and Fillon, M. 1996. 'On the Generalization of Thermohydrodynamic Analyses for Journal Bearings,' *ASME Journal of Tribology*, Vol. 118, pp. 571–579.

Khonsari, M.M. and Jang, J.Y. 1997. 'Design and Analysis of Hydrodynamic Thrust Bearings,' *Tribology Data Handbook*. CRC Press, Boca Raton, Fla, pp. 680–690.

Khonsari, M.M. and Wang, S. 1991. 'Maximum Temperature in Double-Layered Journal Bearings,' *ASME Journal of Tribology*, Vol. 113, pp. 464–469.

Missana, A., Booser, E.R., and Ryan, F.D. 1971. 'Performance of Tapered Land Thrust Bearings for Large Steam Turbines,' *STLE Transactions*, Vol. 14, pp. 301–306.

Neale, P.B. 1970. 'Analysis of the Taper-Land Bearing Pad,' *Journal of Mechanical Engineering Science*, Vol. 12, pp. 73–84.

Raimondi, A.A., Boyd, J., and Kufman, H.N. 1968. 'Analysis and Design of Sliding Bearings,' *Standard Handbook of Lubrication Engineering*,' O'Connor, J. and Boyd, J. (editors), McGraw-Hill, New York.

Raimondi, A.A. 1960. 'The Influence of Longitudinal and Transverse Profile on the Load Capacity of Pivoted Pad Bearings,' *STLE Transactions*, Vol. 3, pp. 265–276.

Raimondi, A.A. and Szeri, A.Z. 1984. 'Journal and Thrust Bearings,' *Handbook of Lubrication*, Vol. 2. CRC Press, Boca Raton, Fla, pp. 413–462.

Rohde, S. and McAllister, G. 1976. 'On the Optimization of Fluid Film Bearings,' *Proceedings of the Royal Society of London. A. Mathematical and Physical Sciences*, Vol. 351, p. 481.

Sneck, H.J. and Vohr, J.H. 1984. 'Hydrodynamic Lubrication,' *Handbook of Lubrication*, Vol. 2. CRC Press, Boca Raton, Fla, pp. 69–91.

Vohr, J.H. 1968. 'An Experimental Study of Taylor Vortices and Turbulence in Flow between Eccentric Rotating Cylinders,' *ASME Journal of Lubrication Technology*, pp. 285–296.

Vohr, J.H. 1981. 'Prediction of the Operating Temperature of Thrust Bearings,' *Journal of Lubrication Technology, Transactions of the ASME*, Vol. 103, p. 97.

Wilcock, D.F. and Booser, E.R. 1957. *Bearing Design and Application*. McGraw-Hill Book Co., New York.

Yang, P. and Kaneta, M. 2013. 'Thermal Wedge in Lubrication,' *Encyclopedia of Tribology*, Y.W. Chung and Q.J. Wang (editors), Springer Science, New York, pp. 3617–3623.

8

Journal Bearings

8.1 Introduction

A journal bearing consists of an approximately cylindrical body around a rotating shaft, used either to support a radial load or simply as a guide for smooth transmission of torque. It involves a stationary sleeve (or bushing) with a complete 360° arc or various arrangements of a partial arc or arcs in a housing structure, as shown in Figure 8.1. The inner surface is commonly lined with a soft bearing material such as lead or tin babbitt, bearing bronze, or a plastic (see Chapter 4).

Plain circular bearings with a 360° arc are the main focus of this chapter, since they lend themselves to general analyses from which good physical insights can be gained. Lubricant supply arrangements range from a simple inlet hole to axial, circumferential, and helical grooves for efficient lubricant distribution (Figure 8.2).

Plain journal bearings are susceptible to a form of instability known as self-excited oil whirl. Over the years, a series of bearing designs have been developed for suppressing such vibration problems. Examples include bearings with tilting-pad, elliptical, pressure dam, and offset split bearings shown in Figure 8.3, with typical applications given in Table 8.1. Since inclusion of oil feed grooves in the bearing bore often does not greatly alter hydrodynamic action in the load zone, the 360° degree analyses which follow also provide useful approximations for many other designs.

Figure 8.4(a) shows the cross-section of a journal bearing at rest with the journal surface in contact with the inner surface of the bushing. When motion is initiated (Figure 8.4(b)), the shaft first rolls up the wall of the sleeve in the direction opposite to rotation due to metal-to-metal friction between the steel shaft and the bearing bore. With an adequate lubricant supply, a supporting wedge-shaped film of lubricant is almost immediately formed to lift the journal into its steady-state position (Figure 8.4(c)).

Once the shaft is in a steady state, it assumes a position in the bearing clearance circle whose coordinates are eccentricity e and attitude angle $\emptyset$ (Figure 8.5). Eccentricity is simply the displacement of the journal center O_j from the bushing center O_b. Attitude angle is the angle between the load line and the line between these centers. Eccentricity decreases and attitude angle commonly increases with more vigorous oil-film pumping action by the journal at higher speeds and with increased lubricant viscosity.

Applied Tribology: Bearing Design and Lubrication, Third Edition. Michael M. Khonsari and E. Richard Booser.
© 2017 John Wiley & Sons Ltd. Published 2017 by John Wiley & Sons Ltd.
Companion Website: www.wiley.com/go/Khonsari/Applied_Tribology_Bearing_Design_and_Lubrication_3rd_Edition

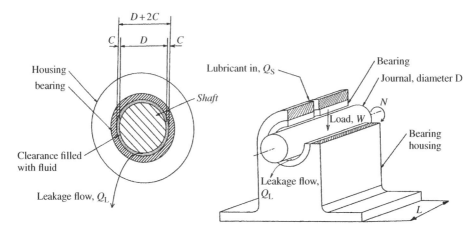

Figure 8.1   Journal bearing geometry and nomenclature.

In many applications, rotor weight and any external load are supported on two identical bearings located on either end of a shaft. A journal may also run centered in its bearing with zero eccentricity if its shaft is positioned vertically so that gravity does not contribute to the load. As in vertical pumps and motors, these lightly loaded bearings are particularly susceptible to troublesome oil whirl instability, as the unloaded journal tends to gyrate in the rotating oil film.

Film Thickness Profile

The geometry of a cylindrical bearing is shown in Figure 8.5, with its circumferential coordinate θ measured directly from the line of centers, i.e. the maximum film thickness. Film thickness h is described by the following approximation, which is used in almost all analyses:

$$h = C + e \cos \theta \qquad (8.1)$$

where C is the *radial* clearance, $C = R_b - R$.

Fluid-film pressure rises rapidly from $\theta = 0$ to reach a peak value before the minimum film thickness, h_{min}, as the fluid is pumped by shaft rotation into the convergent region. After passing the minimum film thickness, the fluid enters a diverging region, then recirculates and mixes with a fresh oil supply at the inlet. As the shaft rotates, it displaces a considerable portion of the feed lubricant axially out from the sides of the converging load-bearing region. This side leakage fluid is collected in a sump or reservoir, then filtered, cooled, and pumped in a closed loop back to the bearing for distribution in the bearing bore through a variety of inlet-hole and distributing groove arrangements (see Figure 8.2 and Section 8.14).

The oil supply may be provided simply by gravity feed at zero gauge pressure. A supply pressure in the 10–100 psi range is often used to enhance flow circulation and distribution, particularly for enhanced cooling when axial or circumferential grooves are used. Much higher pressures in the range up to 2000–3000 psi may be employed for hydrostatic operation (see

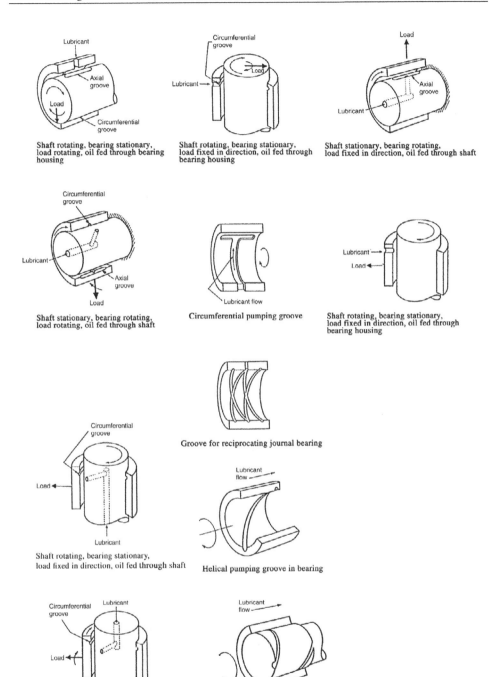

Figure 8.2 Supply lubricant arrangements. *Source*: Kaufman and Boyd, 1968. Reproduced with permission of McGraw-Hill.

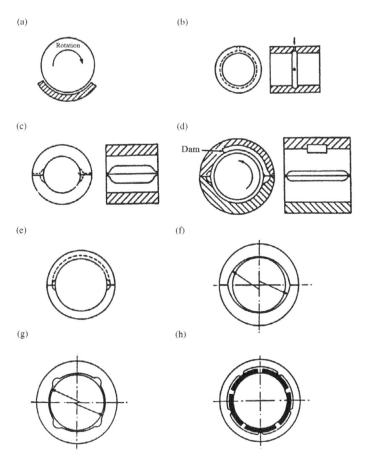

Figure 8.3 Several types of journal bearings: (a) partial arc; (b) circumferential groove; (c) cylindrical bearing-axial groove; (d) pressure dam; (e) cylindrical overshot; (f) elliptical; (g) multilobe; (h) tilting pad. *Source*: Kennedy *et al.*, 1997. Reproduced with permission of Taylor & Francis.

Table 8.1 Journal bearing designs

Type (see Figure 8.3)	Typical loading	Applications
(a) Partial arc	Unidirectional load	Shaft guides, dampers
(b) Circumferential groove	Variable load direction	Internal combustion engines
Axial groove types		
(c) Cylindrical	Medium to heavy unidirectional load	General machinery
(d) Pressure dam	Light loads, unidirectional	High-speed turbines, compressors
(e) Overshot	Light loads, unidirectional	Steam turbines
(f) Multilobe	Light loads, variable direction	Gearing, compressors
(g) Preloaded	Light loads, unidirectional	Minimize vibration
(h) Tilting pad	Moderate variable loads	Minimize vibration

Source: Kennedy *et al.*, 1997. Reproduced with permission of CRC Press.

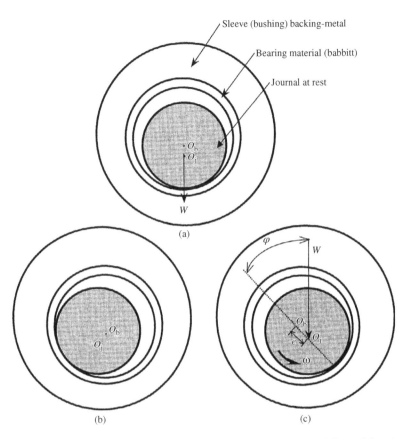

Figure 8.4 Sequence of journal motion from start-up to steady state: (a) journal bearing at rest; (b) journal bearing at the moment of start-up; (c) steady-state position.

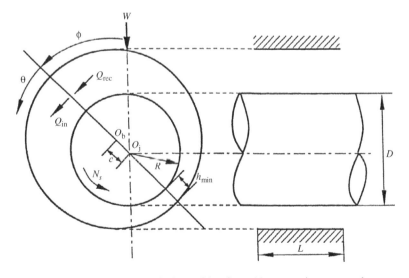

Figure 8.5 Cross-section of a journal bearing with appropriate nomenclature.

Chapter 10). With gravity feed or low supply pressures, the active region of the bearing is limited to the convergent gap and the divergent film section plays a minor role. To avoid unnecessary power loss in inactive portions of the bearing bore, partial arc bearings are used at low to moderate speeds; wide overshot grooves are employed in the upper (unloaded) bearing half with high-speed turbine bearings (Figure 8.3).

8.2 Full-Arc Plain Journal Bearing with Infinitely Long Approximation (ILA)

Physical insight can be gained by examination of analytical governing equations and their solutions. For this purpose, we shall confine our attention to an infinitely long plain journal bearing with a full 360° arc. We begin by referring to the Reynolds equation with ILA, which is reasonably accurate when $L/D > 2$. Referring to Equation (6.52),

$$\frac{dP}{dx} = 6\mu U \frac{h - h_m}{h^3} \tag{8.2}$$

where h_m is the film thickness corresponding to the location where pressure reaches its peak pressure P_{max}.

Letting $x = R\theta$, $U = R\omega$, and rearranging Equation (8.2),

$$\frac{dP}{d\theta} = 6\mu\omega \left(\frac{R}{C}\right)^2 \left[\frac{1}{(1 + \varepsilon \cos\theta)^2} - \frac{h_m}{C(1 + \varepsilon \cos\theta)^3}\right] \tag{8.3}$$

where $h_m = C(1 + \varepsilon \cos\theta_m)$ and θ_m represents the circumferential angle at which $dP/d\theta = 0$.

Equation (8.3) suggests that in addition to speed and viscosity, the ratio of R/C affects the pressure. This ratio turns out to have a pronounced influence on almost all journal-bearing performance parameters. Normally, bearing diametral clearance is chosen to be on the order of 0.001–0.002 per inch of diameter to give R/C values of 500–1000.

Defining a dimensionless pressure $\overline{P} = P(C/R)^2/\mu\omega$ and integrating Equation (8.3) yields

$$\overline{P} = 6\int \frac{d\theta}{(1 + \varepsilon \cos\theta)^2} - 6(1 + \varepsilon \cos\theta_m)\int \frac{d\theta}{(1 + \varepsilon \cos\theta)^3} + A \tag{8.4}$$

where the integration constants A and θ_m are unknowns to be determined by applying the appropriate boundary conditions. Equation (8.4) may be rewritten as follows

$$\overline{P} = 6I_2 - 6(1 + \varepsilon \cos\theta_m)I_3 + A$$

Expressions I_2 and I_3 fall into a class of integrals that can be conveniently evaluated using the Sommerfeld substitution, defined as follows:

$$\cos\gamma = \frac{\varepsilon + \cos\theta}{1 + \varepsilon \cos\theta} \tag{8.5}$$

It can be shown that

$$I_2 = \int \frac{d\theta}{(1 + \varepsilon \cos \theta)^2} = \int \frac{1 - \varepsilon \cos \gamma}{(1 - \varepsilon^2)^{3/2}} d\gamma = \frac{\gamma - \varepsilon \sin \gamma}{(1 - \varepsilon^2)^{3/2}} \tag{8.6}$$

$$I_3 = \int \frac{d\theta}{(1 + \varepsilon \cos \theta)^3} = \int \frac{(1 + \varepsilon \cos \gamma)^2}{(1 - \varepsilon^2)^{5/2}} d\gamma$$

$$= \frac{\gamma - 2\varepsilon \sin \gamma + \varepsilon^2 (\gamma/2) + (\varepsilon^2 \sin 2\gamma)/4}{(1 - \varepsilon^2)^{5/2}} \tag{8.7}$$

Equation (8.4) now reads:

$$\overline{P} = 6 \frac{\gamma - \varepsilon \sin \gamma}{(1 - \varepsilon^2)^{3/2}} - 6 \left(\frac{1 - \varepsilon^2}{1 - \varepsilon \cos \gamma_m} \right) \left(\frac{\gamma - 2\varepsilon \sin \gamma + \varepsilon^2 \gamma 2 + (\varepsilon^2 \sin 2\gamma)/4}{(1 - \varepsilon^2)^{5/2}} \right) + A$$

$$= \frac{6}{(1 - \varepsilon^2)^{3/2}} \left[\gamma - \varepsilon \sin \gamma - \frac{\gamma(1 + \varepsilon^2/2) - 2\varepsilon \sin \gamma + (\varepsilon^2 \sin 2\gamma)/4}{(1 - \varepsilon \cos \gamma_m)} \right] + A \tag{8.8}$$

where γ_m corresponds to θ_m.

8.3 Boundary Conditions

Consider the unwrapped schematic of the film shape shown in Figure 8.6. Let us assume that $\overline{P} = 0$ at $\theta = 0$. Since at $\theta = 0$, $dh/d\theta = -\varepsilon\, C \sin \theta = 0$ and there is no physical wedge at that point, pressure is theoretically zero at $\theta = 0$, where $h = h_{max}$. In most applications, however, this is not the case since the attitude angle is a function of load. In terms of the pressure generation and the load-carrying capacity prediction, this assumption is generally accepted. In applications where the supply pressure is generally greater than the atmospheric pressure,

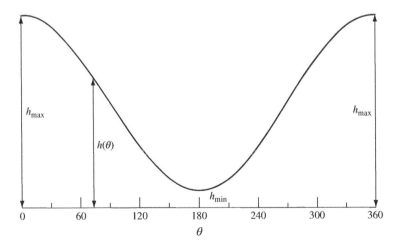

Figure 8.6 Film thickness in an unwrapped journal bearing.

one can define the dimensionless pressure as $\overline{P} = (P - P_s)(C/R)^2/\mu\omega$. Equation (8.8) then remains unchanged, and we can proceed with the derivation.

In what follows we shall assume that at $\theta = 0$, $\overline{P} = 0$, which gives $A = 0$. Evaluation of the second unknown γ_m requires considerably more detailed description.

8.4 Full-Sommerfeld Boundary Condition

One useful boundary condition is periodicity, i.e. $P(\theta) = P(\theta + 2\pi)$, which requires that pressure at $\theta = 2\pi$ is $\overline{P} = 0$. The Sommerfeld substitution allows convenient implementation of this cyclic condition. Examination of $\cos\gamma = (\varepsilon + \cos\theta)/(1 + \varepsilon\cos\theta)$ shows that when $\theta = 0$, π, or 2π, parameter $\gamma = 0$, π, or 2π, respectively. Therefore, applying the condition $\overline{P} = 0$ when $\gamma = 2\pi$ in Equation (8.8) yields $\cos\gamma_m = -\varepsilon/2$. This implies that the peak pressure occurs when $\cos\theta_m = -3\varepsilon/(2 + \varepsilon^2)$. The final expression for the pressure distribution thus becomes

$$\overline{P} = \frac{6\varepsilon\sin\theta(2 + \varepsilon\cos\theta)}{(2 + \varepsilon^2)(1 + \varepsilon\cos\theta)^2} \tag{8.9}$$

In dimensionless form, oil pressure distribution around the bearing circumference is a function of eccentricity ratio ε. Hence, ε is a parameter which directly influences the bearing load-carrying capacity. Figure 8.7 shows the pressure distribution variation for an assumed ε. Note that Equation (8.9) predicts that $P = 0$ at $\theta = \pi$. The magnitude of the peak pressure is

$$\overline{P}_{max} = \frac{3\varepsilon(4 - \varepsilon^2)(4 - 5\varepsilon^2 + \varepsilon^4)^{0.5}}{2(1 - \varepsilon^2)^2(2 + \varepsilon^2)}$$

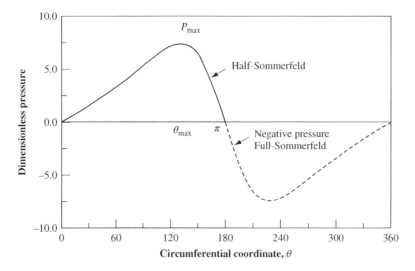

Figure 8.7   Long bearing full- and half-Sommerfeld pressure distributions.

which occurs at

$$\theta_{\max} = \cos^{-1}\left(\frac{-3\varepsilon}{2+\varepsilon^2}\right)$$

Figure 8.7 shows that pressure distribution varies like a sine curve and that from $\theta = \pi$ to 2π, the pressure becomes negative. Physically large negative pressures can be encountered only in the presence of high ambient pressures such as those in a nuclear pressurized water system or a deep submerged submarine. Conventional lubricants cannot withstand negative pressure, and liquid oil in the divergent region will cavitate with gases dissolved in the oil coming out of solution. While the full Sommerfeld boundary condition does not take cavitation into consideration, insight can be gained into bearing performance factors such as load-carrying capacity and attitude angle by accepting the full Sommerfeld solution as given by Equation (8.9).

Load-Carrying Capacity Based on Full-Sommerfeld Condition

Figure 8.8 shows the components of pressure projected along the line of centers (radial direction W_X and tangential direction perpendicular to the line of centers W_Y):

$$W_X = L\int_0^{2\pi} PR\cos\theta\,d\theta \tag{8.10}$$

$$W_Y = L\int_0^{2\pi} PR\sin\theta\,d\theta \tag{8.11}$$

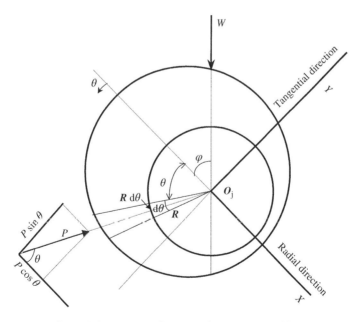

Figure 8.8   Radial and tangential components of load.

where L is the bearing length that is assumed to be much larger than the shaft diameter, in accordance with the ILA.

Substituting pressure from Equation (8.9) into Equations (8.10) and (8.11) and evaluating the integrals yields

$$\overline{W}_X = 0; \qquad \overline{W}_Y = \frac{12\pi\varepsilon}{(1-\varepsilon^2)^{1/2}(2+\varepsilon^2)} \tag{8.12}$$

where the dimensionless load $\overline{W}$ is given by

$$\overline{W} = \frac{P_L}{\mu\pi N_s}\left(\frac{C}{R}\right)^2 \tag{8.13}$$

with $P_L = W/(2LR)$ representing the projected load and N_S denoting the shaft speed in rev/s. The total load-carrying capacity is simply the resultant of the $\overline{W}_X$ and $\overline{W}_Y$ vectors:

$$\overline{W} = \sqrt{\overline{W}_X^2 + \overline{W}_Y^2} = \frac{12\pi\varepsilon}{(1-\varepsilon^2)^{1/2}(2+\varepsilon^2)} \tag{8.14}$$

The attitude angle is given by $\tan\phi = -\overline{W}_Y/\overline{W}_X = \infty$, which implies that

$$\phi = \frac{\pi}{2} \tag{8.15}$$

If the shaft eccentricity ratio is specified, then Equation (8.14) gives the load-carrying capacity. The limiting cases are: as $\varepsilon \to 0$, $\overline{W} \to 0$ and as $\varepsilon \to 1$, $\overline{W} \to \infty$. In practice, a bearing operating with $0 \le \varepsilon \le 0.5$ is considered lightly loaded, which results in large power losses. In a typical design, a range of $\varepsilon = 0.5$–0.85 is common.

Equation (8.15) gives a curious result: no matter what the shaft eccentricity is (or how severe the load) the shaft attitude angle is always 90°. Therefore, full-Sommerfeld solution predicts that upon loading, the shaft center travels along the line of centers which is perpendicular to the load line. See Figure 8.9, where three eccentricity positions are shown. This physically unreasonable prediction stems from neglecting cavitation, and the resulting consequence that $\overline{W}_X = 0$ as the positive and negative components projected along the radial direction X cancel each other.

8.5 Definition of the Sommerfeld Number

To determine the load-carrying capacity and other bearing performance parameters, one must first determine the eccentricity ratio. This can be done by introducing a very useful dimensionless parameter known as the *Sommerfeld number*, S, defined as follows:

$$S = \frac{\mu N_s}{P_L}\left(\frac{R}{C}\right)^2 \tag{8.16}$$

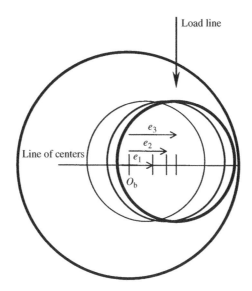

Figure 8.9   Shaft trajectory with a 90° attitude angle (infinitely long solution).

Relating the dimensionless load $\overline{W}$ of Equation (8.14) to the Sommerfeld number, $\overline{W} = 1/(\pi S)$, and solving for S yields

$$S = \frac{(1 - \varepsilon^2)^{1/2}(2 + \varepsilon^2)}{12\pi^2\varepsilon} \tag{8.17}$$

Equations (8.16) and (8.17) allow a convenient method for determining ε. For given bearing specifications such as load, lubricant, and operating speed, Equation (8.16) gives S, which can be used in Equation (8.17) to determine ε. Solution of Equation (8.17) requires a nonlinear equation solver which calls for a numerical solution. Any nonlinear root finder, such as the Newton-Raphson scheme or the bisection search method, can be used to solve for ε.

8.6 Half-Sommerfeld Boundary Condition

Clearly, the results predicted by the full-Sommerfeld condition are objectionable, primarily because the bearing is assumed to run free of cavitation and with the lifting of the journal by negative pressure in the diverging clearance section. The subject of cavitation and the formulation of the appropriate boundary conditions is covered in Section 8.7. At this stage, we shall concentrate on a solution that neglects entirely the negative pressures predicted for $0 \le \pi \le 2\pi$. Known as the *half-Sommerfeld solution*, this analysis essentially limits the pressure to the convergent region, as shown in Figure 8.7.

An analysis similar to the one in the previous section can be carried out to determine the load-carrying capacity and attitude angle. With integration of pressure limited to π, rather

Table 8.2 Comparison of full- and half-Sommerfeld solutions for infinitely long journal bearings

$$\overline{P} = 6\frac{\varepsilon \sin \theta (2 + \varepsilon \cos \theta)}{(2 + \varepsilon^2)(1 + \varepsilon \cos \theta)^2}$$

$$\overline{P}_{max} = \frac{3\varepsilon(4 - \varepsilon^2)(4 - 5\varepsilon^2 + \varepsilon^4)^{0.5}}{2(1 - \varepsilon^2)^2(2 + \varepsilon^2)} \text{ at } \theta_{max} = \cos^{-1}\left(\frac{-3\varepsilon}{2 + \varepsilon^2}\right)$$

$$\omega = 2\pi N_s; \quad P_L = \frac{W}{LD}; \quad P = \mu\omega\left(\frac{R}{C}\right)^2 \overline{P}; \quad W = \mu R\omega L\left(\frac{R}{C}\right)^2 \overline{W};$$

$$S = \frac{\mu N_s}{P_L}\left(\frac{R}{C}\right)^2; \quad \overline{W} = \frac{1}{\pi S}$$

Full-Sommerfeld	Half-Sommerfeld
Solution domain $0 \le \theta \le 2\pi$	Solution domain $0 \le \theta \le \pi; \overline{P} = 0$ elsewhere
$\overline{W}_X = 0$	$\overline{W}_X = \dfrac{12\varepsilon^2}{(1 - \varepsilon^2)(2 + \varepsilon^2)}$
$\overline{W}_Y = \dfrac{12\pi\varepsilon}{(1 - \varepsilon^2)^{0.5}(2 + \varepsilon^2)}$	$\overline{W}_Y = \dfrac{6\pi\varepsilon}{(1 - \varepsilon^2)^{0.5}(2 + \varepsilon^2)}$
$\overline{W} = \dfrac{12\pi\varepsilon}{(1 - \varepsilon^2)^{0.5}(2 + \varepsilon^2)} = \dfrac{1}{\pi S}$	$\overline{W} = \dfrac{6\varepsilon}{(1 - \varepsilon^2)(2 + \varepsilon^2)}[\pi^2 - \varepsilon^2(\pi^2 - 4)]^{0.5} = \dfrac{1}{\pi S}$
$\phi = \dfrac{\pi}{2}$	$\phi = \tan^{-1}\left[\dfrac{\pi}{2\varepsilon}(1 - \varepsilon^2)^{0.5}\right]$
$S = \dfrac{(1 - \varepsilon^2)^{0.5}(2 + \varepsilon^2)}{12\pi^2\varepsilon}$	$S = \dfrac{(2 + \varepsilon^2)(1 - \varepsilon^2)}{6\pi\varepsilon}\left[\dfrac{1}{\pi^2(1 - \varepsilon^2) + 4\varepsilon^2}\right]^{0.5}$
$F = \dfrac{\mu\omega RL}{\pi}\left(\dfrac{R}{C}\right)\left(\dfrac{1 + 2\varepsilon^2}{3\varepsilon}\right)\dfrac{1}{S}$	
$f\left(\dfrac{R}{C}\right) = \dfrac{1 + 2\varepsilon^2}{3\varepsilon}$	$f\left(\dfrac{R}{C}\right) = \dfrac{\varepsilon \sin \phi}{2} + \dfrac{2\pi^2 S}{(1 - \varepsilon^2)^{0.5}}$

than 2π,

$$\overline{W}_Y = \frac{6\pi\varepsilon}{(1 - \varepsilon^2)^{1/2}(2 + \varepsilon^2)} \tag{8.18}$$

$$\overline{W}_X = \frac{12\varepsilon^2}{(1 - \varepsilon^2)(2 + \varepsilon^2)} \tag{8.19}$$

Note that $\overline{W}_X \ne 0$ and $\overline{W}_Y$ is precisely one half of that predicted with the full-Sommerfeld condition. The total load-carrying capacity and the attitude angle are,

$$\overline{W} = \frac{6\varepsilon}{(2 + \varepsilon^2)(1 - \varepsilon^2)}[\pi^2 - \varepsilon^2(\pi^2 - 4)]^{0.5} \tag{8.20}$$

$$\phi = \tan^{-1}\left[\frac{\pi}{2\varepsilon}(1 - \varepsilon^2)^{0.5}\right] \tag{8.21}$$

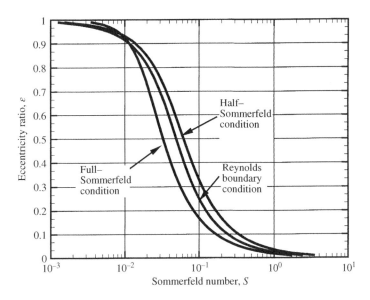

Figure 8.10   Long bearing eccentricity ratio as a function of the Sommerfeld number.

The attitude angle ranges from $\phi = 90°$ when $\varepsilon = 0$ to $\phi = 0°$ when $\varepsilon = 1$. With a given bearing specification the shaft will assume an attitude angle, according to Equation (8.21), between these two extremes.

Finally, Equation (8.20) can be put in terms of the Sommerfeld number as defined in Equation (8.16).

Table 8.2 summarizes the appropriate equations for both full- and half-Sommerfeld conditions. Numerical solutions for ε as a function of S are shown in Figure 8.10. Corresponding attitude angles are shown in Figure 8.11.

Example 8.1   A pump has a horizontal rotor weighing 3200 lb supported on two plain journal bearings, one on either side of the pump impeller. The specifications of the bearing and lubricant are

$$D = 4 \text{ in.}; \qquad L = 4 \text{ in.}; \qquad C = 0.002 \text{ in.}; \qquad N = 1800 \text{ rpm}; \qquad \mu = 1.3 \times 10^{-6} \text{ reyns}$$

Use the full-Sommerfeld and half-Sommerfeld solutions to determine (a) position of the shaft center, (b) minimum film thickness, and (c) location and magnitude of P_{max}.

In this problem $L/D = 1$ and the ILA approximation will *not* be accurate. We shall address this issue in Section 8.12 in depth. For the purpose of illustration, however, the infinitely long solutions follow for a lightly loaded bearing. The load on each bearing is $W = 3200/2 = 1600$ lb. The projected load is $P_L = 1600/(4.4) = 100$ psi.

From the definition of the Sommerfeld number:

$$S = \frac{\mu_i N_s}{P_L}\left(\frac{R}{C}\right)^2 = \frac{(1.3 \times 10^{-6})(1800/60)}{100}\left(\frac{2}{2 \times 10^{-3}}\right)^2 = 0.39$$

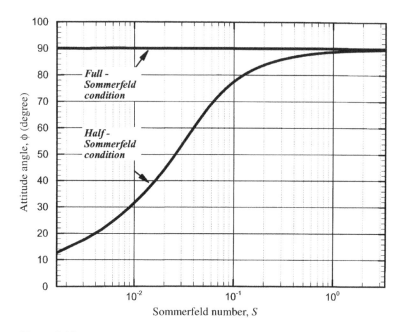

Figure 8.11 Long bearing attitude angle for a specified Sommerfeld number.

1. *Full Sommerfeld.*
    (a) Position of shaft center is determined by using Figures 8.10 and 8.11. With $S =$ 0.39, the steady-state eccentricity ratio and attitude angle are $\varepsilon_0 \cong 0.05$, and $\phi_0 = \pi/2$. The equilibrium shaft position is $(\varepsilon_0, \phi_0) = (0.05, 90°)$. The value of $P_L = 100\,\text{psi}$ is 'light' but fairly common for electric motors and various related machines (see Table 8.11).
    (b) Minimum film thickness $h_{min} = c(1 - \varepsilon_0) = 0.002(1 - 0.05) = 0.0019\,\text{in}.$
    (c) Referring to Table 8.2, the maximum pressure is

$$\overline{P}_{max} = \frac{3\varepsilon_0 \left(4 - \varepsilon_0^2\right)\left(4 - 5\varepsilon_0^2 + \varepsilon_0^4\right)^{0.5}}{2\left(1 - \varepsilon_0^2\right)^2\left(2 + \varepsilon_0^2\right)} = 0.3$$

$$P_{max} = \mu\omega\left(\frac{R}{C}\right)^2 \overline{P}_{max} = 2\pi N_s\mu\left(\frac{R}{C}\right)^2 \overline{P}_{max}$$

$$P_{max} = 2\pi\left(\frac{1800}{60}\right)(1.3 \times 10^{-6})\left(\frac{2}{2 \times 10^{-3}}\right)^2 0.3 = 74\,\text{psi}$$

The location of P_{max} is

$$\theta = \cos^{-1}\left(\frac{-3\varepsilon_0}{2 + \varepsilon_0^2}\right) \cong 105°$$

2. *Half Sommerfeld.*
    (a) Position of shaft center. Using Figures 8.10 and 8.11, we have $\varepsilon_0 \cong 0.1$ and $\phi_0 = 85°$.
    (b) Minimum film thickness $h_{min} = C(1 - \varepsilon_0) = 0.0018$ in.
    (c) Maximum pressure $\overline{P}_{max} = 0.6$, which translates into $P_{max} = 148$ psi.
    The location of P_{max} is $\theta \cong 114°$.

Note: Complete solution to this problem based on numerical simulations of the Reynolds equation with $L/D = 1$ shows that the true equilibrium shaft position is $(0.3\ 68°)$, $h_{min} = 0.0014$ in., $P_{max} \cong 200$ psi, and $\theta_{max} = 126°$. See Section 8.12.

8.7 Cavitation Phenomena

Two types of cavitation are encountered in fluid-film bearings which are related analytically to establishing oil film boundaries: relatively innocuous *gaseous* cavitation and *vapor* cavitation, with its fatigue-type damaging effect on bearing surfaces.

Gaseous Cavitation

Lubricating oil contains roughly 10% by volume of dissolved gas when saturated with air (or most other gases). If oil pressure falls below the usual atmospheric saturation pressure, this dissolved air tends to come out of solution as cavity bubbles. In journal bearings that operate with stationary load under steady-state conditions, this cavitation takes place at the subatmospheric pressure commonly encountered in the divergent section of the oil film.

    The pressurized oil from the convergent film gap cannot occupy the entire clearance space it encounters in the divergent section after passing the minimum film thickness. This insufficient volume of oil then ruptures, commonly into filmlets separated by finger-shaped voids. As ambient pressure then drops toward absolute zero, the air dissolved in the oil tends to expand and develop free bubbles. These cavities containing low-pressure air are then conducted by the converging loaded region of the bearing. Here the cavities experience a relatively gradual increase in pressure, diminish in size, and redissolve in the oil, with no damaging effects. Fluid friction losses in this cavitated region are negligible and oil in this zone drops from its maximum temperature as heat is conducted away into the somewhat cooler shaft and bearing surfaces (Wilcock and Booser, 1988).

Vapor Cavitation

If the load on the bearing fluctuates at high frequency, as with the reaction load from the mesh of a gear train, cavities form in the oil film as the bearing surfaces recede, and then the cavities contract as the surfaces approach each other. The result of this collapse is a high-impact pressure which erodes the bearing surface. The erosion damage is quite similar to that which occurs on the surfaces of propellers of high-powered ships, where water vapor cavities form and eventually collapse with great velocity to give an impact pressure exceeding the fatigue strength of the propeller material. This so-called vapor cavitation in oil films is also considered to be a factor which supplements the usual fatigue damage in reciprocating engine bearings.

For dynamically loaded bearings, formation of a cavitation bubble, its migration, and its eventual collapse require comprehensive modeling. The interested reader is referred to the work of Brewe (1986), Brewe and Khonsari (1988), and Brewe *et al.* (1990) for details of a modeling approach with numerical results that show remarkable agreement with a dynamically loaded test bearing.

This chapter focuses on journal bearings where gaseous cavitation is the primary mode and no bearing damage is expected either from normal film rupture or from air bubbles in diverging oil films under a steady load. Next, we shall describe the appropriate boundary conditions for modeling this gaseous cavitation.

8.8 Swift-Stieber (Reynolds) Boundary Condition

The boundary conditions used in the previous Sommerfeld relations had two shortcomings: the full 360° Sommerfeld condition neglected cavitation altogether, and the half-Sommerfeld condition violated the principle of continuity of flow at $\theta = \pi$. Although Sommerfeld solutions were not available to Reynolds in 1886, he was apparently aware of problems associated with the trailing boundary condition and suggested that pressure should smoothly approach zero at the position where cavitation begins, i.e. $\theta = \theta_{cav}$. This condition requires that

$$\frac{d\overline{P}}{d\theta} = 0 \quad \text{at} \quad \theta = \theta_{cav} \tag{8.22}$$

where the location of cavitation angle θ_{cav} is unknown. It was not until 1932 that Herbert Swift and W. Stieber independently reached the same conclusion. Swift (1931, 1937) proved that condition $\overline{P} = d\overline{P}/d\theta = 0$ must be satisfied in terms of both continuity and the principle of minimum potential energy. This boundary condition has gained wide acceptance for stationary loaded journal bearings.

Figure 8.12 shows the pressure distribution using the Reynolds boundary condition. Schematically, the pressure smoothly approaches the cavitation pressure, and film rupture commences at θ_{cav}. Experimental observations with a transparent bushing show that the film breaks down into a series of filmlets and finger-type cavities (Figure 8.12(c)). Effective length of the film L in the cavitation region is reduced to

$$L_{eff} = \frac{h_{min}}{h(\theta)} L \tag{8.23}$$

Velocity profiles in the circumferential direction in Figure 8.12(d) show that the profile is initially parabolic, and precisely at θ_m it becomes linear since $d\overline{P}/d\theta\big|_{\theta_m} = 0$. Immediately after θ_m, the Poiseuille component of θ_m exists and the velocity profile takes on a parabolic form until θ_{cav} occurs, where $d\overline{P}/d\theta = 0$. Again the velocity profile becomes linear in the entire cavitation region until lubricant is supplied at $\theta = 0\,(2\pi)$. The axial leakage flow in the θ direction is controlled primarily by the convergent region and up to θ_{cav}. There is no side leakage in the cavitation region; air is commonly drawn in from the sides by the subambient pressure.

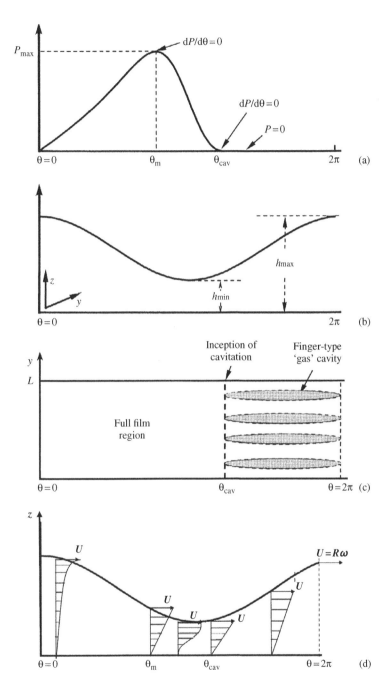

Figure 8.12 Reynolds boundary condition: (a) pressure distribution; (b) unwrapped film profile; (c) illustration of a finger-type cavity; (d) velocity profile.

Accurate prediction of performance parameters, therefore, necessitates the determination of θ_{cav}. Unfortunately, there is no simple analysis that can provide this information, so we must resort to an iterative numerical solution. Referring to Equation (8.8), integration constant A can be determined from an expression for γ_{cav} (or θ_{cav}) at which $\overline{P} = 0$:

$$(1 - \varepsilon \cos \gamma_{cav})(\gamma_{cav} - \varepsilon \sin \gamma_{cav}) - \gamma_{cav}\left(1 + \frac{\varepsilon^2}{2}\right) + 2\varepsilon \sin \gamma_{cav} - \frac{\varepsilon^2}{4}\sin 2\gamma_{cav} = 0 \qquad (8.24)$$

The above expression simplifies to

$$\varepsilon(2\gamma_{cav} - 2\sin\gamma_{cav}\cos\gamma_{cav}) = 4\sin\gamma_{cav} - 4\gamma_{cav}\cos\gamma_{cav} \qquad (8.25)$$

Given an eccentricity ratio, Equation (8.25) can be solved for θ_{cav}, and using the Sommerfeld substitution:

$$\theta_{cav} = \cos^{-1}\left(\frac{\varepsilon - \cos\gamma_{cav}}{\varepsilon\cos\gamma_{cav} - 1}\right)$$

Table 8.3 summarizes the performance parameters for an infinitely long Reynolds solution with the appropriate boundary condition (Pinkus and Sternlicht, 1961). These equations are evaluated numerically for a range of eccentricity ratios from $\varepsilon = 0.05$ to 0.95. The results are given in Table 8.4. Results of numerical solutions for ε as a function of the Sommerfeld are also shown in Figure 8.10.

Table 8.3 Infinitely long journal bearing solution with the Reynolds boundary condition

$$\overline{P} = \frac{6}{(1-\varepsilon^2)^{3/2}}\left\{\gamma - \varepsilon\sin\gamma - \frac{(2+\varepsilon^2)\gamma - 4\varepsilon\sin\gamma + \varepsilon^2\sin\gamma\cos\gamma}{2[1-\varepsilon\cos\gamma_m]}\right\}$$

Cavitation location is $\theta_{cav} = \cos^{-1}\left[\dfrac{(\varepsilon - \cos\gamma_{cav})}{(\varepsilon\cos\gamma_{cav} - 1)}\right]$ where γ_{cav} is obtained from:

$$\varepsilon(2\gamma_{cav} - 2\sin\gamma_{cav}\cos\gamma_{cav}) = 4\sin\gamma_{cav} - 4\gamma_{cav}\cos\gamma_{cav}$$

$$\overline{W}_X = -3\frac{\varepsilon}{(1-\varepsilon^2)}\frac{(1-\cos\gamma_{cav})^2}{(1-\varepsilon\cos\gamma_{cav})} \qquad \overline{W}_Y = \frac{-6}{(1-\varepsilon^2)^{1/2}}\frac{(\gamma_{cav}\cos\gamma_{cav} - \sin\gamma_{cav})}{(1-\varepsilon\cos\gamma_{cav})}$$

$$\overline{W} = \frac{3}{(1-\varepsilon^2)^{1/2}(1-\varepsilon\cos\gamma_{cav})}\left[\frac{\varepsilon^2(1-\cos\gamma_{cav})^4}{1-\varepsilon^2} + 4(\gamma_{cav}\cos\gamma_{cav} - \sin\gamma_{cav})^2\right]^{1/2}$$

$$\phi = \tan^{-1}\left[-\frac{2(1-\varepsilon^2)^{1/2}(\sin\gamma_{cav} - \gamma_{cav}\cos\gamma_{cav})}{\varepsilon(1-\cos\gamma_{cav})^2}\right]$$

$$\left(\frac{R}{C}\right)f = \frac{\varepsilon}{2}\sin\phi + \frac{2\pi^2 S}{(1-\varepsilon^2)^{1/2}}$$

$$S = \frac{(1-\varepsilon^2)(1-\varepsilon\cos\gamma_{cav})}{3\pi[\varepsilon^2(1-\cos\gamma_{cav})^4 + 4(1-\varepsilon^2)(\gamma_{cav}\cos\gamma_{cav} - \sin\gamma_{cav})^2]^{1/2}} = \frac{1}{\pi\overline{W}}$$

Table 8.4 Infinitely long journal bearing solutions with the Reynolds boundary condition

ε	S	ϕ	$\overline{P}_{\max}^{a}$	$\gamma_{\max}$	$\theta_{\max}$	γ_{cav}	θ_{cav}
0.05	0.4759	70	0.4036	103.94	106.7	256.06	253.3
0.1	0.2415	69.03	0.7984	105.31	110.78	254.69	249.22
0.15	0.1631	68	1.1893	106.68	114.77	253.32	245.23
0.2	0.1237	66.9	1.5817	108.03	118.67	251.97	241.33
0.25	0.0999	65.72	1.982	109.37	122.49	250.63	237.51
0.3	0.0838	64.46	2.398	110.71	126.22	249.29	233.78
0.35	0.072	63.11	2.8396	112.03	129.86	247.97	230.14
0.4	0.0629	61.64	3.3199	113.35	133.42	246.65	226.58
0.45	0.0555	60.04	3.8566	114.66	136.9	245.34	223.1
0.5	0.0493	58.3	4.475	115.97	140.31	244.03	219.69
0.55	0.0439	56.37	5.2124	117.28	143.64	242.72	216.36
0.6	0.0389	54.23	6.1265	118.59	146.92	241.41	213.08
0.65	0.0344	51.83	7.3118	119.89	150.16	240.11	209.84
0.7	0.0299	49.1	8.9343	121.2	153.37	238.8	206.63
0.75	0.0256	45.93	11.3123	122.51	156.58	237.49	203.42
0.8	0.0211	42.18	15.1343	123.83	159.83	236.17	200.17
0.85	0.0165	37.59	22.1776	125.15	163.19	234.85	196.81
0.9	0.0115	31.67	38.5663	126.48	166.8	233.52	193.2
0.95	0.0061	23.18	102.7053	127.82	171.03	232.18	188.9

[a] Multiply by 2π to get corresponding maximum pressures in Table 8.6.

8.9 Infinitely Short Journal Bearing Approximation (ISA)

Another type of solution is the opposite extreme, where the bearing length is short, for which the ISA governing equation is

$$\frac{\mathrm{d}}{\mathrm{d}y}\left(\frac{h^3}{\mu}\frac{\mathrm{d}P}{\mathrm{d}y}\right) = 6U\frac{\mathrm{d}h}{\mathrm{d}x} \tag{8.26}$$

The rationale behind this approximation and the valid range of L/D are described in Chapter 6.

Physically, by Equation (8.26), the flow conditions in the direction of rotation do not influence the pressure. This concept was first suggested by Mitchell (1929). Cardullo (1930) integrated the equation and solved it for pressure. Yet, these papers went unnoticed for nearly a quarter of a century until Dubois and Ocvirk (1953), under a contract research for the National Advisory Committee for Aeronautics (NACA, the predecessor of the National Aeronautics and Space Administration [NASA]) verified its validity experimentally. The *Ocvirk short bearing theory*, as it is often referred to, gained almost instant popularity and wide usage. As explained in Chapter 6 and further illustrated in this chapter, the short bearing theory is useful for length-to-diameter ratios up to $L/D = 1/2$ and gives useful approximations of *trends* in performance up to $L/D = 1$.

Integrating Equation (8.26) twice,

$$P = \frac{3\mu U}{h^3} \frac{dh}{dx} y^2 + C_1 y + C_2 \tag{8.27}$$

Applying the boundary conditions $P = 0$ at $y = \pm L/2$,

$$P = \frac{3\mu U}{h^3} \left(y^2 - \frac{L^2}{4} \right) \frac{dh}{dx} \tag{8.28}$$

Since $x = R\theta$ and $dh/d\theta = -C\varepsilon \sin\theta/R$, the pressure distribution becomes

$$P = \frac{3\mu U}{C^2 R} \left(\frac{L^2}{4} - y^2 \right) \frac{\varepsilon \sin\theta}{(1 + \varepsilon \cos\theta)^3} \tag{8.29}$$

Letting $\overline{h} = h/C$, $\overline{y} = y/L$, and $\overline{P} = (C^2 R/\mu U L^2)P$, Equation (8.29) becomes

$$\overline{P} = 3 \left(-\overline{y}^2 + \frac{1}{4} \right) \frac{\varepsilon \sin\theta}{(1 + \varepsilon \cos\theta)^3} \tag{8.30}$$

Pressure is governed by the product of two terms, one related to axial flow and the other to film thickness. Figure 8.13 shows the variation of pressure for an assumed $\varepsilon = 0.5$. Note that pressure takes on a parabolic shape in the axial direction, where the flow is assumed to occur. However, similar to the infinitely long case, the $\sin\theta$ term gives a positive circumferential pressure from 0 to π and a negative circumferential pressure from π to 2π. Once again, large negative pressures are predicted and must be dealt with by either accepting the solution as is (full-Sommerfeld condition) or rejecting the negative pressure region from π to 2π altogether (half-Sommerfeld condition). Clearly, one cannot implement a Reynolds-type pressure boundary condition $(dP/d\theta|_{\theta_{cav}} = 0)$ since this condition does not fit the solution.

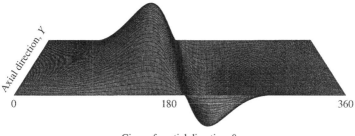

Figure 8.13 Short bearing pressure distribution.

8.10 Full- and Half-Sommerfeld Solutions for Short Bearings (ISA)

Table 8.5 shows a summary of the equations for load components and attitude angle. Again, an attitude angle of 90° is predicted when negative pressures are kept. The definition of the Sommerfeld number that relates the bearing geometry and operating conditions to eccentricity is

$$S = \frac{(1 - \varepsilon^2)^2}{\pi\varepsilon[\pi^2(1 - \varepsilon^2) + 16\varepsilon^2]^{1/2}}\left(\frac{D}{L}\right)^2$$

$$S = \frac{\mu N_S}{P_L}\left(\frac{R}{C}\right)^2 \tag{8.31}$$

Note that eccentricity in Equation (8.31) is determined not by the Sommerfeld number alone, but by a new parameter $S_S = (L/D)^2 S$, sometimes referred to as the *short Sommerfeld number*. Numerical results for ε versus S and ϕ versus S are shown in Figure 8.14 and Figure 8.15, respectively.

Table 8.5 Half and full short bearings solutions

$$\bar{h} = \frac{h}{C}, \quad \bar{y} = \frac{y}{L}, \quad \bar{P} = \frac{R}{\mu U}\left(\frac{C}{L}\right)^2 P, \quad U = R\omega$$

$$\bar{P} = 3\left(-\bar{y}^2 + \frac{1}{4}\right)\frac{\varepsilon \sin\theta}{(1 + \varepsilon\cos\theta)^3} \quad -\frac{1}{2} \le \bar{y} \le \frac{1}{2}$$

$$\bar{P}_{max} = \frac{3}{4}\varepsilon\frac{\sin\theta_m}{(1 + \varepsilon\cos\theta_m)^3} \quad \text{where} \quad \theta_m = \cos^{-1}\left[\frac{(1 - \sqrt{(1 + 24\varepsilon^2)})}{4\varepsilon}\right]$$

$$\bar{Q}_{leak} = Q/(ULC); \quad \bar{F} = FC/(\mu ULR); \quad \bar{W} = \frac{WC^2}{(\mu\omega RL^3)}$$

Full-Sommerfeld	Half-Sommerfeld
$\bar{W}_X = 0$	$\bar{W}_X = \dfrac{-\varepsilon^2}{(1 - \varepsilon^2)^2}$
$\bar{W}_Y = \dfrac{\pi}{2}\dfrac{\varepsilon}{(1 - \varepsilon^2)^{3/2}}$	$\bar{W}_Y = \dfrac{\pi}{4}\dfrac{\varepsilon}{(1 - \varepsilon^2)^{3/2}}$
$\bar{W} = \dfrac{\pi}{2}\dfrac{\varepsilon}{(1 - \varepsilon^2)^{3/2}}$	$\bar{W} = \dfrac{\pi}{4}\dfrac{\varepsilon}{(1 - \varepsilon^2)^2}\sqrt{\left(\dfrac{16}{\pi^2} - 1\right)\varepsilon^2 + 1}$
$\phi = \dfrac{\pi}{2}$	$\phi = \tan^{-1}\left[\dfrac{\pi}{4}\dfrac{\sqrt{1 - \varepsilon^2}}{\varepsilon}\right]$
$S = \dfrac{1}{2\pi^2}\dfrac{(1 - \varepsilon^2)^{3/2}}{\varepsilon}\left(\dfrac{D}{L}\right)^2$	$S = \dfrac{(1 - \varepsilon^2)^2}{\pi\varepsilon[\pi^2(1 - \varepsilon^2) + 16\varepsilon^2]^{0.5}}\left(\dfrac{D}{L}\right)^2$
	$\bar{Q}_{leak} = \varepsilon$
	$\bar{F} = \dfrac{2\pi}{(1 - \varepsilon^2)^{1/2}}$

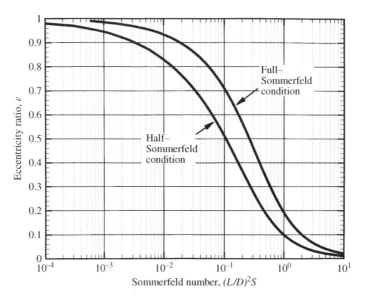

Figure 8.14   Short bearing eccentricity ratio versus Sommerfeld number.

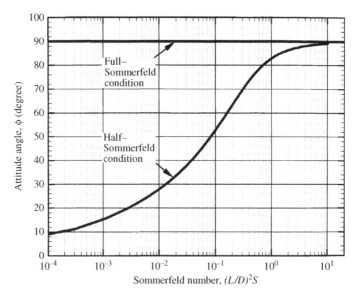

Figure 8.15   Short bearing attitude angle versus Sommerfeld number.

8.11   Bearing Performance Parameters

Leakage Flow Rate and Friction Coefficient

Leakage flow exits both sides of the bearing in the axial direction and is equal to the makeup oil that must be supplied to the bearing. Axial volumetric flow exiting from both sides of the bearing is given by the following equation, as represented in the corresponding term of the Reynolds Equation (6.23(a)):

$$Q_y = 2 \int_0^\pi R \frac{h^3}{12\mu} \frac{dP}{dy}\bigg|_{y=L/2} d\theta \tag{8.32}$$

Substituting

$$\frac{dP}{dy} = \frac{-6\mu U}{Rh^3} y\varepsilon C \sin\theta$$

in the above equation and integrating,

$$Q_{\text{leak}} = \varepsilon ULC = \varepsilon \pi DN_s LC \tag{8.33}$$

Friction force F is obtained by integration of the shear stress over the appropriate area:

$$F = \int_A \tau \, dA$$

where $\tau = \mu du/dz$. The u component of the velocity profile is dominated by the Couette flow only since $dP/d\theta$ is relatively small. Therefore, $du/dz = U/h$ and friction force becomes

$$F = L \int_0^{2\pi} \mu \frac{U}{h} LRd\theta = \frac{\mu ULR}{C} \int_0^{2\pi} \frac{d\theta}{1 + \varepsilon \cos\theta} = \frac{\mu ULR}{C} I_1 \bigg|_0^{2\pi} \tag{8.34}$$

where I_1 is determined using the Sommerfeld substitution as

$$I_1 = \frac{\gamma}{(1 - \varepsilon^2)^{1/2}}$$

Therefore

$$F = \frac{\mu ULR}{C} \frac{2\pi}{(1 - \varepsilon^2)^{1/2}} \tag{8.35}$$

8.12 Finite Journal Bearing Design and Analysis

This section focuses on design and performance analysis based on the full solution of the Reynolds equation. The results pertain to the so-called full, plain cylindrical journal bearing whose angular extent covers $360°$ with any specified L/D ratio.

The bearing is assumed to be steadily loaded, and flow is laminar. The fluid is Newtonian incompressible, and transient effects (i.e. squeeze-film action) are neglected. Under these conditions, a dimensionless form of the Reynolds equation reduces to the following:

$$\frac{\partial}{\partial \theta}\left(\frac{\bar{h}^3}{\bar{\mu}}\frac{\partial \bar{P}}{\partial \theta}\right) + \frac{1}{4\lambda^2}\frac{\partial}{\partial \bar{y}}\left(\frac{\bar{h}^3}{\bar{\mu}}\frac{\partial \bar{P}}{\partial \bar{y}}\right) = 12\pi\frac{\partial \bar{h}}{\partial \theta} \tag{8.36}$$

where

$$\theta = \frac{x}{R}; \qquad \lambda = \frac{L}{D}; \qquad \bar{y} = \frac{y}{L}; \qquad \bar{h} = \frac{h}{C}; \qquad \bar{\mu} = \frac{\mu}{\mu_i}; \qquad \bar{P} = \frac{(P - P_s)}{\mu_i N_s}\left(\frac{C}{R}\right)^2$$

Note that N_S(rev/s) is used in nondimensionalized pressure $\bar{P}$, rather than ω(rad/s).

Assuming the hydrodynamic film begins at h_{max}, the dimensionless film thickness profile is

$$\bar{h} = 1 + \varepsilon \cos\theta \tag{8.37}$$

When the film commences elsewhere around the circumference, for example, at an axial feed groove 90° to the load line, the results should be modified as described in Section 8.13. Accordingly the oil flow rates should be modified as described in Section 8.14. Equation (8.36) is solved numerically using the finite difference method described in Chapter 6. Once the pressure field is determined, all the pertinent bearing performance parameters such as load, flow rates, friction coefficients, etc. can be easily calculated.

Figure 8.16 shows numerical solutions for the eccentricity ratio ε versus the Sommerfeld number S for a wide range of L/D ratios. These solutions are based on treatment of the

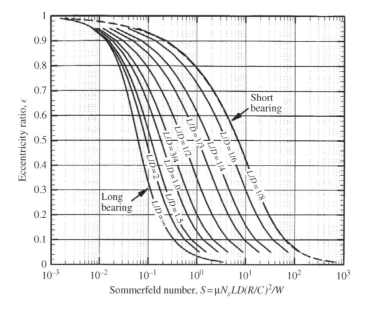

Figure 8.16   Numerical solutions for ε versus S for a wide range of L/D ratios.

Reynolds equation with the proper cavitation boundary condition. Also shown in Figure 8.16 are the corresponding solutions for $L/D = \infty$ (half-Sommerfeld) and the short-bearing solution (half-Sommerfeld) with $L/D = 1/8$.

Tabulated Dimensionless Performance Parameters

Table 8.6 summarizes the results in dimensionless form for a wide range of L/D ratios and $0.05 \le \varepsilon \le 0.95$. Definitions of the dimensionless performance parameters used in Table 8.6 are as follows:

Sommerfeld number:
$$S = \frac{\mu N_s DL}{W}\left(\frac{R}{C}\right)^2 \tag{8.38}$$

Leakage flow rate:
$$Q_L = \overline{Q}_L \left(\frac{\pi}{2}\right) N_s DLC \tag{8.39}$$

Inlet flow rate:
$$Q_i = \overline{Q}_i \left(\frac{\pi}{2}\right) N_s DLC \tag{8.40}$$

Minimum film thickness:
$$h_{min} = C(1 - \varepsilon) \tag{8.41}$$

Friction force:
$$F = fW \tag{8.42}$$

Power loss:
$$E_p = F2\pi RN_s \tag{8.43}$$

Temperature rise:
$$\Delta T = \frac{E_P}{J\rho c_P Q_L} \tag{8.44}$$

where J is the mechanical equivalent of heat if British units are used.

The following examples will illustrate the use of Table 8.6 for design and analysis of journal bearings. To determine that a bearing can function properly, the adequacy of the minimum film thickness, h_{min} (see Equation (8.41)), should be checked with minimum allowable film thickness based on a specified surface finish. Table 8.7 provides a useful guideline based on the surface roughness (ESDU, 1967; Hamrock, 1994). The film thickness values in Table 8.7 are the minimum allowable under clean oil conditions without shaft misalignment taken into consideration. Also, it is assumed that both surface have identical surface roughness. As a rule of thumb, the minimum film thickness should be greater than or at least equal to 20 times the combined conjunction surface roughness values for the surfaces, that is, $h_{min} \ge 10(R_{a,\,shaft}^2 + R_{a,\,bush}^2)^{0.5}$. This can be revealed by referring to Figure 3.17, which shows that hydrodynamic lubrication occurs typically when the film parameter $\Lambda > 10$. At high speeds, it may be necessary to allow for a larger film thickness. At low speeds, satisfactory operation is sometimes possible with as little as 20% of these values. Lastly, adequacy of the minimum film thickness should be verified after the bearing temperature rise is evaluated.

Example 8.2   A large pump has a horizontal rotor weighing 3200 lb supported on two plain $360°$ journal bearings, one on either side of the pump impeller. The specifications of the bearings are as follows

$$R = 2\text{ in.;} \qquad L = 4\text{ in.;} \qquad C = 0.002\text{ in.;} \qquad N = 1800\text{ rpm}$$

The lubricant viscosity $\mu = 1.3 \times 10^{-6}$ reyns (SAE 10 at an inlet temperature of $166\,°F$).

Table 8.6 Numerical solutions for plain 360° journal bearing

L/D	ε	S	$\overline{Q}_L$	$\overline{Q}_i$	$(R/C)f$	$\overline{P}_{max}$	θ_{max}	ϕ	θ_{cav}
2	0.05	1.1279	0.0551	1.0211	22.3219	1.555	102	77.25	234
	0.1	0.5588	0.1088	1.0388	11.1382	3.1471	108	75.17	231
	0.15	0.3678	0.1612	1.0532	7.4197	4.8076	111	73.1	231
	0.2	0.2715	0.2123	1.0643	5.5694	6.565	117	71.03	228
	0.25	0.213	0.2621	1.0724	4.464	8.4574	120	68.94	225
	0.3	0.1733	0.3108	1.0772	3.7285	10.5356	126	66.85	225
	0.35	0.1444	0.3584	1.0793	3.2049	12.8543	129	64.7	222
	0.4	0.1221	0.4049	1.0786	2.8115	15.4881	132	62.49	219
	0.45	0.1043	0.4503	1.0754	2.5033	18.5628	138	60.22	216
	0.5	0.0894	0.4951	1.0692	2.2523	22.2276	141	57.86	216
	0.55	0.0767	0.5389	1.0607	2.0417	26.7118	144	55.37	213
	0.6	0.0656	0.582	1.05	1.8586	32.3784	147	52.72	210
	0.65	0.0556	0.6242	1.0372	1.6943	39.8234	150	49.88	207
	0.7	0.0464	0.6657	1.0241	1.5403	50.0931	153	46.78	204
	0.75	0.0379	0.7065	1.0085	1.39	65.2006	156	43.34	204
	0.8	0.0298	0.7467	0.9916	1.2364	89.4871	159	39.42	201
	0.85	0.0221	0.7861	0.9739	1.0702	134.0831	162	34.82	198
	0.9	0.0145	0.8249	0.9549	0.8769	238.3869	168	29.09	192
	0.95	0.0071	0.8627	0.9361	0.6217	652.069	171	21.16	189
1.5	0.05	1.5448	0.0668	1.0285	30.5612	1.1627	99	79.41	225
	0.1	0.7628	0.1325	1.0547	15.1862	2.3627	105	77.03	225
	0.15	0.4998	0.1971	1.0786	10.0561	3.6295	111	74.66	222
	0.2	0.3667	0.2607	1.1001	7.4864	4.9953	117	72.3	222
	0.25	0.2855	0.3232	1.1194	5.9427	6.4996	120	69.92	219
	0.3	0.2303	0.3848	1.1364	4.9087	8.1877	126	67.53	219
	0.35	0.19	0.4453	1.1513	4.1665	10.1196	129	65.11	216
	0.4	0.1589	0.5051	1.1639	3.6039	12.3702	135	62.64	216
	0.45	0.134	0.5639	1.1745	3.1607	15.0641	138	60.12	213
	0.5	0.1134	0.6221	1.1829	2.7984	18.345	141	57.52	213
	0.55	0.0959	0.6794	1.1893	2.4936	22.4484	144	54.81	210
	0.6	0.0807	0.736	1.1936	2.2301	27.7369	147	51.98	207
	0.65	0.0673	0.7919	1.1961	1.9943	34.8108	150	48.97	207
	0.7	0.0552	0.8471	1.1968	1.7779	44.7305	153	45.74	204
	0.75	0.0442	0.9017	1.1958	1.5727	59.5137	156	42.2	201
	0.8	0.0341	0.9551	1.196	1.3697	83.5183	159	38.26	198
	0.85	0.0246	1.0081	1.1931	1.1609	127.9421	165	33.7	195
	0.9	0.0158	1.0609	1.1873	0.9282	232.444	168	28.12	192
	0.95	0.0075	1.1118	1.1824	0.642	646.358	171	20.5	189
1	0.05	2.7019	0.0807	1.0375	53.4294	0.6823	99	82.18	213
	0.1	1.3302	0.1608	1.0739	26.4433	1.3923	105	79.4	213
	0.15	0.8674	0.2405	1.1092	17.3945	2.1543	111	76.63	213
	0.2	0.6319	0.3196	1.1434	12.8298	2.9958	117	73.86	213
	0.25	0.4876	0.3981	1.1766	10.0627	3.9506	120	71.06	210

Table 8.6   (*Continued*)

L/D	ε	S	$\overline{Q}_L$	$\overline{Q}_i$	(R/C)f	$\overline{P}_{max}$	θ_{max}	φ	θ_{cav}
	0.3	0.3889	0.4763	1.2087	8.1896	5.0587	126	68.25	210
	0.35	0.3164	0.554	1.2396	6.8293	6.3708	129	65.42	210
1	0.4	0.2604	0.6312	1.2696	5.79	7.961	135	62.56	207
	0.45	0.2156	0.708	1.2985	4.9619	9.9301	138	59.65	207
	0.5	0.1786	0.7847	1.3262	4.2824	12.4168	141	56.69	207
	0.55	0.1475	0.8608	1.3529	3.7103	15.6406	144	53.65	204
	0.6	0.1209	0.9368	1.3785	3.2167	19.9477	147	50.52	204
	0.65	0.0979	1.0123	1.403	2.7829	25.9119	150	47.27	204
	0.7	0.0778	1.0877	1.4264	2.3933	34.5708	156	43.83	201
	0.75	0.0601	1.1628	1.4487	2.037	47.8929	159	40.19	198
	0.8	0.0445	1.2377	1.4697	1.7017	70.2402	162	36.2	198
	0.85	0.0308	1.3126	1.4893	1.3785	112.9514	165	31.72	195
	0.9	0.0188	1.3864	1.51	1.0511	215.5045	168	26.4	192
	0.95	0.0084	1.4598	1.5277	0.6896	627.8011	171	19.32	189
3/4	0.05	4.2911	0.0878	1.0421	84.8387	0.4355	99	83.7	207
	0.1	2.112	0.1753	1.0836	41.9534	0.8905	105	80.66	207
	0.15	1.3742	0.2626	1.1246	27.5144	1.3846	111	77.63	207
	0.2	0.9975	0.3496	1.1651	20.1964	1.9389	117	74.59	207
	0.25	0.7658	0.4363	1.2052	15.7341	2.5788	120	71.53	204
	0.3	0.6066	0.523	1.2447	12.6954	3.339	126	68.47	204
	0.35	0.4894	0.6095	1.2837	10.4741	4.2573	132	65.38	204
	0.4	0.3987	0.6958	1.3223	8.7661	5.4	135	62.28	204
	0.45	0.326	0.7821	1.3602	7.4017	6.8439	138	59.14	204
	0.5	0.2663	0.8681	1.3979	6.2802	8.7127	141	55.94	201
	0.55	0.2163	0.9541	1.4349	5.3345	11.1975	144	52.7	201
	0.6	0.174	1.0403	1.4714	4.5232	14.6139	150	49.38	201
	0.65	0.1379	1.1266	1.5072	3.8162	19.4747	153	45.98	201
	0.7	0.1068	1.2126	1.5428	3.1906	26.7257	156	42.42	198
	0.75	0.0803	1.2989	1.5776	2.6305	38.2185	159	38.69	198
	0.8	0.0575	1.3855	1.6119	2.1222	58.1085	162	34.68	195
	0.85	0.0383	1.4723	1.6456	1.6505	97.3573	165	30.27	195
	0.9	0.0223	1.5592	1.6785	1.2019	194.9739	168	25.13	192
	0.95	0.0094	1.6456	1.7107	0.7468	600.1359	171	18.41	189
1/2	0.05	8.7944	0.0941	1.0461	173.8422	0.2154	96	85.06	198
	0.1	4.3273	0.188	1.0921	85.9027	0.442	105	81.74	198
	0.15	2.81	0.2818	1.138	56.1795	0.6916	111	78.43	198
	0.2	2.0322	0.3757	1.1837	41.0417	0.9772	117	75.12	198
	0.25	1.5514	0.4694	1.2292	31.7491	1.3144	123	71.8	198
	0.3	1.22	0.5632	1.2747	25.3869	1.7231	126	68.47	198
	0.35	0.9752	0.657	1.32	20.7103	2.2325	132	65.13	198
	0.4	0.7855	0.7507	1.3652	17.0961	2.8786	135	61.77	198
	0.45	0.6337	0.8446	1.4103	14.1996	3.7169	138	58.4	198
	0.5	0.5093	0.9385	1.4552	11.8145	4.8373	144	54.99	198

(continued)

Table 8.6   (*Continued*)

L/D	ε	S	$\overline{Q}_\mathrm{L}$	$\overline{Q}_\mathrm{i}$	$(R/C)f$	$\overline{P}_\mathrm{max}$	θ_max	ϕ	θ_cav
	0.55	0.4059	1.032	1.5002	9.8131	6.3736	147	51.54	195
	0.6	0.3192	1.126	1.5449	8.1022	8.5489	150	48.03	195
	0.65	0.2463	1.2202	1.5896	6.6262	11.7537	153	44.45	195
	0.7	0.185	1.3146	1.6341	5.3425	16.7182	156	40.78	195
	0.75	0.1338	1.4093	1.6785	4.2197	24.9361	159	36.96	195
	0.8	0.0916	1.5041	1.7228	3.2334	39.8701	162	32.91	192
	0.85	0.0576	1.5995	1.767	2.3621	71.1018	165	28.51	192
	0.9	0.0312	1.6955	1.8111	1.5919	154.0382	168	23.54	192
	0.95	0.0119	1.7916	1.855	0.8916	528.0007	171	17.19	189
1/3	0.05	18.8861	0.0973	1.0482	373.2936	0.1011	99	85.76	192
	0.1	9.2898	0.1944	1.0964	184.3528	0.208	105	82.27	192
	0.15	6.0243	0.2916	1.1445	120.355	0.3271	111	78.8	192
	0.2	4.3465	0.3888	1.1926	87.667	0.4654	117	75.32	192
	0.25	3.3065	0.4859	1.2407	67.5314	0.6315	123	71.83	192
	0.3	2.5882	0.5831	1.2888	53.699	0.8362	129	68.35	192
	0.35	2.0566	0.6803	1.3368	43.4978	1.0948	132	64.86	192
	0.4	1.6445	0.7775	1.3848	35.5971	1.4286	138	61.37	192
	0.45	1.3149	0.8747	1.4328	29.2578	1.8725	141	57.86	192
	0.5	1.0457	0.9719	1.4808	24.0389	2.4756	144	54.33	192
	0.55	0.8228	1.0692	1.5287	19.6628	3.3198	147	50.77	192
	0.6	0.6372	1.1666	1.5766	15.9447	4.5439	150	47.17	192
	0.65	0.4825	1.264	1.6246	12.7583	6.3967	153	43.51	192
	0.7	0.3542	1.3615	1.6725	10.0143	9.3575	156	39.76	192
	0.75	0.2489	1.4593	1.7203	7.6476	14.4421	159	35.87	192
	0.8	0.1641	1.5563	1.7682	5.6151	24.1002	162	31.75	189
	0.85	0.098	1.6543	1.8161	3.8716	45.4776	165	27.31	189
	0.9	0.0492	1.7526	1.8639	2.4024	106.8153	168	22.32	189
	0.95	0.0166	1.8508	1.9117	1.1865	416.307	171	16.14	186
1/4	0.05	32.9983	0.0984	1.049	652.2078	0.058	99	86.01	189
	0.1	16.2271	0.1968	1.0979	321.9788	0.1196	105	82.46	189
	0.15	10.5166	0.2952	1.1469	210.0444	0.1885	114	78.92	189
	0.2	7.5796	0.3936	1.1958	152.8034	0.2691	120	75.37	189
	0.25	5.7573	0.4919	1.2448	117.4957	0.3665	123	71.83	189
	0.3	4.4972	0.5904	1.2937	93.2009	0.4879	129	68.28	189
	0.35	3.5647	0.6887	1.3426	75.278	0.6419	132	64.74	189
	0.4	2.8415	0.7871	1.3915	61.3777	0.8436	138	61.19	189
	0.45	2.2632	0.8856	1.4405	50.2189	1.1131	141	57.63	189
	0.5	1.7913	0.984	1.4894	41.0352	1.4829	144	54.05	189
	0.55	1.4016	1.0824	1.5383	33.3418	2.0064	147	50.45	189
	0.6	1.0778	1.1809	1.5871	26.8159	2.7749	150	46.81	189
	0.65	0.809	1.2795	1.636	21.2381	3.955	153	43.1	189
	0.7	0.5873	1.378	1.6849	16.4563	5.8722	156	39.3	189
	0.75	0.4068	1.4766	1.7338	12.3582	9.2315	159	35.37	189

Table 8.6   (*Continued*)

L/D	ε	S	$\overline{Q}_L$	$\overline{Q}_i$	(R/C)f	$\overline{P}_{max}$	θ_{max}	φ	θ_{cav}
	0.8	0.2631	1.5753	1.7827	8.8657	15.7753	162	31.23	189
	0.85	0.153	1.6741	1.8315	5.9244	30.7464	165	26.78	189
	0.9	0.0736	1.7734	1.8804	3.4979	75.8856	168	21.78	189
	0.95	0.0228	1.8708	1.9293	1.571	324.2043	171	15.55	186
	0.05	73.3037	0.0993	1.0495	1448.799	0.0262	99	86.19	186
	0.1	36.0374	0.1985	1.0991	714.9882	0.054	105	82.6	186
	0.15	23.3433	0.2978	1.1486	466.1314	0.0853	114	79	186
	0.2	16.8097	0.397	1.1982	338.7537	0.1222	120	75.4	186
	0.25	12.7522	0.4963	1.2477	260.0968	0.167	126	71.81	186
1/6	0.3	9.9457	0.5956	1.2972	205.9424	0.2231	129	68.22	186
	0.35	7.8664	0.6949	1.3467	165.9239	0.2951	135	64.64	186
	0.4	6.2546	0.7941	1.3963	134.8851	0.39	138	61.05	186
	0.45	4.9661	0.8934	1.4458	109.962	0.5177	141	57.45	186
	0.5	3.9157	0.9927	1.4953	89.4564	0.6946	144	53.84	186
	0.55	3.0494	1.092	1.5449	72.2868	0.9476	147	50.21	186
	0.6	2.3314	1.1913	1.5944	57.7448	1.3238	150	46.53	186
	0.65	1.7371	1.2906	1.6439	45.3447	1.9096	153	42.79	186
	0.7	1.249	1.3899	1.6934	34.7463	2.8777	156	38.96	186
	0.75	0.8542	1.4892	1.743	25.7086	4.6093	159	35	186
	0.8	0.5427	1.5885	1.7925	18.0628	8.0726	162	30.82	186
	0.85	0.3071	1.6878	1.842	11.6972	16.2889	165	26.33	186
	0.9	0.1411	1.7872	1.8915	6.5543	42.475	168	21.28	186
	0.95	0.0399	1.8865	1.941	2.6472	202.2193	171	15.1	186
1/8	0.05	129.7266	0.0996	1.0497	2563.938	0.0148	99	86.27	186
	0.1	63.7582	0.1992	1.0995	1264.931	0.0305	105	82.65	186
	0.15	41.2921	0.2988	1.1492	824.4794	0.0483	114	79.04	186
	0.2	29.7252	0.3983	1.199	598.9516	0.0693	120	75.43	186
	0.25	22.5403	0.4979	1.2487	459.6426	0.0948	126	71.82	186
	0.3	17.5694	0.5975	1.2984	363.6929	0.1269	129	68.22	186
	0.35	13.8862	0.6971	1.3482	292.7708	0.1682	135	64.62	186
	0.4	11.0306	0.7967	1.3979	237.7464	0.2227	138	61.02	186
	0.45	8.7484	0.8963	1.4476	193.5629	0.2963	141	57.41	186
	0.5	6.8884	0.9959	1.4974	157.209	0.3987	144	53.8	186
	0.55	5.3551	1.0955	1.5471	126.781	0.546	147	50.15	186
	0.6	4.0854	1.1951	1.5969	101.0214	0.7661	150	46.47	186
	0.65	3.0357	1.2947	1.6466	79.074	1.1112	153	42.72	186
	0.7	2.175	1.3943	1.6963	60.3371	1.6862	156	38.89	186
	0.75	1.4802	1.494	1.7461	44.3872	2.725	159	34.91	186
	0.8	0.9339	1.5936	1.7958	30.9285	4.8301	162	30.73	186
	0.85	0.5226	1.6932	1.8455	19.7689	9.9151	165	26.23	186
	0.9	0.2353	1.7929	1.8953	10.8169	26.5901	168	21.15	186
	0.95	0.0631	1.8918	1.945	4.1246	134.5239	171	14.87	183

Determine the following: (a) equilibrium position of the shaft center and location of the film rupture; (b) minimum film thickness; (c) location and magnitude of the maximum pressure; (d) power loss; (e) temperature rise.

Load per bearing is $W = 3200/2 = 1600$ lbf. Equation (8.38) gives the Sommerfeld number as

$$S = \frac{\mu N_S L D}{W} \left(\frac{R}{C}\right)^2 = \frac{1.3 \times 10^{-6} \times 1800/60 \times 4 \times 4}{1600} \left(\frac{2}{2 \times 10^{-3}}\right)^2 = 0.39$$

Using Table 8.6 with $L/D = 1$ and $S = 0.39$:

$$\varepsilon = 0.3; \qquad \overline{Q}_L = 0.4763; \qquad \left(\frac{R}{C}\right)f = 8.1896; \qquad \overline{P}_{max} = 5.0587;$$

$$\theta_{max} = 126; \quad \varphi = 68.25°; \qquad \theta_{cav} = 210°$$

(a) Equilibrium position of the shaft center is $(\varepsilon, \varphi) = (0.3, 68.25°)$. Cavitation begins at $\theta_{cav} = 210°$. This equilibrium position may be modified, since this result corresponds to an assumed viscosity at the inlet temperature.

(b) Minimum film thickness is $h_{min} = C(1 - \varepsilon) = 0.0014$ in. This minimum film thickness of 1400 μin. is far above the 500 μin. minimum allowable film thickness in Table 8.7 for a typical 32 μin. finish.

Table 8.7   Allowable minimum film thickness

Surface finish[a] (center line avg. R_a)		Description of surface	Examples of manufacturing methods	Allowable minimum film thickness	
μm	μin.			μm	μin.
0.1–0.2	4–8	Mirror-like surface without toolmarks	Grind, lap, superfinish	2.5	100
0.2–0.4	8–16	Smooth surface without scratches, close tolerances	Grind and lap	6.2	250
0.4–0.8	16–32	Smooth surface, close tolerances	Grind and lap	12.5	500
0.8–1.6	32–63	Accurate bearing surface without toolmarks	Grind, precision mill, and fine tuning	25	1000
1.6–3.2	63–125	Smooth surface without objectionable toolmarks, moderate tolerances	Shape, mill, grind, and turn	50	2000

[a] *Average* Combined conjunction roughness $R_a = (R_{a,\,shaft}^2 + R_{a,\,bearing}^2)^{0.5}$ (see Chapter 3).
Source: ESDU, 1967 and Hamrock, 1994.

(c) Maximum pressure from the definition of dimensionless parameters given in Equation (8.36) is

$$P_{max} = \mu_i N_s \left(\frac{R}{C}\right)^2 \overline{P}_{max} + P_S$$

$$= (1.3 \times 10^{-6}) \left(\frac{1800}{60}\right) \left(\frac{2}{0.002}\right)^2 (5.0587) = 197.3 \text{ psi}$$

(d) The friction coefficient, friction force, and power loss from Equations (8.42) and (8.43) are

$$f = 0.0082; \qquad F = fW = 13.1 \text{ lbf}$$
$$E_P = F(2\pi R N_s) = (13.1)(2\pi)(2)(30) = 4939 \text{ in. lbf/s} (= 0.7 \text{ hp})$$

(e) To determine the temperature rise, the flow rate is first computed from Equation (8.39):

$$Q_L = (0.4763) \left(\frac{\pi}{2}\right) (30)(4)(4)(0.002) = 0.7182 \text{ in}^3/\text{s}$$

The density ρ and the specific heat c_p of most conventional petroleum oils may be assumed to remain relatively constant. Typically, the following numbers are used for these parameters:

$$\rho = 0.0315 \frac{\text{lbm}}{\text{in}^3} \quad \text{and} \quad c_p = 0.48 \frac{\text{BTU}}{\text{lbm} \,^\circ\text{F}}$$

Therefore, the temperature rise is

$$\Delta T = \frac{4939}{(778)(12)(0.48)(0.0315)(0.7182)} \approx 48.7 \,^\circ\text{F}$$

The effective temperature T_e is related to the inlet temperature, T_{in}, as follows:

$$T_e = T_{in} + \Delta T = 166 + 48.7 = 215 \,^\circ\text{F}$$

This temperature is well above the 166 °F used for the initial viscosity. Next, an updated effective viscosity, μ_e, is computed for the SAE 10 oil at $T_e = 215$ °F. This process should be repeated until the difference in the updated mean temperature between two successive iterations is less than 2–3 °F.

Since a unit load of 100 psi is low compared with usual design practice (see Table 8.11) and the temperature is relatively high, a trial of a modification is indicated with the bearing length reduced by half and the radial clearance doubled. Vibration response and possible interference at shaft seals should then also be checked.

Example 8.3   The performance of a full journal bearing with a minimum film thickness of $h_{min} = 0.0002$ in. is to be evaluated. The specifications for the bearing are as follows:

$$D = 2.0 \text{ in.;} \qquad L = 2.0 \text{ in.;} \qquad C = 0.001 \text{ in.;}$$
$$N = 1000 \text{ rpm;} \qquad \mu = 2.0 \times 10^{-6} \text{ reyns;} \qquad P_S = 40 \text{ psi}$$

Determine the load-carrying capacity of this bearing.

For this h_{min} operating eccentricity ratio $\varepsilon = 1 - h_{min}/C = 0.8$. Utilizing Table 8.6 with $\varepsilon = 0.8$ and $L/D = 1$:

$$\overline{Q}_{leak} = 1.2377; \qquad \left(\frac{R}{C}\right)f = 1.7017; \qquad \overline{P}_{max} = 70.2402;$$
$$\phi = 36.2°; \qquad \theta_{cav} = 198°$$

Equilibrium position of the shaft center at the specified viscosity is

$$(\varepsilon, \varphi) = (0.8, 36.2°)$$

Maximum pressure is

$$P_{max} = \mu_i N_S \left(\frac{R}{C}\right)^2 \overline{P}_{max} + P_S$$
$$= (2 \times 10^{-6}) \left(\frac{1000}{60}\right) \left(\frac{1}{0.001}\right)^2 (70.2402) + 40 \approx 2381 \text{ psi}$$

Position of the peak pressure is predicted to occur at $\theta_{max} = 162°$ ($192°$ from the load line). From the Sommerfeld number in Equation (8.38):

$$S = \frac{2 \times 10^{-6}(1000)(1/60)(2.0)(2.0)}{W} \left(\frac{1}{0.001}\right)^2 = \frac{133.33}{W}$$

Solving for the load-carrying capacity, $W \approx 3000$ lbf. This load-carrying capacity does not take into account the variation of viscosity with temperature and is therefore overpredicted. The mean temperature and viscosity should be predicted to determine a more accurate load-carrying capacity.

8.13 Attitude Angle for Other Bearing Configurations

The numerical results of Table 8.6 can be well represented by the following relationship (Martin, 1998):

$$\phi = \tan^{-1} \left[\left(\frac{\pi}{z\varepsilon}\right) \sqrt{1 - \varepsilon^2} \right] \qquad (8.45)$$

where

$$z = 4 \left[1 + \left(\frac{L}{D} \right) (1 - 1.25\varepsilon^{\delta}) \right] \tag{8.46}$$

which reduces to the corresponding short bearing expression when $L/D \to 0$. Parameter δ pertains to the film's starting position. The results of Table 8.6 are intended for situations where the film starts at h_{max}. For this case, $\delta = 0.18$. If the film begins elsewhere around the circumference – for example when an inlet groove or a hole is placed at a $90°$ angle to the load line – then the equation must be modified. For $L/D \leq 1.0$ and $0.3 \leq \varepsilon \leq 0.9$, Martin (1998) recommends the following. For film starting from h_{max}:

$$\delta = 0.18$$

For film starting from $90°$ to the load line:

$$\delta = 0.505$$

For a feed groove of angle α_g (in degrees) in the range of $90° \leq \alpha_g \leq 120°$:

$$\delta = 0.833 \left(\frac{\alpha_g}{100} \right)^{4.75} \tag{8.47}$$

8.14 Lubricant Supply Arrangement

In this chapter we concentrate on fully flooded lubrication. We focus on the starved lubricant supply as well as dry operation in Chapter 12.

Various methods for distributing lubricant within a journal bearing are available. They range from a supply hole at the inlet to a variety of grooves cut into the bearing's internal surface. The grooves' configuration may be axial, circumferential, or even spiral-shaped, depending on the application.

Supply Hole

A common supply method with small bearings and bushings is to place an inlet port at the bearing midplane opposite to the load line ($\theta = -\phi$), as shown in Figure 8.17(a). Alternatively, a hole may be placed at the maximum film thickness h_{max}, where $\theta = 0$ (Figure 8.17(b)). Clearly, the position of h_{max} depends on the attitude angle, which is a function of load. Therefore, locating its precise location is not practical for all applications. Lubricant is generally fed at a specified supply pressure P_s.

Axial Groove

An axial groove is a single groove of length L_g extending axially from the bearing midplane (opposite to the load line) to either side of the bearing length, leaving a prescribed land segment l at both ends (Figure 8.17(d)). Typically, the total groove length is about 80% of the bearing

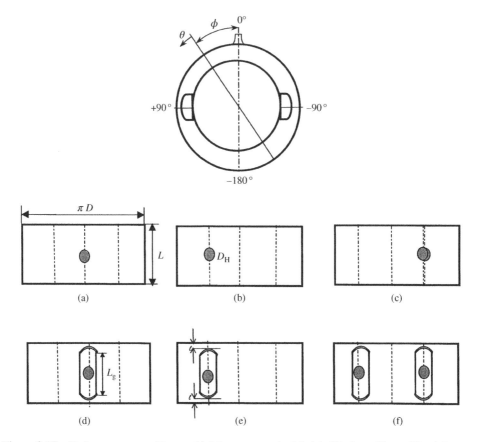

Figure 8.17   Various groove positions and inlet arrangements: (a)–(c) oil hole positions; (d) axial groove at bearing midplane; (e) axial groove at +90° to the load line; (f) double axial grooves at ±90° to the load line.

length, leaving $\ell = 0.4L$ on either side. The depth of the groove and its width w_g vary with different designs. In the absence of any particular specifications, the minimum groove width should be at least one-tenth of the diameter, $w_g = D/10$, with a depth of $D/30$.

   An axial groove is particularly useful for distributing oil over the entire length of the bearing to control its temperature. An alternative location for an axial groove is 90° to the load line, as shown in Figure 8.17(e). In some instances, two axial grooves running parallel at ±90° to the load line are used. This groove configuration (Figure 8.17(f)) is very common for general use in industrial applications, with the axial grooves formed at the split between the halves of a cylindrical bearing.

   If the bearing is steadily loaded, either the supply hole or an axial groove can do the job as long as it is placed in the unloaded portion of the bearing. Clearly, pressure generation is interrupted if a groove or a supply hole is placed in the active region where hydrodynamic pressure is generated. As a guide, if the bearing length is short, then a supply hole is recommended. For a bearing length above about 25–50 mm, an axial groove is more appropriate for better lubricant distribution.

Circumferential Groove

In applications where the direction of load changes over a fairly wide angle of the circumference, such as ±90°, one generally uses a circumferential groove. These grooves are typically seen in dynamically loaded bearings such as engine bearings. A circumferential groove is generally placed centrally, thus dividing the bearing into two identical segments, l'.

Spiral Grooves

In the absence of an oil feed system, pumping action of internal spiral or helical grooves can often be useful in directing oil flow in a sleeve bearing. With wick-oiling and with mist lubrication, similar spiral or diagonal grooving can also minimize end leakage by directing flow inward from the axial ends of a bearing (Khonsari and Booser, 2003).

Three helical grooves, like the one shown in Figure 8.2, represent a typical arrangement for feeding oil from a surrounding oil reservoir through a journal bearing in a vertical drive. Bulk velocity of oil within a groove slanted in the direction of shaft rotation is approximately half the velocity of the shaft in the groove direction. Approximation of this bulk flow Q (cm³/min) in a groove angled at 45° with width w (cm) and depth h (cm) is then given by $Q = wh(\sin 45°)\pi DN/2$ where D is the journal diameter (cm) and N the rotational rpm. Taking into account end effects and geometric differences, groove feed flow will usually be about 80% of this value. Groove width is commonly of the order $D/10$ and depth $D/30$.

Applying this relation for shaft diameter $D = 7.5$ cm, groove width $w = 0.75$ cm, and groove depth $h = 0.25$ cm at shaft speed $N = 900$ rpm:

$$Q = (0.75)(0.25)(\sin 45°)(\pi)(7.5)(900)/2 = 1406 \text{ cm}^3/\text{min}$$

This flow through the bearing introduced by a single groove would then be multiplied by a factor of three for the typical three spiral grooves in such a bearing. The groove dimensions should then be checked and adjusted for temperature rise as this oil flow volume is heated by the bearing power loss (such as calculated by heat balance Equation (8.43) after using Petroff power loss Equation (5.27)).

8.15 Flow Considerations

Holes or Axial Grooves

Side leakage is composed of two components. One is axial flow Q_L, due to the pumping action of an eccentric shaft rotating within the sleeve. The other is pressure-induced flow Q_p, due to the feed (supply) pressure at the inlet hole or a groove.

Axial Flow Due to Rotation

The first component, Q_L, can be predicted from Table 8.6 with a specified Sommerfeld number (or ϵ) and L/D ratio. Recall that in generating the result in Table 8.6, it was assumed that hydrodynamic film begins from the maximum film thickness position. Modification to the flow rates would be needed if the film commences elsewhere around the circumference. Very useful curve fit solutions are available for this purpose. Martin (1998) shows that

$$Q_L = f_L \pi LDN_S C$$

where f_L is a correction factor for the film position as given below.

1. For an oil hole or axial groove positioned in the unloaded section of the bearing opposite to the load line:

$$f_L = \varepsilon f_1$$
$$f_1 = 1 - 0.22 \left(\frac{L}{D}\right)^{1.9} \varepsilon^{0.02} \tag{8.48}$$

2. For an oil hole or axial groove positioned at the maximum film thickness:

$$f_L = 0.965 \varepsilon f_1 \tag{8.49}$$

3. For double axial grooves running parallel at $\pm 90°$ angles to the load line:

$$f_L = \varepsilon f_1 f_2 / f_3$$
$$f_2 = 0.25 + c_1 (1 - \varepsilon)^{c_2}$$
$$f_3 = 10^{c_3}$$
$$c_1 = 0.73 \left(\frac{L}{D}\right)^{-0.035} \tag{8.50}$$
$$c_2 = 0.73 \left(\frac{L}{D}\right)^{0.66}$$
$$c_3 = 0.444 \left(\frac{L}{D}\right)^{0.35} \left(\frac{w_g}{D}\right)^{1.2} (1 + 1.11\varepsilon)$$

where ω_g represents the circumferential width of groove.

4. For a full film starting from the maximum film thickness position, i.e. the results in Table 8.6, we have

$$Q_L = \pi \varepsilon \left[1 - 0.22 \left(\frac{L}{D}\right)^{1.9} \varepsilon^{0.02}\right] LDN_S C \tag{8.51}$$

Equation (8.51) applies to $0.25 \leq L/D \leq 1.0$ and $0.2 \leq \varepsilon \leq 0.9$. In the limiting case when $L/D \to 0$, this expression reduces to $Q_L = \pi LDN_s C\varepsilon$, which is identical to the leakage flow rate expression derived for a short bearing.

Pressure-Induced Flow

To determine the pressure-induced flow Q_p, specification of the position, geometry, and supply pressure at the inlet port is required. Martin (1998) shows:

$$Q_p = f_g \left(\frac{h_g}{C}\right)^3 \frac{P_S C^3}{\mu_i} \tag{8.52}$$

where P_S is the supply pressure and μ_i is the lubricant viscosity at the inlet (or supply) temperature. Parameter f_g is a function of either the hole or the axial groove dimensions, and h_g/C is the film thickness at the inlet port(s). These parameters can be determined as follows (Martin, 1998). Inlet hole of diameter D_H:

$$f_g = 0.675 \left(\frac{D_H}{L} + 0.4\right)^{1.75} \tag{8.53}$$

Axial groove of width w_g and length L_g:

$$f_g = \frac{1.25 - 0.25(L_g/L)}{3\sqrt[3]{(L/L_g) - 1}} + \frac{\omega_g/D}{3(L/D)(1 - L_g/L)} \tag{8.54}$$

1. The film thickness parameter for an oil hole or an axial groove positioned in the unloaded section of the bearing opposite the load line is

$$\left(\frac{h_g}{C}\right)^3 = (1 + \varepsilon \cos \phi)^3 \tag{8.55}$$

2. For an oil hole or axial groove positioned at the maximum film thickness:

$$\left(\frac{h_g}{C}\right)^3 = (1 + \varepsilon)^3 \tag{8.56}$$

3. For double axial grooves running parallel at $\pm 90°$ angles to the load line:

$$\left(\frac{h_g}{C}\right)^3 = (1 + \varepsilon \cos \phi)^3 + (1 - \varepsilon \sin \phi)^3 \tag{8.57}$$

Total Leakage Flow Rate

Based on curve fitting to actual experimental results, Martin (1998) prescribes the following formula for predicting the total leakage flow rate. First, determine a datum flow rate Q_m which calculates the flow rate by ignoring the oil film continuity using the following expression:

$$Q_m = Q_L + Q_P - 0.3\sqrt{Q_L Q_p} \qquad (8.58)$$

Then correct for the film reformation according to the following guidelines. These assume that the oil in the film leaving the converging section will recirculate to join with fresh feed oil at the inlet hole or axial groove:

1. For an oil hole or an axial groove positioned in the unloaded section of the bearing opposite to the load line:

$$Q_{L,\text{total}} = Q_m^{S'} Q_P^{1-S'} \qquad (8.59)$$

$S' = 0.6$ for an oil hole of diameter D_H (in the range of $D_H/L = 0.1$–0.25), and $S' = 0.7(L_g/L)^{0.7} + 0.4$ for an axial groove of length L_g.
2. For an axial groove of length $L_g (L_g/L = 0.3$–$0.8)$ positioned at the maximum film thickness or two axial grooves running parallel at $\pm 90°$ angles to the load line:

$$Q_{L,\text{total}} = Q_m \left(\frac{L_g}{L} \right)^m \qquad (8.60)$$

where $m = 0.27(Q_L/Q_P)^{0.27}$.

The above expressions are grouped together for convenience in Table 8.8.

Example 8.4   Consider a journal bearing with the following specifications that correspond to an actual bearing tested by Dowson *et al.* (1966): $L/D = 0.75$; $R/C = 800$; $D = 0.102$ m, $W = 11\,000$ N, and operating speed is $N_s = 25$ rev/s. An axial groove was cut into the bearing surface in the unloaded portion of the bearing, opposite the load line. The groove width is $\omega_g = 4.76 \times 10^{-3}$ m, and it is $L_g = 0.067$ m long. Lubricant is supplied to the bearing at temperature $T_i = 36.8\,°C$ at a supply pressure of $P_S = 0.276 \times 10^6$ pa(40 psi). The lubricant viscosity is a function of temperature and varies according to $\mu = \mu_i e^{-\beta(T-T_i)}$ with $\mu_i = 0.03$ pa s, and the temperature–viscosity coefficient is estimated to be $\beta = 0.0414$. Lubricant thermal conductivity $k = 0.13$ W/m K, and thermal diffusivity $\alpha_t = 0.756 \times 10^{-7}$ m^2/s. Determine the flow rates, power loss, attitude angle, and maximum pressure.

Using the definition of the Sommerfeld number and Table 8.6, we determine the operating eccentricity ratio with viscosity evaluated at the inlet temperature:

$$S = \frac{\mu N_S L D}{W} \left(\frac{R}{C} \right)^2 = \frac{0.03 \times 25.0 \times 0.0762 \times 0.102}{11000} (800)^2 = 0.338$$

Table 8.8 Flow rate expressions for axial grooves and inlet hole with a specified supply pressure

Inlet specifications	Hydrodynamic leakage flow rate, Q_L	Pressure-induced flow, Q_p	Total leakage flow, $Q_{L,\text{total}}$
Hole D_H ▢ Groove length L_g ▭ Groove width w_g	$Q_L = \pi L D N_S C f_L$	$Q_p = \dfrac{f_g (h_g/C)^3 P_S C^3}{\mu_i}$	$Q_m = Q_L + Q_P - 0.3\sqrt{Q_L Q_p}$
Inlet hole of diameter D_H at opposite load line	$f_L = \epsilon f_1$ $f_1 = 1 - 0.22\left(\dfrac{L}{D}\right)^{1.9}\epsilon^{0.02}$	$f_g = 0.675\left(\dfrac{D_H}{L}+0.4\right)^{1.75}$ $\left(\dfrac{h_g}{C}\right)^3 = (1+\epsilon\cos\phi)^3$	$Q_{L,\text{total}} = Q_m^{S'} Q_p^{1-S'}$ $S' = 0.6$ $\left(\dfrac{D_H}{L} = 0.1 \text{ to } 0.25\right)$
Axial groove of length L_g and width w_g at opposite load line 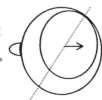	$f_L = \epsilon f_1$ $f_1 = 1 - 0.22\left(\dfrac{L}{D}\right)^{1.9}\epsilon^{0.02}$	$f_g = \dfrac{1.25 - 0.25(L_g/L)}{3\sqrt[3]{(L/L_g) - 1}} + \dfrac{w_g/D}{3\left(\frac{L}{D}\right)(1 - L_g/L)}$ $\left(\dfrac{h_g}{C}\right)^3 = (1+\epsilon\cos\phi)^3$	$Q_{L,\text{total}} = Q_m^{S'} Q_P^{1-S'}$ $S' = 0.75\left(\dfrac{L_g}{L}\right)^{0.7} + 0.4$

Table 8.8 (Continued)

Inlet specifications	Hydrodynamic leakage flow rate, Q_L	Pressure-induced flow, Q_p	Total leakage flow, $Q_{L,\,total}$
Axial groove at h_{max} position	$f_L = 0.965\epsilon f_1$	$f_g = \dfrac{1.25 - 0.25(L_g/L)}{3\sqrt[3]{(L/L_g)}-1} + \dfrac{w_g/D}{3\left(\frac{L}{D}\right)(1-L_g/L)}$ $\left(\dfrac{h_g}{C}\right)^3 = (1+\epsilon)^3$	$Q_{L,\,total} = Q_m\left(\dfrac{L_g}{L}\right)^m$ $m = 0.27\left(\dfrac{Q_L}{Q_p}\right)^{0.27}$ $\left(\dfrac{L_g}{L} = 0.3 \text{ to } 0.8\right)$
Two axial grooves running parallel at $\pm 90°$ to load line	$f_L = \epsilon f_1 f_2 / f_3$ $f_2 = 0.25 + c_1(1-\epsilon)^{c_2}$ $f_3 = 10^{c_3}$ $c_1 = 0.73\left(\dfrac{L}{D}\right)^{-0.035}$ $c_2 = 0.73\left(\dfrac{L}{D}\right)^{0.66}$ $c_3 = 0.444\left(\dfrac{L}{D}\right)^{0.35}\left(\dfrac{w_g}{D}\right)^{1.2}(1+1.11\epsilon)$	$f_g = \dfrac{1.25 - 0.25(L_g/L)}{3\sqrt[3]{(L/L_g)}-1} + \dfrac{w_g/D}{3\left(\frac{L}{D}\right)(1-L_g/L)}$ $\left(\dfrac{h_g}{D}\right)^3 = (1+\epsilon\cos\phi)^3 + (1-\epsilon\sin\phi)^3$	$Q_{L,\,total} = Q_m\left(\dfrac{L_g}{L}\right)^m$ $m = 0.27\left(\dfrac{Q_L}{Q_p}\right)^{0.27}$ $\left(\dfrac{L_g}{L} = 0.3 \text{ to } 0.8\right)$

From Table 8.6, with $L/D = 0.75, \epsilon \cong 0.45$. The table also gives the following predictions for the leakage flow rate (due to the shaft rotation alone), friction coefficient, and attitude angle:

$$\overline{Q} = 0.7821; \qquad (R/C)f = 7.4017; \qquad \phi = 59.14°$$

Therefore

$$Q_L = \left(\frac{\pi}{2}\right) N_S DLC\overline{Q} = 1.51 \times 10^{-5}\,\text{m}^3/\text{s} = 15.1\,\text{cm}^3/\text{s}$$

Alternatively, we can use the curve-fit equations of Table 8.8 to determine the leakage flow:

$$f_1 = 1 - 0.22\left(\frac{L}{D}\right)^{1.9}\epsilon^{0.02} = [1 - 0.22(0.75)^{1.9}(0.45)^{0.02}] = 0.876$$
$$f_L = \epsilon f_1 = 0.45(0.876) = 0.394$$
$$Q_L = \pi LDN_S Cf_L = \pi(0.0762)(0.1)(25)(6.25 \times 10^{-5})(0.394)$$
$$= 1.52 \times 10^{-5}\,\text{m}^3/\text{s}$$

The attitude angle can also be determined using the curve-fit relationship. Since the film is assumed to start from h_{max}, $\delta = 0.18$, and from Equations (8.45) and (8.46) we obtain

$$z = 4\left[1 + \left(\frac{L}{D}\right)(1 - 1.25\epsilon^\delta)\right] = 3.75$$

and

$$\phi = \tan^{-1}\left[(\pi/z\epsilon)\sqrt{(1 - \epsilon^2)}\right] = 59°$$

Next, the pressure-induced flow must be determined. From Table 8.8, determine the groove function and the related film thickness:

$$f_g = \frac{1.25 - 0.25(L_g/L)}{3\sqrt[3]{(L/L_g)} - 1} + \frac{w_g/D}{3(L/D)(1 - L_g/L)}$$

$$= \frac{1.25 - 0.25(0.879)}{3\sqrt[3]{1.137} - 1} + \frac{4.76 \times 10^{-3}/0.102}{3(0.75)(1 - 0.879)} = 0.838$$

$$\left(\frac{h_g}{C}\right)^3 = (1 + \epsilon\cos\phi)^3 - (1 + 0.45\cos 59)^3 - 1.87$$

$$Q_p = f_g\left(\frac{h_g}{C}\right)^3 P_S\frac{C^3}{\mu_i}$$

$$= \frac{0.838(1.87)(2.76 \times 10^5)(6.35 \times 10^{-5})^3}{0.03}$$

$$= 3.69 \times 10^{-6}\,\text{m}^3/\text{s} = 3.69\,\text{cm}^3/\text{s}$$

The total leakage is determined as follows. First, the datum flow rate Q_m is established:

$$Q_m = Q_L + Q_p - 0.3\sqrt{Q_L Q_p}$$
$$= 14.7 + 3.53 - 0.3\sqrt{(14.7)(3.53)}$$
$$= 16.6\,\text{cm}^3/\text{s}(16.6 \times 10^{-6}\,\text{m}^3/\text{s})$$

From Table 8.8, for an axial groove positioned in the unloaded section of the bearing opposite to the load line, we have

$$S' = 0.75\left(\frac{L_g}{L}\right)^{0.7} + 0.4 = 1.085$$

Therefore, the total leakage flow rate becomes

$$Q_{L,\text{total}} = Q_m^{S'} Q_P^{1-S'}$$
$$= (16.6)^{1.085}(3.69)^{-0.085}$$
$$= 18.81\,\text{cm}^3/\text{s}(18.81 \times 10^{-6}\,\text{m}^3/\text{s})$$

The power loss is evaluated using the following expression:

$$E_p = FU = fW(2\pi R N_S)$$
$$= 7.4017\left(\frac{1}{800}\right)(11\,000)(\pi \times 0.102 \times 25)$$
$$= 812\,\text{W}(1.09\,\text{hp})$$

From the definition of thermal diffusivity, we have $\rho C_P = k/\alpha_t = 1.72 \times 10^6\,\text{W.s}/(\text{m}^3\text{K})$. Therefore, the temperature rise is estimated to be

$$\Delta T = \frac{E_P}{(\rho C_P Q_{L,\text{total}})} = \frac{812}{(1.72 \times 10^6 \times 18.81 \times 10^{-6})} = 25.1\,°\text{C}$$

Effective temperature for evaluating the next iteration is (see Section 8.17):

$$T_e = T_i + \Delta T = 36.8 + 25.1 = 61.9\,°\text{C}$$

The corresponding viscosity is

$$\mu = \mu_i e^{-\beta(T-T_i)} = 0.03e^{-0.0414(61.9-36.8)} = 0.011\,\text{pa s}$$

Using this viscosity, a new Sommerfeld number and therefore a new eccentricity ratio must be calculated. The entire set of calculations is then repeated with the new ε. Table 8.9 summarizes the results for several iterations. Note that no attempt has been made to interpolate until the last iteration. After the sixth iteration, the effective temperature is predicted to be 50 °C, which

Table 8.9 Sample iterations for Example 8.4

Iteration No.	T (°C)	μ (pa s)	S	ε	ϕ (deg)	$f(R/C)$	Q_L (cm³/s)	Q_m (cm³/s)	Q_L, total (cm³/s)	E_P (W)	ΔT (°C)
1	36.8	0.03	0.338	0.45	59	7.402	15.1	16.55	18.81	812	25.1
2	61.9	0.011	0.124	0.65	46	3.816	21.8	24.34	27.42	419	8.9
3	45.7	0.021	0.236	0.55	53	5.3	18.42	20.3	23.02	585	14.8
4	51.6	0.016	0.180	0.6	49	4.52	20.1	22.29	25.2	496	11.4
5	48.3	0.019	0.214	0.55	53	5.5	18.42	20.3	23.02	585	14.8
6	51.6	0.016	0.180	0.58	51	4.9	19.3	21.4	24.2	538	13.0

is only 1.5 °C higher than the previous iteration; therefore, no further iterations are needed. The final eccentricity predicted is $\varepsilon \approx 0.58$. From Table 8.6 at this eccentricity, the maximum pressure is $\overline{P}_{max} \cong 13$, which corresponds to

$$
\begin{aligned}
P &= \mu N_S \left(\frac{R}{C}\right)^2 \overline{P}_{max} + P_S \\
&= (0.016)(25)(800)^2(13) + 2.76 \times 10^5 \\
&= 3.60\,\text{MPa (520 psi)}
\end{aligned}
$$

and is predicted to occur at $\theta_{max} = 147°$, which, if measured from the load line, would be $147 + \phi = 198°$.

Table 8.10 shows the experimentally measured values of Dowson *et al.* (1966) and those calculated in this example. The pressure profile, as measured by Dowson *et al.*, is shown in Figure 8.18 for various axial locations. The highest pressure measured at location α corresponds to the bearing midplane. This example shows that the results of the predictions are reasonably close to those measured.

Table 8.10 Measured and predicated bearing performance in Example 8.4

Parameter	Measured (Dowson *et al.*, 1966)	Calculated (Example 8.4)
Load, W (N, lbf)	11 000 N (2500 lbf)	11, 000 N (2500 lbf) given
Maximum pressure, P_{max} (MPa, psi)	3.5 MPa (508 psi)	3.60 MPa (520 psi)
Location of P_{max} (from load line) (degrees)	201°	198°
Outlet drain temperature (°C)	47	50
Leakage $Q_{L, total}$(cm³/s)	26.2	24.2
Attitude angle (degrees)	57	51
Friction force, F (N, lbf)	64.4 N (14.45 lbf)	67.4 N (15.2 lbf)
Shaft torque (N m, ft lb)	3.39 N m (2.5 ft lbf)	3.42 N m (2.52 ft lbf)

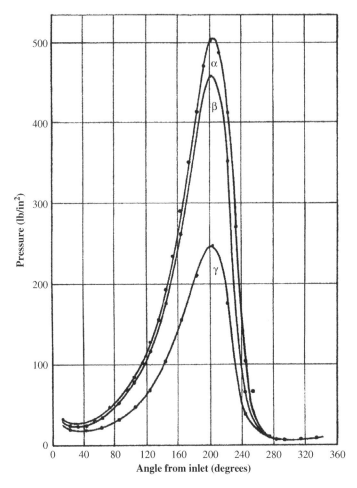

Figure 8.18   Pressure distribution measured by Dowson *et al.* (1966) at various axial locations ($\alpha =$ bearing mid-plane, corresponding to Example 8.6). *Source*: Dowson *et al.*, 1966. Reproduced with permission of Sage Publications.

Circumferential Groove

A fully circumferential groove placed centrally divides the bearing length into two identical short bearings of length $l' = (L - w_g)/2$, where w_g is the width of the groove. In the great majority of applications, $l'/D < 0.5$. Therefore, one is justified in using the short bearing theory to analyze each bearing. For a given load, minimum film thickness is less in each of the shorter bearings than in a bearing without a circumferential groove, and the attitude angle is smaller. This change may suffice to minimize self-excited whirl when it is encountered.

Performance prediction of a circumferentially grooved bearing requires one to estimate the pressure-induced flow exiting the bearing axially in the form of leakage. Assuming that the pressure in the groove remains at P_S and the pressure drops linearly from the groove to the bearing edge, $dP/dy = -P_s/l'$. Substituting this pressure gradient into Equation (8.32) and integrating the result over the bearing circumference yields:

$$Q_c = \frac{\pi RC^3 P_s}{6\mu l'}(1 + 1.5\varepsilon^2) \tag{8.61}$$

where it is assumed that the entire flow is directed outward. Clearly, in the loaded portion of the bearing $P > P_S$, some lubricant may actually flow back into the groove.

The total leakage flow is obtained by superposition of the axial flow due to the turning of the shaft, Q_L, and the pressure-induced flow as given above:

$$Q_{L, total} = Q_L + 2Q_c \tag{8.62}$$

Example 8.5 Consider a plain journal bearing with the following specifications: $D = 8$ in.; $L = 4.00$; $C = 6 \times 10^{-3}$ in.; operating speed $N = 3600$ rpm. The load imposed on the bearing is $W = 4800$ lbf. A narrow circumferential oil feed groove is cut into the bearing at its midlength, and lubricant ($\mu = 10$ cp at $T = 120\,°F$) is supplied at 10 psi. Determine the temperature rise.

With the full-length $L = 4$ divided in half, $l'/D = 0.25$. Accordingly, load on each of the two bearing segments is:

$$W_{l'} = \frac{4800}{2} = 2400\,lbf$$

Projected pressure on each bearing is:

$$P_{l'} = \frac{2400}{(2 \times 8)} = 150\,psi$$

Operating viscosity is:

$$\mu = 10\,cp \quad (1.45 \times 10^{-7})reyns/cp = 1.45 \times 10^{-6}\,reyns$$

Sommerfeld number becomes:

$$S = \frac{\mu N_S(R/C)^2}{P_{L'}} = 0.258$$

Using Table 8.6 or the short bearing theory, operating eccentricity is

$$\varepsilon = 0.8$$

Dimensionless leakage flow rate is

$$\overline{Q}_L = 1.5753$$

Friction coefficient is

$$f\left(\frac{R}{C}\right) = 8.8657$$

In dimensional form, the leakage flow rate due to shaft rotation is

$$Q_L = \bar{Q}_L \left(\frac{\pi}{2}\right) N_S D l' C = 14.25 \text{ in.}^3/\text{s} \quad (14.25 \times 60/231 = 3.70 \text{ gpm})$$

Pressure-induced flow is

$$Q_c = \frac{\pi R C^3 P_s}{6\mu l'}(1 + 1.5\varepsilon^2) = 3.06 \text{ in.}^3/\text{s} \ (0.79 \text{ gpm})$$

Total leakage flow becomes

$$Q_{L,\text{total}} = Q_L + 2Q_c = 20.37 \text{ in.}^3/\text{s} \quad (5.28 \text{ gpm})$$

Power loss for each half-length bearing segment is

$$E_p = fW_{l'}\pi D N_S = 4.81 \times 10^4 \text{ in. lbf/s} \quad (6.83 \text{ hp})$$

Predicted temperature rise is

$$\Delta T = \frac{2(4.81 \times 10^4)}{(778)(12)(0.48)(0.0315)(20.37)} = 33.4\,^\circ\text{F}$$

Mean (outlet) temperature is

$$T_0 = 120 + 33.4 = 153\,^\circ\text{F}$$

This procedure should be repeated with the viscosity at 153 °F and iterations continued as in Example 8.8 until the initially assumed viscosity matches that of the oil at the effective bearing (drain) temperature. See Section 8.17 for discussion on the calculation of effective temperature.

8.16 Bearing Stiffness, Rotor Vibration, and Oil-Whirl Instability

The following three concerns are related to dynamic behavior of journal bearings: (1) avoiding bearing–rotor system natural frequencies (critical speeds) near operating speeds, (2) limiting forced and unbalance vibration amplitude to acceptable levels, and (3) avoiding self–excited oil film *half-frequency* whirl or *oil whip*. For any bearing–rotor system where these become significant questions, reference should be made to a detailed analysis such as those of Sneck and Vohr (1984) and Gross *et al.* (1980).

To illustrate the principles involved, Figure 8.19 represents an idealized rotor of weight $2W$ supported on two bearings. Under steady load W, each bearing center is displaced to a steady

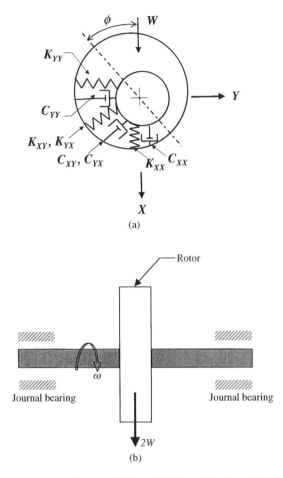

X

(a)

Rotor

Journal bearing   Journal bearing

$2W$

(b)

Figure 8.19 (a) Stiffness and damping coefficients of a journal bearing. (b) Idealized rotor of weight $2W$ supported on two journal bearings.

operating position as shown. While the equivalent spring constant for the vertical displacement is not linear over any extended range, a value K_{XX} can be taken as the differential displacement in the vertical (X) direction with a differential change in W (as with a rotor imbalance or varying external force). With no rotor flexibility, negligible damping, and all of the vertical flexibility in the bearing oil film, the bearing vertical natural frequency becomes, as for any simple spring-mass system:

$$f_1 = \frac{1}{2\pi} \left(\frac{K_{XX}}{W/g} \right)^{0.5}$$

The first subscript of K_{XX} indicates that the weight W gives a force in the X direction; the second subscript indicates that this K is for motion in the X direction. Thus, K_{YX} is the spring constant related to displacement in the Y direction under a force in the X direction.

Since a different spring constant K_{YY} is encountered in the horizontal direction (softer since the load W in the X direction gives a thinner, stiffer film), the bearing horizontal natural frequency becomes

$$f_2 = \frac{1}{2\pi} \left(\frac{K_{YY}}{W/g} \right)^{0.5}$$

Completion of the equations of motion for the rigid rotor system also requires damping coefficients C_{XX} and C_{YY} in the vertical and horizontal directions.

For other than pivoted-pad bearings, cross-coupling coefficients must also be introduced – K_{XY}, for instance, as a spring constant for motion in the Y direction, with variation in vertical load W in the X direction. The forces induced by the cross-coupling coefficients tend to drive the rotor into an orbit. Typically for light loads in fixed arc bearings with their cross-coupling, this orbit frequency is very close to half of the journal rotational speed. If sufficient damping and high enough load are not provided to suppress this cross-coupling force, the rotor will break violently into an uncontrolled whirl orbit, or rotor *whip*, with an amplitude closely matching the full internal clearance in the bearing.

Stability characteristics of a rotor carried on a full 360° journal bearing, based on the half short bearing solution (ISA) of Section 8.9 (with 180° oil film extent), can be estimated from Figure 8.20 (Raimondi and Szeri, 1984). Greater shaft flexibility (lower K_s) tends to decrease stability. K_s can be determined approximately from simple beam theory by dividing the weight of the rotor by the midspan deflection resulting from that weight.

Example 8.6 Determine whirl stability for a horizontal rotor and its bearings with the following characteristics: $D = 2R = 5$ in.; $L = 2.5$ in.; $C = 0.005$ in.; $N_s = 90$ rev/s (5400 rpm); $K_s = 5 \times 10^6$ lb/in. rotor stiffness.

$$W = 5000\,\text{lb rotor weight} \quad (m = W/g = 5000/386 = 13.0\,\text{lb s}^2/\text{in. rotor mass})$$

and

$$\mu = 2 \times 10^{-6}\,\text{lb s/in.}^2 \quad \text{(reyns) viscosity}$$

Using Figure 8.20, the unit bearing load is

$$P = W/(DL) = (5000/2)/(5 \times 2.5) = 200\,\text{psi}$$

The Sommerfeld number is

$$S = (2.5/0.005)^2 \times [(2 \times 10^{-6})(90)/200] = 0.225,$$

The characteristic bearing number is

$$S(L/D)^2 = 0.225(2.5/5)^2 = 0.0563,$$

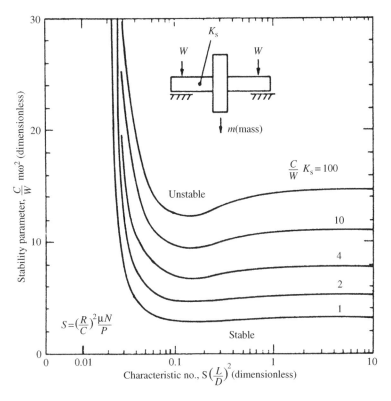

Figure 8.20 Stability of a single mass on a 360° journal bearing. *Source*: Raimondi and Szeri, 1984. Reproduced with permission of Taylor & Francis.

Also

$$(C/W)K_s = (0.005/5000) \cdot (5 \cdot 106) = 10;$$

and

$$(C/W)M\omega^2 = (0.005/2500) \times 13.0 \times (2\pi \times 90)^2 = 8.28$$

Since the coordinate point $S(L/D)^2 = 0.0563$, $(C/W)M\omega^2 = 8.28$ lies below the curve for $(C/W)K_s = 10$, the rotor will be free of oil whip instability.

If the bearing had been unstable, a shift to the stable operating region could have been obtained by increasing the unit loading by shortening the bearing length L, by adding a central circumferential groove, or by modifying other factors shown. A change to one of the more stable designs shown in Figure 8.3 and Table 8.1 would also be an alternative. Pivoted-pad journal bearings do not introduce cross-coupling response coefficients and are free of oil whirl instability, but their higher cost, greater complexity, and higher power loss should also be considered in making a final selection.

A more detailed analytical background and oil-film response coefficients for fixed-arc and pivoted-pad bearings are provided by Raimondi and Szeri (1984) and Sneck and Vohr (1984).

A method for deriving the appropriate linearized dynamic (stiffness and damping) coefficients in polar coordinates is given by Wang and Khonsari (2006a). Exact analytical expressions for a short bearing subject to the half-Sommerfeld boundary conditions for both laminar and turbulent conditions are also available (Wang and Khonsari, 2006e). Using the Routh-Hurwitz stability criterion, one can obtain the threshold of instability based on the linearized dynamic coefficients. Holmes' study of turborotors, based on the short bearing theory, reveals that the assumption of linearity is valid for peak-to-peak vibration of about one-third of the radial clearance at $\varepsilon = 0.7$ (Holmes, 1970, 1975) and cannot accurately predict the response of the journal if the shaft vibration is large. A nonlinear approach requires one to perform a transient, nonlinear study by solving the equations of motion to determine the locus of the shaft center as it orbits within the bearing clearance circle (Hashimoto *et al.*, 1987; Khonsari and Chang, 1993).

Another approach is the application of Hopf bifurcation theory (HBT) to characterize the behavior of rotor bearing systems (Myers, 1984; Hollis and Taylor, 1986). These studies lead to the introduction of the concepts of subcritical and supercritical bifurcation into the instability of a rotor–bearing system. When the running speed is crossing the threshold speed, if supercritical bifurcation occurs, then the journal will gradually lose its stability and the amplitude of the whirl will ramp up. On the other hand, if subcritical bifurcation takes place, the journal will suddenly lose stability and amplitude of the whirl will 'jump up' into a large whirl orbit. Application of HBT provides significant additional insight into the problem (Wang and Khonsari, 2006a–f). Wang and Khonsari (2005), for example, used HBT to predict the occurrence of a hysteresis phenomenon associated with oil whirl, which was experimentally observed. To this end, the effect of oil viscosity (and temperature) is particularly significant. Another application of HBT reveals that drag force can have an important influence on the dynamic performance of a rotor bearing system when the journal operates close to the threshold speed (Wang and Khonsari, 2005).

8.17 Tilting Pad Journal Bearings

Tilting pads are a common choice for optimum rotor stability in high-speed machines. Despite their greater cost and more complexity than cylindrical and elliptical sleeve bearings, tilt-pad bearings of 0.5–50 in. diameter are used in a wide variety of compressors, machine tools, electric motors and generators, marine drives, and high-speed gearing.

The main objective of an ideal tilting pad bearing is to eliminate the destabilizing cross-coupling forces that exist in a plain journal bearing. Figure 8.21 illustrates the position of the shaft center sinking directly below the bearing center in two types of tilting pads, where in one case the load is directly pointed to within one of the pads, and the other where it is directed in between two pads.

Referring to Figure 8.21, in a centrally pivoted $\alpha = \phi/B = 0$. However, as discussed in Chapter 7, a pivot is typically offset 55–60% (i.e., $\alpha = \phi/B = 0.55$–0.6) to improve the load-carrying capacity.

A design with load directed in between pivots typically yields more symmetric stiffness and damping coefficients. This result in a circular shaft orbit and smaller vibration amplitude

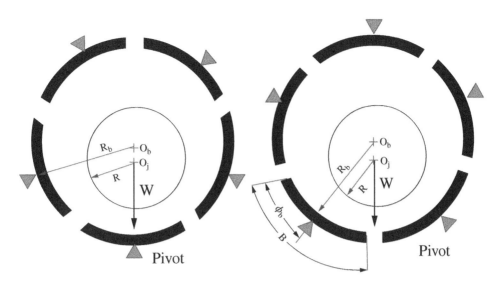

Figure 8.21 Pivoted bearing configurations with load on the pivot (left) and load between the pivots (right).

compared to asymmetric dynamic coefficients that yield an elliptical shaft orbit (Nicholas, 1994, 2012).

Shaft rotation induces a converging pressurized oil wedge over each pad. This gives primary load support from the bottom pads. Each pad is also moved radially inward so that the upper pads can also contribute to shaft centering and rotor damping. To characterize the behavior, a pad preload ($m = 1 - C_b/C_p$) is defined that compares the bearing clearance C_b to the pad clearance C_p. If the pad radius of the curvature is equal to the $C_b = C_p$, the bearing runs with zero preload, $m = 0$ and the shaft center and the pad coincide (Figure 8.22). In a preloaded pad, $C_p > C_d$ and O_p does not coincide with O_j. Therefore, $m > 0$. This fractional displacement inward, often 0.2–0.6 of the otherwise concentric radial clearance, is termed the 'preload ratio.' If, for instance, bearing pads are bored 0.10 in. larger than the shaft diameter and assembled with only

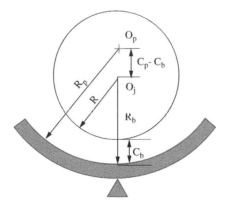

Figure 8.22 Nomenclature of a tilting pad for calculating the preload ratio.

0.005 in. diametral clearance, the preload is 0.005 in. (the preload ratio $0.005/0.010 = 0.5$). Light preload provides oil wedge action and provides effective damping to reduce vibration. Nevertheless, it makes the top pads become susceptible to unloading and thus flutter as they tend to tilt back and forth to reach equilibrium. This causes vibration and is undesirable. As discussed in Section 8.16, tilting pad design then effectively avoids the troublesome half-speed rotor whirl that otherwise plagues high speed rotors in lightly-loaded plain sleeve bearings.

Other tilting pad bearing advantages can include: (1) self-alignment through a double-curvature support surface at the back of each pad, (2) high load capacity with load sharing by two adjacent pads, and (3) quick maintenance. Increased power loss of up to 25% over sleeve bearings at high speeds can be minimized with an oil feed directly to the leading edge of each pad to avoid churning losses in an oil-filled housing. This parasitic loss increases above 10,000 fpm with onset of turbulence (see Chapter 7).

Four and five pad designs are most common with a ratio of axial length to shaft diameter (L/D) in the 0.4–0.75 range. Load is directed on the bottom pad with five pads, between pads with four. Oil feed temperature is commonly in the 120–130 °F range, with oil feed rate set to absorb the total power loss with a drain temperature of 145–160 °F.

In a preliminary evaluation of a pad bearing, its power loss and oil feed requirement can be taken as about 1.25 times that of a plain cylindrical sleeve bearing (see Table 8.6 and Equations 8.38–8.43), and the minimum film thickness would be similar to that in a sleeve bearing. Performance charts for 4-, 5-, 6-, and 8-pad bearings are provided by Raimondi (1983).

8.18 General Design Guides

Effective Temperature

The analysis presented in Section 8.12 represents an approximate method for including thermal effects in bearings. As is typical in bearing performance calculations, a single effective temperature is established as representative of average temperature and the corresponding effective viscosity in the oil film.

A basic approach is to calculate a global effective temperature, T_e, by equating the work done in shearing the oil to the heat carried away by both the oil, E_P, and by conduction and radiation, Φ_R, from the bearing structure and not to the oil stream (Wilcock and Booser, 1957; Martin, 1998):

$$T_e = T_i + \frac{(E_P - \Phi_R)}{J\rho c_P Q_L} \tag{8.63}$$

where, as in Equation (8.44), J is the mechanical equivalent of heat (British, traditional units), ρ is the density of the oil, c_P its specific heat, and Q_L is leakage flow rate. In calculations for bearings larger than about a 2 in. diameter, Φ_R can usually be assumed to be 0. At the other extreme, none of the energy is dissipated through the oil stream for dry and semilubricated bearings, as discussed in Chapter 12.

For small bearings where evaluation of Φ_R is difficult, half of the adiabatic temperature rise is sometimes used:

$$T'_e = T_i + \frac{1}{2}\frac{E_P}{J\rho c_P Q_L} \tag{8.64}$$

The assumption that the average oil-film temperature is equal to the average outlet oil temperature of the bearing is based on test and field observations with many types of industrial bearings. This may be easier to understand if the entire bearing is considered as a mixer. A stream of cool oil is continuously added to the bearing as a mixer, and a stream of oil leaves the bearing at the average temperature in the mixer. The mixer temperature is equal to the temperature of the oil being withdrawn but is above the feed oil temperature by an amount sufficient to carry away heat being generated in the mixer. Discharge oil leaving the ends of axial oil feed grooves has been found to have a measured temperature nearly equal to the overall outlet oil temperature: evidence of the relatively complete mixing in oil feed grooves and of the overall mixer-type action in journal bearings.

Maximum Bearing Temperature

A very important parameter neglected in the analyses of the previous section is the bearing maximum temperature, T_{max}. The bearing temperature normally increases from the inlet to a maximum value in the vicinity of the minimum film thickness and then tends to drop in the divergent region of the bearing. The peak temperature of the bearing and the minimum film thickness must be checked at the design stage to ensure that bearing material wear and thermal failure do not occur. Common babbitt bearing material begins to flow plastically when the local fluid pressure in the oil film reaches the compressive yield strength of the softening babbitt, commonly in the range of 135–150 °C (275–300 °F) (Booser *et al.*, 1970). In design, the maximum bearing temperature should be limited to 120–135 °C, with 110 °C or lower providing more tolerance for operating transients and variations in loading and dimensional factors.

Considerable progress has been made in understanding thermal effects and in predicting performance parameters that take heat effects into consideration (Khonsari, 1987a, b; Khonsari and Wang, 1991a, b, 1992). Recently, design charts have been developed for predicting the maximum temperature in plain bearings (Khonsari et al., 1996). Extension to tilting-pad journal bearings and journal bearings with two axial grooves at 90° angles to the load line have been developed by Fillon and Khonsari (1996) and Keogh *et al.* (1997), respectively. More recently, Jang and Khonsari (2004) provide easy-to-use expressions obtained using thermo-hydrodynamic solutions for both slider and journal bearings. An extension of the analysis to determine the threshold of instability using a simplified thermohydrodynamic solution is presented by Singhal and Khonsari (2005).

A simple and easy-to-apply procedure is presented in the following section for predicting journal bearing maximum temperature as well as effective temperature. The procedure is based on a thermohydrodynamic analysis with realistic simplifications to make the analysis practical for applications without resorting to full numerical simulations.

Monitoring the bearing drain temperature during operation provides a useful measure of the bearing mean temperature. Once this typical temperature is established, a warning indicator can be set for, say, 10 °C higher to provide an indication of any deterioration in bearing performance, loss of oil feed pressure, contamination, or possibly related machine malfunction. An even more sensitive measure of bearing performance is given by a temperature indicator mounted in the bearing bore in the zone of minimum film thickness, where a quicker and higher-amplitude temperature response is provided than in the outlet oil. Oil feed pressure and temperature sensors, usually in the 43–49 °C (110–120 °F) range, are also useful in monitoring the oil supply system.

Maximum Bearing Temperature and Effective Temperature Base on Thermohydrodynamic Analysis

A realistic prediction of the maximum temperature necessities a proper thermohydrodynamic analysis similar to what was presented in dealing with thrust bearings (Section 7.9). For this purpose, we provide a simple procedure that allows one to rapidly estimate the bearing maximum temperature as well as a more realistic effective temperature that can be used for the design of journal bearings (Jang and Khonsari, 2004).

Akin to the temperature rise equations introduced for thrust bearings (see Equation 7.15), we introduce two dimensionless temperature rise parameters (Khonsari et al., 1996) defined as:

$$\kappa_1 = \frac{\alpha \, \mu_i \, \beta \, U}{k \, R} \left(\frac{R}{C} \right)^2, \quad \kappa_2 = U \sqrt{\frac{\mu_i \, \beta}{k}} \tag{8.65}$$

where $\alpha = k/\rho C_p$ is the thermal diffusivity of the lubricant (m/s), k is the lubricant's thermal conductivity (W/m.K), and μ_i is the inlet viscosity (Pa.s), β represents the temperature-viscosity coefficient (1/K), which is based on the fact that the viscosity drops exponentially with temperature increase. This followed the viscosity relation $\mu = \mu_i e^{(T-T_i)}$ described in Chapter 2, where T_i is the inlet temperature.

Once κ_1 and κ_2 are evaluated, then using Table 8.11, one can easily obtain a realistic estimate of the effective bearing temperature, T_{mean}, as well as an estimate of T_{max}. It is interesting to

Table 8.11   Maximum and shaft temperature of journal bearing at $\Lambda = 1$

Valid over the range of $0.001 \leq \kappa_1 \leq 0.5$ $0.01 \leq \kappa_2 \leq 5$			Maximum temperature	Shaft temperature
Eccentricity Ratio	Equation Constant		$\overline{T}_{max} = \exp\left[a + \dfrac{b}{\kappa_1^{0.5}} + \dfrac{c}{\kappa_2^{0.5}} \right]$	$\overline{T}_{shaft} = \left[a + \dfrac{b}{\kappa_1^{0.5}} + \dfrac{c}{\kappa_2^{2}} \right]^{-1}$
0.1 < ε < 0.3 Region I	a		1.63105	0.14092
	b		−0.20250	0.24303
	c		−0.84050	0.10054
0.3 ≤ ε ≤ 0.7 Region II	a		1.68517	0.04734
	b		−0.19962	0.39751
	c		−1.04999	0.11007
0.7 < ε ≤ 0.9 Region III	a		1.79973	0.06200
	b		−0.13627	0.38264
	c		−1.25216	0.10102

Source: Jang and Khonsari, 2004. Reproduced with permissions of Sage Publications.

Table 8.12 Simple expressions for bearing performance parameters

Bearing type	Parameter	Curve-fit performance parameters
	$S = \dfrac{\mu_{eff}N_sDL}{W}\left(\dfrac{R}{C}\right)^2$	$\varepsilon = -0.36667 + \dfrac{0.49737}{\sqrt{\Lambda}} - 0.20986\ln{(S)}$ or $S = \exp\left[-0.85454 + \dfrac{2.83462}{\sqrt{\Lambda}} - 5.38943\sqrt{\varepsilon}\right]$
Journal Bearing	$\left(\dfrac{R}{C}\right)f$	$\left(\dfrac{R}{C}\right)f = \exp\left[2.02994 + \dfrac{2.79622}{\sqrt{\Lambda}} - 4.96338\sqrt{\varepsilon}\right]$
Valid over the range of	$\overline{P}_{max} = \dfrac{C^2\left(P_{max} - P_s\right)}{\mu_{eff}R^2N_s}$	$\overline{P}_{max} = \exp\left[2.25062 - \dfrac{0.39491}{\Lambda} + 4.77721\varepsilon^3\right]$
$0.5 \leq \Lambda \leq 1.5$ $0.1 \leq \varepsilon \leq 0.9$	$\overline{Q}_{leak} = \dfrac{2Q_{leak}}{\pi N_s DCL}$	$\overline{Q}_{leak} = \exp\left[0.72912 - 0.30021\Lambda^{1.5} + 0.97285\ln(\varepsilon)\right]$
	$\overline{Q}_{rec} = \dfrac{2Q_{rec}}{\pi N_s DCL}$	$\overline{Q}_{rec} = \dfrac{1}{0.90566 - 0.16366\sqrt{\Lambda} - \dfrac{0.74706}{\ln{(\varepsilon)}}}$

Source: Jang and Khonsari, 2004. Reproduced with permissions of Sage Publications.

note that in journal bearings, the mean temperature is well represented by the shaft temperature, i.e., $T_{mean} \approx T_{shaft}$. With this mean temperature, Table 8.12 can be used to rapidly predict bearing performance based on curve-fit expressions. Note the results presented in Table 8.11 for the estimation of the shaft and maximum bearing temperatures pertain to $\Lambda = L/D = 1$, the expressions for predicting the bearing performance parameters in Table 8.12 can be used for a wider range of the aspect ratio ranging from $0.5 \leq \Lambda \leq 1.5$.

The procedure is as follows (Jang and Khonsari, 2004):

1. Compute the Sommerfeld number with the viscosity evaluated at the supply temperature.
2. Compute the eccentricity ratio from the expression $\varepsilon = \varepsilon\,(S, \Lambda)$ in Table 8.12.
3. Compute the temperature-rise parameters using the bearing specifications.
4. Compute the maximum and shaft temperatures from the expressions in Table 8.11.
5. Compute the effective viscosity at the determined shaft temperature.
6. Compute the Sommerfeld number again with the effective viscosity.
7. Compute the eccentricity ratio again.
8. Check the region again with the new Sommerfeld number.
9. Compute the bearing performance parameters from Table 8.12 as a function of Λ and ε.

Example 8.7 Consider the following journal bearing specifications: $\Lambda = 0.75$, $R/C = 800$, $L = 0.0762$ m, $\alpha = 0.756 \times 10^{-7}$ m^2/s, $k = 0.13$ W/mK, $\beta = 0.0414$/K, $W = 11000$ N, $N = 1500$ rpm, $T_s = 36.8\,^{\circ}C$, $\mu_i = 0.03$ Pa $\cdot$ s and $P_s = 280$ KPa.

$$R = \frac{L}{2(0.75)} = 0.0508\text{ m}$$
$$U = (0.058)(1500)(1/60)(2\pi) = 7.98\text{ m/s}$$

$$S = \frac{\mu_{eff} N_s DL}{W} \left(\frac{R}{C}\right)^2 = \frac{(0.03)(1500/60)(0.1016)(0.0762)}{11000}(800)^2 = 0.338$$

$$\varepsilon = -0.36667 + \frac{0.49737}{\sqrt{\Lambda}} - 0.20986 \ln(S)$$

$$= -0.36667 + \frac{0.49737}{\sqrt{0.75}} - 0.20986 \ \ln(0.338) = 0.435$$

Therefore, according to Table 8.11, the operation is in Region II.
Now we proceed to evaluate the temperature rise parameters.

$$\kappa_1 = \frac{\alpha \mu_i \beta U}{k R} \left(\frac{R}{C}\right)^2 = \frac{(0.757 \times 10^{-7})(0.03)(0.0414)(7.98)}{(0.13)(0.0508)}(800)^2 = 0.0726$$

$$\kappa_2 = U\sqrt{\frac{\mu_i \beta}{k}} = (7.98)\sqrt{\frac{(0.03)(0.0414)}{0.13}} = 0.78$$

With constants $a = 1.68517$, $b = -0.19962$, and $c = -1.04999$, the maximum dimensionless temperature is:

$$\overline{T}_{max} = \exp\left[a + \frac{b}{\kappa_1^{0.5}} + \frac{c}{\kappa_2^{0.5}}\right] = \exp\left[1.68517 - \frac{0.19962}{0.0726^{0.5}} - \frac{1.04999}{0.78^{0.5}}\right] = 0.783$$

$$T_{max} = \frac{\overline{T}_{max}}{\beta} + T_s = \frac{0.783}{0.0414} + 36.8 = 55.71$$

With constants $a = 0.04734$, $b = 0.39751$, and $c = 0.11007$, the predicted shaft temperature is;

$$\overline{T}_{shaft} = \left[a + \frac{b}{\kappa_1^{0.5}} + \frac{c}{\kappa_2^2}\right]^{-1} = \left[0.04734 + \frac{0.39751}{0.0726^{0.5}} + \frac{0.11007}{0.78^2}\right]^{-1} = 0.587\,°C$$

$$T_{shaft} = \frac{\overline{T}_{shaft}}{\beta} + T_s = \frac{0.587}{0.0414} + 36.8 = 50.98\,°C$$

The above shaft temperature represents an effective (mean) bearing temperature. Therefore, $T_{mean} = T_{eff} \approx 50.98\,°C$.

The effective viscosity can now be calculated at 50.98 °C, and the iteration continued once more with a new Sommerfeld number. The rest of the bearing performance parameters can be then estimated from the expressions in Table 8.12.

$$\mu_{eff} = \mu_i \, e^{-\beta(T_{mean}-T_i)} = 0.017 \, \text{Pa.s}$$

Using this effective viscosity, the updated Sommerfeld and eccentricity ratios are:

$$S = 0.188 \text{ and } \varepsilon = 0.56$$

According to Table 8.11, the operation remains in Region II and no more iterations are needed. We can now proceed to determine the bearing performance parameters – friction coefficient, leakage flow rate, maximum pressure, etc., from Table 8.12.

For example, the friction coefficient is:

$$f = \left(\frac{C}{R}\right) \exp\left[2.02994 + \frac{2.79622}{\sqrt{\Lambda}} - 4.96338\sqrt{\varepsilon}\right] = 0.0059$$

$$\overline{Q}_{leak} = \exp[0.72912 - 0.30021\Lambda^{1.5} + 0.97285\ln(\varepsilon)] = 0.968$$

In dimensional form. the leakage flow is:

$$Q_{leak} = \frac{\pi N_s DCL}{2}\overline{Q}_{leak} = 18.69 \times 10^{-6} \text{ m}^3/\text{s}$$

Supply Temperature and Bearing Whirl Instability

The effect of the supply oil temperature on bearing whirl instability is an extremely important subject and difficult to predict. In fact, as pointed out by Maki and Ezzat (1980), the literature contains apparent disparities and confusion as to the precise effect of supply temperature in whirl instability. Notably, Newkirk, the discoverer of oil whirl instability, reported that instability threshold speed increases if the supply oil temperature is increased (Newkirk and Lewis, 1956). However, opposite to these findings, Pinkus (1956) reported the results of several experiments where the bearing whipped at high inlet oil temperatures, but did not whip when the supply oil was cold. An explanation of these results was offered by an analytical and experimental investigation conducted by Wang and Khonsari (2006b, 2006d), who reported that depending on the operating conditions either heating or cooling can stabilize the bearing. Figure 8.23 shows the whirl instability threshold plotted as a function of the supply temperature. The bearing runs free of whirl instability if it operates below the curve shown and becomes

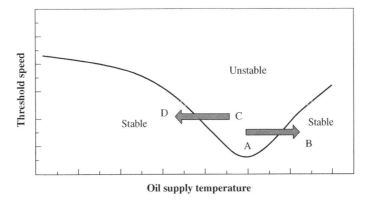

Figure 8.23   Influence of oil supply temperature on threshold speed.

unstable above the threshold. If the operational point falls inside the small dip in the curve, then either cooling or heating can suppress instability. As illustrated in Figure 8.23, the unstable operational point A becomes stable when the supply oil temperature is increased to point B. Alternatively, an unstable point C can be stabilized by reducing the temperature to point D. The interested reader is referred to Wang and Khonsari (2016) for detailed discussions on the nature and treatment of whirl instability and its associated thermal effects.

Turbulent and Parasitic Loss Effects

It must be reemphasized that the results presented in Table 8.6 assume that the fluid is incompressible, is linearly viscous (Newtonian), and remains laminar. Depending on bearing diameter and clearance, operating speed, and lubricant viscosity, the flow may become turbulent. In that case, the data presented in this chapter would have to be modified.

Transition from laminar to turbulent oil film is commonly encountered when the local Reynolds number in the oil film exceeds about 1000–2000 (see Section 6.2). Qualitatively, turbulence increases the apparent viscosity of the film, and hence increases minimum film thickness and power loss above that which would be expected for laminar flow under the same conditions. Note that turbulence influences the bearing temperature field as well. Sneck and Vohr (1984) provide an excellent discussion of the transition to turbulence and the corresponding modification that must be made to the Reynolds equation. Performance charts for cylindrical and pad-type journal bearings with and without turbulence are given by Raimondi and Szeri (1984).

Additional 'parasitic' power losses in large, high-speed bearings may even equal frictional loss in the turbulent bearing oil film itself. These parasitic losses involve churning and turbulence at the high Reynolds numbers in oil grooves and clearance spaces, losses in accelerating feed oil to high surface speed, vortices in feed passages and oil grooves, and general surface drag by oil contacting exposed high-speed surfaces. With journal bearings, these parasitic losses become significant for journals of more than about 250 mm (10 in.) diameter at 3600 rpm. Since larger diameters with their higher surface velocities are normally encountered in the thrust bearings of a machine, Chapter 7 (Sections 7.10 and 7.11) gives details on evaluation of these parasitic losses and some further discussion of turbulence effects on bearing oil films.

Flooded versus Starved Condition

Most predicted bearing performance results in this chapter are based on the flooded inlet condition. That is, the amount of supply lubricant is at least equal to the full hydrodynamic leakage flow rate. When a bearing operates under the so-called fully flooded condition, it is presumed that an excess of supply oil will not result in an improvement in the bearing performance – other than possibly its beneficial cooling effect. Many bearings may operate under starved conditions where the ratio of $Q_{supply}/Q_L < 1$. Dry and starved bearings are discussed in Chapter 12. For analyses including thermal effects with proper consideration of cavitation, the reader is also referred to Artiles and Heshmat (1985), Heshmat and Pinkus (1985), and Pinkas (1990).

Table 8.13 Design journal bearing loads for various applications

Application	Typical mean pressure, P_L (psi)
Steady radial load	
Electric motors	100–250
Turbines and axial compressors	200–300
Railroad car axles	200–350
Water-lubricated bearings	20–40
Dynamic loading	
Automotive engine bearings	2000–3500
Automotive connecting rod	3500–5000
Steel mill roll necks	3500–5000

Bearing Load and Dimensions

A common design criterion for journal bearings is the so-called projected loading $P_L = W/(L \cdot D)$. Typical psi loadings for a variety of applications are available in Table 8.13. In many applications, the shaft diameter D is fixed by other considerations and the designer must select a bearing length L, using Table 8.13 as a starting point. It is important to note that a design with a very small P_L value may be susceptible to vibration, wasteful power loss, and high oil flow in an oversized bearing; and an excessively large P_L may cause overheating. Lastly, the bearing material may impose an additional restriction on the maximum P_L. For example, the maximum mean pressure in a tin-base babbitt should not exceed 1000 psi, whereas a copper–lead alloy bearing can handle three to four times higher mean pressure (Fuller, 1984).

Lift-Off Speed

At start-up and during creep speeds during such operations as windmilling of industrial fans, backflow in pumps, and warm-up of turbines, journal bearings operate at much lower speeds and oil film thickness. As a minimum, film thickness where there is lift-off onto a full oil film under ideal conditions is at least about three times the composite film thickness $\Lambda = (\sigma_j^2 + \sigma_b^2)^{0.5}$ where σ_j and σ_b are, respectively, the rms surface roughnesses of the journal and bearing (Lu and Khonsari, 2005). With further elastic, alignment, and thermal distortions, the minimum lift-off Λ ranges up to about 10 in electric motors and related industrial machines (Elwell and Booser, 1972). See also Section 3.7.

From numerical solution of Reynolds equation, one can vary L/D and h_{min}/C to determine a relationship between the Sommerfeld number, S, and the minimum film thickness, h_{min}. Neglecting thermal effects, the following relationship can be established (Lu and Khonsari, 2005):

$$\frac{h_{min}}{C} = 4.678 \, (L/D)^{1.044} S \qquad (8.66)$$

where C is the radial clearance.

Substituting the definition of Sommerfeld number and solving for N_S, the lift-off speed can be estimated using the following equation:

$$N_S = \frac{P_L h_{min}}{4.678 \, C \, (L/D)^{1.044} \mu (R/C)^2} \tag{8.67}$$

where N_S is rev/s.

Example 8.8 Estimate the lift-off speed N (rpm) and creep-speed minimum oil film thickness h_{min} (μin.) for an industrial electric motor sleeve bearing (diameter $D = 6$ in. and axial length $L = 4.5$ in.) lubricated with a VG-32 turbine oil having a viscosity of 3×10^{-6} reyns at the bearing temperature. Arithmetic average surface finish of the journal is 16 μin., 32 μin. for the bearing bore. The load is $W = 3240$ lb, radial clearance $C = 6 \times 10^{-3}$ in.

For minimum film thickness with $\Lambda = 3$ and with a 1.25 multiplier to convert typical arithmetic average surface finish to rms values:

$$h_{min} = \Lambda \left(\sigma_s^2 + \sigma_b^2 \right)^{0.5} = 3 \cdot [(16 \times 1.25)^2 + (32 \times 1.25)^2]^{0.5} = 134 \, \mu\text{in.}$$

The projected load is

$$P_L = W/(LD) = 120 \, \text{psi}$$

Using Equation (8.67), the lift-speed is estimated to be

$$N_S = \frac{P_L h_{min}}{4.678 \, C \, (L/D)^{1.044} \mu (R/C)^2} = \frac{120 \times 134 \times 10^{-6}}{4.678 \times 6 \times 10^{-3}(4.5/6)^{1.044} \times 3 \times 10^{-6}(3/0.006)^2}$$
$$= 1.033 \, \text{rps} \quad (\times 60 = 62 \, \text{rpm})$$

As break-in of the bearing wears the surfaces to a finer finish, the reducing coefficient of friction in Figure 8.24 reflects a drop in both minimum speed and film thickness at lift-off.

Larger shaft diameters, such as those ranging up to 20–30 in. for rolling-mill drives and large steam turbines, commonly reduce this lift-off speed limit to 2–5 rpm. Conversely, a higher speed is needed to maintain a full film with smaller shafts, especially when operating with very low viscosity gases. Continued running at speeds lower than this lift-off speed will undesirably result in wear and possibly stick–slip vibration with chatter as the journal elastically winds up at higher stick friction before jumping ahead with a lower sliding friction coefficient. The reader interested in the subject of running-in is referred to recent papers by Akbarzadeh and Khonsari (2010a, 2010b, 2013), Mehdizadeh *et al.* (2015) and Mortazavi and Khonsari (2016). Simulations of wear associated with start-up and coast-down operations in journal bearings are reported Sang and Khonsari (2016).

Eccentricity and Minimum Film Thickness

Once the Sommerfeld number is established, the operating eccentricity ratio can be predicted using Table 8.6. The minimum film thickness h_{min} should then be calculated and checked to

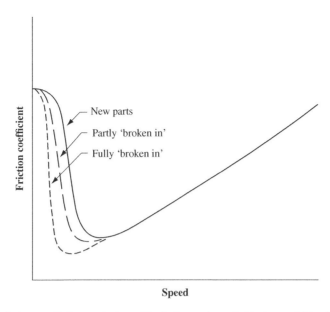

Figure 8.24 Friction coefficient variation with shaft speed. Both friction and lift-off speed decrease as break-in becomes more complete. *Source*: Elwell and Booser, 1972. Reproduced with permission of John Wiley & Sons.

see if it meets the minimum allowable criterion for the combined journal and bearing surface finish given in Table 8.7. Application of typical steady unit loads from Table 8.13 commonly sets the eccentricity ratio in the 0.5–0.85 range, and the corresponding minimum film thickness is given by $h_{\min} = C - e = C(1 - \varepsilon)$.

There is a delicate balance between reliable operation of a bearing and its minimum film thickness. Too small a minimum film thickness leads to excessive bearing temperature and increased susceptibility to wear from contamination and misalignment. Too large a minimum film thickness, on the other hand, might provide poorer vibration stability and would reflect an unnecessarily large bearing with higher than necessary power loss. The typical range for minimum film thickness $h_{\min}$ is 0.020–0.030 mm (0.0007–0.0012 in.) for babbitted bearings under steady loads in electric motors and their driven equipment at medium speeds (500–1500 rpm). At 1500–3600 rpm in large turbine generators, $h_{\min}$ is commonly in the 0.075–0.125 mm (0.003–0.005 in.) range. For small, rigid automotive and reciprocating aircraft engine bearings with very finely finished surfaces, $h_{\min}$ is in the 0.0025–0.005 mm (0.0001–0.0002 in.) range (Fuller, 1984).

Operating Clearance

Bearing clearance can have a significant influence on all performance parameters, particularly the temperature field. Ordinarily, the diametral clearance is chosen to be on the order of 0.002 per inch of diameter.

It is important to make sure that a specified clearance does not lead to excessively high temperatures. From an analytical point of view, a simple trial-and-error study of the bearing mean

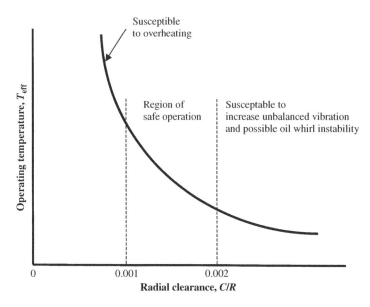

Figure 8.25   Operating temperature as a function of bearing operating clearance.

temperature for a range of clearance values can be helpful. Figure 8.25 shows what happens to the operating temperature when too small a clearance is chosen: it becomes extremely high. This possibility of excessive temperature at small clearances is increased by the differential thermal expansion commonly experienced by the warmer journal (surrounded by its oil film) in a bearing constrained in a cooler housing structure.

Thermally Induced Seizure

Temporary clearance loss of this type can even lead to shaft seizure from an aggravated differential expansion during thermal transients at start-up of a machine. Continued high start-up friction with high contact temperatures before establishment of a full oil film is especially troublesome and there are reported instances where the bearing catastrophically seizes. This mode of failure is often thermally induced seizure (TIS). Although TIS can take place in lubricated bearings, it is predominant when a hydrodynamic bearing happens to operate in the boundary or mixed lubrication regimes (Krishivasan and Khonsari, 2003; Khonsari and Booser, 2004; Jun and Khonsari, 2009a, 2009b, 2011). These conditions are most dangerous if occurring during start-up, with a clogged filter without a bypass system, or in the event of anything else that could cause loss of lubricant. Bearings that have been out of service for a period of time – such as in air conditioning units sitting idle during the winter months or those in spare pumps and compressors – are particularly susceptible because the lubricant may have dried out over a period of time to leave their bearings completely dry at start-up. Therefore, the main contributor to failure is the thermal heating at the interface between the shaft and the bushing and because of high coefficient of friction, which can be in the range of 0.1–0.25 depending on the type of the material properties of the friction pair. This is significantly higher than the typical

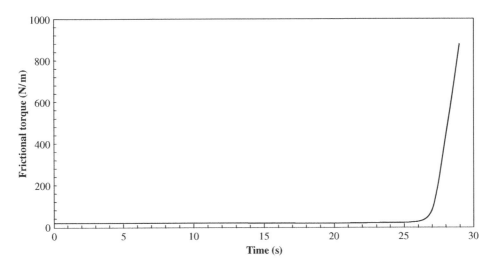

Figure 8.26 Variation of frictional torque during start-up. Note that the initial frictional torque increased 50-fold when seizure occurred. ($W = 4400$ N; $N = 250$ rpm; $R_s = 25.5$ mm; $R_{bo} = 51$ mm; $C = 0.0125$ mm; $L = 51$ mm). *Source*: Krishivasan and Khonsari, 2003. Reproduced with permission of ASME.

fluid-film, hydrodynamic friction with $f = 0.001$–0.008. If the contact is dry, frictional heating is much higher with excessive frictional heating at the contact giving rise to an expansion of the shaft radially outward. Since the bushing is often constrained on the outside by the bearing housing it can only expand inwardly, thus exaggerating the clearance loss even more.

Finite element models of a bearing TIS show that while the shaft expands uniformly, the bushing tends to take on an 'ovalized' shape, generating contact between the top of the shaft and the inner surface of the bushing within a few seconds (Hazlett and Khonsari, 1992a, 1992b; Krishivasan and Khonsari, 2003). This added contact induces further frictional force and heat generation continually increases until seizure takes over. Once this extra contact is formed, torque required to overcome the friction torque increases nonlinearly with time and accelerates rapidly until failure (Figure 8.26). The entire process may take only 20–30 s for a 1 in. shaft running at only 250 rpm.

Based on extensive finite element FEM simulations, the following simple empirical relations will predict the time-to-seizure t_s on start-up of an unlubricated journal bearing (Krishivasan and Khonsari, 2003). Denoting angular speed (ω), load (W), shaft radius (R), radial clearance (C), coefficient of friction (f), coefficient of thermal expansion (α), shaft thermal conductivity (k) and diffusivity (κ), and bearing length (L), the time to seizure (t_s) in seconds can be estimated by:

$$t_s = 90.5 \frac{R^2}{\kappa} \varepsilon_f^{-1.25} \lambda^{-1.65} \quad \text{for } 500 \le R/C \le 1000 \tag{8.68}$$

$$t_s = 1.83 \frac{R^2}{\kappa} \varepsilon_f^{-1.25} \lambda^{-1.03} \quad \text{for } 1000 \le R/C \le 5100 \tag{8.69}$$

where

$$\varepsilon_f = \frac{fW\omega\alpha}{k} \quad \text{and} \quad \lambda = \frac{R^2}{CL}$$

Example 8.9   Consider a journal bearing with the following specifications: $R = 20\,\text{mm}$; $C = 0.008\,\text{mm}$; $L = 50\,\text{mm}$; $W = 5000\,\text{N}$; $N = 1000\,\text{rpm}(104.72\,\text{radians/second})$; $k = 52\,\text{W/m K}$; $\alpha = 0.5 \times 10^{-5}\,\text{m/m K}$, and $\rho C_p = 1.23 \times 10^6$.

Here, $R/C = 2500$, so Equation (8.69) is used. Solving for t_s gives approximately 10.5 s with a start-up friction coefficient of $f = 0.1$. This implies that the system will fail by seizure in about 10 s unless oil gets into the contact, and reduces the coefficient of friction.

These empirical relationships can be used to evaluate changes in pertinent bearing parameters to increase the time to seizure. For example, doubling the radial clearance delays the seizure time to about 22 s. A further increase in the clearance to, say, $R/C = 500$ may be sufficient to avoid seizure because of the reduction in the contact area between the shaft and bearing bore and the related thermal effects as indicated in Figure 8.27. However, the longer period of

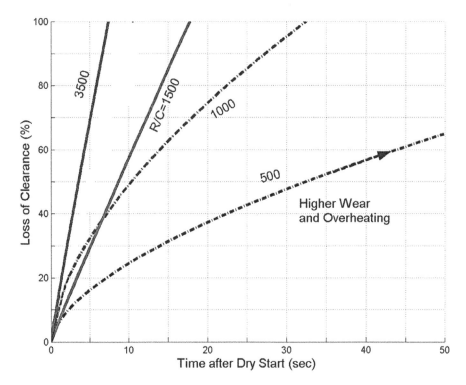

Figure 8.27   Increasing time to loss of clearance and seizure with increasing radial bearing clearance for the 40 mm bore bearing in Example 8.8. *Source*: Khonsari and Booser, 2004 and Elwell and Booser, 1972.

heating while running dry with the larger clearance may lead to excessive wear, overheating and melting of the bearing material, and even catastrophic machine failure. With the increasing clearance, there would also be less control of rotor vibration.

Too large a clearance may not be an acceptable solution because it often results in high susceptibility to oil-whirl instability. The interested reader is referred to Dufrane and Kannel (1989) for simple analytical solutions and experimental verification of TIS; Khonsari and Kim (1989) for TIS analysis with provision for shaft misalignment; and Jang *et al.* (1998a, b), and Pascovici *et al.* (1995) for theoretical and experimental results for seizure in lubricated journal bearings.

Extremely heavily-loaded journal bearings – often referred to as pin-bushing or pin-joint – are used predominately in excavators and construction machinery. These bearings are subject to oscillatory load with large swing angles ranging from $\alpha = \pm 5$ to ± 60 degrees. These bearings are typically grease-lubricated and operate in mixed lubrication with a coefficient of friction in the range of 0.08–0.16 (Lu and Khonsari, 2007a, 2007b; Khonsari *et al.*, 2010). Under normal operating conditions, the temperature initially rises but then reaches a steady level. If the operating conditions are severe and oscillations are continued for a prolonged length of time, the bearing may overheat and fail. Prediction of temperature requires the careful consideration of the moving heat source as illustrated in Figure 8.28, noting that the loads and swing angles of each pin-bushing may be different. The interested reader is referred to Mansouri

Figure 8.28 Pin-bushing bearings in a construction machine with oscillatory heat source.

and Khonsari (2005) for a discussion on heat transfer and to Jun and Khonsari (2009a, 2009b, 2009c; Jun *et al.*, 2011; Aghdam, 2013, 2016) for information on finite element modeling with experimental results on bearing seizure. A comprehensive review of bearing seizure for both journal bearings and rolling element bearings is given by Takabi and Khonsari (2015) and information on modeling and experimental studies on pin-bushing wear is given by Aghdam and Khonsari (2015).

Misalignment and Shaft Deflection

Misalignment due to machinery assembly and installation poses a serious problem in practice. It can occur due to an individual factor or several combinations of factors such as asymmetric bearing loading, shaft deflection, distortion due to shaft thermal growth or bearing housing supports, inaccurate manufacturing tolerances due to machining and assembly defects as well as errors due to improper installation. Shaft misalignment can also occur because of vibration due to the reaction forces generated in the shaft couplings. Thus, the general characterization of shaft misalignment is formidably complex.

A tacit assumption in generating the tabulated journal bearing performance results presented in this chapter (e.g., Table 8.6) is that the system is perfectly aligned so that $h = h(x)$ only. In the case of misaligned shaft, the film thickness varies in the axial direction and $h = h(x,y)$. Figure 8.29 illustrates the geometry of a misaligned shaft depicting the project of the shaft on

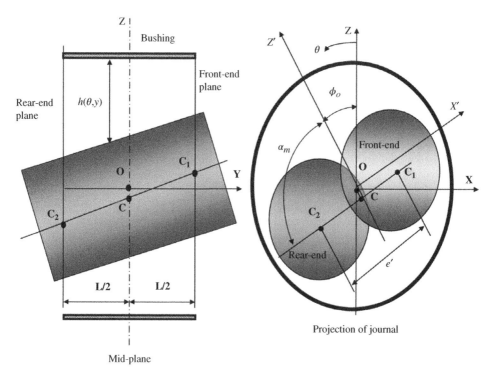

Figure 8.29 Illustration of the shaft misalignment in a plain journal bearing.

the front-end and the rear-end planes. It can be shown that the film thickness can be expressed as follows (Jang and Khonsari, 2010, 2016).

$$h = C + e_o \cos\left(\theta - \varphi_o\right) + e'\left(\frac{y}{L} - \frac{1}{2}\right)\cos\left(\theta - \alpha - \varphi_o\right) \tag{8.70}$$

where C is the clearance, R is the journal radius and L is the bearing width, ϕ_o represents the attitude angle between the load line and the line of centers, and e_o is the eccentricity at the bearing mid-plane. The parameter e' is the magnitude of the projection of the axis of the misaligned journal on the mid-plane of the bearing. The *misalignment angle* α is the angle between the line of centers and the rear center of the misaligned journal). Thus, the *misalignment eccentricity ratio* can be defined as $\varepsilon' = e'/C = D_M \varepsilon'_{max}$ where D_M represents the *degree of misalignment* which varies from 0 to 1, and ε'_{max} is simply the maximum possible ε'. Therefore, to specify the position of a misaligned journal, its eccentricity ratio and attitude angle at its mid-plane and two additional parameters – the degree of misalignment D_M and misalignment angle α – are needed.

Figure 8.30 compares the film thickness and pressure distribution in an aligned and misaligned bearing. Misalignment tends to drastically reduce the minimum film thickness and the resulting pressure distribution in the axial direction is no longer symmetrical. In fact, the pressure distribution in a misaligned journal bearing shows two peaks in the axial direction if D_M is large (Jang and Khonsari, 2010). Further complications in the analytical prediction of performance parameters arise due to the fact that the simple Reynolds cavitation boundary condition may be inadequate and the Reynolds equation should be reformulated via a mass-conservative algorithm to properly account for the film rupture and reformation (Brewe *et al.*, 1990a, 1990b). The interested reader is referred to Elrod (1981), Vijayaraghavan and Keith (1989a, 1998b), Qiu and Khonsari (2009) and Fesanghary and Khonsari (2011, 2013) for details on the implementation of the algorithm and the different numerical treatment approaches available in the literature.

Shaft deflection aggravates the film condition at one edge of the bearing with the load pulling downward on a shaft supported between two journal bearings. As a first approximation, one may estimate the deflection from simple beam theory. Typically, a deflection of 0.002 in./in. is assumed for shaft deflection over half of the bearing length (EDSU, 1967). An adequate minimum film thickness at the bearing edge should be half of the h_{min} established for the aligned bearing. This is clearly only a simple approximation and requires a detailed analysis before being implemented in practice. Especially important is the situation at the bearing edge with high speeds and particular contamination: the high operating temperature may lead to low viscosity, a severe reduction in the film thickness, and local melting (Khonsari *et al.*, 1999).

A more challenging problem is to estimate the temperature rise at the design stage. Pinkus and Bupara (1979) provide some pertinent design information. More recently Jang and Khonsari (2010) present a three-dimensional thermohydrodynamic model with a mass-conservative cavitation formulation for the prediction of misaligned bearing performance. They show that as the degree of misalignment increases, the maximum pressure, the maximum temperature, the bearing load, the leakage flow-rate and the moment increase because of the reduction of the minimum film thickness. They provide illustrative examples for predicting bearing performance based on THD simulations. Additional information is available from Bouyer and Fillon (2002), who show excellent experimental results on the effect of misalignment on hydrodynamic plain journal bearing experimentally. Pierre *et al.* (2002) also

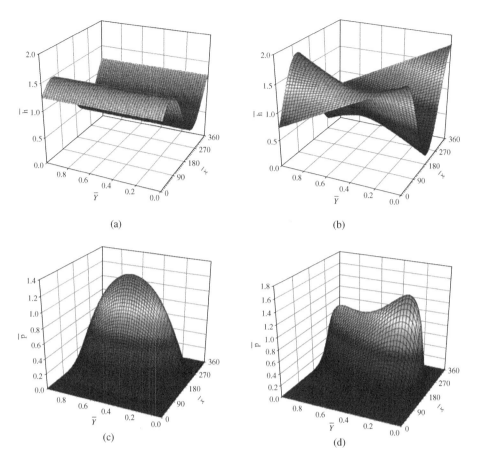

Figure 8.30 Illustration of the film thickness and pressure in a misaligned bearing compared to an aligned bearing. (a) Film thickness in an aligned bearing; (b) misaligned film thickness with considerable axial variation; (c) pressure variation in an aligned bearing with the peak at the bearing mid-plane; and (d) pressure distribution in a misaligned bearing with two peaks in the axial direction when the degree of misalignment is large ($D_M = 0.8$).

studied the thermohydrodynamic behavior of misaligned plain journal bearings with theoretical and experimental approaches under steady-state conditions. More recently, Sun and Gui (2005) have studied the effect of misalignment caused by shaft deformation.

It is interesting to note that despite the deleterious effects of misalignment on steady-state performance, it has been shown that misalignment can improve the stability margin of all journal bearing systems. Experimental results verify that journal bearings are relatively more stable under misalignment (Tieu and Qiu, 1995) and the stability is even more enhanced when the load is higher (Banwait *et al.*, 1998). Extension of the THD results with consideration of turbulence has recently become available by Xu *et al.* (2014). A comprehensive review of the literature survey dealing with various aspects of misalignment is reported by Jang and Khonsari (2015).

Problems

8.1 A journal bearing has the following specifications:

Shaft surface roughness:	RMS: 10×10^{-6} in.
Bearing surface roughness:	RMS: 40×10^{-6} in.
Radial clearance:	0.001 in.
Bearing length:	5.0 in.
Journal diameter	2.5 in.
Projected load	520 psi
Effective oil viscosity	1×10^{-6} reyns

(a) Can this bearing operate safely at 600 rpm? Support your answer with appropriate calculations.
(b) Determine the temperature rise ΔT if the speed is increased to 3600 rpm.

8.2 A journal bearing has the following specifications:

$$D = L = 60\,\text{mm}; \qquad R/C = 1000; \qquad N = 1200\,\text{rpm}$$

This bearing consumes 137 W of power due to friction when lubricated with an oil whose viscosity is $\mu = 18.8$ mpa s. Determine the load being carried on this bearing.

8.3 As a consultant, you are to evaluate the performance of a journal bearing with the following specifications:

$$D = 2\,\text{in.}; \qquad L = 4.0\,\text{in.}; \qquad C = 0.001; \qquad N = 1000\,\text{rpm};$$
$$\mu = 2.0 \times 10^{-6}\,\text{reyns}; \qquad P_s = 100\,\text{psi}$$

Determine the temperature rise, maximum pressure, and location of the shaft center under the operating conditions specified. On coasting down, at what speed would the minimum film thickness drop to 0.0002 in.? Assume that the unit load is 520 psi and use the same roughness information as given in Problem 8.1.

8.4 You are asked to examine the feasibility of modifying the bearing in Problem 8.1 by cutting a full circumferential groove into the bushing. Assume that the width of the groove is negligibly small.
(a) Determine the performance of this bearing, assuming that the total load remains the same as in Problem 8.1. (There is no need to iterate.)
(b) Comment on the performance of the bearings in Problems 8.1 and 8.4. Do you recommend inserting a groove?

8.5 A full journal bearing is circumferentially grooved at the midplane. The thermal performance of the bearing is to be investigated under the most severe operating conditions where the load is 3100 lbf and the operating speed is 3600 rpm, with a radial clearance of $C = 0.001$ in.

Bearing length = 2.4 in.; shaft radius = 0.5 in., width of the circumferential groove = 0.6 in. The inlet oil temperature is 110 °F, and the outlet temperature is not to exceed 140 °F. Assume a constant oil viscosity of $\mu = 2 \times 10^{-6}$ reyns (light turbine oil).
(a) What is the minimum supply pressure that you recommend for this bearing?
(b) Compute the maximum pressure in the bearing.

8.6 As a consulting engineer, you have been asked to evaluate the performance of a new, improved lubricant to be used in a circumferentially grooved bearing. Experiment shows that the viscosity of the lubricant varies according to the following equation:

$$\mu = 6895 \; \exp \left[A \; \exp \left(\frac{63.3}{T_K} \right) + B \right]$$

where T_K = temperature in Kelvin, μ = viscosity in pa s, $A = 53.2$, and $B = -77.14$.
The bearing is a circumferentially grooved, full journal bearing and has the following specifications:

$$D = 100 \, \text{mm}; \qquad l' = 50 \, \text{mm}; \qquad N = 3000 \, \text{rpm}; \qquad C = 0.05 \, \text{mm}; \qquad W = 30 \, \text{kN}$$

The lubricant supply temperature $T = 40 \,°\text{C}$, and the supply pressure $P_S = 20 \, \text{N/mm}^2$.
(a) Determine the eccentricity ratio, oil flow rate, exit oil temperature, friction coefficient, and power loss, assuming that the isoviscous solution prevails, i.e. that the lubricant viscosity remains fixed at the supply temperature $T = 40 \,°\text{C}$.
(b) Take into account the viscosity variation with temperature. For this purpose, iterations are needed and you are to write a computer program to do this task. Tabulate the final, converged results in (b) and compare it to the isoviscous solutions of (a).

8.7 Consider the bearing specifications given in Example 8.8. Write a computer program to determine the bearing performance for the load range of $W = 500, 1000, \ldots 3000 \, \text{lbf}$. Compare your results with the following experimental data:

Load (lbf)	500	1000	1500	2000	2500	3000
Mean temperature (°C)	52	49	48	47	47	47.5
Shaft torque (lbf.ft)	1.75	2.0	2.25	2.35	2.48	2.51
Flow rate (gram/min)	400	800	1000	1200	1300	1400

References

Aghdam, A.B. and Khonsari, M.M. 2013. 'Prediction of Wear in Reciprocating Dry Sliding Via Dissipated Energy and Temperature Rise,' *Tribology Letters*, Vol. 50, pp. 365–378.
Aghdam, A.B. and Khonsari, M.M. 2015. 'Prediction of Wear in Grease-Lubricated Oscillatory Journal Bearings Via Energy-Based Approach,' *Wear*, Vol. 318, pp. 188–201.
Aghdam, A.B. and Khonsari, M.M. 2016. 'Application of a Thermodynamically Based Wear Estimation Methodology,' *ASME Journal of Tribology*. Vol. 208.
Akbarzadeh, S. and Khonsari, M.M. 2010a. 'On the Prediction of Running-In Behavior in Mixed-Lubrication Line Contact,' *ASME Journal of Tribology*, Vol. 132 / 032102:1-11.

Akbarzadeh, S. and Khonsari, M.M. 2010b. 'Experimental and Theoretical Investigation of Running-in,' *Tribology International*, Vol. 22, pp. 92–100.

Akbarzadeh, S. and Khonsari, M.M. 2013. 'On the Optimization of Running-In Operating Conditions in Applications Involving EHL Line Contact,' *Wear*, Vol. 303, pp. 130–137.

Artiles, A. and Heshmat, H. 1985. 'Analysis of Starved Finite Journal Bearings,' *ASME Journal of Tribology*, Vol. 107, pp. 1–13.

Banwait, S.S., Chandrawat, H.N., and Adithan, M. 1998. 'Thermohydrodynamic Analysis of Misaligned 716 Plain Journal Bearing,' *Proceedings of 1st Asia International Conference on Tribology*, Beijing, pp. 35–40.

Boncompain, R., Fillon, M., and Frene, J. 1986. 'Analysis of Thermal Effects in Hydrodynamic Bearings,' *Journal of Tribology*, Vol. 108, pp. 219–224.

Booser, E.R., Ryan, F.D., and Linkinhoker. 1970. 'Maximum Temperature for Hydrodynamic Bearings under Steady Load,' *Lubrication Engineering*, Vol. 26, pp. 226–235.

Bouyer, J. and Fillon, M., 2002. 'An Experimental Analysis of Misalignment Effects,' *ASME Journal of Tribology*, Vol. 124, pp. 313–319.

Brewe, D.E. 1986. 'Theoretical Modeling of the Vapor Cavitation in Dynamically Loaded Journal Bearings,' *ASME Journal of Tribology*, Vol. 108, pp. 628–638.

Brewe, D.E. and Khonsari, M.M. 1988. 'Effect of Shaft Frequency on Cavitation in a Journal Bearings for Non-Centered Circular Whirl,' *STLE Tribology Transactions*, Vol. 31, pp. 54–60.

Brewe, D., Ball, J., and Khonsari, M.M. 1990a. 'Current Research in Cavitating Fluid Films: Part I. Fundamental and Experimental Observation,' *STLE Special Publication on Cavitation*, SP-28.

Brewe, D., Ball, J., and Khonsari, M.M. 1990b. 'Current Research in Cavitating Fluid Films: Part II. Theoretical Modeling,' *STLE Special Publication on Cavitation*, SP-28.

Cardullo, F.E. 1930. 'Some Practical Deductions for the Theory of Lubrication of Cylindrical Bearings,' *Transactions ASME*, Vol. 52, pp. 143–153.

Dowson, D., Hudson, J.D., Hunter, B., and March, C.N. 1966. 'An Experimental Investigation of the Thermal Equilibrium of Steadily Loaded Journal Bearings,' *Proceedings of Institution of Mechanical Engineers*, Vol. 181, Part 3B, pp. 70–80.

Dubois, G.B. and Ocvirk, F.W. 1953. 'Analytical Derivation and Experimental Evaluation of Short Bearing Approximation for Full Journal Bearings,' NACA Report 1157, Washington, DC.

Dufrane, K. and Kannel, J. 1989. 'Thermally Induced Seizures of Journal Bearings,' *ASME Journal of Tribology*, Vol. 111, pp. 288–292.

Elewll, R.C. and Booser, E.R. 1972. 'Low-Speed Limit of Lubrication,' *Machine Design*, 15 June, pp. 129–133.

Elrod, H.G. 1981. 'A Cavitation Algorithm,' *ASME Journal of Lubrication Technology*, Vol. 103, pp. 350–354.

Engineering Science Data Unit (ESDU). 1967. '*General Guide to the Choice of Thrust Bearing Type*,' Item 67033. Institution of Mechanical Engineers, London.

Fesanghary, M. and Khonsari, M.M. 2011. 'A Modification to the Elrod Cavitation Algorithm,' *ASME Journal of Tribology*, Vol. 133, 0245501:1–4.

Fesanghary, M. and Khonsari, M.M. 2013. 'Cavitation Algorithm,' *Encyclopedia of Tribology*, Y.W. Chung and Q.J. Wang (editors), Springer Science, New York, pp. 334–336.

Fillon, M., Frene, J., and Bligoud, J.C. 1992. 'Experimental Study of Tilting-Pad Journal Bearings—Comparison with Theoretical Thermoelastohydrodynamic Results,' *ASME Journal of Tribology*, Vol. 114, pp. 579–587.

Fillon, M. and Khonsari, M.M. 1996. 'Thermohydrodynamic Design Charts for Tilting-Pad Journal Bearings,' *ASME Journal of Tribology*, Vol. 118, pp. 232–238.

Fuller, D. 1984. *Theory and Practice of Lubrication for Engineering*, 2nd edn. Wiley Interscience, New York.

Gross, W.A., Matsch, L.A., Castelli, V., Eshel, A., Vohr, J.H., and Wildmann, M. 1980. *Fluid Film Lubrication*. John Wiley & Sons, New York.

Hamrock, B. 1994. *Fundamentals of Fluid Film Lubrication*. McGraw Hill Book Co., New York.

Harnoy, A. and Khonsari, M.M. 1996. 'Hydro-Roll: A Novel Bearing Design with Superior Thermal Characteristics,' *STLE Tribology Transactions*, Vol. 39, pp. 4555–4619.

Hashimoto, H., Wada, S., and Ito, J. 1987 'An Application of Short Bearing Theory to Dynamic Characteristics Problems of Turbulent Journal Bearing,' ASME *Journal of Lubrication Technology*, Vol. 109, p. 307.

Hazlett, T. and Khonsari, M.M. 1992a. 'Finite Element Model of Journal Bearings Undergoing Rapid Thermally Induced Seizure,' *Tribology International*, Vol. 25, pp. 177–182.

Hazlett, T. and Khonsari, M.M. 1992b. 'Thermoelastic Behavior of Journal Bearings Undergoing Seizure,' *Tribology International*, Vol. 25, pp. 178–183.

Heshmat, H. and Pinkus, O. 1985. 'Performance of Starved Journal Bearings with Oil Ring Lubrication,' *ASME Journal of Tribology*, Vol. 107, pp. 23–31.

Hollis, P. and Taylor, D.L. 1986. 'Hopf Bifurcation to Stability Boundaries in Fluid Film Bearings,' *ASME Journal of Tribology*, Vol. 108, pp. 184–189.

Holmes, R. 1970. 'Nonlinear Performance of Turbine Bearing,' *Journal of Mechanical Engineering Science*, Vol. 12, p. 377.

Holmes, R. 1975. 'Vibration and its Control in Rotating Systems,' *Dynamic of Rotors.* Springer-Verlag Publication, Berlin.

Jang, J., Khonsari, M.M., and Pascovici, M. 1998a. 'Modeling Aspect of a Rate-Controlled Seizure in an Unloaded Journal Bearing,' *STLE Tribology Transactions*, Vol. 41, pp. 481–488.

Jang, J., Khonsari, M.M., and Pascovici, M. 1998b. 'Thermohydrodynamic Seizure: Experimental and Theoretical Evaluation,' *ASME Journal of Tribology*, Vol. 120, pp. 8–15.

Jang, J.Y. and Khonsari, M.M. 2004. 'Design of Bearings Based on Thermohydrodynamic Analysis,' *Journal of Engineering Tribology*, Proceedings of Institution of Mechanical Engineers, Part J, Vol. 218, pp. 355–363.

Jang, J.Y. and Khonsari, M.M. 1997a. 'Performance Analysis of Grease-Lubricated Journal Bearings Including Thermal Effects,' *ASME Journal of Tribology*, Vol. 119, pp. 859–868.

Jang, J.Y. and Khonsari, M.M. 1997b. 'Thermohydrodynamic Design Charts for Slider Bearings,' *ASME Journal of Tribology*, Vol. 119, pp. 733–740.

Jang, J.Y. and Khonsari, M.M. 2001. 'On the Thermohydrodynamic Analysis of a Bingham Fluid in Slider Bearings,' *Acta Mechanica*, Vol. 148(1–4), pp. 165–185.

Jang, J.Y. and Khonsari, M.M. 2002. 'Thermohydrodynamic Analysis of Journal Bearings Lubricated with Multigrade Oils,' *Computer Modeling in Engineering & Sciences (CMCS)*, Vol. 3, No. 4, pp. 455–464.

Jang, J.Y. and Khonsari, M.M. 2008. 'Linear Squeeze Film with Constant Rotational Speed,' *STLE Tribology Transactions*, Vol. 51, pp. 361–371.

Jang, J.Y. and Khonsari, M.M. 2010. 'On the Behavior of Misaligned Journal Bearings Based on Mass-Conservative THD Analysis,' *ASME Journal of Tribology*, Vol. 132, 011702:1–13.

Jang, J.Y. and Khonsari, M.M. 2015. 'On the Characteristics of Misaligned Journal Bearings,' *Lubricant*, Vol. 3, pp. 27–53.

Jang, J.Y. and Khonsari, M.M. 2016. 'On the Relationship Between Journal Misalignment and Web Deflection in Crankshafts,' *ASME Journal of Engineering for Gas Turbines and Power*, Vol. 138, 122501:1–10.

Jun, W. and Khonsari, M.M. 2007. 'Analytical Formulation of Temperature Profile by Duhamel's Theorem for Bodies Subjected to Oscillatory Heat Source,' *ASME Journal of Heat Transfer*, Vol. 129, pp. 236–240.

Jun, W. and Khonsari, M.M. 2009a. 'Thermomechanical Coupling in Oscillatory Systems with Application to Journal Bearing Seizure' *ASME Journal of Tribology*, Vol. 131, 021601: 1–14.

Jun, W. and Khonsari, M.M. 2009b. 'Themomechanical Effects on Transient Temperature in Non-Conformal Contacts Experiencing Reciprocating Sliding Motion,' *International Journal of Heat and Mass Transfer*, Vol. 52, pp. 4390–4399.

Jun, W. and Khonsari, M.M. 2009c. 'Transient Heat Conduction in Rolling/Sliding Components by a Dual Reciprocity Boundary Element Method,' *International Journal of Heat and Mass Transfer*, Vol. 52, pp. 1600–1607.

Jun, W., Khonsari, M.M., and Hua, D. 2011. 'Three-Dimensional Heat Transfer Analysis of Pin-bushing System with Oscillatory Motion: Theory and Experiment,' *ASME Journal of Tribology*, Vol. 133, 011101:1–10.

Kaufman, H.N. and Boyd, J. 1968. 'Sliding Bearings,' *Standard Handbook of Lubrication Engineering*, J.J. O'Conner and J. Boyd (eds). McGraw-Hill Book Co., New York.

Kennedy, F.E., Booser, E.R., and Wilcock, D.F. 1997. 'Tribology, Lubrication and Bearing Design,' *CRC Handbook of Mechanical Engineering*, Vol. 3. CRC Press, Boca Raton, Fla, pp. 128–169.

Keogh, P., Gomiciaga, R., and Khonsari, M.M. 1997. 'CFD Based Design Techniques for Thermal Prediction in a Generic Two-Axial Groove Hydrodynamic Journal Bearing,' *ASME Journal of Tribology*, Vol. 119, pp. 428–436.

Khonsari, M.M. 1987a. 'A Review of Thermal Effects in Hydrodynamic Bearings, Part I: Slider and Thrust Bearings,' *ASLE Transactions*, Vol. 30, pp. 19–25.

Khonsari, M.M. 1987b. 'A Review of Thermal Effects in Hydrodynamic Bearings, Part II: Journal Bearings,' *ASLE Transactions*, Vol. 30, pp. 26–33.

Khonsari, M.M. and Booser, E.R. 2003. 'Grooves Move More Lube,' *Machine Design*, 22 May, pp. 71–72.

Khonsari, M.M. and Booser, E.R. 2004. 'Bearings that Don't Seize,' *Machine Design*, pp. 82–83.

Khonsari, M.M. and Booser, E.R. 2013. 'Ensuring Bearings Run Trouble Free,' *Machine Design*, Vol. 85, pp. 52–54.

Khonsari, M.M., Booser, E.R., and Miller, R. 2010. 'How Grease Keep Bearings Rolling,' *Machine Design*, Vol. 82, pp. 54–59.

Khonsari, M.M. and Chang, Y.J. 1993. 'Stability Boundary of Non-Linear Orbits Within Cleance Circle of Journal Bearings,' *ASME Journal of Vibration and Acoustics*, Vol. 115, pp. 303–307.

Khonsari, M.M., Jang, J., and Fillon, M. 1996. 'On the Generalization of Thermohydrodynamic Analysis for Journal Bearings,' *ASME Journal of Tribology*, Vol. 118, pp. 571–579.

Khonsari, M.M. and Kim, H. 1989. 'On the Thermally Induced Seizure in Journal Bearings,' *ASME Journal of Tribology*, Vol. 111(4), pp. 661–667.

Khonsari, M.M., Pascovici, M., and Kuchinschi, B. 1999. 'On the Scuffing Failure of Hydrodynamic Bearings in the Presence of an Abrasive Contaminant,' *ASME Journal of Tribology*, Vol. 121, pp. 90–96.

Khonsari, M.M. and Wang, S. 1991a. 'On the Maximum Temperature in Double-Layered Journal Bearings,' *ASME Journal of Tribology*, Vol. 113, pp. 464–469.

Khonsari, M.M. and Wang, S. 1991b. 'On the Fluid-Solid Interaction in Reference to Thermoelastohydrodynamic Analysis of Journal Bearings,' *ASME Journal of Tribology*, Vol. 113, pp. 398–404.

Khonsari, M.M. and Wang, S. 1992. 'Notes on the Transient THD Effects in Lubricating Film,' *STLE Tribology Transactions*, Vol. 35, pp. 177–183.

Krishivasan, R. and Khonsari, M.M. 2003. 'Thermally Induced Seizure in Journal Bearings during Start up and Transient Flow Disturbance,' *ASME Journal of Tribology*, Vol. 125, pp. 833–841.

Lu, X. and Khonsari, M.M. 2005. 'On the Lift-Off Speed in Journal Bearings,' *Tribology Letters*, Vol. 20 (3–4), pp. 299–305.

Lu, X. and Khonsari, M.M. 2007a. 'An Experimental Analysis of Grease-Lubricated Journal Bearings,' *ASME Journal of Tribology*, Vol. 129, pp. 640–646.

Lu, X. and Khonsari, M.M. 2007b 'An Experimental Study of Grease-Lubricated Journal Bearings Undergoing Oscillatory Motion,' *ASME Journal of Tribology*, Vol. 129, pp. 84–90.

Mansouri, M. and Khonsari, M.M. 2005. 'Surface Temperature in Oscillating Sliding Interfaces,' *ASME Journal of Tribology*, Vol. 127, pp. 1–9.

Martin, F.A. 1998. 'Oil Flow in a Plain Steadily Loaded Journal Bearings: Realistic Predictions Using Rapid Techniques,' *Proceedings of the Institution of Mechanical Engineers, Part J, Journal of Engineering Tribology*, Vol. 212(J4), pp. 413–425.

Mehdizadeh, M., Akbarzadeh, S., Shams, K., and Khonsari, M.M. 2015. 'Experimental Investigation on the Effect of Operating Conditions on the Running-in Behavior of Lubricated Elliptical Contacts,' *Tribology Letters*, Vol. 59, pp. 1–13.

Mitchell, A.G.M. 1929. 'Progress in Fluid Film Lubrication,' *Transactions ASME*, Vol. 51, pp. 153–163.

Mortazavi, V. and Khonsari, M.M. 2016. 'On the Predication of Transient Wear,' to appear in *ASME Journal of Tribology*.

Myers, C.J. 1984. 'Bifurcation Theory Applied to Oil Whirl in Plain Cylindrical Journal Bearings,' *ASME Journal of Applied Mechanics*, Vol. 51, pp. 244–250.

Pascovici, M., Khonsari, M.M., and Jang, J. 1995. 'On the Modeling of a Thermomechanical Seizure,' *ASME Journal of Tribology*, Vol. 117, pp. 744–747.

Pierre, I., Bouyer, J., and Fillon, M. 2002. 'Thermohydrodynamic Study of Misaligned Plain Journal Bearings – Comparison between Experimental Data and Theoretical Results,' *Proceedings of 2nd World Tribology Congress*, Vienna, Austria, pp. 1–4.

Pierre, I., Bouyer, J., and Fillon, M. 2004. 'Thermohydrodynamic Behavior of Misaligned Plain Journal Bearings: Theoretical and Experimental Approaches,' *STLE Tribology Transactions*, 47, pp. 594–604.

Pinkus, O. 1990. *Thermal Aspects of Fluid-Film Tribology*. ASME Press, New York.

Pinkus, O. and Bupara S.S. 1979. 'Adiabatic Solutions for Finite Journal Bearings,' *ASME Journal of Lubrication Technology*, Vol. 101, pp. 492–496.

Pinkus, O. and Sternlicht, B. 1961. *Theory of Hydrodynamic Lubrication*. McGraw-Hill Book Co., New York.

Qiu, Y. and Khonsari, M.M. 2009. 'On the Prediction of Cavitation in Dimples Using a Mass- Conservative Algorithm,' *ASME Journal of Tribology*, Vol. 131, 041702-1:1–11.

Raimondi, A. and Szeri, A.Z. 1984. *CRC Handbook of Lubrication*, Vol. II, E.R. Booser (ed.). CRC Press, Boca Raton, Fla, pp. 413–462.

Sang, M.C. and Khonsari, M.M. 2016. 'Wear simulation for the journal bearings operating under aligned shaft and steady load during start-up and coast-down conditions,' *Tribology International*, Vol. 97, pp. 440–466.

Singhal, S. and Khonsari, M.M. 2005. 'A Simplified Thermohydrodynamic Stability Analysis of Journal Bearings,' *Journal of Engineering Tribology, Proceedings of Institution of Mechanical Engineers*, Part J, Vol. 229, pp. 225–234.

Sneck, H.J. and Vohr, J.H. 1984. 'Fluid Film Lubrication,' *CRC Handbook of Lubrication*, Vol. II, E.R. Booser (ed.). CRC Press, Boca Raton, Fla, pp. 69–91.

Sun, J., Gui, C.L., and Li, Z.Y. 2005. 'Influence of Journal Misalignment Caused by Shaft Deformation under Rotational load on Performance of Journal Bearing,' *Proceedings of Institution of Mechanical Engineers., Journal of Engineering Tribology*, Vol. 219, pp. 275–283.

Swift, H.W. 1931. 'The Stability of Lubricating Films in Journal Bearings,' *Proceedings of the Institution of Civil Engineers*, Vol. 233, pp. 267–288.

Swift, H.W. 1937. 'Fluctuating Loads in Sleeve Bearings,' *Journal of the Institute of Civil Engineers*, Vol. 5, pp. 151–159.

Takabi, J. and Khonsari, M.M. 2015. 'On the Thermally Induced Seizure in Bearings: A Review,' *Tribology International*.

Tieu, A.K. and Qiu, Z.L. 1995. 'Stability of Finite Journal Bearings – from Linear and Nonlinear Bearing Forces,' *STLE Tribology Transactions*, Vol. 38, pp. 627–635.

Vijayaraghavan, D. and Keith, T.G. 1989a, 'Development and Evaluation of a Cavitation Algorithm,' *STLE Tribology Transactions*, Vol. 32, pp. 225–233.

Vijayaraghavan, D. and Keith, T.G. 1989b, 'Effect of Cavitation on the Performance of a Grooved Misaligned Journal Bearing,' *Wear*, Vol. 134, pp. 377–397.

Wang, J. and Khonsari, M.M. 2005. 'Influence of Drag Force on the Dynamic Performance of Rotor-Bearing Systems,' *Journal of Engineering Tribology, Proceedings of the Institution of Mechonical Engineers*, Part J, Vol. 219, pp. 291–295.

Wang, J. and Khonsari, M.M. 2006a. 'A New Derivation for Journal Bearings Stiffness and Damping Coefficients in Polar Coordinates,' *Journal of Sound and Vibration*, Vol. 290, pp. 500–507.

Wang, J. and Khonsari, M.M. 2006b. 'Prediction of the Stability Envelop of Rotor-Bearing Systems,' *ASME Journal of Vibration and Acoustics*, Vol. 128, pp. 197–202.

Wang, J. and Khonsari, M.M. 2006c. 'On the Hysteresis Phenomenon Associated with Instability of Rotor–Bearing Systems,' Vol. 128(1), pp. 188–196.

Wang, J. and Khonsari, M.M. 2006d. 'Influence of Inlet Oil Temperature on the Instability Threshold of Rotor–Bearing Systems,' *ASME Journal of Tribology*, Vol. 128, pp. 319–326.

Wang, J.K., and Khonsari, M.M. 2006e. 'Application of Hopf Bifurcation Theory to the Rotor–Bearing System with Turbulent Effects,' *Tribology International*, Vol. 39(7), pp. 701–714.

Wang, J.K. and Khonsari, M.M. 2006f. 'Bifurcation Analysis of a Flexible Rotor Supported by Two Fluid Film Journal Bearings,' *ASME Journal of Tribology*, Vol. 128, pp. 594–603.

Wang, J. and Khonsari, M.M. 2009. 'On the Temperature Rise of Bodies Subjected to Unidirectional or Oscillating Frictional Heating and Surface Convective Cooling,' *STLE Tribology Transactions*, Vol. 52, pp. 310–322.

Wang, J. K. and Khonsari, M.M. 2016. *Thermohydrodynamic Instability in Fluid-Film Bearings*, John Wiley and & Sons, Chichester.

Wilcock, D.F. and Booser, E.R. 1957. *Bearing Design and Application*. McGraw-Hill Book Co., New York.

Wilcock, D.F. and Booser, E.R. 1988. 'Temperature Fade in Journal Bearing Exit Regions,' *STLE Tribology Transactions*, Vol. 31, pp. 405–410.

Xu, G., Zhou, J., Geng, H., Lu, M., Yang, L., and Yu, L. 2014. 'Research on the Static and Dynamic Characteristics of Misaligned Journal Bearing Considering the Turbulent and Thermohydrodynamic Effects,' *ASME Journal of Tribology*, Vol. 137, 024504:1–8.

9

Squeeze-Film Bearings

9.1 Introduction

Hydrodynamic pressure developed on a normal approach between two surfaces is resisted by a viscous fluid. As a result, it takes a finite length of time for the fluid to be squeezed out from the sides. Therefore, the fluid offers a certain amount of useful cushioning – that is, load-carrying capacity – even in the absence of sliding motion or a physical wedge effect. By this means, a useful load-carrying capacity can be generated between two parallel plates or disks when one surface is pressed against the other.

Squeeze-film action plays an important role in many mechanical components. The following are important components for which analytical tools are covered in this chapter:

- *Clutch pack in automotive transmission.* Illustrated in Figure 9.1, automatic transmission fluid (ATF) serves as the squeeze-film fluid during the engagement of a set of separator disks and friction disks as they approach each other and then lock for full speed operation (see Example 9.1).
- *Engine piston pin bearings.* Cushioning by squeeze-film action of the engine oil during the power stroke provides an oil film despite lack of the significant relative velocity usually needed to avoid metal-to-metal contact (see Examples 9.2 and 9.4).
- *Human knee joint.* Impact between the femur and tibia in the configuration of Figure 9.8 is cushioned by the body's synovial fluid during aerobic exercises (see Example 9.3).
- *Damper film for jet engine ball bearings.* Since ball bearings have almost no damping capacity, an annular lubricant film enveloping the fixed outer bearing ring provides damping by squeeze-film action to minimize rotor vibration (Figure 9.2).
- *Piston rings.* Both sliding and squeeze-film components exist. At the top and bottom of the stroke in an internal combustion engine, ring sliding velocity drops to zero and squeeze-film action provides the needed cushioning (see Section 9.7).
- *Peeling of a flexible tape.* The hydrodynamics of peeling a flexible strip (e.g. a tape) from a surface attached by a layer of viscous fluid (e.g. adhesive) involves *negative squeeze* analysis in the presence of surface tension. See, for example, the paper by McEwan and Taylor (1966) which describes both theoretical and visual observation of the cavitation. Analyses of

Applied Tribology: Bearing Design and Lubrication, Third Edition. Michael M. Khonsari and E. Richard Booser.
© 2017 John Wiley & Sons Ltd. Published 2017 by John Wiley & Sons Ltd.
Companion Website: www.wiley.com/go/Khonsari/Applied_Tribology_Bearing_Design_and_Lubrication_3rd_Edition

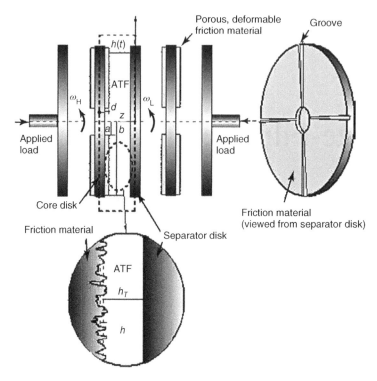

Figure 9.1   Wet clutch system of an automotive transmission. *Source*: Jang and Khonsari, 1999. Reproduced with permission of ASME.

fluid-film meniscus instability of negative squeeze are given by Koguchi *et al.* (1988) and Koguchi and Yada (1990).

Squeeze-film configurations considered in this chapter include planar geometries such as circular, elliptical, triangular, and rectangular sections, as well as circular sectors and concentric annuli. A number of nonplanar squeeze actions such as a nonrotating shaft moving vertically down toward the bushing and piston ring motion are also investigated. The primary parameters covered are the bearing load-carrying capacity, i.e. the contribution of the pressure integrated over the acting area and the time of approach Δt needed for the film thickness to drop from an initial value h_0 to a final value of h.

9.2 Governing Equations

The standard Reynolds equation (6.23) governs the pressure distribution, repeated below, in both the Cartesian and polar coordinate systems, assuming incompressible fluid.

Planar Cartesian coordinate system:

$$\frac{\partial}{\partial x}\left(\frac{h^3}{\mu}\frac{\partial P}{\partial x}\right) + \frac{\partial}{\partial y}\left(\frac{h^3}{\mu}\frac{\partial P}{\partial y}\right) = 6U\frac{\partial h}{\partial x} - 12V_s \tag{9.1}$$

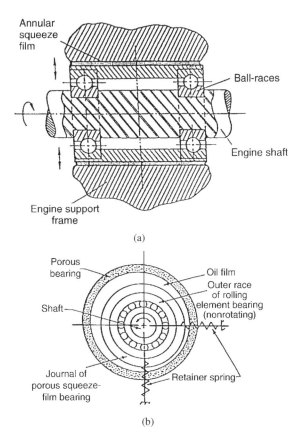

(a)

(b)

Figure 9.2 (a) Squeeze-film damper for absorbing vibration in jet engines. *Source*: Moore, 1972. Reproduced with permission of Pergamon Press. (b) Schematic diagram of a porous bearing squeeze-film damper. *Source*: Cusano and Funk, 1977. Reproduced with permission of ASME.

Polar-cylindrical coordinate system:

$$\frac{\partial}{\partial r}\left(\frac{rh^3}{\mu}\frac{\partial P}{\partial r}\right) + \frac{1}{r}\frac{\partial}{\partial \theta}\left(\frac{h^3}{\mu}\frac{\partial P}{\partial \theta}\right) = 6U\frac{\partial h}{\partial \theta} - 12rV_s \tag{9.2}$$

where $V_s = -dh/dt$ represents the squeeze velocity or the velocity of approach.

9.3 Planar Squeeze Film

In this section, several specific examples of planar squeeze film are covered. The results are then generalized for a variety of configurations.

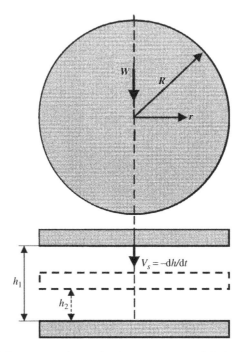

Figure 9.3   Squeeze-film action between two parallel circular disks.

Two Parallel Circular Disks

This is a classical problem in fluid mechanics often referred to as the *Stefan problem*. The primary objective is to determine the time Δt required for the separation distance between the two disks to be reduced from an initial gap of h_1 to h_2 (Figure 9.3).

In the absence of inertia effects, pressure varies radially if $\partial h / \partial \theta = 0$, so that Equation (9.2) reduces to the following ordinary differential equation:

$$\frac{\mathrm{d}}{\mathrm{d}r} \left(\frac{rh^3}{\mu} \frac{\mathrm{d}P}{\mathrm{d}r} \right) = -12 r V_s \tag{9.3}$$

Integrating the above equation twice yields

$$P = -6\mu V_s \int \frac{r}{h^3} \mathrm{d}r + C_1 \frac{\mu}{h^3} \ln r + C_2 \tag{9.4}$$

where C_1 and C_2 are the integration constants.

Since hydrodynamic pressure must be finite at $r = 0$, it may be argued that $C_1 = 0$. The second integration constant is evaluated using the pressure boundary condition at $r = R$, where $P = 0$ (gage pressure). If $h \neq h(r)$, $C_2 = 3\mu r^2 / h^3$ and

$$P = \frac{3\mu V_s}{h^3} (R^2 - r^2) \tag{9.5}$$

The load-carrying capacity is

$$W = \int_A P \, dA = \int_0^R 2\pi r P \, dr = \frac{3\pi\mu R^4}{2h^3} V_s \tag{9.6}$$

Using $V_s = -dh/dt$, Equation (9.6) can easily be integrated with the appropriate limits:

$$\int_{t_1}^{t_2} dt = -\frac{3\pi\mu R^4}{2W} \int_{h_1}^{h_2} \frac{dh}{h^3} \tag{9.7}$$

$$\Delta t = \frac{3\pi\mu R^4}{4W}\left(\frac{1}{h_2^2} - \frac{1}{h_1^2}\right) \tag{9.8}$$

where $\Delta t = t_2 - t_1$ is the time of approach for the film gap to reduce from h_2 to h_1. This equation predicts that $\Delta t \to \infty$ as $h_2 \to 0$: theoretically, it takes an infinite amount of time to squeeze out all the fluid.

Shape Variation: Elliptical Disks

It is possible to obtain a closed-form analytical solution for a variety of other planar squeeze applications with uniform film thickness between the approaching surfaces. To illustrate the solution technique, consider squeeze-film lubrication between two parallel, elliptically shaped disks. Referring to Figure 9.4, the Reynolds equation (9.1) reduces to the Poisson equation:

$$\frac{\partial^2 P}{\partial x^2} + \frac{\partial^2 P}{\partial y^2} = -12\mu \frac{V_s}{h^3} \tag{9.9}$$

It can be directly verified that the solution to Equation (9.9) should be of the form $P(x, y) = f(x^2/a^2 + y^2/b^2 - 1)$, where the function f is to be determined by substitution of the solution into Equation (9.9). The result is

$$f = -6\mu\frac{a^2 b^2}{a^2 + b^2}\frac{V_s}{h^3}$$

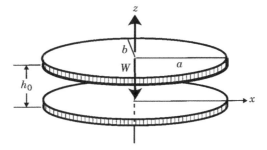

Figure 9.4   Elliptical disk squeeze.

so that the pressure distribution becomes

$$P = -6\mu \frac{a^2 b^2}{a^2 + b^2} \frac{V_s}{h^3} \left(\frac{x^2}{a^2} + \frac{y^2}{b^2} - 1 \right) \tag{9.10}$$

A circular disk of radius R is a special case of the above solution with $a = b = R$, i.e.

$$P = -3\mu \frac{V_s}{h^3} (x^2 + y^2 - R^2) \tag{9.11}$$

which can be converted to the polar coordinate solution using the transformation identities $x = r \cos(\theta)$ and $y = r \sin(\theta)$ to get Equation (9.5).

Load-carrying capacity W can then be evaluated by integrating pressure P over the appropriate area:

$$W = \int_A P dA = \iint P dx dy \tag{9.12}$$

or $W = 2\pi \int_r P r \, dr$ for cylindrical shapes. When the pressure distribution P is substituted into Equation (9.12), the time Δt required for the film thickness to decrease to height h_2 at time t_2 can be obtained.

For elliptically shaped disks, the load-carrying capacity is

$$W = \frac{3\pi \mu a^3 b^3 V_s}{(a^2 + b^2)h^3}$$

and the approach time is

$$\Delta t = \frac{3\pi \mu a^3 b^3}{2(a^2 + b^2)W} \left(\frac{1}{h_2^2} - \frac{1}{h_1^2} \right)$$

9.4 Generalization for Planar Squeeze Film

For planar squeeze-film problems, the time of approach has the following form (Moore, 1993):

$$\Delta t = K \frac{\mu A^2}{W} \left(\frac{1}{h_2^2} - \frac{1}{h_1^2} \right) \tag{9.13}$$

where K is the shape function. Taking the time derivative of Equation (9.13), the surface approach velocity can be obtained:

$$V_s = \frac{1}{2K} \frac{h^3 W}{\mu A^2} \tag{9.14}$$

where A is the plate area and h is the squeeze-film thickness.

Table 9.1 Types of planar squeeze

Type	Ratio, r	Configuration	Constant, K
Circular section	–		$\dfrac{3}{4\pi}$
Elliptical section	$\mathfrak{R} = \dfrac{b}{a}$		$\dfrac{3\mathfrak{R}}{2\pi(1+\mathfrak{R}^2)}$
Rectangular section	$\mathfrak{R} = \dfrac{B}{L}$		$\dfrac{1}{2\mathfrak{R}}\left[1 - \dfrac{192}{\pi^5\mathfrak{R}}\sum_{n=1,3,5,\dots}^{\infty}\dfrac{\tanh(n\pi\mathfrak{R}/2)}{n^5}\right]$
Triangular section	–		$\dfrac{\sqrt{3}}{10}$
Circular sector	$\mathfrak{R} = \dfrac{\alpha}{2\pi}$		$\sum_{n=1,3,5,\dots}^{\infty}\dfrac{24}{n^2\pi^3\mathfrak{R}\left[2+\left(\dfrac{n}{2\mathfrak{R}}\right)\right]^2}$
Concentric annulus	$\mathfrak{R} = \dfrac{D_i}{D_0}$		$\dfrac{3}{4\pi}\left[\dfrac{\ln\mathfrak{R} - \mathfrak{R}^4\ln\mathfrak{R} + (1-\mathfrak{R}^2)^2}{(1-\mathfrak{R}^2)^2\ln\mathfrak{R}}\right]$

Forms of the function K for a series of planar squeeze-film geometries are shown in Table 9.1 and plotted in Figure 9.5 for convenience (Khonsari and Jang, 1997). Using either the table or the figure, one can readily evaluate K. Then, for a given load, W, Equation (9.13) gives the time of approach for the film thickness to drop from an initial h_1 to a final h_2.

Example 9.1 Estimate the time of approach in a wet clutch system modeled as two rigid concentric annuli with $R_i = 0.047$ m and $R_o = 0.06$ m submerged in a lubricant with viscosity of $\mu = 0.006$ Pa s. Hydraulic pressure is $P = 1.25$ MPa, and the initial separation gap is $h_1 = 25 \times 10^{-6}$ m. Estimate the initial squeeze velocity and the length of time necessary for the film thickness to drop to $h_2 = 5 \times 10^{-6}$ m.

The shape factor K for the concentric annulus can be evaluated from Table 9.1:

$$A = \pi\left(R_o^2 - R_i^2\right) = 0.0044 \text{ m}^2$$

$$\mathfrak{R} = \dfrac{R_i}{R_o} = \dfrac{0.047}{0.06} = 0.783$$

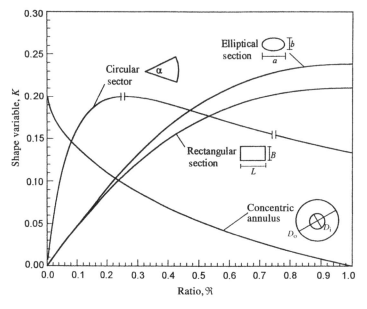

Figure 9.5 Variation of constant K with shape ratio $\Re$.

$$K = \frac{3}{4\pi} \frac{\ln(\Re) - \Re^4 \ln \Re + (1 - \Re^2)^2}{(1 - \Re^2)^2 \ln (\Re)}$$

$$= \frac{3}{4\pi} \frac{\ln(0.783) - (0.783)^4 \ln(0.783) + (1 - 0.783^2)^2}{(1 - 0.783^2)^2 \ln(0.783)} = 0.019$$

Using Equation (9.13), the time of approach is

$$\Delta t = K \frac{\mu A^2}{W} \left(\frac{1}{h_2^2} - \frac{1}{h_1^2} \right) = K \frac{\mu A}{P} \left(\frac{1}{h_2^2} - \frac{1}{h_1^2} \right)$$

$$= \frac{0.019(0.006)(0.0044)}{1.25 \times 10^6} \left[\frac{1}{(5 \times 10^{-6})^2} - \frac{1}{(25 \times 10^{-6})^2} \right] = 0.02 \, \text{s}$$

The squeeze velocity can be estimated using Equation (9.14):

$$V_s = \frac{1}{2K} \frac{h_1^3 W}{\mu A^2} = \frac{1}{2K} \frac{h_1^3 P}{\mu A}$$

$$= \frac{1}{2(0.019)} \frac{(25 \times 10^{-6})^3 1.25 \times 10^6}{(0.006)(0.0044)} = 0.02 \, \text{m/s}$$

The time of approach predicted above is a representation of the first stage of the engagement duration while the clutch operates in the hydrodynamic regime. This engagement begins when pressure is applied hydraulically by means of a piston and hydrodynamic pressure is developed in the ATF as a result of squeeze action which supports most of the applied load. Since the surfaces are separated by a relatively thick film of fluid, behavior of the clutch is governed by the theory of hydrodynamic lubrication. This period lasts only 0.02 s.

During engagement, the fluid film thickness drops to the extent that surface asperities come into contact. As a result, contact pressure at the asperity level begins to support a major portion of the imposed load, significantly influencing the behavior of the wet clutch. The film thickness is further reduced as the friction-lining material is compressed and deforms elastically. The surfaces are subsequently pressed together and 'locked' when their relative speed drops to zero. The timescale of the engagement process is typically of the order of 1 s, during which the squeeze action is of paramount importance. It follows, therefore, that in a typical engagement cycle, the lubrication regime undergoes a transition from hydrodynamic to mixed or boundary lubrication. This shift in the lubrication regime has important implications on the signature of the total torque, i.e. the combination of the viscous torque and contact torque. The total torque reaches a peak value when the film thickness drops to a minimum. After this peak value, the torque initially remains relatively flat for a short period of time and then begins to rise gradually as the relative speed between the clutch disks decreases. This increase in the torque is a direct consequence of the change in the coefficient of friction as a function of speed, in accordance with the Stribeck friction curve. The most interesting torque signature is a highly undesirable sudden spike or 'rooster's tail' toward the end of the engagement. Its occurrence can be predicted analytically (Jang and Khonsari, 1999) and can be treated to minimize its effect by altering the friction behavior.

Note that the friction material is rough, porous, and deformable. Also, the large disk diameters used in wet clutch systems may require consideration of the centrifugal forces. These elements can affect the engagement time and torque-transfer characteristics, as well as the temperature field in the ATF and on the surface of the separator. The interested reader may refer to Jang and Khonsari (1999, 2002, 2011, 2013) for detailed analysis of automotive wet clutch systems. More recent studies of wet clutches have included parametric analysis of variance and experimental results (Mansouri et al., 2001, 2002; Marklund et al., 2007; Li et al., 2014, 2015, 2016, 2017).

9.5 Nonplanar Squeeze Film

Sphere Approaching a Plate

Consider the impact of a perfectly spherical ball approaching a plate covered with a layer of lubricating film of thickness h_1, as shown schematically in Figure 9.6. Film thickness is given by $h = h_0 + R - \sqrt{R^2 - r^2}$ where h_0 is the minimum film thickness at $r = 0$.

The Reynolds equation in the polar-cylindrical coordinate system must be used to analyze this problem. The solution is given by Equation 9.4. Using dh instead of dx to evaluate the integral in Equation (9.4),

$$\frac{dh}{dr} = \frac{r}{\sqrt{R^2 - r^2}} = \frac{r}{(h_0 + R - h)}$$

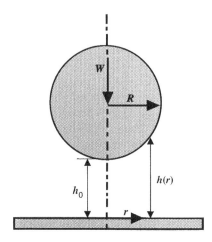

Figure 9.6   Squeeze of a sphere toward a flat plate.

so that

$$P = -6\mu V_s \int \frac{h_0 + R - h}{h^3} dh + C_2 \tag{9.15}$$

Equation (9.15) is integrated to give

$$P = -6\mu V_s \left(\frac{h_0 + R}{2h^2} + \frac{1}{h} \right) + C_2 \tag{9.16}$$

The boundary condition is $P = 0$ at $h = h_0 + R$. Evaluating the integration constant yields $C_2 = 3\mu V_s/(h_0 + R)$. The pressure distribution is therefore

$$P = \frac{3\mu V_s}{h_0 + R} \left(1 - \frac{h_0 + R}{h} \right)^2 \tag{9.17}$$

Integrating this pressure distribution ($W = \int_0^{\pi/2} 2\pi r P R \cos\theta d\theta = \int_0^R 2\pi r P\, dr$) then gives the load-carrying capacity, from which the time of approach can be determined.

Expressions for Several Nonplanar Squeeze-Film Geometries

Load-carrying capacity W and approach time Δt for a sphere approaching a plate and several other nonplanar configurations are given in Table 9.2. The expressions are exact for conical surfaces and hemispherical ball-in-the-socket configurations. However, for the nonplanar cylindrical geometries, such as the journal bearing, it is assumed that $L/D \rightarrow \infty$, so that no side leakage is allowed. In other words, the infinitely long bearing approximation (ILA) is applied to treat the problem as one-dimensional so that $\partial/\partial y$ in Equation (9.1) vanishes. These solutions should be sufficiently accurate when $L/D > 2$. Because L/D may be finite in many practical applications, i.e. $L/D < 2$, treatment requires a numerical solution scheme. See Section 9.6 for squeeze-film solution of bearings with nonrotating journals.

Table 9.2 Closed form solutions for nonplanar squeeze films

Type	Configuration	Solution
Infinitely long journal bearing with 180° arc bushing	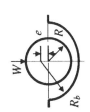	$W = 12\mu L V_s \left(\dfrac{R}{C}\right)^3 \left[\dfrac{\varepsilon}{1-\varepsilon^2} + \dfrac{2}{(1-\varepsilon^2)^{3/2}} \tan^{-1}\left(\sqrt{\dfrac{1+\varepsilon}{1-\varepsilon}}\right)\right]$ $$\Delta t = \frac{24\mu RL}{W}\left(\frac{R}{C}\right)^2 \left[\frac{\varepsilon_2}{\sqrt{1-\varepsilon_2^2}}\tan^{-1}\left(\sqrt{\frac{1+\varepsilon_2}{1-\varepsilon_2}}\right) - \frac{\varepsilon_1}{\sqrt{1-\varepsilon_1^2}}\tan^{-1}\left(\sqrt{\frac{1+\varepsilon_1}{1-\varepsilon_1}}\right)\right]$$
Infinitely long journal bearing with 360° arc bushing	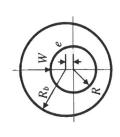	$W = \dfrac{12\pi\mu L V_s}{(1-\varepsilon^2)^{3/2}}\left(\dfrac{R}{C}\right)^3$ $$\Delta t = \frac{12\pi\mu RL}{W}\left(\frac{R}{C}\right)^2 \left[\frac{\varepsilon_2}{\sqrt{1-\varepsilon_2^2}} - \frac{\varepsilon_1}{\sqrt{1-\varepsilon_1^2}}\right]$$
Spherical ball in the socket		$W = 6\pi\mu R V_s \left(\dfrac{R}{C}\right)^3 \left[\dfrac{1}{\varepsilon^3}\ln(1-\varepsilon) + \dfrac{1}{\varepsilon^2(1-\varepsilon)} - \dfrac{1}{2\varepsilon}\right]$ $$\Delta t = \frac{3\pi\mu R^2}{W}\left(\frac{R}{C}\right)^2 \left[\frac{\varepsilon_2 - \varepsilon_1}{\varepsilon_2\varepsilon_1} + \left(\frac{1+\varepsilon_1^2}{\varepsilon_1^2}\right)\ln(1-\varepsilon_1) - \left(\frac{1+\varepsilon_2^2}{\varepsilon_2^2}\right)\ln(1-\varepsilon_2)\right]$$

(continued)

Table 9.2 (*Continued*)

Type	Configuration	Solution
Infinitely long cylinder approaching a plane surface	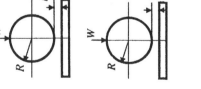	$W = \dfrac{3\pi\mu RLV_s}{h}\sqrt{\dfrac{R}{h}}$ $\Delta t = \dfrac{6\sqrt{2}\,\pi\mu RL}{W}\left[\sqrt{\dfrac{R}{h_2}} - \sqrt{\dfrac{R}{h_1}}\right]$

$W = 6\pi\mu V_s h\left[\dfrac{3}{2}\left(\dfrac{R}{h}-1\right) - \left(1+\dfrac{R}{h}\right)\ln\left(1+\dfrac{R}{h}\right) + \dfrac{1}{2(1+R/h)}\left\{1+2\left(1+\dfrac{R}{h}\right)^3\right\}\right]$

Spherical ball approaching a plane surface	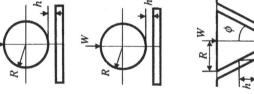	$\approx 6\pi\mu V_s h\left(\dfrac{R}{h}\right)^2$ for $\dfrac{R}{h}\gg 1$ $\Delta t = \dfrac{6\pi\mu R^2}{W}\ln\left(\dfrac{h_1}{h_2}\right)$
Conical surfaces		$W = \dfrac{3\pi\mu R^4 V_s}{2h^3\sin^4\phi}$ $\Delta t = \dfrac{3\pi\mu R^4}{4W\sin^4\phi}\left(\dfrac{1}{h_2^2}-\dfrac{1}{h_1^2}\right)$
Truncated conical surfaces		$W = \dfrac{3\pi\mu V_s}{2h^3\sin^4\phi}\left[R_2^4 - R_1^4 - \dfrac{(R_2^2-R_1^2)^2}{\ln(R_2/R_1)}\right]$ $\Delta t = \dfrac{3\pi\mu}{4W\sin^4\phi}\left[R_2^4 - R_1^4 - \dfrac{(R_2^2-R_1^2)^2}{\ln(R_2/R_1)}\right]\left(\dfrac{1}{h_2^2}-\dfrac{1}{h_1^2}\right)$

Source: Khonsari and Jang, 1997. Reproduced with permission of Taylor & Francis.

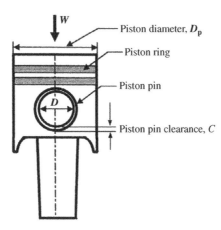

Figure 9.7   Schematic of a piston pin.

Example 9.2   A piston pin (Figure 9.7) of an internal combustion engine is lubricated with SAE 20 oil at an effective engine operating temperature of $150°$ F ($\mu = 2.1 \times 10^{-6}$ reyns). The pin diameter, length, and radial clearance are $D = 1.5$ in., $L = 1.5$ in., and $C = 0.001$ in., respectively. During the power stroke, the top of the piston, whose diameter is 4 in., experiences an effective pressure of 100 psi produced by the explosion of the fuel/air mixture. Assume that side leakage can be neglected.

(a) Determine the time required for the film thickness at the top of the bushing to reduce from 0.0005 in. to 0.0001 in.
(b) Determine the squeeze velocity at the onset of the power stroke.
(c) Assuming that the minimum required film thickness is 0.0001 in., comment on the danger of metal-to-metal contact for two operating speeds: 750 and 1500 rpm.

A piston pin without a sliding velocity can be idealized as a nonrotating journal bearing with a full bushing. The load imposed on the bearing during the power stroke is

$$W = P_{eff}A_P = P_{eff}\left(\pi D_p^2/4\right) = \pi(100)(4)^2/(4) = 1257\,\text{lbf}$$

The minimum film thickness is $h_{min} = C(1 - \varepsilon)$, where C is the radial clearance and ε is the eccentricity ratio, i.e. $e/(R_b - R)$. Therefore, $\varepsilon_1 = 0.5$ at $h_{min} = 0.0005$ in. and $\varepsilon_2 = 0.9$ at $h_{min} = 0.0001$ in., respectively.

(a) From Table 9.2 we have

$$\Delta t = \frac{12\pi\mu RL}{W}\left(\frac{R}{C}\right)^2\left[\frac{\varepsilon_2}{\sqrt{1 - \varepsilon_2^2}} - \frac{\varepsilon_1}{\sqrt{1 - \varepsilon_1^2}}\right] = 0.059\,\text{s}$$

(b) Velocity at the onset of power stroke is

$$V_s = \frac{W\left(1 - \varepsilon_1^2\right)^{3/2}}{12\pi\mu L} \left(\frac{C}{R}\right)^3 = 0.016 \text{ in./s}$$

(c) For the operating speed of 1500 rpm, the power stroke duration is the interval of a half-revolution: $\Delta t_s = (60)/(2)(1500) = 0.02$ s. At 750 rpm, the duration of the power stroke is $\Delta t_s = (60)/(2)(750) = 0.04$ s. The duration of the power stroke Δt_s is less than the time of approach, $\Delta t = 0.059$ s, for both 1500 and 750 rpm. Therefore, the cushioning provided by the lubricant is sufficient to handle these operating conditions, and there is no danger of metal-to-metal contact. The validity of this conclusion is checked in Section 9.6 for the full journal bearing where leakage flow is considered. It may be noted that surface roughness, which is not considered in this example, may have a pronounced effect on the results.

Example 9.3   A human knee joint undergoes significant stresses during aerobic exercise. Typically, loads several times greater than body weight are imposed on the joint during an impact. The upper and lower bones (femur and tibia) shown in Figure 9.8 are to be modeled. The knee joint has poor geometrical conformity (Dowson, 1966). As an idealization, represent the configuration with a rigid cylinder approaching a plane.

Determine the maximum pressure, squeeze velocity, and time of approach required for the film thickness to drop from an initial film thickness of $h_0 = 3.5 \times 10^{-4}$ in. to $h_1 = 3.5 \times 10^{-5}$ in. Assume that the load on each joint is $W = 100$ lbf and take the lubricant viscosity (i.e. the synovial fluid) as $\mu = 1.5 \times 10^{-6}$ reyns, using $R = L = 2$ in.

The cylinder – plane configuration represented by the parabolic film thickness expression $h = h_0 + (x^2/2)R$ is selected for the analysis. The pressure distribution is

$$P = 6\mu \frac{V_s R}{h_0^2} \frac{1}{[1 + x^2/(2Rh_0)]^2}$$

The maximum pressure $P_{max} = 6\mu V_s R/h_0^2$ occurs at $x = 0$.

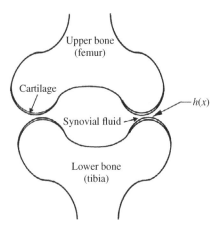

Figure 9.8   Human knee joint.

Referring to Table 9.2, the squeeze velocity can be expressed in terms of the load as follows:

$$V_s = \frac{Wh}{3\pi\mu RL\sqrt{R/h_0}} = \frac{(100)(3.5\times10^{-4})}{3\pi(1.5\times10^{-6})(2)(2)\sqrt{2/(3.5\times10^{-4})}} \approx 8\,\text{in./s}$$

Using this value of V_s, the time of approach and the maximum pressure can readily be evaluated as follows:

$$\Delta t = \frac{6\sqrt{2\pi}\mu RL}{W}\left(\sqrt{\frac{R}{h_1}} - \sqrt{\frac{R}{h_0}}\right)$$

$$= \frac{6\sqrt{2\pi}(1.5\times10^{-6})(2)(2)}{100}\left(\sqrt{\frac{2}{3.5\times10^{-5}}} - \sqrt{\frac{2}{3.5\times10^{-4}}}\right)$$

$$= 2.6\times10^{-4}\,\text{s}$$

$$P_{\text{max}} = \frac{6\mu V_s R}{h_0^2} = 1200\,\text{psi}$$

Fairly large stresses at the minimum film thickness and the small duration of the approach time give an indication of the severity of impact. Fortunately, coupled with the synovial fluid's cushioning effect, a layer of soft cartilage covering the bones provides additional protection (Dowson *et al.*, 1970).

The cartilage is soft and deformable, yet rough and porous. According to Higginson and Unsworth (1981), it is about 2 mm thick in a healthy adult and much thinner in joints suffering from arthritis. The surface roughness of cartilage is much greater than that of typical metal bearings, on the order of $2\,\mu$ m for a healthy adult. Its porosity is high, but with a high water content – roughly 75% under a fully saturated state – it has a low permeability, of the order of $10^{-18}\,\text{m}^2$. The cartilage is a highly elastic membrane whose effective modulus of elasticity varies with time – initially very high upon the application of load and dropping with time (McCutchen, 1967) – giving of a mean value of $E = 10^7\,\text{dyn/cm}^2$.

Synovial fluid also has fascinating characteristics. Involving a dispersion of a long chain polymer known as *hyaluronic acid*, it exhibits a remarkable degree of shear thinning. According to Higginson and Unsworth's citation of Cook's dissertation (Cook, 1974), the apparent viscosity of synovial fluid in a healthy adult drops from $\mu = 50\,\text{Pa s}$ at a shear rate of $\dot{\gamma} \approx 0.1\,\text{s}^{-1}$ to $0.05\,\text{Pa s}$ at $\dot{\gamma} \approx 103\,\text{s}^{-1}$. All these factors may play a role in determining the characteristics of a knee joint, thus requiring careful assessment in a serious modeling attempt. A ball-and-socket configuration is most logical for modeling a hip joint since it must provide stability and load-carrying capacity. The interested reader is referred to an article by Meyer and Tichy (1999).

9.6 Squeeze Film of Finite Surfaces

Finite Planar Squeeze Film

Consider a finite rectangular surface approaching a flat plate with a uniform film thickness h (Figure 9.9). Governing Equation (9.9) is a nonhomogeneous, elliptical partial differential equation which does indeed lend itself to an analytical solution.

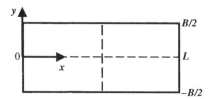

Figure 9.9   Squeeze of finite size, rectangular plates.

Referring to Figure 9.9, the appropriate boundary conditions are

$$
\begin{cases} P(0, y) = 0 \quad \text{(I)} \\ P(x, B/2) = 0 \quad \text{(III)} \end{cases} \quad \text{and} \quad \begin{cases} \dfrac{\partial P}{\partial x}(L/2, y) = 0 \quad \text{(II)} \\ \dfrac{\partial P}{\partial y}(x, 0) = 0 \quad \text{(IV)} \end{cases}
\tag{9.18}
$$

where boundary conditions (II) and (IV) take advantage of the symmetry of the problem.

The method of separation of variables can be applied to solve the problem. The pressure solution is represented by the product of two functions separated in the x and y directions:

$$
P(x, y) = X(x) \cdot Y(y)
\tag{9.19}
$$

Substituting this product solution in the homogeneous part of the governing equation yields

$$
X''Y + Y''X = 0
\tag{9.20}
$$

Dividing through by XY results in the following relationship:

$$
\frac{X''}{X} = -\frac{Y''}{Y} = -\beta^2
\tag{9.21}
$$

where β is a constant due to the fact that left-hand side of the first equality is not a function of Y and the right-hand side side is independent of X. Parameter β represents the eigenvalues of the problem, which are yet to be determined.

Rewriting Equation (9.21) as two sets of equations, we have

$$
X'' + \beta^2 X = 0
$$

and

$$
Y'' - \beta^2 Y = 0
\tag{9.22}
$$

Solution of these equations will determine the eigenvalues and eigenvectors of the problem. From boundary condition (I), we have $X(0) = 0$. Therefore, $C_1 = 0$. Using boundary condition (II), we have $X'(L/2) = 0$. This implies that

$$\cos \beta \frac{L}{2} = 0 = \cos \frac{n\pi}{2} \tag{9.23}$$

The eigenvalues are, therefore, $\beta = n\pi/L$ for $n = 1, 3, 5, \ldots$ (odd integers).
  Solution of the X equation is

$$X_n = \sum_{n=1,3,5,\ldots}^{\infty} C_n \sin \frac{n\pi}{L} x \tag{9.24}$$

and the pressure distribution is of the form

$$P(x, y) = X_n(x)Y(y) = \left(\sum_{n=1,3,5,\ldots}^{\infty} C_n \sin \frac{n\pi}{L} x \right) Y(y) \tag{9.25}$$

where values of C_n are yet to be determined.
  Substituting Equation (9.25) into Equation (9.9) yields the following equation, which must be solved for $Y(y)$:

$$-\sum C_n \left(\frac{n\pi}{L} \right)^2 \sin \frac{n\pi}{L} x \cdot Y + \sum C_n \sin \frac{n\pi}{L} x \cdot \frac{d^2 Y}{dy^2} = -\frac{12\mu V_s}{h^3} \tag{9.26}$$

where $\sum \equiv \sum_{n=1,3,5,\ldots}^{\infty}$ for convenience of notation.
  The right-hand side of Equation (9.26) – a constant quantity – can be expanded in a Fourier series by noting that

$$1 = \frac{4}{\pi} \sum \frac{1}{n} \sin \frac{n\pi x}{L}$$

Therefore

$$-\frac{12\mu V_s}{h^3} = -\frac{12\mu V_s}{h^3} \frac{4}{\pi} \sum \frac{1}{n} \sin \frac{n\pi x}{L} \tag{9.27}$$

and Equation (9.26) becomes

$$-\sum C_n \left(\frac{n\pi}{L} \right)^2 \sin \frac{n\pi}{L} xY + \sum C_n \sin \frac{n\pi}{L} x \frac{d^2 Y}{dy^2} = -\frac{48\mu V_s}{h^3} \sum \frac{1}{n} \sin \frac{n\pi x}{L} \tag{9.28}$$

which simplifies to

$$C_n \frac{d^2 Y}{dy^2} - \left(\frac{n\pi}{L} \right)^2 C_n Y = -\frac{48}{\pi} \frac{\mu V_s}{h^3} \frac{1}{n} \tag{9.29}$$

Equation (9.29) is a nonhomogeneous ordinary differential equation whose solution is

$$Y(y) = Y_H(y) + Y_P \tag{9.30}$$

where Y_H and Y_P are the homogeneous and particular solutions, respectively.
   The homogeneous solution is

$$Y_H(y) = D_1 \sinh\left(\frac{n\pi y}{L}\right) + D_2 \cosh\left(\frac{n\pi y}{L}\right) \tag{9.31}$$

The particular solution is simply a constant which satisfies Equation (9.29). It is determined from

$$-\left(\frac{n\pi}{L}\right)^2 C_n Y_P = -\frac{48}{\pi}\frac{\mu V_s}{h^3}\frac{1}{n} \tag{9.32}$$

Solving for Y_P yields

$$Y_P = \frac{48}{\pi^3}\frac{\mu V_s}{h^3}\frac{L^2}{n^3}\frac{1}{C_n} \tag{9.33}$$

Therefore, the complete solution is

$$P(x, y) = \sum C_n \sin\frac{n\pi}{L}x\left[D_n\sinh\left(\frac{n\pi y}{L}\right) + E_n\cosh\left(\frac{n\pi y}{L}\right) + \frac{1}{C_n}\frac{1}{n^3}\frac{48}{\pi}\frac{\mu V_s L^2}{h^3}\right] \tag{9.34}$$

Combining constants: $C_n D_n = F_n$ and $C_n E_n = G_n$ simplifies the above equation to the following:

$$P(x, y) = \sum \sin\frac{n\pi}{L}x\left[F_n\sinh\left(\frac{n\pi y}{L}\right) + G_n\cosh\left(\frac{n\pi y}{L}\right) + \frac{1}{n^3}\frac{48}{\pi}\frac{\mu V_s L^2}{h^3}\right] \tag{9.35}$$

The unknown constants F_n and G_n are determined using boundary conditions (III) and (IV).
Using (IV), $F_n = 0$. Using (III), we have

$$G_n = -\frac{48}{\pi^3}\frac{\mu L^2 V_s}{h^3}\frac{1}{n^3}\frac{1}{\cosh(n\pi B/2L)}$$

Therefore, the solution for pressure distribution is

$$P(x, y) = -\frac{48}{\pi^3}\frac{\mu L^2 V_s}{h^3}\sum_{n=1,3,5,\ldots}^{\infty}\left[\frac{1}{n^3}\frac{\cosh(n\pi y/L)}{\cosh(n\pi B/2L)} - 1\right]\sin\left(\frac{n\pi x}{L}\right) \tag{9.36}$$

Load-Carrying Capacity

Integrating pressure over the appropriate area yields

$$W = 2 \int_0^{(B/2)L} \int_0^L P(x,y)\,dxdy = \frac{192\mu L^3 V_s}{\pi^4 h^3} \sum_{n=1,3,5,\ldots}^{\infty} \left(\frac{B}{2n^4} - \frac{L}{n^5 \pi} \tanh \frac{n\pi B}{2L} \right) \tag{9.37}$$

Time of Approach

Using $V_s = -dh/dt$, the time of approach to reduce the film thickness from h_1 to h_2 is

$$\Delta t = \frac{96\mu L^3}{\pi^4 W} \left[\left(\frac{1}{h_2^2} - \frac{1}{h_1^2} \right) \sum_{n=1,3,5,\ldots}^{\infty} \left(\frac{B}{2n^4} - \frac{L}{n^5 \pi} \tanh \frac{n\pi B}{2L} \right) \right] \tag{9.38}$$

Dimensionless expressions for the load and time of approach are

$$\overline{W} = \frac{\pi^4 h^3}{\mu L^3 B V_s} W = 192 \sum_{n=1,3,5,\ldots}^{\infty} \left(\frac{1}{2n^4} - \frac{L}{n^5 \pi B} \tanh \frac{n\pi B}{2L} \right) \tag{9.39}$$

$$\overline{\Delta t} = \frac{\pi^4 W}{\mu L^3 B} \left(\frac{h_1^2 h_2^2}{h_1^2 - h_2^2} \right) \Delta t = 96 \sum_{n=1,3,5,\ldots}^{\infty} \left(\frac{1}{2n^4} - \frac{L}{n^5 \pi B} \tanh \frac{n\pi B}{2L} \right) \tag{9.40}$$

Note that the expressions for the summation in Equation (9.40) for $\overline{W}$ and $\overline{\Delta t}$ are identical. Therefore, $\overline{W} = 192\,\mathbf{S}$ and $\overline{\Delta t} = 96\,\mathbf{S}$ where

$$\mathbf{S} = \sum_{n=1,3,5,\ldots}^{\infty} \left(\frac{1}{2n^4} - \frac{L}{n^5 \pi B} \tanh \frac{n\pi B}{2L} \right) \tag{9.41}$$

Figure 9.10 shows the computed value for the summation $\mathbf{S}$ as a function of L/B. This plot is convenient for determining the time of approach and the load-carrying capacity. For example, for an infinitely long rectangular plate when $L/B \to 0$, $\mathbf{S} \to \pi^4 192 = 0.5074$. Therefore,

$$\Delta t = 0.5 \frac{\mu L^3 B}{W} \left(\frac{h_1^2 - h_2^2}{h_1^2 h_2^2} \right)$$

The exact analytical solution for an infinitely long bearing using the ILA yields the identical result. Similar analytical expressions have been derived by Archibald (1956, 1968), Pinkus and Sternlicht (1961), Hays (1961a), Gross (1980), Booker (1984), and Hamrock (1994).

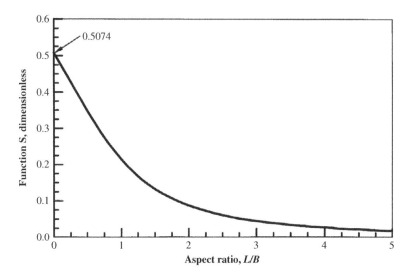

Figure 9.10   Summation term **S** for various L/B ratios.

Squeeze Film of Finite Nonplanar Bodies

Analysis of most nonplanar squeeze-film problems with leakage consideration requires treatment using numerical schemes, although semianalytical solutions are possible in some geometries. See, for example, Hays (1961b) for a squeeze-film solution in a finite journal bearing with a fluctuating load. In many practical situations, both sliding and squeeze components are present. In this section, we shall focus on two examples of nonrotating, nonplanar finite bodies undergoing squeeze motion. The results presented follow the development of Khonsari and Jang (1997).

Partial-Arc and Full Journal Bearing

A full (360°) and a partial arc (180°) journal bearing are shown in Figure 9.11. In the absence of journal rotation, i.e. with $U = 0$, the shaft movement is directly along the line of centers and the minimum film thickness h_{min} is located on the bottom of the shaft. It is therefore convenient to place the origin directly at that position. Film thickness is given by

$$h = C - e \cos \theta \tag{9.42}$$

where C is the radial clearance and e is a measure of eccentricity. Alternatively, one can use the definition of a dimensionless eccentricity ratio, $\varepsilon = e/C$, which varies from $0 \leq \varepsilon < 1$. Differentiating Equation (9.42) with respect to time yields

$$\frac{\partial h}{\partial t} = -\dot{e} \cos \theta = V_s \cos \theta \tag{9.43}$$

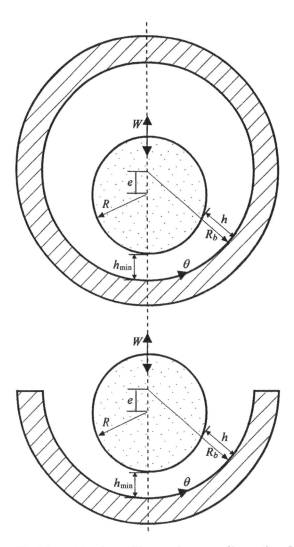

Figure 9.11 Full- and half-journal bearings with normal squeeze-film motion. *Source*: Khonsari and Jang, 1997. Reproduced with permission of John Wiley & Sons.

With $U = 0$, Reynolds Equation (9.1) simplifies to a partial differential equation whose source term (right-hand side) is the squeeze component: $-12\mu V_s$. A similar numerical scheme, i.e. successive over-relaxation, can be used to solve this equation for the pressure distribution.

A typical pressure distribution for a finite journal bearing with $L/B = 1$ and $\varepsilon = 0.5$ is shown in Figure 9.12. It is symmetric about θ, as expected. Note that the pressure for a 180° arc bearing is confined to $-\pi/2$ to $\pi/2$, and there are no complications in terms of the occurrence of negative pressures. When a complete arc is considered, negative pressures are predicted. The full-arc solutions without negative pressures are generated by setting any negative pressure equal to zero during computation of the Reynolds equation. With a sufficiently fine

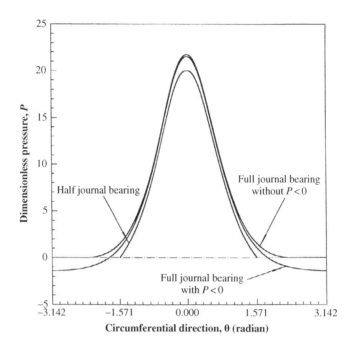

Figure 9.12 Typical pressure distribution for finite bearing squeeze. *Source*: Khonsari and Jang, 1997. Reproduced with permission of John Wiley & Sons.

mesh, this approach closely approximates the Reynolds (Swift–Stieber) boundary condition ($\partial P/\partial\theta = 0$); see Chapter 8 for a detailed explanation of the cavitation boundary conditions.

Figures 9.13 and 9.14 show dimensionless load-carrying capacity and the leakage flow rate in a journal bearing with 180° arc for various aspect ratios, $\Lambda = L/D$. Figures 9.15 and 9.16 show the load-carrying capacity and the leakage flow rate in a full (360°) journal bearing. The leakage flow rate in Figure 9.16 pertains to the volume of lubricant discharged from the convergent area under the shaft ($\theta = 0$ to 180°) displaced as a result of the downward motion.

An application of these charts is illustrated in Example 9.4. Briefly, with a specified load and eccentricity, the approach velocity, $V = C\dot{\varepsilon} = C^3 W/\mu R^3 L\overline{W}$, and the leakage flow rate, $Q = LCV\overline{Q}$, can easily be evaluated.

Example 9.4 Consider the piston pin problem described in Example 9.2. Determine whether there is danger of metal-to-metal contact when leakage is taken into consideration. Also determine the leakage flow rate.

With a known aspect ratio Λ and a given eccentricity ratio ε one can determine the dimensionless load capacity $\overline{W}$ from Figure 9.15 as a function of the time rate of change of eccentricity ratio $\dot{\varepsilon}$. The approach velocity is simply $V = Cd\varepsilon/dt = C\dot{\varepsilon}$. To determine the approach velocity, we may construct a table for eccentricity ratios ranging from the initial velocity of $\varepsilon_1 = 0.5$ to $\varepsilon_2 = 0.9$ at relatively small intervals (see Table 9.3 with $\Delta\varepsilon = 0.05$). At each eccentricity ratio, the corresponding $\overline{W}$ is directly determined from Figure 9.15. The

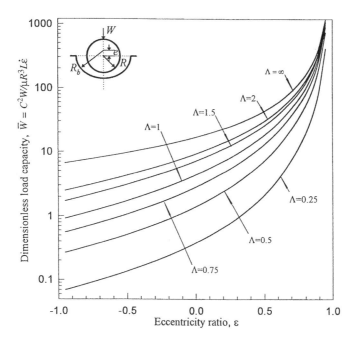

Figure 9.13   Load-carrying capacity for a 180° bearing. *Source*: Khonsari and Jang, 1997. Reproduced with permission of John Wiley & Sons.

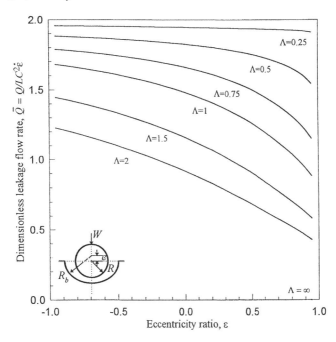

Figure 9.14   Leakage flow rate for a 180° bearing. *Source*: Khonsari and Jang, 1997. Reproduced with permission of John Wiley & Sons.

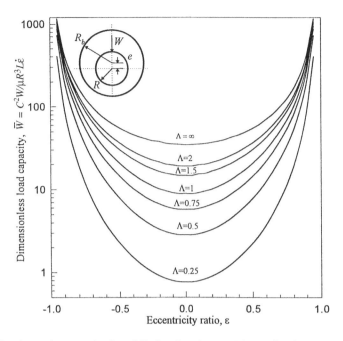

Figure 9.15 Load-carrying capacity for a 360° bearing. *Source*: Khonsari and Jang, 1997. Reproduced with permission of John Wiley & Sons.

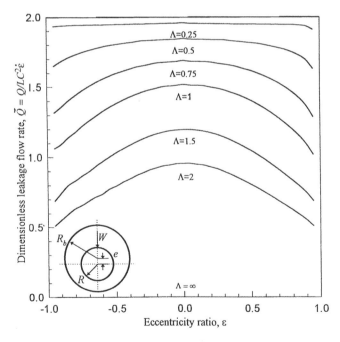

Figure 9.16 Leakage flow rate for a 360° bearing. *Source*: Khonsari and Jang, 1997. Reproduced with permission of John Wiley & Sons.

Table 9.3 Numerical results for finite piston pin calculation using Figures 9.15 and 9.16

ε	$\overline{W}$	V (in./sec)	Time (ms)	$\overline{Q}$	Q($\times 10^{-6}$ in.^{3}/s)
0.50	19.41	0.0487	0.000	1.54	112.5
0.55	23.73	0.0399	1.129	1.51	90.4
0.60	29.40	0.0322	1.388	1.48	71.5
0.65	37.58	0.0252	1.744	1.46	55.2
0.70	49.63	0.0191	2.261	1.43	41.0
0.75	68.18	0.0139	3.037	1.39	29.0
0.80	100.48	0.0094	4.295	1.36	19.2
0.85	163.37	0.0058	6.578	1.33	11.6
0.90	319.57	0.0030	11.430	1.29	5.8
Total	–	–	31.862	–	–

approach velocity

$$V = C\dot{\varepsilon} = \frac{W}{\mu L}\left(\frac{C}{R}\right)^3 \frac{1}{\overline{W}}$$

is computed at every $\overline{W}$, as shown in Table 9.3. To determine the time of approach, we note that $dt = (C/V)d\varepsilon$. Integrating both sides gives $\Delta t \approx C(\varepsilon_2 - \varepsilon_1)/V_{avg}$, where $V_{avg} = (V_2 + V_1)/2$. For example, from $\varepsilon_1 = 0.5$ to $\varepsilon_2 = 0.55$, we have

$$\Delta t \approx 2C(\varepsilon_2 - \varepsilon_1)/(V_1 + V_2) = \frac{(2)(0.001)(0.55 - 0.5)}{(0.0487) + (0.0399)} = 1.129 \times 10^{-3}\,\text{s} \quad (1.129\,\text{ms})$$

The volume of oil displaced by the action of the shaft vertical motion is evaluated from Figure 9.16.

Total approach time is $\Delta t \approx 0.032$ s, which is smaller than the 0.055 s predicted by the infinitely long journal bearing where the leakage flow rate is zero. According to the finite bearing solution, less cushioning time is available than predicted by the simple formulae of Table 9.2. The finite bearing analysis shows that the power stroke for 1500 rpm is still safe, since $\Delta t_s < \Delta t$. However, the squeeze-film action for 750 rpm does not provide sufficient cushioning time, and there is a danger of metal-to-metal contact. Recall that with side leakage neglected, according to Example 9.2, operating at 750 rpm was found to be safe.

Combined Squeeze and Rotational Motion

In the previous section, only the pure squeeze action was considered. We now extend the results to investigate the effect of shaft rotation in a plain journal bearing. Figures 9.17 (a)–(d) show how the dimensionless load-carrying capacity varies with combined rotational and squeeze motion in a plain journal bearing with different aspect ratios. In these figures, $\chi = \frac{\dot{e}}{\omega}$, where $\dot{e} = C\dot{\varepsilon}$ represents the rate of change of eccentricity with time, i.e., the squeeze velocity. The variation of the bearing attitude angle – defined as the angle between the line of center and the load line – is also shown. These results were obtained by a numerical solution of

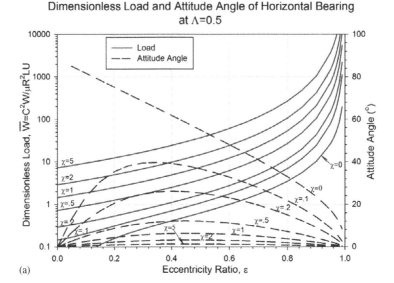

(a)

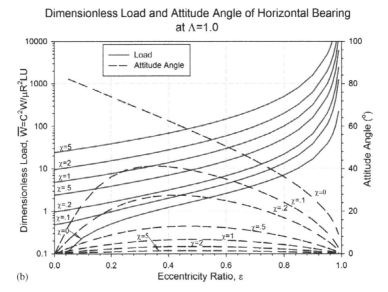

(b)

Figure 9.17 Load-carrying capacity and attitude angles for a radially-loaded plain journal bearing experiencing combined squeeze and rotational motion in a bearing with aspect ratio of $\Lambda = 0.5, 1.0, 1.5,$ and 2.

Reynolds equation without consideration of the thermal effects. Note that the dimensionless load increases with the eccentric ratio and the rate of increase becomes substantially large at eccentricity values $\varepsilon > 0.8$. An illustrative example covered in the next section will be given to show how these charts can be applied.

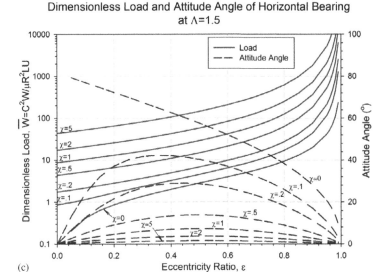

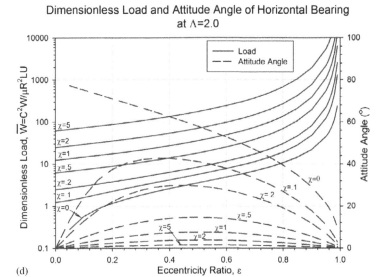

Figure 9.17   *(Continued)*

Vertical Bearing Configuration

The analysis and performance prediction of pumps, mixers, and agitators with a vertical bearing configuration are of considerable interest due to their diverse uses in the field. Surprisingly, however, detailed design information for these applications is not readily available. In this section, we present some results for a vertical shaft that, in addition to rotational motion, may also experience a squeeze motion in the radial direction.

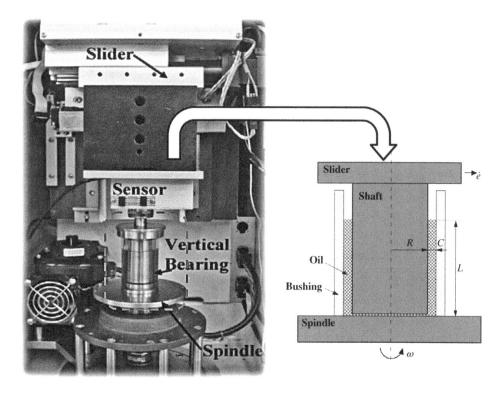

Figure 9.18 Testing apparatus for vertical bearing configuration. *Source*: Jang and Khonsari, 2008. Reproduced with permission of Taylor & Francis.

Figure 9.18 shows a simple laboratory testing apparatus designed to investigate both rotation and squeeze motion. A transparent bushing mounted inside a tribometer is shown in which the spindle produces rotational motion ω and the slider can provide a specified squeeze velocity $\dot{e}$. Figure 9.19 shows the results of measurements and prediction of the load plotted as a function of the eccentricity ratio. These results are obtained using the Reynolds equation with a mass-conservative algorithm and experimental values that correspond to SAE 50 oil. As shown, the theory yields reasonably accurate results compared to experimental measurements. The details of the formulation and numerical simulations are given in Jang and Khonsari (2017).

Figure 9.20 shows the load-carrying capacity of a vertical bearing of various aspect ratios undergoing pure squeeze motion in the radial direction without rotational motion. These results are extended in Figures 9.21 to predict the dimensionless load, the eccentricity ratio, and the attitude angle for a vertical bearing that experiences both rotational and squeezing motion for bearings with different aspect ratios, Λ. In these figures, χ represents the ratio of the squeeze to rotational speed, $\chi = \frac{\dot{\varepsilon}}{\omega}$. Note that thermal effects are not considered in these simulations. It is also interesting to note that the load-carrying capacity increases drastically beyond $\varepsilon \approx 0.9$. Therefore, if the squeeze load were maintained at large eccentricities, it might cause bushing deformation. This is illustrated in the following example.

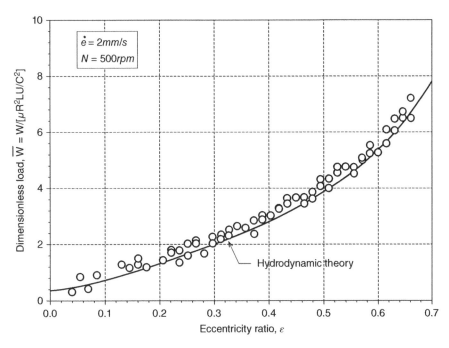

Figure 9.19 Dimensionless load as a function of the eccentricity ratio with comparison with experimental results. *Source*: Jang and Khonsari, 2008. Reproduced with permission of Taylor & Francis.

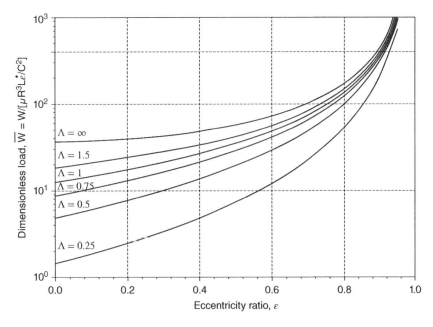

Figure 9.20 Dimensionless load-carrying capacity for a vertical bearing experiencing pure squeeze action. *Source*: Jang and Khonsari, 2008. Reproduced with permission of Taylor & Francis.

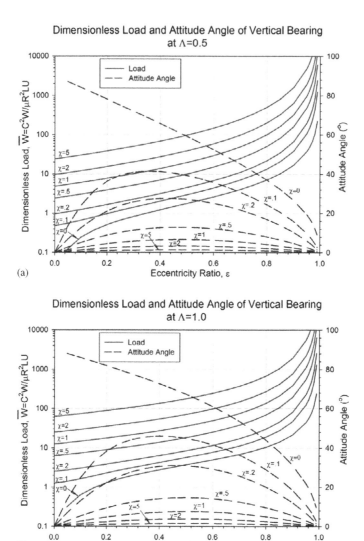

(a)

(b)

Figure 9.21 Load-carrying capacity and attitude angles for a vertically oriented journal bearing experiencing combined squeeze and rotational motion in a bearing with aspect ratio of $\Lambda = 0.5$, 1.0, 1.5, and 2.

Example 9.5   A vertically-oriented journal bearing has an aspect ratio $\Lambda = 0.5$, shaft radius $R = 0.03$ m, and clearance ratio $R/C = 0.001$. The bearing is steadily operating at 200 rpm with SAE 10 oil at 50°C ($\mu = 0.02$ Pa.s). During an abnormal incident, the bearing experiences a sudden radial shock load of 7 KN that lasts for 0.2 s. Safety requirements dictate that the bearing must operate below $\varepsilon < 0.9$; otherwise, the bearing must be replaced. Make a recommendation whether the bearing is safe to operate after the disturbance.

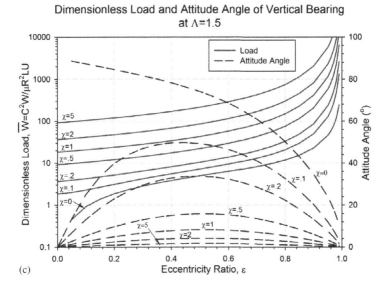

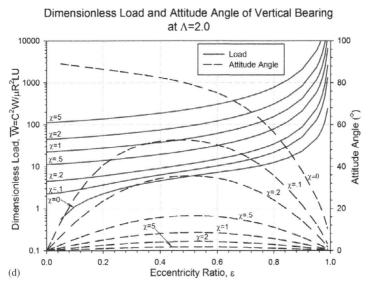

Figure 9.21 (*Continued*)

The first step is to determine the dimensionless load based on the bearing specifications and the lubricant viscosity.

$$\overline{W} = \frac{C^2 W}{\mu R^2 LU} = 18.57$$

Entering Figure 9.21 with $\overline{W} = 18.57$, we read that at $\varepsilon = 0$, $\chi = 3.85$.

Table 9.4 Eccentricity ratio

ε	Read χ From Fig. 9.21a	$\dot{e}$ [cm/s] $\dot{e} = \dfrac{\pi NC\chi}{30}$	$\dot{e}_{avg}$ [cm/s] $\dot{e}_{avg} = \dfrac{\dot{e}_{avg}^n + \dot{e}_{avg}^{n-1}}{2}$	Δt [ms] $\Delta t = \dfrac{e^n - e^{n-1}}{\dot{e}_{avg}}$
0.0000	3.85	0.2418	–	–
0.1000	3.09	0.1941	0.2178	1.377
0.2000	2.41	0.1513	0.1727	1.737
0.3000	1.84	0.1156	0.1335	2.247
0.4000	1.35	0.0848	0.1000	3.000
0.5000	0.95	0.0597	0.0723	4.149
0.6000	0.62	0.0389	0.0493	6.035
0.7000	0.36	0.0226	0.0308	9.774
0.8000	0.16	0.0100	0.0163	18.405
0.9000	0.01	0.0006	0.0053	56.604
Total Time = 103 ms = 0.1 s				

From the definition of $\chi = \dfrac{\dot{\varepsilon}}{\omega}$, we can determine $\dot{e}$ from the following equation

$$\dot{e} = \frac{\pi NC\chi}{30} = 0.2418 \text{ cm/s}$$
$$\text{At } \varepsilon = 0.1, \chi = 3.09 \quad and \quad \dot{e} = 0.1941.$$

Similar to Example 9.4, we compute the average $\dot{e}_{avg} = \dfrac{\dot{e}_{avg}^n + \dot{e}_{avg}^{n-1}}{2} = 0.2178$ which yields $\Delta t = \dfrac{e^n - e^{n-1}}{\dot{e}_{avg}} = 1.377$ ms. Therefore, the shock load causes the shaft to reach the eccentricity ratio of $\varepsilon = 0.1$ in only 1.377 ms.

The calculations should be continued until the limiting eccentricity ratio of 0.9. The result is shown in Table 9.4. It shows that due to disturbance, in roughly $\Delta t = 0.1$ s the shaft reaches its limiting eccentricity level of $\varepsilon = 0.9$, as dictated by the safety requirements. Since the disturbance duration is specified to be 0.2 s, it exceeds the limit and according to the statement of the problem, the bearing must be replaced.

9.7 Piston Rings

A piston ring is considered to be a vital part of an internal combustion engine, for it prevents loss of the pressure that exists above the piston head during the compression and power strokes. That is to say, it provides a 'sealing' mechanism. Clearly, a piston cannot be designed to have a tight fit with its cylinder liner because this would impede the motion of the cylinder and lead to unacceptable wear in a short time. Moreover, additional thermal expansion of a tightly fit piston at normal engine running temperature could then lead to damage of the rod, the crankshaft, or even the piston itself (Crouse, 1971). Overall, 20–40% of an engine friction is attributed to the piston ring assembly (Jeng, 1992).

A piston ring assembly includes compression and oil-control rings (see Figure 9.22). An oil-control ring scrapes off excessive oil from the cylinder liner and returns it to the oil pan. This is

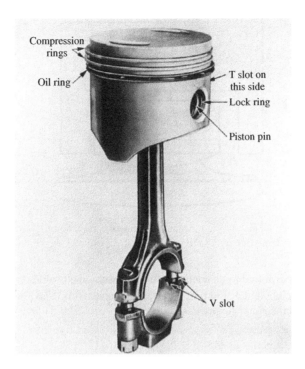

Figure 9.22   Crankshaft showing piston ring assembly. *Source*: Crouse, 1971. Reproduced with permission of McGraw-Hill Book Co., New York.

important because otherwise oil may enter the combustion chamber and burn prematurely. The compression ring, by contrast, is responsible primarily for sealing against combustion taking place above the piston. This is a crucial task, particularly during the power stroke as the piston is forced downward rapidly.

During the intake stroke, the compression ring tilts and scrapes any oil left over by the oil-control ring. However, normally a layer of lubricant remains on the cylinder wall, allowing hydrodynamic pressure development to provide lubricated sliding of the ring akin to that of a slider bearing. The fact that hydrodynamic lubrication prevails between the ring and the cylinder wall during most of the stroke cycle has been demonstrated by friction measurements (Jeng, 1992). If the film thickness falls below the composite surface roughness of the piston and cylinder liner, then mixed film or boundary lubrication may result in intimate contact at the asperity level and affect the instantaneous friction force. As piston velocity drops toward the end of its stroke, hydrodynamic action diminishes and squeeze-film support becomes the only component present when the sliding velocity is zero, i.e. when the piston reaches its top dead center or bottom dead center.

Minimum film thickness with a piston ring is a function of the crank angle and thus varies with time. Various theoretical models and numerical solution schemes have been developed. See, for example, Rhode *et al.* (1979), Ting and Mayer (1974), Sun (1990), and Jeng (1991, 1992). In this section, we closely follow the work of Radakovic and Khonsari (1997).

Figure 9.23 presents a schematic of a piston ring. The piston ring essentially behaves as a reciprocating slider bearing subject to cyclic variations in velocity and pressure. Normally, an

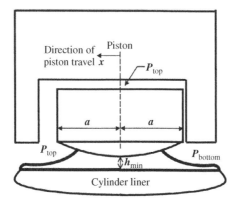

Figure 9.23   Schematic of a piston ring modeled as a slider bearing. *Source*: Radakovic and Khonsari, 1997. Reproduced with permission of ASME.

off-centered parabola is found to be adequate for describing the profile of the ring. Let δ_0 and a denote the crown height and piston ring half-width, respectively. The film thickness between the cylinder wall and the ring is expressed as

$$h(x, t) = h_m(t) + \delta_0(x/a)^2 \tag{9.44}$$

where h_m is the minimum film thickness, which is a function of time (or crank angle). It is assumed that the ring does not tilt in its groove; thus, it will move only radially as a unit under normal loading.

Most piston ring analyses neglect circumferential flow and use the one-dimensional Reynolds equation with an ILA to determine the pressure distribution in the fluid film between the cylinder wall and the piston ring. A generalized form of the Reynolds equation that allows one to incorporate shear thinning of the lubricant as well thermohydrodynamic effects is (cf. Khonsari and Hua, 1994)

$$\frac{\partial}{\partial x}\left(F_2\frac{\partial P}{\partial x}\right) = \frac{\partial}{\partial x}\left(\frac{F_1}{F_0}U\right) + \frac{\partial h}{\partial t} \tag{9.45}$$

where

$$F_0 = \int_0^h \frac{1}{\mu^*}dz; \qquad F_1 = \int_0^h \frac{z}{\mu^*}dz; \qquad F_2 = \int_0^h \frac{z}{\mu^*}\left(z - \frac{F_1}{F_0}\right)dz$$

The above generalized form of the Reynolds equation possesses the proper components for a thermohydrodynamic solution for non-Newtonian fluids since the variation of viscosity across the film is taken into consideration (cf. Khonsari and Beaman, 1986; Paranjape, 1992). The term on the left-hand side of Equation (9.45) involves the hydrodynamic pressure gradient in the sliding direction. The first term on the right-hand side is the physical wedge term, which depends upon the shape of the ring face, for example as prescribed by Equation (9.44). The

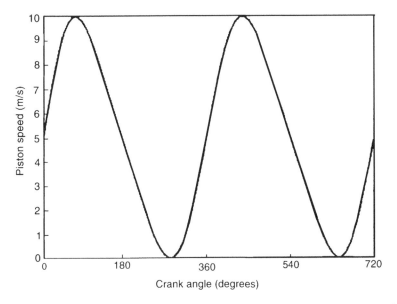

Figure 9.24 Typical sliding velocity as a function of crank angle. *Source*: Jeng, 1992. Reproduced with permission of Taylor & Francis.

second term on the right-hand side describes the squeeze motion of the piston ring toward or away from the cylinder wall.

The slider–crank mechanism imposes a sinusoidal sliding motion on the piston. Typical sliding velocity U as a function of crank angle is shown in Figure 9.24, where the zero degree location is defined as the top dead center of the power stroke.

The boundary conditions are

$$P(x = -a) = P_{\text{top}}$$
$$Pt(x = +a) = P_{\text{bot}}$$

(9.46)

where P_{top} and P_{bot} represent the upper and lower boundary pressures above and below the piston, respectively. It is reasonable to assume that the pressure above the ring, P_{top}, is equal to the combustion chamber pressure and that the pressure below is half of that pressure (Jeng, 1992). Figure 9.25 shows an approximation of a typical gas pressure distribution in the combustion chamber. Again, the zero degree is defined as the start of the engine cycle when the piston is at the top dead center position of the power stroke.

For numerical simulations it is important to nondimensionalize the governing equations. Using the overbar notation to define the following dimensionless qualities

$$\bar{h} = \frac{h}{h_{\text{ref}}}; \qquad \bar{\mu}^* = \frac{\mu^*}{\mu_{i,1}}; \qquad \bar{x} = \frac{x}{b}$$

$$\bar{u} = \frac{u}{R\omega}; \qquad \bar{t} = t\omega; \qquad \bar{h}_m = \frac{h_m}{h_{\text{ref}}}; \qquad \bar{P} = \frac{Ph_{\text{ref}}^2}{\mu_{i,1}R\omega b}$$

(9.47)

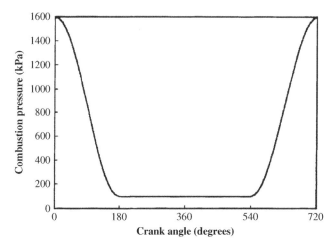

Figure 9.25 Typical pressure distribution in a combustion chamber. *Source*: Radakovic and Khonsari, 1997. Reproduced with permission of ASME.

we arrive at the following equation:

$$\frac{\partial}{\partial \bar{x}} \left(\bar{F}_2 \bar{h}^3 \frac{\partial \bar{P}}{\partial \bar{x}} \right) = \bar{u} \frac{\partial}{\partial \bar{x}} \left(\bar{h} \frac{\bar{F}_1}{\bar{F}_0} \right) + \frac{b}{R} \frac{\partial \bar{h}}{\partial \bar{t}}$$

(9.48)

where R is the crank radius and

$$\bar{F}_0 = \int\limits_0^1 \frac{1}{\bar{\mu}^*} d\bar{z}; \qquad \bar{F}_1 = \int\limits_0^1 \frac{\bar{z}}{\bar{\mu}^*} d\bar{z}; \qquad \bar{F}_2 = \int\limits_0^1 \frac{\bar{z}}{\bar{\mu}^*} \left(\bar{z} - \frac{\bar{F}_1}{\bar{F}_0} \right) d\bar{z}$$

The finite difference form of Equation (9.48) is (Radakovic and Khonsari, 1997)

$$(\bar{F}_2)_i \bar{h}_i^3 \left(\frac{\bar{P}_{i-1} - 2\bar{P}_i + \bar{P}_{i+1}}{\Delta \bar{x}^2} \right) + \left[3(\bar{F}_2)_i \bar{h}_i^2 \left(\frac{d\bar{h}}{d\bar{x}} \right)_i + \bar{h}_i^3 \left(\frac{d\bar{F}_2}{d\bar{x}} \right)_i \right] \left(\frac{\bar{P}_{i+1} - \bar{P}_{i-1}}{2\Delta \bar{x}} \right)$$

$$= \bar{u} \left[\bar{h}_i \left(\frac{d\bar{F}_3}{d\bar{x}} \right)_i + (\bar{F}_3)_i \left(\frac{d\bar{h}}{d\bar{x}} \right)_i \right] + \frac{b}{R} \frac{\partial \bar{h}}{\partial \bar{t}}$$

(9.49)

The film thickness $\bar{h}_i$ is based on an initial assumption of minimum film thickness. The unknown quantities are the pressure $\bar{P}_i$ and squeeze term $d\bar{h}_m/d\bar{t}$, which sum to $(n + 1)$ unknowns. Equation (9.49), however, describes only n equations. The additional equation is obtained by assuming that the piston ring can be modeled as a thin-walled cylinder subject to internal and external pressure. The equation that describes the deflection of the ring is

$$E\bar{h}_m = (\bar{P}_m - \bar{P}_{top})$$

(9.50)

where $\overline{P}_m$ is the mean external hydrodynamic pressure in the fluid film; $\overline{h}_m$ describes the deflection of the ring, equivalent to the minimum film thickness; and $\overline{E}$ is a dimensionless form of the ring's Young modulus of elasticity. Mean pressure $\overline{P}_m$ is obtained through integration of pressure distribution in the film. Using the trapezoidal integration rule, it becomes:

$$\overline{P}_m = \frac{1}{b}\left[\frac{\Delta x}{2}\left(\overline{P}_{top} + 2\sum_i \overline{P}_i + \overline{P}_{bot}\right) + \overline{P}_{div}l_{div}\right] \tag{9.51}$$

where b is the total width of the piston ring, Δx is the distance between nodes, and l_{div} is the length of the divergent portion of the gap between the ring and the cylinder wall where the pressure in the lubricant is equal to the pressure at the trailing edge of the ring. $\overline{P}_{div}$ can be equal to $\overline{P}_{top}$ or $\overline{P}_{bot}$, depending on the direction of motion of the piston. Combining Equations (9.50) and (9.51), the unknown pressures in the film can be expressed as

$$\sum_i \overline{P}_i = \frac{1}{\Delta x}[b(\overline{E}\overline{h}_m + \overline{P}_{top}) - \overline{P}_{div}l_{div}] - \frac{(\overline{P}_{top} + \overline{P}_{bot})}{2} \tag{9.52}$$

Upon simplification, Equations (9.49) and (9.52) can be expressed as

$$\overline{P}_{i+1}B_i + \overline{P}_iA_i + \overline{P}_{i-1}C_i + D_i\frac{\partial\overline{h}}{\partial\overline{t}} = R_i, \qquad \sum_i \overline{P}_i = \alpha$$

where

$$A_i = \frac{-2(\overline{F}_2)_i\overline{h}_i^{-3}}{\Delta\overline{x}^2}; \qquad D_i = b/R$$

$$B_i = \frac{(\overline{F}_2)_i\overline{h}_i^{-3}}{\Delta\overline{x}^2} + \frac{-3(\overline{F}_2)_i\overline{h}_i^{-2}}{2\Delta\overline{x}}\left(\frac{\partial\overline{h}}{\partial\overline{x}}\right)_i + \frac{\overline{h}_i^{-3}}{2\Delta\overline{x}}\left(\frac{d\overline{F}_2}{d\overline{x}}\right)_i$$

$$C_i = \frac{(\overline{F}_2)_i\overline{h}_i^{-3}}{\Delta\overline{x}^2} - \frac{-3(\overline{F}_2)_i\overline{h}_i^{-2}}{2\Delta\overline{x}}\left(\frac{\partial\overline{h}}{\partial\overline{x}}\right)_i + \frac{\overline{h}_i^{-3}}{2\Delta\overline{x}}\left(\frac{d\overline{F}_2}{d\overline{x}}\right)_i \tag{9.53}$$

$$R_i = \overline{u}\left[\overline{h}_i\left(\frac{d\overline{F}_3}{d\overline{x}}\right)_i + (\overline{F}_3)_i\left(\frac{d\overline{h}}{d\overline{x}}\right)_i\right]$$

$$\alpha = \frac{1}{\Delta x}[b(\overline{E}\,\overline{h}_m + \overline{P}_{top}) - \overline{P}_{div}l_{div}] - \frac{(\overline{P}_{top} + \overline{P}_{top})}{2}$$

Details of the solution scheme and implementation of the Reynolds boundary condition, $\partial\overline{P}/\partial\overline{x} = 0$, are described in Radakovic and Khonsari (1997). Briefly, a time-marching technique is used to solve for new values of the minimum film thickness. The squeeze term $d\bar{h}_m/d\bar{t}$, a product of the solution to the Reynolds equation, can be written as

$$\frac{d\bar{h}_m(\bar{t})}{d\bar{t}} = f(\bar{t}, \bar{h}_m) \tag{9.54}$$

Equation (9.54) can be integrated in time using an ordinary differential equation solver such as the Runge-Kutta method. With the new value for minimum film, the time (or crank angle) is incremented by $\Delta\bar{t}$ and the procedure is repeated. The minimum film thickness is calculated continuously until a value of $\bar{h}_m$ is found at the end of the engine cycle. This value is compared to the initially assumed value. If the error is less than a prescribed convergence tolerance, the $\bar{h}_m(\bar{t})$ is taken as the solution; otherwise, a new initial guess is made and the process continues until the solution converges. Typically, the process converges in slightly more than one engine cycle.

Friction Force and Power Loss

The viscous drag force acting on the piston ring can be computed using the following expression:

$$F(t) = \int_A \left(\frac{h}{2} \frac{dP}{dx} - \frac{\mu U}{h} \right) dA \tag{9.55}$$

where A is the area of the lubricated region. If boundary lubrication is present, the friction force is calculated using Jeng (1992):

$$F(t) = \mu_c \, (BW(t)) \, |U| \tag{9.56}$$

where μ_c is the friction coefficient and $W(t)$ is the load-carrying capacity of the oil film per unit length represented by

$$W(t) = \int P(x,t)dx \tag{9.57}$$

Power loss is given as

$$E_P(t) = F(t)|U| \tag{9.58}$$

When hydrodynamic lubrication is present between the piston ring and cylinder wall, the power loss is

$$P(t) = \int_A \left(\frac{\mu}{h} \right) U^2 dA + \int_A \left(\frac{h^3}{12\mu} \right) |\nabla P|^2 dA \tag{9.59}$$

This equation automatically includes the power loss due to squeeze motion.

Table 9.5   Relevant engine parameters

Parameters used by Jeng (1992)			
Engine speed			$S = 2000$ (rpm)
Cylinder bore			$B = 88.9$ (mm)
Bore radius			$r = 44.45$ (mm)
Crank radius			$R = 40.0$ (mm)
Connecting rod length			$L = 141.9$ (mm)
Composite roughness			$\sigma = 0.50$ (μm)
Piston ring width			$b = 1.475$ (mm)
Ring thickness			$t_r = 3.8$ (mm)
Ring modulus			$E = 70.0$ (GN/m^2)
Ring crown height			$\delta_0 = 14.9$ (μm)
Friction coefficient			$f = 0.08$
Lubricant viscosity			$\mu = 0.00689$ (Pa s)

Lubricant properties (non-Newtonian, shear thinning)[a]

Low shear rate viscosity			
100 °C	0.0111 pa s		

Non-Newtonian properties			
100 °C	κ(pa)	μ_1(pa s)	μ_2(pa s)
	1500	0.0111	0.0063
Temperature–viscosity coefficients			
100 °C	$\beta_1(°K^{-1})$	$\beta_2(°K^{-1})$	$\beta_3(°K^{-1})$
	0.0214	0.0214	0.0270

[a]See Radakovic and Khonsari, 1997 and Chapter 2 for discussion on shear thinning fluids.

Relevant engine parameters, similar to those simulated by Jeng (1992), are shown in Table 9.5. The corresponding calculated minimum film thickness, friction force, and power loss are shown in Figures 9.26, 9.27, and 9.28, respectively. Note that the minimum film thickness is highest at crank positions where the piston velocity is greatest, i.e. around crank angles 90°, 270°, 450°, and 630°. Instantaneous power loss (and friction force) have a trend similar to that of the minimum film thickness.

Computations of Radakovic and Khonsari (1997) for shear thinning oil representing multigrade oils with consideration of thermal effects indicated that at the zero degree, the minimum film thickness for the non-Newtonian simulations falls below the composite roughness σ of the piston ring and cylinder wall. It is here where wear patterns are most evident on cylinder liners from piston ring scuffing (Ting and Mayer, 1974). At this time during the engine cycle (onset of the power stroke), the piston velocity approaches zero and there is a loss of hydrodynamic action. This, coupled with the high combustion pressure, increases the possibility of boundary lubrication. Boundary lubrication is also evident in Figure 9.27, which shows large spikes in friction force at the zero degree. The friction force resulting from boundary lubrication is significantly higher than the friction force where the ring is hydrodynamically lubricated. Similarly, this effect is evident in Figure 9.28, showing the power loss during the engine cycle. Recent research shows that by engraving carefully designed textures into to surface of the piston rings via a laser, it may be possible to reduce the friction coefficient and reduce power loss (Shen and Khonsari, 2016).

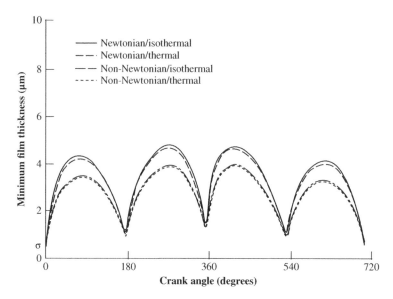

Figure 9.26 Minimum film thickness in a piston as a function of crank angle. *Source*: Radakovic and Khonsari, 1997. Reproduced with permission of ASME.

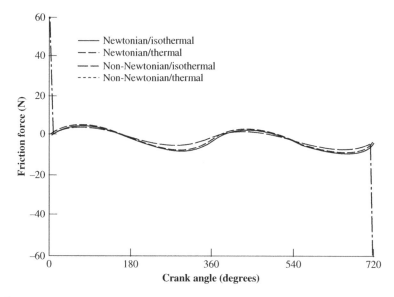

Figure 9.27 Friction force in a piston ring. *Source*: Radakovic and Khonsari, 1997. Reproduced with permission of ASME.

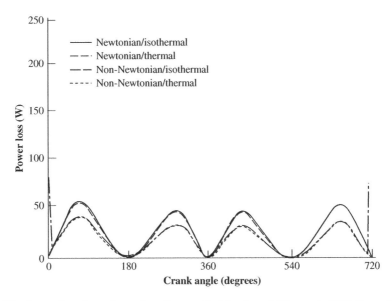

Figure 9.28   Power loss in a piston ring. *Source*: Radakovic and Khonsari, 1997. Reproduced with permission of ASME.

Problems

9.1   Compare the time of approach and the approaching velocity for the following:
   (a) A square 2 in. on each side.
   (b) A circular disk of radius $R = 1.25$ in.
   (c) An elliptical section with major and minor dimensions of $a = 3$ in. and $b = 1.5$ in.
   (d) Concentric annulus with inner and outer radii of $R_i = 1.0$ in. and $R_o = 2.0$ in.
      Assume that the load imposed on each of the above configurations is $W = 100$ lbf and that the lubricating oil is SAE 30 at $80\,°F$ ($\mu = 2.8 \times 10^{-5}$ reyns). Determine the time of approach required for the initial film gap of $h_0 = 0.02$ in. to drop fivefold.

9.2   Repeat Example 9.2, but assume that the piston pin is modeled using an infinitely long journal bearing with a 180° arc.

9.3   Repeat Problem 9.2, but use the finite bearing results.

9.4   Show that the squeeze-film pressure distribution for a nonrotating, full-journal bearing when the journal is forced downward is given by the following expression:

$$P = \frac{6\mu V}{R} \left(\frac{R}{C}\right)^3 \frac{\cos\theta\,(2 - \varepsilon\cos\theta)}{(1 - \varepsilon\cos\theta)^2}$$

where θ is measured from h_{min}. Integrate the above pressure, and verify the load-carrying capacity expression in Table 9.2. (*Hint*: $W = 2\int_0^\pi P\cos\theta R d\theta$.)

9.5   Formulate the squeeze-film pressure distribution, load-carrying capacity, and time of approach of a 180° journal bearing, assuming that the ISA applies.

9.6   Determine the time of approach in Problem 9.5 from an initial eccentricity ratio ε_1 to ε_2. Using your derivation, analyze the piston pin in Example 9.2, and comment on the possibility of metal-to-metal contact with the same specifications and operating conditions.

9.7   In nonconformal geometries, there is a significant increase in the lubricant viscosity due to the rise in pressure in accordance with $\mu = \mu_o e^{\alpha P}$, where α is the pressure viscosity coefficient. Formulate the pressure distribution, load-carrying capacity, and time of approach for a sphere approaching a plate. Verify your solution by letting $\alpha = 0$.

9.8   Repeat Problem 9.7 for a cylinder approaching a plate and verify your results using Table 9.2.

9.9   Two infinitely wide flat plates approach each other with normal velocities of V_1 and V_2, respectively. Show that the pressure distribution in the fluid can be computed from the following equation:

$$\frac{dP}{dx} = 12\mu(V_2 - V_1)\left(\frac{x - x_m}{h^3}\right)$$

where x_m represents the position where the pressure is maximum.

9.10   The top surface of an infinitely wide step bearing approaches the bottom plate at a normal velocity V_s. Assume that there is no sliding motion. State the boundary conditions for this problem and show that the pressure at the step can be determined from the following expression:

$$P_C = \frac{6\mu V_s(L_1 + L_2)L_1 L_2}{h_1^3 L_2 + h_2^3 L_1}$$

References

Archibald, F.R. 1956. 'Load Capacity and Time Relations for Squeeze Film,' *ASME Transactions*, Vol. 78, pp. 29–35.

Archibald, F.R. 1968. 'Squeeze Films,' *Standard Handbook of Lubrication Engineering*, J.J. O'Conner and J. Boyd (eds). McGraw-Hill Book Company, New York, pp. 7.1–7.8.

Booker, J.F. 1984. 'Squeeze Films and Bearing Dynamics,' *CRC Handbook of Lubrication*, Vol. II, E.R. Booser, (ed.). CRC Press, Boca Raton, Fla, pp. 121–137.

Cook, A. 1974. 'A Study of the Rheological Characteristics of Some Fluids and Tissues in Synovial Joint Lubrication,' Ph.D. thesis, University of Leeds, UK.

Crouse, W.H. 1971. *Automotive Fuel, Lubricating, and Cooling Systems*. McGraw-Hill Book Co., New York.

Cusano, C. and Funk, P.E. 1977. 'Transmissibility Study of a Flexibly Mounted Rolling Element Bearing in a Porous Bearing Squeeze-Film Damper,' *Journal of Lubrication Technology*, Vol. 99, pp. 50–56.

Dowson, D. 1966. 'Modes of Lubrication in Human Joints,' *Proceedings of Society of Mechanical Engineers, London*, Vol. 101, Pt. 3J, pp. 45–54.

Dowson, D. Unsworth, A., and Wright, V. 1970. 'Analysis of "Boosted Lubrication" in Human Joints,' *Journal of Mechanical Engineering Sciences*, Vol. 12, pp. 364–369.

Gross, W. 1980. *Fluid Film Lubrication*. Wiley Interscience, New York.

Hamrock, B. 1994. *Fundamentals of Fluid Film Lubrication*. McGraw-Hill Book Co., New York.

Hays, D.F. 1961a. 'Squeeze Film for Rectangular Plates,' *ASME Journal of Basic Engineering*, Vol. 85, pp. 243–246.

Hays, D.F. 1961b. 'Squeeze Films: a Finite Journal Bearing with a Fluctuating Load,' *ASME Transactions, Journal of Basic Engineering*, Vol. 83, pp. 579–588.

Higginson, G.R. and Unsworth, T. 1981. 'The Lubrication of Natural Joints,' *Tribology of Natural and Artificial Joints*, J. Dumbleton (ed.). Elsevier.

Jang, J. and Khonsari, M.M. 1999. 'Thermal Characteristics of a Wet Clutch,' *ASME Journal of Tribology*, Vol. 121, pp. 610–618.

Jang, J.Y. and Khonsari, M.M. 2002. 'On the Formation of Hot Spots in Wet Clutch Systems,' *ASME Journal of Tribology*, Vol. 124, pp. 336–345.

Jang, J.Y. and Khonsari, M.M. 2008. 'Linear Squeeze Film with Constant Rotational Speed,' *STLE Tribology Transactions*, Vol. 51, pp. 361–371.

Jang, J.Y. and Khonsari, M.M. 2013. 'Wet Clutch,' *Encyclopedia of Tribology*, Y.W. Chung and Q.J. Wang (editors), Springer Science, New York, pp. 4102–4108.

Jang, J.Y., Khonsari, M.M. and Maki, R. 2011. 'Three-Dimensional Thermohydrodynamic Analysis of a Wet Clutch with Consideration of Grooved Friction Surfaces,' *ASME Journal of Tribology*, Vol. 133, 011703:1–12.

Jeng, Y.R. 1991. 'Theoretical Analysis of Piston Ring Lubrication. Part II – Starved Lubrication and Its Application to a Complete Ring Pack,' *Tribology Transactions*, Vol. 35, pp. 707–714.

Jeng, Y.R. 1992. 'One Dimensional Analysis of Piston Ring Lubrication: Part I – Fully Flooded Lubrication,' *Tribology Transactions*, Vol. 35, pp. 696–705.

Khonsari, M.M. and Beaman, J. 1986. 'Thermohydrodynamic Analysis of Laminar Incompressible Journal Bearings,' *ASLE Transactions*, Vol. 29, pp. 141–150.

Khonsari, M.M. and Hua, Y. 1994. 'Thermal Elastohydrodynamic Analysis Using a Generalized Non-Newtonian Formulation with Application to Bair-Winer Constitutive Equation,' *ASME Journal of Tribology*, Vol. 116, pp. 37–46.

Khonsari, M.M. and Jang, J.Y. 1997. 'Squeeze-Film Bearings,' *Tribology Data Handbook*, E.R. Booser (ed.). CRC, Boca Raton, Fla, pp. 691–707.

Koguchi, H., Okada, M., and Tamura, K. 1988. 'The Meniscus Instability of a Thin Liquid Film,' *Journal of Applied Mechanics*, Vol. 55, pp. 975–980.

Koguchi, H. and Yada, T. 1990. 'The Meniscus Instability in Non-Newtonian Negative Squeeze Film,' *Journal of Applied Mechanics*, Vol. 57, pp. 769–775.

Li, M., Khonsari, M.M., McCarthy, D.M.C., and Lundin, J. 2014. 'Parametric Analysis of a Paper-Based Wet Clutch with Groove Consideration,' *Tribology International*, Vol. 80, pp. 222–233.

Li, M., Khonsari, M.M., McCarthy, D.M.C., and Lundin, J. 2015. 'On the Wear Prediction of the Friction Material in a Wet Clutch,' *Wear*, Vol. 334–335, pp. 56–66.

Li, M., Khonsari, M.M., Lingesten, N., Marklund, P., McCarthy, D.M.C., Lundin, J. 'Model Validation and Uncertainty Analysis in the Wear Prediction of a Wet Clutch,' *Wear*, Vol. 364–365, pp. 112–121.

Li, M., Khonsari, M.M., McCarthy, D.M.C., and Lundin, J. 2017. 'Parametric Analysis of Wear Factors of a Wet Clutch Friction Material with Different Groove Patterns,' to appear in Institution of Mechanical Engineers, Part J, *Journal of Engineering Tribology*.

McCutchen, C.W. 1967. 'Physiological Lubrication,' Symposium in Lubrication and Wear in Living and Artificial Human Joints, Institution of Mechanical Engineers, London.

McEwan, A.D. and Taylor, G.I. 1966. 'The Peeling of a Flexible Strip Attached by a Viscous Adhesive,' *Journal of Fluid Mechanics*, Vol. 26, pp. 1–15.

Mansouri, M. Holgerson, M., Khonsari, M.M. and Aung, W. 2001. 'Thermal and Dynamic Characterization of Wet Clutch Engagement with Provision for Drive Torque,' *ASME Journal of Tribology*, Vol. 123 (2), pp. 313–323.

Mansouri, M., Khonsari, M.M., Holgerson, M., and Aung, W. 2002. 'Application of Analysis of Variance to Wet Clutch Engagement,' *Journal of Engineering Tribology*, Proceedings of Institution of Mechanical Engineers, Part J, V. 216, pp. 117–126.

Marklund, P, Maki, R. Larsson, R., Hoglund, E, Khonsari, M.M. and Jang, J. Y. 2007. 'Thermal Influence on Torque Transfer of Wet Clutches in Limited Skip Differential Applications,' *Tribology International*, Vol. 40, pp. 876–884.

Meyer, D. and Tichy, J.A. 1999. 'Lubrication Model of an Artificial Hip Joint: Pressure Profile versus Inclination Angle of the Acetabular Cup,' *ASME Journal of Tribology*, Vol. 121, pp. 492–498.

Moore, D.F. 1972. *The Friction and Lubrication of Elastomers*. Pergamon Press, Ltd. Oxford.

Moore, D.F. 1993. *Viscoelastic Machine Elements: Elastomers and Lubricants in Machine Systems*. Butterworth-Heinemann, Ltd. Oxford.

Paranjape, R.S. 1992. 'Analysis of Non-Newtonian Effects in Dynamically Loaded Finite Journal Bearings Including Mass Conserving Cavitation,' *ASME Journal of Tribology*, Vol. 114, pp. 1–9.

Pinkus, O. and Sternlicht, B. 1961. *Theory of Hydrodynamic Lubrication*. McGraw-Hill Book Co., New York.

Radakovic, D. and Khonsari, M.M. 1997. 'Heat Transfer in a Thin-Film Flow in the Presence of Squeeze and Shear-Thinning: Application to Piston Rings,' *ASME Journal of Heat Transfer*, Vol. 119, pp. 249–257.

Rhode, S.M., Whitaker, K.W., and McAllister, G.T. 1979. 'A Study of the Effects of Piston Ring and Engine Design Variables on Piston Ring Friction,' *Energy Conservation through Fluid Film Lubrication Technology: Frontiers in Research and Design*. ASME Press, New York, pp. 117–134.

Shen, C. and Khonsari, M.M. 2016a. 'The Effect of Laser Machined Pockets on the Lubrication of Piston Ring Prototypes,' *Tribology International*, Vol. 101, pp. 273–283.

Shen, C. and Khonsari, M.M. 2016b. 'Tribological and Sealing Performancxe of Laser Pocketed Piston Rings in a Diesel Engine,' *Tribology Letters*, Vol. 64, pp. 26–35.

Sun, D.C. 1990. 'A Subroutine Package for Solving Slider Lubrication Problems,' *ASME Journal of Tribology*, Vol. 112, pp. 84–92.

Ting, L.L. and Mayer, J.E. 1974. 'Piston Ring Lubrication and Cylinder Bore Wear Analysis, Part 1 – Theory,' *ASME Journal of Lubrication Technology*, Vol. 96, pp. 305–314.

10

Hydrostatic Bearings

10.1 Introduction

Hydrodynamic bearings covered in previous chapters are categorized as self-acting. In other words, pressure is self-generated within the bearing. In contrast, *hydrostatic* bearings are externally pressurized by means of the lubricating fluid fed under pressure from an external source – usually a pump. If pressurized lubricant is supplied continuously and without interruption, separation between bearing surfaces with an adequate film thickness can be maintained even when the speed is nil. Thus, it is possible to achieve extremely low operating friction coefficients allowing heavily loaded machinery to undergo rotational motion with very little effort. Both hydrodynamic and hydrostatic bearings operate with a relatively thick film separating the surfaces. The term *hydrostatic* is introduced to differentiate its operating mechanism from the self-acting hydrodynamic lubrication, where speed and wedge effects are the main requirements.

Hydrostatic bearings commonly involve more complex lubrication systems, and require specialized design and application practices. Nevertheless, they find applications ranging from small precision machine tools to some of the largest and heaviest equipment in use. Following are some primary reasons for their applications:

1. Friction is essentially zero at zero speed and during starting and stopping, and is quite low at low speeds that would not provide a full hydrodynamic film.
2. These bearings avoid mechanical contact and wear, even at start-up and at low speeds.
3. Poor lubricants such as air, water, and liquid metals are usable. With low-viscosity gases like air, very high speeds are possible.
4. Very high loads can be supported on small bearing areas.

Despite this broad range of use, liquid-lubricated hydrostatic bearings are typically associated with low-speed operations such as those described below.

Heavily Loaded Precision Machinery

Hydrostatic bearings are employed to support enormous structures such as telescopes, observatory domes, and large radio antennas where weight requirements often range from 250 000

Applied Tribology: Bearing Design and Lubrication, Third Edition. Michael M. Khonsari and E. Richard Booser.
© 2017 John Wiley & Sons Ltd. Published 2017 by John Wiley & Sons Ltd.
Companion Website: www.wiley.com/go/Khonsari/Applied_Tribology_Bearing_Design_and_Lubrication_3rd_Edition

Figure 10.1 Magellan telescope being assembled for installation in Chile. *Source*: Courtesy of L&F Industries.

to over 1 million lbf. For example, the Magellan telescope for the observatories of the Carnegie Institution of Washington was designed by L&F Industries to support 320 000 lbf of rotating weight, which includes an enormous 6.5 m diameter mirror (Figure 10.1). The system utilizes 18 hydrostatic bearing pads to support the weight of the azimuth and elevation axes, allowing them to rotate merely by gentle pressure from the operator's fingertip.

Hydrostatic Oil Lifts

Excessive bearing wear associated with start-up of large hydroelectric generators and turbine generators can be virtually eliminated with hydrostatic bearings. An external pump supplies pressurized oil to maintain an adequate film thickness during startup, during shutdown, and when undergoing direction reversal. In addition to wear reduction, a hydrostatic oil lift minimizes the start-up torque in steam turbine-generator systems when they are put on turning gear at about 3–10 rpm.

Severe Operating Conditions

In corrosive environments as well as in applications where the bearing temperatures are high and surface-to-surface contact cannot be tolerated, a hydrostatic bearing helps to maintain a separation gap to avoid scuffing failure, surface welding, and abrupt breakdown.

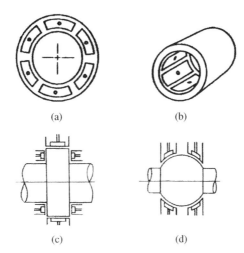

(a)   (b)

(c)   (d)

Figure 10.2   Several types of hydrostatic bearings: (a) multirecess, annular thrust; (b) journal bearing; (c) combined (hybrid) journal and thrust; (d) spherical shape. *Source*: Rowe, 1983. Reproduced with permission of Elsevier.

Stadium Mover/Converter

The entire grandstand seating section of a football stadium can be moved for conversion to a baseball field using a water-lubricated hydrostatic bearing. The Mile High Stadium in Denver applies multiple rubber-pad bearing pads 1.22 m (4 ft) in diameter to lift and move this nearly 20 million kg (4500 ton) load (Fuller, 1984).

10.2 Types and Configurations

Hydrostatic bearings are designed to carry normal loads (thrust bearings), radial loads (journal bearings), and/or combined thrust and radial hybrid loads, allowing shaft rotation with axial constraint. Figure 10.2 shows a schematic of several hydrostatic bearings. Figures 10.2(a) and 10.2(b) show typical thrust and journal bearing configurations, respectively. Figure 10.2(c) shows a hybrid journal and thrust bearing which is suitable when an axial constraint is needed. The spherical bearing shown in Figure 10.2(d) is useful when misalignment is of particular concern.

Each hydrostatic bearing is equipped with a recess (pocket) whose depth is several times greater than the film thickness. Pressurized oil is fed into the bearing, either directly from the pump or through a compensating element such as an orifice or a capillary. Depending on the application, one or more pads may be needed with multiple recesses (typically four to six in journal bearings) to provide different pocket pressures in supporting asymmetric loads. Without a compensating element, a constant volume pump would be needed to feed each recess. With a restrictor in place, a single pump can supply two or more of the pockets.

A significant amount of thrust load with a reasonable tolerance to misalignment can be achieved by using one of the annular thrust configurations shown in Figure 10.3. In Figure 10.3(a), a series of feed holes are used with no recess. Often referred to as

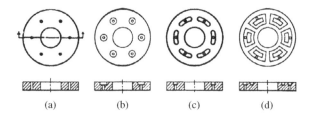

Figure 10.3 Annular thrust bearings: (a) film-compensated; (b) circular recess; (c) circumferential recess; (d) circumferentially grooved. *Source*: O'Conner and Boyd, 1968. Reproduced with permission of McGraw-Hill.

film-compensated, this configuration offers the least load-carrying capacity. The grooved configuration of Figure 10.3(d) offers the greatest load-carrying capacity (stiffness).

Dynamic characteristics and stability of bearings are also important considerations, particularly when gas is used as the lubricating agent. Generally, introducing a recess improves both load and stiffness, while damping is enhanced by greater land area. The film-compensated configuration without a recess in Figure 10.3(a), therefore, turns out to be the most stable and the grooved configuration, Figure 10.3(d), offers the least stability. Circular and longitudinal grooves (Figures 10.3(b), (c)) offer intermediate stiffness and stability (O'Conner and Boyd, 1968). The interested reader is also referred to Wilcock and Booser (1957), Pinkus and Sternlicht (1964), and Anderson (1964) for additional details.

10.3 Circular Step Thrust Bearings

Circular step bearings are the most common type of hydrostatic thrust bearings. They are particularly useful for carrying large vertical shafts in turbogenerators. Analyses of performance parameters for both constant flow and compensated bearings are presented under separate headings.

Figure 10.4 shows a simple flat plate thrust bearing with a single lubricant supply at feed pressure P_S. For the purpose of analysis, our attention will be restricted to liquid-lubricated bearings.

Pressure Distribution and Load

The governing equation for the pressure distribution is the Reynolds equation in cylindrical coordinates (Equation 6.25). For this hydrostatic action in Figure 10.4, the wedge and velocity terms on the right-hand side of the equation drop to zero, as does the $\partial h / \partial \theta$ term. This then gives the following:

$$\frac{d}{dr} \left(r \frac{h^3}{12\mu} \frac{dP}{dr} \right) = 0 \tag{10.1}$$

The boundary conditions are

$$P = P_S \quad \text{at} \quad r = R_1 \tag{10.2}$$

$$P = 0 \quad \text{at} \quad r = R_2 \tag{10.3}$$

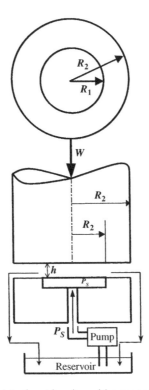

Figure 10.4   Flat plate thrust bearing without a compensating element.

The solution to Equation (10.1) subject to the above boundary condition is

$$P = P_S \frac{\ln(r/R_2)}{\ln(R_1/R_2)} \tag{10.4}$$

The pressure profile is shown in Figure 10.5. Note that pressure distribution in the pocket remains at the supply pressure, and it drops to ambient pressure in a nonlinear fashion over the land.

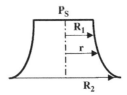

Figure 10.5   Pressure distribution.

Load-Carrying Capacity

Integration of pressure over the bearing area yields the following load capacity result:

$$W = \pi R_1^2 P_S + \int_{R_2}^{R_1} 2\pi r P \, dr = \frac{\pi P_S \left(R_2^2 - R_1^2 \right)}{2 \ln(R_2/R_1)} \tag{10.5}$$

Flow Rate Requirement

The volumetric flow rate that must be supplied to the bearing is

$$Q = (2\pi r) \left(-\frac{h^3}{12\mu} \frac{dP}{dr} \right) = \frac{\pi h^3 P_S}{6\mu \ln(R_2/R_1)} \tag{10.6}$$

Equations (10.5) and (10.6) show that the flow rate and load-carrying capacity are related as shown:

$$W = \frac{3\mu Q \left(R_2^2 - R_1^2 \right)}{h^3} \tag{10.7}$$

These equations can be written in the following forms:

$$W = A_{\text{eff}} P_S \tag{10.8}$$

and

$$W = C_1 \frac{Q}{h^3} \tag{10.9}$$

where A_{eff} is the effective area defined as $A_{\text{eff}} = \pi(R_2^2 - R_1^2)/(2 \ln(R_2/R_1))$ and $C_1 = 3\mu(R_2^2 - R_1^2)$.

Equation (10.9) implies that for a specified bearing, the load-carrying capacity varies simply with the flow rate and the film thickness. Solving for film thickness:

$$h = \sqrt[3]{\frac{C_1 Q}{W}} \tag{10.10}$$

Equation (10.10) implies that if the flow rate is maintained constant, doubling the load results in only a 38% reduction in film thickness, which indicates remarkable film stiffness. As the load is increased, the system automatically adjusts its film thickness to reestablish an operating equilibrium.

Bearing Stiffness

Stiffness is one of the most important parameters in hydrostatic bearings. Its definition is

$$K = -\frac{dW}{dh} \tag{10.11}$$

where the negative sign indicates that the stiffness decreases as the film thickness increases. Using Equation (10.7), the stiffness of a constant flow hydrostatic bearing becomes

$$K = \frac{3W}{h} \tag{10.12}$$

Friction Torque

Friction torque due to the integrated effect of oil film shear stresses is

$$\mathbf{T} = \int_A Fr dA \tag{10.13}$$

which reduces to the following expression assuming Couette flow prevails:

$$\mathbf{T} = \frac{\pi \mu \omega}{2h} \left(R_2^4 - R_1^4 \right) \tag{10.14}$$

System Power Loss

Total power loss is a summation of the power loss due to friction and that due to pumping losses:

$$E_{total} = E_f + E_P$$

Film Power Loss

Power loss associated with fluid-film friction is

$$E_f = \omega \mathbf{T} = \frac{\pi \mu \omega^2}{2h} \left(R_2^4 - R_1^4 \right) \tag{10.15}$$

In applications where the operating speed is low, typical of a great majority of applications, this power loss is insignificant in comparison to the losses associated with pumping power requirement.

Pumping Power Loss

This is the most significant source of power loss in a hydrostatic bearing, contributing substantially to the operating cost. Let η_P denote the efficiency of a pump. Then by definition, pumping power loss is given by the following expression:

$$E_P = \frac{P_S Q}{\eta_P} \tag{10.16}$$

Substituting for flow rate Q from Equation (10.6),

$$E_P = \frac{\pi h^3 P_S^2}{6 \eta_P \mu \ln(R_2/R_1)} \tag{10.17}$$

Note that power loss rises rapidly with increasing film thickness. Yet, it is important to keep in mind that too small a film thickness makes the bearing very susceptible to failure due to particulate contamination in the lubricant.

Optimization

Pumping power loss often presents the greatest concern in the design of constant flow hydrostatic bearings, and one normally seeks to minimize it. In most applications, the outer radius R_2 is fixed by space limitations. However, R_1 is a parameter which the designer can control. To determine the condition in which power loss is minimum, we differentiate Equation (10.17) with respect to R_1 and set it equal to zero:

$$\frac{dE_P}{dR_1} = 0 \tag{10.18}$$

The following transcendental equation arises:

$$\ln \left(\frac{R_2}{R_1} \right) = 0.25 \left(\frac{R_2^2}{R_1^2} - 1 \right) \tag{10.19}$$

Equation (10.19) can be easily solved by a numerical scheme. The result is

$$\frac{R_1}{R_2} = 0.53 \tag{10.20}$$

By choosing the inner diameter of the step as 53% of the OD, pumping power will be minimum.

Thermal Effects and Typical Operating Conditions

High-precision applications such as those used for pointing mechanisms, telescopes, etc. often demand unusual accuracy. For example, operation of a modern telescope requires the entire telescope – including the mirror – to be maintained at a uniform temperature, often ambient

Table 10.1 Typical design and operating parameters for the vertical azimuth of Magellan 6.5 m telescope

Operating pressure	1100 psi
Lubricant type and viscosity	Mobil DTE 13 M (viscosity = 56 cSt at 80 °F)
Oil film thickness	0.0024 in.
Oil flow rate	0.48 gpm
Film stiffness	1.0×10^8 lbf/in.
Recess pressure	570 psi
Bearing load	80 000 lbf
Bearing type	Rectangular pad with central rectangular recess

Source: Chivens, 2000. Reproduced with permission of John Wiley & Sons.

temperature. This is required to minimize the optical distortions caused by convection currents in the air of the light path. Therefore, during the day, the entire telescope is maintained at the temperature anticipated during the night, when observation takes place.

While the metal thickness and internal ventilation system must be considered, the temperature of oil exiting the bearing pads must also be controlled in such a manner that it remains at the ambient temperature. For telescope applications, thermal control of the oil dictates that the hydraulic system be maintained at unusually low oil temperatures. Consideration must first be given to the heating of oil by friction and pressurizing the oil as it passes through the hydraulic pump. For a supply pressure of 1100 psi, the oil pump temperature must be kept at roughly 9 °F below ambient temperature at all times (Chivens, 2000). Table 10.1 gives flow rates and film stiffness for the large vertical azimuth hydrostatic bearing used in the Magellan telescope (Chivens, 2000).

If one assumes that all power dissipates into heat, then the temperature difference between the oil leaving the bearing and that of reservoir can be estimated as $E_P/(\rho C_P Q)$. This temperature rise, if significant, could result in a hot spot on the bearing surface. Thermal warping and associated distortion also may be a possibility (Elwell, 1984). A complete heat transfer analysis must include convection and radiation losses in particular because of the large surface areas involved.

Example 10.1 A circular step hydrostatic thrust bearing ($OD = 16$ in.) is to be designed to carry a load of 107 700 lbf. The recommended film thickness based on the surface finish is $h = 0.003$ in. and the lubricant viscosity is $\mu = 7.85 \times 10^{-6}$ reyns.

(a) What recess size would you recommend?
(b) Determine the corresponding pressure and flow rates.
(c) Determine the pumping power requirement, assuming that the pump efficiency is 80%.
(d) Determine the temperature difference between the leakage oil and the reservoir.
(e) Determine the bearing stiffness.

Solution

(a) For optimum stiffness, $R_1/R_2 = 0.53$. Therefore, the recommended recess radius is

$$R_1 = 0.53R_2 = 4.24 \text{ in.}$$

(b) The supply pressure is obtained by solving Equation (10.5) for pressure:

$$P_S = \frac{2W \ln(R_2/R_1)}{\pi (R_2^2 - R_1^2)} = 946 \, \text{psi}$$

The flow rate is (Equation 10.6)

$$Q = \frac{\pi h^3 P_S}{6\mu \, \ln(R_2/R_1)}$$

$$= \frac{\pi (0.003)^3 (946)}{6(7.85 \times 10^{-6}) \ln(1.887)}$$

$$= 2.68 \, \text{in}^3/\text{s} \quad (\times 60/231 = 0.7 \, \text{gal/min})$$

(c) Pumping power is

$$E_P = \frac{P_S Q}{\eta} = \frac{946 \times 2.68}{0.8} = 3169 \, \text{in. lb/s} \quad (0.48 \, \text{hp})$$

(d) Using typical values for density and specific heat, the temperature rise is estimated to be

$$\Delta T = \frac{E_P}{(\rho C_p Q)} = 8 \, ^\circ\text{F}$$

(e) Stiffness is $K = 3W/h = 3(107\,700)/0.003 = 1.08 \times 10^8 \, \text{lbf/in}$.

10.4 Capillary-Compensated Hydrostatic Bearings

Compensating elements allow the supply of two or more bearings pads with a single pump. Lubricant flows from the supply pump into a pressure control valve and then passes through the restrictor and into the bearing recess (Figure 10.6). The pressure in the recess P_r is therefore not equal to the supply pressure P_S exiting the pump. Capillary and orifice restrictors are the most commonly used compensating elements.

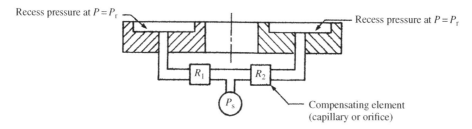

Figure 10.6 Hydrostatic bearing with a compensating element (capillary or orifice) where one pump supplies lubricant to several bearings.

Governing Equations

Simply replace supply pressure P_S with P_r in Equations (10.4)–(10.6) to determine the pressure distribution, load-carrying capacity expression, and flow rate equation, respectively. Determination of P_r requires an analysis of flow through the restrictor.

Governed by the Hagan-Poiseuille equation, the flow rate in a capillary tube Q_C is a function of the shearing action of the fluid flow and therefore depends on the viscosity. For a given fluid, the pressure difference ΔP across the capillary controls the volumetric flow rate, which in terms of the capillary constant K_C is

$$Q_C = K_C \Delta P \tag{10.21}$$

where $\Delta P = P_S - P_r$ and $K_C = \pi R_C^4/(8\mu l_C)$.

Continuity requires the flow through the capillary and the bearing to be the same, i.e.

$$Q_C = Q_B \tag{10.22}$$

From Equation (10.6), $Q_B = C_2 h^3 P_r$ where $C_2 = \pi/(6\mu \ln(R_2/R_1))$. Therefore, Equation (10.22) becomes

$$C_2 h^3 P_r = K_C P_S \left(1 - \frac{P_r}{P_S} \right) \tag{10.23}$$

Solving Equation (10.23) for the pressure ratio:

$$\frac{P_r}{P_S} = \frac{K_C}{C_2 h^3 + K_C} \tag{10.24}$$

The load-carrying capacity is

$$W_C = A_{eff} P_r = A_{eff} P_S \frac{K_C}{C_2 h^3 + K_C} \tag{10.25}$$

Stiffness and Optimization

Using the definition of stiffness given by Equation (10.11), the stiffness of a capillary-compensated bearing is:

$$K_{cap} = -\frac{dW}{dh} = \frac{3 A_{eff} P_r}{h[(P_S/P_r)/(P_S/P_r - 1)]} \tag{10.26}$$

The stiffness of a hydrostatic bearing with a restrictor generally is much less than that of a constant flow bearing. For this reason, it is important to design for optimum stiffness by selecting the capillary parameters properly. Letting $m = P_r/P_S$, Equation (10.26) is

$$K_{cap} = \frac{3 A_{eff} P_S (m - m^2)}{h} \tag{10.27}$$

Evaluating $dK_{cap}/dm = 0$ reveals that optimum stiffness occurs when $m = 1/2$. Therefore, the capillary should be chosen so that the recess pressure is

$$P_r = 0.5P_S \tag{10.28}$$

It follows, therefore, that the total load-carrying capacity and the stiffness are simply

$$W_C = 0.5A_{eff}P_S \quad \text{and} \quad K_{cap} = 0.5\left(\frac{3W}{h}\right) \tag{10.29}$$

which reveals that when optimized for stiffness, the load-carrying capacity and stiffness of a capillary-compensated bearing are reduced to only 50% of those of a constant flow bearing, without a restrictor in line.

10.5 Orifice-Compensated Bearings

Flow through an orifice is governed by

$$Q_O = \pi R_O^2 C_D \sqrt{\frac{2}{\rho}(P_S - P_r)} \tag{10.30}$$

where R_O is the orifice radius and C_D is the discharge coefficient. Parameter C_D is a function of the Reynolds number, Re. It typically varies in a nonlinear fashion from $C_D \approx 0.3$ for $Re \approx 2$ to $C_D \approx 0.7$ at $Re \approx 100$, and drops to about $C_D \approx 0.6$ for higher Re values.

Flow continuity requires

$$Q_O = Q_B \tag{10.31}$$

Therefore,

$$\frac{P_r}{P_S} = \frac{k_O^2}{C_2^2 h^6 P_r + k_O^2} \tag{10.32}$$

where $k_O = \pi R_O^2 C_D \sqrt{2/\rho}$.

The load in terms of the effective area is given by

$$W = A_{eff}\left[\frac{\left(-k_O^2 + k_O\sqrt{k_O^2 + 4C_2^2 h^6 P_S}\right)}{2C_2^2 h^6}\right] \tag{10.33}$$

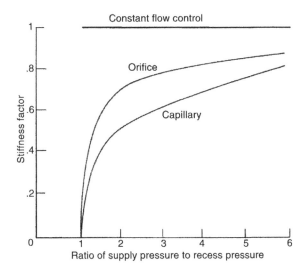

Figure 10.7 Comparison of the stiffness factor for constant flow rate with orifice- and capillary-compensated bearings. *Source*: Malanoski and Loeb, 1960. Reproduced with permission of ASME.

Stiffness and Optimization

The stiffness is obtained using Equation (10.11):

$$K_O = \frac{3P_S A_{\text{eff}}}{h} \left(\frac{2m(1-m)}{2-m} \right) \tag{10.34}$$

which gives an optimum K_O when $m = 0.586$. Under this condition, the load and the maximum stiffness are

$$W = 0.586 A_{\text{eff}} P_r \quad \text{and} \quad K_O = 0.586 \left(\frac{3W}{h} \right) \tag{10.35}$$

An orifice-compensated bearing often has relatively better stiffness than a capillary, yet its stiffness is much less than that of a constant flow system. Figure 10.7 compares the performance of these bearings.

Consideration must be given to parameters other than stiffness as well. For example, an orifice is temperature sensitive. That is, oil delivered to the recess may become too hot for a particular application where the position control is of great concern. An orifice is also susceptible to dirt and contamination in the lubricant, and this, together with the possibility of pressure surges, may cause damage to the orifice plate. A capillary, on the other hand, is a very simple and inexpensive device, but care must be taken in the selection of the type since the capillary may become excessively long. Figure 10.8 shows a schematic of an adjustable-length capillary that is particularly suitable for such applications.

It should be noted that all the derivations are restricted to a laminar flow assumption. Hence, it is crucial to examine the validity of the analysis by computing the Reynolds number and restricting it to $Re < 2500$.

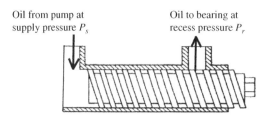

Figure 10.8   Adjustable capillary-compensating device.

10.6 Design Procedure for Compensated Bearings

1. Given the dimensions and the load, compute the recess pressure P_r.
2. Unless specified otherwise, choose the pressure ratio m for maximum stiffness.
3. Determine the supply pressure and flow rates and select an appropriate pump.
4. Using the flow rate information, determine the capillary parameter.
5. Choose a reasonable capillary diameter and compute the appropriate capillary length l_c. Make certain that l_c is reasonable. Select an adjustable-length capillary if needed.
6. Compute the Reynolds number for flow in the capillary to ascertain that the flow is laminar, i.e. $Re < 2500$.

Example 10.2   Design a capillary-compensated, circular step bearing with the following specifications: $W = 40\,000$ lbf, $R_2 = 5$ in.; $R_1 = 2.5$ in.; $N = 100$ rpm $\mu = 7 \times 10^{-6}$ reyns and the operating film thickness $h = 0.003$ in. Take a capillary of radius $R_c = 0.03$ in.

First, the recess pressure must be predicted:

$$P_r = \frac{2W \ln(R_2/R_1)}{\pi\left(R_2^2 - R_1^2\right)} = \frac{2(40\,000)\ln(2)}{\pi(5^2 - 2.5^2)} = 941 \text{ psi}$$

For optimum stiffness, the supply pressure should be selected such that

$$P_S = \frac{P_r}{0.5} = 1882 \text{ psi}$$

The flow rate through the bearing is

$$Q_B = \frac{\pi h^3 P_r}{6\mu \ln(R_2/R_1)} = 2.74 \text{ in.}^3/\text{s}$$

which must be identical to the flow rate through the capillary tube. Making use of Equation (10.21) and solving for the capillary constant yields

$$K_C = \frac{Q_C}{\Delta P} = 0.003$$

$$l_c = \frac{\pi R_c^4}{8\mu K_c} = 15 \text{ in.}$$

To check the Reynolds number, first determine the mean flow velocity:

$$U_{\text{mean}} = \frac{Q_C}{\pi R_c^2} = 969 \, \text{in./s}$$

Taking the density of oil to be $\rho = 0.03 \, \text{lbm/in.}^2$ gives the kinematic viscosity $v = \mu/\rho = 0.085 \, \text{in.}^3/\text{s}$. Therefore, the Reynolds number is $Re = U_{\text{mean}}D/v = 680$, which indicates that the flow is laminar.

10.7 Generalization to Other Configurations

The simple step hydrostatic bearing presented above is only one possible configuration, albeit an excellent starting point for analysis of more complicated ones such as multirecess thrust bearings of either square or circular types. Loeb and Rippel (1958) used the step-bearing formulations as a building block and came up with convenient design charts for other configurations. Their methodology is based upon defining three new parameters called the *pressure factor* P_f, *flow factor* q_f, and *power loss factor* H_f, presented below.

Pressure Factor P_f

Performance factors for various configurations are shown in Figure 10.9. The definitions of these factors are given as follows. Referring to Equation (10.8), the expression for supply pressure can be written as

$$P_S = \frac{W}{A_{\text{eff}}} \frac{1/A}{1/A} \tag{10.36}$$

where A is the bearing projected area, $A = \pi R^2$ for a circular step. Let $a_f = A_{\text{eff}}/A$ and define a so-called pressure factor parameter such that

$$P_f = \frac{1}{a_f} = \frac{A}{A_{\text{eff}}} \tag{10.37}$$

Therefore

$$P_S = P_f \frac{W}{A}$$

Flow Factor q_f

Similarly, the flow rate equation can be rewritten in the following form:

$$Q = \frac{CP_S h^3}{12\mu} = C\left(P_f \frac{W}{A}\right) \frac{h^3}{12\mu} \tag{10.38}$$

$$P_S = P_f \times \frac{W}{16\ell^2} \text{psi}$$

$$Q = q_f \times \frac{W}{16\ell^2} \times \frac{h^3}{12\mu} \text{ in.}^3/\text{s/pad}$$

$$E_P = H_f \times \left(\frac{W}{16\ell^2}\right) \times \left(\frac{h^3}{12\mu}\right) \text{ in. lb/s/pad}$$

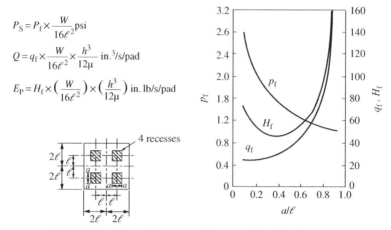

(a) Performance factor for a square, four-recess thrust bearing

$$P_S = P_f \times \frac{W}{(12+\pi)\ell^2} \text{ psi}$$

$$Q = q_f \times \frac{W}{(12+\pi)\ell^2} \times \frac{h^3}{12\mu} \text{ in.}^3/\text{s/pad}$$

$$E_P = H_f \times \left[\frac{W}{(12+\pi)\ell^2}\right]^2 \times \frac{h^3}{12\mu} \text{ in. lb/s/pad}$$

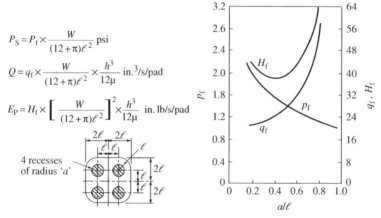

(b) Performance factor for a square, four-recess thrust bearing with rounded corners

Figure 10.9 Performance factors for various configurations: (a) square, four-recess thrust bearing; (b) square, four-recess thrust bearing with rounded corners; (c) circular, four-recess thrust bearing; (d) partial, 120° journal bearing. (*Note*: to obtain flow/recess, divide Q by 4 in (a)–(c)). *Source*: Loeb and Rippel, 1958. Reproduced with permission of John Wiley & Sons.

or

$$Q = q_f \frac{W}{A} \frac{h^3}{12\mu} \tag{10.39}$$

where

$$q_f = CP_f \tag{10.40}$$

$$P_S = P_f \times \frac{W}{81\pi\ell^2/16} \text{ psi}$$

$$Q = q_f \times \left[\frac{W}{81\pi\ell^2/16}\right] \times \frac{h^3}{12\mu} \text{ in.}^3/\text{s/pad}$$

$$E_P = H_f \times \left[\frac{W}{81\pi\ell^2/16}\right]^2 \times \frac{h^3}{12\mu} \text{ in. lb/s/pad}$$

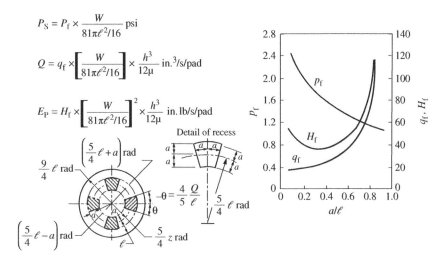

(c) Performance of circular, four-recess thrust bearing

$$P_S = P_f \times \frac{W}{\sqrt{3}D^2/2} \text{ psi}$$

$$Q = q_f \times \frac{W}{\sqrt{3}D^2/2} \times \frac{h^3_m}{12\mu} \text{ in.}^3/\text{s/pad}$$

$$E_P = H_f \times \left[\frac{W}{\sqrt{3}D^2/2}\right]^2 \times \frac{h^3_m}{12\mu} \text{ in. lb/s/pad}$$

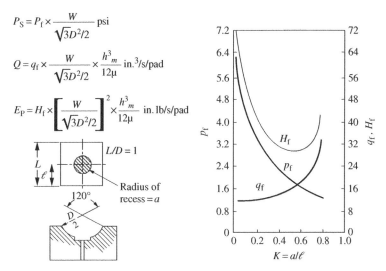

(d) Performance factors for a 120-degree partial journal bearing

Figure 10.9   (*Continued*)

Power Loss Factor, H_f

Similarly, a power loss factor can be defined in the following manner:

$$E_P = H_f \left(\frac{W}{A}\right)^2 \frac{h^3}{12\mu} \tag{10.41}$$

where

$$H_f = \frac{P_f q_f}{\eta} \tag{10.42}$$

Note that parameters P_f, q_f, and H_f are dimensionless. For a circular step bearing, the pressure factor, flow factor, and power factor become

$$P_f = \frac{2\ln(R_2/R_1)}{[1 - (R_1/R_2)^2]}; \qquad q_f = \frac{4\pi}{[1 - (R_1/R_2)^2]}; \qquad H_f = \frac{8\pi \ln(R_2/R_1)}{[1 - (R_1/R_2)^2]^2} \tag{10.43}$$

Example 10.3   Determine the flow rate, pressure supply, and power loss expected to support a 120° partial journal bearing with 1000 lb load on a 2 in. shaft with 2 in. long bearing. The recess diameter is 0.4 in. and $l = 1$ in. Assume that the minimum film thickness is $h_m = 0.003$ in. and the oil viscosity is 7.85×10^{-6} reyns.

Referring to Figure 10.9 (d), $K = a/l = 0.2$ and the following coefficients can be read from the performance factors chart:

$$q_f = 12; \quad P_f = 3.5, \quad \text{and} \quad H_f = 43$$

Therefore

$$P_S = P_f \frac{W}{(\sqrt{3}/2)D^2} = (3.5)\frac{1000}{(\sqrt{3}/2)2^2} = 1.01 \times 10^3 \text{ psi}$$

$$Q = q_f \left(\frac{W}{(\sqrt{3}/2)D^2}\right) \frac{h_m^3}{12\,\mu} = (12)\frac{1000}{(\sqrt{3}/2)2^2} \frac{0.003^3}{12(7.85 \times 10^{-6})} = 0.993 \text{ in}^3/s$$

$$E_f = H_f \left(\frac{W}{(\sqrt{3}/2)D^2}\right)^2 \frac{h_m^3}{12\,\mu} = 43 \left(\frac{1000}{(\sqrt{3}/2)2^2}\right)^2 \frac{0.003^3}{12(7.85 \times 10^{-6})} = 1.027 \times 10^3 \text{ in.lb/s}$$

10.8 Hydraulic Lift

Externally pressurized oil lifts are useful in the bearings of turbines, generators, and other large machines to eliminate both high starting torque and bearing wear during slow rotation in the range up to about 10–20 rpm before establishing a full hydrodynamic film.

Several high-pressure oil lift patterns are illustrated in Figure 10.10. Groove patterns (a), (b), and (c) generally provide somewhat lower breakaway pressure, while the pocket pattern in (d) allows formation of a more complete hydrodynamic oil film at normal running speeds. The oil pressure needed to start lifting action commonly exceeds somewhat the elastic contact pressure between the bearing material and journal around the lift pocket edges. Once this required feed pressure is reached, oil is forced between the static bearing–journal contact surfaces to start the lifting action (Booser and Khonsari, 2011). Equilibrium two-dimensional lift flow $Q\,(\text{in}^3/s)$

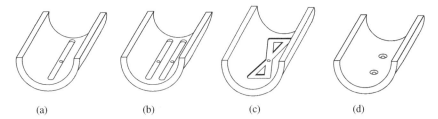

(a)   (b)   (c)   (d)

Figure 10.10 Representative pressurized oil lift grooves and pockets in journal bearings: (a) steel mill drive motor; (b) electric generator; (c and d) large steam turbines.

circumferentially from the pressurized oil zone along the bottom center line of the bearing is given by Fuller (1984).

$$Q = \frac{W\,C}{6\,\mu}(C/R)^2\frac{(1-\varepsilon)^2}{(2-\varepsilon)} \tag{10.44}$$

and the equilibrium feed pressure P (psi) following the lift is given by

$$P = \frac{2\,W}{B\,D}\frac{\frac{\varepsilon}{2}(4-\varepsilon^2) + \frac{2+\varepsilon^2}{(1-\varepsilon^2)}\tan^{-1}\left(\frac{1+\varepsilon}{(1-\varepsilon^2)^{1/2}}\right)}{(1-\varepsilon^2).(2-\varepsilon)} \tag{10.45}$$

where W = total load on bearing (lbf), C = radial clearance (in.), R = bearing bore radius (in.), D = bearing bore (in.), ε = eccentricity ratio, and B = effective axial length of oil lift zone: distance between axial ends of groove or pocket pattern + $^1/_2$ remaining bearing length.

The break-away pressure commonly ranges from about 1.2 to 4 times the equilibrium feed pressure P as total area of lift grooves or pockets drops from 10% to 1–2% of the total projected bearing area bearing area. For the lift arrangement shown in Figure 10.11, the breakaway

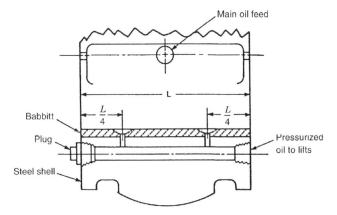

Figure 10.11 Pressurized oil lifts showing two elliptical oil pockets are incorporated along the axis of a journal bearing. *Source*: Elwell, 1984. Reproduced with permission of Taylor & Francis.

pressure can be approximated from the following (Elwell, 1984):

$$P_{BA} = \frac{k_{BA}\, W}{D n \sqrt{A}} \tag{10.46}$$

where n = number of pockets along the bottom center line of the bearing, A = area of a single oil lift pocket, and the breakaway parameter k_{BA} is determined experimentally (typically, $k_{BA} = 3$).

Example 10.4   A bearing with oil lift pockets such as that illustrated in Figure 10.11 is to carry a load of 100 000 lbf with diameter $D = 20$ in., length $L = 18$ in. and radial clearance ratio $C/R = 0.016/10$ in./in. With three lift pockets, each having an area $A = 6$ in^2. and with an oil viscosity $\mu = 7 \times 10^{-6}$ reyns. Determine the breakaway pressure and the oil feed required at equilibrium for a vertical lift of 0.005 in.

Applying Equation (10.46) gives the breakaway lift pressure P_{BA}:

$$P_{BA} = \frac{3(100\,000)}{(20)(3)(\sqrt{6})} = 2041\,\text{psi}$$

Once the journal is lifted 0.005 in., this breakaway pressure in the pockets will drop approximately in half to the value given by Equation (10.45). Lift oil feed rate needed to maintain this lift (with $\varepsilon = 1 - h/C = 1 - 0.005/0.016 = 0.688$) is then given by Equation (10.44):

$$Q = \frac{100\,000(0.016)(0.016/10)^2(1 - 0.688)^2}{6\,(7 \times 10^{-6})\,(2 - 0.688)} = 7.24\,\text{in}^3/\text{s} \quad (7.24 \times 60/231 = 1.9\,\text{gpm})$$

Problems

10.1   A hydrostatic bearing is being considered for lifting the grandstand seating section of a stadium, and initial draft calculations are needed to determine its feasibility. Assume that the entire section weighs 100 000 lbf and that water is used as the lubricating agent (take $\mu = 2 \times 10^{-9}$ reyns). For a preliminary design, take inner and outer diameters of $D_i = 10$ and $D_O = 30$ in. Assume a supply pressure of 20 psig. Determine the following:
   (a) How many pads are required?
   (b) What is the operating film thickness if a pump capable of supplying 5000 in.^{3}/s is used?
   (c) Compute the pumping power loss, assuming 80% efficiency.
   (d) Optimize the bearing for minimum power loss. What pumping power would be saved when optimized?

10.2   A circular hydrostatic bearing of the thrust type has a recess radius of 3.0 cm and a pad radius of 7.0 cm. The pad is designed to support a mass of 3000 kg with a uniform clearance of 0.04 mm. The required rotational speed is 200 rpm. The lubricant is incompressible, with dynamic viscosity of 0.1 poise, and is supplied to the pad at a pump pressure of 5×10^6 N/m^2 through a capillary restrictor.

(a) Determine the volumetric flow rate.
(b) Determine the pad stiffness.
(c) What is the maximum stiffness that can be achieved if this system is optimized?
(d) Determine the coefficient of friction.

10.3 Design a circular thrust bearing capable of carrying 20 000 lbf. An oil pump is available which has the maximum pressure, and flow ratings are 1000 psi and 1.0 gal/min with $\mu = 7 \times 10^{-6}$ reyns. Select the bearing size, pumping power, and flow rate requirement.

10.4 A hydrostatic thrust bearing of the rectangular type has a recess with length b and width L. Let l denote the land on either side of the pocket and let $L \gg l$.
(a) Starting with the Reynolds equation, show that the pressure distribution is given by
$$P = P_r(1 - x/l).$$
(b) Show that the load-carrying capacity is given by $W = P_r L(b + l)$.
(c) Show that the recess pressure is related to the flow rate according to the expression
$$P_r = 6\mu Q l / L h^3.$$
(d) Show that the stiffness of the bearing is given by

$$K = \frac{18\mu l}{h^4} \left(\frac{Q}{L}\right) L(b + l)$$

(e) If the bearing is fitted with a capillary tube, what would the stiffness be? How can it be optimized?
(f) Repeat part (e) for an orifice-compensated bearing.

10.5 A constant flow, rectangular hydrostatic bearing supports a large lathe in a steel forging plant. The pocket length is $b = 4.0$ in., and the land on each side is $l = 1.0$ in. A pump ($\eta_P = 0.7$) delivers SAE 10 oil ($\mu = 4.35 \times 10^{-6}$ reyns at the reservoir temperature of 95 °F) at a supply pressure of 4000 psi. Estimate the power loss and the temperature rise between the fluid leaving the bearing and the reservoir for an operating film thickness ranging from $h = 0.002$ to 0.004 in.

References

Anderson, W.J. 1964. 'Hydrostatic Lubrication,' *Advanced Bearing Technology*, E.E. Bisson and W.J. Anderson (eds.). NASA Publication SP-38, NASA, Washington, DC.

Booser, E.R. and Khonsari, M.M. 2011. 'Giving Bearings a Lift,' *Machine Design*, Vol. 83, pp. 48–51.

Chivens, D. 2000. Private communication. L&F Industries, San Luis Obispo, CA 93401.

Elwell, R.C. 1984. 'Hydrostatic Lubrication,' *CRC Handbook of Lubrication*, Vol. II, E.R. Booser (ed.). CRC Press, Boca Raton, Fla, pp. 105–120.

Fuller, D. 1984. *Theory and Practice of Lubrication for Engineering*, 2nd edn. Wiley Interscience, New York.

Hamrock, B. 1994. *Fundamentals of Fluid Film Lubrication*. McGraw-Hill Book Co., New York.

Loeb., A. and Rippel, H.C. 1958. 'Determination of Optimum Properties for Hydrostatic Bearings,' *ASLE Tribology Transactions*, Vol. 1, pp. 241–247.

Malanoski, S.B. and Loeb, A.M. 1960. 'The Effect of the Method of Compensation on Hydrostatic Bearing Stiffness,' ASME Paper No. 60-LUB-12.

O'Conner, J.J. and Boyd, J. (eds). 1968. *Standard Handbook of Lubrication Engineering*. McGraw-Hill Book Co., New York.

Pinkus, O. and Sternlicht, B. 1964. *Theory of Hydrodynamic Lubrication*. McGraw-Hill Book Co., New York.

Raimondi, A.A., Boyd, J., and Kauffman, H.N. 1968. 'Analysis and Design of Sliding Bearings,' *Standard Handbook of Lubrication Engineering*, J.J. O'Conner and J. Boyd (eds). McGraw-Hill Book Co., New York, pp. 5-1, 5-118.

Rowe, W.B. 1983. *Hydrostatic and Hybrid Bearing Design*. Butterworths & Co., Ltd, London.

Wilcock, D.F. and Booser, E.R. 1957. *Bearing Design and Application*. McGraw-Hill Book Co., New York.

11

Gas Bearings

Gas bearings are utilized in a variety of applications ranging from small instruments such as flying heads in read/write devices to large turbomachinery. Of the two broad categories of gas bearings, self-acting hydrodynamic bearings and those pressurized from an external source, this chapter will concentrate primarily on self-acting bearings. Typical examples of self-acting thrust bearings are shown in Figure 11.1.

Gas bearings are particularly attractive for use in high-speed applications because of the low viscosity, which results in much lower power losses than occur in their oil-lubricated counterparts. Also noteworthy is the increase in gas viscosity with increased temperature, which minimizes thermal issues. Gas bearings have minimal problems with cavitation, sealing, and leakage that often plague oil-lubricated bearings. With these attributes, gas bearings can operate without difficulty at extremely high speeds. For example, air-lubricated journal bearing dental drills operate smoothly at speeds of over 600 000 rpm (Grassam and Powell, 1964). Fuller (1984) states a practical limit of 700 000 rpm for gas bearings with no cooling requirements.

The load-carrying capacity of gas bearings is, however, orders of magnitude lower than that of oil-lubricated bearings, and vibrational stability and gas compressibility introduce complex analytical questions.

Gas bearings represent a specialized field, and many books are devoted to their development. Two of the most comprehensive reference books are by Grassam and Powell (1964) and Gross (1980). Tang and Gross (1962), Wilcock (1972), and O'Connor and Boyd (1968) cover externally pressurized gas bearings. Rowe (1983) covers hybrid hydrostatic and hydrodynamic bearings. The material presented in this chapter is based extensively on a more detailed article by Khonsari *et al.* (1994).

11.1 Equation of State and Viscous Properties

Equation of State

For a perfect (ideal) gas, physical properties follow a functional relationship known as the *ideal gas law*:

$$PV = nRT \tag{11.1}$$

Applied Tribology: Bearing Design and Lubrication, Third Edition. Michael M. Khonsari and E. Richard Booser.
© 2017 John Wiley & Sons Ltd. Published 2017 by John Wiley & Sons Ltd.
Companion Website: www.wiley.com/go/Khonsari/Applied_Tribology_Bearing_Design_and_Lubrication_3rd_Edition

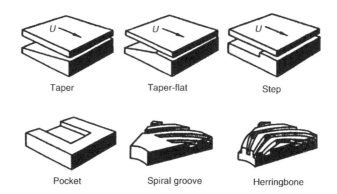

Taper   Taper-flat   Step

Pocket   Spiral groove   Herringbone

Figure 11.1   Various thrust bearing configurations. *Source*: Ausman, 1959.

where V is the volume occupied by n kg mol of a gas at the absolute temperature T in Kelvin and R is the universal gas constant: $R = 8.3143$ kJ/kg mol K.

Viscous Properties

In general, the viscosity of a fluid is governed by the combined effect of intermolecular forces and momentum transfer. While the viscous properties of a liquid are primarily dominated by intermolecular forces, momentum transfer is the dominant factor in a gas. Table 11.1 shows the viscous properties of several gases at atmospheric pressure (Wilcock, 1984).

The most important property as far as hydrodynamic lubrication is concerned is viscosity. Pressure tends to increase the viscosity of fluids (liquids and gases). But for most gases, this effect is realized only beyond a certain pressure. At a given temperature, the pressure effect on the viscosity of a gas is mild up to a certain pressure. For example, for nitrogen at 300 K, viscosity remains unchanged at roughly 18 mPa s up to roughly 100 bar pressure, after which it increases rapidly to roughly 50 mPa s at 1000 bar. Data and relationships are available from Wilcock (1997) for determining the viscosity of several gases as a function of pressure.

In a liquid, intermolecular forces tend to weaken with increasing temperature, and thus viscosity drops. In a gas, however, a rise in temperature brings about an increase in momentum transfer, and thus viscosity tends to increase. Several equations are available for characterizing the effect of temperature on viscosity. The simplest one is

$$\mu = \mu_{\mathrm{r}}(T/T_{\mathrm{r}})^{n} \tag{11.2}$$

where μ_{r} is the reference viscosity evaluated at the reference temperature T_{r} (in Kelvin). Parameter n is unique to a given gas and is typically $n \approx 0.7$ (see Table 11.1). Exceptions such as ammonia ($n = 0.128$) and sulfur dioxide ($n = 1.027$) do exist. Table 11.2 shows the viscosity of many different gases at various temperatures.

11.2 Reynolds Equation

Pressure distribution in gas-lubricated bearings can be obtained by solving the appropriate form of the Reynolds equation, with provision for compressibility. Assuming (1) that

Table 11.1 Viscous properties of gases at atmospheric pressure

Gas	Chemical formula	Molecular weight	Boiling point at P-760 mm Hg K (1.01 bar)	Heat capacity, C (kJ/kg deg)		Viscosity (N s/m² × 10⁶)		Speed of sound (m/s)[a]		Specific volume (m³/kg)[a]		Viscosity exponent, n
				300 K	400 K	300 K	400 K	300 K	400 K	300 K	400 K	
Air	$N_2H_7O_2$	28.96	78.8	1.007	1.014	18.5	23.0	—	—	0.861	1.148	0.757
Ammonia	NH_3	17.03	239.6	2.158	2.287	15.76[b]	16.35[b]	—	—	1.451	1.947	0.128
Argon	Ar	39.94	87.3	0.522	0.521	22.7	28.9	322.6	372.6	0.624	0.833	0.839
Carbon dioxide	CO_2	44.01	194.6[c]	0.851	0.942	14.9	19.4	269.4	307.3	0.564	0.754	0.918
Carbon monoxide	CO	28.01	81.6	1.043	1.049	17.7	21.8	353.1	407.3	0.890	1.188	0.724
Fluorine	F_2	38.00	85.2	—	—	24.0	30.0	—	—	0.656	0.876	0.776
Freon 12	CF_2Cl_2	120.92	243.3	—	—	12.4	15.3	—	—	0.203	0.273	0.731
Freon 13	CF_3Cl	104.47	191.6	—	—	—	—	—	—	0.237	0.318	—
Freon 21	$CHFCl_2$	102.92	282.0	—	—	11.4	—	—	—	0.236	0.320	—
Freon 11	$CFCl_3$	137.39	296.8	—	—	11.5	13.7	—	—	0.176	0.239	0.608
Freon 113	$C_2F_3Cl_3$	187.39	320.8	0.155	0.178	10.3	—	—	—	—	—	—
Helium	He	4.003	4.2	5.192	5.192	19.81	24.02	—	—	6.23	8.31	0.670
Hydrogen	H_2	2.016	20.3	14.31	14.48	8.96	10.8	1319.3	1519.6	12.38	16.50	0.634
Krypton	Kr	83.8	119.8	0.2492	0.2486	25.6	33.1	—	—	0.297	0.397	0.893
Nitrogen	N_2	28.02	77.4	1.041	1.045	17.9	22.1	353.1	407.4	0.890	1.187	0.733
Neon	Ne	20.183	27.09	—	—	31.8	38.8	—	—	1.24	—	0.692
Oxygen	O_2	32.00	90.2	0.920	0.942	20.7	25.8	329.6	379.0	0.779	1.039	0.766
Sulfur dioxide	SO_2	64.07	263.0	0.622	0.675	12.8[d]	17.2	—	—	0.369	0.495	1.027
Water	H_2O	18.02	373.0	4.20[d]	2.00	858.0[a]	13.2	—	—	—	—	—
Xenon	Xe	131.3	160.05	0.160	0.159	23.3	30.8	—	—	0.189	0.253	0.970

[a] At 1 bar.
[b] At 30 bar.
[c] Sublimes.
[d] Liquid.

Source: Wilcock, 1984. Reproduced with permission of Taylor & Francis.

Table 11.2   Effect of temperature on gas viscosity

	Viscosity in micropascal seconds (μPa s)					
	100 K	200 K	300 K	400 K	500 K	600 K
Air	7.1	13.3	18.6	23.1	27.1	30.8
Argon	8.0	15.9	22.9	28.8	34.2	39.0
Boron trifluoride	—	12.3	17.1	21.7	26.1	30.2
Hydrogen chloride	—	—	14.6	19.7	24.3	—
Sulfur hexafluoride ($P = 0$)	—	—	15.3	19.8	23.9	27.7
Hydrogen ($P = 0$)	4.2	6.8	9.0	10.9	12.7	14.4
Deuterium ($P = 0$)	5.9	9.6	12.6	15.4	17.9	20.3
Water	—	—	10.0	13.3	17.3	21.4
Deuterium oxide	—	—	11.1	13.7	17.7	22.0
Helium ($P = 0$)	9.7	15.3	20.0	24.4	28.4	32.3
Krypton ($P = 0$)	8.8	17.1	25.6	33.1	39.8	45.9
Nitric oxide	—	13.8	19.2	23.8	28.0	31.9
Nitrogen ($P = 0$)	—	12.9	17.9	22.2	26.1	29.6
Nitrous oxide	—	10.0	15.0	19.4	23.6	27.4
Neon ($P = 0$)	14.4	24.3	32.1	38.9	45.0	50.8
Oxygen ($P = 0$)	7.5	14.6	20.8	26.1	30.8	35.1
Sulfur dioxide	—	8.6	12.9	17.5	21.7	—
Xenon ($P = 0$)	8.3	15.4	23.2	30.7	37.6	44.0
Carbon monoxide	6.7	12.9	17.8	22.1	25.8	29.1
Carbon dioxide	—	10.0	15.0	19.7	24.0	28.0
Chloroform	—	—	10.2	13.7	16.9	20.1
y Methane	—	7.7	11.2	14.3	17.0	19.4
Methanol	—	—	—	13.2	16.5	19.6
Acetylene	—	—	10.4	13.5	16.5	—
Ethylene	—	7.0	10.4	13.6	16.5	19.1
Ethane	—	6.4	9.5	12.3	14.9	17.3
Ethanol	—	—	—	11.6	14.5	17.0
Propane	—	—	8.3	10.9	13.4	15.8
n-Butane	—	—	7.5	10.0	12.3	14.6
Isobutane	—	—	7.6	10.0	12.3	14.6
Diethyl ether	—	—	7.6	10.1	12.4	—
n-Pentane	—	—	6.7	9.2	11.4	13.4
n-Hexane	—	—	—	8.6	10.8	12.8

Note: Unless otherwise noted, the viscosity values are for a pressure of 1 bar. The notation $P = 0$ indicates the low-pressure limiting value.
Source: Wilcock, 1997. Reproduced with permission of Taylor & Francis.

continuum holds, i.e. that the molecular mean free path of the gas is negligibly small compared to the film gap and (2) that the ideal gas laws hold for the working fluid, the Reynolds equation takes on the following form (Gross, 1980):

$$\frac{\partial}{\partial X}\left(H^3 P \frac{\partial P}{\partial X}\right) + \frac{\partial}{\partial Y}\left(H^3 P \frac{\partial P}{\partial Y}\right) = \Lambda \frac{\partial}{\partial X}(PH) + \sigma \frac{\partial}{\partial T}(PH) \tag{11.3}$$

where H, P, and T are dimensionless film thickness, pressure, and time, respectively, defined as follows:

$$X = \frac{x}{L_{ref}}; \qquad Y = \frac{y}{L_{ref}}; \qquad H = \frac{h}{h_{ref}}; \qquad T = \omega t \qquad (11.4)$$

In Equation (11.4), reference parameters are indicated by the subscript ref. The appropriate form of the reference parameters depends on the type of the bearing, i.e. thrust or journal. For thrust bearings, typically, the minimum (outlet) film thickness h_{min} is used as the characteristic dimension. For journal bearings, clearance C is commonly chosen. For thrust bearings, $L_{ref} = B$, the bearing length in the direction of motion. For journal bearings, $x = R\theta$ and the appropriate characteristic length is $L_{ref} = R$, the shaft radius.

Parameter Λ in Equation (11.3) represents the bearing number or compressibility number, defined for slider bearings as

$$\Lambda = \frac{6\mu_a UB}{p_a h_{min}^2} \qquad (11.5)$$

where U is the sliding velocity and p_a and μ_a are the ambient pressure and viscosity, respectively. For a journal bearing application, the bearing number is defined as

$$\Lambda = \frac{6\mu_a \omega}{p_a}\left(\frac{R}{C}\right)^2 \qquad (11.6)$$

where R is the shaft radius, C is the radial clearance, and ω is the rotational speed (rad/s). The final parameter to be defined is the bearing squeeze number:

$$\sigma = \frac{12\mu_a B}{p_a[h(t=0)]^3}\frac{\partial h}{\partial t} = \frac{12\mu_a \omega B^2}{p_a[h(t=0)]^2} \qquad (11.7)$$

where $h(t=0)$ is reference film thickness at $t = 0$.

Restrictions and Limitation

Equation (11.3) assumes that continuum holds, i.e. the mean free path of the molecules is small compared to the film thickness. Whether continuum holds or not can be determined by calculating the Knudsen number, $Kn = \lambda/h$, where λ is the molecular mean free path. The continuum assumption holds when $Kn < 0.01$. Physically, λ is a measure of the distance that a gas particle travels between collisions. Therefore, if the gap size is very small – as in the case of a hydrodynamic bearing – then it is likely that the mean free path becomes of the same order as the gap size. When $0.01 < Kn < 15$, slip flow (between the gas and the bearing surfaces) occurs, and for $Kn > 15$, molecular flow is thought to become fully developed (Gross, 1980). When slip flow occurs, a reduction in pressure, akin to a reduction in viscosity, can be expected (Burgdorfer, 1959; Eshel, 1970; Sereny and Castelli, 1979).

Slip flow can become an issue in dealing with gas bearings. For example, consider an air-lubricated bearing with a minimum film thickness of 20 μ in.(0.5 μ m). Given that the molecular free path of air at atmospheric condition is 2.52 μ in.(0.064 μ m), the Knudsen number becomes $Kn = 0.064/0.54 = 0.119$, which signifies the importance of slip flow. Under this condition, the Reynolds equation must be modified.

The need to consider slip flows is much more crucial in hard-disk technology. Current practice in magnetic recording devices is to push the separation gap to a minimum. The read/write head used in magnetic recording devices is essentially a slider bearing. By 'flying' over the disk, which rotates at speeds of up to 10 000 rpm, it accesses information recorded on the disk as magnetic bits. The separation gap in current disk drives is about 500 Å (see e.g. Bhushan, 1996; Mate and Homola, 1997) as the head tends to fly over the disk at speeds of the order of 10 m/s. This is analogous to an aircraft flying at the speed of 360 000 km/h at an altitude of only 0.5 mm (Gellman, 1998). This flying instrument must maintain its separation gap without crashing into the hard disk. Needless to say, surface roughness effects are also a major concern (Bhushan, 1999).

Significant progress has been made in understanding the nature and the modeling aspects of molecular gas film lubrication with application to micro-electro-mechanical systems (MEMS). The fundamental theoretical development is based on the kinetic theory of gases applied to ultra-thin clearances. To this end, the Boltzmann equation (see e.g. Patterson, 1956) can be applied to model the flow at an *arbitrary* Knudsen number. Gans (1985) was the first to apply the linearized Boltzmann equation and demonstrate its application to lubrication problems. A detailed account of microscopic equations for lubrication flow known as the *molecular gas film lubrication (MGL) theory* is presented by Fukui and Kaneko (1995).

Interesting recent advances have been made in modeling the flow of powders within the context of the lubrication theory by applying the kinetic theory of gases (Dai *et al.*, 1994; McKeague and Khonsari, 1996a,b; Zhou and Khonsari, 2000; Pappur and Khonsari, 2003; ElKholy and Khonsari, 2007, 2008a,b; Iordanoff *et al.*, 2008, Khonsari, 2013). The objective of the model, known as *powder lubrication*, is to lubricate bearings by injecting a suitable powder into the clearance space. The feasibility of powder lubrication for use in thrust and journal bearings has been demonstrated by extensive experimental tests reported by Heshmat (1992) and Heshmat and Brewe (1995). Powder lubrication is intended for high-temperature applications where conventional lubricants are ineffective. It is possible to derive a modified Reynolds equation that could predict the behavior of a powder lubricant as it flows within the clearance space of a bearing (Jang and Khonsari, 2005).

To account for the slip-flow regime, as a first approximation valid when $0 < Kn < 1$, Burgdorfer (1959) suggests treating the problem as the conventional continuum with modified boundary conditions, which take slip velocities into account. At the runner surface, for example, the boundary condition with slip is

$$u\Big|_{z=0} = U + \lambda \frac{\partial u}{\partial z}\Big|_{z=0} \tag{11.8a}$$

and at the stationary surface, the boundary condition for the gas in contact with the surface is:

$$u\Big|_{z=h} = -\lambda \frac{\partial u}{\partial z}\Big|_{z=h} \tag{11.8b}$$

Similarly, the boundary conditions for the velocity component in the leakage direction are

$$v\Big|_{z=0} = \lambda \frac{\partial v}{\partial z}\Big|_{z=0} \tag{11.8c}$$

and

$$\left. v \right|_{z=h} = - \lambda \frac{\partial v}{\partial z}\bigg|_{z=h} \tag{11.8d}$$

Using these boundary conditions, Burgdorfer (1959) derived the following modified form of the Reynolds equation, which takes slip flow into account and is valid for very small Knudsen numbers in the range of $0 < Kn \ll 1$:

$$\frac{\partial}{\partial x}\left(h^3 P (1 + 6Kn)\frac{\partial P}{\partial x} \right) + \frac{\partial}{\partial y}\left(H^3 P (1 + 6Kn)\frac{\partial P}{\partial y} \right) = 6\mu U \frac{\partial}{\partial x}(Ph) \tag{11.9}$$

This equation can be nondimensionalized and extended to include transient effects similar to those shown in Equation (11.3). The result is as follows:

$$\frac{\partial}{\partial X}\left[H^3 P \left(1 + \frac{6Kn^*}{PH} \right) \frac{\partial P}{\partial X} \right] + \frac{\partial}{\partial Y}\left[H^3 P \left(1 + \frac{6Kn^*}{PH} \right) \frac{\partial P}{\partial Y} \right] = \Lambda \frac{\partial}{\partial X}(PH) + \sigma \frac{\partial}{\partial T}(PH) \tag{11.10}$$

where $Kn^* = \lambda_{ref}/h_{ref}$. The appropriate reference dimension, for example for a slider bearing, can be taken as the minimum film gap, h_{min}.

Limiting Cases

Referring to Equation (11.3), the following limiting cases can be deduced:

1. *Incompressible solution.* When $\Lambda \to 0$, and steady-state performance prevails, the effect of compressibility diminishes and the solution to the Reynolds equation reduces to the incompressible case. Pressure with subscript zero, P_0, will be used to designate this limiting case.
2. *Compressible solution.* When Λ is large and in the limit as $\Lambda \to \infty$, a limiting solution for compressible flow can be obtained by solving the Reynolds equation. The subscript ∞ for the pressure solution, P_∞, will be used to denote this solution.

These limiting cases are asymptotes that bound the actual pressure (and therefore the load-carrying capacity) that lies between these limiting solutions. This is illustrated in Figure 11.2 for a slider bearing with a film thickness ratio of $H_1 = h_1/h_2 = 2$. The influence of the film thickness ratio and the compressibility number on the load-carrying capacity is shown in Figure 11.3. Note that $\Lambda \propto H_1$, so that as the slider's inclination angle becomes sharper ($H_1 \geq 4$), the compressible asymptote is delayed (Gross, 1980).

It is important to note that depending on the operating conditions – particularly speed – the compressibility number can vary widely.

Example 11.1   Consider a journal bearing working with air as the lubricating fluid. Let $R/C = 3 \times 10^3$. Determine the range of the compressibility number variation over operating speeds ranging from 3600 rpm (at operating temperature of 300 K) to 360,000 rpm (at 400 K).

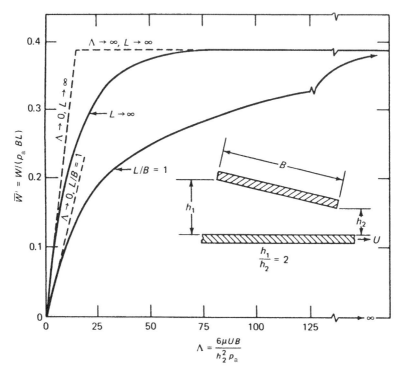

Figure 11.2 Effect of film thickness ratio and bearing load on an infinitely long, plane slider bearing illustrating the incompressible and compressible asymptotes. *Source*: Gross, 1980. Reproduced with permission of John Wiley & Sons.

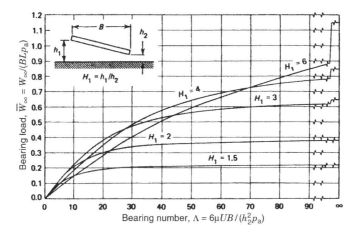

Figure 11.3 Effect of compressibility number on slider bearing load-carrying capacity based on the infinitely long solution. *Source*: Gross, 1980. Reproduced with permission of John Wiley & Sons.

For the top speed and maximum temperature:

$$\Lambda = \frac{6\mu_a\omega}{p_a}\left(\frac{R}{C}\right)^2 = \frac{6(18.5\times 10^{-6})(360\,000\times 2\pi/60)}{101\,325}(3\times 10^3)^2 = 370$$

For the minimum speed and temperature $\Lambda = 4.6$, which is the incompressible range.

When Λ is small, the incompressible theory is used to obtain a solution, but with a question about the inclusion of cavitation. According to the simulations of Shapiro and Colsher (1968), the cavitating incompressible theory yields results closer to those of the compressible solutions based on a series of numerical experiments with thrust bearings with a parabolically shaped film thickness. They attribute this anomaly to the fact that a cavitating pad with liquid lubricant produces a greater load-carrying capacity than a noncavitating pad. Up to $\Lambda = 50$, the incompressible and compressible results are close. This upper limit for Λ does depend on the film thickness. The permissible range for the incompressible solution drops with a reduction in film thickness. A similar conclusion can be interpreted from Figure 11.3.

11.3 Closed-Form Solutions

Closed-form analytical solutions for various bearing types and configurations are available from Gross (1980). These relationships assume that side leakage is negligible. The following broad categories of these solutions are adopted here from that work.

Infinitely Long Tapered-Step and Slider Bearings

The theoretical load-carrying capacity of an infinitely long slider bearing is predicted to be optimum if the shape is designed to form a combination tapered-step bearing (Gross, 1980). Shown in Table 11.3 is the predicted performance of a simple slider bearing. Performance parameters, including load and friction force, can be evaluated from the expressions provided in the table. Performance results for simple geometric configurations such as taper (plane wedge), taper-flat, and step bearings are available from Gross (1980).

Full numerical solutions for finite bearings indicate that for bearings with some geometric shapes, such as Rayleigh step bearings, the theoretical load-carrying capacity is considerably reduced when side leakage is considered, since there is a large pressure gradient at the step where the pressure is maximum. A practical method for improving performance is to reduce side leakage by applying shrouds along the sides of the bearing to form a pocket, as shown in Figure 11.1.

Infinitely Long Journal Bearings

For a full 360° journal bearing, similar limiting expressions are summarized in Table 11.4. Parameters W_1 and W_2 are dimensionless components in the vertical and horizontal directions, respectively. The total load $\overline{W}$ is the resultant of the two components: $\overline{W} = \sqrt{\overline{W}_1^2 + \overline{W}_2^2}$. The following Sections 11.4, 11.5 and 11.6 provide finite length solutions of the Reynolds equation for a variety of thrust and journal bearing designs.

Table 11.3 Limiting solutions for infinitely long slider bearings

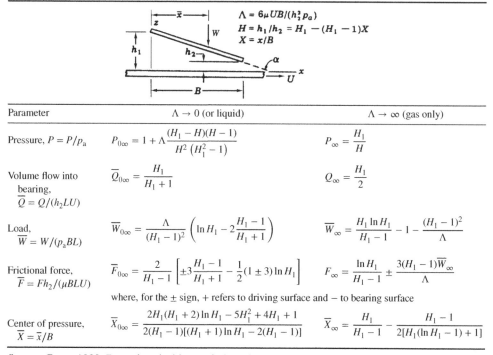

Parameter	$\Lambda \to 0$ (or liquid)	$\Lambda \to \infty$ (gas only)
Pressure, $P = P/p_a$	$P_{0\infty} = 1 + \Lambda \dfrac{(H_1 - H)(H - 1)}{H^2\,(H_1^2 - 1)}$	$P_\infty = \dfrac{H_1}{H}$
Volume flow into bearing, $\overline{Q} = Q/(h_2 LU)$	$\overline{Q}_{0\infty} = \dfrac{H_1}{H_1 + 1}$	$Q_\infty = \dfrac{H_1}{2}$
Load, $\overline{W} = W/(p_a BL)$	$\overline{W}_{0\infty} = \dfrac{\Lambda}{(H_1 - 1)^2}\left(\ln H_1 - 2\dfrac{H_1 - 1}{H_1 + 1}\right)$	$\overline{W}_\infty = \dfrac{H_1 \ln H_1}{H_1 - 1} - 1 - \dfrac{(H_1 - 1)^2}{\Lambda}$
Frictional force, $\overline{F} = Fh_2/(\mu BLU)$	$\overline{F}_{0\infty} = \dfrac{2}{H_1 - 1}\left[\pm 3\dfrac{H_1 - 1}{H_1 + 1} - \dfrac{1}{2}(1 \pm 3)\ln H_1\right]$	$F_\infty = \dfrac{\ln H_1}{H_1 - 1} \pm \dfrac{3(H_1 - 1)\overline{W}_\infty}{\Lambda}$
	where, for the $\pm$ sign, $+$ refers to driving surface and $-$ to bearing surface	
Center of pressure, $\overline{X} = \bar{x}/B$	$\overline{X}_{0\infty} = \dfrac{2H_1(H_1 + 2)\ln H_1 - 5H_1^2 + 4H_1 + 1}{2(H_1 - 1)[(H_1 + 1)\ln H_1 - 2(H_1 - 1)]}$	$\overline{X}_\infty = \dfrac{H_1}{H_1 - 1} - \dfrac{H_1 - 1}{2[H_1(\ln H_1 - 1) + 1]}$

Source: Gross, 1980. Reproduced with permission of John Wiley & Sons.

11.4 Finite Thrust Bearings

Rectangular Thrust Bearings

Load-carrying capacity as a function of the bearing number for a square-shaped slider is presented in Figure 11.4. For comparison purposes, various solutions for different tilt angles are shown for slenderness ratios for $L' = L/B = 1$ and ∞. This figure is useful for assessment of the accuracy of infinitely long solutions as well as the limiting solution when $\Lambda \to \infty$.

Sector-Pad Thrust Bearings

A simple approach for analyzing a sector-pad thrust bearing is to replace the geometry by that of a rectangular pad whose width is equal to the mean value of the sector circumference and whose length is equal to the difference between the outer radius r_0 and the inner radius r_i. This, however, distorts the sector shape geometry. Starting with the Reynolds equation written in polar coordinates, Etsion and Fleming (1977) provide accurate solutions for flat sector-shape thrust bearings over a wide range of compressibility numbers. When a pivot (assumed frictionless) is used, the film thickness will depend both on the pitch (tilting about a radial line) and on the roll (tilting about a tangent line). If the film thickness is decreased substantially, the perfect plane pivoted-load bearing surfaces become parallel, thereby leading to a loss of load-carrying capacity (Gross, 1980). For this reason, and because of difficulties with starting perfectly plane-pivoted bearings, a crown is commonly machined into the surface.

Table 11.4 Limiting solutions for infinitely long journal bearings

$$\Lambda = \frac{6\mu U r}{C^2 p_a} \qquad \epsilon = \frac{e}{C}$$

$$U = rw \qquad \theta^*, P_M \quad \text{(position and magnitude of peak pressure)}$$

Parameter	$\Lambda \to 0$	$\Lambda \to \infty$
Pressure, $P = P/p_a$	$1 + \Lambda\epsilon \sin\theta \dfrac{2 + \epsilon\cos\theta}{(2 + \epsilon^2)(1 + \epsilon\cos\theta)^2}$	$\dfrac{\left(1 + \frac{3}{2}\epsilon^2\right)^{1/2}}{1 + \epsilon\cos\theta}$
$\theta^*, \left.\dfrac{dP}{d\theta}\right|_{\theta^*} = 0$	$\pm\cos^{-1}\dfrac{-3\epsilon}{2 + \epsilon^2}$	$\pm\pi/2$
$P_M - P_{\min}$	$\dfrac{\Lambda\epsilon}{2(2 + \epsilon^2)}\left(\dfrac{4 - \epsilon^2}{1 - \epsilon^2}\right)^{3/2}$	$\dfrac{2}{1 - \epsilon^2}\left(1 + \dfrac{3\epsilon^2}{2}\right)^{1/2}$
$\overline{W}_1 = \dfrac{W_1}{p_a DL}$	0	$\dfrac{\pi}{\epsilon}\left(1 + \dfrac{3\epsilon^2}{2}\right)^{1/2}\dfrac{1 - (1 - \epsilon^2)^{1/2}}{(1 - \epsilon^2)^{1/2}}$
$\overline{W}_2 = \dfrac{W_2}{p_a DL} = \dfrac{W_2}{p_a DL}$	$\dfrac{\pi\epsilon\Lambda}{(2 + \epsilon^2)(1 - \epsilon^2)^{1/2}}$	0
Attitude angle, ϕ	$\pi/2$	0
$\overline{M}_j = M_j c/(\mu L r^3 \omega)$	$\dfrac{4\pi(1 + 2\epsilon^2)}{(2 + \epsilon^2)(1 - \epsilon^2)^{1/2}}$	$2\pi/(1 - \epsilon^2)^{1/2}$
Friction coefficient, f	$\dfrac{1 + 2\epsilon^2}{3\epsilon}\dfrac{C}{r}$	$\dfrac{\mu U}{p_a C}\dfrac{\left(1 + \frac{3}{2}\epsilon^2\right)^{1/2}}{1 - (1 - \epsilon^2)^{1/2}}$

Source: Gross, 1980. Reproduced with permission of John Wiley & Sons.

Performance curves and appropriate design procedures for sector thrust bearings with crowned surfaces (parabolic, cylindrical, and spherical) are covered by Shapiro and Colsher (1968). Solutions are based on the Reynolds equation in polar coordinates as follows:

$$\frac{\partial}{\partial R}\left(\frac{RH^3}{2}\frac{\partial Q}{\partial R}\right) + \frac{1}{R}\frac{\partial}{\partial\theta}\left(\frac{H^3}{2}\frac{\partial Q}{\partial\theta}\right) = \Lambda R\left[\frac{\omega}{\omega_1}\left(\sqrt{Q}\frac{\partial H}{\partial\theta} + \frac{H}{2Q}\frac{\partial Q}{\partial\theta}\right) + \sqrt{Q}\frac{\partial H}{\partial T} + \frac{H}{2\sqrt{Q}}\frac{\partial Q}{\partial T}\right]$$

$$(11.11)$$

where $Q = P^2$ and T is the dimensionless time $= \omega_1 t/2$. Parameter ω_1 is a reference speed $= 2T/t$.

The following design information is extracted from Shapiro and Colsher (1968) for a tilting-pad bearing with cylindrical crowns, as shown in Figure 11.5. The parameters of interest for design are position of pivot, optimum height, and bearing performance prediction. Location of the pivot is defined by the radius r_{pv} and the angle γ measured from the pad's leading edge.

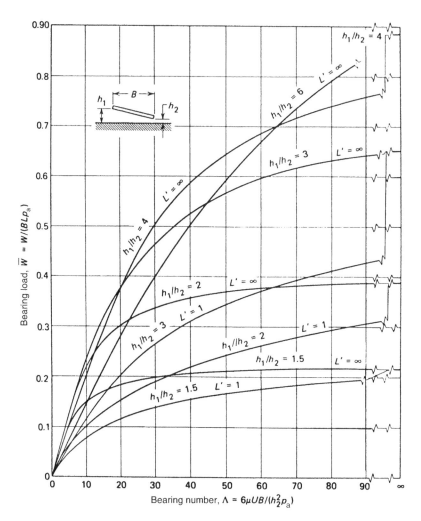

Figure 11.4 Effect of bearing number on slider bearing load. *Source*: Gross, 1980. Reproduced with permission of John Wiley & Sons.

The pitch is represented by $\alpha_r = \alpha$ and the roll by α_θ. The influence of the crown is governed by two parameters: the crown height δ and clearance of the aligned pad measured from the peak point of the crown C. The angular extent of the pad θ_p is assumed to be 45° (0.7854 rad).

When the pad is approximated as a rectangular shape, the pivot location is normally taken to be along the mean radius. Proper formulation of the problem shows that the actual position is slightly outside the mean radius for a zero roll angle. Although it is a function of δ/C and also varies with Λ, a value of $r_{pv}/r_o \approx 0.76$ represents a typical pivot position (Shapiro and Colsher, 1968). For representative numbers of $\Lambda = 25$ (incompressible) and $\Lambda = 250$ (compressible), this radial position of the pivot is about 1–2% beyond the mean radius and is rather insensitive to crown height, inclination angle, and compressibility number. When the dimensionless minimum film thickness $H_{min} = h_{min}/C$ is very small, the load coefficient $c_L = W/p_a r_o^2$ does indeed become sensitive to the pivot position.

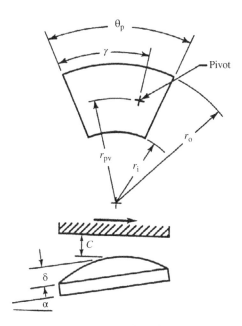

Figure 11.5 Nomenclature of a crowned tilting-pad thrust bearing. *Source*: Shapiro and Colsher, 1968. Reproduced with permission of John Wiley & Sons.

Figures 11.6 and 11.7 illustrate the importance of the dimensionless crown height δ/C for two representative compressibility numbers $\Lambda = 25$ and $\Lambda = 250$, respectively. An optimum load can be achieved by letting $\delta/C = 1$. This results in a significant improvement over a flat shape ($\delta/C = 0$). Figures 11.8 and 11.9 present field plots as a function of the pivot position for the representative bearing numbers. Figures 11.10 and 11.11 present the variation of the load coefficient and friction as a function of Λ, respectively.

Design Procedure for Tilting-Pad Thrust Bearings

The following steps provide a representative design procedure:

1. In most applications, load and speed are specified as design constraints and r_o is fixed. Therefore, the first step is to compute the load coefficient.
2. If the crown parameters C and δ are available, determine the bearing number and select the appropriate design curves in this chapter. If the crown parameters are not known, start by selecting a representative Λ.
3. Compute H_{min} from Figures 11.6–11.7 or Figures 11.8–11.9, as appropriate.
4. Determine C using the definition of the bearing number.
5. Choose $\delta/C = 1$ for the maximum loading condition and determine the crown height.
6. Having determined H_{min} and C, evaluate the dimensional minimum film thickness and check its reasonableness by comparing it to the bearing surface roughness. Table 8.7 in Chapter 8 provides a useful guide for this purpose.
7. Determine the friction moment factor.
8. Determine the pivot position from the appropriate field plots.

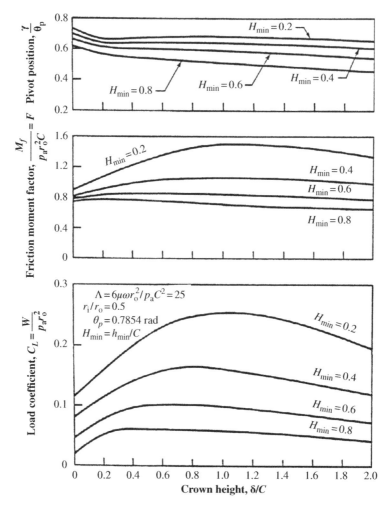

Figure 11.6 Performance of tilting-pad thrust bearing versus crown height ($\Lambda = 25$) (incompressible). *Source*: Shapiro and Colsher, 1968. Reproduced with permission of John Wiley & Sons.

In a thorough evaluation of a design, it may be necessary to perform an off-design performance analysis by maintaining a constant pivot position and making use of the field plots. This step requires repetition of the calculations for various δ/C and load values to determine the adequacy of the film thickness and the reasonableness of the friction moment (Shapiro and Colsher, 1968).

Example 11.2 Evaluate the performance of an air-lubricated tilting-pad thrust bearing with the following specifications (Khonsari *et al.*, 1994).

Total load and speed requirements:

$$W_T = 450 \text{ lb} \quad (2000 \text{ N}); \qquad \omega = 20\,000 \text{ rpm} \quad (2094 \text{ rad/s})$$

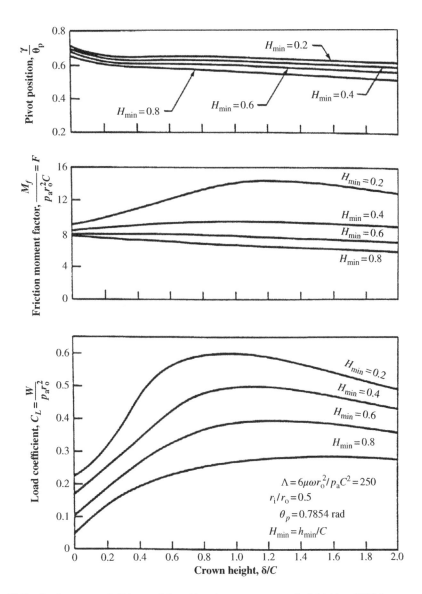

Figure 11.7   Performance of tilting-pad thrust bearing versus crown height ($\Lambda = 250$) (compressible).
Source: Shapiro and Colsher, 1968. Reproduced with permission of John Wiley & Sons.

Dimensions:

$$r_i = 2 \text{ in.} \quad (5.08 \times 10^{-2} \text{ m});$$
$$r_o = 4 \text{ in.} \quad (1.02 \times 10^{-1} \text{ m}); \qquad \theta_p = 45\,^\circ$$

Ambient pressure:

$$p_a = 15 \text{ psi} \quad (103 \text{ kp}_a)$$

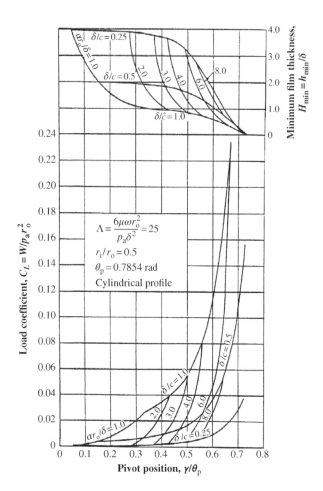

Figure 11.8 Load coefficient and minimum film thickness versus pivot position, constant crown height, varying film thickness ($\Lambda = 25$). *Source*: Shapiro and Colsher, 1968. Reproduced with permission of John Wiley & Sons.

Viscosity:

$$\mu = 3.19 \times 10^{-9}\, \text{reyns} \quad (2.2 \times 10^{-5}\, \text{Pa s})$$

Select $n = 6$ pads, so that the load per pad becomes: $W_p = 75\, \text{lb}(334\, \text{N})$.

1. The load coefficient, based on the maximum load per pad, is

$$C_L = \frac{W}{p_a r_o^2} = 0.3125$$

2. Since crown parameters C and δ are not known *a priori*, the bearing number cannot be computed directly. Therefore, begin by selecting as a representative bearing number $\Lambda = 250$.

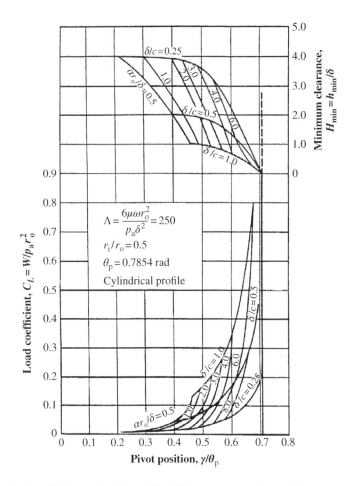

Figure 11.9 Load coefficient and minimum film thickness versus pivot position, constant crown height, varying film thickness ($\Lambda = 250$). *Source*: Shapiro and Colsher, 1968. Reproduced with permission of John Wiley & Sons.

3. Use Figure 11.7 to evaluate $H_{min} = 0.732$.
4. Compute C using the definition of the bearing number:

$$C = r_0 \sqrt{\frac{6\mu\omega}{p_a\Lambda}} = 0.414 \times 10^{-3} \text{ in.} \quad (1.052 \times 10^{-5} \text{ m})$$

5. For an optimum load-carrying capacity, choose $\delta = C = 0.414 \times 10^{-3}$ in.
6. Compute the dimensional minimum film thickness:

$$h_{min} = H_{min}C = 0.303 \times 10^{-3} \text{ in.} \quad (7.7 \times 10^{-6} \text{ m})$$

Referring to Table 8.7, this minimum film thickness requires a smooth surface finish with close tolerances, with $R_a = 8$–16 in. $(0.2$–$0.4\,\mu\text{m})$.

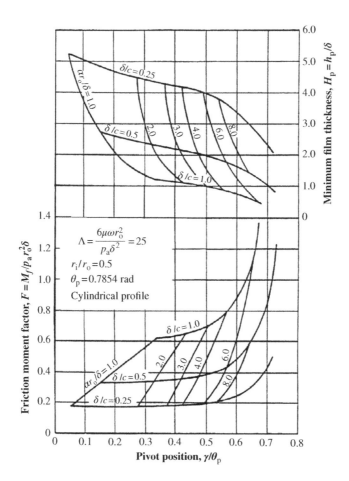

Figure 11.10   Friction moment factor and pivot film thickness versus pivot position, constant crown height, varying film thickness ($\Lambda = 25$). *Source*: Shapiro and Colsher, 1968. Reproduced with permission of John Wiley & Sons.

7. Compute the friction moment factor:

$$F = \frac{M_f}{p_a r_o^2 C} = 6.75$$

$$M_f = 6.75 p_a r_o^2 C = 0.671 \text{ in. lb/pad} \quad (7.58 \times 10^{-2} \text{ N m/pad})$$

$$M_f(6 \text{ pads}) = 4.02 \text{ in. lb} \quad (0.455 \text{ N m})$$

$$E_P = \frac{M_f N}{63000} = 1.276 \text{ hp} (0.952 \text{ kW})$$

8. Determine the pivot position. Use Figure 11.9 with $\delta/C = 1.0$, and read $\gamma/\theta_p = 0.6$. Since $\theta_p = 45°, \gamma = 27°$. Take the radius of the pivot as 3.05 in. (1.5% greater than the 3 in. mean radius).

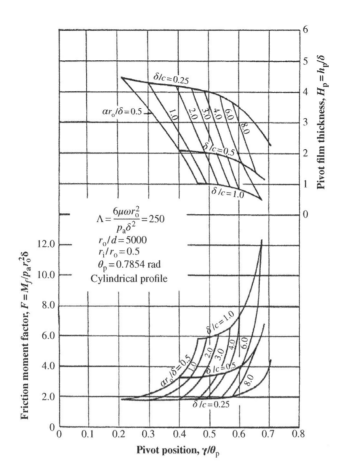

Figure 11.11 Friction moment factor and pivot film thickness versus pivot position, constant crown height, varying film thickness ($\Lambda = 250$). *Source*: Shapiro and Colsher, 1968. Reproduced with permission of John Wiley & Sons.

The influence of the bearing compressibility number on the load coefficient and friction moment for $\theta_p = 45°$ and $\delta/C = 1.0$ (optimum loading) is shown in Figures 11.12 and 11.13, respectively. Also shown are results based on the incompressible fluid assumption, which closely approximate those of the compressible assumption up to roughly $\Lambda = 50$ (Shapiro and Colsher, 1968).

11.5 Finite Journal Bearings

Steady-State Performance

Raimondi (1961) has developed design charts for full journal bearings of finite length with various L/D ratios based on numerical solutions of the Reynolds equation. Results for $L/D = 1$ are given in Figure 11.14. The design procedure is straightforward. With a specified load and operating conditions, compute $\Lambda = 6\mu\omega(r/c)^2/p_a$ and $P/p_a = W/(2rLp_a)$. Figure 11.14(a) then gives operating eccentricity ε. Entering Figures 11.14(b) and 11.14(c) with compressibility

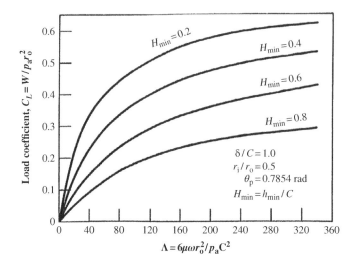

Figure 11.12 Variation of load coefficient with Λ corresponding to Example 11.2. *Source*: Shapiro and Colsher, 1968. Reproduced with permission of John Wiley & Sons.

and ε then gives friction force (hence, power loss) and attitude angle, respectively. Alternatively, in a particular design, the minimum film thickness may be specified. The eccentricity ratio can then be directly computed from $h_{min} = C(1 - \varepsilon)$, and the load-carrying capacity can be predicted from Figure 11.14(a). Note that the compressibility number has a major influence on the attitude angle.

Note in Figure 11.14 that nonlinearity sets in when $\Lambda/6 \geq 1$; that is, the incompressible solutions deviate significantly from the compressible solutions when $\Lambda \geq 6$. In contrast, Shapiro and Colsher (1968) recommend a limit of $\Lambda = 50$ for the compressibility number with tilting-pad thrust bearings.

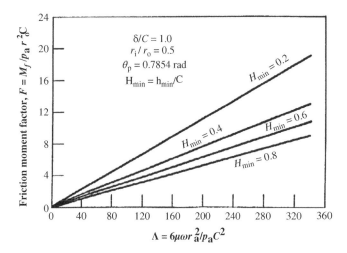

Figure 11.13 Variation of friction moment with Λ. *Source*: Shapiro and Colsher, 1968. Reproduced with permission of John Wiley & Sons.

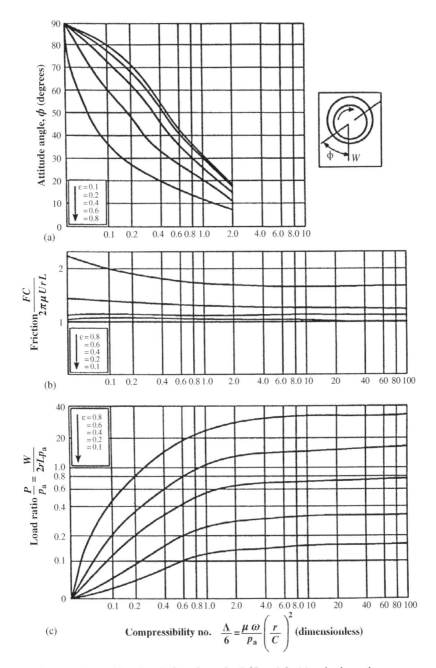

Figure 11.14 Plain journal bearing design charts for $L/D = 1.0$: (a) attitude angle versus compressibility number; (b) dimensionless friction force variable versus compressibility number; (c) dimensionless load ratio versus compressibility number. *Source*: Raimondi, 1961. Reproduced with permission of Taylor & Francis.

Example 11.3   A small blower operating at 3600 rpm at room temperature has an air-lubricated journal bearing. The following specifications are given:

$$W = 0.5 \, \text{lbf}; \qquad L = D = 0.6 \, \text{in.}; \qquad C = 0.0003 \, \text{in.}$$

Recommend a surface finish for this bearing and estimate the power loss requirement. The angular speed and linear speed are

$$\omega = \frac{3600 \times 2\pi}{60} = 377 \, \text{rad/s}$$
$$U = r\omega = 113 \, \text{in./s}$$

From the viscosity table, $\mu = 18.5 \, \text{Pa.s} \times 1.45 \times 10^{-4} \, \text{reyns/(Pa.s)} = 2.68 \times 10^{-9} \, \text{reyns}$.

$$\frac{\Lambda}{6} = \frac{\mu U r}{C^2 p_a} = \frac{(2.68 \times 10^{-9})(113)(0.3)}{(0.0003)^2(14.7)} = 0.07$$

The load ratio is $W/(2rLp_a) = 0.09$. From Figure 11.14, we estimate an operating eccentricity of $\varepsilon \approx 0.4$. The minimum film thickness is $h_{min} = C(1 - \varepsilon) = 180 \, \mu \text{in}$. From the allowable film thickness (Table 8.7), a surface finish of $R_a = 8 - 10 \, \mu \text{in}$. is recommended.

  To determine the power loss, first determine the friction force. From Figure 11.14, the friction force variable is estimated to be $FC/(2\pi\mu UL) \approx 1.13$. Solving for friction force yields $F = 0.004 \, \text{lbf}$. The friction moment is therefore $M = F \cdot r = 0.004 \times 0.3 = 0.0012 \, \text{in. lbf}$. and power loss is remarkably low as $E_p = F \cdot U = 0.004 \times 113 = 0.487 \, \text{in lbf/s} = 7.3 \times 10^{-5} \, \text{hp}$.

Angular Stiffness and Misalignment Torque

A system involving a number of gas-lubricated journal bearings may require one to take into account the effect of possible misalignment and angular stiffness. Ausman (1959) provides a design chart, Figure 11.15, for estimating the *moment-carrying capacity*. For a given Λ, this

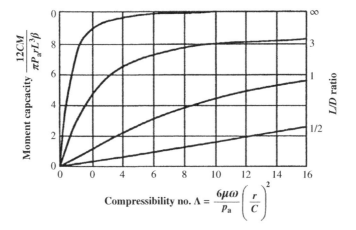

Figure 11.15   Moment-carrying capacity as a function of compressibility number. *Source*: Ausman, 1959.

figure relates misalignment torque T_m and misalignment angle β. Once a maximum misalignment angle is estimated, one can calculate the maximum allowable moment. The results shown in Figure 11.15 are based on a perturbation method and are valid for small eccentricity ratios (up to 0.5). Extensions to higher ε values are provided by Rice (1965).

Example 11.4   Estimate the friction moment corresponding to Example 11.3 if the shaft is subject to a misalignment angle of $\beta = 0.0005\,\text{rad}(0.057\,°)$.

Entering Figure 11.15 with $\Lambda = 0.07 \times 6 = 0.42$ and $L/D = 1$, we estimate the friction moment capacity to be roughly 0.1. Solving for the moment capacity:

$$M = \frac{0.1\pi p_a r L^3 \beta}{(12C)} = \frac{0.1\pi(14.7)(0.3)(0.6)^3(0.0005)}{(12 \times 0.0003)} = 0.042\,\text{in. lbf}$$

This value of the moment is substantially greater than the 0.0012 in. lbf predicted in the previous example. This example illustrates the significance of misalignment.

Whirl Instability

Plain journal bearings lubricated with gas are uniquely susceptible to troublesome self-excited whirl instability. The threshold of instability depends on the eccentricity ratio and the L/D ratio. When the compressibility number is known, Figure 11.16 for $L/D = 1$ can be used to determine the critical speed above which instability is likely to set in (Cheng and Pan, 1965; O'Connor and Boyd, 1968); the higher the eccentricity ratio, the higher the threshold of instability. Stability of plain journal bearings may be improved (Gross, 1980) by incorporating multiple sections or lobed-shaped bore geometries. See Figure 11.17.

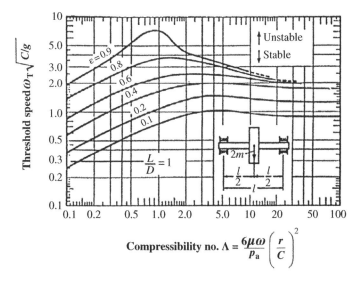

Figure 11.16   Threshold of whirl instability as a function for $L/D = 1$. *Source*: O'Connor and Boyd, 1968. Reproduced with permission of McGraw-Hill.

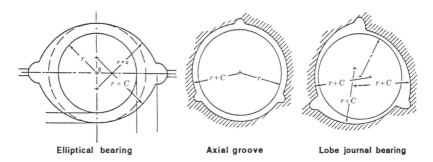

Elliptical bearing   **Axial groove**   **Lobe journal bearing**

Figure 11.17 Multiple section and elliptical bearings with improved stability characteristics. *Source*: Gross, 1980. Reproduced with permission of John Wiley & Sons.

Example 11.5 Determine the critical whirl instability threshold of the bearing specified in Example 11.3.

Entering Figure 11.16 with $\Lambda = 0.07 \times 6 = 0.42$ and $\varepsilon = 0.4$, predicts that the dimensionless critical speed parameter is $\omega_T \sqrt{C/g} \approx 1.3$, where g is the gravitational constant: $g = 32.2 \, \text{ft/s}^2 = 386 \, \text{in./s}^2$.

Solving for critical speed gives $\omega_T = 1.3 \sqrt{386/0.0003} = 1475 \, \text{rad/s}(14\,080 \, \text{rpm})$. Therefore, this bearing operates well within the stable region.

11.6 Tilting-Pad Journal Bearings

Tilting-pad journal bearings offer excellent stability and tolerance for misalignment. Extensive design curves are provided by Shapiro and Colsher (1970). Their design charts are presented here for three-pad journal bearings, with the load directed toward one of the pads.

Figure 11.18 defines the geometry of a three-pivoted-pad journal bearing with pivot film thickness values of h_{p1}, h_{p2}, and h_{p3}. Parameters α and β denote the pad angle and angle from the load vector to the pivot, respectively. Two 'clearances' are involved: one is the machined clearance C, and the other is the pivot clearance C'. The ratio C'/C, the so-called preload factor, plays an important role in bearing performance and stability.

Complete design charts with different compressibility numbers, and with the load vector directed either toward a pad or between the two pads, are available from Shapiro and Colsher (1970, 1976). These authors also provide design charts for spring-loaded pads that may be needed to accommodate thermal expansion of the shaft.

This chapter considers a pivoted-pad journal bearing with rigidly supported pads and with load directed toward a pad. Figure 11.19 shows the variation of the bearing coefficient $C_{LT} = W/p_a RL$ versus bearing eccentricity ratio $\varepsilon' = e'/C'$ for $R/D = 0.5$ and $\Lambda = 1.5$. For a given C_{LT} and C'/C, one can predict the eccentricity ratio ε' and the dimensionless film thickness at the pivot points. As shown, $H_{p,1} = H_{p,2}$ for all cases.

Steady-State Performance

Steady-state performance can be conveniently evaluated using the charts presented (Shapiro and Colsher, 1970; Khonsari *et al.*, 1994). The design procedure is illustrated in the following example.

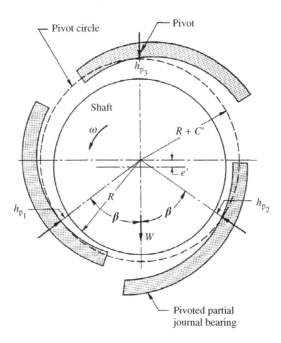

Figure 11.18 Nomenclature of a pivoted-pad journal bearing. *Source*: Shapiro and Colsher, 1970. Reproduced with permission of John Wiley & Sons.

Example 11.6   Design a miniature cryogenic bearing lubricated with nitrogen gas for a turbocharger with the following specifications:

$$R = 0.25 \text{ in. } (6.35 \times 10^{-3}\text{m}); \qquad p_a = 15 \text{ psia } (103 \text{ kp}_a); \qquad N = 200\,000 \text{ rpm};$$
$$T = 703\,°\text{F } (373\,°\text{C}); \qquad W = 2.5 \text{ lb}(11 \text{ N});$$
$$\mu = 1.01 \times 10^{-9} \text{ reyns } (6.97 \times 10^{-6} \text{ Pa.s})$$

1. Since bearing length and clearance are not specified, both are left to the designer's discretion. Typically, the start-up load under its own weight should be limited to $W/DL \le 3$ psi, while $W/DL \le 10$ psi is recommended for maximum loading under normal operating conditions. Roll stability should also be considered. As a guide, keep $L/D < 1.5$.

$$\frac{W}{DL} \le 10 \Rightarrow L_{\min} = 0.5 \text{ in. } \quad (1.27 \times 10^{-2}\text{m})$$

   and choose $L/D = 1$.
2. The bearing clearance must be established. Since the compressibility number cannot be predicted, begin by choosing compressibility number $\Lambda = 1.5$ and compute 'machined-in' clearance C:

$$\Omega = \frac{N\pi}{30} = 2.0944$$

$$C = R\sqrt{\frac{6\mu\Omega}{p_a\Lambda}} = 5.937 \times 10^{-4} \text{ in. } \quad (1.508 \times 10^{-5} \text{ m})$$

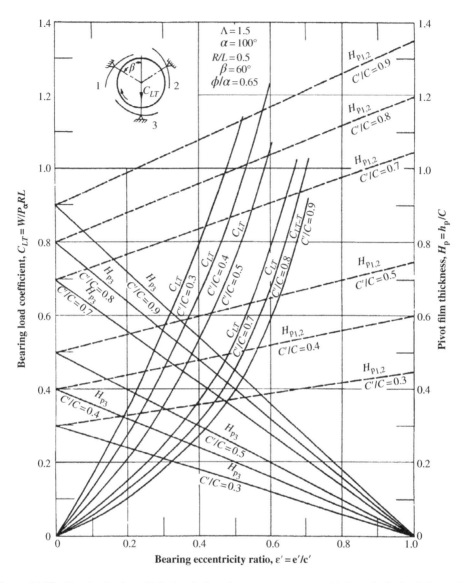

Figure 11.19 Bearing load coefficient and pivot clearance versus eccentricity ratio for load directed at a pad with all pads rigidly supported ($\Lambda = 1.5, R/L = 0.5$). *Source*: Shapiro and Colsher, 1970. Reproduced with permission of John Wiley & Sons.

3. Compute the load coefficient:

$$C_{\text{LT}} = \frac{W}{p_a RL} = 1.333$$

4. Use Figure 11.19 with this value of C_{LT} and tabulate the results in relating eccentricity and pivot film thickness for selected values of preload factors C'/C.

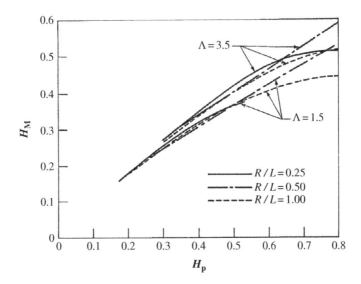

Figure 11.20 Pad minimum film thickness as a function of pivot clearance. *Source*: Shapiro and Colsher, 1970. Reproduced with permission of John Wiley & Sons.

(a) Evaluate the bearing eccentricity ratio ε'. Move horizontally to intersect a selected C'/C curve and read ε' at the intersection point. For example, for $C'/C = 0.3, \varepsilon' = 0.575$ (extrapolation may be needed).

(b) Evaluate pivot film thickness. At the established eccentricity and appropriate C'/C curves (= 0.3 in this example), determine the dimensionless film thickness values:

$$H_{p1} = H_{p2} = 0.39 \quad \text{(dashed lines)}; \qquad H_{p3} = 0.12 \quad \text{(solid lines)}$$

(c) Evaluate the dimensional film thicknesses by multiplying H_{pi} by C; i.e. $h_{pi} = C \times H_{pi}$ for each pad.

5. Use Figure 11.20 to determine the minimum film thickness for each pad. In this example, pad 3 has the smallest minimum film thickness at $H_M \cong 0.1$. Dimensionalize the smallest minimum film thickness:

$$H_M = CH_M = 0.59 \times 10^{-4} \text{ in.} \quad (1.27 \times 10^{-6} \text{ m})$$

6. Determine the pad load coefficient using Figure 11.21 and pad coefficient of friction $F = F_f/p_a CL$ for each pad by utilizing Figure 11.22:
with $H_{p_1} = H_{p_2} = 0.39 \Rightarrow F_1 = F_2 = 0.95$; $\quad F_3 = 1.7 \quad$ for $\quad H_{P_3} = 0.12$ Solve for friction force on each pad and sum them to get the total:

$$\sum_{i=1}^{3} F_{fi} = p_a CL \sum_{i=1}^{3} F_i = p_a CL(2F_1 + F_3) = 1.73 \times 10^{-2}$$

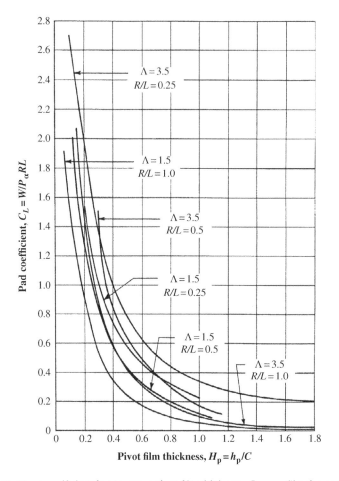

Figure 11.21 Pad load coefficient factor versus pivot film thickness. *Source*: Shapiro and Colsher, 1970. Reproduced with permission of John Wiley & Sons.

7. Compute the frictional power loss:

$$E_{\mathrm{P}} = \frac{746RN \sum\limits_{i=1}^{3} F_{\mathrm{fi}}}{63\,000} = 10.2\,W \quad (= 1.37 \times 10^{-2}\,\mathrm{hp})$$

8. Repeats steps 4 through 7 for various preload factors, tabulate the results, and plot the dimensional minimum film thickness h_{M} versus the preload factor C'/C, as well as the power loss P_f, as a function of the preload factor to determine the optimum preload factor.

9. A small C'/C value (tight film thickness) results in a large power loss. A large C'/C value, on the other hand, may cause vibration due to excessive pad loading and is susceptible to pad lock-up at the *leading edge*. Therefore, the recommended range for the preload factor is $0.3 \le C'/C \le 0.8$. For this problem, minimum film thickness approaches a limiting maximum value approximately at $C'/C = 0.5$, which also gives a reasonable power loss.

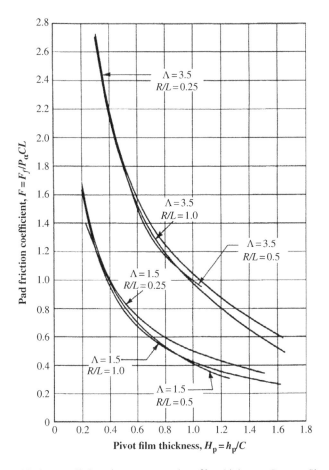

Figure 11.22 Pad friction coefficient factor versus pivot film thickness. *Source*: Shapiro and Colsher, 1970. Reproduced with permission of John Wiley & Sons.

11.7 Foil Gas Bearings

The application of foil bearings in gas turbine engines has been identified as a revolutionary concept: a practical approach to realize a significant improvement in speed, efficiency, and reliability. Properly designed foil bearings can offer significant advantages over conventional bearings, ranging from operational safety since they do not require external pressurization, considerable weight reduction due to elimination of lubricant supply arrangements, and substantial savings with reduction in maintenance costs (Heshmat *et al.*, 1982). In addition, air is an environmentally benign lubricant, capable of accommodating operation at elevated temperatures. In contrast, conventional oil-lubricated bearings are susceptible to failure because with increasing temperature, the oil viscosity drops exponentially, thus reducing the film thickness to a dangerously low level.

Historically, the use of foil gas bearings has been in lightly loaded applications ranging from the magnetic disk drive read/write heads in PCs to the air cycle machines (ACMs) in aircraft. However, new generations of foil journal bearings have opened many opportunities for advanced rotating machinery with their improvement in load-carrying capacity through soft metal foil coatings (Heshmat and Shapiro, 1984) and advances in compliant foil structure (Heshmat, 1994). In fact, an advanced design of a foil bearing achieved a breakthrough load-carrying capacity of $670\,000\,p_a$ (97 psi) at 59 700 rpm over a decade ago (Heshmat, 1994). Researchers at NASA Glenn Research Center have exercised extraordinary leadership in taking this technology to the next level (Dellacorte, 1997).

In the early 1980s, major advances were made by Heshmat *et al.* (1983a,b) on the characterization of the behavior of bump-type foil bearings – both journal and thrust types. These considered foil compliance and the coupling of elasticity with the hydrodynamics of flow – akin to elastohydrodynamic analysis. Since then, several approaches have been developed by different authors to predict the load performance of foil bearings. Dellacorte and Valco (2000) estimated foil bearing load-carrying capacity by an empirical formula based on first principles and experimental data available in the literature. The effects of different parameters, including compliance, nominal radial clearance, bearing speed, etc., on bearing performance have also been reported (Heshmat, 1994; Faria and San Andres, 2000).

Experimental and theoretical investigations have shown that unique structural stiffness and damping characteristics of foil bearings enable better stability than conventional rigid gas bearings. Both Ku and Heshmat (1992) and Heshmat (1994) have reported the modeling of structural damping and stiffness characteristics of the bump foils. Peng and Carpino (1993) and Carpino and Talmage (2006) formulated appropriate equations for determining dynamic (four stiffness and four damping) coefficients.

However, the overall elastohydrodynamic behavior of foil bearings is still not well understood, due in part to their structural and hydrodynamic nonlinearity (Dellacorte and Valco, 2000). While full elastohydrodynamic analyses that couple the hydrodynamics of compressible air flow with elasticity of the foil structure are rare, a noteworthy publication on this type is available on a thrust foil bearing (Heshmat *et al.*, 2000).

A schematic of a foil journal bearing and its compliant structure is shown in Figure 11.23. It is comprised of a cylindrical shell (sleeve) lined with corrugated bumps (bump foil) topped with a flat top foil. In this particular configuration, the leading edge of the top foil is attached to the bearing sleeve and the bump foils support the rest of the top foil. The top and bump foils are typically made from a nickel-based superalloy coated with soft films that abrade easily during operation to accommodate nonoptimum foil geometry. For high-temperature operations, foils are not coated, but rather the shaft is treated with high-temperature solid lubricant coatings (DellaCorte, 2000).

Coupling between Hydrodynamics and Structure

A foil gas bearing is self-acting with air drawn between the top foil and the shaft to completely separate them when the 'lift-off' speed is exceeded. With an air film maintained between the rotating and stationary surfaces, adequate pressure is generated to support the applied load. The bump foil supports the top foil and its compliance allows the top foil to deform under the action of hydrodynamic pressure. Therefore, successful operation relies on a coupling between a thin film of lubricating gas and an elastic structure. Modeling and simulation of foil bearing are very complex since conventional bearing parameters such as clearance and eccentricity ratio do not directly apply. Additional complications arise because of large displacements and

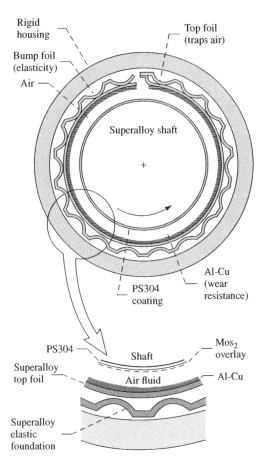

Figure 11.23 Geometry and nomenclature of a foil bearing. *Source*: DellaCorte *et al.*, 2004. Reproduced with permission of ASME.

the contribution of friction, both between the bumps and the top foil as well as between the bumps and support structure. Without these considerations, models tend to overestimate the dynamic stiffness (Le Lez *et al.*, 2007).

Analysis

One of the many challenges in numerical analysis of foil journal bearings lies in modeling the operational film thickness and the boundary conditions that establish the *volume of air flowing in and out of a foil bearing*. Determination of this film thickness requires a solution coupling the air flow hydrodynamics with the bearing structural compliance.

Breakthroughs in this type of analysis were made by a number of researchers such as Heshmat and his colleagues (1983a, 2000), Carpino and Peng (1991), Carpino and Talmage (2003), etc. Recently, Peng and Khonsari (2004a) have devised a model for characterizing the foil bearing film thicknesses and the interrelated hydrodynamic performance such as eccentricities and attitude angles. Analysis of the limiting load-carrying capacity of foil bearings under high speeds reported by Peng and Khonsari (2004b) has been extended by Kim and San Andres (2006) to include dynamic analysis.

Assuming that the ideal gas law holds, the dimensionless compressible Reynolds equation governing the hydrodynamic pressure takes on the following form:

$$\frac{\partial}{\partial \theta}\left(\overline{P}\,\overline{h}^3\frac{\partial \overline{P}}{\partial \theta}\right) + \left(\frac{D}{L}\right)^2 \frac{\partial}{\partial \overline{y}}\left(\overline{P}\,\overline{h}^3\frac{\partial \overline{P}}{\partial \overline{y}}\right) = \Lambda\frac{\partial}{\partial \theta}(\overline{P}\,\overline{h}) \tag{11.12}$$

where

$$\overline{y} = \frac{y}{L/2}, \qquad \overline{P} = \frac{P}{p_a}, \qquad \overline{h} = \frac{h}{C}, \quad \text{and} \quad \Lambda = \frac{6\mu_0\omega}{p_a}\left(\frac{R}{C}\right)^2$$

Equation (11.12) assumes the shaft is perfectly aligned within its sleeve and takes advantage of the axial symmetry.

The dimensionless film thickness as a function of pressure is given by (Heshmat *et al.*, 1983a)

$$\overline{h} = 1 + \varepsilon_f \cos \theta + \alpha(\overline{P} - 1) \tag{11.13}$$

where ε_f is the eccentricity ratio of a foil bearing and third term on the right-hand side represents the change in the film thickness due to the deformation of the bump foil with pressure acting on it. The parameter α is given by the following expression:

$$\alpha = \frac{2p_a s}{CE}\left(\frac{l}{t_B}\right)^3(1 - v^2) \tag{11.14}$$

Referring to Figure 11.24, the parameters s, l, and t_B denote the bump foil pitch, the half length, and the thickness, respectively. The parameters E and v are the bump foil modulus of elasticity and Poisson's, respectively.

While always less than 1 for a rigid bearing, the eccentricity ratio, ε_f, can be greater than 1 in a foil bearing while still maintaining a positive minimum film thickness. This is so since hydrodynamic pressure deforms the top foil, enlarges the original clearance, and thus provides additional room for the shaft to move towards the minimum film thickness, $h_{\min}$.

Peng and Khonsari (2004a) developed a numerical algorithm that applies the finite difference method to Equation (11.12) along with Equations (11.13) and (11.14) to predict the hydrodynamic pressure and the eccentricity ratio ε_f, based on given minimum film thicknesses. The results of the load-carrying capacity are shown in Figure 11.25 with comparison to an

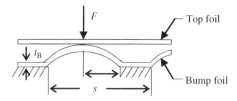

Figure 11.24 Geometry of bump foil.

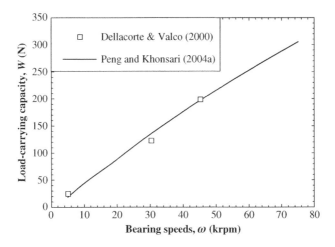

Figure 11.25 Foil bearing load-carrying capacity as a function of shaft speed.

expression developed by Dellacorte and Valco (2000). These empirical and computational results establish that the load-carrying capacity increases linearly with increasing speed up to moderately high speeds. There exists, however, a certain limiting speed after which the load-carrying capacity reaches a plateau as explored in the following section.

Limiting Speed Analysis

Based on the following analysis (Peng and Khonsari, 2004b), foil bearings tend to reach their limiting load-carrying capacity at much larger bearing speeds than do their rigid counterparts, signifying better performance at high speeds. With the dimensions shown in Table 11.5 for example, the limiting load-carrying capacity for the rigid bearing occurs at about 7×10^5 rpm, whereas the foil bearing reaches its plateau at speeds exceedingly 2×10^6 rpm.

Referring to Equation (11.12), as the shaft speed $\omega \to \infty$, $\Lambda \to \infty$. For the pressure field to remain bounded as $\Lambda \to \infty$, it is necessary that

$$\frac{\partial}{\partial \theta}(\overline{P}\,\overline{h}) \to 0 \tag{11.15}$$

Table 11.5 Data used in the simulations of a foil bearing

Shaft radius (D)	35×10^{-3} m
Diametrical clearance (C)	66×10^{-6} m
Bearing length (L)	62×10^{-3} m
Bump foil thickness (t_B)	0.076×10^{-3} m
Bump pitch (s)	3.17×10^{-3} m
Bump length ($2l$)	2.54×10^{-3} m
Young's modulus of bump foil (E)	$200 \times 10^9\ P_a$
Poisson's ratio of bump foil (v)	0.31

implying that $\overline{P}\,\overline{h}$ must be constant so that

$$\overline{P} = \frac{\widetilde{A}}{\overline{h}} \tag{11.16}$$

At $\theta = 0, \overline{P} = 1$ and $\widetilde{A} = 1 + \varepsilon$. The dimensionless pressure reads:

$$\overline{P} = \frac{1+\varepsilon}{\overline{h}} \tag{11.17}$$

Note that this result identically satisfies zero hydrodynamic pressure at $\varepsilon = 0$.

Therefore, when $\omega \to \infty$, the limiting pressure of a rigid gas journal bearing is simply

$$\overline{P} = \frac{1+\varepsilon_r}{1+\varepsilon_r \cos\theta} \tag{11.18}$$

and the maximum pressure is

$$\overline{P}_{max} = \frac{1+\varepsilon_r}{1-\varepsilon_r} \tag{11.19}$$

whose magnitude is a function only of eccentricity ratio ε for a rigid bearing.

The same expression should remain valid for the limiting pressure of a foil bearing with the appropriate film thickness of Equation (11.13). That is:

$$\overline{P} = \frac{1+\varepsilon_f}{1+\varepsilon_f \cos\theta + \alpha(\overline{P}-1)} \tag{11.20}$$

Solving Equation (11.20) for $\overline{P}$, we obtain

$$\overline{P} = \frac{-(1+\varepsilon_f \cos\theta - \alpha) + \sqrt{(1+\varepsilon_f \cos\theta - \alpha)^2 + 4\alpha(1+\varepsilon_f)}}{2\alpha} \tag{11.21}$$

The corresponding limiting load-carrying capacity based on the infinitely long approximation (ILA) is determined using the following equation:

$$W = p_a LR \sqrt{\left(\int_0^{2\pi} \overline{P}\cos\theta d\theta\right)^2 + \left(\int_0^{2\pi} \overline{P}\sin\theta d\theta\right)^2} \tag{11.22}$$

Comparison between the limiting load-carrying capacity of a foil bearing and a similar rigid bearing is shown in Figure 11.26, based on the simulations of Peng and Khonsari (2004b). These results clearly indicate that foil bearings have a much greater load-carrying capacity than their rigid counterparts. Further, a foil bearing saturation speed for reaching the load plateau occurs at a much greater speed than that of a rigid bearing.

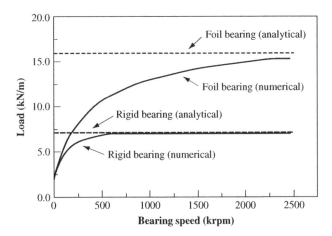

Figure 11.26 Limiting load-carrying capacity for a foil bearing and a similar sized rigid bearing. *Source*: Peng and Khonsari, 2004b. Reproduced with permission of ASME.

Problems

11.1 Formulate the Rayleigh step-bearing equations for a gas-lubricated bearing with inlet and outlet pressures maintained at p_a. The inlet and outlet film thicknesses are h_1 and h_2, respectively, which are constant over the lengths L_1 and L_2. Show that the pressure distribution can be determined from the following equation:

$$P + (6Kn^* + C_1)\ln(P - C_1) = \Lambda X + C_2$$

where C_1 and C_2 are integration constants to be determined from the boundary conditions for each region. Plot the pressure distribution for $Kn^* = 0, 0.05$, and 0.1 for $L_1/L_2 = 1$ and $h_1/h_2 = 2$.

11.2 A micromotor is constructed in the shape of step bearing with the following dimensions:

$$L_1 = 4\,\mu m; \qquad L_2 = 10\,\mu m; \qquad h_1 = 0.2\,\mu m; \qquad h_2 = 0.1\,\mu m$$

(a) The rotor weight is 0.3 nN. Determine the critical velocity needed for the rotor to fly.

(b) Suppose that rotor spins at 5 m/s. What is the minimum film gap?

11.3 Analyze the performance of a tilting-pad thrust bearing lubricated with air at room temperature carrying a load of 4000 N, $r_i = 4$ cm, $r_o = 8$ cm, and $\theta_P = 45°$ operating at 3600 and 36 000 rpm.

11.4 Determine the operating eccentricity, attitude angle, minimum film thickness, power loss, and critical whirl instability threshold for a journal bearing operating lubricated with krypton at 400 K with the following specifications:

$$L = D = 5\,cm; \qquad C = 0.0015\,cm; \qquad U = 10\,000\,rpm$$

Determine the friction moment if the shaft is subjected to a misalignment of $0.04°$.

11.5   Determine the maximum load that can be carried by a 1 in. diameter, 1 in. long journal bearing lubricated with argon at room temperature operating at 15 000 rpm. Let the radial clearance be $C = 300\,\mu$ in.

11.6   Design a tilting-pad journal bearing with three shoes to meet the following specifications:

$$D = L = 8\,\text{cm}; \quad N = 17\,000\,\text{rpm}; \qquad \mu = 0.018\,\text{cP};$$
$$C = 0.09\,\text{mm}; \quad W(\text{per pad}) = 150\,\text{N}$$

References

Ausman, J. 1959. 'Theory and Design of Self-Acting Gas-Lubricated Journal Bearings Including Misalignment Effects,' *Proceedings of the First International Symposium on Gas Lubricated Bearings*, Report ONR, ACR-49. US Government Printing Office, Washington, DC, p. 161.

Bhushan, B. 1996. *Tribology and Mechanics of Magnetic Storage Devices*, 2nd edn. Springer-Verlag, New York.

Bhushan, B. 1999. *Principles and Applications of Tribology*. Wiley Interscience, New York.

Burgdorfer, A. 1959. 'The Influence of the Molecular Mean Free Path on the Performance of Hydrodynamic Gas Lubricated Bearings,' *ASME Journal of Basic Engineering*, Vol. 81, pp. 41–100.

Carpino, M. and Peng, J.P. 1991. 'Theoretical Performance of Foil Journal Bearings,' *Proceedings, AIAA/SAE/ASME/ASEE 27th Joint Propulsion Conference*, AIAA-91-2105, Sacramento, calif.

Carpino, M. and Talmage, G. 2003. 'A Fully Coupled Finite Element Formulation for Elastically Supported Foil Journal Bearings,' *Tribology Transactions*, Vol. 46, pp. 560–565.

Carpino, M. and Talmage, G. 2006. 'Prediction of Rotor Dynamic Coefficients in Gas Lubricated Foil Journal Bearings with Corrugated Sub-Foils,' *STLE Tribology Transactions*, Vol. 49, pp. 400–409.

Cheng, H. and Pan, C. 1965. 'Stability Analysis of Gas-Lubricated Self-Acting Plain Cylindrical Bearing of Finite Length,' *Transactions of the ASME, Journal of Basic Engineering*, Vol. 87, p. 185.

Dai, F., Khonsari, M.M., and Liu, Z. 1994. 'On the Lubrication Mechanism of Grain Flows,' *STLE Tribology Transactions*, Vol. 53, pp. 516–524.

Dellacorte, C. 1997. 'A New Foil Air Bearing Test Rig for Use to 700 °C and 70,000 rpm,' NASA Technical Memorandum 107405.

DellaCorte, C. 2000. 'The Evaluation of a Modified Chrome Oxide Based High Temperature Solid Lubricant Coating for Foil Gas Bearing,' *Tribology Transactions*, Vol. 43, pp. 257–262.

DellaCorte, C., Antonio, D., Zaldana, R., and Radil, K.C. 2004, 'A Systems Approach to the Solid Lubrication of Foil Air Bearings for Oil-Free Turbomachinery,' *ASME Journal of Tribology*, Vol. 126, pp. 200–207.

DellaCorte, C. and Valco, M.J. 2000, 'Load Capacity Estimation of Foil Air Journal Bearings for Oil-Free Turbo-Machinery Applications,' *Tribology Transactions*, Vol. 43 (4), pp. 795–801.

Elkholy, K. and Khonsari, M.M. 2007. 'Granular Collision Lubrication: Experimental Investigation and Comparison with Theory,' *ASME Journal of Tribology*, Vol. *129*, pp. 923–932.

ElKholy, K. and Khonsari, M.M. 2008a. 'An Experimental Investigation of Stick-Slip Phenomenon in Granular Collision Lubrication,' *ASME Journal of Tribology*, Vol. *130*, pp. 021302-1, 021302-7.

Elkholy, K. and Khonsari, M.M. 2008b. 'On the Effect of Enduring Contact on the Flow and Thermal Characteristics in Powder Lubrication,' *Proceedings of the Institution of Mechanical Engineers, Journal of Engineering Tribology*, Vol. 222, pp. 741–759.

Eshel, A. 1970. 'On Controlling the Film Thickness in Self-Acting Foil Bearings,' *ASME Transactions, Journal of Lubrication Technology*, Vol. 92, p. 630.

Etsion, I. and Fleming, D. 1977. 'An Accurate Solution of the Gas Lubricated, Flat Sector Thrust Bearing,' *Transactions of the ASME, Journal of Lubrication Technology*, Vol. 99, p. 82.

Faria, M.T. and San Andres, L. 2000, 'On the Numerical Modeling of High-Speed Hydrodynamic Gas Bearing,' *Transactions of the ASME*, Vol. 122, pp. 124–130.

Fukui, S. and Kaneko, R. 1995. 'Molecular Gas Film Lubrication (MGL),' *Handbook of Micro/Nanotribology*, B. Bhushan (ed.). CRC Press, Boca Raton, Fla, pp. 559–604.

Fuller, D.D. 1984. *Theory and Practice of Lubrication for Engineers*. John Wiley & Sons, New York.

Gans, R.F. 1985. 'Lubrication Theory at Arbitrary Knudsen Number,' *ASME Journal of Tribology*, Vol. 107, pp. 431–433.

Gellman, A.J. 1998. 'Lubricants and Overcoats for Magnetic Storage Media,' *Current Opinion in Colloid and Interface Science*, Vol. 3, pp. 368–372.

Grassam, N. and Powell, J. (eds). 1964. *Gas Lubricated Bearings*. Butterworths, London.

Gross, W. 1980. *Fluid Film Lubrication*. Wiley Interscience, New York.

Heshmat, H. 1992. 'The Quasi-Hydrodyamic Mechanism of Powder Lubrication. Part 2. Lubricant Film Pressure Profile,' *Lubrication Engineering*, Vol. 48, pp. 373–383.

Heshmat, H., 1994, 'Advancements in the Performance of Aerodynamic Foil Journal Bearings: High Speed and Load Capacity,' *Journal of Tribology*, Vol. 116, pp. 287–295.

Heshmat, H. and Brewe, D.E. 1995. 'Performance of Powder-Lubricated Journal Bearings with MoS$_2$ Powder: Experimental Study of Thermal Phenomena,' *ASME Journal of Tribology*, Vol. 117, pp. 506–512.

Heshmat, H. and Shapiro, W. 1984. 'Advanced Development of Air-Lubricated Thrust Bearings,' *ASLE Lubrication Engineering*, Vol. 40, p. 21.

Heshmat, H., Shapiro, W., and Gray, S. 1982, 'Development of Foil Journal Bearings for High Load Capacity and High-Speed Whirl Stability,' *ASME Journal of Lubrication Technology*, Vol. 104 (2), pp. 149–156.

Heshmat H., Walowit J.A., and Pinkus O. 1983a. 'Analysis of Gas-Lubricated Foil Journal Bearings,' *ASME Journal of Lubrication Technology*, Vol. 105, pp. 647–655.

Heshmat H., Walowit J.A., and Pinkus O. 1983b. 'Analysis of Gas-Lubricated Compliant Thrust Bearings,' *Journal of Lubrication Technology*, Vol. 105, pp. 638–646.

Heshmat, C.A., Xu, D., and Heshmat, H. 2000. 'Analysis of Gas Lubricated Foil Thrust Bearings Using Coupled Finite Element and Finite Difference Methods,' *ASME Journal of Tribology*, Vol. 122, pp. 199–204.

Iordanoff, I., Elkholy, K. and Khonsari, M.M. 2008. 'Effect of Particle Size Dispersion on Granular Lubrication Regimes,' *Proceedings of the Institution of Mechanical Engineers, Journal of Engineering Tribology*, Vol. 222, pp. 725–739.

Jang, J.Y. and Khonsari, M.M. 2005. 'On the Granular Lubrication Theory,' *Proceedings of the Royal Society of London*, Series A, Vol. 461, pp. 3255–3278.

Khonsari, M.M. 2013. 'Granular Lubrication,' *Encyclopedia of Tribology*, Y. W. Chung and Q. J. Wang (editors), Springer Science, New York, pp. 1546–1549.

Khonsari, M.M., Lee, A.M., and Shapiro, W. 1994. 'Gas Bearings,' *CRC Handbook of Lubrication*, Vol. III, E.R. Booser (ed.). CRC Press, Boca Raton, Fla, pp. 553–575.

Kim, T.H. and San Andres, P. 2006. 'Limits for High speed Operation of Gas Foil Bearings,' *ASME Journal of Tribology*, Vol. 128, pp. 670–673.

Ku, C.-P. R. and Heshmat, H. 1992. 'Compliant Foil Bearing Structural Stiffness Analysis: Part I – Theoretical Model Including Strip and Variable Bump Foil Geometry,' *Journal of Tribology*, Vol. 114, pp. 394–400.

Le Lez, S., Arghir, M., and Frene, J. 2007. 'Static and Dynamic Characterization of a Bump-Type Foil Bearing Structure,' *ASME Journal of Tribology*, Vol. 129, pp. 75–83.

McKeague K. and Khonsari, M.M. 1996a. 'An Analysis of Powder Lubricated Slider Bearings,' *ASME Journal of Tribology*, Vol. 118, pp. 206–214.

McKeague, K. and Khonsari, M.M. 1996b. 'Generalized Boundary Interaction for Powder Lubricated Couette Flows,' *ASME Journal of Tribology*, Vol. 118, pp. 580–588.

Mate, C.M. and Homola, A.M. 1997. 'Molecular Tribology of Disk Drives,' *Micro Nanotribology and Its Applications*, B. Bhushan (ed.). Kluwer Academic Publishers, Netherlands, pp. 647–661.

O'Connor, J.J. and Boyd, J. 1968. *Standard Handbook of Lubrication Engineering*. McGraw-Hill Book Co., New York.

Pappur, M. Vol. and Khonsari, M.M. 2003. 'Flow Characterization and Performance of Powder Lubricated Slider Bearings,' *ASME Journal of Tribology*, Vol. 125 (1), pp. 135–144.

Patterson, G.N. 1956. *Molecular Flow of Gases*. John Wiley & Sons, New York.

Peng, J.P. and Carpino, M. 2003. 'Calculations of stiffaess and Damping Coefficients for Elastically Supported Gas Foil Bearings,' *ASME Journal of Tribology*, Vol. 115, pp. 20–27.

Peng, Z-C. and Khonsari, M.M. 2004a. 'Hydrodynamic Analysis of Compliant Foil Bearing with Compressible Air Flow,' *ASME Journal of Tribology*, Vol. 126, pp. 542–546.

Peng, Z-C. and Khonsari, M.M. 2004b. 'On the Limiting Load Carrying Capacity of Foil Bearings,' *ASME Journal of Tribology*, Vol. 126, pp. 117–118.

Raimondi, A.A. 1961. 'A Numerical Solution for the Gas Lubricated Full Journal Bearing of Finite Length,' *Transactions of the ASLE*, Vol. 4, p. 131.

Rice, J. 1965. 'Misalignment Torque of Hydrodynamic Gas-Lubricated Journal Bearings,' *Transactions of the ASME, Journal of Basic Engineering*, Vol. 87, p. 193.

Rowe, W.B. 1983. *Hydrostatic and Hybrid Bearing Design*. Butterworths, Ltd, London.

Sereny, A. and Castelli, V. 1979. 'Experimental Investigation of Slider Gas Bearings with Ultra-Thin Films,' *Transactions of the ASME, Journal of Lubrication Technology*, Vol. 101, p. 510.

Shapiro, W. and Colsher, R. 1968. 'Analysis and Performance of the Gas-Lubricated Tilting Pad Thrust Bearing,' Report I-A2049-30. Franklin Institute Research Laboratories, Philadelphia.

Shapiro, W. and Colsher, R. 1970. 'Analysis and Design of Gas-Lubricated Tilting Pad Journal Bearings for Miniature Cryogenic Turbomachinery,' Report AFFOL-TR-70-99. Air Force Flight Dynamics Laboratory, Wright-Patterson Air Force Base, Ohio.

Shapiro, W. and Colsher, R. 1976. 'Computer-Aided Design of Gas-Lubricated, Tilting-Pad Journal Bearings,' *Computer-Aided Design of Bearings and Seals*. American Society of Mechanical Engineers, New York, p. 67.

Tang, I.C. and Gross, W. 1962. 'Analysis and Design of Externally Pressurized Gas Bearings,' *ASLE Transactions*, Vol. 5, p. 261.

Wilcock, D. (ed.). 1972. *Gas Bearing Design Manual*. Mechanical Technology, Inc., Latham, NY.

Wilcock, D. 1984. 'Properties of Gases,' *CRC Handbook of Lubrication*, Vol. II, E.R. Booser (ed.). CRC Press, Boca Raton, Fla, pp. 290–300.

Wilcock, D. 1997. 'Gas Properties,' *Tribology Data Handbook*, E.R. Booser (ed.). CRC Press, Boca Raton, Fla, pp. 169–176.

Zhou, L. and Khonsari, M.M. 2000. 'Flow Characteristics of a Powder Lubricant,' *ASME Journal of Tribology*, Vol. 122, pp. 147–155.

12

Dry and Starved Bearings

While previous chapters have dealt primarily with bearings supplied with a full film of lubricant, millions of bearings operate dry or with limited lubrication at low loads and low speeds. A variety of plastics, porous metals, and specially selected materials provide good performance under conditions of poor or nonexistent lubrication where at least some of the load is carried by solid-to-solid contact during sliding.

Operating characteristics under dry and starved lubrication unfortunately cannot be defined easily by clear physical relations and formal mathematical expressions. Instead, empirical correlations based on model tests, laboratory evaluations, and field experience are commonly used. Final applications should normally be made only after prototype testing or careful comparison with past experiences under very similar conditions.

Guidelines for selecting and applying dry and semilubricated bearings follow. Coverage is then given to starved lubrication. While the focus of this chapter is primarily on journal bearings, corresponding principles will provide performance characteristics of thrust bearings when operating with less than full lubrication.

12.1 Dry and Semilubricated Bearings

Plastics, porous bronze and porous iron, carbon-graphite, rubber, and wood are widely used for bearings operating dry or with boundary lubrication (Booser, 1992; Khonsari and Booser, 2005a). With the availability of new composite materials and broadening experience, use of this class of bearings has been rapidly growing for design simplicity and to avoid the higher cost of ball bearings or fluid-film lubrication. Supplementary property information and application limits for these types of bearing materials are given in Chapter 4.

Plastics

Most commercial plastics find some use, both dry and lubricated, in slow-speed bearings and at light loads. The most commonly used thermoplastics are PTFE, nylon, and acetal resins. Thermosetting resins used for bearings include phenolics, polyesters, and polyimides. Table 12.1

Applied Tribology: Bearing Design and Lubrication, Third Edition. Michael M. Khonsari and E. Richard Booser.
© 2017 John Wiley & Sons Ltd. Published 2017 by John Wiley & Sons Ltd.
Companion Website: www.wiley.com/go/Khonsari/Applied_Tribology_Bearing_Design_and_Lubrication_3rd_Edition

Table 12.1 Typical operating limits for nonmetallic bearings

Material	Maximum temperature (°C)	PV limit (MN/(m s)	Maximum pressure, $P(MN/m^2)^a$	Maximum speed, V (m/s)
Thermoplastics				
Nylon	90	0.90	5	3
Filled	150	0.46	10	—
Acetal	100	0.10	5	3
Filled	—	0.28	—	—
PTFE	250	0.04	3.4	0.3
Filled	250	0.53	17	5
Fabric	—	0.88	400	0.8
Polycarbonate	105	0.03	7	5
Polyurethane	120	—	—	—
Polysulfone	160	—	—	—
Thermosetting				
Phenolics	120	0.18	41	13
Filled	160	0.53	—	—
Polyimides	260	4	—	8
Filled	260	5	—	8
Others				
Carbon-graphite	400	0.53	4.1	13
Wood	70	0.42	14	10
Rubber	65	—	0.3	20

^{a} To convert MN/m^2 to psi, multiply by 145.
Source: Kennedy *et al.*, 2005. Reproduced with permission of Taylor & Francis.

compares characteristic limits for the use of plastic bearings with those of carbon-graphite, wood, and rubber, which are used in similar applications.

The four operating limits listed in Table 12.1 are normally used to help define the useful range of operation of plastics: (1) maximum temperature above which physical deterioration or abnormal softening is to be expected; (2) maximum load at low speed, which reflects the compressive yield strength; (3) a PV load–speed limit at intermediate speeds; and (4) maximum surface speed when running at light loads, probably limited by melting and other local distortions at asperities. With a given coefficient of friction, this PV is the product of unit load P (MPa, psi) multiplied by surface velocity V (m/s, ft/min), which gives a measure of surface frictional heating and temperature rise. P is simply the total force divided by the bearing area for a thrust bearing or any two flat surfaces. For a journal bearing, P is taken as the load on the bearing divided by the projected area: journal diameter D× length of the bearing L. A further example of the relation of temperature rise to PV follows in Equation (12.3) in the discussion of porous metal bearings.

PV values also give a measure of wear rate. Many tests have shown that the total volume of material worn away is approximately proportional to the normal load and the distance traveled (see Archard Equation 4.4). This gives the relation for wear depth

$$W_d = k(PV)t \qquad (12.1)$$

Table 12.2 Wear factors for plastic bearings[a]

Material	Wear factor, k (10^{-15} m^3/N m)		Kinetic friction coefficient	
	No filler	Filled[b]	No filler	Filled[b]
Nylon 6,6	4.0	0.24[c]	0.61	0.18[c]
PTFE	400	0.14[d]	0.05	0.09[d]
Acetal resin	1.3	0.49	0.21	0.34
Polycarbonate	50	3.6	0.38	0.22
Polyester	4.2	1.8	0.25	0.27
Poly(phenylene oxide)	60	4.6	0.39	0.27
Polysulfone	30	3.2	0.37	0.22
Polyurethane	6.8	3.6	0.37	0.34

[a] See Blanchet, 1997 and Polymer Corp., 1992.
[b] With 30 wt.% glass fiber unless otherwise noted.
[c] Twenty percent PTFE.
[d] Fifteen percent glass fiber.

where t is the operating time in seconds. Representative values of wear factor k are given in Table 12.2 for operation with low-carbon steel (generally Rc hardness of 20 with 12–16 μ in. surface finish) at room temperature.

Example 12.1 How much radial wear will be experienced by a nylon bushing supporting a 10 mm shaft running at 900 rpm (0.47 m/s) under 0.5×10^6 N/m^2 (70 psi) load?

The PV of 0.235×10^6 N/m^2.m/s (6510 psi fpm) and $k = 0.24 \times 10^{-15}$ m^2/N for filled nylon in Table 12.2 gives a wear rate of

$$\dot{W}_d = kPV = (0.24 \times 10^{-15})(0.235 \times 10^6)(0.47) = 5.64 \times 10^{-11} \text{ m/s}$$

In 1000 h, wear depth then becomes

$$(1000 \cdot 60 \cdot 60) \text{ s times } 5.64 \times 10^{-11} = 0.00020 \text{ m } (0.20 \text{ mm})$$

Increased ambient temperature reduces the tolerable temperature rise related to the PV. Generally, the PV limit drops by 70% as ambient temperature approaches the limiting temperatures in Table 12.1 (Polymer Corp., 1992). Shaft hardness above about 50 Rc usually provides minimum wear. Wear rate increases with soft stainless steel by a factor of up to 5 and by up to 20- to 50-fold with soft aluminum alloys (Theberge, 1970). Shaft surface roughness in the 12–16 μ in. range is usually optimum; adhesion appears to promote higher wear at a finer finish; abrasion appears to promote more wear for rougher surfaces.

Since wear rates calculated from k values represent only very broad estimates, prototype tests are highly desirable for any planned application with no more than moderate acceleration by employing somewhat higher than expected loads and surface speeds. As a broad generalization for unlubricated conditions near the maximum allowable PV, an estimated wear rate of approximately 0.020–0.040 in./1000 h of running can occur with thermoplastic bearings, compared to 0.080–0.160 with acetal (Polymer Corp., 1992).

Added fillers can reduce wear factor k by a factor of 10–1000 or more (Blanchet, 1997). Common fillers include inorganic powders such as clay, glass fibers, graphite, molybdenum disulfide, and powdered metal. Silicone fluid at a concentration of about 2% is also beneficial as an internally available lubricant for reducing the friction and wear of plastics. The wear rate of polymer composites is strongly dependent on the loading level of the filler (Blanchet, 1997). Dependence of composite wear rate k_c on matrix volume fraction x_m and filler volume fraction x_f is typically related by an inverse rule of mixtures:

$$\frac{1}{k_c} = \frac{x_m}{k_m} + \frac{x_f}{k_f} \qquad (12.2)$$

where k_m and k_f are the wear factors of the matrix and filler, respectively. For a given system, values of k_m and k_f can be determined from Equation (12.2) from wear data at any two different loading levels (one of which may be the unfilled case). At high filler volume fractions ($x_f >$ 0.35), particulates may not be well dispersed. Increasing weak particle-to-particle interfaces may eventually result in an increased wear rate with increasing volume fraction, deviating from the pattern predicted by Equation (12.2).

Porous Metal Bearings

Porous metal bearings made of compressed and sintered powders of bronze, iron, and aluminum alloys are very common because of their simplicity, self-contained lubrication, and low cost. Millions of these bearings for shaft sizes ranging from about 1.6 to 150 mm are produced daily for small electric motors, household appliances, machine tools, automotive accessories, farm and construction equipment, and business machines (Morgan, 1984; Cusano, 1994). Material standards for porous metal bearings are given by ASTM B328, B438, B439, B612, and B782 and in the standards of the Metal Powder Industries Federation (MPIF, 1991–1992).

Production consists of five primary steps (Cusano, 1994): (1) metal powders are uniformly blended, sometimes with a solid lubricant; (2) this blend is compacted in dies at room temperature under high pressure; (3) the compact is sintered below the melting point of the base metal to bond the metal particles metallurgically; (4) the sintered bearings are repressed with sizing tools for improved dimensional accuracy and surface finish; and (5) oil is introduced into the interconnecting pores.

Traditional powder metal bearings consist of 90% copper and 10% tin (Table 12.3). Porous iron bearings are used for lower cost, often with some copper or graphite added for higher load capacity at low speeds. Up to 40% added 90-10 bronze powder provides many of the characteristics of porous bronze bearings together with the lower cost of iron. Porous aluminum containing 3–5% copper, tin, and lead finds limited use because of its cooler operation, better conformability, and lower weight.

The main limitation of porous metal bearings is that they commonly operate with only boundary or mixed film lubrication. A typical pore volume of 20–30% is commonly impregnated with a VG 68 turbine petroleum oil, although the ISO viscosity grade ranges from 32 to 150. Figure 12.1 provides a general guide for the selection of lower-viscosity oil with increasing shaft surface speed. Synthetic oils such as those listed in Table 12.4 are employed to meet extreme temperature conditions and other special application requirements. To promote

Table 12.3 Operating limits for porous metal bearings

Porous metal	Nominal composition (wt%)	Pressure limit, P (MN/m^2)		Speed limit, V (m/s)	PV limit (MN/(m s))
		Static	Dynamic		
Bronze	Cu 90, Sn 10	59	28	6.1	1.8[a]
Iron		52	25	2.0	1.3
Iron–copper	Fe 90, Cu 10	140	28	1.1	1.4
Iron–copper–carbon	Fe 96, Cu 3, C 0.7	340	56	0.2	2.6
Bronze–iron	Fe 60, Cu 36, Sn 4	72	17	4.1	1.2
Aluminum		28	14	6.1	1.8

Note: To convert MN/m^2 to psi, multiply by 145.
[a] Approximately equivalent to 50 000 psi ft/min limit often quoted by US suppliers.
Source: Kennedy *et al.*, 2005. Reproduced with permission of Taylor & Francis.

oil-film formation, high porosity with its high oil content is used for higher speeds, often with an oil wick or grease providing a supplementary lubricant supply. Lower porosity with up to 3.5% added graphite is used for lower speeds and oscillating motion where oil-film formation is difficult.

When received from the manufacturer, a porous metal bearing is about 90% saturated with oil. If conditions favor hydrodynamic pressure generation in the loaded portion, oil recirculates from the loaded to the unloaded regions of the bearing, as shown in Figure 12.2. Oil is then lost by evaporation, creepage, and throw-off radially at the ends of the bearing. The extent of the oil film drops as the oil content in the pores decreases, as in the typical pattern shown in Figure 12.3, until asperity contact and finally boundary lubrication exists. Morgan (1984) suggests that before the oil content falls below 65%, it should be replenished by means of wicks, oil reservoirs, or oil cups.

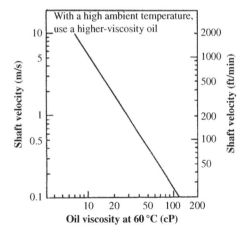

Figure 12.1 Guide to oil viscosity as a function of shaft surface speed. *Source*: Morgan, 1980. Reproduced with permission of Taylor & Francis.

Table 12.4   Advantages of synthetic lubricants for porous metal bearings

Base fluid	Temperature range	Advantages	Limitations
Polyalphaolefins	−40 °C to 179 °C (−40 °F to 355 °F)	Plastic and elastomer compatibility, low evaporation rates	Additive solubility
Diesters	−54 °C to 204 °C (−65 °F to 400 °F)	Detergency, price, good lubricating characteristics	Elastomer and plastic incompatibility
Polyolester	−57 °C to 279 °C (−70 °C to 535 °F)	Thermal stability, low evaporation rates	Higher price, elastomer and plastic incompatibility
Silicone	−73 °C to 279 °C (−100 °F to 535 °F)	Chemical resistance, high viscosity index	Poor lubricity, high price, generally replaced by low-viscosity synthetic hydrocarbons

Source: Cusano, 2007. Reproduced with permission of Taylor & Francis.

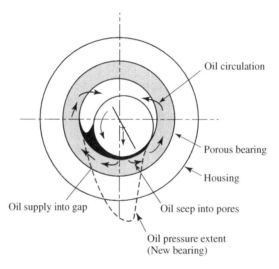

Figure 12.2   Mechanism of oil flow in a porous metal bearing. *Source*: Cusano, 1994. Reproduced with permission of Taylor & Francis.

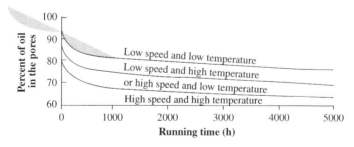

Figure 12.3   Typical oil loss curves for porous metal bearings. *Source*: Cusano, 1994. Reproduced with permission of Taylor & Francis.

Table 12.3 gives approximate operating limits for porous metal bearings. Maximum PV values for sleeve bearings range up to 1.8 MN/(m s)(50 000 psi ft/min). For long-term, continuous service and for thrust bearings, PV levels should generally not exceed about 20% of these values. Detailed property and application data are available from manufacturers of porous metal bearings, from the Metal Powder Industries Association, and from Cusano (1994, 1997).

The coefficient of friction with the usual boundary lubrication depends on load, speed, material and lubricant properties, and temperature (Cusano, 1994). Typical values range from 0.05 to 0.15. For operation near the PV limit, a coefficient of friction of about 0.1 is a reasonable value. Heat generated in the bearing is then proportional to this coefficient of friction f:

$$E_P = fWV; \qquad \frac{E_P}{A} = fPV \tag{12.3}$$

where E_P is the heat generated in watts, W is load (N), and V is the surface velocity of the shaft in m/s. This equation also gives the heat generated per unit of projected bearing area E_P/A (and the rise in bearing temperature T_b above its surrounding temperature T_s) as being generally proportional to the PV operating factor. That is

$$T_b - T_s = K_t(fPV) \tag{12.4}$$

where K_t is heat flow resistance to the surroundings. Morgan (1980) suggested performing experimental evaluation of K_t by using an electric heater to simulate bearing frictional heating while measuring bearing temperature with an embedded thermocouple. Values of K_t commonly fall in the range of 1.5×10^{-4} to 4×10^{-4}.

Operation under hydrodynamic conditions with the load fully supported by an oil film is not the usual case for porous metal bearings. Analyses for both copious and starved lubricant feed are summarized by Cusano (1994, 1997) while considering hydrodynamic action at the shaft–bearing interface together with permeability of the bearing material.

Porous metal bearings should be mounted with a press fit light enough to give acceptable ID close in. While low internal running clearance is more likely to cause problems of seizure, misalignment, and high temperature, internal clearances should be kept low for a low noise level and good bearing life. For small shafts up to about 0.5 in. in diameter, diametrical clearance is commonly held at 0.0003–0.0012 in. For larger shafts, the ratio of diametrical clearance to shaft diameter typically ranges from 0.0005 to 0.0015. Shafts should have a surface finish of 0.4 μm or better and an R_b hardness at least 30 above the hardness of particles making up the bearing. With a bronze bearing particle hardness of 65 R_b, shaft hardness should then be 95 or more (Cusano, 1997).

12.2 Partially Starved Oil-Film Bearings

While many bearings depend for normal operation on a relatively thick oil film for complete separation of moving and stationary surfaces, this frequently involves a smaller lubricant supply than is needed to be fully flooded. Oil is supplied to many such journal bearings by metered feed from a centralized system or by self-contained devices such as oil rings, chains, wicks, disks, pumping holes, and capillary feed. Disks lubricate some thrust bearings; wicks and rings are generally not used.

Table 12.5 Realms of fluid-film bearing operation

Realm	Supply methods	Typical applications	Approximate minimum oil feed rate, % of full-film flow rate
Full feed	Pressure fed, flooded	Turbines, compressors, superchargers	100
Partially starved	Ring-oiled, chain, disk, metered feed	Electric motors, pumps, fans, line shafts, miscellaneous large machines	10
Minimal feed	Wick, drop feed, mist, air-oil	Appliances, conveyors, pumps, various small and low-speed units	< 1

Three general realms are suggested in Table 12.5 for the degree of lubricant feed for fluid-film operation: (1) *full feed*, with all the lubricant being fed that can be accommodated for full hydrodynamic behavior; (2) *partially starved* to provide an incomplete hydrodynamic fluid film; and (3) *minimal feed* to create a fluid film similar in thickness to the surface roughness.

Full feed is the choice in high-speed rotating machinery such as turbines and compressors. When the lubricant supply to bearings in these machines drops below about 80–90% of full flow, trouble can be expected, with an excessive temperature rise and increased vibration accompanying the lower damping and other adverse dynamic film effects.

Satisfactory operation is often possible in lower-speed equipment with partially starved operation involving low power loss. This can be carried to an extreme, with minimal lubricant being supplied to give a film having a thickness no greater than several times the surface roughness of the bearing–shaft interface. This roughness is commonly defined by the surface finish of a harder journal when operating with a readily polished, softer bearing material such as babbitt.

These partially starved oil supply systems commonly offer simplicity and reliability, smaller lubricant supplies, and less need for auxiliary controls, instrumentation, and related lubrication system components.

Oil-Ring Bearings

Journal bearings lubricated with oil rings (Figure 12.4) have been used widely for over a century in electric motors, pumps, fans, and other medium-sized and large machines at surface speeds of up to about 10–14 m/s (33–46 ft/s) (Elwell, 1997). For lower cost and design simplicity, ring-oiled bearings are frequently replaced by ball and roller bearings for shaft diameters below about 75–100 mm (3–4 in.). Following are common design rules for oil-ring bearings:

- Bearing length/diameter ratio: $L/D = 0.6–1.1$ (based on net bearing working length)
- Ring bore/shaft diameter ratio: $D/d = 1.5–1.9$
- Clearance ratio (bearing bore–shaft journal diameter)/shaft journal diameter: $C_d/d = C/R = 0.0020–0.00275$
- Journal surface finish: commonly 0.4 μm (16 μ in.) CLA, 0.8 μm (32 μ in,) on slow-speed units.

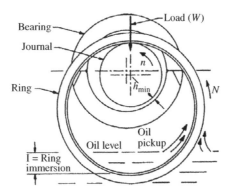

Figure 12.4 Typical elements of an oil-ring bearing. *Source*: Elwell, 1997. Reproduced with permission of Taylor & Francis.

Common oil-ring dimensions are indicated in Table 12.6. Trapezoidal rings in bores up to about 170 mm are often zinc die castings or are machined from brass or bronze tubing. Larger rings are machined from brass or bronze. T-rings, generally formed from rolled brass sheet, are used in large rings that can be split and hinged in two segments for convenient disassembly during bearing maintenance. Both of these cross sections minimize drag by the sides of the ring on the walls of the ring slot in the upper half of the bearing.

The typical bearing shown in Figure 12.5 is split horizontally for mounting in a split housing. The bearing casing then forms an oil sump beneath the bearing. Extending the babbitt from the bore to provide grooved thrust faces at one or both ends of the journal bearing will enable light

Table 12.6 Typical oil-ring dimensions

Typical trapezoidal ring dimensions (mm, in.)				
d	D	A	B	C
53–67 (2.1–2.6)	100 (3.94)	6 (0.24)	6 (0.24)	1 (0.04)
71–90 (2.8–3.5)	135 (5.31)	7 (0.28)	7 (0.28)	1 (0.04)
87–110 (3.4–4.3)	165 (6.50)	8 (0.31)	8 (0.31)	2 (0.08)
105–133 (4.1–5.2)	200 (7.87)	16 (0.63)	13 (0.51)	2 (0.08)

Typical T-ring dimensions (mm, in.)					
d	D	E	F	G	H
105–133 (4.1–5.2)	200 (7.87)	16 (0.63)	5 (0.20)	11 (0.43)	10 (0.39)
176–223 (6.9–8.8)	335 (13.2)	21 (0.83)	6 (0.24)	13 (0.51)	14 (0.55)
308–390 (12.1–15.4)	585 (23.0)	25 (1.0)	7 (0.28)	14 (0.55)	20 (0.79)
437–553 (17.2–21.8)	830 (32.7)	32 (1.3)	8 (0.31)	16 (0.63)	27 (1.1)

Source: Elwell, 1997. Reproduced with permission of Taylor & Francis.

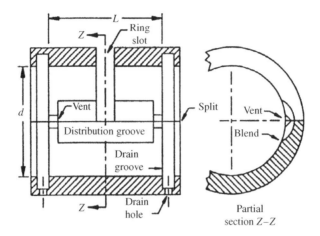

Figure 12.5 Cross section of a typical oil-ring bearing. *Source*: Elwell, 1997. Reproduced with permission of Taylor & Francis.

axial loads to be carried. The outboard ends of the oil distribution grooves are vented to release any accumulated air, which might otherwise block the small oil flows along the grooves.

Ring Speed and Oil Delivery

The ring, rotated by its contact at the top of the rotating shaft, picks up a sheath of oil from the lower sump. Oil wiped from the bore of the ring onto the shaft at the top of its travel is then carried to the downside oil distribution groove to form the bearing oil film.

General patterns of ring speed and oil delivery with increasing shaft speed are illustrated in Figure 12.6. At low shaft speeds with 'no-slip' operation, the driving friction of the shaft keeps the ring surface velocity equal to that of the shaft. Ring speed (revolutions/second, N) is then related to shaft rps (n) and their corresponding diameters D and d as follows:

$$N = \frac{n \cdot d}{D} \qquad (12.5)$$

This no-slip range ends at N_S when viscous drag on the portion of the ring immersed in the oil reservoir beneath exceeds the driving force from the rotating journal, which experimentally was found to be (Lemmon and Booser, 1960)

$$N_S = \frac{0.00048 \times W}{\mu D^2} \qquad (12.6)$$

where the slip rps is proportional to the ring weight W (lb), and is lower with higher oil viscosity μ (lb s/in.^{2}) and larger ring diameter D (in.). This relation held for both trapezoidal and T-rings, with a normal immersion of 15% of the ring diameter. The no-slip speed limit can be increased several fold by grooving the internal surface of the ring. This grooving correspondingly raises the ring speed and oil delivery in large, low-speed machinery. Greater immersion of the ring in the oil reservoir increases viscous drag and lowers N_S.

Following a transition with the ring being driven by a partial oil film on the shaft, full-film drive is reached when ring speed N reaches about 2.9 times N_S. Once full-film drive

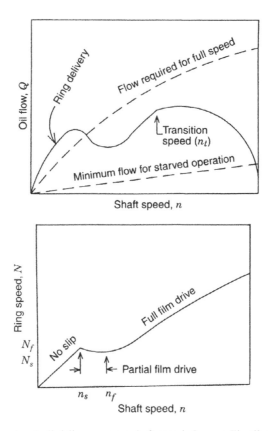

Figure 12.6 Ring speed and oil delivery versus shaft speed. *Source*: Elwell, 1997. Reproduced with permission of Taylor & Francis.

is established, ring speed N (rps) reflects a balance between viscous drive and viscous drag forces in the following relation. In this region, ring speed becomes relatively independent of geometric ring grooving or other geometric features that are useful with no-slip or partial-film operation for increasing driving force on the ring:

$$N = \frac{1.67 \times v^{0.2}(nd^2)^{0.8}}{D^2} \tag{12.7}$$

where v is kinematic viscosity of the oil (in.^{2}/s). This ring speed drops significantly when the rotating ring contacts the wall on either side of the ring slot as it gradually oscillates back and forth during normal operation.

Oil rings in most small machines operate in the full-film region, with the ring rpm being of the order of one-tenth of shaft rpm. No-slip or partial-film drive of rings may be expected in large, lower-speed units such as steel rolling-mill drive motors and ship line shaft bearings.

Oil delivery Q (in.^{3}/s) from a ring for most of its low and normal operating speeds is given by the following equation (Lemmon and Booser, 1960):

$$Q = 0.14Av^{0.65}(DN)^{1.5} \tag{12.8}$$

where A (in.) is the axial width at the bore of the ring from which oil is delivered to the shaft. This has the same general form as the following theoretical maximum lift of fluid on a rising surface, which reflects a balance of viscous lift and gravity forces (Deryagin and Levi, 1964):

$$Q = \frac{2}{3}A(U)^{1.5}\left(\frac{\mu}{g}\sin\theta\right)^{0.5} \tag{12.9}$$

where U is velocity of the rising surface (in./s), μ is fluid viscosity (lb. s/in.^{2}), g is acceleration of gravity (386 in./s^2), and θ is the angle of the rising surface above horizontal.

As indicated in Figure 12.6, a transition speed is eventually reached with increasing ring speed beyond which oil flow no longer increases according to the pattern of Equation (12.8). The probable cause of this decreased oil delivery is increased centrifugal action, which first reduces availability to the journal of oil exuded between the ring and the journal. Eventually, centrifugal throw-off from the ring and increased windage from shaft rotation reduce oil flow to unusable quantities. This generally occurs when the journal surface velocity approaches about 16 m/s (46 ft/s).

Example 12.2   Calculate the oil delivery to a 5 in. diameter, 5 in. long ($D = L = 5$ in.) journal bearing at 1500 rpm fed with ISO VG-68 oil with an oil sump temperature of 131 °F (55 °C) by a 200 mm (7.87 in.) bore bronze oil ring with dimensions taken from Table 12.6 (weight = 2.4 lb). Determine the speed at which ring slip is likely to occur and the corresponding oil delivery rate.

With oil viscosity of 4.2×10^{-6} lb s/in.^{2} (35 cSt), Equation (12.6) gives the following top ring speed N_S for operation at the same surface speed as the shaft with no slip:

$$N_S = \frac{0.00048W}{(\mu D^2)} = \frac{0.00048 \times 2.4}{[(4.2 \times 10^{-6})(7.87)^2]} = 4.43\,\text{rps}$$

and the corresponding limiting shaft speed for no-slip operation follows from Equation (12.5):

$$n_S = \frac{N_S D}{d} = \frac{4.43 \times 7.87}{5} = 6.97\,\text{rps}$$

With the shaft speed of $1500/60 = 25$ rps being more than 2.9 times the slip speed limit of 6.97 rps, ring speed N is given by Equation (12.6) for full-film operation. With oil density of 0.0307 lb/in.^{3} and kinematic viscosity $v = 35$ cSt [$35 \times (1.55 \times 10^{-3}) = 0.0543$ in.^{2}/s], the ring speed is

$$N = \frac{1.67 \times v^{0.2}(nd^2)^{0.8}}{D^2}$$

$$= \frac{1.67(0.0543)^{0.2}[(1500/60)(5)^2]^{0.8}}{(7.87)^{0.2}} = 2.60\,\text{rps}$$

Oil delivery given by Equation (12.8) then becomes the fluid transferred to the shaft from the bore of the ring whose axial width A in Table 12.6 is 0.63 in.

$$Q = 0.14Av^{0.65}(DN)^{1.5}$$

$$= 0.14(0.63)(0.0543)^{0.65}(7.87 \times 2.60)^{1.5} = 1.23\,\text{in.}^3/\text{s}$$

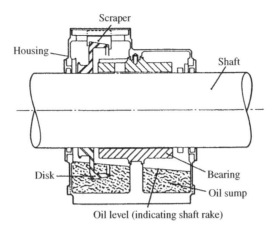

Figure 12.7 Disk lubricated marine line shaft bearing assembly. *Source*: Gardner, 1977. Reproduced with permission of ASME.

Disk-Oiled Bearings

Disks usually require more axial space and involve significant power loss at higher speeds but offer the following advantages over oil rings: more positive oil supply at very low speeds, relatively large oil feed under some pressure at high speed, and better operation on tilted shafts in marine and earth-moving machinery. The disk can also serve as a thrust runner to operate with a substantial thrust bearing for axial loads (Elwell, 1997).

Disk oiling for a low-speed marine line shaft bearing is illustrated in Figure 12.7. Oil delivery from the scraper floating on top of the large disc ($D = 0.94$ m) to grooving in the bearing is proportional to the following disk flow parameter (*DFP*) (Gardner, 1977). This resembles Equation (12.8) for oil lift by a rising surface:

$$DFP = 1455 \times 10^9 \, B(Dn)^{1.5} \mu^{0.5} \qquad (12.10)$$

where B is the axial breadth (m) of the disk being scraped, n is rotational rps, and μ is viscosity (kg s/m^2). Gardner's experiments showed that oil delivery was proportional to *DFP* up to 2×10^9 where oil feed was 112×10^{-6} m^3/s (1.8 gpm). At high speeds, oil flow dropped below proportionality with *DFP* under the influence of centrifugal throw-off and windage effects. At *DFP* of 6×10^9, for instance, measured flow was about 75% of the theoretical level.

Centrifugal pumping disk lubrication, shown in Figure 12.8, has been developed for lubrication at higher speeds (Kaufman *et al.*, 1978). Oil metered through an orifice hole into the bottom of the hollow disk is carried by rotation to the top, where a scoop diverts the oil to a trough to feed the bearing. The shroud at the bottom minimizes power loss from churning by the disk. While centrifugal pumping action increases oil delivery with increasing speed, excess oil circulation at high speed with splashing and leakage is avoided by control of oil circulation with orificing oil feed to the bottom of the disk. At low speed in the absence of centrifugal pumping, a dam or scraper on or close to the outside disk surface can supply oil from the top of the disk in the manner shown in Figure 12.7.

The disk for supplying oil can also be completely enclosed in a close-fitting oil-circulating shroud, sometimes referred to as a *viscous pump*, which surrounds the disk. This disk may also

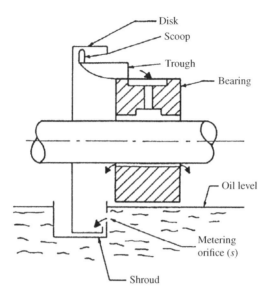

Figure 12.8 Centrifugal pumping disk assembly. *Source*: Elwell, 1997. Reproduced with permission of Taylor & Francis.

serve as a thrust collar for an axial bearing. After entering the viscous pump through a small opening at the bottom in the oil reservoir, oil is carried by rotation of the collar rim in a shallow circumferential pumping groove. At the end of the oil travel, oil velocity provides pressure at a dam in the pumping groove for sending the oil to either journal or thrust bearings.

12.3 Partially Starved Bearing Analysis

When lubricant feed is less than that needed for a full oil film, the circumferential extent of the film is reduced to the degree needed to match the inlet supply with the calculated oil leakage from the sides of the bearing. Charts for carrying out performance analyses under these starved conditions are given in Figures 12.9 and 12.10 (Connors, 1962). Although not readily adaptable for general cases, more detailed analyses can be made with current computer programs for simultaneous solution of the Reynolds equation and energy relations in the oil film for evaluating temperature effects with specific oils (Artiles and Heshmat, 1985).

The effect of a reduced oil supply on minimum film thickness h_{min} can be estimated from Figure 12.9. For a given bearing application, shaft radius R, radial clearance C, speed N, viscosity μ, and unit loading P are used to establish the dimensionless bearing characteristic number S (Sommerfeld number). The 'classical input flow' line gives both the minimum film thickness and the dimensionless feed rate for this value of S. Supplying greater amounts of oil to flood the bearing has essentially no effect, but reducing the feed rate below this classical value reduces minimum film thickness between the bearing and its journal with a lower value of h_{min}/C. This is the region where ring-oiled, disk, and other self-lubricated bearings typically operate.

The corresponding effect of reduced oil supply on friction and power loss is provided in Figure 12.10. For a given bearing Sommerfeld number S, the classical input flow line again

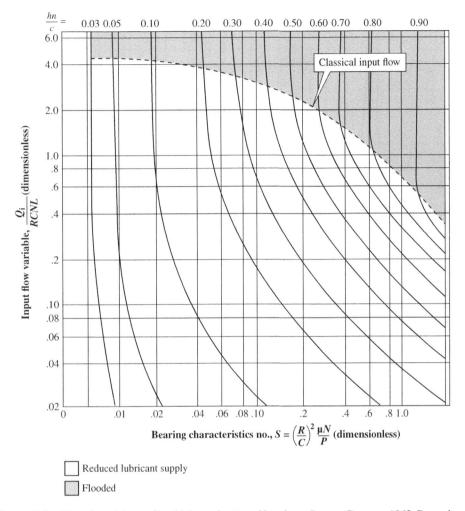

Figure 12.9 Chart for minimum film thickness in starved bearings. *Source*: Connors, 1962. Reproduced with permission of Taylor & Francis.

differentiates the operating regimes of flooded and reduced lubricant supplies. Feeding more oil than the classical input has no influence on the calculated power loss: the additional oil does not enter the hydrodynamic film. With a reduced oil supply, power loss often drops with the diminished circumferential extent of the oil film. With ring-oiled bearings in electric motors, for instance, bearing power loss may drop by half, as the ring supplies less than a full oil film. On the other hand, the influence of the reduced film extent may be more than matched by the rising friction introduced by the increased shear rate in a thinning oil film Booser and Khonsari (2011).

Example 12.3   For the ring-oiled 5 in. diameter ($R = 2.5$) bearing in Example 12.5 with a 5 in. axial length L, shaft rotational speed $n = 25$ rps, radial clearance $C = 0.005$ in., and radial load $P = 150$ psi, assume that 1.23 in.^{3}/s oil fed at 131 °F will rise to 170 °F in passing through the

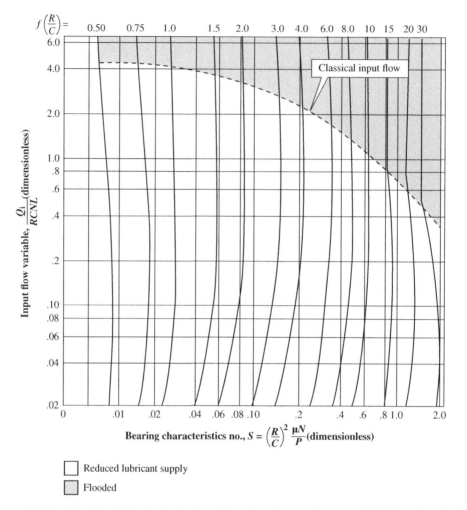

Figure 12.10 Chart for friction factor in starved bearings. *Source*: Connors, 1962. Reproduced with permission of Taylor & Francis.

bearing, with a corresponding viscosity μ at this bearing temperature of 1.95×10^{-6} lb s/in.^{2}. Determine power loss and oil film temperature rise.

Referring to Figure 12.10, bearing characteristic number is

$$S = \left(\frac{R}{C}\right)^2 \left(\frac{\mu n}{P}\right) = \left(\frac{2.5}{0.005}\right)^2 \left[\frac{(1.95 \times 10^{-6})(25)}{150}\right] = 0.081$$

Input flow variable is

$$\frac{Q_i}{RCnL} = \frac{1.23}{2.5 \times 0.005 \times 25 \times 5} = 0.79$$

This is 23% of the classical input flow (with $Q_i/RCnL = 3.5$ in Figure 12.10) needed to supply the bearing fully. This 23% should be adequate since it is more than twice the usual minimum of 10% required for satisfactory performance of ring-oiled bearings. (Two rings might be used for an additional safety margin.) From Figure 12.10:

$$f\left(\frac{R}{C}\right) \qquad = 1.90; \qquad f = 0.0038$$

Load W $\qquad = P \times \text{Area} = 150 \times (5 \times 5) = 3750\,\text{lb}$

Friction torque $\mathbf{T} = f \times W \times R = 0.0038 \times 3750 \times 2.5 = 35.6\,\text{lb in.}$

Power loss $E_p \qquad = 2\pi n\mathbf{T} = 2\pi(25)(35.6) = 5592\,\text{lb in.}/\text{s}(0.85\,\text{hp})$

With oil heat capacity of 4535 in. lb/°F lb(0.49 Btu/°F lb):

$$\text{Oil temperature rise} = \frac{\text{power loss}}{\text{oil flow} \times \text{density} \times \text{heat capacity}}$$

$$\Delta T = \frac{E_p}{Q_i \rho c_p}$$

$$= \frac{5592}{1.23 \times 0.0307 \times 4535} = 33\,°\text{F} \qquad (12.11)$$

Effective bearing temperature $T_b = T_{in} + \Delta T = 131 + 33 = 164\,°\text{F}$. Since this is a questionable match on the initially assumed 170 °F, a new assumption of temperature rise should be made and the calculation repeated until the calculated bearing temperature matches the initially assumed temperature. Figure 12.11 and Section 12.5 provide further guidelines for estimation of bearing operating temperature.

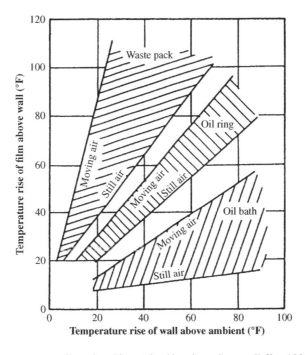

Figure 12.11 Temperature gradients in self-contained bearings. *Source*: Fuller, 1984. Reproduced with permission of John Wiley & Sons.

To obtain minimum oil film thickness with the above conditions, $S = 0.081$ and input flow variable of 0.79 in Figure 12.9 gives $h_{min}/C = 0.27$. From this, $h_{min} = 0.27 \times 0.005 = 0.00135$ in., which represents about 80% of the minimum film thickness for full oil feed.

12.4 Minimum Oil Supply

A limit in reducing the oil feed rate is reached for fluid-film lubrication when the film becomes so thin as to allow contact between asperities of the shaft and the bearing surface. This limit commonly represents a minimum film thickness in the general range of 3–10 times the rms surface roughness of the harder journal material: soft bearing material such as babbitt is quickly polished in operation.

Two examples drawn from experience illustrate how surprisingly little lubrication will serve with even sizable bearings at low to moderate speeds. A 16 in. journal bearing was run at 3600 rpm with a feed rate of VG 32 oil reduced to just that required to limit the babbitt bearing temperature to 250 °F following 15 min of operation under load. As the speed was reduced in steps, the oil feed was cut correspondingly. At 600 rpm, oil feed was eliminated in a final step without reaching the limiting temperature. In another case in which a finely finished 9 in. journal of a railway propulsion motor was run in its wick-oiled bearing at 600 rpm, operation continued for 1 h without failure after the lubricating wick was removed. While operation in these two examples of extreme starvation was unreliable at best, they demonstrate operation with a very limited oil supply at surface speeds up to about 7–10 m/s (23–33 ft/s).

Two factors may contribute to these low lubrication needs: (1) With the very low film thickness and circumferentially short oil film, very little axial flow takes place from the ends of the bearing. (2) Much of the small amount of oil displaced axially toward the ends of the bearing is then returned by air being sucked into the diverging journal-bearing clearance downstream of the load zone. With starvation, these effects are accompanied by gradual retreat of the oil film axial boundaries inward from the ends of the bearing, and the load support is provided in a pressurized 'puddle'. Essentially all oil-film action then occurs within the envelope of the journal-bearing clearance space, as suggested in Figure 12.12.

Based on reported experience, Fuller (1984) concludes that minimum inlet flow Q_m in.^{3}/s required by a journal bearing just to maintain fluid-film conditions is often given by the relation

$$Q_m = K_m U C_r L \tag{12.12}$$

where U is the surface velocity (in./s), C_r is radial clearance (in.), and K_m is a proportionality characteristic which depends on the bearing L/D, surface finish, and unit loading P in psi on the projected area. The following value was found to be representative of tests with a variety of speeds, loads, radial clearance, and oils with a 360° journal bearing:

$$K_m = 0.0043 + 0.0000185P \tag{12.13}$$

Similar values of 0.0043 and 0.0045 were deduced for K_m from other literature reports. Influence of surface finish, not indicated for these test results, would also directly influence K_m. These K_m values generally represent about 2–3% of the feed rate called for in Chapter 8 to form a full hydrodynamic film.

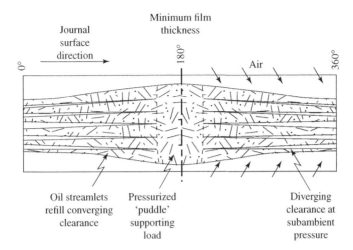

Figure 12.12 Unwrapped journal bearing oil film with minimal feed. Essentially no make-up oil is needed, as film action is confined within the bearing axial length.

Elastohydrodynamic lubrication effects may further reduce minimum oil feed requirements. Once significant elastic deflections are encountered in EHL film formation at low speeds and high loads, the thin oil film involved remains relatively uniform in thickness throughout the load zone so as to induce very limited oil loss from the ends of the bearing. The circumferential extent of the oil film, film thickness, and other details can then be developed from the EHL relations given in Chapter 15.

Friction coefficients in EHL contacts are commonly in the 0.03–0.06 range rather than approximately 0.001, as normally encountered with full oil feed to hydrodynamic journal and thrust bearings.

Whatever the mechanism, needed oil flow is surprisingly low at modest speeds, where power loss is insufficient to cause destructive temperature rise. Factors for drop feeding, wick oiling, and mist lubrication, which take advantage of these minimal requirements, are discussed below.

Drop-Feed Oiling

Drop-feed oilers are used in a variety of low-speed industrial applications such as pumps and mill equipment ranging up to fairly large sizes. Generally, 1 drop of oil in the viscosity range of SAE 10 to SAE 30 has a volume of $0.033\,\mathrm{cm}^3$ ($0.0020\,\mathrm{in.}^3$) and this drop size increases somewhat for feed rates above 60 drops/min (Fuller, 1984).

Wick Oiling

Wicks are commonly used for lifting oil less than about 5 cm (2 in.) in small machines such as electric motors, fans, pumps, and office machines; wicks can, however, raise oil up to 13 cm (5 in.) or more in applications like railway journal bearings. While earlier units used primarily wool felt (or waste packing in railroad bearings), synthetic fibers and injectable oil/fiber

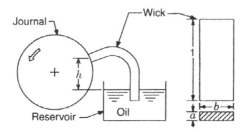

Figure 12.13 Bottom-fed wick lubricated bearing. *Source*: Elwell, 1997. Reproduced with permission of Taylor & Francis.

wicking mixtures have come into use. Although generally confined to surface speeds below 4 m/s (13 fps) by the inability of the low oil flows to carry away heat, surprisingly heavy loads can be carried with proper attention to surface finish and manufacturing details (Elwell, 1994).

Arranged as in Figure 12.13, mineral oil delivery $Q(cm^3/min)$ by a typical wick can be approximated by the following relation:

$$Q = \frac{k_w A F_0 (h_u - h)}{(\mu L)} \tag{12.14}$$

where A is the cross section area of the wick (cm^2) through which oil is carried for distance L(cm). F_0 is the volume fraction of oil in the saturated wick (often about 0.75); h_u is the ultimate wicking height, about 18 cm for SAE Grade F-1 felt; h is the height above the reservoir surface to which oil is to be delivered (cm); and μ is centipoise oil viscosity at the wick temperature. The empirical constant k_w, involving the capillary spacing in the wick and the oil surface tension, is approximately 4.6 for mineral oil with SAE Grade F-1 felt.

Comparison of wick oil delivery with full-film flow is of interest. For a railway traction motor wick 2.5 cm thick, 10 cm wide, and 14 cm long from the reservoir to the journal 4.5 cm above the reservoir oil level, Equation (12.14) gives the following wick delivery for oil of 58 centipoise viscosity at 40 °C oil temperature:

$$Q = \frac{4.6(2.5 \times 10)(0.75)(18)}{(58 \times 14)} = 1.91\,ml/min \quad (0.117\,in.^3/min)$$

Corresponding full hydrodynamic film flow of 1.4 gpm (320 in.^{3}/min) is approximately 2700 times this value from analysis in Chapter 8 for a 9.0 in. diameter traction motor bearing 4.75 in. in axial length at 700 rpm with a 5400 lb radial load.

This wick delivery is far less than the oil supplied by rings or disks, as discussed in the previous sections, and inlet flow with wicks is so low that the charts in Figures 12.9 and 12.10 are no longer applicable. Since severe oil starvation gives a circumferentially short oil film in which elastic deformations of the bearing and journal surfaces play a major role, minimum film thickness can be estimated from EHL theory covered in Chapter 15 for its more usual application with rolling bearings.

Developments in recent years have allowed closer integration of wick-oiled bearings with machines. Figure 12.14 shows an end shield for a fractional horsepower motor with a simple oil reservoir filled with either oil-soaked felt or an injectable oil–fiber wicking mixture. The porous metal wall limits passage of oil to the working bearing film.

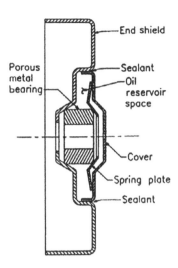

Figure 12.14 Small bearing in FHP motor end shield using an oil–fiber wicking mixture. *Source*: Elwell, 1994. Reproduced with permission of Taylor & Francis.

Performance characteristics of porous metal bearings were covered in Section 12.1. Further characteristics of wick lubrication for small bearings are given by Connors (1966).

Mist and Air–Oil Feed

While mist systems are predominantly used with rolling-element bearings, sleeve bearings can also be lubricated for relatively mild operating conditions (Khonsari and Booser, 2005b). Oil supplied as mist particles having diameters less than 6 or 7 μm (0.00024–0.00028 in.) are supplied in an airstream from a mist generator up to several hundred feet away (Bornarth, 1994). For air velocities of less than 24 ft/s, oil particles are carried smoothly in laminar flow through their supply line to be metered to each lubrication point. Condensing fittings are used at supply points for sliding bearings to reclassify the fine mist particles into larger drops for feeding into the bearings through grooving such as that shown in Figure 12.15.

Usual recommendations for mist feed to sleeve bearings with a standard density of about 0.5 in.^{3} of oil per hour per scfm (standard cubic feet per minute) of mist flow are as follows (Rieber, 1997):

$$\text{Moderate service:} \qquad \text{scfm} = LD/100 \, (\text{in.}^3 \, \text{oil/h} = 0.005LD)$$
$$\text{Heavy service:} \qquad \text{scfm} = LD/60 \, (\text{in.}^3 \, \text{oil/h} = 0.008LD) \qquad (12.15)$$
$$\text{Heavy service, high loss:} \quad \text{scfm} = LD/30 \, (\text{in.}^3 \, \text{oil/h} = 0.017LD)$$

where L is the bearing axial length in inches and D is shaft diameter in inches. Moderate service involves horizontal shafts with a unidirectional load, bearings in any position with oil retained by seals, or porous metal or other 'self-lubricated' bearing materials. Heavy service includes oscillating bearings, and unsealed bearings subject to shock load with a constantly shifting load zone; but boundary lubrication is possible, as in kingpins and spring pins on trucks. More feed is needed in the high-loss category with crank arm and crankshaft bearings and large

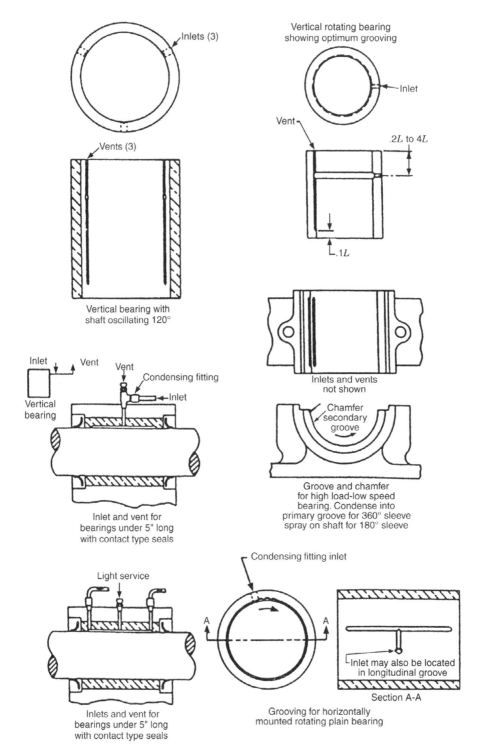

Figure 12.15 Mist lubrication of plain bearings. *Source*: Rieber, 1997. Reproduced with permission of Taylor & Francis.

bearings without seals on nonhorizontal shafts. Special considerations are needed in light of related experience or supplier information for surface speeds of over 600 ft/min.

Air–oil delivery systems are replacing oil mist in many applications to avoid environmental problems with escaping mist. This type of feed uses an airstream to move evenly spaced oil droplets to the bearing with a required quantity of oil $Q(\text{cm}^3/\text{h})$ (Simpson, 1997):

$$Q = 2 \times 10^{-4} LDk_m \qquad (12.16)$$

where L is the bearing axial length in millimeters, D is journal diameter in millimeters, and k_m is 1.0 at 100 rpm. Variation of k_m with speed N (rpm) modifies Equation (12.16) to the following:

$$Q = 5 \times 10^{-5} LDN^{0.3} \qquad (12.17)$$

With fluid grease, Q can generally be reduced 20–40%, assuming that the lubricating properties of the fluid grease are similar to those of oil.

These approximate delivery rates in Equations (12.16) and (12.17) are rather similar for oil mist, air–oil feed, and fluid grease. Each feed rate represents approximate hourly replacement of an oil layer 0.1 mm thick between a sleeve bearing and its journal, far less than the full-film feed analyzed in Chapter 8. Since the above rates are not related directly to bearing load and only marginally to bearing speed, the oil feed primarily serves only to replace minor end leakage while the oil film behaves in the pattern of Figure 12.12.

Grease Lubrication

Grease lubrication finds wide use for pins, bushings, and slow-speed sliding surfaces up to 10–20 ft/min. Grease provides a thicker lubricant film than oil and minimizes wear during boundary lubrication with slow speeds, shock loads, frequent stops and starts, and reversing direction (Glaeser and Dufrane, 1978). Oscillatory operation in heavily loaded construction equipment can be especially troublesome with a tendency for introducing wear particles, early galling and scuffing, sudden temperature rise, and seizure. Oscillation at very low amplitudes and high frequency can also restrict replenishment of grease into a contact to cause failure from fretting and false brinelling (Booser and Khonsari, 2007).

NLGI Grade 2 grease is typically employed with a base oil viscosity of 150–220 cSt at 40 °C. Higher–viscosity base oils, and extreme-pressure (EP) and solid additives are used for low speeds, high loads, and high temperatures. Higher temperatures increase the rate of oil loss from the grease structure by oxidation, creep, and evaporation. Figure 12.16 gives a recommended operating time for relubrication of sleeve bearings in farm and construction machinery as a function of bearing temperature. Above 10–20 ft/min, continuous feed systems using semifluid NLGI 00, 0, or 1 grades (see Table 2.11) can apply 20–40% less grease than the oil volume indicated in Equations (12.16) and (12.17) to coat the contact zone with a 0.07 mm grease layer hourly. Section 15.10 covers the generally different grease application and life factors for ball and roller bearings.

Example 12.4 Estimate the minimal oil feed rate to be used with a 3 in. long sleeve bearing on a 4 in. diameter shaft running under a steady radial load of 1000 lb at 150 rpm.

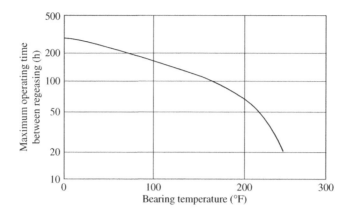

Figure 12.16 Guideline for sleeve bearing regreasing intervals with mild operating conditions. *Source*: Glaeser and Dufrane, 1978. Reproduced with permission of Machinery Lubrication.

Because this operation falls in the moderate service category, Equation (12.15) calls for $0.005LD = 0.005 \times 3 \times 4 = 0.06$ in.^{3}/h (0.98 cm^3/h) oil delivery as oil mist. Equation (12.17) gives

$$5 \times 10^{-5}LDN^{0.3} = 5 \times 10^{-5}(2.54 \times 3)(2.54 \times 4)(150) = 1.04\,\text{cm}^3/\text{h}$$

Hourly replacement of a layer 0.1 mm thick between the bearing bore area and the journal would correspond to 0.77 cm^3/h. The range from 0.77 to 1.04 cm^3/h probably represents the general uncertainty involved. A likely approach would be to set the input feed rate to a some-what higher value than 1.04 cm^3/h and then reduce the feed rate gradually if trouble develops with undesirable oil leakage to the surroundings, or raise the feed rate upon any evidence of mechanical or thermal distress.

12.5 Temperature of Starved Bearings

The characteristic temperature of partially starved bearings with self-contained lubrication can be related approximately to the housing wall temperature using Figure 12.11 (Fuller, 1984). With oil-bath and disk lubrication, the agitated oil is effective in promoting heat transfer and keeps temperature gradients much lower than with oil-ring or oil wicking in waste-packed bearings. These approximations suggest that the wall structure around the bearing (and the reservoir of oil beneath an oil-ring bearing) is about halfway between ambient and the bearing oil-film temperature for oil-ring bearings. Oil agitation with bath lubrication is very effective in transporting heat from the bearing to the housing wall, while the temperature rise is much higher with minimal lubrication in wick and waste-packed bearings.

In considering the factors involved, the calculated rise above the oil-feed temperature from a self-contained oil sump (commonly similar to the bearing wall temperature) at the end of Section 12.3 assumes that all frictional power loss in the bearing goes into heating the oil entering the bearing film. Adjustment should be made in the net power loss for heat passing to or from the bearing surroundings – for instance, for thermal conductance along the shaft from hotter machine elements or away to cooler surfaces. Since starved bearings operate at mild loadings and moderate speeds, a simple heat balance such as reflected in Equation (12.18) and Example 12.5 is usually sufficient to characterize bearing performance.

Temperature rise ΔT_h from ambient air to the surface of the bearing housing can be roughly approximated from the surface dissipation heat transfer factor $K_h(\text{Btu}/(h \times ft^2 \times° \text{ F})$ in the following relation:

$$\Delta T_h = \frac{E_P}{K_h \times A} \qquad (12.18)$$

where E_P is the power loss (Btu / h), K_h varies from 2.1 for still air to 5.9 for a velocity of 500 fpm. Housing surface area A commonly varies from 10 to 15 times the product of bearing length times diameter $(L \times D)$ for a simple pillow block bearing and up to 18–25 times $(L \times D)$ for pillow blocks with separate shells.

Example 12.5 Consider the 5 in. diameter ring-oil bearing of Example 12.3. Estimate the temperature rise.

Take the housing area as $25(L \times D) = 25(5 \times 5) = 625$ in.^{2} $(4.34$ ft$^2)$. Calculated power loss is $E_p = 0.85$ hp $(0.85 \times 2545 = 2163$ Btu/h$)$. The bearing housing surface temperature rise using Equation (12.18) is

$$\Delta T_h = \frac{2163}{5.9 \times 4.34} = 84\,°\text{F}$$

With ambient air at 70 °F, the temperature of the housing surface (and approximately the oil sump temperature) becomes $70 + 84 = 154$ °F. Since a lower sump temperature of only 131 °F was assumed in Example 12.3, the above calculation routine would be repeated while using an intermediate assumed sump temperature (such as 145 °F), and the process would be repeated until a satisfactory check is reached with the starting assumptions of sump and bearing temperatures.

Problems

12.1 What oil feed rate do you suggest for a sleeve bearing on a horizontal shaft at 200 rpm with length $L = 3$ in., diameter $D = 4$ in., diametral clearance $= 0.012$ in., and load $P = 150$ psi? Consider (a) a full classical hydrodynamic oil feed, (b) an oil ring supplying oil at 35% of the full flow rate, and (c) a centralized oil–air system. The bearing will operate at 122 °F (50 °C) with VG 68 oil. What rate of drop feed from a drop-feed oiler would be equivalent to the delivery from the centralized oil–air system?

12.2 Calculate the power loss for the bearing in Problem 12.1 with (a) full oil feed, (b) an oil ring supplying 35% of the full feed rate, and (c) a porous bronze bearing operating in boundary lubrication with oil fed by an air–oil centralized system.

12.3 Friction loss of 0.2 hp is encountered in a 2 in. long × 2 in. diameter sleeve bearing. If half of the heat loss is conducted out through a 0.030 in. thick liner of bearing material, calculate the temperature drop across the liner if the bearing material is tin babbitt, porous bronze, or nylon.

12.4   Determine the minimum cross section for a 6 in. long oil wick to deliver VG 68 mineral oil to a line shaft bearing 3 in. above the reservoir oil level where the oil is at 50 °C. Apply Equation (12.17) to establish the oil required for this 5 in. long by 4 in. diameter bearing running at 450 rpm.

12.5   A trapezoidal cross section oil ring is to be applied in an electric motor to supply VG 32 oil at 50 °C to a 2.5 in. long by 2.5 in. diameter sleeve bearing under 150 psi load at 1800 rpm with a 0.005 in. diametral clearance. Determine the full-flow hydrodynamic oil feed requirements of the bearing running at 60 °C and establish an appropriate geometry of an oil ring to supply 50% of full feed flow. Calculate the power loss and minimum oil-film thickness for the bearing.

12.6   Complete the analysis of the sump and bearing temperature in Example 12.5. Suggest three possible features to lower the bearing temperature. Approximate the temperature reduction in each case.

References

Artiles, A. and Heshmat, H. 1985. 'Analysis of Starved Journal Bearings Including Temperature and Cavitation Effects,' *Journal of Tribology, Transactions of the ASME*, Vol. 107(1), p. 1.

Blanchet, T.A. 1997. 'Friction, Wear, and PV Limits of Polymers and Their Composites,' *Tribology Data Handbook*. CRC Press, Boca Raton, Fla, pp. 547–562.

Booser, E.R. 1992. 'Bearing Materials,' *Encyclopedia of Chemical Technology*, 4th edn, Vol. 4. John Wiley & Sons, New York, pp. 1–21.

Booser, E.R. and Khonsari, M.M. 2007. 'Systematically Selecting the Best Grease for Equipment Reliability,' *Machinery Lubrication*, Jan–Feb, pp. 19–25.

Bornarth, D.M. 1994. 'Oil Mist Lubrication,' *Handbook of Tribology and Lubrication*, Vol. 3. CRC Press, Boca Raton, Fla, pp. 409–422.

Booser, E.R. and Khonsari, M.M. 2011. 'Should you Flood or Starve a Sleeve Bearing,' *Machine Design*, Vol. 83, pp. 40–44.

Connors, H.J. 1962. 'An Analysis of the Effect of Lubricant Supply Rate on the Performance of the 360° Journal Bearings,' *STLE Transactions*, Vol. 5(2), pp. 404–417.

Connors, H.J. 1966. 'Fundamentals of Wick Lubrication for Small Sleeve Bearings,' *Transactions of the ASLE*, Vol. 9(2), p. 299.

Cusano, C. 1994. 'Porous Metal Bearings,' *Handbook of Lubrication and Tribology*, Vol. 3. CRC Press, Boca Raton, Fla, pp. 491–513.

Cusano, C. 1997. 'Porous Metal Bearings,' *Tribology Data Handbook*. CRC Press, Boca Raton, Fla, pp. 719–733.

Deryagin, B.V. and Levi, S.M. 1964. *Film Coating Theory*. Focal Press, London.

Elwell, R.C. 1994. 'Self-Contained Bearing Lubrication: Rings, Disks, and Wicks,' *Handbook of Lubrication and Tribology*, Vol. 3. CRC Press, Boca Raton, Fla, pp. 515–533.

Elwell, R.C. 1997. 'Ring- and Wick-Oiled Starved Journal Bearings,' *Tribology Data Handbook*. CRC Press, Boca Raton, Fla, pp. 708–718.

Fuller, D.D. 1984. *Theory and Practice of Lubrication for Engineers*, 2nd edn. John Wiley & Sons, New York.

Gardner, W.W. 1977. 'Bearing Oil Delivery by Disk-Scraper Means,' *Journal of Lubrication Technology, Transactions of the ASME*, Vol. 99(2), p. 174.

Glaeser, W.A. and Dufrane, K.F. 1978. 'New Design Methods for Boundary-Lubricated Sleeve Bearings,' *Machine Design*, 6 April, pp. 207–213.

Kaufman, H.N., Szeri, A., and Raimondi, A.A. 1978. 'Performance of a Centrifugal Disk-Lubricated Bearing,' *Transactions of the STLE*, Vol. 21(4), p. 315.

Kennedy, F.E., Booser, E.R., and Wilcock, D.F. 2005. 'Tribology, Lubrication, and Bearing Design,' *The CRC Handbook of Mechanical Engineering*, 2nd edn, F. Kreith (ed.). CRC Press, Boca Raton, Fla, pp. 3.129–3.170.

Khonsari, M.M. and Booser, E.R. 2005a. 'A Dry Subject. These Bearings Don't Need Liquid Lubricants,' *Machine Design*, 6 Nov., pp. 96–102.

Khonsari, M.M. and Booser, E.R. 2005b. 'Guidelines for Oil Mist Lubrication.' *Machinery Lubrication*, Sept.–Oct., pp. 58–63.

Lemmon, D.C. and Booser, E.R. 1960. 'Bearing Oil-Ring Performance,' *Transactions of the ASME*, Series D, Vol. 82(2), pp. 327–334.

Metal Powder Industries Federation. 1991–1992. *Material Standards for P/M Self-Lubricating Bearings*. MPIF Standard 35. MPIF, Princeton, NJ.

Morgan, V.T. 1980. 'Self-Lubricating Bearings, 2. Porous-Metal Bearings,' *Engineering*, Vol. 220(8), Tech. File No. 80.

Morgan, V.T. 1984. 'Porous Metal Bearings and Their Application,' MEP-213. Mechanical Engineering Publications, Workington, UK.

Polymer Corp. 1992. *Plastic Bearing Design Manual*. Polymer Corp., Reading, Pa.

Rieber, S.C. 1997. 'Oil Mist Lubrication,' *Tribology Data Handbook*. CRC Press, Boca Raton, Fla, pp. 396–403.

Simpson, J.H., III. 1997. 'Centralized Lubrication for Industrial Machines,' *Tribology Data Handbook*. CRC Press, Boca Raton, Fla, pp. 385–395.

Theberge, J.E. 1970. 'A Guide to the Design of Plastic Gears and Bearings,' *Machine Design*, 5 February, pp. 114–120.

Part III

Rolling Element Bearings

13

Selecting Bearing Type and Size

Chapter 1 summarized the advantages and disadvantages of ball and roller bearings in comparison with those of fluid-film and sliding bearings. While performance and cost are usually the major considerations, other selection factors commonly include physical size, lubrication needs, reliability, and maintenance.

Rolling element bearings consist basically of two hardened steel rings, a set of hardened balls or rollers, and a separator. They are generally classified by the type of rolling element, i.e. ball, roller, or needle, and by whether they are designed primarily for radial or thrust loads. Figure 13.1 illustrates components of common types of rolling bearings, and the capabilities of different classes are characterized in Table 13.1.

The following sections describe some typical types of ball and roller bearings listed in Table 13.1, their functions, bearing nomenclature, and load–life relations. Details of the composition and properties of lubricants and bearing materials for rolling element bearings are presented in Chapters 2 and 4.

13.1 Ball Bearing Types

Deep-groove single-row, or Conrad, type shown at the upper-left-hand side of Figure 13.1 is low in cost and able to carry both radial and thrust loads. They are usually given first consideration in most new machine designs.

The number of balls is kept small enough that they can be bunched into slightly more than one-half of the circumference to allow their assembly between the inner and outer rings. During manufacture, the balls are distributed to give uniform spacing, and the separator is then installed to maintain this ball distribution.

The inner ring is usually fastened to the rotating shaft, with the groove on its outer diameter providing a circular ball raceway. The outer ring, incorporating a similar ball raceway groove in its bore, is mounted in the bearing housing. These raceway grooves usually have curvature radii of 51.5–53% of the ball diameter. Smaller raceway curvatures are generally avoided because of high rolling friction from the tight conformity of the balls and raceways and more restricted tolerance for misalignment. Higher curvatures undesirably shorten the fatigue life from increased stress in the smaller ball-race contact area. While the inner ring normally carries

Applied Tribology: Bearing Design and Lubrication, Third Edition. Michael M. Khonsari and E. Richard Booser.
© 2017 John Wiley & Sons Ltd. Published 2017 by John Wiley & Sons Ltd.
Companion Website: www.wiley.com/go/Khonsari/Applied_Tribology_Bearing_Design_and_Lubrication_3rd_Edition

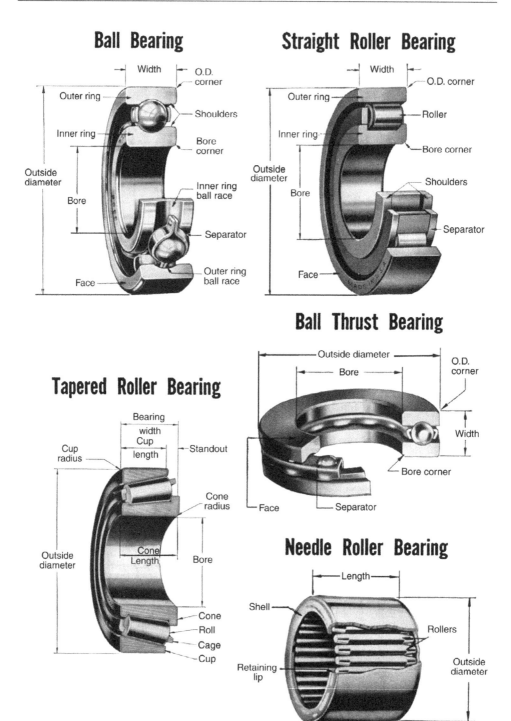

Figure 13.1 Components of common types of rolling element bearings. *Source*: Courtesy of ABMA, Manual 100.

Table 13.1 Relative performance of ball and roller bearing types

Type		Size range in inches		Average relative ratings				Available with			Dimensions	
				Capacity		Limiting speed	Permissible misalignment	Shields	Seals	Snap rings	Metric	Inch
		Bore	O.D.	Radial	Thrust							
Ball bearings	Conrad type	0.1181 to 41.7323	0.3750 to 55.1181	Good	Fair ↔	Conrad is basis for comparison 1.00	±0° 8′ Std. Radial Clearance. ±0° 12′ C3 Clear	X	X	X	X	X
	Maximum type	0.6693 to 4.3307	1.5748 to 8.4646	Excellent	Poor ↔	1.00	±0° 3′	X		X	X	
	Angular contact 15°/40°	0.3937 to 7.4803	1.0236 to 15.7480	Good	Good (15°) Excellent (40°) ←	$\frac{1.00}{0.70}$	±0° 2′				X	
	Angular contact 35°	0.3937 to 4.3307	1.1811 to 9.4488	Excellent	Good ←	0.70	0°				X	
	Self aligning	0.1969 to 4.7244	0.7480 to 4.7244	Fair	Fair ↔	1.00	±4°				X	
Cylindrical roller bearings	Separable inner ring nonlocating	0.4724 to 19.6850	1.2598 to 23.3465	Excellent	0	1.00	±0° 4′				X	
	Separable inner ring ore directions locating	0.4724 to 12.5984	1.2598 to 22.8346	Excellent	Poor ↓	1.00	±0° 4′				X	
	Self-contained two directions locating	0.4724 to 3.9370	1.4567 to 8.4646	Excellent	Poor ↔	1.00	±0° 4′				X	

(continued)

Table 13.1 (Continued)

Type			Size range in inches		Average relative ratings				Available with			Dimensions	
					Capacity		Limiting speed	Permissible misalignment	Shields	Seals	Snap rings	Metric	Inch
			Bore	O.D.	Radial	Thrust							
Tapered roller bearings	Separable		0.6205 to 6.0000	1.5700 to 10.0000	Good	Good ⟶	0.60	±0° 2′				X	X
	Self-aligning		0.9843 to 12.5984	2.0472 to 22.8346	Good	Fair ↕	0.50	±4°				X	
Spherical roller bearings	Self-aligning		.9843 to 35.4331	2.0472 to 46.4567	Excellent	Good ⟵	0.75	±1°				X	
Needle bearings	Complete bearing with or without locating rings and lubricating groove		0.2362 to 14.1732	0.6299 to 17.3228	Good	0	0.60	±0° 2′		X		X	X
	Drawn cup		0.1575 to 2.3622	0.3150 to 2.6772	Good	0	0.30	±0° 2′				X	X
Thrust bearings	Single direction ball grooved race		0.2540 to 46.4567	0.8130 to 57.0866	Poor	Excellent ↑	0.30	0°				X	X
	Single direction eye roller		1.1811 to 23.6220	1.8504 to 31.4960	0	Excellent ↑	0.20	0°				X	
	Self-aligning spherical roller		3.3622 to 14.1732	4.3307 to 22.0472	Poor	Excellent ↑	0.50	±3°				X	

the rotating element of a machine, an application may require that the inner ring be stationary, with the outer ring rotating.

Filling-notch, or maximum-capacity, ball bearings have a slot machined in the side wall of both the inner and outer ring grooves to permit insertion of additional balls for higher radial load capacity than is available with the Conrad type. The balls induce some increase in noise in passing over these filling notches. Because the filling slots disrupt the shoulders of the ball grooves, these bearings are not recommended for thrust load.

Angular-contact ball bearings are especially designed to carry a heavy thrust load in one direction. This is accomplished by providing a high shoulder on one side of the raceway, and the raceways are arranged so that the thrust load is transmitted from one raceway to the other under a certain contact angle. A bearing having a high contact angle will support a heavier thrust load. While 40° is common for high loads, bearings are made with lower contact angles such as 15° or 25°. Since these bearings have almost no raceway shoulder to carry thrust in the opposite direction, angular-contact bearings are usually mounted in pairs, with free end play removed or preloaded against each other to stiffen the assembly in the axial direction. Bearings may also be mounted in tandem for higher thrust load capacity.

Instrument and *miniature* ball bearings are small units which are usually fabricated from stainless steel, frequently under clean room conditions. Made in a variety of configurations in addition to the deep groove, these bearings are available in surprisingly small sizes with bore diameters down to the 1 mm range. The bore diameter of instrument bearings generally ranges from 5 to 15 mm; smaller sizes are frequently called *miniature bearings*. Ball raceway curvatures commonly fall in the 52–58% range.

Seals and shields are provided with many pregreased ball bearings to minimize the entrance of dirt and other contaminants. An integral seal generally consists of an elastomer partially encased in a steel or plastic carrier. The seal at each side face of the bearing is normally attached to the outer ring and rubs lightly on a land or recess at the outer edge of the inner ring. Since friction from the rubbing of the seal on the inner ring is often much greater than that within the ball assembly itself, use of seals is commonly restricted to smaller and low-speed bearings. Where important, low interference can be specified for the rubbing fit of the seal at the inner ring to avoid high friction and overheating.

To avoid the friction involved with a seal, many packaged-type bearings employ noncontacting shields that are usually similar in construction to seals but have a small clearance at their interface with the inner ring.

When filled about one-third full with a premium grease, sealed or shielded bearings provide long life and offer distinct simplicity in handling and mounting arrangements. For dirty and hot applications requiring regreasing, a grease cavity with a regreasing fitting can be designed for the open outboard side of the bearing. In such a case, a single integral shield can be used to protect its inboard side.

13.2 Roller Bearing Types

Roller bearings are usually employed for higher loads than can be accommodated with ball bearings.

Cylindrical roller bearings, with their cylindrical rollers running in cylindrical raceways, have quite low friction, high radial load capacity, and high-speed capability. The usual design is free to float axially, with roller guiding flanges on both sides of one ring and none on the other. This enables the bearing to accommodate thermal expansion of a rotor when used in combination with the fixed location of a ball bearing at the opposite end. If a guiding flange is added on one side of the opposing ring, thrust load can be supported in one direction, and an appended second flange is occasionally added for two-directional thrust capacity.

The rollers usually have a length not much greater than their diameter to minimize the tendency of the rollers to skew. The rollers are slightly crowned to prevent high edge stresses and to minimize the detrimental effects of slight misalignment.

Tapered roller bearings use rollers in the shape of truncated cones. In addition to unidirectional thrust capacity, the radial load capacity is similar to that of a cylindrical roller bearing of the same size. The contact angle in the majority of tapered roller bearings ranges from 10° to 16°, but a steeper contact angle of approximately 30° is used for higher thrust load capacity. With different nomenclature than for other bearing types, the outer ring is called a *cup* and the inner ring the *cone*.

Since single-row tapered roller bearings generally cannot accept a pure radial load and since their rings are separable, they are generally mounted in pairs and one bearing is adjusted against the other. In large bearings, two or four rows of tapered rollers are commonly combined in a single unit for heavy-duty applications such as those in rolling mills and railroad axle bearings. Because of the taper, a force component is carried on the larger roller face as it slides against its guide flange. While the face is generally slightly crowned to induce lubrication at this interface, the geometry involved in tapered roller–race contacts generally leads to higher friction, a lower limiting speed, and a need for more lubrication than with cylindrical roller bearings.

Spherical roller bearings typically employ two rows of barrel-shaped rollers running in two raceways ground on the inner ring and on a continuous spherical surface ground on the inner diameter of the outer ring. This spherical raceway enables operation with some misalignment. With barrel roller profiles that closely match the profiles of the raceways, spherical roller bearings are very robust and provide the highest load capacity of all rolling bearings in rolling mills, paper mills, stone crushers, and marine applications.

The geometry involved in the roller–raceway contacts leads to higher friction than with cylindrical roller bearings and corresponding limitations for high-speed operation. While significant thrust loads can be carried, a high ratio of thrust load to radial load leads to wear at roller–raceway contacts.

Needle roller bearings use elongated cylindrical rolling elements, needles of small diameter, for use in applications where radial space is limited. The diameter/length ratio for the needles varies between 1/2.5 and 1/10. Since the rollers cannot be guided accurately, their friction is relatively high and their use is commonly restricted to low speeds and oscillating motions. Needle bearings may be made with cages for improved retaining and guiding of the needles.

To conserve space, the needles may run directly on a hardened shaft, or an independent needle and cage assembly can be run directly between a hardened shaft and a hardened housing.

13.3 Thrust Bearing Types

Thrust capacity of ball and roller bearings of varying geometry can be extended beyond that of the more common series of ball and roller bearings discussed above by increasing the contact angle up to a maximum of 90° to the rotor axis. The final three illustrations in Table 13.1 show designs for much higher thrust load capability.

Roller thrust bearings require sliding within the roller–race contact to accommodate the surface speed variation with varying diameter across the contact zone. While the large amount of sliding with cylindrical thrust bearings limits their use to slow speeds, this sliding is reduced somewhat by using several short rollers in each cage pocket. This sliding is reduced with barrel-shaped rollers in the spherical roller thrust bearing or with tapered rollers, but only with a resulting force component being applied to sliding-type contacts between faces of the rollers and a raceway flange. While very high loading is possible with roller thrust bearings, lubrication, wear, and limiting speed require careful attention.

A ball thrust bearing consists of two grooved flat plates with a set of balls between them. These bearings also involve sliding action in the ball–race contacts, which is increased at high speeds by centrifugal force on the balls. As indicated in general terms in Table 13.1, a cylindrical roller thrust bearing is limited to about 20% of the speed of its radial bearing counterpart, a ball thrust bearing limited to 30% of the speed, and a spherical roller bearing limited to 50%.

Because sliding action in the contacts of these ball and roller thrust bearings tends to introduce wear, higher friction, and lubrication problems, the first preference for many thrust load conditions is given primarily to radial bearings such as angular contact ball bearings, and tapered and spherical roller bearings with their less demanding sliding motions. Nevertheless, spherical roller thrust bearings and the other types discussed above find use for very high thrust loads in heavy-duty worm gears, large vertical electric motors, ship propeller thrust blocks, large cranes, and heavy equipment engineering.

13.4 Nomenclature

Rolling bearings are usually designated by a three-part basic code that indicates their type of construction and dimensions in the following form:

[code for bearing type] [code for bearing cross-section] [code for bore size]

This basic designation is often supplemented by suffixes for specifying nonbasic features of internal clearance and internal design, external design features such as seals and shields, cage design, dimensional precision, vibration class, and heat stabilization for high-temperature use. Figure 13.2 gives an example of typical nomenclature. Catalogs from individual bearing manufacturers should be consulted for specific details ranging beyond this universal basic code.

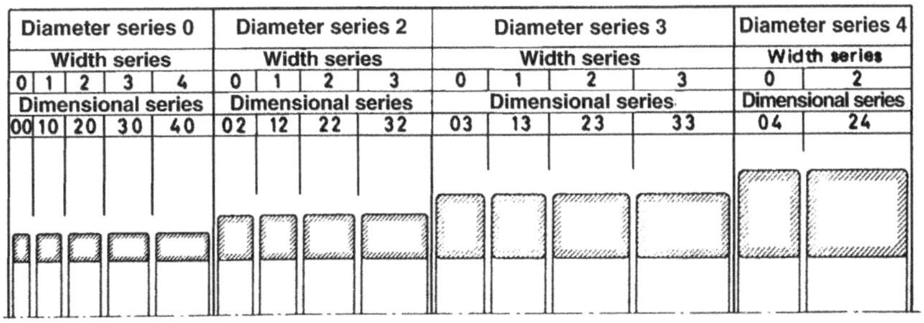

Figure 13.2 Example of diameter and width series for radial bearings. For 120 mm bearing bore from ANSI/ABMA Std. 20. *Source*: Eschmann, 1985. Reproduced with permission of John Wiley & Sons.

Bearing Type Code

While variations among different manufacturers and different countries exist, the following preliminary numbers or letters commonly indicate the bearing type:

1. Self-aligning ball bearing
2. Spherical roller bearing
3. Tapered roller bearing or double-row, angular-contact ball bearing
5. Thrust ball bearing
6. Deep-groove ball bearing
7. Single-row, angular-contact ball bearing
8. Cylindrical roller thrust bearing
N Line Cylindrical roller bearing or needle roller bearing
T Metric tapered roller bearings according to ISO 355.

This initial designation may be combined with several other letters to indicate construction details. For a cylindrical roller bearing, for instance, the letter N alone usually signifies two flanges on the inner ring with none on the outer ring, NU has two flanges on the outer ring with none on the inner, and NJ has two flanges on the outer ring with a single flange on one side of the inner ring to accommodate thrust in one direction.

Bearing Bore Code

Bore code designations are relatively universal. For all rolling bearings having a bore diameter from 20 to 480 mm, except needle, tapered roller, and double-acting thrust bearings, this is a standardized two-digit bore-code number: the bore diameter divided by 5. For example, the bore diameter of 30 mm divided by 5 gives the bore code number 06; the bore diameter of 280 mm gives bore code number 56.

For small-bore diameters from 10 to 17 mm, a 10 mm bore is designated by bore number 00, a 12 mm bore by 01, a 15 mm bore by 02, and a 17 mm bore by 03. Below 10 mm, the usual two-digit bore code is commonly replaced by one digit which gives the actual bore diameter in millimeters. A 624 bearing, for instance, has a 4 mm bore. As shown in the 'bearing type' code, the '6' indicates a single-row ball bearing. The '2' shows the small-diameter series, discussed in the next section.

For 500 mm and larger bores, the bore number is given by the actual diameter in millimeters preceded by a slash; for example, a 710 mm bore has the bore number /710. For smaller bore bearings from 32 mm down to 0.6 mm having bores intermediate between those in the original boundary dimension plan, the bore number also uses a slash before the actual diameter in millimeters. A 60/2.5 bearing, for instance, has a 2.5 mm bore diameter.

Bearing Cross Section Code

This portion of the bearing designation contains either one or two numbers in the pattern shown in Figure 13.2. The final number indicates the outside diameter relative to the bearing bore, 0 originally being the smallest and 4 the largest. Subsequent need for bearings with even smaller radial dimensions resulted in the present ISO expanded Diameter Series 7, 8, 9, 0, 1, 2, 3, and 4 (in order of ascending size). For general machinery applications, the '2' diameter series is often

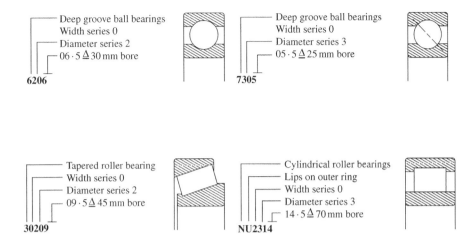

Figure 13.3   Example of nomenclature for rolling element bearings. *Source*: FAG Bearings Corp., 1995. Reproduced with permission of Schaeffler Group.

given first consideration. Larger cross-section bearings would then be considered primarily if higher load capacity were needed; '1' and even smaller diameter series would be considered for light loads where space is limited.

Within each diameter series, Width Series 8, 0, 1, 2, 3, 4, 5, 6, and 9 have been established in order of increasing width. A 0 width is common, and no designation is given in the bearing series code for this standard width. For any other width, the width series number is inserted directly before the diameter series number. This commonly gives the following sequence in a bearing designation: first, a number or letter giving the bearing type; second, a number for the section height (outside diameter) series; and finally, two digits for the bore size.

The American Bearing Manufacturers Association (ABMA) has standardized this dimensional plan covering width and diameter series. This provides a wide range of load-carrying capacities for a given size of shaft or housing in standard sizes of interchangeable radial bearings, as illustrated in Figure 13.2.

Examples of standardized nomenclature for ball and roller bearings are given in Figure 13.3. The number from the basic pattern just outlined is sometimes supplemented by added designations of nonstandard bearing features. Commonly added details in an appendix to the basic bearing code include the designation of seals or shields at the faces of the bearing and any off-standard 'C' clearance range.

Extensions of the Dimensional Plan

For approximations in extending the basic dimensional plan, ANSI/ABMA Std. 20 gives the following formulas for calculating approximate bearing dimensions. These values are commonly modified and rounded; bearing manufacturers' catalogs should be used as a reference for dimensional details of any specific bearing. The approximate bearing outside diameter D can be calculated from the bore diameter d by the formula:

$$D = d + f_D \, d^{0.9} \text{ mm} \qquad (13.1)$$

Values of factor f_D are:

Diameter Series	7	8	9	0	1	2	3	4
f_D	0.34	0.45	0.62	0.84	1.12	1.48	1.92	2.56

Width of the bearing ring B is calculated from the formula

$$B = \frac{f_B(D-d)}{2} \tag{13.2}$$

Values of factor f_B are:

Width Series	0	1	2	3	4	5	6	7
f_B	0.64	0.88	1.15	1.5	2	2.7	3.6	4.8

13.5 Boundary Dimensions

Outer dimensions of ball and roller bearings have been standardized on a worldwide basis both for simplicity in machine design and for obtaining spares and replacements. Although internal design details such as the number and size of rolling elements are not standardized, manufacturers have generally arrived at similar practices with similar load capacities for individual series of bearings with the same boundary dimensions.

Table 13.2 outlines dimensions abstracted from ISO 15 for common series of rolling element bearings. This dimensional plan is not restricted to specific designs. For example, a cylindrical roller bearing can be employed for a higher load capacity to replace a deep-groove ball bearing without the need to alter the mounting space. In making this change, all boundary dimensions of Table 13.2 and the tolerances on these dimensions in Table 13.3 would be unaltered.

Expanded versions of Table 13.2 are available from the American Bearing Manufacturers Association (1996), along with similar dimensional tabulations covering tapered roller bearings (ISO 355), and single-acting and double-acting thrust bearings (ISO 104). Bearing manufacturers' catalogs should also be consulted since up to six or more cross-sections may be available for one given bore diameter, rather than just the common cross-section series shown in Table 13.2.

13.6 Chamfer Dimensions

Figure 13.4 illustrates the corner dimensions of ball and roller bearing rings which must be accommodated at mating shaft and housing shoulders. While bearing rings should be closely fitted to these shoulders, they must not foul the shoulder filet radii. Consequently, the maximum fillet radius of the mating part must be smaller than the minimum corner r_{smin} listed in Table 13.2 for metric radial bearings having the basic dimensions given in the table. Radially, r_{smax} along the face of the bearing is usually about twice r_{smin}; r_{smax} in the axial direction

Table 13.2 Standard boundary dimensions for radial ball and roller bearings[a]

Bearing bore		19XX Series			01XX Series			02XX Series			03XX Series		
Bore No.	mm	OD	Width	$r_{s\,min}$	OD	Width	$r_{s\,min}$	OD	Width	$r_{s\,min}$	OD	Width	$r_{s\,min}$
5	5	13	4	0.2	14	5	0.2	16	5	0.3	19	6	0.3
6	6	15	5	0.2	17	6	0.3	19	6	0.3	22	7	0.3
7	7	17	5	0.3	19	6	0.3	22	7	0.3	26	9	0.3
8	8	19	6	0.3	22	7	0.3	24	8	0.3	28	9	0.3
9	9	20	6	0.3	24	7	0.3	26	8	0.3	30	10	0.6
00	10	22	6	0.3	26	8	0.3	30	9	0.6	35	11	0.6
01	12	24	6	0.3	28	8	0.3	32	10	0.6	37	12	1
02	15	28	7	0.3	32	9	0.3	35	11	0.6	42	13	1
03	17	30	7	0.3	35	10	0.3	40	12	0.6	47	14	1
04	20	37	9	0.3	42	12	0.6	47	14	1	52	15	1.1
05	25	42	9	0.3	47	12	0.6	52	15	1	62	17	1.1
06	30	47	9	0.3	55	13	1	62	16	1	72	19	1.1
07	35	55	10	0.6	62	14	1	72	17	1.1	80	21	1.5
08	40	62	12	0.6	68	15	1	80	18	1.1	90	23	1.5
09	45	68	12	0.6	75	16	1	85	19	1.1	100	25	1.5
10	50	72	12	0.6	80	16	1	90	20	1.1	110	27	2
11	55	80	13	1	90	18	1.1	100	21	1.5	120	29	2
12	60	85	13	1	95	18	1.1	110	22	1.5	130	31	2.1
13	65	90	13	1	100	18	1.1	120	23	1.5	140	33	2.1
14	70	100	16	1	110	20	1.1	125	24	1.5	150	35	2.1
15	75	105	16	1	115	20	1.1	130	25	1.5	160	37	2.1
16	80	110	16	1	125	22	1.1	140	26	2	170	39	2.1
17	85	120	18	1.1	130	22	1.1	150	28	2	180	41	3
18	90	125	18	1.1	140	24	1.5	160	30	2	190	43	3
19	95	130	18	1.1	145	24	1.5	170	32	2.1	200	45	3
20	100	140	20	1.1	150	24	1.5	180	34	2.1	215	47	3
21	105	145	20	1.1	160	26	2	190	36	2.1	225	49	3
22	110	150	20	1.1	170	28	2	200	38	2.1	240	50	3
24	120	165	22	1.1	180	28	2	215	40	2.1	260	55	3
26	130	180	24	1.5	200	33	2	230	40	3	280	58	4
28	140	190	24	1.5	210	33	2	250	42	3	300	62	4
30	150	210	28	2	225	35	2.1	270	45	3	320	65	4

[a]Selected size range from ANSI/ABMA Std. 20-1996. Does not cover tapered roller bearings.

along the bearing bore or outside diameter is usually about three times r_{smin} for bearings with a bore diameter of 280 mm or smaller.

The radial height of mating shaft and housing shoulders must sufficiently exceed r_{smax} to provide an adequate abutment surface to support axial loads applied to the bearing. The tolerance range for chamfers at ring corners is relatively loose to ease the bearing manufacturing process. Some bearing manufacturers hold a closer radial tolerance on r_{smax} than listed in Table 13.2, which may enable the use of smaller-diameter bar stock in manufacturing a shaft.

Table 13.3 Tolerances for dimensions and runouts (ABEC-1, RBEC-1 tolerances for radial bearings in micrometers)

Inner ring

Bore diameter (mm)		Mean bore diameter deviation		Single-bore diameter variation[a] — Diameter Series			Mean bore diameter variation	Radial runout	Single-width deviation		Width variation
Over	Incl.	High	Low	9	0, 1	2, 3, 4			High	Low	
				Max.			Max.	Max.			Max.
2.5	10	0	−8	10	8	6	6	10	0	−120	15
10	18	0	−8	10	8	6	6	10	0	−120	20
18	30	0	−10	13	10	8	8	13	0	−120	20
30	50	0	−12	15	12	9	9	15	0	−120	20
50	80	0	−15	19	19	11	11	20	0	−150	25
80	120	0	−20	25	25	15	15	25	0	−200	25
120	180	0	−25	31	31	19	19	30	0	−250	30
180	250	0	−30	38	38	23	23	40	0	−300	30

Outer ring

Outside diameter (mm)		Mean outside diameter deviation		Single outside diameter variation[a] — Open bearings Diameter Series			Sealed and shielded bearings[b]	Mean outside diameter variation	Radial runout	Width deviation and variation
Over	Incl.	High	Low	9	0, 1	2, 3, 4				
				Max.			Max.	Max.	Max.	
6	18	0	−8	10	8	6	10	6	15	
18	30	0	−9	12	9	7	12	7	15	Identical
30	50	0	−11	14	11	8	16	8	20	to those
50	80	0	−13	16	13	10	20	10	25	of inner
80	120	0	−15	19	19	11	26	11	35	ring of
120	150	0	−18	23	23	14	30	14	40	same
150	180	0	−25	31	31	19	38	19	45	bearing
180	250	0	−30	38	38	23	—	23	50	
250	315	0	−35	44	44	26	—	26	60	
315	400	0	−40	50	50	30	—	30	70	

[a]No values established for Diameter Series 7 and 8.
[b]No values established for Diameter Series 9, 0 and 1.
Source: American Bearing Manufacturers Association, 1995. Reproduced with permission of American Bearing Manufacturers Association.

This enables correspondingly less machining to generate an adequate shoulder. ANSI/ABMA Std. 20-1996 covers standards for chamfer dimensions of radial bearings, along with requirements for some types of cylindrical roller bearings and various other special cases.

Internal Clearance

Radial internal clearance, the distance the inner and outer rings can be moved relative to each other when the bearing is not mounted, is normally of Class C0 when bearings are shipped. This

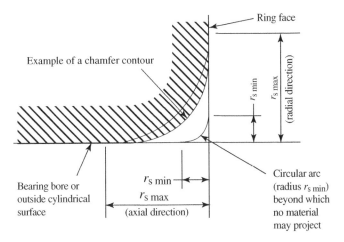

Figure 13.4   Chamfer dimensions for bearing rings.

normal range results in a minimum of essentially zero internal clearance when the bearing is mounted on a shaft with the usually recommended fits and under normal operating conditions. Special internal clearance ranges are identified as follows:

C2	Clearance less than normal
C3	Clearance greater than normal
C4	Clearance greater than C3
C5	Clearance greater than C4

Variations in the normal C0 radial clearance with the bore size of deep-groove ball bearings, cylindrical roller bearings, and double-row cylindrical roller bearings are given in Table 13.4. Roller bearings have a larger internal radial clearance to accommodate heavier interference fits on a shaft and also generally more severe loading.

The span of clearance increases just slightly for each larger clearance range. A 50 mm-bore deep-groove, single-row ball bearing, for instance, has the following clearances (in micrometers):

Clearance group:	C2	Normal	C3	C4	C5
Minimum clearance:	1	6	18	30	45
Minimum clearance:	11	23	36	51	73

Bearings in the C0 range bear no markings to indicate their normal clearance. An internal clearance of C3 has become the usual choice with ball bearings for electric motors and similar equipment operating with grease lubrication at moderate speeds. This increased clearance results in significantly fewer bearing failures by providing an increased thrust load capacity, as well as more tolerance both for misalignment and for differential thermal expansion of the shaft. With heavy press fits and other demanding conditions, looser clearances than C3 may be necessary. Tighter than normal clearance in the C2 range would be used for more precise location of a rotor or machine tool.

Table 13.4   Normal radial internal clearance[a]

Bore diameter (mm)		Radial contact ball bearings		Cylindrical roller bearings		Spherical roller bearings	
Over	Incl.	Min.	Max.	Min.	Max.	Min.	Max.
2.5	10	2	13				
10	18	3	18	20	45		
18	24	5	20	20	45	20	35
24	30	5	20	20	45	25	40
30	40	6	20	25	50	30	45
40	50	6	23	30	60	35	55
50	65	8	28	40	70	40	65
65	80	10	30	40	75	50	80
80	100	12	36	50	85	60	100
100	120	15	41	50	90	75	120
120	140	18	48	60	105	95	145
140	160	18	53	70	120	110	170
160	180	20	61	75	125	120	180
180	200	25	71	90	145	130	200
200	225	28	82	105	165	140	220
225	250	31	92	110	175	150	240

[a]Values in micrometers from ABMA Std. 20-1996.

The axial motion or end play Δ_e in a deep-groove ball bearing is several times (often about 10 times) the radial play Δ_r. This end play is also a function of the raceway conformities, as given by the following equation:

$$\Delta_e = \left[4d\,\Delta_r(f_i + f_o - 1) - \Delta_r^2 \right]^{0.5} \tag{13.3}$$

where f_i and f_o are the ratios of the inner and outer raceway curvature radii to the ball diameter, respectively.

Angular misalignment β that can be freely tolerated by a ball bearing (without inducing contact deflections) is directly related to end play by the following equation:

$$\tan \beta = \frac{2\Delta_e}{D_1 + D_2} \tag{13.4}$$

where D_1 and D_2 are the inner and outer bearing diameters, respectively. These end play relations are also affected by contact deformations resulting from loads on a bearing. Common limits for misalignment and the influence on internal clearance of shaft and housing mounting fits are covered in the next chapter.

Precision Classifications

Normal dimensional tolerances for the ring surfaces and balls of ball bearings are covered as ISO Class Normal and as the identical ABEC 1 for ball bearings and RBEC 1 for roller bearings (American Bearing Manufacturers Association, 1996). These standard dimensional tolerances are abstracted in Table 13.3. Closer precision ABEC 3 (ISO Class 6), ABEC 5 (ISO Class 5),

ABEC 7 (ISO Class 4), and ABEC 9 (ISO Class 2) provide for increasing degrees of precision for ball bearings. RBEC 1, RBEC 3, RBEC 5, RBEC 7, and RBEC 9 (corresponding to ISO Classes Normal, 6, 5, 4, and 2, respectively) are the precision grades of roller bearings. Closer than normal ISO Classes 6 and 5 are referred to as *super-precision*, and Classes 4 and 2 are used as ultra-precision bearings where the quietest and smoothest operation is desired. Detailed tolerances for these various degrees of precision are given by ABMA and in manufacturers' catalogs.

Unground bearings find some use in slow-speed applications such as conveyor rollers, heavy doors, and drawers where accuracy and precision are not required. Made to much looser tolerances than for ABEC 1, the rings and raceways are turned rather than ground and are then hardened after turning.

13.7 Shaft and Housing Fits

Earlier sections reviewed tolerance details for major dimensions of ball and roller bearings. It is particularly important to apply suitable fits between the inner ring of the bearing and the shaft, and between the outer ring and the housing.

Rotating Inner Ring with a Stationary Load

For the common case of a rotating shaft with a stationary load, as in an electric motor belt driving a fan, it is important that the bearing inner ring have a tight fit on the bearing shaft. With a loose fit, rotation of the load with respect to the bearing inner ring will cause the ring to rotate slowly on its shaft. This slow sliding motion commonly results in a great deal of wear after relatively few hours of operation.

The only way to prevent this motion is to fit the bearing ring to the shaft so tightly that no clearance exists and none can develop because of the action of the load. Higher than normal interference fits would then be required for heavy radial loads, hollow shafts, abnormal bending of the shaft, and any unusual temperature rise within the bearing that might induce loss of fit.

A bearing locknut to provide axial clamping and adhesive cements cannot normally be relied upon to prevent this relative motion and wear between the bearing bore and its shaft surface. Keying the bearing bore to the shaft with suitable key ways in both surfaces will avoid rotation, but intolerable fretting may be encountered at the key way. If a loose fit is necessary, lubrication of the shaft surface with grease or a solid lubricant coating will minimize wear for mild operating conditions with smaller bearings.

Stationary Inner Ring with a Rotating Load

Whichever bearing ring will be rotating with respect to the load must be fitted tightly to its mating member. If the shaft remains stationary, as in the front wheel of a car, and the load is always directed toward the same portion of the inner ring raceway, then the outer ring must be mounted with an interference fit in its housing.

Normally one bearing ring, usually the outer, is stationary with respect to the load. In a normal application, therefore, the outer ring is fitted slightly loose in the housing both to facilitate assembly and disassembly and to accommodate thermal expansion.

Indeterminate or Variable Load Direction

If the load direction is variable or indeterminate, both the inner and outer rings should be tightly fitted to the shaft and housing, respectively. If more than one ball bearing is involved in the design, care must be taken to avoid high internal axial loading of the bearings from thermal expansion or from improper axial tolerances.

Vibration is one possible source of variable load. With normally steady deadweight loading, as with the weight of an electric motor rotor, a tight fit of the inner bearing ring is common, along with a loose fit of the outer ring in its housing bore. When rotor imbalance results in excessive vibration, however, the outer ring will slowly rotate in the housing, causing wear and early failure. In horizontal 3600 rpm motors, this outer ring rotation has been observed to be approximately 1 rpm and was commonly initiated when vibration amplitude exceeded $0.5g$, i.e. when the vibrational force was greater than about one-half the gravitational pull on the rotor. This effect is particularly troublesome with the inherent vibration in a rotor system employing a diesel or other internal combustion engine drive.

Mounting Tolerances

Since rolling bearing manufacturers make bearings to standardized ISO tolerances, proper fits can be obtained only by selecting proper tolerances for the mating shaft and housing surfaces. Figure 13.5 illustrates the zones of fit involved in the various ISO tolerance grades. While these tolerance grades form useful recommendations, the final fits selected for a given machine design should reflect specific manufacturing processes, application demands, and field experience.

Normal rotating shaft fits for ball bearings with a stationary load direction usually involve the k5 range, with a light interference fit ranging from approximately line-to-line (zero interference with the smallest shaft diameter and the largest bearing bore) to a maximum interference with the largest shaft diameter and the smallest bearing bore. This shaft interference is commonly increased to the m5 range for roller bearings. Corresponding normal tolerance of H7 for the housing fit would range from a minimum line-to-line fit to a maximum looseness, with the smallest bearing outside diameter assembled within the largest housing bore. In unusually demanding cases, selective assembly may even be required where, for instance, a nearly line-to-line fit could be obtained in a housing by having the bearing manufacturer group bearing outer diameters into narrow size ranges within their normal tolerance range for fitting to housing bores which are similarly grouped by the machine manufacturer.

Surface finish of shaft and housing surfaces also influences bearing fits: the rougher the surface, the greater the loss of interference fit (Eschmann *et al.*, 1985). For a unidirectional load with a rotating shaft, the shaft surface should be ground to an R_a finish of 0.8 μm (32 μin.) or better, while the less critical surface in the housing for a unidirectional load can normally be a turned finish of 1.6–3.2 μm (63–125 μin.).

13.8 Load–Life Relations

The size and type of a ball or roller bearing to be used for an application are selected to provide the required load-carrying capacity in relation to the life required. The life is defined as the number of revolutions, or operating time, which the bearing can endure under ideal conditions

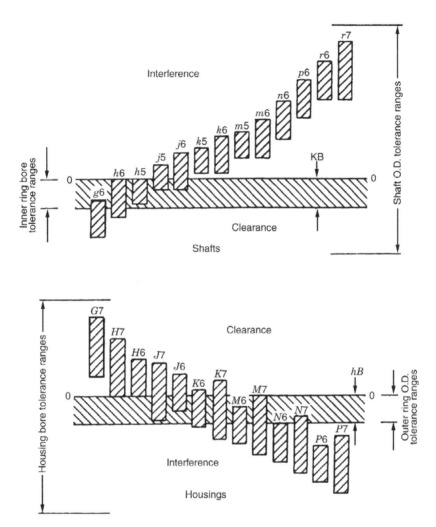

Figure 13.5 Graphic representation of shaft and housing fits. *Source*: Harris, 1991. Reproduced with permission of John Wiley & Sons.

of cleanliness and lubrication before the first sign of fatigue flaking on a raceway or rolling element appears. Most fatigue cracks then begin at a depth beneath the surface where the alternating shear stress reaches a maximum (Eschmann *et al.*, 1985).

The basic dynamic load rating C is used in load–life calculations for bearings rotating under load. This basic load rating C represents the static load, which the bearing can carry for 1 000 000 inner ring revolutions with only a 10% probability of fatigue failure (i.e. 90% reliability). The basic bearing life L_{10} in millions of revolutions for 90% reliability can then be determined for any other bearing load P by the following relationship:

$$L_{10} = \left(\frac{C}{P}\right)^P \tag{13.5}$$

where the load–life exponent $p = 3$ for ball bearings and $P = 10/3$ for roller bearings, and P is the equivalent bearing load. For bearings operating at a constant speed n rpm, the 10% basic fatigue life in hours L_h becomes

$$L_h = \frac{1000000(C/P)^P}{60n} \quad \text{or} \quad C = P\left(\frac{60nL_n}{1000000}\right)^{1/P} \tag{13.6}$$

For road and rail vehicle bearings, especially wheel hub and axle bearings, the basic rating life may preferably be expressed by the following L_{10s} basic rating life in millions of kilometers travel (SKF, 1999):

$$L_{10s} = \frac{\pi D L_{10}}{1000} \tag{13.7}$$

where D is the outside wheel diameter in meters.

To use these relations in selecting bearings, the required life of the application must be known. When previous experience is lacking, Tables 13.6 and 13.7 can serve as a guide for establishing design life. After the required minimum basic load rating is calculated from Equation (13.6) or (13.7), Table 13.5 will give the required bearing sizes for common series of ball and roller bearings. Because of variations in construction details and in the size and number of rolling elements used by different manufacturers, reference should be made to individual bearing catalogs for specific values and for considering possible bearing series other than the ones commonly used, listed in Table 13.5.

Example 13.1   For a 70 mm diameter shaft running at $n = 1000$ rpm, compare the 10% fatigue life with 20 000 N (4494 lb) radial load P being carried on 6214 and 6314 deep-groove ball bearings and N214 and N314 cylindrical roller bearings. Which bearing should be selected to meet a specification requirement of 10 000 h?

As a first step, the *basic dynamic load rating* for each of the bearings is read from Table 13.5, giving the values indicated in the following tabulation. Fatigue life L_h is then calculated from Equation (13.6) to give the 10% life in hours.

Bearing	Basic Dynamic Load Rating, N C	Calculated 10% Life, hr L_{10}
Deep-groove ball bearings, 6214	61400	480
6314	104500	2380
Cylindrical roller bearings, N214	119000	6360
N314	204000	38360

While the N314 bearing meets the life requirement, the means for axial location of the shaft should be checked since none is provided by a conventional cylindrical roller bearing. Use of additional flanges for this purpose could be called for on the cylindrical roller bearing.

Table 13.5 Typical basic dynamic load ratings of ball and roller bearings (N)

Bearing bore		Deep-groove ball bearings				Angular-contact ball bearings		Cylindrical roller bearings	
Bore No.	mm	19XX Series	01XX Series	02XX Series	03XX Series	72XX Series	73XX Series	N2XX Series	N3XX Series
5	5			1 220	2 100				
6	6			2 130					
7	7		2 130	3 260					
8	8		3 260						
9	9		3 700	4 580					
00	10	1 950	4 580	5 520	8 100	6 000			
01	12	2 250	5 070	6 910	9 800	7 260	10 700		
02	15	4 030	5 570	7 750	11 400	8 400	13 200	12 600	19 400
03	17	4 360	6 000	9 580	13 500	10 400	16 100	14 400	22 800
04	20	6 370	9 400	12 800	16 600	13 400	19 000	26 300	31 200
05	25	6 630	10 700	14 000	22 400	14 900	26 300	28 900	40 600
06	30	7 280	13 000	19 500	28 600	22 000	33 600	38 500	51 400
07	35	9 560	16 100	25 400	33 300	28 900	39 100	49 100	64 700
08	40	13 800	16 900	29 900	42 000	34 000	49 600	53 600	81 100
09	45	14 000	20 400	32 100	53 000	36 600	60 000	60 700	98 400
10	50	14 600	21 100	35 700	62 000	38 000	72 700	64 000	111 000
11	55	15 900	28 300	43 600	74 000	47 500	82 600	83 500	137 000
12	60	16 500	29 400	52 600	81 600	56 400	93 200	93 900	150 000
13	65	17 400	30 200	57 800	92 400	65 000	106 000	107 000	182 000
14	70	23 800	38 100	61 400	104 500	69 800	117 000	119 000	204 000
15	75	24 200	39 900	66 500	113 700	70 800	130 000	130 000	241 000
16	80	25 100	47 800	71 400	123 000	81 600	144 000	139 000	257 000
17	85	31 900	49 600	83 000	130 000	92 300	153 000	164 000	284 000
18	90	33 200	58 600	96 000	138 000	107 000	164 000	182 000	317 000
19	95	33 800	60 700	108 000	149 000	121 000	178 000	219 000	337 000
20	100	42 300	60 100	123 000	168 000	134 000	199 000	250 000	385 000
21	105	44 200	72 000	133 000	178 000	146 000	209 000	262 000	440 000
22	110	43 600	81 000	144 000	196 000	158 000	228 000	292 000	468 000
24	120	55 300	84 000	145 000	210 000	166 000	246 000	337 000	527 000
26	130	65 000	105 000	161 000	230 000	185 000	268 000	357 000	601 000
28	140	66 300	110 000	171 000	252 000	189 000	287 000	388 000	675 000
30	150	88 400	124 000	176 000	280 000	210 000	315 000	445 000	769 000

Combined Radial and Thrust Load

Contributions from the combination of a radial load P_r and a thrust load P_t to the equivalent bearing load P can be calculated from the following equation:

$$P = XP_r + YP_t \tag{13.8}$$

If the resulting P from Equation (13.8) is less than P_r for a radial bearing, use $P = P_r$. If P is less than P_t for a usual ball or roller thrust bearing, use $P = P_t$.

Table 13.6   Typical rolling bearing design life

Machine or application type	Life (h)
Blowers	30 000
Compressors	50 000
Continuous-duty cranes	30 000
Continuous-duty shipboard equipment	150 000
Conveyors and elevators	10 000
Door actuators	500
Farm machinery	4 000
Foundry and assembly cranes	8 000
General-purpose gearboxes	20 000
Machine tools	30 000
Mine hoists	50 000
Mine pumps	150 000
Public power stations	200 000
Public pumping-station equipment	200 000
Pulp and paper machines	100 000
Pumps	50 000
Single-screw extrusion equipment	100 000
Stationary electric motors	20 000
Twin and multiple-screw extruders	50 000

Source: Power Transmission Design, 1990. Reproduced with permission of Power Transmission Design.

Table 13.8 gives representative values of X and Y. Since multiplying factors X and Y vary with the relative magnitude of the thrust and radial load, internal clearance, contact angles, and related internal geometry set by the bearing manufacturer, bearing catalogs or analytical sources (Derner and Pfaffenberger, 1984; Eschmann *et al.*, 1985; Harris, 1991) should be consulted for details of other bearing types not given in Table 13.8, such as double-row, spherical, and tapered roller bearings.

Table 13.7   Guide to requisite rating life L_{10s} for road and rail vehicles

Vehicle use	L_{10s} (millions of km)
Wheel hub bearings for road vehicles	
Cars	0.3
Commercial vehicles, buses	0.6
Axlebox bearings for rail vehicles	
Railroad freight cars	0.8
Commuter trains, street cars	1.5
Railroad passenger cars	3
Traction motors	3–4
Railroad diesel and electric locomotives	3–5

Source: SKF, 1999/2005. Reproduced with permission of John Wiley & Sons.

Table 13.8 *X and Y load factors for equivalent fatigue load P^a*

	Radial ball bearings										
Contact angle, deg.	0–5		10		15		20		30		40	
Internal clearance	Normal C0		C3		C4							
Fa/C	X	Y	X	Y	X	Y	X	Y	X	Y	X	Y
0.025	0.56	2	0.46	1.75	0.44	1.42	0.43	1	0.39	1	0.35	1
0.04	0.56	1.8	0.46	1.61	0.44	1.36	0.43	1	0.39	1	0.35	1
0.07	0.56	1.6	0.46	1.46	0.44	1.27	0.43	1	0.39	1	0.35	1
0.13	0.56	1.4	0.46	1.30	0.44	1.16	0.43	1	0.39	1	0.35	1
0.25	0.56	1.2	0.46	1.14	0.44	1.05	0.43	1	0.39	1	0.35	1
0.5	0.56	1	0.46	1	0.44	1	0.43	1	0.39	1	0.35	1

	Thrust ball bearings						
Contact angle, deg.	45		60		75		90
	X		Y	X	Y	X	Y	X		Y

Contact angle, deg.   45   60   75   90

X	Y	X	Y	X	Y	X	Y
0.66	1	0.92	1	1.66	1	—	1

Cylindrical roller bearings

Contact angle, deg.   0   90

X	Y	X	Y
1	—	—	1

[a] Where $P = XP_r + YP_t$. If $P < P_r$, use $P = P_r$.
Source: Harris, 1991. Reproduced with permission of John Wiley & Sons.

For deep groove ball bearings, X varies from 0.56 for standard radial clearance, to 0.46 for a typical contact angle of 10° for C3 larger than normal internal clearance, to 0.44 with the larger 15° ball–raceway contact for even larger C4 radial internal clearance. Y generally varies from 1.0 for a high thrust load to 2.0 for a low thrust load. The Y value of 1.0 at a high thrust load reflects the change when load support shifts from a limited number of contacts under a purely radial load to half or more of the rolling elements carrying the thrust load (prerequisite for the basic load rating) (SKF, 1999). Contact angles of 20° or more become independent of the load with $Y = 1.0$.

Similar criteria for combined loads with radial roller bearings are given at the bottom of Table 13.8. Caution must be observed in using double-row spherical roller thrust bearings under axial load. With high axial load F_a and low radial load $F_r(F_a/F_r < 2)$, rollers in the loaded row skew to produce troublesome sliding action, and these loaded rollers may even roll over the edge of the outer ring raceway. Damaging sliding may also be encountered with unloading of the row of rollers opposite to the axial load.

These combined load effects change only marginally for outer ring rotation. Equation (13.8) should be used for either outer or inner ring rotation.

Example 13.2 A thrust load P_t of 7000 N (2247 lb) combined with a radial load P_r of 3000 N (1348 lb) is to be carried on a 200 series ball bearing of 70 mm bore at 1000 rpm. Determine the equivalent radial load to be used for calculating the fatigue life, and then the life to be expected with a 6214 single-row, deep-groove ball bearing in comparison with that for a 7214 angular contact ball bearing with its nominal contact angle of 30°.

First, the basic dynamic load rating C (Newtons) for each bearing is taken from Table 13.5. Table 13.8 then provides appropriate X and Y multipliers for Equation (13.8) $(P = XP_r + YP_t)$ with normal internal clearance and a 0–5° contact angle for the 6214 bearing. The 10% fatigue life L_h follows from Equation (13.6) to provide the following comparison:

Bearing	Basic Load Rating C, N	P_t/C	X	Y	P (N)	L_h (hr)
6214	61,400	0.11	0.56	1.5	12,180	2,135
7214	69,800	—	0.39	1.0	8,170	10,390

Equivalent load P for the 6214 deep-groove bearing is 20% of its basic load rating, which is in the undesirably 'heavy' load range (see Chapter 14). If the fatigue life is insufficient with the 7214 angular contact bearing, 6314 and 7314 bearings with larger cross sections should be considered.

Varying Load

If the magnitude of the load fluctuates, an equivalent mean constant load P_m which would have the same influence on fatigue life as the actual fluctuating load must first be determined. A constantly fluctuating load can be approximately resolved into a number of constant single forces, and mean load P_m becomes as an approximation:

$$P_m = \left[\frac{(P_1^3 U_1 + P_2^3 U_2 + P_3^3 U_3 + \cdots)}{U} \right]^{1/3} \tag{13.9}$$

where P_m = equivalent constant mean load (N); P_1, P_2 = constant load during $U_1, U_2 \ldots$ revolutions (N) and U = total number of revolutions $(U_1 + U_2 + \cdots)$ during which loads $P_1, P_2, \ldots$ act. While exponent 3 is for ball bearings and should be 10/3 for roller bearings, the difference has very little influence on the final results, and 3 is commonly used for both bearing types (Harris, 1991).

With constant bearing speed and load direction, but with load constantly fluctuating between a minimum value P_{min} and a maximum value P_{max}, the mean load is

$$P_m = 1/3 P_{min} + 2/3 P_{max} \tag{13.10}$$

If the load on a bearing consists of both a constant unidirectional load P_1 (like the weight of a rotor) and a rotating unbalance load P_2, an approximate mean equivalent load for $P_2 > P_1$ is given by Eschmann *et al.* (1985):

$$P_m = P_2 \left[1 + 0.5 \left(\frac{P_1}{P_2} \right)^2 \right] \tag{13.11}$$

For $P_1 > P_2$:

$$P_m = P_1 \left[1 + 0.5 \left(\frac{P_2}{P_1} \right)^2 \right] \tag{13.12}$$

For $P_1 = P_2$, this mean equivalent load $P_m = 0.75P_1$. Expressions for P_m for various other cases of fluctuating loads have been derived by Harris (1991).

Example 13.3   With the 214 size deep-groove ball bearing in Example 13.2 with its basic dynamic load rating $C = 61400\,N$ from Table 13.5, calculate the 10% fatigue life to be expected when operating in the following 1 h cycle:

Condition	Radial load, P (N)	Speed (rpm)	Time (min)	Revolutions, U (rpm × min)
1	8 000	500	10	5 000
2	4 000	1 200	30	36 000
3	7 000	800	20	16 000

Applying the three speed conditions to obtain the total revolutions/h, U:

$$U = U_1 + U_2 + U_3 = 5000 + 36\,000 + 16\,000 = 57\,000\,\text{rev/h}$$

Equivalent constant load P_m from Equation (13.9):

$$P_m = \left[\frac{(8000)(3)(5000) + (4000)(3)(36\,000) + (7000)(3)(16\,000)}{57\,000} \right]^{1/3} = 5663\,N$$

Basic 10% fatigue life in one million revolutions from Equation (13.5) then becomes

$$L = (C/P_m)^3 = (61\,400/5663)^3 = 1275 \text{ million revolutions}$$
$$1275/0.057 \text{ million rev/h} = 22\,360\,h$$

Minimum Load and Preloading

To avoid skidding of rolling elements, a bearing must be subjected to some minimum load. As a rule of thumb, SKF (1999) states that a minimum load of $0.02C$ should be imposed on roller bearings and $0.01C$ on ball bearings (where C is the basic load rating). If the external force plus the weight of components supported by the bearing do not supply this minimum load, added preload should be supplied. For deep-groove ball bearings under a purely radial load, NTN Bearing Company recommends an added axial spring preloading force (newtons) of four to eight times the bearing bore diameter in millimeters.

Two methods are normally used to apply any required preload: (1) fixed position preload, which increases rigidity, or (2) constant pressure preload, as obtained with coil springs or wavy washers. Fixed position preload with spacers or threaded screw adjustment is applied with precision angular-contact ball bearings, tapered roller bearings, thrust ball bearings and angular-contact ball bearings. This technique aids in maintaining the accuracy of location and gives increased rigidity in grinders, lathes, milling machines, and wheel axles.

Spring preloads are used with deep-groove ball bearings, angular-contact ball bearings, precision tapered roller bearings, and spherical roller thrust bearings. This spring preloading produces relatively constant axial force with changing temperatures and loads to give less vibration and noise in electric motors and with high-speed shafts in small machines. Relatively flat wavy spring washers have become common in industrial electric motors. Such a washer provides the preload by being compressed between the housing shoulder and the floating outer ring of the ball bearing at one end of the motor, with the reaction force also preloading the restrained opposite end bearing. This preload then avoids cage vibration with possible bearing squeaking and squealing as the balls would otherwise change speed, skid, and oscillate back and forth in their cage pockets as they pass through the unloaded sector of the bearing.

13.9 Adjusted Rating Life

With standard ball and roller bearings used in conventional applications, L_{10} life calculations based on the basic rating life from Equation (13.5) are adequate for representing fatigue life experience based on the usual speed, load, and lubrication factors.

For considering other factors influencing bearing life, ABMA introduced the following revised life equation (Zaretsky, 1992; Jendzurski and Moyer, 1997):

$$L_{na} = a_1 a_2 a_3 (C/P)^p \quad \text{or simply} \quad L_{na} = a_1 a_2 a_3 L_{10} \tag{13.13}$$

where

L_{na} = adjusted rating life, millions of revolutions (index n represents the difference between requisite reliability and traditional 90% reliability).

a_1 = life adjustment factor for reliability. This is used to determine fatigue lives, which are attained with a greater than 90% probability. Values are given in Table 13.9.

a_2 = life adjustment factor for material. A factor of 1.0 is used for current vacuum-degassed steels. The approximate range for special and premium steels is 0.6-10.

a_3 = life adjustment factor for operating conditions. The following factors are used: 1.0 for lubricant film thickness similar to surface roughness; < 1.0 for DN < 10 000 and for low-lubricant viscosity; > 1.0 for favorable lubrication conditions.

Figure 13.6 gives an example of the influence of lubricant film thickness (see also Chapter 15) on fatigue life. The loss of life for low film thickness with Λ less than 1.0 can sometimes be minimized with gear-type lubricants containing extreme pressure or antiwear additives. Since some of these additives chemically attack bearing steels, with a corresponding reduction in fatigue life, suppliers should be consulted for a suitable product. Contaminants (liquid or

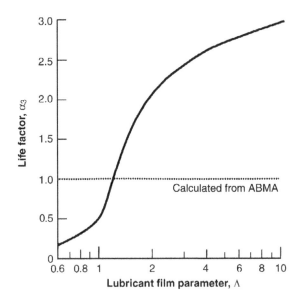

Figure 13.6 Life adjustment factor as function of lubricant film parameter Λ, ratio of film thickness to composite surface roughness. *Source*: Jendzurski and Moyer, 1997. Reproduced with permission of Taylor & Francis.

solid), alignment, and surface chemistry may also influence life. Most manufacturers have detailed programs to help select bearings and appropriate lubricants to meet various operating requirements (Jendzurski and Moyer, 1997). For the generally accepted reliability of 90%, for the usual bearing steels to which the basic dynamic load rating C corresponds, $a_1 = a_2 = a_3 = 1$ for normal operating conditions, and the basic and adjusted rating lives are identical.

Recent practical experience and analysis indicate much longer fatigue life under ideal conditions than predicted by traditional life calculations (SKF, 2005). With special steels and appropriate lubrication in some lightly loaded aerospace units, an adjusted rating life of the order of 30 times the basic rating life has been found. With full separation of rolling surfaces by a lubricant film at light loads and in the absence of significant contamination, even 'infinite' fatigue life now appears to be available and SKF includes in its catalog a fatigue load capacity limit P_u. For loads below P_u, infinite fatigue life is to be expected in the absence of contamination and with a lubricant film of sufficient thickness. Although the details are open to question, infinite fatigue life has been attributed to ball bearings for Hertz maximum contact stresses below about 200 000 psi, 150 000 psi for roller bearings, or alternatively for lubricant film thickness well above (approximately six times or more) composite surface roughness in the rolling contact zone.

Based partly on the concept of this load limit P_u below which fatigue life becomes infinite (Ioannides and Harris, 1985), SKF has generated a new life factor a_{SKF}. This replaces the combined effect of factors a_2 and a_3 of Equation (13.13) in estimating fatigue life as operating conditions vary from severe to ideal (SKF, 2005):

$$L_{naa} = a_1 a_{SKF} L_{10}$$

(13.14)

Table 13.9   Values of life adjustment
reliability factor a_l

Reliability %	a_1
90	1
95	0.62
96	0.53
97	0.44
98	0.33
99	0.21

Source: SKF, 1999/2005. Reproduced
with permission of John Wiley & Sons.

where

L_{naa} = rating life adjusted for lubrication conditions and contamination
a_1 = life adjustment factor for reliability (see Table 13.9)
a_{SKF} = life adjustment factor for lubricant film thickness, for loading relative to fatigue load
          limit P_u, and for contamination

Figure 13.7 relates the factors involved in a_{SKF} for radial ball bearings (SKF, 1999). Factor κ in this figure is the ratio of the lubricant kinematic viscosity (cSt) at the bearing temperature to the minimum viscosity required for the oil film thickness to match the composite surface roughness in the ball–raceway contact (from Figure 15.5). Factor η_c is an adjustment factor for contamination as applied to P_u/P, the ratio of maximum load for infinite life P_u to the applied radial bearing load P. Table 13.10 gives suggested values for these η_c contamination factors (SKF, 1999).

For heavily contaminated lubricant and inadequate lubricant film thickness to separate the balls from their raceways, a_{SKF} drops to a limiting value of 0.1. With clean oil (contamination factor $\eta_c = 1$), light loads, and lubricant viscosity three to four times that required to just match the composite surface roughness in the loaded rolling contacts ($\kappa = 3$–4), a_{SKF} rises to very high values. SKF recommends that the resulting life estimates be used only as supplementary guides in estimating bearing fatigue life and that a_{SKF} values above 50 should not be used. Charts similar to Figure 13.7 are also available for roller and thrust bearings (SKF, 1999/2005).

A detailed ASME review concluded that the many parameters affecting bearing fatigue life are interrelated and should be represented by a single integrated stress-life factor (ASME, 2003). This study continued to focus on the traditional C/P data given in bearing catalogs, but with further attention to contact effects that create surface microspalls that then progress to pits, cracks, and ultimate bearing failure. This contrasts with previous life rating theory that placed all fatigue initiation sites below the surface in the zone of maximum shear stress.

Interesting examples in this ASME review show significantly shortened fatigue life from surface chemical effects by extreme-pressure additives in gear oils. While 0.5% water reduced fatigue life by a factor of 3 with one ISO VG-220 mineral oil, water had minimal effects with some other lubricant compositions. (In experience with greased ball bearings in industrial electric motors, the authors of this present book observed no significant loss of life with

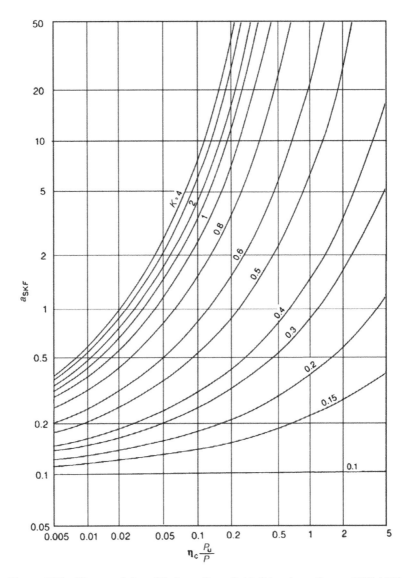

Figure 13.7   The a_{SKF} fatigue life factor for radial ball bearings. *Source*: SKF, 1999.

minor amounts of water contamination.) Oil cleanliness and friction effects at loaded contacts are included in this ASME review. A detailed computer program for making overall
fatigue life calculations is provided with the ASME 2003 publication (ASME Life Program,
Version 1).

Example 13.4   Calculate the 10% fatigue life for a 210 deep-groove ball bearing (50 mm
bore, 90 mm OD with 70 mm mean diameter) running at 5000 rpm under a 5000 N (1124 lb)

radial load (1) by the traditional fatigue life Equation (13.5), (2) added adjustment as in Equation (13.14) for the SKF life factor for loading relative to the fatigue load limit. Assume fatigue load limit of $P_u = 980\,\text{N}$ as the maximum load for infinite fatigue life (SKF, 1999).

1. For 1 million revolutions $C = 35\,700\,\text{N}$ (8026 lb) from Table 13.5. Using Equation (13.5), the basic load rating is

$$L_{10} = (C/P)^3 = (35\,700/5000)^3 = 364 \text{ million revs}$$

2. Introduce lubricant film parameter $\kappa = 20/8 = 2.5$ in Figure 13.7. With fatigue load limit $P_u = 980\,\text{N}$ as the maximum load for infinite fatigue life (SKF, 1999), $P_u/P = 980/5000 = 0.20$. Using $\eta_c = 1$ for super clean conditions then gives SKF life factor $\alpha_{SKF} = 20$ in Figure 13.7. For 10% probability of failure, $\alpha_1 = 1$ from Table 13.9. Then Equation (13.14) gives the following adjusted fatigue life:

$$\text{SKF } L_{10} = \alpha_1\,\alpha_{SKF}\,L_{10} = 1 \times 20 \times 364 = 7280 \text{ million revs}$$

Using $\eta_c = 0.5$ for normal cleanliness with double-shielded, greased-for-life bearings in Table 13.10, Figure 13.7 drops α_{SKF} to 5 and then SKF $L_{10} = 1 \times 5 \times 364 = 1820$ million revs.

While these adjustment factors for use with traditional fatigue life are likely oversimplified, they do indicate the extremely long fatigue life available with lightly loaded bearings having good lubrication and in the absence of contamination. Lubricant viscosity and elastohydrodynamic film thickness also influence the expected fatigue life of ball and rolling element bearings. These factors are covered in Chapter 15.

Life Reduction Due to Particle Contamination

As illustrated in Example 13.4 and Figure 13.7, particle contamination has been proposed as a major adjustment factor η_c in fatigue life calculations. These η_c multipliers range down to 0.1

Table 13.10   Adjustment factor η_c for different degrees of contamination

Condition	η_c
Very clean: Debris size of the order of lubricant film thickness	1
Clean: Typical of bearings greased for life and sealed	0.8
Normal: Typical of bearings greased for life and shielded	0.5
Contaminated: Typical without integral seals; with coarse lubricant filters. and/or particle ingress from surroundings	0.5–0.1
Heavily contaminated: Under extreme contamination, can be more severe reduction in life than predicted from Equation (13.14) or Figure 13.7	0

Note: Values of η_c refer only to solid contaminants. Contamination by water or other detrimental fluids is not included.
Source: SKF, 1999/2005. Reproduced with permission of John Wiley & Sons.

and lower for smaller ball bearings having mean diameters below about 50 mm when running in a severely contaminated circulating oil system (SKF, 2005; ASME, 2003). The downward adjustment is much lower for larger ball bearings. For mean diameters above 500 mm, η_c rises to the 0.8–1.0 range for normal levels of oil contamination and with lubricant film thickness parameters of 1.0 or more from Figure 15.5.

Some questions exist as to the general need for these major downward contamination adjustments for general applications of ball and roller bearings. In examination of hundreds of greased ball bearings ranging up to 100 mm mean diameter returned from field problems in industrial electric motors, less than 10% displayed fatigue damage. Those that did commonly reflected early fatigue damage from abnormally high contact stress induced at installation by misalignment or motor assembly problems. Instead, after years of service most of the damaged bearings suffered not from fatigue, but from worn or dented ball and raceway surfaces involving dried grease, various contaminants, or occasional rusting.

Other bearing manufacturers have shared their observation that the η_c factor appears to reflect test results primarily with hard particles and that they encounter no such generally shortened fatigue life with the softer contaminant particles more common in industrial and commercial service. Bearing manufacturers should be consulted for their advice where significant questions of this type are encountered in matching bearings to operating demands in new machine designs or in the case of operating troubles.

13.10 Static Load Capacity

Where static loads are encountered with bearings that are not rotating, for intermittent shock loads, and at very low speeds, resistance to plastic indentation becomes important. While load–life Equations (13.6) and (13.7) based on material fatigue would indicate infinite load capacity for a bearing at zero speed, a limiting static capacity is commonly defined as the load resulting in plastic deformation of 1/10 000 of the rolling element diameter. This criterion corresponds to a Hertzian contact pressure of approximately 4200 N/mm^2 (610 000 psi) at the center of the contact area for standard radial ball bearings, and this limiting pressure varies less than 10% for most other ball and roller bearing types (Harris, 1991). Larger deformations may result in objectionable noise, vibration, and short fatigue life during normal running. Where these effects and short fatigue life are tolerable, stationary loads up to four times the static capacity may be carried. Loads of about eight times the static capacity may fracture the bearing steel.

For combined radial and thrust loads, P_r and P_t, the equivalent static bearing load P for radial contact single row ball bearings becomes

$$P = 0.6P_r + 0.5P_t \qquad (13.15)$$

For angular-contact ball bearings, the thrust multiplier drops: from 0.5 for radial contact to 0.33 for a 30° contact angle (SKF, 1999).

As shown in Table 13.11, the static capacity for deep-groove ball bearings commonly ranges from about half or less of the basic dynamic load rating C for quite small bearings up to near equality for larger bearings of the common series. For large roller thrust bearings, the static capacity may range up to over twice the dynamic rating.

Table 13.11 Typical CO/C ratios of static capacity to basic dynamic load rating

| Bearing bore (mm) | Deep-groove ball bearings | | | | Angular-contact ball bearings | | Cylindrical roller bearings | |
	19XX Series	01XX series	02XX series	03XX series	72XX series	73XX series	N2XX series	N3XX series
5			0.34	0.36				
7		0.36	0.42					
10	0.38	0.42	0.46	0.42	0.48			
15	0.51	0.51	0.48	0.47	0.54	0.52	0.82	0.79
20	0.57	0.53	0.52	0.49	0.59	0.55	0.88	0.84
30	0.63	0.62	0.57	0.57	0.66	0.61	0.96	0.94
50	0.71	0.74	0.66	0.61	0.78	0.69	1.08	1.02
70	0.77	0.82	0.74	0.65	0.84	0.76	1.15	1.18
100	0.98	0.89	0.75	0.81	0.90	0.94	1.21	1.26
150	1.05	1.00	0.95	1.03	1.15	1.21	1.24	1.37

Problems

13.1 A cylindrical 1500 lb horizontal rotor is centered between two supporting bearings 30 in. apart. The rotor is driven at 900 rpm by a gear located 9 in. outboard of the support bearing at the drive end. This drive applies a 400 lb horizontal force on the gear. An L_{10} life of 25 000 h is required. Determine the minimum size N2XX series cylindrical roller bearing to use at the drive end and the 62XX single-row, deep-groove ball bearing at the opposite end.

13.2 A 209 single-row, deep-groove ball bearing running at 900 rpm carries a combined force involving 500 lb radial load and 1000 lb thrust. What is the fatigue life for C0 normal internal clearance and for C3 and C4 greater than standard internal clearance? Repeat for 500 lb radial load and 200 lb thrust.

13.3 A vertical centrifuge rotor weighing 200 lb runs at 900 rpm with a typical radial imbalance force of 500 lb which is to be carried equally on a lower 72XX angular contact ball thrust bearing with a 40° contact angle and an upper N62XX cylindrical roller bearing. What size angular-contact ball bearing should be used to provide a 5000 h fatigue life? Suggest the mounting fits to be used on the shaft and in the housing. What size roller bearing should be used? Suggest its mounting fits.

13.4 The load on a ball bearing running at 500 rpm is composed of a steady component of 600 lb with an added sinusoidal varying load of 300 lb. What is the equivalent effective load to be used in establishing fatigue life?

13.5 With 1000 lb radial load and 1000 rpm speed, determine the smallest 6200 and 6300 single-row ball bearings, a 7200 angular-contact ball bearing with a 40° contact angle, and a N200 cylindrical roller bearing for a 20 000 h fatigue life. What is the volume within the envelope of each?

13.6 What precautions should be taken when mounting ball bearings in aluminum housings for low-temperature service? For high-temperature service?

13.7 A pump impeller is to operate at 6500 rpm with a thrust load of 175 lb. The impeller weighs 25 lb and overhangs 2 in. The shaft weighs 10 lb and is directly connected to its drive motor. Bearings are to be spaced 6 in. apart. Assuming a minimum shaft diameter of 3/4 in., select rolling bearings from the range of those given in Table 13.5 for a 10% fatigue life of 25 000 h.
   (a) Compute the load on the front and rear bearings.
   (b) Would you use a roller bearing? If so, would you place it near the impeller or the coupling? Why?
   (c) What bearing sizes will give you the smallest outer diameter? The narrowest bearings?

13.8 The radial load on a 6310 ball bearing is unidirectional but varies sinusoidally. Peak load is 1000 lb. For the purpose of computing bearing fatigue life, what is the equivalent mean load? Compute the bearing life at 3600 rpm and compare it with the life if (a) the maximum load had been assumed and (b) the average load had been assumed.

References

American Bearing Manufacturers Association. 1995. *Shaft and Housing Fits for Metric Radial Ball and Roller Bearings (Except Tapered Roller Bearings) Conforming to Basic Boundary Plan*, Std. 70. American National Standards Institute/American Bearing Manufacturers Association, Washington, DC.

American Bearing Manufacturers Association. 1996. *Radial Bearings of Ball, Cylindrical Roller and Spherical Roller Types – Metric Design*, ANSI/ABMA Std. 20. American National Standards Institute/American Bearing Manufacturers Association, Washington, DC.

ASME. 2003. *Life Ratings for Modern Rolling Bearings*, Trib-Vol. 14, ASME, New York.

Derner, W.J. and Pfaffenberger, E.E. 1984. 'Rolling Element Bearings,' *Handbook of Lubrication*, Vol. II. CRC Press, Boca Raton, Fla, pp. 495–537.

Eschmann, P., Hasbargen, L., and Weigand, K. 1985. *Ball and Roller Bearings – Theory, Design, and Application*. John Wiley & Sons, New York. (See also the 3d edn, 1999.)

FAG Bearings Corp. 1995. *FAG Rolling Bearings*, Catalog WL41-520ED. FAG Bearings Corp., Danbury, Conn.

Harris, T.A. 1991. *Rolling Bearing Analysis*, 3rd edn. John Wiley & Sons, New York.

Ioannides, E. and Harris, T.A. 1985. 'A New Fatigue Life Model for Rolling Bearings,' *Journal of Tribology, Transactions of the ASME*, Vol. 107, pp. 367–378.

Jendzurski, T. and Moyer, C.A. 1997. 'Rolling Bearings Performance and Design Data,' *Tribology Data Handbook*. CRC Press, Boca Raton, Fla, pp. 645–668.

Power Transmission Design. 1990. 'Rolling Bearing Design Life,' Sept. issue.

SKF. 1999/2005. *General Catalog*. SKF USA, King of Prussia, Pa.

Zaretsky, E.V. (ed.). 1992. *Life Factors for Rolling Bearings*. Society of Tribologists and Lubrication Engineers, Park Ridge, Ill.

14

Principles and Operating Limits

Chapter 13 provided details on the dimensions and fatigue life of ball and roller bearings. With that background, this chapter covers internal details of local contact areas and stresses, deflections under load, and relative speed of the rolling elements. Limits on operating parameters such as temperature, load, and speed encountered with ball and roller bearings are also covered in this chapter.

While the usual applications of ball and roller bearings treat them as rather simple mechanisms, features of their internal geometry and highly stressed contacts become quite complex, as discussed by Harris (1991) and Eschmann *et al.* (1985). These details, normally considered and studied primarily by the bearing manufacturers, involve such factors as relative conformities of rolling elements and raceways (osculations), angles of contact between balls and raceways under thrust load, pitch diameter and speed of rolling elements, curvature and fit of pockets in the separator, and especially the uniquely demanding stress levels at a point contact (between balls and raceways) or line contact (between rollers and raceways).

A number of complex computer programs have been developed to analyze details of the stresses and motions in rolling bearings for demanding applications such as those in jet engines and various aerospace units (ASME, 2003).

14.1 Internal Geometry

Despite the overall relative simplicity of ball and roller bearings, the details of their internal geometry are quite complex. With an ordinary single-row, deep-groove ball bearing, for instance, even a steady radial load will give different and varying loads on individual balls. With the application of added axial loading, the contact angle with sides of the raceway grooves will vary around the bearing circumference and the ball speed will fluctuate. While the principles and examples presented here will provide insight into the basic factors involved in general machine applications, insight into demanding applications requires detailed coverage (Jones, 1946; Eschmann *et al.*, 1985; Harris, 1991; ASME, 2003).

Applied Tribology: Bearing Design and Lubrication, Third Edition. Michael M. Khonsari and E. Richard Booser.
© 2017 John Wiley & Sons Ltd. Published 2017 by John Wiley & Sons Ltd.
Companion Website: www.wiley.com/go/Khonsari/Applied_Tribology_Bearing_Design_and_Lubrication_3rd_Edition

Point and Line Contact

So-called *point contact* characterizes all ball bearings. An unloaded ball will contact its race-way at a point to produce point contact when the radius of the ball is smaller than the radius of its groove-shaped raceway. Under load, elastic deformation expands this point into an ellip-tical contact area with characteristic dimensions and stresses, which will be covered in the following sections.

As an alternative with cylindrical or tapered rollers, the rollers and their raceway have the same curvature in their axial plane and the unloaded contact is a line. Under load, this *line contact* then expands to a rectangular area for cylindrical rollers and to a trapezoid for tapered rollers.

Barrel-shaped rollers in spherical roller bearings have greater curvature than their raceway and correspondingly produce point contact when unloaded. While a truly elliptical contact area is generated at light loads, at medium and high loads with a barrel roller bearing the contact area extends to the ends of the rollers, cutting off the ends of an otherwise elliptic shape.

Curvature and Ball Contact Geometry

Several conventions are used to describe general curvature relations, such as those shown in Figure 14.1 when applied to the mating contact of a rolling element with its raceway in a bearing ring. Curvature ρ is the reciprocal of the radius of curvature r. Convex curvature is taken as positive, with the center of curvature being within the body. Concave curvature is negative, with the center of curvature being outside the body. Subscripts involving two indices qualify each curvature; the first number refers to the body and the second to the principal plane. For example, ρ_{12} indicates the curvature of body 1 in the principal plane 2.

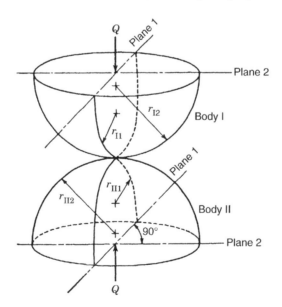

Figure 14.1   Contact of two bodies and their principal planes of curvature. *Source*: Harris, 1991. Repro-duced with permission of John Wiley & Sons.

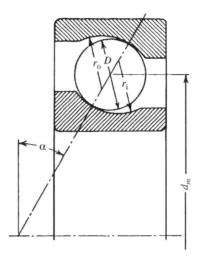

Figure 14.2   Ball bearing geometry. *Source*: Harris, 1991. Reproduced with permission of John Wiley & Sons.

The following factors describe curvature relations in the contact position:

Curvature sum:
$$\Sigma\rho = \frac{1}{r_{11}} + \frac{1}{r_{12}} + \frac{1}{r_{21}} + \frac{1}{r_{22}} \tag{14.1}$$

Curvature difference:
$$F(\rho) = \frac{(\rho_{11} - \rho_{12}) + (\rho_{21} - \rho_{22})}{\Sigma\rho} \tag{14.2}$$

Planes 1 and 2 are always selected to give a positive value for this curvature difference. Alternatively, if the calculation gives a negative sign, the value of curvature difference from Equation (14.2) should simply be set positive.

For a single-row ball bearing, radii for use in Equation (14.1) become the following for the ball as body 1 and raceway as body 2, as shown in Figure 14.2, with the contact radius on the raceway becoming a function of the contact angle α:

At the inner ring:

$$r_{11} = r_{12} = \frac{D}{2}$$

$$r_{21} = \frac{d_i}{2} = \left(\frac{1}{2}\right)\left(\frac{d_m}{\cos\alpha} - D\right)$$

$$r_{22} = f_i D$$

At the outer ring:

$$r_{11} = r_{12} = \frac{D}{2}$$

$$r_{21} = \frac{d_o}{2} = \left(\frac{1}{2}\right)\left(\frac{d_m}{\cos\alpha} + D\right)$$

$$r_{22} = f_o D$$

$$(14.3)$$

Two additional performance factors are the internal clearance in the bearing and curvature ratios f_i and f_o, respectively. With ball bearings, f_i and f_o represent the ratios of inner and outer raceway curvature radii to ball diameter D.

For a cylindrical roller bearing, the above relations are simplified, with only a single radius (and curvature) each for both the roller and its individual raceways. With a spherical roller bearing and its barrel-shaped rollers, two unequal curvatures are involved with the rolling

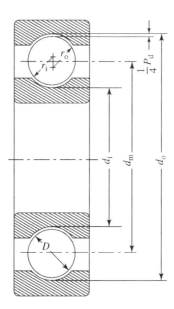

Figure 14.3 Radial ball bearing showing diametral clearance. *Source*: Harris, 1991. Reproduced with permission of John Wiley & Sons.

elements: one in the rolling axis and the other in the axial plane of the roller. The diameter D of the roller is taken at its midlength. The mean diameter of a roller can approximate a tapered roller bearing. The force applied by the end of the roller on the guide flange of a tapered roller bearing can be determined from the total load on a roller and the contact angle involved.

Radial (Diametral) Internal Clearance

Standard ranges of diametral internal clearance are covered in Chapter 13, along with factors involved in their selection. Diametral clearance P_d is shown in Figure 14.3 to be

$$P_d = d_o - d_i - 2D \tag{14.4}$$

where d_i and d_o are the diameters of the inner and outer ring raceways and D is the ball (or roller) diameter. While clearance is always taken on a diameter, the measurement is made in a radial plane. As is typically the case, the terms *radial* and *diametral clearance* for rolling element bearings are used interchangeably in this textbook.

This clearance is diminished by any interference fits used in mounting the inner ring on its shaft or the outer ring in its housing. Analytical relations allow calculation of this reduction of clearance for important applications (see e.g. Harris, 1991). Experience and calculations show that about 80% of the interference fit in mounting either ring of a light series bearing shows up as loss of internal diametral play, 70% for medium series, and 60% for heavy series.

Internal diametral play may be reduced further from an increase in the temperature and thermal expansion of the inner ring above those of the outer ring, and some increase in internal clearance may result from a heavy radial or axial load. Particular attention is required for differential thermal expansion effects when aluminum housings with the usual steel rolling element bearings are used. With all-steel systems and for industrial machinery generally, no interference fit is common in housings, and any increase in the temperature of the inner ring above that of the outer ring is only of the order of 5–10 °C.

Axial Clearance

Axial clearance or end play P_e in a deep-groove ball bearing is several times (often of the order of 10 times) the diametral play P_d. This end play is a function of the raceway conformities, as given by the following equation:

$$P_e = \left[4DP_d(f_i + f_o - 1) - P_d^2\right]^{0.5} \tag{14.5}$$

where f_i and f_o are the ratios of inner and outer raceway curvature radii to ball diameter. This internal axial looseness is also affected by many of the same mounting and load influences involved in diametral internal clearance.

Angular Misalignment

Angular misalignment β that can be freely tolerated by a ball bearing (without inducing elastic contact deflections; see Figure 14.3) is related to end play by the following equation:

$$\tan \beta = \frac{P_e}{\text{mean diameter}} \tag{14.6}$$

As covered in Section 14.13, 10–15 min is a common misalignment limit in applying deep-groove ball bearings, and 5 min is common with cylindrical roller bearings.

Example 14.1   A 6210 deep-groove ball bearing involves the following dimensions in Figure 14.3. Calculate the ball–raceways contact curvatures involved under radial load, diametral and axial internal clearance, and maximum misalignment angle.

Bore	50 mm
Outside diameter	90 mm
Ball diameter, D	12.7 mm (0.5 in.) ($R = 6.35$ mm)
Number of balls, Z	10
Inner raceway diameter, d_i	57.291 mm (radius = 28.65 mm)
Outer raceway diameter, d_o	82.709 mm (radius = 41.355 mm)
Inner groove ball track radius, r_i	6.6 mm ($f_i = 6.6/12.7 = 0.52$)
Outer groove ball track radius, r_o	6.6 mm ($f_o = 6.6/12.7 = 0.52$)

Curvature sums at the inner ring and outer ring raceways are from Equation (14.1):

$$\Sigma \rho_i = \frac{1}{6.35} + \frac{1}{6.35} + \frac{1}{-6.6} + \frac{1}{28.65} = 0.198 \, \text{mm}^{-1}$$

$$\Sigma \rho_o = \frac{1}{6.35} + \frac{1}{6.35} + \frac{1}{-6.6} + \frac{1}{-41.355} = 0.139 \, \text{mm}^{-1}$$

and curvature differences at the inner and outer ring raceways are from Equation (14.2):

$$F(\rho)_i = \frac{[(1/6.35) - (1/6.35)] + [(1/-6.6) - (1/28.65)]}{0.198} = 0.942$$

$$F(\rho)_o = \frac{[(1/6.35) - (1/6.35)] + [(1/-6.6) - (1/-41.355)]}{0.139} = 0.916$$

(Positive sign required for $F(\rho)$; see Section 14.1.)
  Diametral clearance P_d from Equation (14.4):

$$P_d = d_o - d_i - 2D$$
$$= 82.709 - 57.291 - 2 \times 12.7$$
$$= 0.018\,\text{mm} \quad (0.0007\,\text{in.})$$

Axial play is given by Equation (14.5) as follows:

$$P_e = \left[4DP_d(f_i + f_o - 1) - P_d^2\right]^{0.5}$$
$$= [4(12.7)(0.018)(0.52 + 0.52 - 1) - (0.018)^2]^{0.5}$$
$$= 0.19\,\text{mm} \quad (0.0075\,\text{in.})$$

where f_i and f_o are the ratios of inner and outer raceway curvature radii to ball diameter.
  Maximum angle of misalignment β without elastic deflection of ball–raceway contacts is given by Equation (14.6):

$$\tan \beta = \frac{P_e}{[1/2 \times (\text{OD} + \text{bore})]} = \frac{0.19}{[1/2(90 + 50)]} = 0.0027$$
$$\beta = 9\,\text{min}$$

14.2 Surface Stresses and Deformations

Calculation of stresses and corresponding deformation at contact points between rolling elements and raceways employs elasticity relations established by Hertz in 1881. Hertz's theory considers the contact of two bodies with curved surfaces under force Q (Figure 14.1).

Ball–Raceway Contacts

The contact area between two curved bodies in point contact has an elliptical shape (Figure 14.4). Using the effective modulus of elasticity E' and the compressive force Q gives the following a and b as the axes of the elliptical area of contact, and deformation δ as relative approach of the two bodies involving the sum of deflections of the two individual surfaces:

$$a = a^* \left(\frac{3Q}{\Sigma \rho E'}\right)^{1/3} \tag{14.7}$$

$$b = b^* \left(\frac{3Q}{\Sigma \rho E'}\right)^{1/3} \tag{14.8}$$

$$\delta = \delta^* \left(\frac{3Q}{\Sigma \rho E'}\right)^{2/3} \left(\frac{\Sigma \rho}{2}\right) \tag{14.9}$$

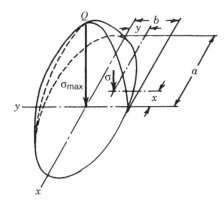

Figure 14.4 Ellipsoidal stress distribution in point contact. *Source*: Harris, 1991. Reproduced with permission of John Wiley & Sons.

Hertz dimensionless contact coefficients a^*, b^*, and δ^* for any particular case are extracted by interpolation from Table 14.1 based on curvature difference values $F(\rho)$ from Equation (14.2) (Harris, 1991). Effective modulus of elasticity E' for ball–raceway contact involves the individual modulus of elasticity E and the Poisson ratio v of the two contacting materials:

$$\frac{2}{E'} = \left(\frac{1 - v_1^2}{E_1}\right) + \left(\frac{1 - v_2^2}{E_2}\right) \tag{14.10}$$

Table 14.2 gives values of these elasticity parameters for steel-on-steel contacts and several other material combinations which might be encountered in rolling contacts. E' has a value of $219 \times 10^3 \, \text{N/mm}^3$ for either through-hardened or case-hardened steels in raceway contacts with balls or rollers.

The maximum compressive stress (1.5 times the mean stress $Q/\pi ab$) occurs at the geometric center of this elliptical contact area with the magnitude

$$\sigma_{max} = \frac{3Q}{2\pi ab} \tag{14.11}$$

and the stress at other points within the contact area is given by the following ellipsoidal distribution, as shown in Figure 14.4:

$$\sigma = \sigma_{max} \left[1 - \left(\frac{x}{a}\right)^2 - \left(\frac{y}{b}\right)^2\right]^{1/2} \tag{14.12}$$

Example 14.2 The 6210 deep-groove ball bearing in Example 14.1 is to carry a radial load F_r of 3000 N (674 lb). Calculate the radial load, Hertz contact stresses, and deflection for the most heavily loaded ball.

From Equation (14.21) in Section 14.4, load Q_{max} on the most heavily loaded ball becomes

$$Q_{max} = \frac{5.0 F_r}{Z} = \frac{5.0(3000)}{10} = 1500 \, \text{N}$$

Table 14.1 Dimensionless elastic contact parameters

$F(\rho)$	a^*	b^*	δ^*
0	1	1	1
0.10	1.07	0.938	0.997
0.20	1.15	0.879	0.991
0.30	1.24	0.824	0.979
0.40	1.35	0.771	0.962
0.50	1.48	0.718	0.938
0.60	1.66	0.664	0.904
0.70	1.91	0.607	0.859
0.80	2.30	0.544	0.792
0.85	2.60	0.507	0.745
0.90	3.09	0.461	0.680
0.91	3.23	0.450	0.664
0.92	3.40	0.438	0.646
0.93	3.59	0.426	0.626
0.94	3.83	0.412	0.603
0.95	4.12	0.396	0.577
0.955	4.30	0.388	0.562
0.960	4.51	0.378	0.546
0.965	4.76	0.367	0.530
0.970	5.05	0.357	0.509
0.975	5.44	0.343	0.486
0.980	5.94	0.328	0.459
0.982	6.19	0.321	0.447
0.984	6.47	0.314	0.433
0.986	6.84	0.305	0.420
0.988	7.25	0.297	0.402
0.990	7.76	0.287	0.384
0.992	8.47	0.275	0.362
0.994	9.46	0.260	0.336
0.996	11.02	0.241	0.302
0.997	12.26	0.228	0.279
0.998	14.25	0.212	0.249
0.9985	15.77	0.201	0.230
0.9990	18.53	0.185	0.207
0.9995	23.95	0.163	0.171
1		0	0

Source: Eschmann *et al.*, 1985, Harris, 1991.

Curvature sums $\Sigma\rho$ and differences $F(\rho)$ at the inner ring and outer ring raceways from Equations (14.1) and (14.2) as calculated in Example 14.1:

$$\Sigma\rho_i = 0.198 \text{ mm}^{-1}; \quad \Sigma\rho_o = 0.139 \text{ mm}^{-1}$$
$$F(\rho)_i = 0.942; \quad\quad F(\rho)_o = 0.916$$

Table 14.2   Elastic properties and density of rolling bearing materials

Material	Elastic modulus $(N/mm^2 \times 10^3, psi \times 10^6)$	Poisson's ratio	Density (g/cm^3)
AISI 5200 through hardened	202 (29)	0.28	7.8
AISI 8620 carburized	202 (29)	0.28	7.75
AISI 440C stainless	200 (29)	0.28	7.5
AISI M50 tool steel	193 (28)	0.28	7.6
Si_3N_4 silicon nitride ceramic	310 (45)	0.26	3.2

From these values of curvature difference to enter Table 14.1, the following are the contact parameters at the inner and outer ring raceways:

	Inner raceway	Outer raceway
a^*	3.88	3.33
b^*	0.409	0.443
δ^*	0.598	0.653

From Equations (14.7)–(14.9) with $Q = Q_{max} = 1500\,N$ and $E' = 219 \times 10^3\,N/mm^2$:

	Inner raceway	Outer raceway
$a = a^* \left(\dfrac{3Q}{\Sigma \rho E'} \right)^{1/3}$	1.823 mm	1.761 mm
$b = b^* \left(\dfrac{3Q}{\Sigma \rho E'} \right)^{1/3}$	0.192 mm	0.243 mm
$\delta = \delta^* \left(\dfrac{3Q}{\Sigma \rho E'} \right)^{1/3} \left(\dfrac{\Sigma \rho}{2} \right)$	0.0278 mm	0.0240 mm

Maximum Hertz elastic compressive stress in the contacts from Equation (14.11) becomes:

	Inner	Outer
$\sigma_{max} = \dfrac{3Q}{2\pi ab}$	$2044\,N/mm^2$ (296 000 psi)	$1736\,N/mm^2$ (252 000 psi)

As is usually the case, this contact stress on the inner ring raceway is higher than the stress on the outer ring because of the poorer conformity of the inner ring raceway curvature with the balls. Since more contact stress cycles are encountered in the stationary loaded zone of the outer ring raceway with a steady radial load and a rotating inner ring, however, the likelihood of fatigue failure on either raceway is similar.

Roller–Raceway Line Contacts

In this section, relations for the contact of infinitely long cylinders are developed. They may be applied, however, to both cylindrical and tapered rollers. With cylindrical rollers, the cylindrical contour used for bearing elements is normally modified with a slight crown and with relieved roller ends to provide a uniform stress pattern along the roller length. This eliminates a peak contact stress of the order of 1.5 times the mean, which otherwise would be encountered at the ends of a cylinder. For tapered rollers, calculations can be based on the mean outer ring (cone) diameter.

For the contact of a cylindrical roller with its raceway in a roller bearing, curvatures in the axial plane for both bodies become zero, and the semiwidth b of the contact area becomes

$$b = \left(\frac{8Q}{\pi L \Sigma \rho E'} \right)^{1/2} \tag{14.13}$$

Compressive stress distribution in the contact area degenerates to the semicylindrical form shown in Figure 14.5, with the maximum stress at the center line of the pressure rectangle being

$$\sigma_{max} = \frac{2Q}{\pi b L} \tag{14.14}$$

Effective length L is the total roller length minus possible recesses in the raceways and the two corner radii normally provided at the ends of the rollers. This maximum stress is 1.273 times the mean stress Q/bL. Compressive stress acting at any distance y from the center line of the contact area is given by the following equation:

$$\sigma = \sigma_{max} \left[1 - \left(\frac{y}{b} \right)^2 \right]^{1/2} \tag{14.15}$$

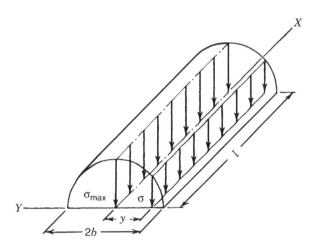

Figure 14.5 Semicylindrical surface compressive stress distribution of ideal line contact. *Source*: Harris, 1991. Reproduced with permission of John Wiley & Sons.

Deflection δ for line contact of a cylindrical roller with its raceway (approach of the roller involving the sum of the elastic deformations of both the roller and raceway surfaces) is given by the following equation for steel on steel:

$$\delta = \frac{3.84 \times 10^{-5} Q^{0.9}}{L^{0.8}} \tag{14.16}$$

For other material combinations, load Q should first be multiplied by the ratio of the effective modulus E' for steel on steel to that of the alternative material combination. The units for Q and L are N and mm, respectively.

Example 14.3   A 209 cylindrical roller bearing with the following dimensions is to carry a 6000 N radial load. Calculate the load, maximum Hertz elastic compressive contact stresses, and deflections for the most heavily loaded roller.

Inner raceway diameter, d_i	54.991 mm (2.165 in.)
Outer raceway diameter, d_o	75.032 mm (2.954 in.)
Roller diameter, D	10 mm (0.3937 in.)
Roller effective length, L	9.601 mm (0.3780 in.)
Number of rollers, Z	14

Curvature sums for the inner and outer raceway contacts are as follows from Equation (14.1) (simplified for curvature only in the rolling direction for an ideal line contact):

$$\Sigma\rho = \frac{2}{d} + \frac{2}{D}; \qquad \Sigma\rho_i = \frac{2}{54.991} + \frac{2}{10} = 0.236 \, \text{mm}^{-1}$$

$$\Sigma\rho_o = \frac{2}{75.032} + \frac{2}{10} = 0.227 \, \text{mm}^{-1}$$

From Equation (14.21), which follows, load Q_{max} on the most heavily loaded roller becomes

$$Q_{\text{max}} = \frac{5.0 F_r}{Z} = \frac{5.0(6000)}{14} = 2143 \, \text{N}$$

Using this roller load $Q = Q_{\text{max}} = 2143 \, \text{N}$ and $E' = 219 \times 10^3 \, \text{N/mm}^2$, half-width b of the rectangular contact areas is given as follows from Equation (14.13):

$$b = \left(\frac{8Q}{\pi L \Sigma \rho E'} \right)^{1/2}$$

$$b_i = \left[\frac{8 \times 2143}{(3.1416 \times 9.601 \times 0.236)(219 \times 10^3)} \right]^{1/2} = 0.1049 \, \text{mm}$$

$$b_o = \left[\frac{8 \times 2143}{(3.1416 \times 9.601 \times 0.227)(219 \times 10^3)} \right]^{1/2} = 0.1069 \, \text{mm}$$

Maximum elastic compressive stresses at the inner and outer raceway contacts from Equation (14.14) become

$$\sigma_{max} = \frac{2Q}{\pi bL}$$

$$\sigma_{imax} = \frac{2 \times 2143}{\pi 0.1049 \times 9.601} = 1355 \, \text{N/mm}^2 \quad (196\,000\,\text{psi})$$

$$\sigma_{omax} = \frac{2 \times 2143}{\pi 0.1069 \times 9.601} = 1329 \, \text{N/mm}^2 \quad (193\,000\,\text{psi})$$

Finally, elastic deflection in each of the inner and outer raceway contacts is given by Equation (14.16):

$$\delta = \frac{3.84 \times 10^{-5}Q^{0.9}}{L^{0.8}} = \frac{3.84 \times 10^{-5}(2143)^{0.9}}{(9.601)^{0.8}} = 0.0063 \, \text{mm}$$

14.3 Subsurface Stresses

In addition to compressive surface stresses, stresses within the bearing material just below the contact surface are of interest. Examination of bearings which have suffered surface fatigue failures under high load indicates that the fatigue cracking which initiates a surface pit or spall commonly originates in zones below the surface.

With surface stress being maximum at the Z axis at the center line of the surface contact area, related principal stresses σ_x, σ_y, and σ_z acting on a subsurface element on the Z axis are illustrated in Figure 14.6. Details on the evaluation of these stresses are summarized by Harris (1991).

Figure 14.7 illustrates how these stresses change with increasing depth z below the center of an ideal line contact. Compressive stress σ_z slowly decreases from its maximum at the surface, where it equals the contact pressure. At the same time, stress σ_y in the direction of contact area width $2b$ decreases more rapidly with z. The main shear stress $\tau_{yz} = (\sigma_z - \sigma S_y)/2$ reaches its maximum when depth z below the surface is $0.786b$ for a line contact. For simple point contact with $a = b$, maximum shear stress occurs at depth $z = 0.467b$. For a typical ball bearing contact with $b/a = 0.1$, maximum shear stress occurs approximately at $z = 0.76b$ (Harris, 1991). During cycling with the passage of a rolling element, these fluctuating stresses increase until they reach their maximum value at the Z axis and then decrease.

Quite different is the case of the shear stress τ_{yz} that occurs in a plane orthogonal to the external load, i.e. parallel to the contact surface and also parallel to the Y axis. During cycling, τ_{yz} acts in the zone before the Z axis in the same direction as the rolling motion. At the Z axis, τ_{yz} drops to zero and then acts opposite to the direction of motion. These peak values for line contacts are $+0.25\sigma_{max}$ to $-0.25\sigma_{max}$ and give a double amplitude of $0.5\sigma_{max}$ at depth $z = 0.5b$. For a typical ball–raceway contact with contact ellipse $b/a \sim 0.1$, double amplitude is $0.50\sigma_{max}$ at a depth of $0.49b$; for a circular contact the value is $0.43\sigma_{max}$ at a depth of $0.35b$. Harris (1991) summarizes observations indicating which subsurface stress factors are major contributors to fatigue failures.

Example 14.4 Determine the amplitude of the maximum shear stress and the maximum orthogonal shear stress at the inner and outer ring raceways of the 6210 ball bearing of Example 14.2. Estimate the depth below the ball–raceway contact at which these stresses occur.

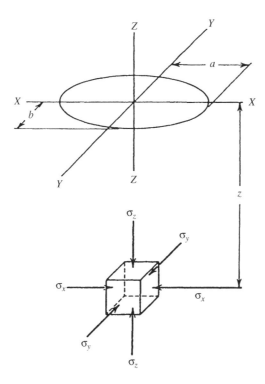

Figure 14.6 Principal stresses occuring on an element on the Z axis below the contact surface. *Source*: Harris, 1991. Reproduced with permission of John Wiley & Sons.

	Inner raceway	Outer raceway
From Example 14.2:		
Max. compressive contact stress, σ_{max}(N/mm^2)	2044	1736
Contact ellipse dimensions (mm)		
a	1.823	1.761
b	0.192	0.243
Following the text of Section 14.3:		
Maximum shear stress		
Magnitude $\tau_{yz\,max} = 0.33\sigma_{max}$(N/mm^2)	675	573
Depth below contact surface, $z = 0.77b$ (mm)	0.148	0.187
Maximum orthogonal shear stress		
Magnitude $\tau_o = 0.5\sigma_{max}$(N/mm^2)	1022	868
Depth below contact surface, $z = 0.49b$ (mm)	0.094	0.119

Harris (1991) gives more details for variation of the numerical constants involved as a function of ellipse dimension ratio b/a. The values used here were taken for $b/a = 0.1$ as typical for ball–raceway contacts, and the variation is minor for the range of ellipticity usually

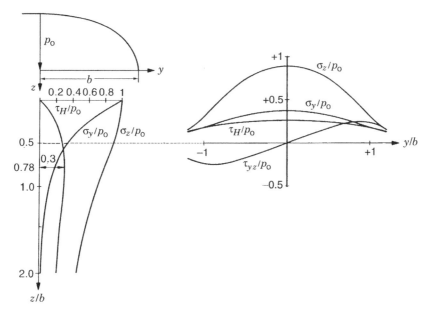

Figure 14.7 Subsurface stresses at ideal line contact. *Source*: Eschmann *et al.*, 1985. Reproduced with permission of John Wiley & Sons.

encountered. For case-hardened steels, the case depth should exceed the depth to the zone of maximum orthogonal shear stress by at least a factor of 3.

14.4 Load Distribution on Rolling Elements

Balls or rollers transmit the external force on a bearing from one ring to the other. This external force, generally composed of a radial force F_r and an axial force F_a, is distributed over a number of supporting rolling elements in a statically indeterminate system.

In a general case, load distribution may be determined by starting with a mutual displacement of the rings and calculating the associated deformation of each rolling element. These deformations correspond to forces on individual rolling elements, as covered in the previous section. The geometric sum of these forces then represents the total load on the bearing. If this calculation is repeated for a range of arbitrary displacements, data are obtained which provide the load distribution and overall bearing deflection for any external load. Harris (1991) and Eschmann *et al.* (1985) give charts for use in hand calculations. The relations discussed below are generally suitable for preliminary approximations.

Radially Loaded Bearings

The load distribution in a radially loaded bearing is represented in Figure 14.8. Total radial force F_r on the bearing equals the sum of the support provided by the individual rolling elements.

$$F_r = Q_1 + 2Q_2 \cos \beta + 2Q_3 \cos 2\beta + \cdots \tag{14.17}$$

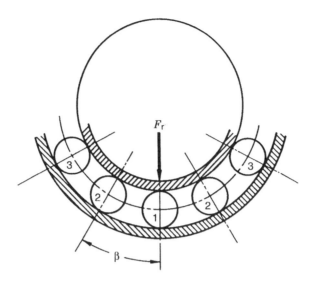

Figure 14.8   Distribution of ball load under radial force F_r.

With a number of balls Z, $\beta = 360/Z$ and Q is the force supported on an individual ball or roller.

For a ball bearing, Equation (14.9) gives $Q_2/Q_1 = (\delta_2/\delta_1)^{3/2}$. Then $Q_2/Q_1 = (\cos \beta)^{3/2}$ and Equation (14.17) becomes

$$F_r = Q_1[1 + 2(\cos \beta)^{5/2} + 2(\cos 2\beta)^{5/2} + \cdots] \qquad (14.18)$$

For the line contacts in roller bearings, Q is proportional to $(\delta)^{1/0.9}$ from Equation (14.16). For roller bearings, Equation (14.17) becomes

$$F_r = Q_1[1 + 2(\cos \beta)^{1.9/0.9} + 2(\cos 2\beta)^{1.9/0.9} + \cdots] \qquad (14.19)$$

The load-distribution quantities in brackets in Equations (14.18) and (14.19) depend only on the number of rolling elements Z. For bearings with zero internal clearance under purely radial load and with zero contact angles for the rolling elements, the following equations give the maximum force $Q_{max} = Q_1$:

$$\begin{array}{cc} \text{Ball bearings} & \text{Roller bearings} \\ Q_{max} = \dfrac{4.37 \times F_r}{Z} & Q_{max} = \dfrac{4.06 \times F_r}{Z} \end{array} \qquad (14.20)$$

The numerical constants vary for bearings having radial clearance or preload and also with the total radial load. In calculations for the usual applications with either ball or roller bearings under radial load and having nominal internal clearance, a factor of 5.0 is commonly used for estimating the load on the most heavily loaded ball or roller (Harris, 1991):

$$Q_{max} = \dfrac{5.0 \times F_r}{Z} \qquad (14.21)$$

Examples 14.2 and 14.3 include applications of this relation. The factor 5.0 increases with increasing radial clearance and for light loads when fewer balls are involved in sharing the total radial load. As an extreme example, a single ball or roller will carry the total radial load while involving the Hertz elastic contact parameters given in Equations (14.7)–(14.9) for ball contact and Equations (14.13)–(14.16) for a roller. After decreasing moderately for minor preloads, the factor 5.0 again increases for heavy preloads when preload force adds to the loading from the radial load (Eschmann *et al.*, 1985).

With a contact angle α less than 90° for the rolling elements in a loaded bearing, Equation (14.21) becomes

$$Q_{max} = \frac{5.0 \times F_r}{(Z \cos \alpha)} \tag{14.22}$$

Thrust Loaded Bearings

Under thrust load, ball and roller bearings have the total load F_a equally distributed among the rolling elements to give normal load Q on each rolling element:

$$Q = \frac{F_a}{(Z \sin \alpha)} \tag{14.23}$$

where α is the contact angle in the loaded bearing. For thrust ball bearings whose contact angles are less than 90°, the contact angle increases under load. In general, however, operation will involve a contact angle close to a design value such as 40°. Bearing manufacturers can supply details for such cases, and analytical relations are available (Harris, 1991).

Combined Radial and Thrust Loads

When both radial and axial loads are applied to a bearing inner ring, the ring generally will be displaced a distance δ_a in the axial direction and δ_r in the radial direction. At any angular position ψ measured from the most heavily loaded rolling element, approach of the rings becomes (Harris, 1991)

$$\delta_\psi = \delta_a \sin \alpha + \delta_r \cos \alpha \cos \psi \tag{14.24}$$

This relation is illustrated in Figure 14.9, where the maximum displacement is encountered at $\psi = 0$. For equilibrium, summation of rolling element forces in the radial and axial directions must equal the applied loads in each direction F_r and F_a. This balance can be given as follows in terms of radial and thrust integrals $J_r(\varepsilon)$ and $J_a(\varepsilon)$ as related to the normal load Q_{max} on the most heavily loaded ball:

$$F_r = ZQ_{max}J_r(\varepsilon)\cos\alpha; \qquad F_a = ZQ_{max}J_a(\varepsilon)\sin\alpha \tag{14.25}$$

Table 14.3 gives values of J_r, J_a, and ε versus $F_r \tan\alpha/F_a$ for point (ball) and line (roller) contacts. There ε is a load distribution factor, which represents the fraction of the projected diameter over which the rolling elements come under load, as shown in Figure 14.9.

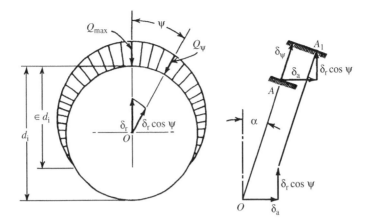

Figure 14.9   Rolling bearing displacements due to combined radial and axial loading. *Source*: Harris, 1991. Reproduced with permission of John Wiley & Sons.

In a pattern analogous to that of Figure 14.8 for a radial load on a ball bearing, as given in Equation (14.18), normal force Q_n on a ball other than that at Q_{max} becomes for ball bearings:

$$Q_n = Q_{max}\left[1 - \frac{(1 - \cos\psi)}{(2\varepsilon)}\right]^{3/2} \tag{14.26}$$

Table 14.3   Radial and thrust integrals for single-row bearings

	Point contact			Line contact		
	$\dfrac{F_r\tan\alpha}{F_a}$			$\dfrac{F_r\tan\alpha}{F_a}$		
ε	$\dfrac{F_r\tan\alpha}{F_a}$	$J_r(\varepsilon)$	$J_a(\varepsilon)$	$\dfrac{F_r\tan\alpha}{F_a}$	$J_r(\varepsilon)$	$J_a(\varepsilon)$
0	1	$1/z$	$1/z$	1	$1/z$	$1/z$
0.2	0.9318	0.1590	0.1707	0.9215	0.1737	0.1885
0.3	0.8964	0.1892	0.2110	0.8805	0.2055	0.2334
0.4	0.8601	0.2117	0.2462	0.8380	0.2286	0.2728
0.5	0.8255	0.2288	0.2782	0.7939	0.2453	0.3090
0.6	0.7835	0.2416	0.3084	0.7480	0.2568	0.3433
0.7	0.7427	0.2505	0.3374	0.6999	0.2636	0.3766
0.8	0.6995	0.2559	0.3658	0.6486	0.2658	0.4098
0.9	0.6529	0.2576	0.3945	0.5920	0.2628	0.4439
1	0.6000	0.2546	0.4244	0.5238	0.2523	0.4817
1.25	0.4338	0.2289	0.5044	0.3598	0.2078	0.5775
1.67	0.3088	0.1871	0.6060	0.2340	0.1589	0.6790
2.5	0.1850	0.1339	0.7240	0.1372	0.1075	0.7837
5	0.0831	0.0711	0.8558	0.0611	0.0544	0.8909
∞	0	0	1	0	0	1

Source: Harris, 1991. Reproduced with permission of John Wiley & Sons.

The exponent shifts from 3/2 for ball bearings to 1/0.9 for roller bearing line contacts, analogous to Equation (14.19).

Example 14.5   A 7220 angular-contact ball bearing is to carry a load of 24 000 N (5393 lb) axially and 20 000 N (4494 lb) radially. What is the load on each of the 16 balls? Assume that the nominal 40° contact angle remains constant.

For entering Table 14.3:

$$\frac{F_r \tan \alpha}{F_a} = \frac{20\,000 \tan 40°}{24\,000} = 0.699$$

giving

$$\varepsilon = 0.799; \qquad J_r(\varepsilon) = 0.256; \qquad J_a(\varepsilon) = 0.366$$

Applying Equation (14.22) with number of balls $Z = 16$, the normal load on the heaviest loaded ball becomes

$$Q_{max} = Q_0 = \frac{F_r}{ZJ_r(\varepsilon)\cos 40°} = \frac{20\,000}{16 \times 0.256 \times 0.766} = 6374\,N$$

Alternatively,

$$Q_{max} = \frac{F_a}{ZJ_a(\varepsilon)\sin 40°} = \frac{24\,000}{16 \times 0.366 \times 0.643} = 6374\,N$$

Angular spacing of balls:

$$\Delta_\psi = \frac{360}{Z} = \frac{360}{16} = 22.5°$$

Load on subsequent balls from Equation (14.26):

$$Q_n = Q_{max}\left\{1 - \frac{[1 - \cos(n \times 22.5)]}{2\varepsilon}\right\}^{3/2}$$

$$Q_1 = 6374\left\{1 - \frac{[1 - 0.9239]}{2 \times 0.799}\right\}^{3/2} = 5924\,N$$

Ball No., n	Angle ψ (degrees)	$\cos(n \times 22.5°)$	Normal load on ball (N)
0	0	1	6374
1	22.5	0.9239	5924
2	45	0.7071	4704
3	67.5	0.3827	3064
4	90	0	1459
5	112.5	−0.3827	315
6	135	−0.7071	0
7	157.5	−0.9239	0
8	180	−1	0

14.5 Speed of Cage and Rolling Elements

While the speed of the balls and cage can be defined for many common geometries and operating conditions, sliding within rolling contacts, frictional drag in cage pockets, lubricant friction, gyroscopic motions, and centrifugal forces may modify relations expected with no gross slip at rolling contacts. Although the no-slip speeds which follow are typical with most bearings, wide variations may be encountered since the driving force required to maintain these speeds is often through a low-friction EHL fluid film (traction coefficient typically ~0.5) between a few loaded rolling elements and their rotating raceway. High drag by a viscous lubricant in cage pockets and at any other cage locating surfaces then tends to slow rotation of the complement of rolling elements below their no-slip speed. During regreasing, for instance, drag by the grease commonly reduces the cage speed by half until excess grease is dispelled from the path of the rolling elements. With low winter temperatures resulting in stiffened grease, the balls and their cage may even be brought to a standstill.

Velocity of the ball center and the cage, v_m in Figure 14.10, becomes the mean of the velocity at the two contact surfaces:

$$v_m = \frac{v_i + v_o}{2} \tag{14.27}$$

For a general case with various degrees of angular contact α of the rolling elements and with rotational speeds n_i and n_o (rpm) for the inner and outer rings:

$$v_m = \left(\frac{\pi d_m}{120}\right)\left\{n_i\left[1 - \left(\frac{D}{d_m}\right)\cos\alpha\right] + n_o\left[1 + \left(\frac{D}{d_m}\right)\cos\alpha\right]\right\} \tag{14.28}$$

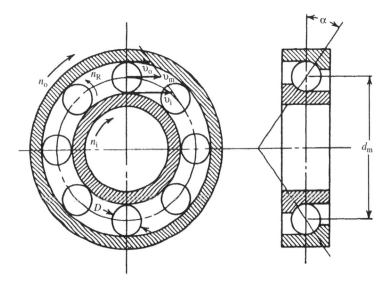

Figure 14.10   Relation of rolling speeds and velocities. *Source*: Harris, 1991. Reproduced with permission of John Wiley & Sons.

With no gross slip at the inner or outer ring raceways, ball or roller velocity n_R equals the raceway velocity at the point of contact to give the following rolling element rotational speed n_R:

$$n_R = \left(\frac{d_i}{D}\right) n_i \quad \text{or} \quad n_R = \left(\frac{d_o}{D}\right) n_o \tag{14.29}$$

From either of these relations:

$$n_R = 0.5 \left(\frac{d_m}{D}\right) (n_o - n_i) \left[1 - \left(\frac{D}{d_m}\right) \cos\alpha\right] \left[1 + \left(\frac{D}{d_m}\right) \cos\alpha\right] \tag{14.30}$$

For the inner ring rotation alone:

$$n_R = 0.5 \left(\frac{d_m}{D}\right) n_i \left[1 - \left(\frac{D}{d_m}\right)^2 \cos^2\alpha\right] \tag{14.31}$$

And for a thrust bearing or a deep-groove ball bearing where the contact angle is 90° with $\cos\alpha = 0$, Equation (14.30) becomes

$$n_R = 0.5 \left(\frac{d_m}{D}\right) (n_o - n_i) \tag{14.32}$$

14.6 Cage Considerations

The primary function of a cage (also called a *separator* or *retainer*) is to maintain the proper spacing of the rolling elements while producing the least possible friction through sliding contacts.

Riveted ball-riding cages pressed from low-carbon steel strip, with their low weight and high elasticity, are usually the most suitable for ball bearings at normal speeds. Several bearing manufacturers attribute to brass and bronze cages an increased speed limit of about 15%. Machined bronze cages, commonly centered on the shoulders of one of the bearing rings, have been traditional for many roller bearings and for high-speed ball bearings. A variety of metal and plastic cage materials in current use are reviewed in Chapter 4.

The stresses and demands on the cages are slight in many bearings at low and moderate speeds. Figure 14.11 illustrates the following primary forces acting on a cage; unbalance force F_{UB} and centrifugal body force F_{CF} become significant at high rotational speeds.

1. Rolling element force on the cage pocket F_{CP}. The direction depends on whether the cage is driving the rolling element or vice versa. Circumferential clearance in the cage pocket should be sufficient to accommodate minor changes in rolling element speed during each revolution. With the common limit of 10–15 minutes of misalignment for a deep-groove ball bearing, for instance, this pocket clearance should be sufficient to accommodate the variation in ball speed as it changes its angular contact with the raceway. Otherwise, cage forces, bearing vibration, and wear may be encountered. Any greater looseness of the cage, however, may result in poor cage guidance, cage vibration, and objectionable cage rattle and squealing.

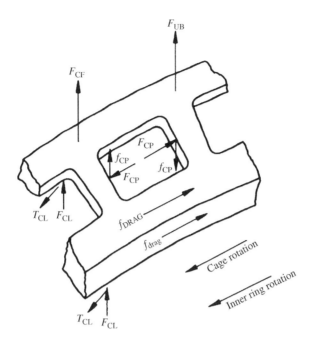

Figure 14.11 Cage forces. *Source*: Harris, 1991. Reproduced with permission of John Wiley & Sons.

2. Viscous drag f_{CP} by lubricant in the cage pocket. This tends to slow rotation of the rolling element. As an extreme example, at winter temperatures below -15 to $-20\,°C$, conventional industrial grease may apply sufficient drag to immobilize the cage and rolling elements despite shaft and inner ring rotation at their normal full speed.
3. Friction force at the cage–land contact F_{CL}. For cage location on the shoulders of a rotating inner ring, this will be a driving force that tends to accelerate the cage. With a cage location on a stationary outer ring, F_{CL} friction will tend to retard rotation. Swelling of plastic cages by a synthetic lubricant or by moisture may cause loss of this cage–land clearance.

Geometric details of the cage pocket should be selected to promote lubricant film formation. The edges of each pocket, for instance, should be shaped and rounded to help rotating ball or roller surfaces draw lubricant into the pocket. The radius of curvature for the pocket should also be slightly greater than that of its rotating element to produce a converging wedge-shaped clearance to promote formation of a lubricant film.

While forces and strains on cages are light under normal operating conditions, cages are commonly sensitive to poor lubrication. Any wear from inadequate lubrication first develops where the cage slides against the rolling elements or rings. As this wear develops, the cage may eventually break and fail the bearing. In cage failures, lubrication faults such as dried grease should always be checked first.

Cage surfaces should be selected from materials that need minimum lubrication for special applications. Silver plating is commonly used for jet engine bearing cages to enable short-time running in the absence of lubricant feed, and for ball bearings without lubrication in the vacuum of X-ray tubes. Self-lubricating composites and bonded solid-film lubricants also find use in vacuum and space applications. Polyimide and other plastic cages help minimize lubricant needs. Cotton-filled phenolics, with their ability to retain oil, are useful for high-speed grease lubrication.

14.7 Vibration

Trueness of shaft rotation depends on the fit and runout (radial and axial) of the bearing and its mounting surfaces, as well as on the elastic behavior of bearing components and related machine elements, as influenced by rotor unbalance.

Effects of bearing irregularities are illustrated in Figure 14.12 for a machine tool spindle such as in a lathe (Eschmann *et al.*, 1985). Despite the inner ring raceway runout in case (a), the workpiece axis of rotation will be the axis of the inner ring raceway. A stationary cutting tool therefore produces a round workpiece. If the tool were rotating against a plane surface, however, the radial runout would take effect to produce a wavy, scalloped surface.

For case (c) of Figure 14.12, with a concentric but elliptical inner ring raceway, the rotation axis is not fixed and the lathe spindle transfers the elliptical raceway shape to the rotating workpiece. For case (b), the irregular surface of the raceway does not take full effect because

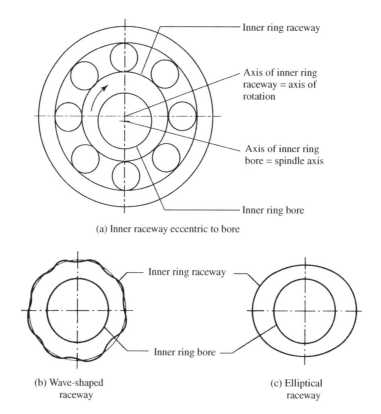

(a) Inner raceway eccentric to bore

(b) Wave-shaped raceway

(c) Elliptical raceway

Figure 14.12   Representative inner ring raceway irregularities. *Source*: Eschmann *et al.*, 1985. Reproduced with permission of John Wiley & Sons.

of some compensation in elastic deformation at rolling contact areas. In practice, the individual deviations idealized in Figure 14.12 are generally superimposed on one another.

Since the relatively thin rings of rolling bearings adapt themselves significantly to the shapes of their mating parts, these parts must have the same form accuracy as the rolling bearings. Further factors to consider in selecting housing and shaft fits are covered in Chapter 13.

While the runout values for rolling bearings are also influenced somewhat by diameter differences in rolling elements, these differences are extremely small for quality bearings. The slight change in deflection occurring as balls or rollers pass through the loaded zone for a radially loaded bearing are also usually negligibly small and are largely compensated for by elastic deformation.

Bearing Frequencies

Fundamental passing frequencies of rolling elements are useful in establishing bearing waviness test speeds and filter bands. Knowledge of these frequencies is also useful in monitoring machinery condition. Based on the relations discussed in Section 14.5 and assuming no skidding of rolling elements, the following relations apply to a stationary outer ring and a rotating inner ring, which is most often the case in machinery applications (Harris, 1991). Rotational speed of the cage is

$$f_c = \left(\frac{N}{2}\right)\left[1 - \left(\frac{D}{d_m}\right)\cos\alpha\right] \tag{14.33}$$

where N is the rotational speed of the inner ring, d_m is the bearing pitch diameter, and D is the rolling element diameter. The rotational speed f_{ci} at which a fixed point on the inner ring passes a fixed point on the cage is

$$f_{ci} = \left(\frac{N}{2}\right)\left[1 + \left(\frac{D}{d_m}\right)\cos\alpha\right] \tag{14.34}$$

The rate f_{bpor} at which balls pass a point in the outer ring raceway (outer raceway defect frequency) is

$$f_{bpor} = Zf_c \tag{14.35}$$

The rate f_{bpir} at which a point in the inner ring raceway passes balls (inner raceway defect frequency) is

$$f_{bpir} = Zf_{ci} = Z(N - f_c) \tag{14.36}$$

The rate of rotation of a ball or roller about its own axis becomes

$$f_r = \left(\frac{Nd_m}{2D}\right)\left\{1 - \left[\left(\frac{D}{d_m}\right)\cos\alpha\right]^2\right\} \tag{14.37}$$

A single defect on a ball or roller in one revolution would contact both raceways to give a defect frequency of $2f_r$.

14.8 Bearing Elasticity

Since bearings deflect when subject to load, they have the properties of mechanical springs, which play a vital role in the dynamic behavior of a rotor system (Gargiulo, 1980). This overall elasticity is a complex function of the stiffness and shape of the housing, support of bearing rings on their mating parts, internal bearing clearance, any bearing preload, and the position of rolling elements during their rotation.

Usually, however, deflection under a radial load can be related primarily to elastic contact deflection for the most heavily loaded ball or roller, which sets the approach between the inner and outer rings. The normal load Q_{max} on this rolling element is approximately related to the total radial force on the bearing F_r and the number of rolling elements Z by the following equation:

$$Q_{max} = \frac{5.0 \times F_r}{Z \cos \alpha} \tag{14.38}$$

Thrust ball and roller bearings have the load equally distributed on the rolling elements:

$$Q_{max} = \frac{F_a}{Z \sin \alpha} \tag{14.39}$$

where α is the contact angle between a line through the rolling element contact and the radial plane of the bearing.

After Q_{max} has been determined for an individual ball bearing from Equation (14.38) or (14.39), the corresponding ball contact deflections at the inner and outer ring raceways can be obtained from Equation (14.9) in the form $\delta \sim Q_{max}^{2/3}$. Displacement between the inner and outer rings δ_s is the sum of the individual contact deflections at the inner and outer ring raceways. Bearing stiffness is then determined by $dF/d\delta_s = dQ_{max}/d\delta_s = kQ_{max}^{1/3}$. Stiffness of a radially loaded line-contact roller bearing is similarly available through the contact deflection relation of Equation (14.16) in the pattern $dF_r/d\delta_s \sim F_r^{1/0.9}$.

Table 14.4 gives the approximate deflection and stiffness for several common types of rolling element bearings (for steel rolling elements and raceways) under a specified radial or axial external load (Gargiulo, 1980). For more complicated bearing systems or where greater accuracy is required, the bearing manufacturer should be consulted in order to obtain values reflecting the geometry of a particular bearing design.

Example 14.6   Using Table 14.4, determine the deflection and spring constant for a 208-size deep-groove ball bearing carrying a 300 lb radial load (contact angle $\alpha = 0°$). The bearing has $Z = 8$ balls, and their diameter D is 5/8 in.

Applying the relation in Table 14.4 for deflection δ:

$$\delta_r = 46.2 \times 10^{-6} \left(\frac{F_r^2}{DZ^2 \cos^5 \alpha} \right)^{1/3}$$

$$= 46.2 \times 10^{-6} \left(\frac{300^2}{5/8 \times 8^2 \times 1^5} \right)^{1/3}$$

$$= 0.00060 \text{ in.}$$

Table 14.4 Deflection and stiffness of rolling bearings under a given external force

Bearing type	Radial deflection, δ_r (in.)	Radial bearing stiffness, K_r (lb/in.)	Axial deflection, δ_a (in.)	Axial bearing stiffness, K_a (lb/in)
Deep-groove or angular-contact radial ball	$45.2 \times 10^{-6}\sqrt[3]{\dfrac{F_r^2}{DZ^2\cos^5\alpha}}$	$0.0325 \times 10^6\sqrt[3]{DF_rZ^2\cos^5\alpha}$	$15.8 \times 10^{-6}\sqrt[3]{\dfrac{F_a^2}{Z^2D\sin^5\alpha}}$	$0.0949 \times 10^6\sqrt[3]{DF_aZ^2\sin^2\alpha}$
Self-aligning ball	$74.0 \times 10^{-6}\sqrt[3]{\dfrac{F_r^2}{DZ^2\cos^5\alpha}}$	$0.0203 \times 10^6\sqrt[3]{DF_rZ^2\cos^5\alpha}$	$25.3 \times 10^{-6}\sqrt[3]{\dfrac{F_a^2}{Z^2D\sin^5\alpha}}$	$0.0593 \times 10^6\sqrt[3]{DF_aZ^2\sin^5\alpha}$
Thrust ball			$19.0 \times 10^{-6}\sqrt[3]{\dfrac{F_a^2}{Z^2D\sin^5\alpha}}$	$0.0789 \times 10^6\sqrt[3]{DF_aZ^2\sin^5\alpha}$
Spherical roller	$14.5 \times 10^{-6}\sqrt[4]{\dfrac{F_r^3}{\ell^2Z^3\cos^7\alpha}}$	$0.0921 \times 10^6\sqrt[4]{F_r\ell^2Z^3\cos^7\alpha}$	$4.33 \times 10^{-6}\sqrt[4]{\dfrac{F_a^3}{\ell^2Z^3\sin^7\alpha}}$	$0.308 \times 10^6\sqrt[4]{F_a\ell^2Z^3\sin^7\alpha}$
Cylindrical or tapered roller	$3.71 \times 10^{-6}\dfrac{F_r^{0.9}}{\ell^{0.8}Z^{0.9}\cos^{1.9}\alpha}$	$0.300 \times 10^6 F_r^{0.1}Z^{0.9}\ell^{0.8}\cos^{1.9}\alpha$	$0.871 \times 10^{-6}\dfrac{F_a^{0.9}}{\ell^{0.8}Z^{0.9}\sin^{1.9}\alpha}$	$1.28 \times 10^6 F_a^{0.1}Z^{0.9}\ell^{0.8}\sin^{1.9}\alpha$

D = rolling element diameter (in.).
F = external force (lb.).
ℓ = effective roller length (in.).
Z = number of rolling elements.
α = contact angle (radians).

Source: Gargiulo, 1980. Reproduced with permission of John Wiley & Sons.

Radial stiffness K_r is given from Table 14.4 as:

$$K_r = 0.0325 \times 10^6 [DF_r Z^2 \cos^5 \alpha]^{1/3}$$
$$= 0.0325 \times 10^6 [(5/8)(300)(8)^2 \cos^5 0]^{1/3}$$
$$= 0.74 \times 10^6 \, \text{lb/in.}$$

Since the low stiffness of a lightly loaded bearing can lead to low natural frequencies and undesirable runout under unbalanced loads, preload of bearings to obtain greater internal stiffness is often desirable in applications such as machine tool spindles. Axial preloads are common for ball bearings as obtained with spring loading or by specifying suitably matched 'duplex' pairs of bearings. Roller bearings are most often radially preloaded, either by a tapered adapter on the shaft or by press fit between a bearing ring and a housing or shaft.

Lack of any significant damping in ball and roller bearings makes their use troublesome in operation near rotor critical speeds, in passing through criticals during start-up, or while coasting during shutdown. Oil squeeze-film dampers (see Chapter 6) are commonly used to introduce needed damping in the bearing housings of aircraft gas turbines and other high-speed machines (Cookson and Kossa, 1981).

14.9 Noise

Rolling bearing noise is commonly of little significance in industrial, transportation, and construction machinery operating at relatively high noise levels. Analysis of noise spectra is, nevertheless, useful in quality control evaluation of new bearings, in monitoring machinery for onset of bearing failures, and in the diagnosis of operating problems in many types of applications. Quiet operation is essential in many applications of small ball bearings, such as in electric motors running in hospitals, homes, and offices.

Noise consists of airborne vibrations in a frequency and amplitude range which is detected by the human ear and is usually disagreeable. These air vibrations may originate from mechanical vibration within bearings themselves, or from shocks or vibrations of surrounding machine elements induced by the rotating bearings. Rolling bearings may also be induced to vibrate by the action of related machine components such as imbalance, gear teeth, belt drives, or various other mechanical or electrical alternating forces. Particularly noisy operation may be encountered when an exciting frequency coincides with a natural frequency.

Cage Noise

The greatest source of noise in rolling bearing applications at moderate speeds in the usual industrial applications is the *rattle* and *squeal* of their metal cages. Occasionally, a defect in a rolling element may produce some noise by contacting one or both sides of the cage pocket with a frequency given by Equation (14.33), but this is not a usual source of noise or vibration observed external to the bearing.

General cage noise, which may be classified loosely as rattling, is somewhat typical of common pressed steel ribbon cages. With a loose fit of the ball pockets and with sparse lubrication providing insufficient damping of cage motions, the cage may rattle as it moves erratically and experience shocks with changing speed of the balls when entering and leaving the load zone or when located in a misaligned bearing. With a loose-fitting cage and with insufficient lubricant in the cage pockets for damping, natural frequency bending of the cage may also be excited to give a disturbing squeal or bird-chirping sound.

The following practices commonly minimize noise with pressed steel ribbon cages in ball bearings:

1. Set ball pocket geometry to provide a small axial clearance with the balls that is just sufficient to accommodate the usual standard of holding misalignment to a maximum of 15 min with single-row, deep-groove bearings. With a 6216 80 mm ball bearing, for instance, this misalignment involves 0.2 mm axial displacement (at the bearing mean diameter) of a ball back and forth as it travels around the circumference of its raceway. The ball pocket curvature should also be sufficiently open to give circumferential looseness of about two and a half to four times the 0.2 mm axial looseness. This clearance range will minimize rubbing contact and bending forces on the cage.

   Larger clearance will leave the cage with less support. At low speeds up to 500–1000 rpm, the cage will repeatedly rise with the rotating balls and then drop to rattle as its support shifts on the rotating ball complement. At higher speeds, the cage will run eccentrically, and may squeal or squeak as it vibrates at its natural frequency. The first bending natural frequency of the cage ranges from 3000 Hz for a 6205 25 mm bore bearing to about 1500 Hz for a 6312 (60 mm) bore bearing. Other vibration modes generally give higher frequencies.

2. Use a cage pocket curvature 5–10% larger than the ball curvature, and pilot the cage at the center of the ball pocket to minimize contact and wear on the pocket edges. This piloting gives the lowest possible rubbing velocity between the ball and cage, along with a converging clearance shape for the rotating ball to draw lubricant to this pilot point.

3. Use reasonably soft grease that will feed into the cage pockets. NLGI Grade 2 greases commonly produce somewhat less noise and lower friction than Grade 3. Greases using a mineral oil with a viscosity greater than about VG 100 (100 cSt at 40 °C) generally contribute to noise. Oil lubrication, rather than grease, involves a minimum of cage noise questions.

4. Apply a thrust washer to give a light axial preload to the bearing. This thrust preload should be just sufficient to keep the motion of a ball relatively constant at its nonskid velocity as it travels around the bearing and in and out of the load zone. An axial spring force equivalent to 1–2% of the specific dynamic capacity of the bearing will commonly suffice. Assuming a coefficient of friction of about 0.35, this spring force should be enough to induce the outer bearing ring to slip in its housing to reapply correct preload following thermal transients or changes in machine operation which shift the relative location of shaft and housing seats.

Many of the same factors are involved in roller bearing noise. Formed steel cages and machined bronze and steel cages tend to encounter similar stimuli and natural vibration frequencies to produce squeaks, squeals, chirps, and howling similar to those encountered with ball bearing cages. When operating with a light bearing load, frictional drag on the roller complement to slow it below its nonskid velocity has sometimes been observed to induce this noise. Guiding the cage on the rotating inner ring tends to avoid this noise problem by keeping the cage up to its nonskid speed. Recirculation of a relatively soft, oily grease around the cage and over the flat ends of rollers also helps. The influence of lubrication traction coefficient on the cage angular velocity is reported by Takabi and Khonsari (2014).

Table 14.5 Approximate normal DN speed limits for rolling element bearings

Bearing type	Normal speed limit (DN)[a]		Highest DN attained under ideal conditions
	Grease or oil bath	Circulating oil	
Radial or angular contact groove ball bearings	300K[b]	500K	3500K
Cylindrical roller bearings	300K	500K	3500K
Tapered roller bearings	150K	300K	3500K
Spherical roller bearings	150K	300K	1000K

[a] DN = bore (D), mm × speed (N), rpm.
[b] K indicates × 1000 (e.g. 300K = 300 000).
Source: Jendzurski and Moyer, 1997. Reproduced with permission of Taylor & Francis.

14.10 Speed Limit

The mechanics of operation covered in the previous sections give a background for defining several limiting factors in applying rolling bearings. Table 14.5 gives common speed limits. Typical DN (mm bore × rpm speed) operating ranges for various type bearings shown in Figure 14.13 give a further view of the maximum speed range with the various bearing types. At speeds above these usual limits, early failures can be expected due to factors such as lubricant starvation from centrifugal throw-off from the inner ring and cage, overheating, skidding (involving some sliding rather than pure rolling) of rolling elements, centrifugal forces on rolling elements, and centrifugal expansion of components (Jendzurski and Moyer, 1997).

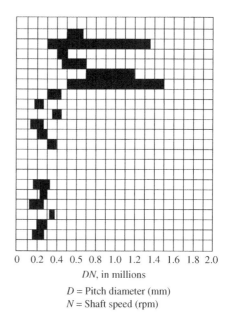

Radial bearings

Aligning ball bearing
Angular-contact ball bearing, 1-row
Angular-contact ball bearing, 2-row
Cylindrical roller bearing, industrial
Cylindrical roller bearing, precision
Deep-groove ball bearing, 1-row
Deep-groove ball bearing, 2-row
Journal roller bearing
Needle roller bearing, caged
Needle roller bearing, full-complement
Spherical roller bearing
Tapered roller bearing

Thrust bearings

Angular-contact ball bearing
Ball thrust bearing
Cylindrical roller bearing
Needle roller bearing
Spherical roller thrust bearing
Tapered roller thrust bearing

0 0.2 0.4 0.5 0.8 1.0 1.2 1.4 1.6 1.8 2.0
DN, in millions

D = Pitch diameter (mm)
N = Shaft speed (rpm)

Figure 14.13 Typical operating speed ranges for rolling element bearings. *Source*: Jendzurski and Moyer, 1997. Reproduced with permission of Taylor & Francis.

These upper speed limits are not absolute values, but they do reflect speeds typical of those in bearing catalogs below which ordinary lubricants and lubrication will give satisfactory performance with standard bearings. Special attention to bearing components and lubrication will usually enable operation up to about twice these typical catalog values. Three to four times these limits are reached under laboratory conditions, in jet engines and various aerospace applications, and with special components such as low-density ceramic balls.

Physical factors involved in these limits are generally not clearly established, and prototype testing is usually appropriate before applications at higher speeds. With grease lubrication, centrifugal throw-off of the lubricant is probably the primary speed-limiting factor. From observation of results in test fixtures and machine applications, this throw-off limit appears to be encountered on cage surfaces. When surface tension of the fluid in the grease and the channeling nature of the thickener are no longer able to maintain an adequate lubricant separating film between the cage and rolling element at rising speeds, failure quickly follows. With silicone grease, this speed limit drops in proportion to the lower surface tension of its fluid. With cotton-filled phenolic cages, the lubricant-soaked cage enables much higher speeds.

Centrifugal force generated by rotation of balls or rollers is a serious dynamic limitation for high-speed operation. This centrifugal loading may become so high that added stress at contacts on the outer ring raceway may materially reduce bearing fatigue life. This effect generally becomes significant in the general speed range above about 5000–20 000 rpm or so for common bearing sizes. Centrifugal force F_c acting on a ball of mass m is given by

$$F_c = \frac{1}{2} m d_m \omega_m^2 \qquad (14.40)$$

where ω is the orbital speed (rad/s) of the ball at its orbit diameter d_m. To minimize this centrifugal ball force, silicon nitride ceramic balls, which have lower density (42% of steel), can be employed. Table 14.2 gives typical densities of rolling bearing materials.

Because of the influence on contact angle in angular-contact bearings, possible ball skidding, and differences in elastohydrodynamic oil film thickness, a simple definition of the effect of centrifugal ball loading and other high-speed effects becomes very difficult. Harris (1991) presents the analytical relations involved in fatigue life at high speeds, and bearing suppliers can apply computer programs for the complex calculations involved.

Simple overheating is the most common problem at high speeds, and some bearing catalogs give thermal speed limits for each bearing type. These are simply approximate speeds, below which ordinary lubricants and lubrication methods can be expected to give satisfactory service.

Example 14.7   A 110 mm diameter shaft is to be run at 5000 rpm. What types of rolling element bearings should be considered? What type of lubrication?

Assuming a 150 mm pitch diameter for the 110 mm bore bearings, this operation involves a DN speed factor in Figure 14.13 of $150 \times 5000 = 750\,000$. Figure 14.13 indicates that both angular contact and cylindrical roller bearings would be typical. Table 14.5, using the bearing bore diameter for a DN of $110 \times 5000 = 550\,000$, would suggest circulating oil lubrication and consideration of precision bearings since this DN well exceeds the limit for grease and is even above the normal limit for circulating oil.

Low and Ultra-low Speeds

At low operating speeds, the surface roughness can influence the dynamic performance of roller bearings, particularly at large contact loads. Different lubrication regimes may become active in the loaded and unloaded sections of the bearing. Careful assessment is necessary to predict the operating film thickness, heat generation and the wear rate (Takabi and Khonsari, 2015a).

Considerably more complex is the characterization of ultra-low speeds ball bearings used in precision sensing mechanisms such as those used for attitude control in spacecrafts. As discussed more fully in Chapter 15, such instruments often run near zero rotational speeds for extended periods with frequent start-stop and reversal of rotational direction. Typically, the when slowly rotating from rest, the balls tend to exhibit a steadily increasing friction torque from zero to their "free-running" torque asymptotically. Then upon reversal, the friction torque yields a hysteresis loop characterized by what is generally known at the Dahl friction (1968). This phenomenon is sometimes referred to as the ball jitter. Lovell *et al.* (1992, 1993a, 1993b, 1994, 1996a) provide extensive testing and analysis of Dahl friction parameters in both liquid-lubricated and dry contacts. Solid-lubricated ball and roller bearings with thin layers of MoS_2 also operating at low speeds are analyzed using the finite element method (Lovell *et al.*, 1996c, 1997a, 1997b). The durability of the coated ball bearing performance operating under slow speeds in oscillatory motion is reported by Mesgarnejad and Khonsari (2010) and Kahirdeh and Khonsari (2010).

14.11 Load Limit

The basic fatigue load ratings given in catalogs and in Chapter 13 are extremely high and not normally applied in practice. Table 14.6 defines several relative load ranges as percentages of this basic load rating. With proper fitting, lubrication, and design practice, satisfactory operation can be expected throughout the range from light to heavy loading.

For extremely light loads, providing very little driving force to rotate the rolling elements, especially at high speeds and with high-viscosity lubricants giving high cage drag, rolling elements sometimes skid rather than roll. As reviewed in detail by Harris (1991), this problem of skidding is a common concern with gas turbine engine bearings.

At the other end of the scale, with very heavy loads and especially at low speeds, boundary lubrication is encountered together with failures originating at the surface rather than caused by the more usual subsurface fatigue. Under extremely heavy loads of 25–30% of the basic

Table 14.6 Load ranges for rolling element bearings

Load	Ball bearings	Roller bearings
Light	Up to 7% C	Up to 8% C
Normal	7 to 15% C	8 to 18% C
Heavy	Over 15% C	Over 18% C

C = basic dynamic load rating.
Source: Derner and Pfaffenberger, 1984. Reproduced with permission of Taylor & Francis.

load rating C, heavy press fit on the shaft in combination with a locking nut to avoid turning of the inner ring on its shaft fit can add to stresses from the high load force itself to cause premature failure with bulk cracking.

14.12 Temperature Limit

The upper temperature limit for ball and roller bearings depends on the materials used in bearing components, on auxiliary materials used in cages and seals, and on the lubricants. Table 14.7 gives typical temperature limits. Special reference should be made to bearing manufacturers and lubricant suppliers for operation at temperatures approaching these limits. For common bearings of 52100 steel using mineral oil or mineral oil greases for lubrication, caution should be used when approaching any operation or heating during assembly that involves a temperature above about 125 °C. Softening and lowered density of the steel can lead to both decreased fatigue life and to lower internal clearance at any higher temperature unless specially heat-stabilized bearings are employed.

Table 14.7 Temperature limits for rolling bearing components

	Temperature limit	
	°C	°F
Bearing materials		
AISI 51200 through hardened	140–175	285–350
AISI 8620 carburized	150	300
AISI 440C stainless	170	340
AISI M50 tool steel	315	600
Hybrid Si_3N_4-M50	425	800
All ceramic Si_3N_4	650	1200
Seal materials		
Felt	95	200
Buna N	110	225
Polyacrylic	150	300
Viton	260	500
Nonmetallic cages		
Heat-stabilized nylon 66	120	250
Phenolic	120	250
Fiberglass-reinforced nylon 66	150	300
Polyphenylene-sulfide	230	450
Lubricants		
Standard petroleum greases	110	225
Special petroleum greases	150	300
Mineral oils	150	300
Synthetic greases and oils	230	450
Solid films	Bearing material limit	

Source: Derner and Pfaffenberger, 1984. Reproduced with permission of Taylor & Francis.

Table 14.8 Recommended misalignment limits for ball and roller bearings

Bearing type	Minutes′/degrees°
Deep-groove ball	12–16′
Cylindrical roller	3–4′
Tapered roller	3–4′
Spherical roller	0.5–2°
Spherical roller thrust	0.5–2°
Self-aligning ball	4°

Source: Eschmann *et al.*, 1985 and Harris, 1991.

14.13 Misalignment Limit

Table 14.8 gives typical limits for the maximum misalignment to avoid significant reduction in the fatigue life of rolling bearings (Eschmann *et al.*, 1985; Harris, 1991). These misalignments arise when mounting pillow blocks; when shaft seats, bearing bores, and supporting shoulders are not machined in one operation; and from shaft and housing deformations during operation. While these limits are within the range of conventional accuracy, further accommodation of misalignment is available with self-aligning ball bearings, spherical roller bearings, and flexible or spherically shaped housing surfaces. Even with these self-aligning types, misalignment must be limited to avoid edge loading with contact surfaces of rolling elements projecting beyond the raceway width.

With misalignment of a single-row ball bearing, ball contact shifts from one shoulder of its raceway to the other during each revolution. Each ball will, therefore, alter its speed four times during each bearing revolution while tending each time to approach or leave behind the adjacent balls. This action will shorten the fatigue life of the raceways and cause stress and wear in cage pockets. Sensitivity of thrust ball bearings to angular misalignment may by alleviated by a spherical housing washer and a seating ring.

In cylindrical and tapered roller bearings, roller motions are restricted by the cylindrical raceway tracks and lateral flanges. While angular misalignment leads to nonuniform loading over the roller length, the added stresses are tolerable within the misalignment limits listed in Table 14.8. The conventionally crowned form of rollers or raceways also aids accommodation of this misalignment, and slight crowning of the ends of rollers and lateral mating flanges helps promote the formation of a beneficial lubricant film in their contact zone.

Problems

14.1 A 12.7 mm (0.5 in.) diameter ball rolls under load in a raceway having a maximum diameter of 52.29 mm (2.059 in.). Raceway curvature is 52% of the ball diameter, and axial length of the contact ellipse is 0.10 in. How many more revolutions would the ball make in rolling completely around the raceway at the ends of the ellipse rather than at the center? Assuming that the ball actually rolls halfway between these two extremes, how much sliding occurs in inches per inch of travel?

14.2 Using a steel 210-size ball bearing to carry a 2200 N (494 lb) radial load, compute the reduction in hours of fatigue life in going from 1000 rpm with negligible centrifugal ball loading to 5000, 10 000, and 20 000 rpm. (See Example 14.1 for bearing dimensions.) Use as fatigue life criteria the contact stress at the outer raceway for the most heavily loaded ball. What is the reduction in life from centrifugal loading of the balls? Table 13.5 gives 35 700 N as the dynamic load rating for a fatigue life of 1 million revolutions with no centrifugal force effects.

14.3 Repeat the life calculations of Problem 14.2 for 20 000 rpm with a 210-size ball bearing using silicon nitride ceramic balls with M50 tool steel rings. Material properties for the balls and rings are given in Table 14.2.

14.4 For 2200 and 6000 N (494 and 1348 lb) radial loads, calculate the maximum Hertz contact stress on the steel outer ring raceway (with diameter $d_o = 80.4$ mm) for a 210 cylindrical roller bearing with $Z = 14$ steel rollers, $D = 11.0$ mm roller diameter, and $L = 11.30$ mm effective roller length. Compare this with the maximum stress under these loads for a 210 deep-groove ball bearing with the dimensions given in Example 14.1.

14.5 Calculate the spring constant for a 210-size ball bearing (of 52100 steel) at 200, 2200, and 4000 N radial loads based on Equations (14.21) and (14.9), together with the bearing dimensions given in Example 14.1. Compare these results with those from the approximate relation given in Table 14.4.

14.6 For the cylindrical roller bearing of Example 14.3 running at 10 000 rpm, estimate the minimum radial load needed to keep the roller complement from skidding at 40 °C with VG 100 oil (13×10^{-6} reyns of viscosity). Loaded contact of the rollers with the inner raceway exerts a traction coefficient of 0.04 (ratio of driving friction force to radial load) while involving just a minor roller/raceway slip velocity. Each of the 14 cylindrically shaped cage pockets of the roller–riding cage has a diametral clearance of 0.20 mm with the circumference of its roller. Viscous shearing of the oil film between the cage pockets, each of which encloses 25% of its roller outer diameter, and the circumference of the rollers represent 50% of the drag force on the roller–cage assembly.

Repeat for 0 °C where the oil viscosity becomes 200×10^{-6} reyns. Suggest three ways of reducing this load needed to avoid roller skidding.

References

ASME. 2003. *Life Ratings of Modern Rolling Bearings.* ASME International, New York.

Cookson, R.A. and Kossa, S.S. 1981. 'The Vibration Isolating Properties of Uncentralized Squeeze-Film Damper Bearings Supporting a Flexible Rotor,' *Transactions of the ASME Journal of Engineering for Power,* Vol. 103, pp. 781–787.

Dahl, P.R. 1968. 'A Solid Friction Model,' AFO 4695-67-C-D158, Aerospace Corporation.

Derner, W.J. and Pfaffenberger, E.E. 1984. 'Rolling Element Bearings,' *Handbook of Lubrication,* Vol. II. CRC Press, Boca Raton, Fla, pp. 495–537.

Eschmann, P., Hasbargen, L., and Weigand, K. 1985. *Ball and Roller Bearings–Theory, Design, and Application.* John Wiley & Sons, New York. (See also the 3d edn., 1999.)

Gargiulo, E.P. 1980. 'A Simple Way to Estimate Bearing Stiffness,' *Machine Design,* 24 July, pp. 107–110.

Harris, T.A. 1991. *Rolling Bearing Analysis*, 3rd edn. John Wiley & Sons, New York.

Jendzurski, T. and Moyer, C.A. 1997. 'Rolling Bearing Performance and Design Data,' *Tribology Data Handbook*. CRC Press, Boca Raton, Fla, pp. 645–668.

Jones, A.B. 1946. *Analysis of Stresses and Deflections*. New Departure, Bristol, Conn.

Kahirdeh, A. and Khonsari, M.M. 2010. 'Condition Monitoring of Molybdenum Disulphide Coated Thrust Ball Bearings Using Time-Frequency Signal Analysis,' *ASME Journal of Tribology*, Vol. 132, 041606:1–11

Lovell, M., Khonsari, M.M., and Marangoni, R. 1992. 'Evaluation of Ultra-Low Speed Jitter in Rolling Balls,' *ASME Journal of Tribology*, Vol. 114, pp. 589–594.

Lovell, M., Khonsari, M.M., and Marangoni, R. 1993a. 'Ultra Low-Speed Friction Torque on Balls Undergoing Rolling Motion,' *STLE Tribology Transactions*, Vol. 36, pp. 290–296.

Lovell, M., Khonsari, M.M., and Marangoni, R. 1993b. 'The Response of Balls Undergoing Oscillatory Motion: Crossing from Boundary to Mixed Lubrication,' *ASME Journal of Tribology*, Vol. 115, pp. 261–266.

Lovell, M., Khonsari, M.M., and Marangoni, R. 1994. 'Experimental Measurements of the Rest-Slope and Steady Torque on Ball Bearings Experiencing Small Angular Rotation,' *STLE Tribology Transactions*, Vol. 37, pp. 261–268.

Lovell, M., Khonsari, M.M., and Marangoni, R. 1996a. 'Comparison of the Ultra-Low Speed Frictional Characteristics of Silicon Nitride and Steel Balls Using Conventional Lubricants,' *ASME Journal of Tribology*, Vol. 118, pp. 43–51.

Lovell, M., Khonsari, M.M., and Marangoni. R. 1996b. 'A Finite Element Analysis of Frictional Forces Between A Cylindrical Bearing Element and MoS_2 Coated and Uncoated Surfaces,' *Wear*, Vol. 194, pp. 60–70.

Lovell, M., Khonsari, M.M., and Marangoni, R. 1996c. 'Dynamic Friction Measurements of MoS_2 Coated Ball Bearing Surfaces,' *ASME Journal of Tribology*, Vol. 118, pp. 858–864.

Lovell, M., Khonsari, M.M., and Marangoni, R. 1997a. 'Parameter Identification of Hysteresis Friction for Coated Ball Bearings Based on Three-Dimensional FEM Analysis,' *ASME Journal of Tribology*, Vol. 119, pp. 462–470.

Lovell, M., Khonsari, M.M., and Marangoni, R. 1997b. 'Frictional Analysis of MoS_2 Coated Ball Bearings: A Three-Dimensional Finite Element Analysis,' *ASME Journal of Tribology*, Vol. 119, pp. 754–763.

Mesgarnejad, A. and Khonsari, M.M. 2010. 'On the tribological behavior of MoS_2-coated thrust ball bearings operating under oscillating motion,' *Wear*, Vol. 269, pp. 547–556.

Takabi, J. and Khonsari, M.M. 2013. 'Experimental Testing and Thermal Analysis of Ball Bearings,' *Tribology International*, Vol. 60, pp. 93–103.

Takabi, J. and Khonsari, M.M. 2014. 'On the Influence of Traction Coefficient on the Cage Angular Velocity in Roller Bearings,' *Tribology Transactions*, Vol. 57, pp. 793–805.

Takabi, J. and Khonsari, M.M. 2015a. 'On the dynamic performance of roller bearings operating under low rotational speeds with consideration of surface roughness,' *Tribology International*, Vol. 86, pp. 62–71.

Takabi, J. and Khonsari, M.M. 2015b. 'On the Thermally Induced Seizure in Bearings: A Review,' *Tribology International*, Vol. l91, pp. 118–130.

Takabi, J. and Khonsari, M.M. 2015c. 'On the Thermally–Induced Failure of Rolling Element Bearings,' *Tribology International*, Vol. 94, pp. 661–674.

15

Friction and Elastohydrodynamic Lubrication

This chapter initially covers friction and wear in ball and roller bearings, areas of primary concern in their application and maintenance. Guidelines are then presented for lubricant selection and related performance factors.

A relatively predictable friction coefficient of about 0.001–0.002 is involved in run-of-the-mill applications of ball and roller bearings. While this level of friction is similar to that for oil-film bearings while running, rolling element bearings offer a decided advantage in having very low starting torque requirements.

Wear is a major question in many industrial, agricultural, and transportation vehicle applications. While the load rating procedures covered in Chapter 13 form the basis for machine design practices to avoid general fatigue problems, wear commonly determines the service life of many ball and roller bearings.

Operating temperature and lubrication are considered in the final portion of this chapter. Successful operation of ball and roller bearings requires the presence of a lubricating film in areas of moving contact: between rolling elements and raceways, and on separator surfaces contacting either rolling elements or any guiding surfaces on inner or outer bearing rings. Without lubrication, high friction will develop, with wear, elevated temperature, skidding and fretting at rolling element contacts, and early cage breakage with a jammed bearing.

15.1 Friction

Frictional resistance to motion in a rolling bearing arises from various sources; the following commonly predominate (Eschmann *et al.*, 1985; Harris, 1991):

1. *Rolling friction:* Elastic hysteresis and deformation at raceway contacts;
2. *Sliding friction:* Sliding from unequal curvatures in contact areas, sliding contact of the cage with rolling elements and guiding surfaces, sliding between the ends of rollers and ring flanges, and seal friction;
3. *Lubricant friction:* Viscous shearing on rolling element, cage, and raceway surfaces; churning and working of lubricant dispersed within the bearing cavity.

Applied Tribology: Bearing Design and Lubrication, Third Edition. Michael M. Khonsari and E. Richard Booser.
© 2017 John Wiley & Sons Ltd. Published 2017 by John Wiley & Sons Ltd.
Companion Website: www.wiley.com/go/Khonsari/Applied_Tribology_Bearing_Design_and_Lubrication_3rd_Edition

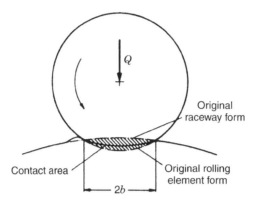

Figure 15.1   Deformation of a rolling element and raceway in the direction of rolling.
Source: Eschmann *et al.*, 1985. Reproduced with permission of John Wiley & Sons.

Rolling contact friction between loaded rolling elements and their raceways results partly
from elastic hysteresis. The ball and raceway material just ahead of the contact in the circumferential direction undergo distortion and compression, while the material just behind the
contact is relieved of stress. Energy loss in this stress reversal is a traditional source of friction.
According to Harris (1991), however, this is a small factor that involves a rolling coefficient
of friction as low as 0.0001.

Within the loaded rolling contact, microslip occurs in the rolling direction as the roller in
Figure 15.1 is depressed and the raceway stretches: deformations which produce friction and
cause the roller to go forward a distance minutely less than its circumference in one revolution.
Additional sliding results in a ball bearing with the contact area in the raceways being curved
transverse to the rolling direction (Figure 15.2). Since the rolling diameter varies throughout
the contact zone, surface speed also varies. The middle portion of the ball slides opposite to
the rolling direction, and the outer portion slides in the rolling direction. Frictional forces are
balanced to result in pure rolling at points D and D'.

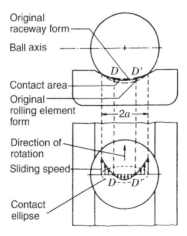

Figure 15.2   Sliding motion resulting from the curvature of the contact area. *Source*: Eschmann *et al.*,
1985. Reproduced with permission of John Wiley & Sons.

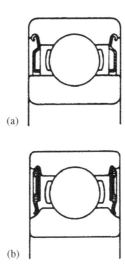

(a)

(b)

Figure 15.3 Enclosed deep-groove ball bearings with (a) noncontacting shields and (b) rubbing seals. *Source*: SKF, 1999/2005.

Spinning friction must be added to these frictional effects when there is angular contact between a ball and its raceway under thrust load. In such cases, *gyroscopic* action results in a tendency of the ball to rotate about an axis perpendicular to the ball–raceway contact, with resulting *macrosliding* in the contact (Harris, 1991).

Sliding motion at cage surfaces (see Section 14.6 and Figure 14.11) and at any seals is encountered in addition to friction components in the rolling element contacts with their raceway. With adequate lubrication at normal operating conditions, this sliding resistance at cage surfaces is small. This friction may increase considerably, however, with inadequate lubrication, with high lubricant viscosity at cold temperatures, with contamination, and particularly at high speeds. Seal friction can be quite high during initial run-in of a new bearing, so high as to cause overheating damage to the bearing and its lubricant and to create difficulty in assessing the efficiency of an electric motor or other machine during manufacture. Careful control of seal tightness in its rubbing contact with the inner ring shoulder and wear during the early life of a machine can be expected to reduce this friction to low values, for which estimates are given below. In many applications where contamination problems are not severe, noncontacting seals or shields are employed to avoid this component of friction (Figure 15.3).

Viscous drag depends mainly on the amount and viscosity of the lubricant dispersed in the bearing cavity and on bearing speed.

15.2 Friction Moments

For most normal operating conditions, the total frictional moment can be estimated with sufficient accuracy as load dependent using a constant coefficient of friction (Jendzurski and Moyer, 1997; SKF, 2005):

$$M = 0.5\mu Pd \qquad (15.1)$$

Table 15.1   Coefficient of friction for rolling element bearings

Bearing type	Coefficient of friction, μ
Deep-groove ball bearings	0.0015[a]
Self-aligning ball bearings	0.0010[a]
Angular-contact ball bearings	
Single row	0.0020
Double row	0.0024[a]
Four-point contact ball bearings	0.0024
Cylindrical roller bearings	
With cage	0.0011[b]
Full complement	0.0020[a,b]
Needle roller bearings	0.0025[a]
Spherical roller bearings	0.0018
Taper roller bearings	0.0018
Thrust ball bearings	0.0013
Cylindrical roller thrust bearings	0.0050
Needle roller thrust bearings	0.0050
Spherical roller thrust bearings	0.0018

[a] Applies to unsealed bearings.
[b] No appreciable axial load ($F_a \approx 0$).
Source: SKF, 1999/2005.

where

M = bearing frictional moment, calculated at the bearing bore radius of $0.5d$ (N mm)
μ = coefficient of friction for the bearing (Table 15.1)
P = bearing load (N)
d = bearing bore diameter (mm)

Note that designating this coefficient of friction as μ is common practice for ball and roller bearings rather than the parameter f used elsewhere in this text for friction coefficient in sliding contacts and fluid-film bearings.

The starting coefficient of friction can generally be taken as being about 60% higher than the running values given in Table 15.1. More accurate calculations of bearing friction are available which account for variations in the coefficient of friction with relative bearing load, bearing size, and cross section series (SKF, 1999/2005). Details on friction of tapered roller bearings are provided by Aihara (1987) and Witte (1973).

Example 15.1   Estimate the friction power loss in a 306 deep-groove ball bearing running at 3600 rpm under a 712 N (160 lb) radial load.

With its 30 mm bore, the friction moment from Equation (15.1) with a coefficient of friction $\mu = 0.0015$ from Table 15.1 becomes

$$M = 0.5\mu\, Pd = 0.5 \times 0.0015 \times 712 \times 30 = 16\, \text{N mm} \quad (0.14\, \text{in. lb})$$

Power loss $= 2\pi(n/60)M = 2\pi(3600\,\text{rpm}/60)(16) = 6030\,\text{N mm/s} = 6.03\,\text{W}$. This compares with 5–10 W measured in many grease life test bearings under conditions simulating those of electric motor service.

Quantitative evaluation of friction including the influence of individual friction elements becomes complex. A more complete calculation can be made by considering that friction and its frictional moment M include the following: (1) load-dependent moment M_P resulting from rolling and sliding friction in the loaded rolling contacts, which predominates with slow rotation and heavy loads; (2) combination M_L of lubricant viscous friction plus related cage friction in rolling element pockets, and at cage guiding surfaces, which dominates at high speeds and light loads; and (3) rubbing seal friction M_S in sealed bearings. Summing these friction moments gives the following (Harris, 1991; SKF, 1999):

$$M = M_P + M_L + M_S \tag{15.2}$$

The load-dependent component M_P can be approximated from Equation (15.1). Details for making a better match to the bearing type and operating conditions are given by SKF (1999).

The lubricant friction moment M_L should be added for high-speed, lightly loaded bearings. This fluid friction factor is independent of bearing load and depends on lubricant viscosity, rolling velocity, and quantity of lubricant. This moment can be calculated as follows (SKF, 1999):

$$M_L = 10^{-7}f_L(vn)^{2/3}d_m^3 \quad vn > 2000 \tag{15.3}$$

$$M_L = 160 \times 10^{-7}f_L d_m^3 \quad vn < 2000 \tag{15.4}$$

where

$M_L =$ load-independent moment representing lubricant losses (N mm)
$d_m =$ mean diameter of bearing, 0.5 (bore + outer diameter) (mm)
$n =$ bearing speed (rpm)
$f_L =$ a factor depending on bearing type and method of lubrication
$v =$ kinematic viscosity at bearing operating temperature (mm²/s)

For grease this becomes the base oil viscosity at the operating temperature.

With grease, oil mist, oil spot, and other limited forms of lubrication which offer minimum friction, f_L in Table 15.2 varies from about 1 for radial ball bearings and light series cylindrical roller bearings, to the order of 3 with tapered roller and spherical roller bearings, and 6–14 with needle bearings (SKF, 1999). In the last column, 'oil jet' pressurized feed through adjacent nozzles is common for aircraft turbine engine bearings with the oil stream velocity overcoming windage resistance around a high-speed ball or roller complement.

Where a bearing is fitted with rubbing seals, frictional losses from the seals may exceed those from the bearing itself. While subject to considerable variation, frictional moment M_S for seals on both sides of a bearing can be estimated from the following empirical equation (SKF, 1999):

$$M_S = \left(\frac{d+D}{f_1}\right)^2 + f_2 \tag{15.5}$$

Table 15.2 Lubrication friction factor f_L for Equations (15.3) and (15.4)

Bearing type	Grease lubrication	Oil spot lubrication	Oil bath lubrication	Oil bath with vertical shaft, oil jet
Deep-groove ball bearing	0.75–2	1	2	4
Self-aligning ball bearing	1.5–2	0.7–1	1.5–2	3–4
Angular-contact ball bearing				
Single row	2	1.7	3.3	6.6
Double row, bearing pair	4	3.4	6.5	13
Cylindrical roller bearing	0.6–1	1.5–2.8	2.2–4	2.2–4[a]
Needle roller bearings	12	6	12	24
Spherical roller bearings	3.5–7	1.75–3.5	3.5–7	7–14
Taper roller bearings				
Single row	6–3	6	8	8–10[a]
Paired single row	12	6	12	16–20[a]
Thrust ball bearings	5.5	0.8	1.5	3
Cylindrical roller thrust bearings	9		3.5	7
Needle roller thrust bearings	14		5	11
Spherical roller thrust bearings			2.4–5	5–10

Note: When a range of values is indicated, small values apply to light series bearings and large values to heavy series bearings.
[a] Value given for oil jet lubrication. Double for oil bath lubrication with a vertical shaft.
Source: SKF, 1999/2005.

where

M_S = friction moment of two seals (N mm)
d, D = bearing bore diameter, outside diameter (mm)
f_1, f_2 = factors from Table 15.3

Example 15.2   Estimate the frictional power loss for a 7220 angular-contact ball bearing running at 10 000 rpm under a 26 700 N (6000 lb) thrust load when jet lubricated by synthetic ester jet engine oil having a viscosity of 6 mm²/s (cSt) at operating temperature. Following are

Table 15.3 Friction factors for bearing seals in Equation (15.5)

Bearing (design)	Factors	
	f_1	f_2
Deep-groove ball bearings (2RS1), self-aligning ball bearings (2RS1), angular-contact ball bearings (2RS), Y-bearings (series 17262(00)-2RS1 and 17263(00)-2RS1)	20	10
Y-bearings (all other series), needle roller bearings (2RS)	20[a]	25[a]
Cylindrical roller bearings, full complement (2LS)	10	50

[a] Does not apply to Y-bearings with special fingers.
Source: SKF, 1999/2005.

the characteristics of the bearing:

$d_i = 100$ mm bore
$d_o = 180$ mm outer diameter
$D = 23.81$ mm (0.9375 in.) ball diameter
$Z = 16$ balls
$\alpha = 40°$ nominal contact angle

Applying Equation (15.1) for the load-dependent portion M_P of the friction moment and using 0.002 as friction factor μ from Table 15.1:

$$M_P = 0.5\,\mu\,Pd_i = 0.5 \times 0.002 \times 26\,700 \times 100 = 2670\,\text{N mm} \quad (23.62\,\text{in. 1b}) \quad (15.6)$$

Using Equation (15.4) for lubricant frictional drag in the high-speed range (with $vn = 60\,000$) and $f_L = 6.6$ from Table 15.2 for an angular-contact ball bearing with jet lubrication:

$$M_L = 10^{-7}f_L(vn)^{2/3}d_m^3 = 10^{-7}(6.6)(60,000)^{2/3}(140)^3 = 2777\,\text{N mm} \quad (24.58\,\text{in. 1b})$$

Total friction moment from Equation (15.2) with seal friction $M_S = 0$ then gives

$$M = M_P + M_L + M_S = 2670 + 2777 + 0 = 5447\,\text{N mm} \quad (48.21\,\text{in. 1b})$$

More detailed SKF analyses modify Equation (15.2) to provide the overall friction moment M as the sum of individual friction moments involving the following four factors: (1) rolling, (2) sliding, (3) seals, and (4) lubricant drag, churning and splashing (SKF, 2005).

15.3 Wear

Wear severe enough to impair bearing performance should not occur even after long service with a bearing which is well lubricated, perfectly sealed, and running at moderate load and speed. Conditions found in practice are almost always less favorable, however, and wear along with fatigue is commonly a factor in establishing bearing life in the usual industrial applications.

Investigations carried out with 100 000 bearings used in diverse machinery, vehicles, and manufacturing showed life-limiting wear (Eschmann *et al.*, 1985). This wear was encountered at contact surfaces of the rolling elements and rings, at cage sliding surfaces, and on the lips and roller faces of roller bearings. The initial phase of the damage involved a dull appearance and roughening of the rolling path. The cause of wear was predominantly abrasion from foreign particles entering the bearing through insufficient or worn seals. However, wear debris from related machine elements and from bearing surfaces themselves was also found to play a role. Another cause of wear was corrosion from water contamination or condensation during temperature changes, from corrosive liquids and fumes, and from chemically aggressive EP lubricant additives.

Inadequate lubrication often leads to wear through metal-to-metal contact at bearing surfaces. High temperatures and severity of operating conditions influence grease life, with drying of grease becoming a problematic factor. Increased wear rates are also commonly encountered with high speeds and rapid speed changes, marginal lubrication, and decreased seal effectiveness.

Wear increases the bearing clearance, noise and vibration increase, local surface stresses increase from raceway surface changes, and rolling contact surfaces sometimes undergo early fatigue damage. By measuring the increase in radial clearance, the extensive study reported by Eschmann et al. (1985) encountered patterns of increasing wear with accumulating service hours. The following limiting radial wear factor f_v was established as a standard of comparison for bearings of various sizes:

$$f_v = \frac{V}{e_0} = \frac{V}{0.46d^{2/3}}$$

(15.7)

where V is the total wear at bearing failure in micrometers, as reflected in the increased radial clearance measured in an unmounted bearing. Size factor e_0 in the denominator is approximately related to bearing bore diameter d (mm) by $e_0 = 0.46 \cdot d^{2/3}$.

As an approximation from the Eschmann study, wear (and wear factor f_v) increases linearly with time until it reaches a limiting value beyond which the bearing operation becomes unsatisfactory. Table 15.4 lists representative values of limiting wear factor f_v. These range from very little tolerable wear for machine tool spindles with their need for precision, to the much greater leeway in agricultural and construction machines.

The wide range in wear life T_L listed in Table 15.4 reflects the broad range experienced in various applications. The shorter wear life values in each case would be expected with environmental coefficients k in the lower end of their range given in the final column with $T_L = kf_v$. These lower environmental coefficients involve such factors as high speed, rapid speed changes, decreased lubricant efficiency, decreased seal effectiveness, high ambient temperature, and environmental contamination. With subsequent improvement in seals, lubrication, and maintenance practices, wear life values from this study are likely being exceeded currently (Harris, 1991). Nevertheless, the general pattern, and particularly the limits observed in wear, can serve as useful guides for evaluating troublesome application and service problems.

Table 15.4 Representative limits for wear and wear life[a]

Application	Limiting Value of wear factor, f_v[b]	Range of wear life, T_L(h)	Environmental factor, $k = T_L/f_v$[c]
Agricultural machines	8–25	1 600–25 000	100–1 500
Construction machinery	6–12	1 900–7 000	200–800
Crushers	8–12	7 000–21 000	700–2 100
Electric motors	3–5	9 000–50 000	2 200–12 500
Gears, general engineering	3–8	6 000–57 000	1 100–10 400
Machine tool spindles	0.5–1.5	5 000–45 000	5 000–45 000
Motor vehicles	3–5	1 000–4 000	250–1 000
Paper–making machines	5–10	50 000–200 000	6 700–27 000
Plastic processing	8–12	11 000–70 000	1 100–7 000
Pumps, fans, compressors	3–5	3 500–27 000	900–6 800
Rail vehicles	6–12	13 000–120 000	1 400–13 300
Textile machines	2–8	6 000–200 000	1 200–40 000

[a] Derived from Eschmann et al., 1985. Reproduced with permission of John Wiley & Sons.
[b] Limiting increase in radial clearance V by wear (μm): $V_L = f_v/(0.46d^{2/3})$ with bearing bore d in mm.
[c] Environmental factor $k = T_L/f_v$ using mean value of f_v.

In an unrelated evaluation of 100 bearings taken randomly from medium-sized industrial electric motors returned to a repair shop for various problems, the authors of this book found internal wear to be by far the predominant bearing fault. The wear factor measured for the used bearings generally exceeded significantly the 3–5 μm limiting range of f_v for electric motors given in Table 15.4, and this excessive wear indicated that almost none of the bearings would have been satisfactory for continued service. Despite this general wear damage, the bearings showed essentially no fatigue.

Example 15.3 A 3600 rpm motor using 210-size deep-groove ball bearings (with bore diameters $d = 50$ mm) is to be used in a dusty cement plant. What would be the limiting increase in radial internal clearance V and the expected wear life T_L?

The dusty environment and high motor speed would set environmental factor $k = 3000$ in Table 15.4, at the low end of the range for electric motors. Assuming a mean value of 4 for the limiting wear factor f_v, limiting radial wear is

$$V = f_v/(0.46\,d^{2/3}) = 4/[0.46 \times (50)^{2/3}] = 0.64\ \mu m \quad (25\ \mu\,in.)$$

Wear life is

$$T_L = k \times f_v = 3000 \times 4 = 12\,000\,h$$

15.4 Bearing Operating Temperature

Operating temperature bears important relations to bearing and seal friction, design of the bearing assembly, and especially lubrication considerations, to be covered in the next section. Since most usual industrial rolling bearing applications involve only a small amount of frictional heating, their operating temperature rise is seldom more than 5–10 °C above that of their surrounding structure. Representative operating temperatures observed in a wide variety of machinery applications are given in Table 15.5 (Eschmann *et al.*, 1985).

In considering a new bearing application using a self-contained lubricant supply of grease or oil, the temperature rise ΔT of the bearing above ambient can be estimated from the following relation:

$$\Delta T = T_{brg} - T_{amb} = \frac{E_p}{K_h \times A} \tag{15.8}$$

where the frictional power loss E_p in the bearing obtained from Section 15.2 is transmitted as heat to adjacent components and to ambient air through surface area A.

K_h is a characteristic heat dissipation factor for the system. Harris (1991) gives $K_h = 0.006$ W/cm² C as representative of natural convection over a bearing housing, and this rises fourfold in a 50 ft/s airstream. A reference value of heat dissipation factor K_h could also be established from Equation (15.8) from the calculated bearing power loss together with measurement of bearing temperature rise and surface area in similar existing equipment. Palmgren (1959) gives the outer cooling area A for a rolling element bearing pillow block as $A = \pi H(B + H/2)$ where H is the vertical height and B the axial breadth of the pillow block. While examples exist for defining the temperature distribution within a bearing assembly (Harris, 1991), details of the heat dissipation area and the heat dissipation factor for specific cases involve large degrees of uncertainty (Eschmann *et al.*, 1985).

Table 15.5 Representative bearing temperatures in various machines

Operating temperature (°C)	Application
40–45	Cutter shaft of planing machine, bench drill, horizontal boring spindle, circular saw shaft, blooming and slabbing mill
50–55	Lathe spindle, vertical turret lathe, wood cutter spindle, calendar roll of paper-making machine, backup rolls of hot strip mills, face-grinding machine
60–65	Jaw crusher, axle bearings of locomotives and passenger cars, hammer mill, wire mill
70	Vibratory motor, rope-stranding machine
80	Vibrating screen, impact mill, ship propeller thrust bearing
90	Locomotive traction motor, hot gas fan, vibrating road roller
120	Water pump in automobile engine, turbocompressors, dryer rolls of paper-making machines
180	Calendar for plastic materials
200–300	Wheel bearings of kiln trucks

Source: Eschmann *et al.*, 1985. Reproduced with permission of John Wiley & Sons.

Equation (15.8) can be applied usefully, however, in comparing the expected temperature rise ΔT_{new} for a new design with reference temperature rise ΔT_{ref} observed in similar applications. Applying Equation (15.8) to both cases:

$$\frac{\Delta T_{new}}{\Delta T_{ref}} = \frac{(E_{p-new}/E_{p-ref})}{(A_{new}/A_{ref})} \tag{15.9}$$

If the power loss for a new application, for instance, were calculated from the relations in Section 15.2 to be twice that for a reference case where ΔT was observed to be 15 °C and with new area A 50% larger, the estimated bearing temperature rise for the new assembly becomes $15(2/1.5) = 20$ °C. A representative area ratio can commonly be taken from the product of the bearing axial width times its mean diameter for each case. Further details should be evaluated for significant changes in airflow velocity or for a different pattern in the heat transfer path.

A reference value of heat dissipation factor K_h could also be established from Equation (15.8) from the calculated bearing power loss together with measurement of bearing temperature rise and surface area in similar existing equipment.

Accurate prediction of the ball bearing thermal field is quite complex and the procedure described above represents a simplified estimation method of the bearing effective temperature. For a more complete treatment, Tu and Stein (1998) divide the entire process into several inter-connected stages and recommend an iterative procedure. Figure 15.4 shows a block diagram of the four calculations stages in a recent study reported by Takabi and Khonsari (2013). The heart of the calculation involves the determination of the total heat generation. This can be obtained in the first stage by considering the appropriate geometrical parameters, lubrication type, initial preload, friction coefficient, and the desired operational variables such the load and speed. The next stage is to determine how heat is transferred to the shaft and housing from the balls to the inner and outer rings. For this purpose, one has to consider the viscous effects, heat conduction, and convection into the air by convection (and radiation if warranted) as well as the thermal constriction resistances between the ball and the inner race as well as the ball and

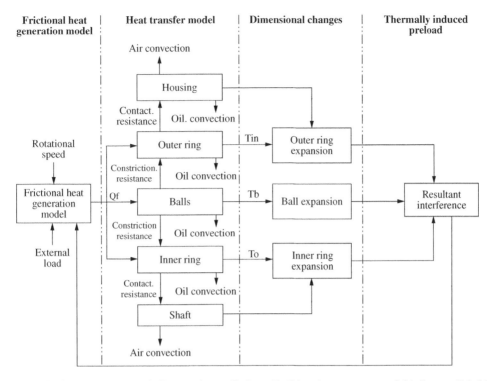

Figure 15.4   Four-stage block diagram for prediction of ball bearing temperature field. *Source*: Takabi and Khonsari, 2013. Reproduced with permission of Taylor & Francis.

the outer race. Any dimensional changes due to the ball expansion or the inner-and-outer ring thermal expansion/contraction should be considered in the third stage. The Hertzian contact theory can be used for this purpose and the change in the clearance can be determined along with thermal expansion. Finally, thermally-induced preload must be taken into account (Stein and Tu, 1994). The changes should subsequently feedback into the first stage and the procedure repeated with an updated preload and viscosity the temperature changes in the successive cycles become negligible and the calculations converge. Takabi and Khonsari (2013) show that predictions using this approach yield close agreement with laboratory measurements. At high rational speeds thermally-induced preload can become large and one has to guard against bearing seizure. Considerations should be given to the influence of traction coefficient and surface roughness. The interested reader is referred to recent reports by Takabi and Khonsari (2014, 2015a, 2015b, 2015c).

15.5 Rolling Bearing Lubrication

The primary need for lubrication is in the heavily stressed load-carrying contact between rolling elements and their raceways. With a ball bearing under simple radial load, some sliding occurs in the elliptical contact area, as illustrated in Figure 15.2. To balance frictional forces in the contact with the varying diameter and surface velocity, sliding in the direction opposite to that of rolling must occur in the center of the contact area while sliding in the same direction occurs at the outer areas of the contact ellipse. Lines of pure rolling between these two areas and wear in the sliding areas show up clearly in ball bearings when dryness of aged grease has led to less than full-film separation in this loaded area. With thrust loading of a ball bearing, this sliding motion in the contact zone is more complex, reflecting the shifted axis of ball rotation.

In addition to minimizing friction and wear in the ball track, lubrication is needed for sliding of the ball surfaces within each pocket of the separator. While sliding action in the load zone is minimal with the uniform diameter of a cylindrical roller in a roller bearing, elastic deflection, misalignment and skewing of the rollers, flat ends of the rollers, and separator surfaces again require lubrication to avoid rapid wear and early failure. Similar needs are present for all other types of rolling bearings.

Secondary functions are almost always required from lubricants for rolling bearings. These may include (1) cooling, (2) preventing corrosion, (3) providing protection from contaminants, and (4) minimizing maintenance. Such secondary functions commonly determine the choice of lubricant and the means of its application. In the absence of the need for cooling at high loads and high speeds, only a very small amount of lubricant is needed to supply a film to cover surface roughness of the balls or rollers, their raceways, and their separator. A single drop of oil, for instance, is adequate to provide lubrication for a ball bearing of 20–40 mm bore size for many days of running under moderate load at 3600 rpm (Wilcock and Booser, 1957).

15.6 Elastohydrodynamic Lubrication (EHL) of Rolling Contacts

Hydrodynamic lubrication relations developed in earlier chapters for sliding bearings can be applied to lightly loaded surfaces of rolling bearings such as cage contacts and roller face contacts with guiding lips on races. For the highly stressed contacts between rolling elements and their raceways, however, one must also consider the vital role of elastic deformation of surfaces and the dramatic increase in oil viscosity of up to 10 000 times or more its base value under the high contact pressure.

Line Contact

Figure 15.5 provides a schematic representation of the lubricant film and pressure distribution between a roller and its raceway under EHL conditions. As lubricant is drawn into the leading edge of the film by the roller surface, the viscosity of the oil increases rapidly and the lubricant film pressure rises to match the elastic Hertz contact pressure. After passing the center of the contact area, pressure of the lubricant film falls until it reaches the outlet zone. Here the decreasing contact pressure relieves the compression of the roller surface, the gap narrows to h_0, and a distinctive pressure spike is reached at the film exit. As speed decreases or load increases, this pressure spike moves further toward the exit and eventually disappears for very heavily loaded contacts (Dowson, 1970).

From simultaneous numerical solutions of Hertz elastic deflection in the contact area, Reynolds solutions for oil-film behavior, and the effect of pressure on raising oil viscosity, the following minimum film thickness h_0 is generated in a roller–raceway line contact (Dowson, 1970):

$$\frac{h_0}{R} = \frac{2.65 G^{0.54} U^{0.7}}{W^{0.13}} \qquad (15.10)$$

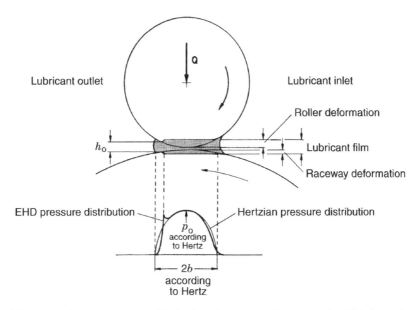

Figure 15.5 EHL film and pressure distribution. *Source*: Eschmann *et al.*, 1985. Reproduced with permission of John Wiley & Sons.

involving the following dimensionless parameters:

$$G = \alpha E' \quad \text{(material parameter)} \qquad (15.11)$$

$$U = \frac{\mu_0 u}{E' R} \quad \text{(speed parameter)} \qquad (15.12)$$

$$W = \frac{w}{E' L R} \quad \text{(load parameter)} \qquad (15.13)$$

where α is the pressure coefficient of viscosity (mm^2/N), generally in the 0.01–0.03 range (see Chapter 2); μ_0 is the dynamic viscosity at atmospheric pressure (N s/mm^2); u is the entrainment velocity in the contact with $u = (u_1 + u_2)/2$ (mm/s) and for pure rolling $u = u_1 = u_2$; $R = R_1 R_2/(R_1 + R_2)$ involving R_1 and R_2 radii of the roller and its raceway (mm); w is the total load (N) on effective roller length L (mm); and E' is the equivalent elastic modulus of the contact $E/(1 - v^2)$, where E is the modulus (N/mm^2) and v is Poisson's ratio for contacting materials that are the same and for which typical properties are given in Table 14.2. (See Equation (14.10) for contact of different materials.)

Point Contact

For a ball bearing, minimum oil-film thickness h_0 for 'point' contact of a ball with its raceway is given by the following (Hamrock and Dowson, 1977; Cheng, 1984):

$$\frac{h_0}{R_x} = 3.63 U^{0.68} G^{0.49} W^{-0.073} (1 - e^{-0.68k}) \qquad (15.14)$$

involving the following dimensionless parameters corresponding to Equations (15.10)–(15.13):

$$U = \mu_0 u / E' R_x \qquad \text{(speed parameter)} \tag{15.15}$$

$$W = \frac{w}{E' R_x^2} \qquad \text{(load parameter } w \text{ is normal load on ball (N))} \tag{15.16}$$

$$k = 1.03 \left(\frac{R_y}{R_x} \right)^{0.64} \qquad \text{(elliptical parameter)} \tag{15.17}$$

where

$$R_x = \frac{R_{x1} R_{x2}}{R_{x1} + R_{x2}} \qquad \begin{array}{l} \text{(conjunction radii of ball and raceway (} R_{x1} \text{ and } R_{x2} \text{)} \\ \text{in rolling direction (mm))} \end{array} \tag{15.18}$$

$$R_y = \frac{R_{y1} R_{y2}}{R_{y1} + R_{y2}} \qquad \begin{array}{l} \text{(conjunction radii of ball and raceway (} R_{y1} \text{ and } R_{y2} \text{)} \\ \text{in plane perpendicular to rolling (mm))} \end{array} \tag{15.19}$$

R_x reflects the sum of the curvatures in the direction of ball rolling; G is as defined in Equation (15.11).

Equations (15.10) and (15.14) indicate for both ball and roller bearings that EHL oil-film thickness is strongly dependent on both bearing speed and oil viscosity and is very slightly related to load. While increasing load increases the area in the contact zone according to Hertz elasticity relations, film thickness diminishes only as the 0.13 power of the load for roller contact and as the 0.073 power for ball contact.

To avoid wear and related problems, minimum film thickness should generally be in the range of one to three or more times the composite surface finish at the loaded rolling contacts. Oils with suitable extreme pressure or antiwear additives should be considered for lower film thicknesses.

Example 15.4   The 6210 deep-groove ball bearing of Examples 14.1 and 14.2 has balls with a 12.7 mm diameter, and the maximum normal ball load is 1500 N. Oil bath lubrication uses industrial VG 32 viscosity grade mineral turbine oil at 65 °C, at which its kinematic viscosity is 13.5 cSt (mm^2/s), its dynamic viscosity is 1.13×10^{-8} N s/mm^2, and its pressure–viscosity coefficient is 0.018 mm^2/N(1.8×10^{-8} m^2/N) from Example 2.1. Calculate the minimum oil-film thickness at 900 rpm for the most heavily loaded ball contact (on the inner ring raceway).

Lubricant parameter G from Equation (15.11) uses the following effective modulus of elasticity E' from Equation (14.10) for the steel-on-steel contact:

$$E' = \frac{E}{1 - v^2} = \frac{202 \times 10^3 \text{ N/mm}^2}{1 - 0.28^2} = 219\,000 \text{ N/mm}^2$$

Then

$$G = \alpha E' = (0.018 \text{ mm}^2/\text{N})(219\,000 \text{ N/mm}^2) = 3940$$

Example 14.1 gives the ball radius $R = R_{x1} = R_{y1} = 6.35$ mm, inner raceway radius in the rolling direction $R_{x2} = 28.35$ mm, and the ball groove radius perpendicular to the rolling plane $R_{y2} = 6.6\,\text{mm}$. These give the following conjunction radii from Equations (15.18) and (15.19):

$$R_x = \frac{R_{x1}R_{x2}}{R_{x1} + R_{x2}} = \frac{6.35 \times 28.35}{6.35 + 28.35} = 5.188\,\text{mm}$$

$$R_y = \frac{R_{y1}R_{y2}}{R_{y1} + R_{y2}} = \frac{6.35 \times 6.6}{6.35 + 6.6} = 3.236\,\text{mm}$$

and the elliptical parameter in Equation (15.17) becomes

$$k = 1.03 \left(\frac{R_y}{R_x}\right)^{0.64} = 1.03 \left(\frac{3.236}{5.188}\right)^{0.64} = 0.761$$

Entrainment velocity u in the contact with pure rolling of the balls on their inner raceway with its radius $R_{x2} = 28.35$ is

$$u = u_i = 2\pi R_{x2}n_i = 2\pi(28.35)(900/60) = 2672\,\text{mm/s}$$

Using these values gives the following dimensionless parameters in Equations (15.12) and (15.13):

$$U = \frac{\mu_0 u}{E' R_x} = \frac{(1.13 \times 10^{-8})(2672)}{(219\,000)(5.188)} = 2.655 \times 10^{-11}$$

$$W = \frac{w}{E' R_{x2}} = \frac{1500}{(219\,000)(5.188)^2} = 2.543 \times 10^{-4}$$

Applying these values to obtain dimensionless film thickness in Equation (15.14):

$$\frac{h_0}{R_x} = 3.63 U^{0.68} G^{0.49} W^{-0.073}(1 - e^{-0.68k})$$

$$= 3.63(2.655 \times 10^{-11})^{0.68}(3940)^{0.49}(2.543 \times 10^{-4})^{-0.073}(1 - e^{-[0.68 \times 0.761]})$$

$$= 9.985 \times 10^{-6}$$

$$h_0 = (9.985 \times 10^{-6})R_x = (9.985 \times 10^{-6})(5.188)$$

$$= 51.8 \times 10^{-6}\,\text{mm} = 0.0518\,\mu\text{m}(0.0518 \times 39.37\,\text{in./m} = 2.039\,\mu\text{in.})$$

This minimum film thickness at the inner raceway contact is smaller than at the outer raceway contact because of the smaller effective radius at the inner raceway. Other factors such as the entrainment velocity are the same at both the inner and outer ring raceways. Lubricant starvation and any thermal heating of the lubricant at the conjunction would reduce h_0 calculated by the above procedure (Khonsari and Hua, 1997).

Assuming $\sigma_1 = 0.012\,\mu\text{m}$ finish for the balls and $\sigma_2 = 0.025\,\mu\text{m}$ for the raceway grooves, the dimensionless lambda ratio Λ of the minimum oil film thickness h_0 to composite surface roughness σ is given as follows:

$$\Lambda = \frac{h_0}{\sigma} = \frac{h_0}{(\sigma_1^2 + \sigma_2^2)^{1/2}} = \frac{0.0518}{(0.012^2 + 0.025^2)^{1/2}} = 1.9$$

As discussed in Section 3.7, full EHL with undiminished fatigue life is expected when Λ exceeds 3. Given that in this example the film parameter is below 3, for greater assurance of full-film lubrication and to help accommodate any unexpected contamination, the next higher viscosity grades of turbine mineral oil (VG 46 or VG 68) should be considered.

Lubrication Regimes

While the dominating lubrication regime in rolling element bearings is elastohydrodynamic, it is important to realize that other regimes may exist depending on the materials properties, lubricant viscosity or operating conditions. These are briefly reviewed below and references are given for the details (e.g., Johnson, 1970; Winer and Cheng, 1980; Hamrock and Dowson, 1981). Application examples for determining the regimes based on specifications of properties and operating conditions are available in Khonsari and Hua (1997), Khonsari & Kumar (2013), Kumar and Khonsari (2013) and more recently Takabi and Khonsari (2015a).

(i) *Isoviscous-rigid.* In this regime, the elastic deformation and variation of viscosity with pressure are negligibly small and can be safely neglected. A familiar example is the thrust bearings covered in Chapter 7, where the hydrodynamic pressure is typically low and elastic deformation of a pad does not appreciably affect the bearing performance.

(ii) *Piezoviscous-rigid.* In this regime, the hydrodynamic pressure becomes high enough to have a major influence on the viscosity while the surface deformation remains negligible. An example is the lubrication of a piston ring or a moderately-loaded roller bearing.

(iii) *Isoviscous-elastic (soft EHL).* In this regime, the effect of pressure on viscosity is negligible. However, the elastic deformation of one or both of the surfaces is large and must be taken into consideration. Commonly referred to as soft EHL, this is typically the case when one deals with a material with a low elastic modulus that even low pressures—e.g. on the order of 1 MPa—can cause significant deformation. A lip seal, water-lubricated rubber bearing, the bone-cartilage interaction in a human knee joint (see discussion in Example 9.3) are examples of soft EHL.

(iv) *Piezoviscous-elastic (hard EHL).* In this regime, typically referred to as EHL, one deals with heavily-loaded concentrated contacts where the maximum pressure ranges from 0.5 and 3 GPa. Therefore, both the elastic deformation of solids as well as the variation of viscosity with pressure must be considered in the analysis. This is the primary mode of lubrication in ball and roller bearings and gears.

Mixed-Film Lubrication

Heavily loaded bearings or those that operate at relatively low speed may not generate adequate film thickness. If the dimensionless film parameter Λ drops below 3, the surfaces comes into contact at the asperity level and, according to the Stribeck curve, the mixed-film lubrication prevails (Zhu and Wang, 2011, 2012, 2013; Zhu et al., 2015; Lu et al., 2006). Under these conditions, the load is supported by both the hydrodynamic pressure as well as surface asperities through what is commonly referred to as the load-sharing concept introduced by Johnson et al. (1972).

Figure 15.6 shows remarkable results of computer simulations for the prediction of the pressure distribution and the film thickness in an elliptical contact with transition from smooth

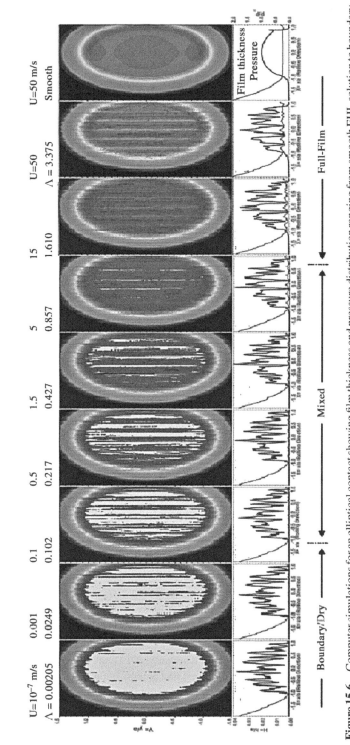

Figure 15.6 Computer simulations for an elliptical contact showing film thickness and pressure distribution ranging from smooth EHL solution to boundary lubrication. *Source*: Zhu *et al.*, 2015. Reproduced with permission of ASME.

EHL solution at U = 50 m/s to mixed lubrication in the range of U = 15 to 0.1 m/s ($\Lambda = 1.610$ to 0.102) and to boundary lubrication and dry condition when the speed is reduced to 0.001 and 10^{-7} m/s ($\Lambda = 0.0249$ and 0.00205), respectively (Zhu et al., 2015). Surface-roughness pattern—isotropic, longitudinal, and transverse (see Chapter 6, Section 6.3)—also affect the traction and film thickness. Expressions of estimating film thickness and traction coefficient in mixed lubrication regime are available for both line and point contact (Masjedi and Khonsari, 2014a, 2014b; 2015a, 2015b). They show that the non-dimensional film thickness is:

$$H_{mixed} = H_{smooth} f(\overline{\sigma}, \overline{V}, W, U, G, k) \tag{15.20}$$

where f is a function of dimensionless parameters for roughness $\overline{\sigma}$, hardness $\overline{V}$, load W, speed U, material parameter G and the ellipticity parameter k for treating point contact problems.

Lubricant Starvation

An assumption that is generally made in deriving the film thickness equations is that the contact's inlet is *fully flooded* with lubricant. If the lubricant quantity is insufficient to fill the inlet, then the contact is said to be starved. *Lubricant starvation* reduces the film lubricant film and increases the tendency of the asperity-to-asperity to come into intimate contact. This can result in excessive and surface damage if the unit is allowed to operate in this mode for a prolong length of time. It is thus no surprise that the subject has attracted the attention of many researchers in the field (Hamrock and Dowson, 1977; Chevalier et al., 1998; Damiens et al., 2004; Cann et al., 2004, Faran and Schipper, 2007; Yang et al., 2004; Kumar and Khonsari, 2008; Ali et al., 2013a, 2013b; Svoboda et al., 2013).

The central and the minimum film thickness in fully flooded regime $H_{C(fully\,flooded)}$ and $H_{min(fully\,flooded)}$ are reduced by a starvation factor Φ_{H_C} and $\Phi_{H_{min}}$ obtained as:

$$\begin{aligned} H_{C(starved)} &= \Phi_{H_C} \times H_{C(fully\,flooded)} \\ H_{min(starved)} &= \Phi_{H_{min}} \times H_{min(fully\,flooded)} \end{aligned} \tag{15.21}$$

For the line-contact configuration, regression analyses of more than 200 simulation cases in the mixed-EHL regime reveal that the ratio of the starved to fully-flooded central film thickness and minimum film thickness can be defined as follows (Masjedi and Khonsari, 2015):

$$\begin{aligned} \Phi_{H_C} &= 1 - \zeta \\ \Phi_{H_{min}} &= 1 - \zeta^{1.08} \end{aligned} \tag{15.22}$$

where $\zeta = 1 - \frac{\dot{m}_s}{\dot{m}_f}$ in which $\dot{m}_f$ and $\dot{m}_s$ represent the mass flow rate in the fully-flooded and the starved conditions, respectively. Thus, the starvation degree, ζ, is zero in the fully-flooded condition and one in the fully-starved condition.

For point contacts, according to Masjedi and Khonsari (2015), the correction factors depend on the elasticity parameter as well. They showed that a regression analysis of the results of 200 simulations within a specified range of operating conditions and material properties can be described by the following relationships.

$$\begin{aligned} \Phi_{H_C} &= 1 - 1.561\,\zeta^{0.849} k^{-0.214} \\ \Phi_{H_{min}} &= 1 - 0.845\,\zeta^{0.893} k^{0.081} \end{aligned} \tag{15.23}$$

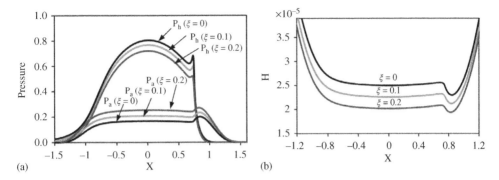

Figure 15.7 Fully-flooded and starved regimes in mixed-EHL line contact with $\overline{\sigma} = \sigma/R = 2 \times 10^{-5}$ (a) Pressure profile and (b) film profiles. *Source*: Masjedi and Khonsari, 2015. Reproduced with permission of John Wiley & Sons.

Figure 15.7 illustrates the contribution of hydrodynamic and asperity pressure to load-carrying capacity and the film thickness profiles in a mixed-EHL line contact for fully-flooded ($\xi = 0$) as well as a starved with $\xi = 0.1$ and $\xi = 0.2$. Here, the combined roughness is assumed to be $\sigma = 0.5$ µm for the effective radius of 1 in., and the load is maintained constant. The bearing material is steel and the maximum Hertzian pressure is roughly 0.64 GPa. Referring to Figure 15.6, with 20% starvation degree ($\xi = 0.2$), the asperity load ratio increases to 34% compared to 23% in the fully flooded case. Accordingly, the fully-flooded contact and central film thickness drops by roughly 20% with $\xi = 0.2$. The film parameter for the fully-flooded case is $\Lambda = 1.15$, but it drops to $\Lambda = 0.97$ for $\xi = 0.2$. Thus, as the degree of starvation increases, the lubrication regime can shift to the boundary lubrication.

Figure 15.8 shows analogue simulation results for circular point contact ($k = 1$). See Masjedi and Khonsari (2015) for details. Here the load ratio increases from 43% for the fully-flooded mixed-EHL to 65% for $\xi = 0.2$, and the reduction in the central is more than that of the line contact. Compared to fully-flooded, the film thickness drops by roughly 30% for $\xi = 0.2$.

Non-Newtonian Effects

As described in Chapter 2, many lubricants exhibit shear-thinning behavior at high shear rates, and the Newtonian fluid model is simply incapable of accurately predicting their behavior. Prediction of the traction coefficient, hence the power loss, is particularly sensitive to the lubricant rheological model. While a critical review of the subject is beyond the scope of this book, it is nevertheless worthwhile to provide a brief exposition of the available modeling schemes and pertinent literature. The interested reader is referred to Kumar and Khonsari (2009c) for more detailed discussion.

The Sinh Law Model (so-called 'Ree-Eyring')

A rheological behavior sometimes referred to as the Sinh law is commonly used to explain the shear-thinning behavior of lubricants. It has the following form.

$$\dot{\gamma} = \frac{\tau_o}{\mu} \sinh\left(\frac{\tau}{\tau_o}\right)$$ (15.24)

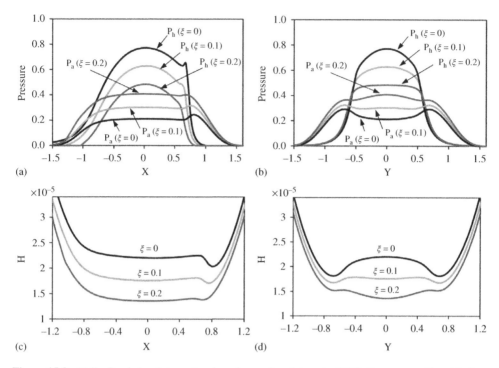

Figure 15.8 Fully-flooded and starved regimes in rough point-contact EHL Pressure profiles: (a) along X in $Y = 0$ plane (b) along Y in $X = 0$ plane film profiles: (c) along X in $Y = 0$ plane (d) along Y in $X = 0$ plane with $k = 1$ and $\overline{\sigma} = \sigma/R_x = 2 \times 10^{-5}$. *Source*: Masjedi and Khonsari, 2015. Reproduced with permission of John Wiley & Sons.

where $\dot{\gamma}$, τ, μ are the lubricants shear rate, shear stress, and the low shear viscosity, respectively. The parameter τ_o represents the Newtonian limit of the lubricant, and is called the Eyring stress. A plot of traction against the sliding velocity for small sliding velocities approximately follows the inverse hyperbolic sine function. The existing literature commonly attributes this rheological equation to Ree-Eyring.

The Generalized Maxwell Model
Proposed by Tanner (1960), a generalized Maxwell model for lubrication for simple shear can be expressed as:

$$\tau + \lambda_m \frac{D\tau}{Dt} = \eta(\tau)\,\dot{\gamma} \tag{15.25}$$

where D/Dt is the convective time derivative and λ_m is a relaxation time. Equation (15.25) is often attributed to a later work by Johnson and Tewaarwerk (1977) who considered $\eta(\tau)$ to be of Sinh law form, as in Equation (15.24). The time-dependent term in Equation (15.2) is often found negligible in many EHL applications. Johnson and Greenwood (1980) examine the behavior of lubricant in the contact with consideration of thermal effects. Addition references include the work of Houpert and Hamrock (1985), Conry *et al.* (1987), and Chang *et al.*

(1989). Sui and Sadeghi (1991) present a full numerical solution to study the combined effect of temperature rise and non-Newtonian Sinh law fluid in EHL line contacts. Salehizadeh and Saka (1991) investigated the present combined effect of thermal and non-Newtonian character in EHL line contacts under pure rolling conditions with Sinh law type of fluid. Results for simple sliding condition using the Sinh law is given by Wang *et al.* (1992). Kumar *et al.* (2002) present a thermal EHL analysis of rolling/sliding line contacts using a mixed rheological model consisting of two flow units corresponding to Newtonian and Sinh law fluids, which follows the mixture theory introduced by Dai and Khonsari (1994).

The Bair-Winer Model
Bair and Winer (1979) proposed the following non-linear constitutive equation based upon extensive laboratory experiments

$$\dot{\gamma} = \frac{1}{G_\infty} \frac{d\bar{\tau}}{dt} + \frac{\tau_L}{\mu} \ln(1 - \bar{\tau})^{-1} \tag{15.26}$$

where $\bar{\tau} = \tau/\tau_L$ and $\tau_L = $ limiting shear strength of the lubricant; See Chapter 2 for discussion. Numerical implementation of this model in a line-contact application with thermal effect and experimental verification is presented by Khonsari and Hua (1994a,b).

The Sinh Law with Limiting Shear Strength
Wang *et al.* (1992) suggest the following rheological model that combines the Sinh law and limiting shear strength of Bair and Winer.

$$\tau = \begin{cases} \tau_o \sinh^{-1}(\mu\dot{\gamma}/\tau_o) & for \ |\mu\dot{\gamma}| \leq \tau_o \sinh(\tau_L/\tau_o) \\ \tau_L & for \ |\mu\dot{\gamma}| \geq \tau_o \sinh(\tau_L/\tau_o) \end{cases} \tag{15.27}$$

Analyses presented by Wang *et al.* (1992) includes provisions for thermal effects.

The Carreau-Yasuda Model
This is the most general form of a power law that describes high shear rate as well as the limiting second Newtonian regime (Bair and Khonsari, 1996) as described in Chapter 2. The Carreau-Yasuda (Yasuda *et al.*, 1981) rheological equation has the following form:

$$\eta = \tau/\dot{\gamma} = \mu_\infty + (\mu - \mu_\infty)[1 + (\lambda\dot{\gamma})^a]^{(n-1)/a} \tag{15.28}$$

where μ_∞ is the second Newtonian viscosity and λ is the time constant, generally expressed as a Maxwell time constant $\lambda = \mu/G_{cr}$ with G_{cr} representing a shear modulus or a critical stress associated with the longest relaxation time. It can be estimated as:

$$G_{cr} \approx \frac{\rho RT}{M} \tag{15.29}$$

where $\rho = $ lubricant density, $R = $ universal gas constant, $T = $ absolute temperature and $M = $ molecular weight. By setting $a = 2$, which is the case for monodisperse liquids, in Equation (15.28), the Carreau viscosity equation (1972) is obtained:

$$\eta = \tau/\dot{\gamma} = \mu_\infty + (\mu - \mu_\infty)[1 + (\dot{\gamma}\mu/G_{cr})^2]^{(n-1)/2} \tag{15.30}$$

The following modified form of Carreau-Yasuda model is often recommended for EHL film thickness calculations:

$$\eta = \tau/\dot{\gamma} = \mu[1 + (\tau/G_{cr})^2]^{(1-1/n)/2} \tag{15.31}$$

Setting $n = 1/3$ in the above Equation (15.31) yields the so-called Rabinowitsch's generalized Newtonian viscosity model (Bair and Qureshi, 2003):

$$\eta = \tau/\dot{\gamma} = \frac{\mu}{[1 + (\tau/G_{cr})^2]} \tag{15.32}$$

Bair and Qureshi (2003) conclude that the shear thinning response can be described by a generalized Newtonian model and that both traction and film thickness are of power-law form as described by the Carreau viscosity model. Mongkolwongrojn et al. (2005) show how the performance characteristics of EHL line contacts changes with varying the Carreau viscosity model parameters. More recently, Jang et al. (2007) presented extensive EHL line-contact simulations using the Carreau model and showed that the results closely agree with published experimental measurements of Dyson and Wilson (1965). In contract, Jang et al. (2007) find that the Sinh law is incapable of describing the experimentally observed film-thinning pattern correctly. This is also in accordance with findings reported by Bair (2004), Kumar and Khonsari (2009a) and Kumar et al. (2009b). For accurate quantification of EHL performance, in addition to the proper rheological relationship, careful attention must be given to the evaluation of the lubricant's piezo-viscous properties as described in Chapter 2, Section 2.4. The interested reader is referred to Bair (2007) who provides a detailed account of high pressure rheology for quantitative EHL analyses.

15.7 Selection of Oil Viscosity

To provide adequate film thickness under normal cleanliness conditions, Figure 15.9 gives a simplified criterion to determine the minimum oil viscosity required to match the general pattern defined in Equations (15.10) and (15.14). In essence, Figure 15.9 relates surface speed and required viscosity to the required EHL minimum film thickness needed to match the composite surface roughness in the loaded contact while neglecting the relatively minor influence of load on the bearing.

Based on this required viscosity from Figure 15.9, the viscosity–temperature chart in Figure 15.10 enables selection of the appropriate ISO grade (mm²/s viscosity at 40°C) of petroleum mineral oil to provide the required viscosity at the bearing operating temperature. Use of Figures 15.9 and 15.10 is also valid for greases; the base oil viscosity needed in the grease at the operating temperature is again given in Figure 15.9. Many industrial ball and roller bearing greases incorporate a mineral oil having a viscosity in the approximate range of 70–130 mm²/s (cSt) at 40°C.

Eschmann et al. (1985) indicate that the limiting viscosity in Figure 15.9 should be increased by a factor of 2 when the load changes from a radial direction to a combined radial–thrust load and by a factor of 3 for an axial load. Since various degrees of sliding motion and different curvature combinations become involved in bearings designed for thrust and combined loads, bearing manufacturers should be consulted on demanding applications.

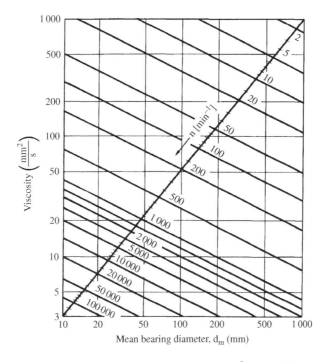

Figure 15.9 Minimum required lubricant kinematic viscosity (mm²/s) for EHL film versus mean bearing diameter (mm) and speed (rpm). *Source*: FAG Bearings Corp., 1995. Reproduced with permission of Schaeffler Group.

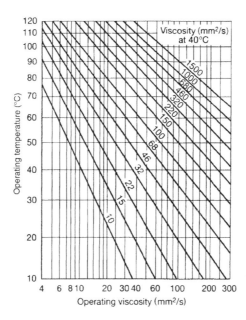

Figure 15.10 Lubricant kinematic viscosity–temperature chart for mineral oils of standard VG viscosity grades at 40 °C. *Source*: FAG Bearings Corp., 1995. Reproduced with permission of Schaeffler Group.

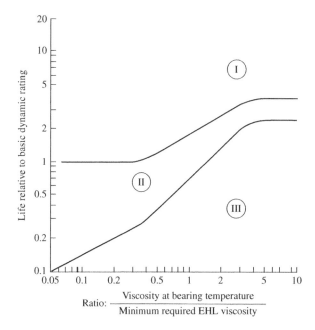

Figure 15.11 Bearing fatigue life factor versus viscosity ratio. Zone I: Utmost cleanliness, not too high load, high speed; Zone II: Customary experience with normal cleanliness; Zone III: Contaminated lubricant, unsuitable lubricant, unfavorable conditions. *Source*: FAG Bearings Corp., 1995. Reproduced with permission of Schaeffler Group.

The viscosity determined by Figure 15.9 represents a minimum value for forming a full EHL film. As the ratio of operating viscosity to this minimum value rises above 1.0, the fatigue life given in Figure 15.11 then increases up to four times the usual fatigue life given in catalogs. (See Chapter 13 for an indication that even fatigue life multipliers up to the 10–30 range are sometimes found in Zone I with high-speed aircraft applications.)

The broad range of diminishing life shown in Figure 15.11 for low-viscosity ratios below 1.0 reflects the increasing influence of other factors, such as wear resulting from contamination and from sliding contact by the rolling elements. Catalog ratings (basic dynamic load ratings) are still achieved even at very low viscosity ratios (as low Λ values give oil-film thickness well below the surface roughness) with suitable antiwear additives in the lubricant, with freedom from contaminants, and at low loads relative to static load capacity. Zone II in Figure 15.11 is generally to be expected with a normal degree of cleanliness. A drop from the lower extreme of the life factor in Zone II down to Zone III occurs with unfavorable lubrication conditions and with heavy contamination (FAG, 1985).

Significantly higher viscosity than indicated from Figure 15.9 is not expedient since it would lead to unnecessarily higher bearing friction and higher bearing temperature. Where a synthetic oil is to be used for extreme temperatures or to fill other special application requirements, the pressure viscosity coefficient of the synthetic should also be compared with that of mineral oil (see Equation (2.13) and Table 2.6). Using Equations (15.10) and (15.14) as guides, oil viscosity needed for a given minimum oil-film thickness is seen to decrease with increasing pressure viscosity coefficient.

Example 15.5 For the 6210 size ball bearing of Example 15.6 with a mean diameter of 70 mm and running at 900 rpm, Figure 15.9 gives a required minimum viscosity of 18 mm^2/s (cSt) for providing a full separating elastohydrodynamic oil film. For the 65 °C operating temperature, Figure 15.10 indicates a need for a minimum VG 46 viscosity grade of oil (having a viscosity of 46 mm^2/s at the 40 °C reference temperature). The more detailed analysis of minimum film thickness for the particular bearing and its load in Example 15.6 indicates that VG 32 would be adequate.

15.8 Oil Application

Rolling element bearings generally require very little oil to supply their needed surface-coating films. Other factors such as carrying away of frictional heat in heavily loaded and high-speed bearings, and the needs of adjacent gearing or other machine elements, may dictate the nature of an oil supply, as covered in Chapter 2. Suitable enclosures and seals are essential in oil lubrication with many rolling bearings. Guidelines follow for selection of common oil lubrication methods (SKF, 1999).

Oil bath supply is the simplest method of oil lubrication for low and moderate speeds. Oil is distributed within the bearing by its rotating components and then drains back to the oil bath. The oil level at standstill, as checked with a simple sight glass, should be maintained just below the center of the lowest ball or roller. Windage and centrifugal effects set the approximate speed limits in Table 14.5 for using oil bath lubrication.

Circulating system supply (Figure 15.12) is preferred at higher speeds to provide the increased lubrication required, to provide a lower operating temperature than is obtained with a simple sump, and for increased reliability under conditions which may otherwise reduce oil life and lead to increased oil loss by leakage. Oil feed to the bearing housing should be located on one side of the bearing and well below the centerline for a horizontal shaft. With a drain on the opposite side of the bearing, all oil fed then passes through the bearing. A drain standpipe can be used to maintain the static oil level just below the center of the bottom rolling element, as with a simple oil bath. With a double-row roller bearing or a pair of single-row bearings, oil feed is best supplied at the center plane and an oil groove may be incorporated in the outer diameter of a double-row bearing for this purpose. A drain with a center feed is made from each side of the housing.

Oil mist from a centralized system or from a separate mist generator is effective in supplying the small amount of oil needed for lubrication of most rolling bearings, while cooling from the associated airstream provides effective cooling. Since the oil mist is discharged into the environment, air–oil systems have been replacing oil mist for many centralized systems for industrial lubrication. With air–oil feed, a continuous supply of compressed air transports small metered amounts of oil at programmed time intervals to bearings at specific locations.

Oil jet feeding is effective for very high speeds. A jet of circulating oil under high pressure is directed at the side of the bearing with a velocity of at least 15 m/s at high *DN* (diameter × rpm) values, sufficient so that some of the oil will penetrate the turbulence surrounding the rotating bearing. The oil feed volume should usually be set so that the temperature rise of the oil does not exceed about 30–50 °C in carrying away frictional heat generated within the bearing.

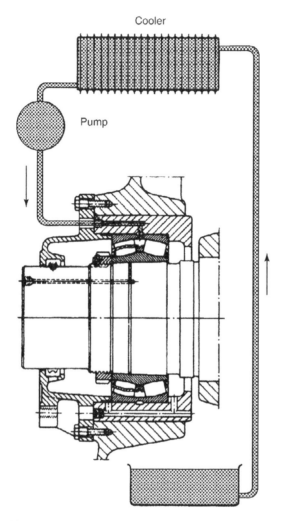

Figure 15.12   Elements of an oil circulating system. *Source*: SKF, 1999/2005.

15.9 Oil Change Intervals

The required frequency for oil change depends both on oil quality and on operating conditions. Chapter 2 gives details on the limiting oxidation life to be expected with different types of oil. General field experience with rolling bearing applications gives further useful guidelines (SKF, 1999).

With oil bath lubrication, yearly change is usually sufficient for bearing temperatures up to 50 °C, where there is little risk of contamination. Higher temperatures call for more frequent change: every three months for temperatures around 100 °C. With circulating oil systems and oil jet lubrication, a suitable oil change interval can generally be determined by regular oil inspections to check both for oil oxidation and for contamination using the background given in Chapter 8 and 17, and with guidance from the lubricant supplier.

15.10 Grease Selection and Application

While oil generally provides superior lubrication, grease is a common first choice for most ball and roller bearings (Booser, 1995; Schrama, 1997). From the viewpoint of lubrication, oil is more effective in carrying away heat, feeds more readily into loaded contact areas, enables higher operating speeds and loads, and flushes out contaminates such as dirt, wear products, and water. Oil also better satisfies the lubrication requirements of any related machine elements such as gears, cams, guides, and sliding bearing surfaces.

Despite the desirable lubricating properties of oils, grease is used in a far larger number of rolling bearing applications. This broad use of grease is dictated by the possibilities for simpler housing designs, less maintenance, less difficulty with leakage, and generally easier sealing against dirt and moisture. For these reasons, greases are now almost universally used in electric motors, household appliances, instruments, transportation wheel bearings, and many other applications where high speed, temperature, or design do not preclude the use of grease.

While some factors involved in grease performance can be determined from oil and thickener composition (see Chapter 2) and laboratory tests, field experience is usually the ultimate guide to grease life, noise, speed and temperature limits, friction, and related operating behavior. Generally, the soap or other gel structure can be viewed as a sponge reservoir for an oil which itself would be appropriate for lubricating the rolling element bearing.

Grease Composition

Mineral oils are employed in greases comprising well over 95% of current usage. To meet the needs of a wide range of applications, use of mineral oil in the viscosity range of 70–130 cSt at 40 °C is common. Although involving higher friction and noise, more viscous oils in the 200–600 cSt range to meet the guidelines of Equations (15.10) and (15.14) (and Figure 15.9) are employed under more demanding EHL conditions of high contact stress, low speed, and high temperature (Booser and Khonsari, 2007).

Lower-viscosity oils down to 25 cSt at 40 °C enable operation at lower temperatures, with easier dispensing and possibly operation at higher speeds, but such greases suffer from more rapid oil evaporation at elevated temperatures. As an extreme upper viscosity limit, the starting torque of an industrial electric motor is commonly insufficient to start turning a ball bearing when the oil viscosity in the grease exceeds 100 000 cSt. This viscosity is reached with conventional ball bearing greases in the −15 to −20 °C range. With sufficient torque to start rotation at these and lower temperatures, the inner ring may initially spin independently within a stationary assembly of rolling elements, which has been immobilized by the very stiff grease. For bearings in small, low-torque equipment at low temperatures, greases must be considered which employ either low-viscosity mineral oils or corresponding synthetic oils.

Lithium soaps are used in a majority of greases as the gelling agent. Lithium soap greases provide reasonable mechanical stability in the presence of the shearing action within a rolling bearing, have generally adequate water resistance, and are usable over a wide temperature range. Nonmelting polyurea powder is finding expanding use for premium bearing greases, especially at elevated temperatures (above about 110 °C for lithium hydroxystearate and 150 °C for complex lithium, calcium, or aluminum soaps). Clay-type thickeners are used in less demanding bearings because of their low cost.

NLGI Grade 2 'soft' stiffness grade is normally given first consideration. Stiffer Grade 3 is used in many ball and roller bearings with larger bore sizes than about 80–100 mm and with vertical shafts to avoid grease slumping, churning, and leakage. Grade 3 greases also find use in some double-sealed and double-shielded ball bearings to avoid mechanical breakdown of the soap structure under the shearing action of the ball–separator assembly. Grade 1 and softer grades provide easier feeding to a variety of machine elements and to multiple-row bearings. The following representative self-supporting heights provide some guidance as to the minimum softness grade of grease for various grease cavity dimensions within a bearing housing to avoid grease slumping and churning problems:

NLGI grease grade	0	1	2	3	4	5
Approximate self-supporting height						
mm	13	18	30	56	132	381
inches	0.5	0.7	1.2	2.2	5.2	15

Chapter 2, Section 2.7, supplies further details on composition, properties, and temperature limits with mineral oil and synthetic greases. The interested reader is also referred to Booser and Khonsari (2013a, 2013b).

15.11 Grease Life: Temperature and Speed Relations

The primary function of grease in a rolling bearing is to serve as a supply of lubricant to coat rolling elements and cage surfaces. A wide range of laboratory tests and observations from industrial, appliance, railroad, and aerospace applications indicate that a grease commonly fails to meet this lubrication demand when half of its initial oil content is lost (Booser, 1997). At this degree of dryness, friction and noise rise to generally intolerable levels; continued operation is accompanied by wear of bearing surfaces and early failure. Loss of half of the initial oil content corresponds to the following percentage soap (thickener) S_f in the grease at failure:

$$S_f = 100 \times \frac{2S_0}{100 + S_0} \tag{15.33}$$

where S_0 is the percentage of thickener in the fresh grease. For a fresh grease containing 10% soap, for instance, failure would be expected at $S_f = 18\%$; for 25% initial soap content, $S_f = 40\%$. Other factors used to monitor remaining grease life include oil content by atomic absorption spectroscopy, grease acidity, antioxidant content, and iron content as a reflection of bearing wear.

Grease life in rolling bearings is related to temperature in much the same pattern as was oil life in Equation (2.31). Under ideal conditions of light load and moderate surface speeds, log of 10% grease life L(h) is generally a function of grease stability term A and the reciprocal of absolute temperature (Kelvin) in the first two terms on the right-hand side of the following equation (Booser, 1974, 1975):

$$\log L = A + \frac{B}{T} - 9.6 \times 10^{-7} k_f D N \tag{15.34}$$

where $T = 273 + °$ C.

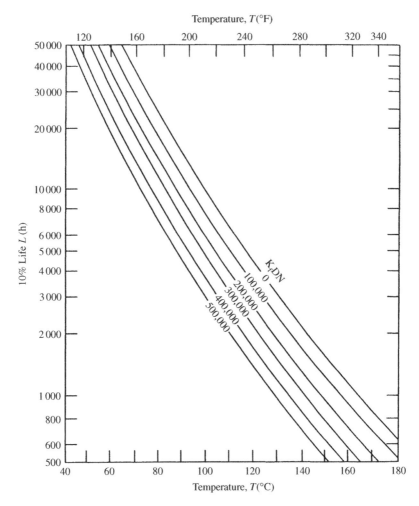

Figure 15.13 Grease life versus rolling bearing temperature and $k_f DN$ speed factor. *Source*: Booser, 1997. Reproduced with permission of Taylor & Francis.

Grease life typically increases by a factor of 1.5 for each 10 °C lower reduction in temperature in the 100 °C range, giving $B = 2450$. Life factor A becomes -2.60 for test lives with a number of premium mineral oil and synthetic greases using various thickeners such as lithium hydroxystearate, complex metal soaps, and nonmelting polyurea powders (Khonsari and Booser, 2003). Reduction in grease life with increasing surface speeds is reflected in the $k_f DN$ term (D mm bore times N rpm) in the above equation and in Figure 15.13. Speed factor life-reduction coefficients k_f in Table 15.6 reflect the lubrication needs of various bearing types with increasing speed (FAG). The higher values for a given bearing type apply to larger cross section bearings (higher load capacity) and smaller values to smaller cross-section bearings.

While this $k_f DN$ velocity term reflects typical experience for premium NLGI Grade 2 greases in open ball and roller bearings at loads experienced in electric motors and related equipment, the pattern of decreasing life with increasing DN varies greatly with different greases. Results more sensitive to DN reported by some bearing suppliers probably reflect

Table 15.6 Speed factor for grease life reduction

Bearing type	Factor k_f
Deep-groove ball bearing	
Single row	0.9–1.1
Double row	1.5
Angular-contact ball bearing	
Single row	1.6
Double row	2
Spindle bearing	
$\alpha = 15°$	0.75
$\alpha = 25°$	0.9
Four-point bearing	1.6
Self-aligning ball bearing	1.3–1.6
Thrust ball bearing	5–6
Angular-contact thrust ball bearings, double row	1.4
Cylindrical roller bearing	
Single row	1.8–2.3
Double row	2
Full complement	25
Cylindrical roller thrust bearing	90
Needle roller bearing	3.5
Tapered roller bearing	4
Barrel roller bearing	10
Spherical roller bearing without lips	7–9
Spherical roller bearing with center lip	9–12

Source: FAG Bearings Corp., 1995. Reproduced with permission of Schaeffler Group.

behavior with somewhat stiffer greases selected to accommodate close confinement in double-sealed bearings and for adaptability to high temperatures.

The effect of temperature shown in Equation (15.34) and Figure 15.13 reflects a variety of life tests in the temperature range from 75 to 175 °C. With ultimate oxidation stability afforded by perfluoroalkylpolyether greases, over 750 h of life were provided at 260 °C (500 °F) in ASTM D3336 tests (DuPont, 1996). Both SKF (1999) and FAG (1995) indicate that grease life becomes relatively independent of temperature as the temperature drops below 70 °C. As bearing temperature drops further into the range below about 30–40 ° C, grease life can be expected to drop as the grease begins to stiffen and bleeds oil less readily. Grease life was found to drop rapidly, for instance, with decreasing temperature in this region of 'cool' operation with a large sample of fractional horsepower motors running at speeds of up to 25 000 rpm with synthetic grease (Smith and Wilson, 1980).

Example 15.6 What grease life is to be expected in a 214 deep-groove ball bearing (with its 70 mm bore diameter D) at 85 °C in a 3600 rpm electric motor?

Applying Equation (15.34),

$$\log L = A + \frac{B}{T} - 9.6 \times 10^{-7} k_f D N$$

Neglecting the final speed factor term $k_f DN$ would give the following life for low-speed operation:

$$\log L = A + \frac{B}{T}$$

with $A = -2.60$ and $B = 2450$

$$\log L = -2.60 + \frac{2450}{(273 + 85)} = 4.244$$

$$L = 17\,539\,\text{h}$$

Attributing life reduction for the DN speed factor ($70\,\text{mm} \times 3600\,\text{rpm}$) with $k_f = 1.0$ for a single-row, deep-groove ball bearing from Table 15.6 for the 10% grease life in this application:

$$\log L = 4.244 - (9.6 \times 10^{-7})k_f DN$$

$$= 4.244 - (9.6 \times 10^{-7})(1.0)(70)(3600) = 4.002$$

$$L = 10\,050\,\text{h}$$

These same life values are provided graphically in Figure 15.13. Regreasing should be considered at half of this interval, approximately yearly, to avoid possible problems with a bearing failure resulting from hardening and drying of the grease.

15.12 Greasing and Regreasing

For initial lubrication, grease is applied to fill about 30–40% of the initial free space within a bearing housing with an open ball or roller bearing or within the enclosed space in a double-sealed or double-shielded bearing. Larger amounts generally have to be dispelled at shaft seals to provide free space for the rotating complement of balls or rollers. Commonly, a temperature spike is encountered during initial running while a new grease charge is distributed. After this run-in period of up to a few hours, bearing friction and temperature rise then commonly drop to only 30–50% of their run-in values.

Ball and roller bearings must be relubricated for continued service if the grease life is less than the life of the bearing itself. This relubrication should always be undertaken while the condition of the grease in the bearing is still satisfactory. Relubrication at half of the 10% grease life from Equation (15.21) is a useful guideline. A more direct guide is provided by operating experience with similar units or by monitoring samples from critical bearings for a drop in oil content, for a rise in acid number by potentiometric titration, or for antioxidant content by chromatography (Booser, 1997). For mild service conditions with an estimated grease life of several years or more, such as in small and medium-sized electric motors and line shafts, omission of grease fittings from the housing is advisable to avoid introducing contaminants and overgreasing.

More frequent regreasing is necessary in contaminated and wet environments, as in paper-making machines where bearing housings are washed with water. For bearings on vertical shafts, intervals should be halved. For large roller bearings with bore diameters of 300 mm or more, more frequent lubrication is also required. Adding small quantities of fresh grease at regular intervals on a plantwide basis is often convenient to avoid reaching critical dryness of the existing grease charge in bearings. Suggested quantities to be added are given by the following equation (SKF, 1999):

$$G = 0.005 \times D \times B \qquad (15.35)$$

where

$G =$ grease quantity to be added (g)
$D =$ bearing outside diameter (mm)
$B =$ bearing total width (total height for thrust bearings) (mm)

For a 207-size deep-groove ball bearing, for instance, Table 13.2 gives outside diameter $D = 72$ mm and width $B = 17$ mm. Grease quantity G to be added by Equation (5.35) is then $0.005 \times 72 \times 17 = 6.1$ g.

Grease must be replenished periodically, and the frequency for regreasing depends on the efficiency of the grease, its thermal stability, and rate at which wear debris is produced. Temperature has a dominating effect of grease life, and operating in higher temperatures increases the rate of oil loss from the grease structure by oxidation, creepage, and evaporation. With this drying, grease hardens and loses its ability to distribute itself in the bearing and replenish boundary films. Figure 15.14 gives recommended time of continuous operation for applications in ball and rolling element bearings along with large sleeve bearings used in farm and construction machinery as a function of bearing temperature (Glaeser and DuFrane, 1978; Khonsari and Booser, 2007; Khonsari et al., 2010). More recently, Kuhn (2010), Rezasoltani and Khonsari (2014, 2016a, 2016b) reported the development of an engineering approach for estimating the mechanical degradation. Physical (mechanical) degradation is primarily due to the loss of the base oil or destruction of the thickener structure during shearing action. Destruction of the thickener structure is either due to long-term use of the grease (mechanical breakdown) or chemical breakdown. When the grease operates below the oxidation temperature, the most important factor for reduction of life is the destruction of thickener structure due to mechanical working of the grease as it is sheared. A review of available method for assessment of grease degradation is given by Rezasoltani and Khonsari (2016c) including both mechanical and chemical degradation. An excellent assessment of the state-of-the art in grease lubrication in rolling element bearings is provided by Lugt (2012).

Another concern with heavily loaded construction machinery is their oscillatory nature of operation, which can be more severe than those operating in continuous, unidirectional speeds typical of most machines. Scuffing failure and seizure can be a concern (Jun and Khonsari, 2009, 2011). Laboratory experiments with oscillatory shaft motion show that friction coefficient can suddenly rise after a number of cycles with little warning. The sudden rise can be induced by many factors, including wear of the protective layer, introduction of wear particles in the contact, or a sudden change from external sources such as a shock. Thus, the type of grease plays a major role. During laboratory experiments, temperature and friction rose rapidly after a few cycles for some greases, whereas with others the temperature and friction both stabilized, and the bearing operated smoothly. Oscillatory speeds under very low amplitudes and high frequency can also restrict replenishment of grease into the contact, causing fretting failure as well as false brinelling.

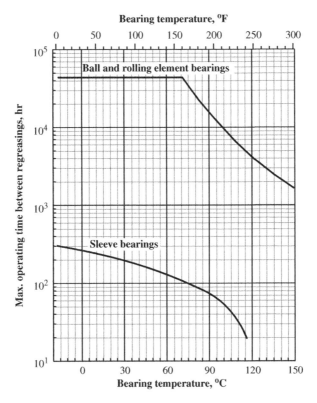

Figure 15.14 Guidelines for bearing regreasing intervals with mild operating conditions.
Source: Booser and Khonsari, 2007. Reproduced with permission of Machinery Lubrications.

15.13 Solid Lubricants

An alternative to oil or grease for severe conditions is solid lubrication (see Chapter 2). Molybdenum disulfide (MoS_2), Teflon, and related self-lubricating solids are used, often incorporated in the cage material, for high temperatures or vacuum conditions where liquid lubricants and greases are impractical. Silver plating of ball bearing surfaces has been used in X-ray tubes. Solid lubricants do not prevent solid–solid contact, however, and bearing life is limited by wear and depletion of the lubricant film. When incorporated in greases for conventional rotating machinery, many solid lubricants cause accelerated wear of rolling bearings, indentations of rolling contact surfaces, and shortened life.

Excellent information on the performance of solid lubricants, particularly MoS_x, in contact with 440C stainless steels disks in terms of friction coefficient, wear, durability in terms of number of cycles to failure, are presented by Miyoshi (2001) based on extensive testing in ultrahigh vacuum for aerospace applications. Such information is vital for control of satellites, telescope and tracking system equipped with precision ball bearings that require very stringent positioning accuracy—in the range of microns—while operating at ultra-low speeds (0.0001 to 1 cm/s). Control of motion requires development of appropriate friction compensation algorithms that can position the device at the exact desired location, keeping in mind that even minute inaccuracies in friction estimation can translate into totally unacceptable displacement in the during tracking of a target's position. The friction behavior of ball bearings operating under ultra-low speeds is characterized by Dahl (1968, 1978) who derived empirical equations—commonly referred to as Dahl's friction—for modeling their behavior during what is known as pre-rolling before pure rolling motion commences. Lovel *et al.* (1992; 1993a, 1993b; 1994; 1996a, 1996b; 1997a, 1997b) and more recently Lovel and Morrow (2006) present extensive information on experimental results and finite element modeling of ball bearings lubricated with oil as well as solid lubricants. Theoretical modeling of these problems is challenging because of the anisotropy of the solid lubricant coatings. Additional experimental results on the investigation of the tribological behavior and durability of MoS_2-coated ball bearings undergoing oscillatory motion are presented by Mesgarnejad and Khonsari (2010) and Kahirdeh and Khonsari (2010).

Another class of lubrication involves ejection of solid lubricants into the clearance spaces of a bearing, rather than directly coating the surfaces. Major experimental work in understanding the nature and application of this concept has been reported in a series of papers by Heshmat (1991; 1992a, 1992b; 1994; 1995) in what he refers to as quasi-hydrodynamic lubrication. This technology also has application in auxiliary bearings particularly for use during start-up periods and has already been demonstrated by Kaur and Heshmat (2002). In addition, successful demonstration of such operation at speeds up to 2×10^6 DN (shaft diameter in mm $\times$ rpm) on a three-pad journal bearing using MoS_2 powder as the lubricant has been reported (Heshmat and Brewe, 1994). Heshmat and Brewe (1996) find that dry tungsten disulfide (TiO_2) powder lubricant is also viable for it provides excellent wear resistance. They report that friction properties of TiO_2 are similar to that of MoS_2 with the added advantage that it can be used at much higher temperatures. Bearing tests at speeds up to 30,000 rpm and loads up to 236 N are reported to yield stable thermal operation (Heshmat and Brewe, 1996).

The need for bearings that can withstand severe operating temperatures has necessitated consideration of alternative means of lubrication since conventional lubricants ordinarily lose their effectiveness beyond 200° C. Thus, the possibility of using solid lubricants with a carrier fluid has been proposed and numerous investigations reported both for hydrodynamic and elastohydrodynamic regimes. Rylander (1952, 1966) reported one of the earliest investigations to study the frictional behavior of lubricating oil including small concentration of MoS_2 in a journal bearing. Khonsari and Esfahanian (1988) examined the thermal effects and Dai and Khonsari (1991, 1992, 1993, 1994) provided a general formulation for analyzing granular material in a lubricating oil from continuum mechanics point of view. Hisakado (1982) describes the physical mechanism of lubrication of solid lubricants in oil and Cusano and Sliney (1983a, 1983b), Wan and Spike (1987), Khonsari *et al.* (1990a, 1990b) and Hua and Khonsari (1996) present results for EHL applications.

If instead of liquid, air (or gas) is used as a carrier fluid for the transport of cohesive micron-size granules—i.e. a powder lubricant—into the clearance space of a bearing, then higher speeds and temperatures can be accommodated, as demonstrated by Heshmat (2010), who gives a general exposition of the subject matter under the heading of tribology of interface layers. Of particular interest is the investigation conducted by Heshmat (1992b), who shows the first documented measured pressure generated by a powder lubricant as it flows in the clearance space of a bearing. The ensuing pressure distribution generated is remarkably similar to that of a liquid lubricant. This sets the stage for exploring the granular flow as a fluid mechanics problem by invoking a pseudo-energy equation to treat the granules' fluctuation velocity as described by Haff (1983), Johnson and Jackson (1987) Elrod (1988), Dai *et al.* (1994), Yu *et al.* (1994, 1996), and Khonsari (1997). Complications occur because, in contrast to the no-slip boundary condition—which prevails in conventional fluids—the granules tend to slip at the boundaries and this requires special analytical treatment (Hui *et al.*, 1984; Jenkin and Richman, 1986; Campbell, 1993; McKeague and Khonsari, 1996a, 1996b). While the majority of these formulations have ben applied to understand the nature of granular flow in a shear cell (Zhu and Khonsari, 2000; Tichy and Higgs, 2004; Sawyer and Tichy, 2001; Elkholy and Khonsari, 2007, 2008), successful application to bearings is available (Pappur and Khonsari, 2002; Jang and Khonsari, 2005; Tsai and Jeng, 2006). Clearly, the analyses and performance prediction of bearings lubricated with powder lubricants and computational and experimental work with granular material—typically characterized by cohesionless hard particles—are of interest to many tribologists and remain a subject of ongoing research from many viewpoints. The interested reader is referred to a comprehensive review given by Wornyoh *et al.* (2007).

Problems

15.1 Noise is detected in the drive end ball bearing of an oil well pump. State five checks that might be made to determine the possible source of trouble without dismantling the unit.

15.2 Overheating and lubrication-type failures are being encountered in 316-size deep-groove ball bearings in a 4200 rpm pump shaft. Initial operation was satisfactory, but noise, overheating, and bearing failures have since been encountered within one month of operation. There is no evidence of misalignment or incorrect mounting, and the bearings are regreased on a three-month schedule. Suggest a likely cause of the trouble and a remedy.

15.3 Based on a limiting viscosity of $100\,000\,\text{mm}^2/\text{s}$ (cSt) for the oil in a grease, determine the lowest tolerable operating temperature for greases compounded with the oils given in the following table. Use Equation (2.3) to obtain the extrapolated low-temperature viscosity.

	Viscosity, cSt	
Oil type	40 °C	100 °C
Mineral oil	100	11
Mineral oil	32	5.2
Synthetic hydrocarbon	63	9.6
Diester	12	3.2

15.4 Radial play in a 210-size ball bearing in an electric motor has increased 0.008 mm (0.0003 in.) in its first 4500 h of operation. Assume that the rate of increase in its f_V in Equation (15.7) remains constant. What bearing wear life in hours can be expected for a limiting $f_V = 8$?

15.5 What power loss can be expected with a 214 cylindrical roller bearing (70 mm bore, 125 mm outside diameter) running at 3600 rpm at 90 °C while under a radial load of 6000 N at 3600 rpm when lubricated with a conventional ball and roller bearing grease? Estimate power loss in watts from Equation (15.1).

Assume that half of this loss is viscous friction from reaction with the grease (which incorporates an oil of 100 cSt viscosity at 40 °C). What loss would be expected if a shift were made to bath lubrication at 65 °C using a VG 32 oil with a viscosity of 32 cSt at 40 °C?

15.6 What grease life would be expected for the 214 cylindrical roller bearing in Problem 15.5 at 110 °C? Suggest possible ways to double this expected life.

References

Aihara, S. 1987. 'A New Running Torque Formula for Tapered Roller Bearings under Axial Load,' *Journal of Tribology, ASME Transactions*, Vol. 109, pp. 471–478.

Ali, F., Krupka, I., and Hartl, M. 2013. 'An Approximate Approach to Predict the Degree of Starvation in Ball-Disk Machine Based on the Relative Friction.' *Tribology Transactions*, Vol. 561, pp. 681–686.

Ali, F., Krupka, I., and Hartl, M. 2013. 'Analytical and experimental investigation on friction of non-conforming point contacts under starved lubrication,' *Mechanica*, Vol. 48, pp. 545–53.

Bair, S. 2004. 'Actual Eyring Models for Thixotropy and Shear-Thinning: Experimental Validation and Application to EHD,' *ASME Journal of Tribology*, Vol. 126, pp. 728–732.

Bair, S. 2007. *High Pressure Rheology for Quantitative Elastohydrodynamics*, Elsevier, Oxford, UK.

Bair, S., and Khonsari, M.M. 1996. 'An EHD Inlet Zone Analysis Incorporating the Second Newtonian,' *ASME Journal of Tribology*, Vol. 118, pp. 341–343.

Bair, S. and Qureshi, F. 2003. 'The Generalized Newtonian Fluid Model and Elastohydrodynamic Film Thickness', *ASME Journal of Tribology*, Vol. 125, pp. 70–75.

Bair, S. and Winer, W.O. 1979. 'A Rheological Model for Elastohydrodynamic Contacts Based in Primary Laboratory Data,' *ASME Journal of Lubrication Technology*, Vol. 101, pp. 258–265.

Booser, E.R. and Khonsari, M.M. 2013a. 'Grease Life,' *Encyclopedia of Tribology*, Y.W. Chung and Q.J. Wang (Editors), Springer Science, New York, NY, pp. 1555–1561.

Booser, E.R. and Khonsari, M.M. 2013b. 'Oil Life,' *Encyclopedia of Tribology*, Y.W. Chung and Q.J. Wang (Editors), Springer Science, New York, NY, pp. 2475–24478.

Booser, E.R. 1974. 'Grease Life Forecast for Ball Bearings,' *Lubrication Engineering*, Vol. 30, pp. 530–540.

Booser, E.R. 1975. 'When to Grease Ball Bearings,' *Machine Design*, 21 Aug., pp. 70–73.

Booser, E.R. 1995. 'Lubricants and Lubrication,' *Encyclopedia of Chemical Technology*, 4th edn, Vol. 15. John Wiley & Sons, New York, pp. 463–517.

Booser, E.R. 1997. 'Life of Oils and Greases,' *Tribology Data Handbook*. CRC Press, Boca Raton, Fla, pp. 1018–1028.

Booser, E.R. and Khonsari, M.M. 2007. 'Systematically Selecting the Best Grease for Equipment Reliability,' *Machinery Lubrication*, Jan–Feb, pp. 19–25.

Campbell, C.S. 1993. 'Boundary Interactions for Two-Dimensional Granular Flow–Part I. Flat Boundaries, Asymmetric Stress and Couple Stress,' *Journal of Fluid Mechanics*, 1993, Vol. 247, pp. 137–156.

Cann, P.M.E., Damiens, B., and Lubrecht, A.A. 2004. 'The transition between fully flooded and starved regimes in EHL,' *Tribology International*, Vol. 37, pp. 859–864.

Carreau, P.J. 1972. 'Rheological Equations from Molecular Network Theories,' *Transactions of the Society of Rheologists*, Vol. 16(1), pp. 99–127.

Cheng, H.S. 1984. 'Elastohydrodynamic Lubrication,' *Handbook of Lubrication*, Vol. II. CRC Press, Boca Raton, Fla, pp. 139–162.

Chang, L., Cusano, C., and Conry, T.F. 1989. 'Effects of Lubricant Rheology and Kinematic Conditions Micro-Elastohydrodynamic Lubrication,' *ASME Journal of Tribology*, Vol. 111, pp. 344–351.

Chevalier, F., Lubrecht, A.A., Cann, P.M.E., Colin, F., and Dalmaz, G. 1998. 'Film Thickness in Starved EHL Point Contacts,' *ASME Journal of Tribology*, Vol. 120, pp. 126–133.

Conry, T.F., Wang, S., and Cusano, C. 1987. 'A Reynolds-Eyring Equation for Elastohydrodynamic Lubrication in Line Contacts,' *ASME Journal of Tribology.*, Vol. 109, pp. 648–658.

Cusano, C. and Sliney, H.E. 1982a. 'Dynamics of Solid Dispersions in Oil During the Lubrication of Point Contacts, Part I—Graphite,' *ASLE Transactions*, Vol. 25, pp. 183–189.

Cusano, C. and Sliney, H.E. 1982b. 'Dynamics of Solid Dispersions in Oil During the Lubrication of Point Contacts, Part II—Molybdenum Disulfide,' *ASLE Transactions*, Vol. 25, pp. 190–197.

Dahl, P.R. 1968. 'A Solid Friction Model,' AFO 4695-67-C-D-158, Aerospace Corporation.

Dahl, P.R. 1978. 'Measurements of Solid Friction Parameters of Ball Bearings,' Aerospace Corporation, pp. 49–60.

Dai, F. and Khonsari, M.M. 1991. 'On the Solution of a Lubrication Problem with Particulate Solids,' *International Journal of Engineering Science*, Vol. 29, pp. 1019–1033.

Dai, F. and Khonsari, M.M. 1992. 'Analytical Solution for the Mixture of a Newtonian Fluid and Granules in Hydro-dynamic Bearings,' *Wear*, Vol. 156, pp. 327–344.

Dai, F. and Khonsari, M.M. 1993. 'A Continuum Theory of a Lubrication Problem with Solid Particles,' *ASME Journal of Applied Mechanics*, Vol. 60, pp. 48–58.

Dai, F. and Khonsari, M.M. 1994a. 'Generalized Reynolds Equation for Solid-Liquid Lubricated Bearings,' *ASME Journal of Applied Mechanics*, Vol. 61, pp. 460–466.

Dai, F. and Khonsari, M.M. 1994b. 'A Theory of Hydrodynamic Lubrication Involving the Mixture of Two Fluids,' *ASME Journal of Applied Mechanics*, Vol. 61, pp. 634–641.

Dai, F., Khonsari, M.M., and Lu, Y.Z. 1994. 'On the Lubrication Mechanism of Grain Flows.' *STLE Tribology Trans-actions*, Vol. 37, pp. 516–524.

Damiens, B., Venner, C.H., Cann, P.M.E., and Lubrecht, A.A. 2004. 'Starved Lubrication of Elliptical EHD Contacts,' *ASME Journal of Tribology*, Vol. 126, pp. 105–111.

Dowson, D. 1970. 'Elastohydrodynamic Lubrication,' *Interdisciplinary Approach to the Lubrication of Concentrated Contacts*, NASA Spec. Pub. SP-237. National Aeronautics and Space Administration, Washington, DC, p. 34.

DuPont Corp. 1996. *Krytox Oils and Greases*. DuPont Bulletin H-58505. DuPont Corp., Wilmington, Del.

Dyson, A. and Wilson, A.R. 1965–66. 'Film thickness in Elastohydrodynamic Lubrication by Silicone Fluids', *Pro-ceedings of the Institute of Mechanical Engineers, Part K: Journal of Multi-body Dynamics*, Vol. 180, pp. 97–105.

Elkholy, K. and Khonsari, M.M. 2007. 'Granular Collision Lubrication: Experimental Investigation and Comparison with Theory,' *ASME Journal of Tribology*, Vol. 129, pp. 923–932.

Elkholy, K. and Khonsari, M.M. 2008. 'On the Effect of Enduring Contact on the Flow and Thermal Characteristics in Powder Lubrication,' *Proceedings of the Institute of Mechanical Engineers Journal of Engineering Tribology*, Vol. 222, pp. 741–759.

Elrod, H.G. 1988. 'Granular Flow as a Tribological Mechanism – A First Look.' Interface Dynamics, Leeds-Lyon Symposium, BHRA, 1988: pp. 75–102.

Eschmann, P., Hasbargen, L., and Weigand, K. 1985. *Ball and Roller Bearings: Theory, Design, and Application*. John Wiley & Sons, New York. (See also the 3d edn, 1999.)

FAG Bearings Corp. 1985. *The Lubrication of Rolling Bearings*, Pub. WL 81 115/2 EC/ED. FAG Bearings Corp., Stratford, Ontario.

FAG Bearings Corp. 1995. *FAG Rolling Bearings*, Catalog WL 41 520 ED. FAG Bearings Corp., Danbury, Conn.

Faraon I.C. and Schipper D.J. 2007. 'Stribeck Curve for Starved Line Contacts,' *ASME Journal of Tribology*, Vol. 129, pp. 181–187.

Glaeser, W.A. and Dufrane, K.F. 1978. 'New Design Methods for Boundary-Lubricated Sleeve Bearings,' *Machine Design*, April 6, pp. 207–213.

Haff, P.K. 1983. 'Grain Flow as a Fluid-Mechanical Phenomenon,' *Journal of Fluid Mechanics*, Vol. 134, pp. 401–430.

Hamrock, B. and Dowson, D. 1977. 'Isothermal Elastohydrodynamic Lubrication of Point Contacts. Part III – Fully Flooded Results,' *ASME Journal of Lubrication Technology*, Vol. 92, pp. 155–162.

Hamrock, B.J., Dowson, D. 1977. 'Isothermal Elastohydrodynamic Lubrication of Point Contacts: Starvation Results,' *ASME Journal of Lubrication Technology*, Vol. 99, pp. 15–23.

Hamrock, B.J., and Dowson, D. 1981. *Ball Bearing Lubrication – The Elastohydrodynamics of Elliptical Contacts*, John Wiley & Sons, Ltd, Chichester.

Harris, T.A. 1991. *Rolling Bearing Analysis*, 3rd edn. John Wiley & Sons, New York.

Heshmat, H. 1991. 'The Rheology and Hydrodynamics of Dry Powder Lubrication,' *STLE Tribology Transactions*, Vol. 34, No. 3, pp. 433–439.

Heshmat, H. 1992a. 'The Quasi-Hydrodynamic Mechanism of Powder Lubrication—Part I: Lubricant Visualization,' *Lubrication Engineering*, Vol. 48, pp. 96–104.

Heshmat, H. 1992b. 'The Quasi-Hydrodynamic Mechanism of Powder Lubrication—Part II: Lubricant Pressure Profile,' *Lubrication Engineering*, Vol. 48, pp. 373–383.

Heshmat, H. 1994. 'On the Theory of Quasi-Hydrodynamic Powder Lubrication with Dry Powder. Application to
Development of High-Speed Journal Bearings for Hostile Environments,' *Dissipative Process in Tribology*, D.
Dowson, C.M. Taylor, T.H.C. Childs, M. Godet, and G. Dalmaz, eds., Tribology Series 27, Elsevier Scientific
Publishing Co., New York, pp. 45–64.

Heshmat, H. 2010. *Tribology of Interface Layers*, CRC Press, Boca Raton, Fl.

Heshmat, H. and Brewe, D. 1994. 'Performance of Powder-Lubricated Journal Bearings with M0S2 Powder: Experimental Study of Thermal Phenomena,' *ASME Journal of Tribology*, Vol. 117, pp. 506–512.

Heshmat, H., Godet, M., and Berthier, Y. 1995. 'On the Role and Mechanism of Dry Triboparticulate Lubrication,'
Lubrication Engineering, Vol. 51, pp. 557–564.

Heshmat, H. and Brewe, D.E. 1996. 'Performance of a Powder Lubricated Journal Bearing With WS_2 Powder: Experimental Study, *ASME Journal of Tribology*, Vol. 118, pp. 484–491.

Higgs, C.F., III and Tichy, J. 2004. 'Granular Flow Lubrication: Continuum Modeling of Shear Behavior.' *ASME
Journal of Tribology*, Vol. 126, pp. 499–510.

Hisakado, T., Tsukizoe, T., and Yoshikawa, H. 1983. 'Lubrication Mechanism of Solid Lubricants in Oils,' *Asme
Journal of Lubrication Technology*, Vol. 105, pp. 245–253.

Houpert, L.G. and Hamrock, B.J. 1985. 'Elastohydrodynamic Lubrication Calculations Used as a Tool to Study Scuf-
fling,' *Proceedings of the 12th Leeds-Lyon Symposium on Tribology*, Tribology Series, Elsevier, Amsterdam,
pp. 146–155.

Hua, D. and Khonsari, M.M. 1996. 'Elastohydrodynamic Lubrication of Powder Slurries,' *ASME Journal of Tribology*,
Vol. 118, pp. 67–73.

Hui, K., Haff, P.K., Ungar, J.E., and Jackson, R. 1984. 'Boundary Conditions for High-Shear Grain Flows.' *Journal
of Fluid Mechanics*, Vol. 145, pp. 223–233.

Jang, J.Y. and Khonsari, M.M. 2005. 'On the Granular Lubrication Theory,' *Proceedings of the Royal Society of
London, Series A*, Vol. 461, pp. 3255–3278.

Jang, J.Y., Khonsari, M.M., and Bair, S. 2007. 'On the Elastohydrodynamic Analysis of Shear-Thinning Fluids,'
Proceedings of the Royal Society of London, London, 463, pp. 3271–3290.

Jendzurski, T. and Moyer, C.A. 1997. 'Rolling Bearings Performance and Design Data,' *Tribology Data Handbook*.
CRC Press, Boca Raton, Fla, pp. 645–668.

Jenkins, J.T. and Richman, M.W. 1986. 'Boundary Conditions for Plane Flows of Smooth, Nearly Elastic, Circular
Disks.' *Journal of Fluid Mechanics*, 1986, Vol. 171, pp. 53–69.

Johnson, K.L. 1970. 'Regimes of Elastohydrodynamic Lubrication,' *Journal of Mechanical Engineering Sciences*,
Vol. 12, pp. 9–16.

Johnson, K.L. and Greenwood, J.A. 1980. 'Thermal Analysis of an Eyring Fluid in EHD Traction,' *Wear*, 61, pp. 353–
374.

Johnson, K.L. and Tevaarwerk, J.L. 1977. 'Shear Behaviour of EHD Oil Films,' *Proceedings of the Royal Society of
London, Series A*, 356, pp. 215–236.

Johnson, K.L., Greenwood, J.A., and Poon, S.Y. 1972. 'Simple Theory of Asperity Contact in Elastohydrodynamic
Lubrication,' *Wear*, Vol. 19, pp. 91–108.

Johnson, P.C. and Jackson, R. 1987. 'Frictional-Collisional Constitutive Relations for Granular Materials with Applications to Plane Shearing,' *Journal of Fluid Mech.*, Vol. 176, pp. 67–93.

Jun, W. and Khonsari, M.M. 2009. 'Thermomechanical Coupling in Oscillatory Systems with Application to Journal
Bearing Seizure,' *ASME Journal of Tribology*, Vol. 131/021601:1–14.

Jun, W., Khonsari, M.M., and Hua, D. 2011. 'Three-Dimensional Heat Transfer Analysis of Pin-bushing System with
Oscillatory Motion: Theory and Experiment,' *ASME Journal of Tribology*, Vol. 133, 011101–1.

Kahirdeh, A. and Khonsari, M.M. 2010. 'Condition Monitoring of Molybdenum Disulphide Coated Thrust Ball Bearings Using Time-Frequency Signal Analysis,' *ASME Journal of Tribology*, Vol. 132, pp. 041606:1–11.

Kaur, R.G. and Heshmat, H. 2002. '100mm Diameter Self-Contained Solid/Powder Lubricated Auxiliary Bearing
Operated at 30,000 rpm,' STLE *Tribology Transactions*, Vol. 45, pp. 76–84.

Khonsari, M.M. and Dai, F. 1992. 'On the Mixture Flow Problem in Lubrication of Hydrodynamic Bearings: Small
Solid Volume Fraction,' STLE *Tribology Transactions*, Vol. 35, pp. 45–52.

Khonsari, M.M. and Esfahanian, V. 1988. 'Thermohydrodynamic Analysis of Solid-Liquid Lubricated Journal Bearings,' *ASME Journal of Tribology*, Vol. 110, No. 2, pp. 367–374.

Khonsari, M.M., Booer, E.R., and Miller, R. 2010. 'How Grease Keep Bearings Rolling,' *Machine Design*, Vol. 82, pp. 54–59.

Khonsari, M.M. and Hua, D. 1994a, 'Generalized Non-Newtonian Elastohydrodynamic Lubrication,' *Tribology International*, Vol. 26, pp. 405–411.

Khonsari, M.M. and Hua, D. 1994b. 'Thermal Elastohydrodynamic Analysis Using a Generalized Non-Newtonian with Application to Bair-Winer Constitutive Equation,' ASME *Journal of Tribology*, Vol. 116, pp. 37–46.

Khonsari, M.M., Wang, S., and Qi, Y. 1989. 'A Theory of Liquid-Solid Lubrication in Elastohydrodynamic Regime,' *ASME Journal of Tribology*, Vol. 111, No. 3, pp. 440–444.

Khonsari, M.M., Wang, S., and Qi, Y. 1990. 'A Theory of Thermo-Elastohydrodynamic Lubrication for Liquid-Solid Lubricated Cylinders,' *ASME Journal of Tribology*, Vol. 112, No. 2, pp. 259–265.

Khonsari, M.M. 1997. 'On the Modeling of Multi-Body Interaction Problems in Tribology,' *Wear*, Vol. 207, pp. 55–62.

Khonsari, M.M. and Booser, E.R. 2003. 'Predicting Lube Life,' *Machine Design*, Jan., pp. 89–92.

Khonsari, M.M. and Hua, D.Y. 1997. 'Fundamentals of Elastohydrodynamic Lubrication,' *Tribology Data Handbook*. CRC Press, Boca Raton, Fla, pp. 611–637.

Kuhn, E. 2010. 'Analysis of a Grease-Lubricated Contact From an Energy Point of View,' *International Journal of Materials and Product Technology*, Vol. 38, pp. 5–15.

Kumar, P. and Khonsari, M.M. 2008. 'Effect of Starvation on Traction and Film Thickness in Thermo-EHL Line Contacts with Shear-Thinning Lubricants,' *Tribology Letters*, Vol. 32, pp. 171–177.

Kumar, P. and Khonsari, M.M. 2009a. 'Traction in EHL Line Contacts Using Free Volume Pressure-Viscosity Relationship with Thermal and Shear Thinning Effects,' *ASME Journal of Tribology*, Vol. 131, 011503:1–8.

Kumar, P. and Khonsari, M.M. 2009b. 'On the Role of Lubricant Rheology and Piezo-viscous Properties in Line and Point Contact EHL,' *Tribology International*, Vol. 42, pp. 1522–1530.

Kumar, P., Jain, S.C., and Ray, S. 2002. 'Thermal EHL of Rough Rolling/Sliding Line Contacts Using a Mixture of Two Fluids at Dynamic Loads,' *ASME Journal of Tribology.*, Vol. 124, pp. 709–715.

Kumar, P., Khonsari, M.M., and Bair, S. 2009. 'Full EHL Simulations using the Actual Ree-Eyring Model for Shear Thinning Lubricants,' *ASME Journal of Tribology*, Vol. 131, 011802:1–6.

Lovell, M. and Morrow, C. 2006. 'Contact Analysis of Anisotropic Coatings,' STLE *Tribology Transactions*, Vol. 49, pp. 33–38.

Lovell, M., Khonsari, M.M., and Marangoni, R. 1992. 'Evaluation of Ultra-Low Speed Jitter in Rolling Balls,' *ASME Journal of Tribology*, Vol. 114, pp. 589–594.

Lovell, M., Khonsari, M.M., and Marangoni, R. 1993a. 'Ultra Low-Speed Friction Torque on Balls Undergoing Rolling Motion,' STLE *Tribology Transactions*, Vol. 36, pp. 290–296.

Lovell, M., Khonsari, M.M., and Marangoni, R. 1993b. 'The Response of Balls Undergoing Oscillatory Motion: Crossing from Boundary to Mixed Lubrication,' *ASME Journal of Tribology*, Vol. 115, pp. 261–266.

Lovell, M., Khonsari, M.M., and Marangoni, R. 1994. 'Experimental Measurements of the Rest-Slope and Steady Torque on Ball Bearings Experiencing Small Angular Rotation,' STLE *Tribology Transactions*, Vol. 37, pp. 261–268.

Lovell, M., Khonsari, M.M., and Marangoni, R. 1996a. 'Comparison of the Ultra-Low Speed Frictional Characteristics of Silicon Nitride and Steel Balls Using Conventional Lubricants,' *ASME Journal of Tribology*, Vol. 118, pp. 43–51.

Lovell, M., Khonsari, M.M., and Marangoni. R. 1996b. 'Finite Element Analysis of Frictional Forces Between a Cylindrical Bearing Element and MoS2 Coated and Uncoated Surfaces,' *Wear*, Vol. 194, pp. 60–70.

Lovell, M., Khonsari, M.M., and Marangoni, R. 1996c. 'Dynamic Friction Measurements of MoS2 Coated Ball Bearing Surfaces,' *ASME Journal of Tribology*, Vol. 118, pp. 858–864.

Lovell, M., Khonsari, M.M., and Marangoni, R. 1997a. 'Parameter Identification of Hysteresis Friction for Coated Ball Bearings Based on Three-Dimensional FEM Analysis,' *ASME Journal of Tribology*, Vol 119, pp. 462–470.

Lovell, M., Khonsari, M.M., and Marangoni, R. 1997b. 'Frictional Analysis of MoS2 Coated Ball Bearings: A Three-Dimensional Finite Element Analysis,' *ASME Journal of Tribology*, Vol. 119, pp. 754–763.

Lu, X.B., Khonsari, M.M., and Gelinck, E.R.M. 2006. 'The Stribeck Curve: Experimental Results and Theoretical Prediction,' *ASME Journal of Tribology*, Vol. 128, pp. 789–794.

Lugt, P.M. 2012. *Grease lubrication in rolling bearings*, John Wiley & Sons, New York, NY.

Masjedi, M. and Khonsari, M.M. 2012. 'Film Thickness and Asperity Load Formulas for Line-Contact Elastohydrodynamic Lubrication with Provision for Surface Roughness.; *ASME Journal of Tribology*, Vol. 134, 011503.

Masjedi, M. and Khonsari, M.M. 2014a. 'Theoretical and Experimental Investigation of Traction Coefficient in Line-Contact EHL of Rough Surfaces,' *Tribology International*, Vol. 70, pp. 179–189

Masjedi, M. and Khonsari, M.M. 2014b. 'Mixed elastohydrodynamic lubrication line-contact formulas with different surface patterns, *Proceedings of the Institute of Mechanical Engineers, Part J: Journal of Engineering Tribology*, Vol. 228, pp. 849–859.

Masjedi, M. and Khonsari, M.M. 2014c. 'On the Effect of Surface Roughness in Point-Contact EHL: Formulas for Film Thickness and Asperity Load,' *Tribology International*, DOI:10.1016/j.triboint.2014.09.010.

Masjedi, M. and Khonsari, M.M. 2015a. 'On the Effect of Surface Roughness in Point-Contact EHL: Formulas for Film Thickness and Asperity Load,' *Tribology International*, Vol. 82, Part A, pp. 228–244.

Masjedi, M. and Khonsari, M.M. 2015b. 'An Engineering Approach for Rapid Evaluation of Traction Coefficient and Wear in Mixed EHL,' *Tribology International*, Vol. 92, pp. 184–190.

McKeague, K. and Khonsari, M.M. 1996a, 'An Analysis of Powder Lubricated Slider Bearings,' *ASME Journal of Tribology*, Vol. 118, pp. 206–214.

McKeague, K. and Khonsari, M.M. 1996b. 'Generalized Boundary Interaction for Powder Lubricated Couette Flows,' *ASME Journal of Tribology*, Vol. 118, pp. 580–588.

Mesgarnejad, A. and Khonsari, M.M. 2010. 'On the Tribological Behavior of Mos2-Coated Thrust Ball Bearings Operating Under Oscillating Motion,' *Wear*, Vol. 269, pp. 547–556.

Miyoshi, K. 2001. *Solid Lubrication: Fundamentals and Applications*, Marcel-Dekker, New York, NY.

Mongkolwongrojn, M., Wongseedakeaw, K., Yawong, S., Jeenkour, P., and Aiumpornsin, C. 2005. 'Effect of Model Parameters on Elastohydrodynamic Lubrication Line Contact with Non-Newtonian Carreau Viscosity Model', *International Journal of Applied Mechanics and Engineering*, Special Issue: ICER 2005, Vol. 10, pp. 255–261.

Palmgren, A. 1959. *Ball and Roller Bearing Engineering*, 3rd edn. SKF Industries, Philadelphia, Pa.

Pappur, M. and Khonsari, M.M. 2002. 'Flow Characterization and Performance of a Powder Lubricated Slider Bearing,' *ASME Journal of Tribology*, Vol. 124: pp. 1–10.

Rezasoltani, R. and Khonsari, M.M. 'On the Correlation Between Mechanical Degradation of Lubricating Grease and Entropy,' *Tribology Letters*, Vol. 56, pp. 197–204.

Rezasoltani, A. and Khonsari, M.M. 2016a. 'An Engineering Model to Estimate Consistency Reduction of Lubricating Grease Subjected to Mechanical Degradation Under Shear,' *Tribology International*, Vol. 103, pp. 465–474.

Rezasoltani, A. and Khonsari, M.M. 2016b. 'Mechanical Degradation of Lubricating Grease in an EHL Line Contact,' *Tribology International*, Vol. 109, pp. 541–551.

Rezasoltani, A. and Khonsari, M.M. 2016c. 'On Monitoring Physical and Chemical Degradation and Life Estimation Models for Lubricating Greases,' *Lubricants*, Vol. 4, pp. 1–24.

Rylander, H. 1952. 'The Effect of Solid Inclusions in the Oil Supply to Sleeve Bearings Oil Supply,' *Mechanical Engineering*, Vol. 74, 1952, pp. 963–966.

Rylander, H. 1966. 'A Theory of Liquid-Solid Hydrodynamic Film Lubrication,' *ASLE Transactions*, Vol. 9, pp. 264–271.

Salehizadeh, H. and Saka, N. 1991. 'Thermal Non-Newtonian Elastohydrodynamic Lubrication of Rolling Line Contacts,' *ASME Journal of Tribology*, 113, pp. 181–191.

Sawyer, G.W. and Tichy, J.A. 2001. 'Lubrication with Granular Flow: Continuum Theory, Particle Simulation, Comparison with Experiment,' *ASME Journal of Tribology*, Vol. 123, pp. 777–784.

Schrama, R.C. 1997. 'Greases,' *Tribology Data Handbook*. CRC Press, Boca Raton, Fla, pp. 138–155.

SKF USA. 1999/2005. *General Catalog*. SKF USA, King of Prussia, Pa.

Smith, R.L. and Wilson, D.S. 1980. 'Reliability of Grease-Packed Ball Bearings for Fractional Horsepower Motors,' *Lubrication Engineering*, Vol. 36, pp. 411–416.

Stein J.L. and Tu, J.F. 1994. 'A State-Space Model for Monitoring Thermally-Induced Preload in Antifriction Spindle Bearings of High-Speed Machine-Tools. *ASME Journal of Dynamic Systems*,' Vol. 116, pp. 372–386.

Sui, P.C. and Sadeghi, F. 1991. 'Non-Newtonian Thermal Elastohydrodynamic Lubrication,' *ASME Journal of Tribology*, Vol. 113, pp. 390–397.

Svoboda, P., Kostal, D., Krupka, I., and Hartl, M. 2013. 'Experimental Study Of Starved EHL Contacts Based on Thickness of Oil Layer in the Contact Inlet,' *Tribology International*, Vol. 67, pp. 140–145.

Takabi, J. and Khonsari, M.M. 2013. 'Experimental Testing and Thermal Analysis of Ball Bearings,' *Tribology International*, Vol. 60, pp. 93–103.

Takabi, J. and Khonsari, M.M. 2014. 'On the Influence of Traction Coefficient on the Cage Angular Velocity in Roller Bearings,' *Tribology Transactions*, Vol. 57, pp. 793–805.

Takabi, J. and Khonsari, M.M. 2015a. 'On the Dynamic Performance of Roller Bearings Operating Under Low Rotational Speeds with Consideration of Surface Roughness,' *Tribology International*, Vol. 86, pp. 62–71.

Takabi, J. and Khonsari, M.M. 2015b. 'On the Thermally Induced Seizure in Bearings: A Review,' to appear in *Tribology International*, 2015.

Takabi, J. and Khonsari, M.M. 2015c. 'On the Thermally–Induced Failure of Rolling Element Bearings,' to appear in *Tribology International*.

Tanner, R.I. 1960. 'Full-Film Lubrication Theory for a Maxwell Liquid,' *International Journal of Mechanical Science*, Vol. 1, p. 206.

Tsai, H. and Jeng, Y. 2006. 'Characteristics of Powder Lubricated Finite-Width Journal Bearings: A Hydrodynamic Analysis,' *ASME Journal of Tribology*, Vol. 128, pp. 351–356.

Tu, J.F. and Stein, J.L. 1998. 'Model Error Compensation for Observer Design,' *International Journal of Control*, Vol. 69, pp. 329–345.

Wan, G.T.Y. and Spikes, H.A. 1987. 'The Behavior of Suspended Solid Particles in Rolling and Sliding Elastohydrodynamic Contacts,' *ASLE Transactions*, Paper No. 87-AM-2F-1.

Wang, S., Conry, T.F., and Cusano, C. 1992. 'Thermal Non-Newtonian Elastohydrodynamic Lubrication of Line Contacts Under Simple Sliding Conditions,' *ASME Journal of Tribology*, Vol. 114, pp. 317–327.

Wilcock, D.F. and Booser, E.R. 1957. *Bearing Design and Application*. McGraw-Hill Book Co., New York.

Winer, W.O. and Cheng, H.S. 1980. 'Film Thickness, Contact Stress and Surface Temperatures,' *Wear Control Handbook*, Peterson, M. and Winer, W. (editors), ASME Publication, New York, NY, pp. 81–141.

Witte, D.C. 1973. 'Operating Torque of Tapered Roller Bearings,' *ASLE Transactions*, Vol. 16, pp. 61–67.

Yang, P., Wang, J., and Kaneta, M. 2006. 'Thermal and non-Newtonian numerical analyses for starved EHL Line Contacts,' *ASME Journal of Tribology*, Vol. 128, pp. 282–290.

Yasuda, K., Armstrong, R.C., and Cohen, R.E. 1981. 'Shear Flow Properties of Concentrated Solutions of Linear and Star Branched Polystyrenes,' *Rheologica Acta*, Vol. 20, pp. 163–178.

Yu, C.M., Craig, K., and Tichy, J. 1994. Granular Collision Lubrication. *Journal of Rheol.*, Vol. 38, pp. 921–936.

Yu, C.M. and Tichy, J. 1996. 'Granular Collisional Lubrication: Effect of Surface Roughness, Particle Size and Solid Fraction.' *STLE Tribology Transactions*, Vol. 67, pp. 537–546.

Zhou, L. and Khonsari, M.M. 2000. 'Flow Characteristics of a Powder Lubricant,' *ASME Journal of Tribology*, Vol. 122, pp. 147–155.

Zhu, D. and Wang, Q. 2011. 'Elastohydrodynamic Lubrication (EHL): A Gateway to Interfacial Mechanics-Review and Prospect,' *ASME Journal of Tribology*, Vol. 133, p. 041001.

Zhu, D. and Wang, Q. 2012. 'On the λ Ratio Range of Mixed Lubrication,' *Proceedings of the Institute of Mechanical Engineers*, Part J, Vol. 226(12), pp. 1010–1022.

Zhu, D. and Wang, Q. 2013. 'Effect of Roughness Orientation on the Elastohydrodynamic Lubrication Film Thickness,' *ASME Journal of Tribology*, Vol. 135, p. 031501

Zhu, D. and Wang, Q.J. 2013. 'Effect of Roughness Orientation on the Elastohydrodynamic Lubrication Film Thickness.' *ASME Journal of Tribology*, Vol. 135, pp. 225–228.

Zhu, D., Wang, J., and Wang, Q.J. 2015. 'On the Stribeck Curves for Lubricated Counterformal Contacts of Rough Surfaces,' *ASME Journal of Tribology*, Vol. 137, p. 021501.

Part IV

Seals and Monitoring

16

Seals Fundamentals

While the main focus of this book is on bearings, overall coverage would be incomplete without the introduction of seals – a vital component and ubiquitous companion to bearings. Fitted around the shaft or its periphery, the seal eliminates or reduces the leakage flow between two regions. Use of bearings and seals is predominant in petrochemical and process industries, aerospace units, and power generation that utilize them in turbines (both water and steam), compressors, pumps, gearboxes, and other auxiliary units (Lebeck, 1991).

Improper design, installation, or maintenance of seals can easily lead to catastrophic failure of a system, endangering the safety of plant operators and posing a significant environmental hazard as a result of pollution associated with leakage from seals. According to Lebeck (1991), the costs associated with repairing bearings and seals in a *single* large refinery exceed $1 million annually. For process plants, Nau (1985) attributes the major portion of the maintenance costs to mechanical seal failures.

Pumps represent a major and critical area for seal applications (see Figure 16.1). A survey by Sabini (1999) indicates that the currently installed population of pumps in chemical processes *alone* exceeds 750 000 units. This includes 40 000 new pumps that go into service annually. For each pump, on average, industry spends $2500 for repairs annually with 80% being related to mechanical seals and bearings. This becomes significant since the industry average mean time between maintenance (MTBM) is only 18 months. This calls for a major improvement in seal life and reliability since the 1990 Clean Air Act calls for an average seal life of three years (Netzel, 1999).

Classification of Seals

Fluid seals fall into two categories of *static* and *dynamic* as shown in the general classification of Figure 16.2. Examples of static seals include gaskets, O-rings, packings and the like used with little or no motion between mating parts. This chapter, on the other hand, will cover primarily dynamic seals that are used between relatively moving surfaces to restrict flow. The following section involves the broad range of *clearance*, or *interstitial*, dynamic seals. This is then followed by coverage of radial and axial *contact*, or *interfacial*, seals with their distinctly different design characteristics.

Applied Tribology: Bearing Design and Lubrication, Third Edition. Michael M. Khonsari and E. Richard Booser.
© 2017 John Wiley & Sons Ltd. Published 2017 by John Wiley & Sons Ltd.
Companion Website: www.wiley.com/go/Khonsari/Applied_Tribology_Bearing_Design_and_Lubrication_3rd_Edition

Figure 16.1 A typical industrial pump. *Source*: Courtesy of Dow Chemical Company.

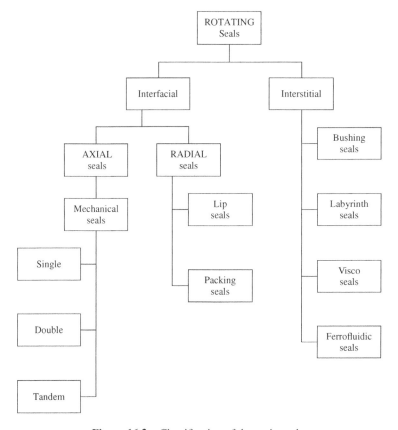

Figure 16.2 Classification of dynamic seals.

16.1 Clearance Seals

These include a broad category of shaft seals whose principle of operation is based on putting an adequate resistance in front of the flow of sealing fluid to reduce leakage to a tolerable amount. Buchter (1979) categorizes clearance seals as bushing, labyrinth, visco, and ferrofluidic. A brief description of these follows.

Bushing Shaft Seals

In bushing seals, the desired pressure drop is created by using the viscous friction losses that occur in a small annular clearance gap around the bushing. The simplest device of this type is referred to as the *fixed clearance bushing seal* (Figure 16.3). The resistance to leakage depends on the bushing length, clearance, and viscosity of the fluid. The smaller the clearance, the lower is the leakage; however, proper selection must ensure that shaft rotation is not impeded by too small a clearance. To this end, thermal expansion of the shaft and the bushing should be considered at the design stage.

Floating Bushing Seals

The next level of sophistication in design is to have the bushing seal float on the shaft without rotating (Figure 16.4). In the simplest form, this uses a bushing pressed against a ring that is allowed to move radially and 'ride' on the shaft, hence the name *floating ring* or *floating bushing seal*. The floating ring clearance is normally smaller than that of the fixed bushing seal. More sophisticated designs including *multiple floating rings* and *segmented floating ring* arrangements are described by Buchter (1979). These seals can be used for containment of either liquid or hot gases. The flow is often choked when using gases.

The rate of leakage can be estimated using standard equations for slit flow, and concentric and eccentric axial flow between the shaft and the sleeve (Figures 16.5(a) and (b)). The leakage mass flow of an incompressible Newtonian fluid of viscosity μ flowing in slit of gap size h,

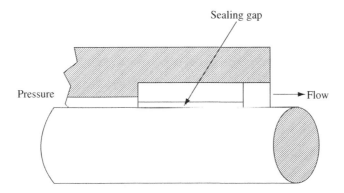

Figure 16.3   Fixed clearance bushing seal.

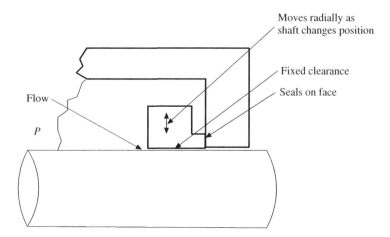

Figure 16.4   Floating ring seal.

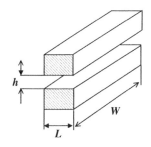

(a) slit flow

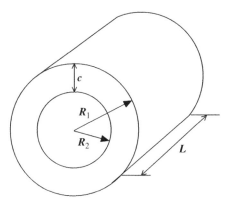

(b) annular flow

Figure 16.5   (a) Slit flow and (b) annular flow.

length L, width w exposed to pressure differential $\Delta P = P_s - P_{atm}$ is (Stair, 1983):

$$\text{Laminar flow:} \quad \dot{m} = \frac{\rho h^3 w}{12 \mu L} \Delta P \tag{16.1}$$

$$\text{Turbulent flow:} \quad \dot{m} = 4.71 w \left(\frac{\rho h^3 \Delta P}{\mu^{1/4} L} \right)^{4/7} \tag{16.2}$$

For axial flow between two concentric cylinders, the leakage flow is obtained by letting $w = 2\pi R$. The result is:

$$\text{Laminar flow:} \quad \dot{m} = \frac{\pi \rho c^3 R}{6 \mu L} \Delta P \tag{16.3}$$

$$\text{Turbulent flow:} \quad \dot{m} = 9.42 \, \pi R \left(\frac{\rho c^3 \Delta P}{\mu^{1/4} L} \right)^{4/7} \tag{16.4}$$

where c is the radial clearance. For eccentric cylinders, the leakage flow can be estimated using Equation (8.61) repeated below:

$$\dot{m} = \frac{\pi \rho R c^3}{6 \mu L} (1 + 1.5 \varepsilon^2) \Delta P \tag{16.5}$$

where ε is the eccentricity ratio. Note that leakage increases with eccentricity. For the same pressure gradient, as $\varepsilon \to 1$ leakage of the fully eccentric case becomes 2.5 times that for the concentric case with $\varepsilon = 0$, resulting in a significant performance deterioration. The critical Reynolds number where the flow transition occurs may be taken from the literature on general pipe flow with, $Re_{crit} = \rho U D_h / \mu$, where D_h is the hydraulic diameter defined as the four times the ratio of the cross sectional area divided by the wetted perimeter. Generally, in pipe flows, flow remains laminar when $Re_{crit} \leq 2000$.

Example 16.1 Consider a stainless steel shaft operating at 3600 rpm. Estimate the leakage flow in a bushing seal used for containing ISO69 oil at $T = 90\,°F$ against 10 psi pressure gradient. The radial clearance is $c/R = 1/1000$, $R = 1$ in., $L = 2$ in. Estimate the leakage flow rate. How does the leakage flow rate change if the oil temperature is raised to $180\,°F$?

The oil viscosity is $\mu = 2 \times 10^{-5}$ reyns and density is $\rho = 0.04\,lbm/in^3$. We first estimate the Reynolds number to determine if the flow is laminar or turbulent. The characteristic length is the hydraulic diameter which is defined based on the cross sectional area and the wetted perimeter. For a concentric tube, the hydraulic diameter is

$$D_h = \frac{4(\pi/4) \left(D_o^2 - D_i^2 \right)}{\pi D_o + \pi D_i} = D_o - D_i = 2c$$

where c is the radial clearance.

$$Re = \frac{\rho R \omega D_h}{\mu} = \frac{(0.04)(1)(377)(2)(1 \times 10^{-3})}{2 \times 10^{-5}} = 1508 < 2000$$

Therefore, the flow is laminar and Equation (16.3) applies.

$$\dot{m} = \frac{\pi \rho c^3 R}{6 \mu L} \Delta P = \frac{\pi.(0.04)(1 \times 10^{-3})^3(1)}{6(2 \times 10^{-5})(2)}(10) = 5.236 \times 10^{-6} \text{ lbm/s} \quad (0.019 \text{ lbm/h})$$

If the oil is heated to 170 °F, the viscosity drops ten fold: $\mu = 2 \times 10^{-6}$ reyns. Assuming the density remains unchanged, the Reynolds number increases ten fold and flow becomes turbulent. Neglecting thermal expansion, Equation (16.4) gives the leakage flow rate as follows:

$$\dot{m} = 9.42 \, \pi \, R \left(\frac{\rho c^3 \Delta P}{\mu^{1/4} L} \right)^{4/7} = 9.42\pi(1.0) \left(\frac{(0.04)(1 \times 10^{-3})^3(10)}{(2 \times 10^{-6})^{1/4}(2)} \right)^{4/7}$$

$$= 5.534 \times 10^{-4} \text{ lbm/s} \quad (1.99 \text{ lbm/h})$$

Note that at this temperature, the leakage is over 100-fold greater than the laminar case. In reality, dimensional changes due to thermal expansion will have to be taken into account for a more realistic assessment of leakage.

Labyrinth Shaft Seals

Labyrinth seals use a series of flow restrictors as illustrated in Figure 16.6 with the steps formed by inserting a series of sleeve rings around the shaft. Variations shown in Figure 16.7 include *straight, high-low, stepped,* and various other tooth arrangements (Stair, 1983). The straight arrangement of teeth is common to accommodate axial displacement. Knife edges are usually preferable as they thermally soften and displace readily during any rubs for minimum damage to their close-fitting shaft surface. While stepped and high-low teeth involve a more complex design, they give moderate improvement in sealing by blocking free flow of the discharge from one stage into the clearance of the next.

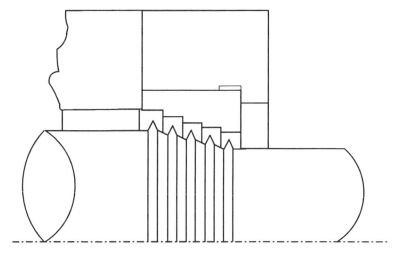

Figure 16.6   Labyrinth seal assembly formed by sleeve rings around shaft.

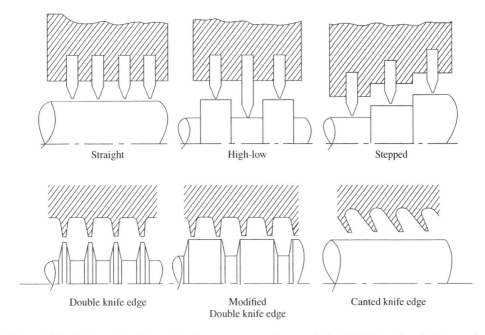

Figure 16.7 Various labyrinth seal teeth arrangements. *Source*: Stair, 1983. Reproduced with permission of Taylor & Francis.

In the selection and design of a labyrinth seal, attention needs to be given to the thermal expansion of the surface to ensure that an adequate clearance exists and that the shaft motion is not impeded. For this reason, labyrinth seals often leak more than other types. However, they generally require very little maintenance and are well suited for high-speed centrifugal machines such as compressors as well as gas and steam turbines.

Flow analysis has generally considered each labyrinth element as an orifice, or as having similarity to turbulent pipe flow. The actual flow pattern involves elements of both. Gas flow is often choked when discharged to the atmosphere. Egli (1935) established appropriate seal leakage equations. The reader is referred to Buchter (1979) for a more comprehensive discussion of this type of seal.

Brush Seals

High sealing losses accompanying the generous radial clearance needed with labyrinth seals has led to varied design details and tooth materials involving abradable and deformable bronze, steel and honeycomb materials. Most promising for many industrial compressors, power-generating turbines and aircraft turbine engines are static wire brush designs that typically reduce by 80% or more the leakage rate of labyrinth seals (Dinc *et al.*, 2002; Chupp *et al.*, 1997).

The static circular brush seal controlling leakage along a rotating shaft contains a bristle pack of many fine, round wires of 0.05–0.15 mm (0.002–0.006 in.) diameter The wire material is selected from superalloys such as cobalt-base Haynes 25 to accommodate high gas

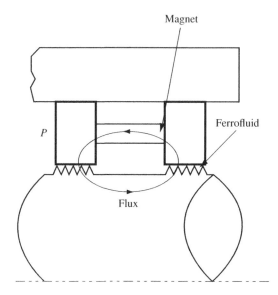

Figure 16.8   Schematic of a ferrofluid seal.

temperatures, wear and abrasion requirements. These bristles are welded at their outer diameter between two plates and inclined at angles up to 40–50 degrees in the direction of rotation. The supporting back plate uses a close clearance of its inner diameter with the rotor to protect the bristles from a bending overload by elevated upstream gas pressures. Minor modification in side-plate arrangement has also allowed two-way sealing with bidirectional flow (Hendricks *et al.*, 1998). In recent years, brush seal applications have been attempted at surface speeds over 400 m/s, temperatures above 650 °C, and at discontinuous contact surfaces at blade tips in a turbine (Dinc *et al.*, 2002).

Ferrofluid Seals

Ferrofluid seals were developed in the 1960s for NASA as a means of controlling rocket fuel in space applications (Buchter, 1979), but their use has extended to other applications including disk drives. A unique feature of a ferrofluid seal is that it functions without rubbing contact with the shaft surface. Hence, there is no wear, and the seal can effectively result in zero leakage.

The clearance gap around the shaft is hermetically closed off by filling the gap with a 'magnetic fluid' exposed to a magnetic field (see Figure 16.8). The magnetic fluid is a random suspension of coated microscopic magnetic particles (roughly 100 Å in diameter) that becomes oriented by the application of an external magnetic field. The magnetized fluid bridges the gap and blocks leakage while still allowing the shaft to rotate freely – even at high speeds – with relatively small viscous friction at the interface. Removing the magnetic field will demagnetize the fluid without concern for an appreciable hysteresis loss. Because there is no contact, there is no wear, and life expectancy is indefinite. However, the ferromagnetic fluid should not be mixed with the process fluid, and despite many attractive features and advantages that

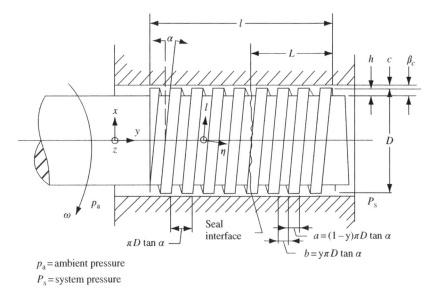

p_a = ambient pressure
P_s = system pressure

Figure 16.9 Geometry of a spiral-shaped visco seal. *Source*: Stair and Hale, 1966.

this type of a seal has to offer, its applications are somewhat limited to special devices. Some pertinent design considerations are presented by Buchter (1979).

16.2 Visco Seals

Operation of visco seal hinges upon the pressure developed as a result of rotation of a shaft with spiral-shaped grooves on its surface (Figure 16.9) or a straight shaft with grooves machined into the ID of the sleeve. Either one would result in a screw pumping action to hydrodynamically seal against pressure in the opposite direction. These seals find use in turbines and compressors regardless of whether the flow is laminar or turbulent.

Analysis of Flow in a Visco Seal: Laminar and Turbulent

(a) Concentric Laminar Flow

Referring to Figures 16.9 and 16.10 and following the work of Stair (1965), the following notation will be used to formulate visco seal performance parameters:

a = axial land width

b = axial groove width

c = radial clearance

d = seal diameter

h = groove depth

h_g = film thickness above grooves = $h + c$

l = axial threaded length of seal

L = active seal length

t = tangent of helix angle α

α = helix angle

x, y, z = Cartesian coordinates

ξ, η, z = groove coordinates system

$\beta = (h + c)/c$

$\gamma = b/(a + b)$

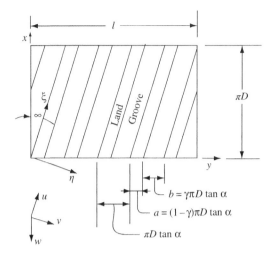

Figure 16.10 Unwrapped geometry of a spiral-shaped visco seal. *Source*: Stair and Hale, 1966.

Visco-Seal Coefficient

Performance of a visco seal depends on a dimensionless parameter known as *seal coefficient*, Λ, which is inversely proportional to the pressure drop between the system pressure and ambient. It is defined as

$$\Lambda = \frac{6\mu\, UL}{c^2 \Delta P} \tag{16.6}$$

For concentric laminar flows, the visco seal coefficient Λ is a function of screw geometry as shown by the following equation (Stair, 1965):

$$\Lambda = \frac{\beta^3(1+t^2) + \gamma\, t^2(1-\gamma)(\beta^3-1)^2}{\gamma\, t(1-\gamma)(\beta^3-1)(\beta-1)} \tag{16.7}$$

Optimization of Seal Head Coefficient

For a specified μ, U, L, and c, the maximum ΔP can be achieved by minimizing Λ. Taking partial derivatives of each one of the terms on the right-hand side of Equation (16.7) and setting the result equal to zero, i.e.

$$\frac{\partial \Lambda}{\partial \gamma} = \frac{\partial \Lambda}{\partial \beta} = \frac{\partial \Lambda}{\partial t} = 0 \tag{16.8}$$

yield the following:

$$\gamma = 0.5; \qquad \beta = 3.65; \qquad t = 0.28; \quad \text{and} \quad \alpha = 15.68° \tag{16.9}$$

Using the above values, the theoretical minimum seal coefficient and the corresponding maximum pressure drop that can be achieved in a visco seal operating in a laminar regime are as follows:

$$\Lambda_{min} = 10.97 \quad \text{and} \quad \Delta P_{max} = 0.55\frac{\mu UL}{c^2} \tag{16.10}$$

Power Loss

For a visco seal operating in the laminar regime:

$$E_P = \Phi \frac{\pi \mu D L U^2}{c} \tag{16.11}$$

where Φ is the so-called total dissipation function, which is the sum of two components: the Couette flow dissipation Φ_c, which depends on angle α, and Poiseuille flow dissipation Φ_p, which is a function of both α and β.

$$\Phi = \Phi_c + \Phi_p \tag{16.12}$$

$$\Phi = \left(1 - \gamma + \frac{\gamma}{3}\right) + 3 \left[\frac{\gamma t^2 (1 - \gamma)(\beta - 1)(1 - \gamma + \gamma \beta^3)}{\beta^3 (1 + t^2) + t^2 \gamma (1 - \gamma)(\beta^3 - 1)^2} \right] \tag{16.13}$$

In terms of the seal coefficient, Λ, the power loss is given by

$$E_P = \frac{1}{6} \Phi \Lambda \pi D U c \Delta P \tag{16.14}$$

The power loss in a visco seal designed for maximum pressure drop (minimum seal coefficient) is

$$E_P = 1.36 \pi D U c \Delta P_{max} \tag{16.15}$$

where the optimized parameters in Equations (16.10) are used.

Letting $U = D\omega/2$, the power loss with the minimum seal coefficient is

$$E_P \approx 0.59 \frac{\mu D^3 \omega^2 L}{c} \tag{16.16}$$

Estimating power loss at the design stage permits prediction of temperature rise and corresponding drop in viscosity if the sealing fluid is oil. While this equation shows that increasing clearance tends to reduce power loss and thus the fluid temperature, this option may unacceptably increase the leakage flow. External cooling may thus be required.

Friction Resistance Coefficient

Similar to laminar flow in pipes, a friction resistance coefficient – sometimes called the friction parameter – can be defined. For laminar flow in a circular pipe $f = 64/Re_D$, where the Reynolds number is determined based on pipe diameter D. In a visco seal operating in a laminar regime, the friction parameter is related to the dissipation function according to the following equation (Stair and Hale, 1966):

$$f_s = \frac{4\pi \Phi}{Re_c} \tag{16.17}$$

where clearance, c, is used as the characteristic length for determining the Reynolds number.

To calculate the performance of a visco seal operating with turbulent flow, it is appropriate to first identify the critical Reynolds number, Re_{crit}, where the flow is likely to undergo transition from laminar to turbulent (Stair and Hale, 1966).

$$Re_{crit} = 41.1 \left[\frac{D/2}{(1-\gamma)c + \gamma\beta c} \right]^{0.5} \tag{16.18}$$

For an ungrooved shaft with $\beta = 1$ and $\gamma = 0$ (or $\gamma = 1$), the above equation reduces to the well-known critical Re_{crit} number for rotating cylinders, i.e.

$$Re_{crit} = 41.1\sqrt{D/2c} \tag{16.19}$$

Using the clearance c as the characteristic length, the flow is considered to be laminar so long as the following condition is satisfied:

$$Re_c = \frac{\rho c U}{\mu} \le Re_{crit} \tag{16.20}$$

(b) Turbulent Seal Coefficient

Analytical derivations are based on a series of functions, some of which are best determined experimentally, and give the following turbulent seal coefficient (Stair and Hale, 1966):

$$\Lambda = \frac{6\mu\, UL}{c^2 \Delta P} = K_4 \left(\frac{I_1 + I_2}{I_4} \right) + K_5 \left(\frac{I_3}{I_4} \right) \tag{16.21}$$

where

$$I_1 = (1-\gamma)t^2$$
$$I_2 = \beta^3 \gamma t^2$$
$$I_3 = \frac{\beta^3}{\gamma + \beta^3(1-\gamma)}$$
$$I_4 = t\left[1 - \gamma + \gamma\beta - \frac{\gamma + \beta^3(1-\gamma) + \gamma(\beta-1)}{\gamma + \beta^3(1-\gamma)} \right]$$

$$K_4 = \frac{3}{2F_\xi}\left[1 - \frac{F_\xi}{10.5F_\xi - 7.5} + \frac{1 - F_\xi}{3.92F_\xi^2 - 1.4F_\xi - 1} \right] \tag{16.22}$$

$$K_5 = \frac{3}{2F_\eta}\left[1 - \frac{F_\eta}{10.5F_\eta - 7.5} + \frac{1 - F_\eta}{3.92F_\eta^2 - 1.4F_\eta - 1} \right]$$

$$F_\xi = 0.00509 \left(2000\cos\alpha \frac{Re_c}{Re_{crit}} \right)^{0.747}$$

$$F_\eta = 0.0101 \left(2000\sin\alpha \frac{Re_c}{Re_{crit}} \right)^{0.754}$$

Note that Equation (16.21) reduces to the laminar case when $K_4 = K_5 = 1$.

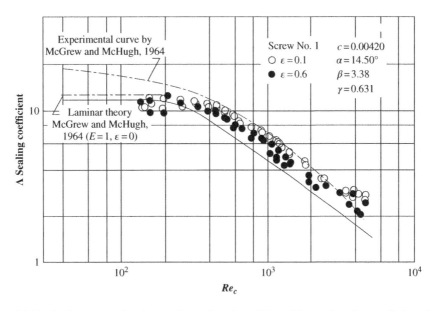

Figure 16.11 Performance of a visco-seal as a function of Reynolds number. *Source*: Stair and Hale, 1966.

Experimental and theoretical sealing coefficients obtained by Stair and Hale (1966) and comparison with the results of McGrew and McHugh (1964) are shown in Figure 16.11. Note that for this particular seal and the operating conditions, the sealing coefficient remains fairly constant at low Re_c values until $Re_c \approx 250$–300, after which it drops with increasing Re_c. Two sets of eccentricity ratios $\varepsilon = 0.1$ and $\varepsilon = 0.6$ are shown. The sealing coefficient is lower for the eccentric case, as expected.

Friction Resistance Parameter
Stair and Hale (1966) recommend the following equation for determining the friction resistance parameter when the flow is turbulent:

$$f = \frac{0.156\pi}{Re_c^{0.43}} \tag{16.23}$$

This relationship is a modified version of the turbulent flow relation observed in journal bearings by Smith and Fuller (1956). The modification in the multiplier brings the Smith and Fuller experiments closer to observations of Stair and Hale with visco seals.

Gas Ingestion

One problem with a high-speed visco seal is the gas ingestion from the low-pressure side of the seal to the high-pressure end and formation of cavitation bubbles (Stair and Hale, 1966). This deteriorates the effectiveness of the seal and opens the seal to potential contamination. The results of experiments reported by Ludwig *et al.* (1966) with a grooved shaft and smooth sleeve revealed that gas ingestion began at $Re = 1200$ and its rate increased with increasing the Reynolds number. On the other hand when the sleeve was grooved and the shaft was straight and smooth, gas ingestion did not occur, even when operating at much higher Reynolds numbers (ranging from $Re = 10\,400$ to 60 500) and for pressures ranging from 1.4 to 68.9 N/cm^2 (2–100 psig). In addition, Ludwig et al. (1966) found that concentricity of the shaft is important to minimize gas ingestion: the greater the eccentricity, the higher the possibility of gas ingestion.

Example 16.2 Consider a visco seal operating with the following specifications:

$$\alpha = 9.67°, \quad t = 0.26, \quad \beta = 6.16, \quad \gamma = 0.504, \quad c = 0.00265 \text{ in. and } D = 1.243 \text{ in.}$$

(a) Determine the seal coefficient for operating condition when $Re_c = 400$ and with $Re_c = 100$.
(b) Determine the friction resistance at both operating Reynolds numbers.

The critical Reynolds number is given by Equation (16.18):

$$Re_{crit} = 41.1 \left[\frac{D/2}{(1-\gamma)c + \gamma\beta c} \right]^{0.5}$$

$$= 41.1 \left[\frac{1.243/2}{(1 - 0.504)(0.00265) + (0.504)(6.16)(0.00265)} \right]^{0.5} = 332$$

Therefore, when operating with $Re_c = 400$, the flow is turbulent. Referring to Equations (16.22), we can determine the remaining parameters needed to calculate the seal coefficient.

$$I_1 = (1 - \gamma)t^2 = (1 - 0.504)(0.26)^2 = 0.033$$
$$I_2 = \beta^3 \gamma t^2 = (6.16)^3(0.504)(0.26)^2 = 7.878$$

$$I_3 = \frac{\beta^3}{\gamma + \beta^3(1-\gamma)} = \frac{(6.16)^3}{0.504 + (6.16)^3(1-0.504)} = 2.007$$

$$I_4 = t\left[1 - \gamma + \gamma\beta - \frac{\gamma + \beta^3(1-\gamma) + \gamma(\beta-1)}{\gamma + \beta^3(1-\gamma)}\right]$$

$$= (0.26)\left[1 - 0.504 + (0.504)(6.16) - \frac{0.504 + (6.16)^3(1-0.504) + 0.504(6.16-1)}{0.504 + (6.16)^3(1-0.504)}\right]$$

$$= 0.667$$

$$F_\xi = 0.00509\left(2000\cos\alpha\frac{Re_c}{Re_{crit}}\right)^{0.747} = 0.00509\left(2000\cos(9.67°)\frac{400}{331.703}\right)^{0.747} = 1.693$$

$$F_\eta = 0.0101\left(2000\sin\alpha\frac{Re_c}{Re_{crit}}\right)^{0.754} = 0.0101\left(2000\sin(9.67°)\frac{400}{331.703}\right)^{0.754} = 0.934$$

$$K_4 = \frac{3}{2F_\xi}\left[1 - \frac{F_\xi}{10.5F_\xi - 7.5} + \frac{1 - F_\xi}{3.92F_\xi^2 - 1.4F_\xi - 1}\right]$$

$$= \frac{3}{2(1.693)}\left[1 - \frac{1.693}{10.5(1.693) - 7.5} + \frac{1 - 1.693}{3.92(1.693)^2 - 1.4(1.693) - 1}\right] = 0.662$$

$$K_5 = \frac{3}{2F_\eta}\left[1 - \frac{F_\eta}{10.5F_\eta - 7.5} + \frac{1 - F_\eta}{3.92F_\eta^2 - 1.4F_\eta - 1}\right]$$

$$= \frac{3}{2(1.051)}\left[1 - \frac{1.051}{10.5(1.051) - 7.5} + \frac{1 - 1.051}{3.92(1.051)^2 - 1.4(1.051) - 1}\right] = 1.051$$

The turbulence seal coefficient is

$$\Lambda_T = K_4\left(\frac{I_1 + I_2}{I_4}\right) + K_5\left(\frac{I_3}{I_4}\right) = 0.662\left(\frac{0.033 + 7.878}{0.667}\right) + 1.051\left(\frac{2.007}{0.667}\right) = 11.019$$

The friction parameter is

$$f = \frac{0.156\pi}{Re_c^{0.43}} = \frac{0.156\pi}{400^{0.43}} = 0.037$$

When $Re_c = 100$, flow is laminar and the seal coefficient can be evaluated from the above equation as well by setting $K_4 = K_5 = 1$:

$$\Lambda_T = K_4\left(\frac{I_1 + I_2}{I_4}\right) + K_5\left(\frac{I_3}{I_4}\right) = (1)\left(\frac{0.033 + 7.878}{0.667}\right) + (1)\left(\frac{2.007}{0.667}\right) = 14.876$$

Prediction of the friction parameter in the laminar flow calls for computing the dissipation function and making use of Equation (16.13):

$$\Phi = \left(1 - \gamma + \frac{\gamma}{3}\right) + 3\left[\frac{\gamma t^2(1 - \gamma)(\beta - 1)(1 - \gamma + \gamma\beta^3)}{\beta^3(1 + t^2) + t^2\gamma(1 - \gamma)(\beta^3 - 1)^2}\right] = \left(1 - 0.504 + \frac{0.504}{3}\right)$$

$$+ 3\left[\frac{0.504(0.26)^2(1 - 0.504)(6.16 - 1)(1 - 0.504 + 0.504(6.16)^3)}{(6.16)^3(1 + 0.26^2) + (0.26)^2(0.504)(1 - 0.504)(6.16^3 - 1)^2}\right] = 0.691$$

$$f = \frac{4\pi\Phi}{Re_c} = \frac{4\pi(0.668)}{100} = 0.087$$

High Pressure Gas Sealing Using Double Screw Pump with a Buffer Fluid

Figure 16.12 shows a schematic of a shaft with threaded left-hand grooves on the system pressure side and right-hand groove on the air side. This is a double screw pump working in the opposite direction to each other. The point at which the sealing fluid enters is on the neutral level and the maximum pressure occurs there. Upon shaft rotation, the pressure-side pump with left-hand screw pumps the fluid from pressure P to the maximum pressure P_{max} and the air-side pump with right-hand screw pumps from P_{atm} to P_{max}. The net effect is practically zero leakage.

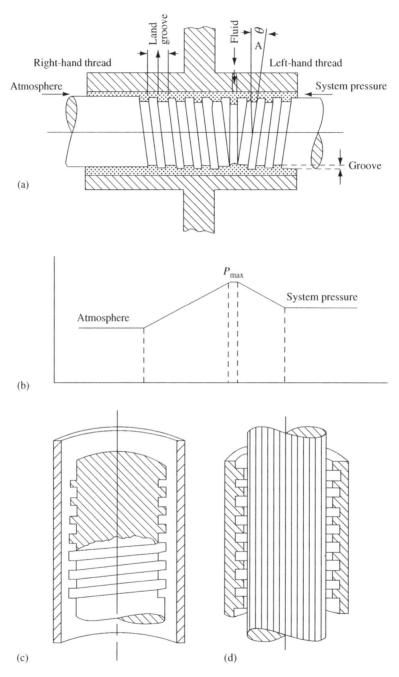

Figure 16.12 (a) High-pressure double screw pump; (b) pressure distribution; (c) screw machined in shaft; (d) screw machined in sleeve. *Source*: Buchter, 1979. Reproduced with permission of John Wiley & Sons.

16.3 Radial Contact Seals

Lip Seals

Elastomeric radial lip seals are the most widely used shaft seals in machinery – particularly rolling element bearings – for retaining oil in and keeping dirt out of the system. They can tolerate shaft misalignment, runout, and varying eccentricity due to the changes in operating conditions. Their low cost and sealing effectiveness make them ideal for use in automotive and truck engines, gearboxes, and transmissions. However, their service life is limited. While their use is predominately in low and moderate pressures, they have been put to use at relatively high pressures and operation speeds (Wheelock, 1981).

Figure 16.13 shows one of the wide varieties of lip seals. The lip is a made of a flexible elastomeric membrane. It is pushed against the shaft using a garter spring to form a 'tight' interface fit between the lip and the rotating shaft, which seals the oil side under pressure against the air side at atmospheric pressure. The other end of the membrane is bonded to a metal casing and affixed to the housing. The principle of pumping action from the air side is somewhat akin to that of a visco seal.

It is important to discuss some key features of lip seal operation and design (Müller and Nau, 1998; Buchter, 1979). The lip, being an elastomer, may react with oil, lose its elasticity, and thus wear. Even under normal conditions, the lip–shaft interface wears from an initial width of about 0.1–0.15 mm to about 0.2–0.3 mm after operating only 500–1000 h in a relatively clean atmosphere. The seal can experience even fivefold increase in width if operating in an abrasive environment.

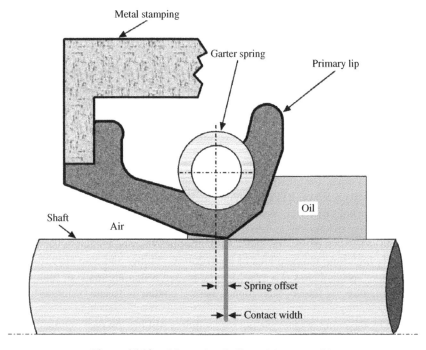

Figure 16.13   Schematic of a lip seal (not to scale).

Average radial pressure for a new seal is typically about 1 N/mm², which translates to a tip load of 0.1–0.15 N/mm. After operation, the tip pressure drops by a factor of two because of the increase in the contact width. However, the lip can effectively maintain its sealing function with a tip load of 0.05 N/mm. The membrane angle exposed to the oil side is normally twice the air side, typically 40°–60° on the oil side and 20°–30° on the air side to prevent oil leakage (Müller and Nau, 1998). The spring offset (see Figure 16.13) is typically about 0.5 mm towards the oil side. Lip seals without a spring are limited to use in moderate speeds.

Life and performance of a lip seal depend on the lubrication characteristics of the contact. It is widely accepted that a very thin seal fluid (of the order of a micron) partially separates the lip from the shaft to reduce wear. Here, the lip asperities play a major role (Jagger, 1957). On the one hand, asperities help to prevent leakage by providing a pumping action in the opposite direction of leakage, i.e. from the air side to the high-pressure liquid side. On the other hand, each asperity is responsible for generating load-carrying capacity since as the shaft rotates, fluid is dragged over the asperities, creating a local pressure that supports the load – as if each asperity were a tiny slider bearing – thereby lifting the lip off from the shaft (Horve, 1992; Salant, 1996, 1999; Salant and Rocke, 2004). At low speeds, the capillary forces at the oil–air interface on the air side counteract the pumping action of the seal. At higher speeds, cavitation is likely to occur in the contact area. This complicates the elastohydrodynamic analysis of lip seals which is necessarily coupled to thermal effects. Each of these factors is known to have a significant impact on the seal performance, particularly at high speeds. Similar to the thermohydrodynamic consideration of bearings, increasing speeds lead to a higher interface temperature, reduced viscosity, and associated thinning of the film thickness. As a result, according to the Stribeck curve, the friction will rise to further aggravate the contact.

Packing Seals

Soft packing or compression packing was the first type of sealing technique ever employed and historically had been the dominant method for sealing rotating shafts – particularly in pumps in chemical and process industries – although recent environmental regulations have largely displaced packing in favor of mechanical face seals.

Implementation of packing is relatively simple and the resulting sealing is effective and tolerable so long as the leakage is not hazardous. Conventional packings are either the *jam* or *automatic* type, with a multitude of geometric shapes (e.g. circular, wedge, cone for jam-type and C-rings, U-rings, V-rings for automatic type), configurations (braided, twisted, woven), and materials (leather, fiber, Viton, PTFE, graphite, and metallic). See Buchter (1979) for details. In selecting the packing material, attention should be given to the temperature limit, thermal conductivity, thermal expansion, and friction characteristics. For example, PTFE is susceptible to fairly large thermal expansion and suffers from low thermal conductivity. While PTFE can resist fairly high fluid temperatures of up to 250 °C, Müller and Nau (1998) recommend limiting its use to speeds below 0.5 m/s. Metallic packings that are made of aluminum foils, which are corrugated and folded, offer higher thermal conductivity and better performance at high speeds.

As implied by the name, a *jam-type packing* requires stuffing the packing material into the pump chamber cavity commonly called the *stuffing box*. Figure 16.14 illustrates the placement of a set of rings (typically four or five) of soft packing materials in the stuffing box, where

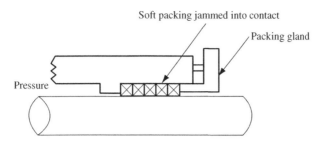

Figure 16.14   Schematic of a packing seal.

they are compressed using a bolt on the adjustable packing gland ring plate. Packing materials are porous – typically with 40–45% available void space – to allow the lubricant to fill the interstitial space. Axial compression by bolt tightening reduces the voids in the packing and shortens the axial length. Simultaneously, pressure against the shaft and the bushing increases as the packing expands radially. This forms a tight sealing against the process fluid.

The axial pressure is maximum at the gland plate and decreases exponentially to a minimum at the last ring at the stuffing box throat. This axial pressure reduction may not be sufficient to ensure adequacy of lubricant to reach the end rings placed near the stuffing box throat. The so-called *lantern ring* design improves sealing performance by dividing the packing in half and introducing the lubricant in the center via a lantern ring to lubricate both halves (Figure 16.15).

Some leakage in a packing seal is considered normal and is expected. In fact, it is necessary to provide some cooling. Overtightening the packing is not recommended, although periodic adjustment as specified by the packing manufacturer may be needed. Gland-side adjustments are normally specified based on leakage. After each adjustment, the packing material tends to undergo creep and redistribution of axial stresses. Therefore, allowance should be given for the settling time which may be of the order of hours or even days (Muller and Nau, 1998). For a water pump, Muller and Nau (1998) show that with a one-ring pack, the leakage can be as low as 15 ml/h (1 drop every 15 s), whereas the manufacturer recommends adjusting the packing to give 2 drops/s (300 ml/h).

The *automatic-type packing* does not require tightening after the system is pressurized. These packings are 'pressure-energized' and their sealing action is accomplished by a lip-type

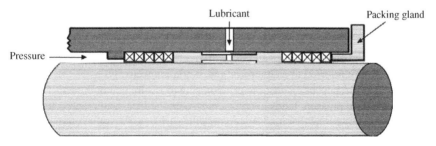

Figure 16.15   Schematic of a lantern ring seal. *Source*: Buchter, 1979. Reproduced with permission of John Wiley & Sons.

configuration installed with small interference. They are used in high-pressure hydraulic and pneumatic fields as well as reciprocating applications. With the lip pointed in the direction of the system pressure, the packing seals against the contact areas. With minute interference, increasing the pressure tends to increase the force on the contact area, and improve sealing (Stair, 1983).

The efficiency of a packing seal in terms of rate of leakage is dependent upon the shaft gyration, i.e. the cumulative effect of shaft runout, misalignment, deflection, and vibration. Buchter (1979) recommends maximum radial gyration of 0.003 in. (0.76 mm) for a pump operating at 3600 rpm, limiting the increase in packing diameter by 0.006 in. (0.152 mm) per shaft revolution. Gyrations beyond this limit are liable to create an annular gap through which leakage may become excessive.

16.4 Mechanical Face Seals

The remainder of this chapter focuses on mechanical seals, or face seals. Consider a container or a vessel which has a rotating shaft extending through its housing. The vessel may contain oil or a hazardous fluid under pressure that must be sealed from the environment. Clearly, upon shaft rotation, oscillation, normal vibration, or even wobble, the fluid will escape around the shaft without a suitable seal, such as a mechanical face seal.

Schoenherr (1965) broadly categorizes mechanical face seals based on their design or the varying arrangements. From design consideration, seals are either *balanced* or *unbalanced*. They are either pusher seals (using one or more springs) or nonpusher that use a *bellows* mechanism to hold the sealing-ring faces together. In terms of arrangements, either single or multiple seals are used. Single seals are mounted on either the inside or the outside of the vessel, generally referred to as *inside* or *outside seals*. Multiple seals use *double seals*, *tandem*, or *combination of double/tandem* arrangements. A brief description of each of these categories follows some elementary concepts.

Elementary Considerations and Terminology

The most elementary design would call for adding a shoulder on the shaft that could form a direct contact around the housing wall opening area where the sealing fluid is expected to leak out. The face of the shoulder area which rubs against the housing would act as a seal, so long as full contact is maintained. This design is, of course, prone to failure and is not a practical solution: surfaces will wear quickly and the contact will be lost, resulting in leakage. Nonetheless, this general concept lays the foundation for the design of practical seals.

The basic elements of a mechanical seal are illustrated in Figure 16.16. Here, the surfaces of two rings form the main faces of the seal. One surface – referred to the *primary ring* – rotates with the shaft and the other – *mating ring* – is affixed to the housing (gland). These two rings are sometimes simply referred to as the rotor and stator, respectively.

A spring mechanism pushes against the primary ring to ensure that both faces are in contact. Nevertheless, the fluid pressure tends to enter in between the faces and create a hydrodynamic pressure as an 'opening force' that tends to separate the surfaces. Two other elements contribute to this opening force: (1) hydrostatic lift resulting from the pressure difference between the fluid pressure in the container and ambient, and (2) centrifugal action caused by rotation of the primary ring.

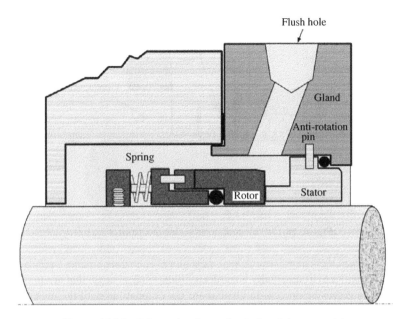

Figure 16.16   Schematic of a mechanical seal (not to scale).

This opening force is opposed by a 'closing force' generated by the hydraulic pressure of the fluid containment as well as the spring. The net pressure acting on the surfaces is a delicate design matter. If the closing force were too large, friction and wear would become excessive. On the other hand, if the opening force dominates, then much of the fluid would leak out. The combination of spring(s) and associated *O-rings* that move axially together are called *pusher seals*. *Nonpusher seals* use elastomeric or metallic elements such as bellows or diaphragms as secondary seals that are not in contact with the shaft. Here the axial movement is simply made possible by 'convolution'.

In the majority of applications, the primary ring is affixed to the shaft by a set screw and the leak path between the shaft and the ring is closed off by an O-ring, which also provides some resilience as it deforms elastically and moves axially with the spring. An O-ring, therefore, functions as a static *secondary sealing element*. Similarly, an O-ring is placed between the mating ring and the gland. The combination of the primary ring, together with the O-ring, spring, and its retainer, is often referred to as the *seal head*. The combination of the mating ring and its associated O-ring is sometimes referred to as the *seat assembly*. While in the majority of applications the seal head rotates with the shaft, there are designs – particularly for high-speed shafts – where the head is stationary and the seat rotates instead.

Figures 16.17 and 16.18 depict configurations of a single *inside seal* and a single *outside seal*. In an inside seal, the pressure acts from the OD of the seal to the ID, where typically the pressure is atmospheric. So, the seal is externally pressurized. An outside seal, on the other hand, is internally pressurized. Inside seals are by far the most common type of single seals in use because they can withstand high pressures and can accommodate different types of flush plans. However, they are best suited to noncorrosive process fluids, unless relatively expensive corrosion-resistant materials are used.

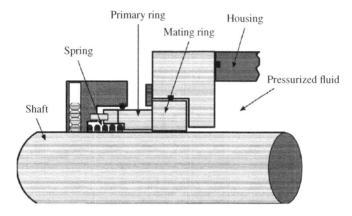

Figure 16.17   Illustration of inside mechanical seal.

When dealing with environmentally hazardous fluids, such as those that are toxic, or in the case of extremely corrosive process fluids that are not compatible with the materials used in a single seal, one may have to resort to a *double (dual) seal* or *tandem seal*. In a double seal, a pressurized barrier liquid or pressurized gas is used. Because of environmental emission regulations, an inert gas is often used when dealing with toxic process fluids. In applications involving carcinogenic fluids with severe environmental and safety regulations, a tandem seal with unpressurized buffer fluid is used.

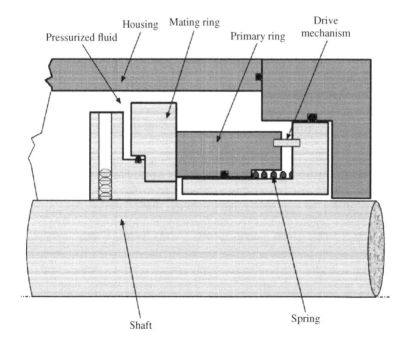

Figure 16.18   Illustration of outside mechanical seal.

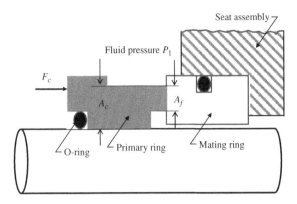

Figure 16.19   Schematic of an unbalanced seal.

Balanced vs Unbalanced Seal

Seal face pressure is critical: if it is too great, the interface temperature will become exceedingly high and the seal face will crack, blister, heat check, and fail. If not enough, the surfaces lose contact and the seal leaks. The delicate job of providing an adequate pressure on the seal face is achieved by simple balancing. Basically, the process involves modification to the seal head ring so that the effect of the hydraulic pressure on the seal face is essentially eliminated from consideration, thus leaving the spring force to counteract the hydrodynamic pressure.

To illustrate the operational concept, consider Figure 16.19, which shows a sketch of the essential parts of an *unbalanced seal*. In this configuration the entire fluid pressure exerted on the back of the primary ring (area A_c) is transmitted to the seal face (area A_f). Note that hydrodynamic pressure generated in between the faces counteracts the closing force and tends to push the surfaces open. This is the opening force acting on A_f. Any increase the fluid pressure directly increases the seal interfacial pressure by the same amount, making the seal susceptible to excessive wear. This pushes the primary (rotating) ring against the mating (stationary) ring. This is called the closing force F_c, and the area on which it acts is the closing area, A_c.

Figure 16.20 shows the schematic of a *fully balanced* seal with two changes: a reduction in area A_c – by providing a shoulder or simply a sleeve over the shaft – and reducing the closing

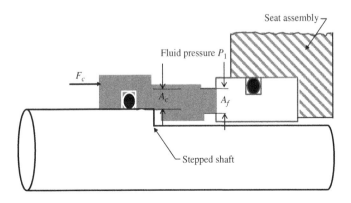

Figure 16.20   Schematic of a balanced seal.

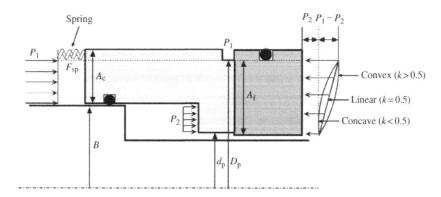

Figure 16.21   Balance ratio nomenclature.

area the seal head face so that hydraulic pressure can fully counteract the pressure on A_c. In this design, the new seal face A_f is practically under no pressure and therefore a change in the fluid pressure does not affect the interfacial pressure. The primary ring essentially 'floats' with the contact pressure controlled by one or more springs in the seal head.

In practice, seals are only partially balanced – typically around 65–75% – and a balance ratio in terms of the areas is defined to specify the magnitude of the partial balancing.

Hydrodynamic Pressure and Balance Ratio

Consider an externally pressurized seal (Figure 16.21), where the pressure at the OD of the face is P_1 and at the ID it is atmospheric (zero gage pressure). Fluid enters the contact and forms a wedge where fluid pressure acts on the face. The average pressure on the face is

$$\overline{P} = \frac{\int\limits_{r=r_1}^{r_2}\int\limits_{\theta=0}^{2\pi} Pr d\theta \, dr}{A_f} \tag{16.24}$$

where A_f represents the friction area. Assuming that the pressure is circumferentially uniform, Equation (16.24) reduces to

$$\overline{P} = \frac{2\pi \int\limits_{r=r_1}^{r_2} Pr \, dr}{A_f} \tag{16.25}$$

For simplicity, the pressure is commonly assumed to vary linearly across the face. In reality, the pressure profile is nonlinear in shape, either concave or convex in shape. Regardless of the shape, this pressure pushes the faces out axially, creating an opening force. The ratio of the closing area, on which the surrounding hydraulic pressure acts, to that of the opening area, where the hydrodynamic pressure is generated, is called the *balance ratio*:

$$b = \frac{\text{closing area}}{\text{opening area}} = \frac{A_c}{A_f} \tag{16.26}$$

For an inside seal, this area ratio is

$$b = \frac{\pi/4 \left(D_p^2 - B^2\right)}{\pi/4 \left(D_p^2 - d_p^2\right)} = \frac{\left(D_p^2 - B^2\right)}{\left(D_p^2 - d_p^2\right)} \tag{16.27}$$

For an outside seal, the balance ratio is

$$b = \frac{\pi/4 \left(B^2 - D_p^2\right)}{\pi/4 \left(D_p^2 - d_p^2\right)} = \frac{\left(B^2 - D_p^2\right)}{\left(D_p^2 - d_p^2\right)} \tag{16.28}$$

The opening force results from the hydrodynamic pressure, P_h, across the gap:

$$F_o = P_h A_f = k(P_1 - P_2)A_f + P_2 A_f \tag{16.29}$$

where k is called the *pressure gradient factor* to account for the shape of the pressure profile and its deviation from linearity. If assumed linear, $k = 0.5$; for convex pressure distribution, $k < 0.5$ and $k > 0.5$ for concave. Note that P_2 is typically atmospheric. The value of 0.5 is most commonly used for water and nonflashing fluids. Typically, k ranges between 0.5 and 0.8.

The closing force is the sum of the face pressures P_1, P_2 and the spring load:

$$F_c = P_1 A_c + P_2 (A_f - A_c) + F_{sp} \tag{16.30}$$

The net force on the face is

$$F_f = F_c - F_o = \Delta P (A_c - k A_f) + F_{sp} \tag{16.31}$$

where $\Delta P = P_1 - P_2$. The *face pressure* can be obtained by dividing Equation (16.30) by A_f. The result is

$$P_f = \Delta P(b - k) + P_{sp} \tag{16.32}$$

where P_{sp} is the spring pressure, typically in the range of 10–40 psi (69–276 kPa).

Equation (16.32) reveals that if pressure profile is linear, i.e. $k = 0.5$, and that the seal is 50% balanced, i.e. balance ratio is $b = 0.5$, then the only pressure acting on the face is theoretically identical to that exerted by the spring(s) in the seal head. If $b < 0.5$, then the first term in Equation (16.32) becomes negative: the face pressure will be less than the spring pressure and the faces pop open and the seal leaks. According to Schoenherr (1965), seals designed with a balance ratio in the range of 55–60% have failed with their faces popping open, since the pressure distribution is convex ($k < 0.5$). Therefore, typically the balance ratio is maintained between 65% to 75%. Note that a balanced seal simply refers to a seal whose face pressure is lower than an unbalanced one. Balanced seals are particularly useful when the pressure in the seal chamber is greater than 200 psi. At very high pressures, the balance ratio may become greater than 100%. Therefore, the PV factor should be carefully examined.

Pressure-Velocity, Heat Generation, and Power Loss

Having discussed the face pressure, we are now in a position to determine the seal performance in terms of the PV factor, which has important implications in terms of the wear, heat generation, and power loss. The PV factor is simply evaluated using the face pressure, P_f, multiplied by the mean velocity of the rotating ring V_m:

$$PV = P_f V_m \qquad (16.33)$$

where the mean velocity is evaluated at the mean diameter, D_m, of the rotating ring (primary), i.e. $D_m = D_p + d_p/2$.

Substituting Equation (16.32) into (16.33) yields

$$PV = \lfloor \Delta P(b-k) + P_{sp} \rfloor . V_m \qquad (16.34)$$

The power loss is given by

$$E_p = f P_f A_f V_m \qquad (16.35)$$

or

$$E_p = f A_f \lfloor \Delta P(b-k) + P_{sp} \rfloor V_m \qquad (16.36)$$

where the face area is

$$A_f = \pi D_m \left(\frac{D_p - d_p}{2} \right) \qquad (16.37)$$

where the terms in brackets give the face width.

The importance of thermal effects in mechanical face seals cannot be over-emphasized. As discussed in depth by Lebeck (1991), with the film gap on the order of a few microns, if the face temperature is high, the fluid may flash to vapor and damage the surfaces. Another problem is associated with non-uniform heating of seal rings that could thermally distort surfaces and cause damage. Excessive thermal distortion is particularly problematic for pump seals that run on hydrocarbons. It therefore comes as no surprise that many seal failures are directly caused by excessive interfacial heat generation leading to hot spotting, blistering, heat checking, and cracking (Lebeck, 1991).

Hot spotting is a form of surface damage, which is not unique to mechanical seals; it is commonly observed in brakes and clutches. It typically shows up as macroscopic dark spots or patches on the surface where intensive localized frictional heating occurs. Its occurrence is thought to be due to concentrated local temperatures that aggravate friction heating that causes large contact pressure and causes a form of instability known as thermoelastic instability (TEI). Akin to many other types of instabilities, it occurs when the speed exceeds a certain value known as the critical (or threshold) speed. There is a rich volume of literature available on the investigation of TEI starting from the pioneering work of Barber (1969), followed by Burton et al. (1973), who analyzed the occurrence of TEI between two sliding friction pairs where one disk is the conductor and the mating disk is the insulator. Lebeck (1980) extended the analysis by considering the effect of ring deflection and heat transfer in mechanical seals. Later developments include taking into account surface roughness and extension of analyses to a two-conductor friction system and its application in mechanical seals (Jang and Khonsari,

2000, 2003, 2004a, 2004b). Of particular interest is the reported number of hot spots measured experimentally by Netzel (1980) and close prediction of the TEI analysis reported by Jang and Khonsari (2003, 2013). According to Peng *et al.* (2003), the maximum number of hot spots is a strong function of film thickness. This is particularly important in face seals where the film thickness between the primary and mating rings can be on the order of several microns.

Method for Reducing Surface Temperature

Significant reduction of the face temperature is possible by innovative approaches, particularly by augmenting the design of the mating ring with a heat exchanger. For example, Winkler *et al.* (2011) reported the design of a mechanical seal assembly with an integrated heat transfer capacity in which the interior of the stationary seal ring is made of a porous material. The design allows coolant to flush through the porous structure and dissipate the interfacial heat generation. Khonsari and Somanchi (2005) designed a so-called double-tier mating ring which possesses a series of radial holes for coolant to enter into a built-in cooling channel inside the mating ring. In another related design, the cooling fluid is manipulated to enter and circulate in the mating ring with a plurality of through-channels to achieve an optimal cooling result (Khonsari and Somanchi, 2007). The cooling fluid can be either a gas or a liquid, separate from the flush fluid. Extensive laboratory tests with prototypes and long-term field testing in a chemical refinery have shown that effective cooling and controlling of the face temperature can be realized with these designs. A review of recent patents is reported by Xiao and Khonsari (2013).

Another methodology that has emerged in recent years is laser surface texturing. By engraving appropriate shaped dimples onto the mating ring face, it is possible to improve the load-carrying capacity (Wang *et al.*, 2001), reduce the coefficient of friction and wear (Etsion and Kligerman, 1999; Etsion *et al.*, 2004), expand the range of hydrodynamic lubrication regime (Kovalchenko *et al.*, 2005), and provide seizure resistance and improved cooling (Wang and Kato, 2003). This technology has received significant attention in the research community and the reader can refer to an extensive and authoritative review by Etsion (2005). The main objective of texturing the seal face(s) is to increase the load-carrying capacity (opening force) and reduce the interfacial friction coefficient. In this regard, textured dimples behave as many tiny bearings or oil reservoirs within which the fluid cavitates. In-depth discussions of dimple shape optimization and modeling aspects are beyond the scope of this book. The interested reader is referred to Qiu and Khonsari (2011a, 2011b, 2011c, 2012), Fesanghary and Khonsari (2011, 2012, 2013), and Shen and Khonsari (2013, 2016).

A direct application of texturing to reduce face temperature is to engrave an array of circumferential grooves on the mating ring side wall rather than the seal's face and take advantage of the added effective area that has been wetted by fluid in the seal chamber (Xiao and Khonsari, 2012; Khonsari and Xiao, 2017). Laboratory tests and numerical simulations reveal that the face temperature can be reduced by a maximum of 10% by appropriately shaped dimples. This approach works well with any type of seal face material and is quite attractive due to its effectiveness, simplicity, and the fact that it does not require any changes to the seal gland design. Thus, the user is not required to change the existing flush plan. If a significantly higher reduction in the face temperature is desired, then one may explore a recent development that incorporates a built-in heat pipe within the mating ring (Xiao and Khonsari, 2014; Khonsari and Xiao, 2017). Due to its extremely high effective thermal conductivity, this design is capable of absorbing significant heat from the seal-contacting face into the mating ring and transferring

Table 16.1 Properties values for different seal materials

Properties	Thermal conductivity, k (W/(m K))[a]	Thermal expansion coefficient, $\alpha \times 10^6$ (cm/cm/K)[b]	Elastic modulus, E (MPa)[c]	Density, ρ (kg/m³)[d]	Hardness, H
Ceramic (Al$_2$O$_3$) 85–99%	14.70–25.08	7.02–7.74	221–345	3405–3792	87 Rockwell A
Tungsten carbide	70.93–83.04	4.55	621	16 331	92 Rockwell A
Silicon carbide	70.93–103.8	3.38	331–393	2879	86-88 R 45N
Carbon (resin-filled)	6.57–20.76	4.14–6.12	17.2–27.6	1771–1910	80-105 Shore
Ni-resist	43.25–48.44	11.7–12.24	72–117	7307–7418	131-183 Brinell
Cast-iron	39.79–50.17	11.88	90–110	7169–7418	217-269 Brinell

[a] Multiply by 0.5778 to convert to Btu/h ft °F.
[b] Multiply by 0.926 to convert to in./in./°F.
[c] Multiply by 0.145 to convert to Mpsi.
[d] Multiply by 3.6×10^{-5} to convert to lb/in³.
Source: Johnson and Schoenherr, 1980. Reproduced with permission of ASME.

it from the outer surface to the flush fluid efficiently. The direct cooling of the face yields a more uniform interface temperature with less thermal distortion and lower wear rate (Xiao and Khonsari, 2015a, 2015b).

Seal Materials and Coefficient of Friction

As is normally the practice in choosing friction pair materials, the primary and mating rings should be selected from dissimilar materials to minimize friction and wear. Buck (2001) recommends using a hard face and soft face with a hardness difference of 20%. Also, the primary and mating rings should possess a high modulus of elasticity difference in order for the stiffer face to run into the softer one and provide a 'tighter' seal. Another element to consider is the coefficient of friction between the two faces, which is responsible for heat generation and thermal expansion.

Pertinent data on a variety of seal materials are presented in Table 16.1 based on the data provided by Johnson and Schoenherr (1980). Additional information is available from Lebeck (1991) and Table 4.8. Table 16.2 presents the coefficient of friction results based on the information provided by Johnson and Schoenherr (1980). The friction coefficient can be three to five times greater than the values in Table 16.2 when the interface is dry. This is important in determining the friction torque at start-up and the suitability of the motor to provide the high breakaway torque. Also crucial is the flatness of seal surfaces. Typically, most face seals are flat to within 2–3 helium light bands (see Chapter 3) with a surface finish of 4–5 µ in.

Seal Flushing Systems

In most industrial applications, the seal rings are cooled by an externally supplied coolant known as flush fluid. Ideally, the flush fluid provides coolant to quench the rings and also to

Table 16.2 Coefficient of friction based on tests with water at 35.03 bar m/s and *PV* limit of 10^5 psi ft/min

Rotating (primary) ring	Stationary (mating) ring	Friction coef. in water[a,b]
Carbon-graphite (resin-filled)	Cast-iron ceramic	0.07
Carbon-graphite (resin-filled)	Tungsten carbide	0.07
Carbon-graphite (resin-filled)	Tungsten carbide	0.07
Carbon-graphite (resin-filled)	Silicon carbide	0.02
Carbon-graphite (resin-filled)	Silicon carbide (converted carbon)	0.015
Silicon carbide	Tungsten carbide	0.02
Silicon carbide (converted carbon)	Silicon carbide (converted carbon)	0.05
Silicon carbide	Silicon carbide	0.02
Tungsten carbide	Tungsten carbide	0.08

[a] For oil, the friction coefficient is 25–50% higher.
[b] For lower *PV* values, the coefficient of friction is 10–20% lower. If hydropads or spiral grooves are used on the face seal, the friction coefficient can be reduced by a factor of 1/3.
Source: Johnson and Schoenherr, 1980. Reproduced with permission of ASME.

vent air or any vapor that may have formed away from the seal rings as it exits the chamber (Figure 16.22).

A variety of flush system standards are provided by the American Petroleum Institute (API) for pump–shaft sealing systems (API 682). For example, API Plan 02 is used for dead-ended seals that do not use flush fluid circulation (Figure 16.23). Instead, these pump seals typically use flow enhancers and taper the seal chamber to enhance flow. Nevertheless, a dead-end seal is primarily used for low pressure, low temperature, and low speeds. In the chemical industry

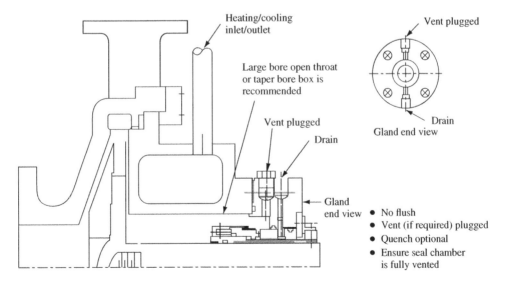

Figure 16.22 Flush Plan 02. *Source*: Buck and Gabriel, 2007a. Reproduced with permission of Cahaba Media Groups.

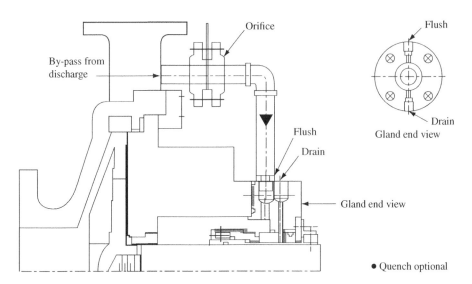

By-pass from discharge

Orifice

Flush

Drain

Gland end view

Flush

Drain

Gland end view

Flush

• Quench optional

Figure 16.23   Seal Flush Plan 11. *Source*: Buck and Gabriel, 2007b. Reproduced with permission of Cahaba Media Groups.

they are used for service involving nonvolatile fluids, often water. In these applications, heat generated at the interface is transferred to the fluid in the chamber, to the metal housing, and to the ambient until equilibrium is reached. This is often referred to as *heat soak* for which API 682 (2005) provides guidelines.

The most common is the flash Plan 11 where a bypass from the pump discharge is injected onto the seal faces to cool the rings. Plan 11 is considered to be the default seal flush plan for single seals. This flush plan calls for routing fluid from the pump discharge into the seal chamber by injection through an inlet port in the gland onto the seal faces to provide cooling and cleaning as well as for venting vapors out of the seal chamber (Figure 16.24). For high-head applications, Flash Plan 13 is often considered. It is also used for vertical pumps where cooling is provided by taking the process fluid from the seal chamber back to the pump suction (Figure 16.25). The interested reader is referred to Buck and Gabriel (2007a, b, c) for detailed information on different types of flush plans and recommendations for single and multiple seal arrangements based on many years of field experience.

Ideally, the heat generated at the interface between the mating and rotating ring is dissipated into the flush fluid. The flow in the chamber may be laminar or turbulent depending on the type of the fluid and the operating conditions. Accurate prediction of the flow and heat transfer characteristics within the seal chamber requires analysis by computational fluid dynamics (CFD) or a similar computational tools capable of taking into account the specific chamber design, fluid properties, operating conditions, and flush rate. The flow induced by primary ring rotation, injection rate and pressure can be either laminar or turbulent, depending on the speed and fluid properties. Recent publications provide the flow pattern for laminar flows in a seal with relatively viscous fluid, and turbulent flow for less viscous fluids such as water (Luan and Khonsari, 2006, 2007a, 2007b). Because of its high specific heat capacity, water is an ideal flush fluid if the design allows its use.

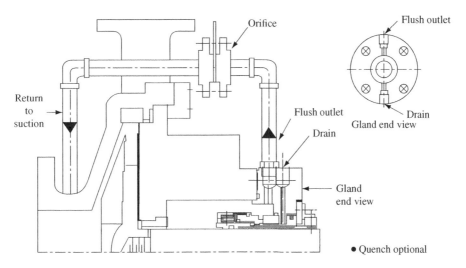

Figure 16.24   Seal Flush Plan 13. *Source*: Buck and Gabriel, 2007b. Reproduced with permission of Cahaba Media Groups.

The amount of coolant required is related to the *PV* factor as well as the friction coefficient. For minimum flush rate the manufacturer's specifications should be consulted. If the flow rate is less than the minimum, then the seal life can be negatively impacted. As such, the seal may not be suitable for the system pressure and should be derated (Buck, 1999). According to Buck, the minimum flush rate can be based on a temperature rise of $\Delta T = 30\,°\mathrm{F}$ for oils, 15 °F for water, and 5 °F for volatile hydrocarbons.

Seal Face Temperature

Accurate prediction of the seal face temperature is complicated by various factors in addition to unknowns associated with hydrodynamic and asperity interaction between the mating and

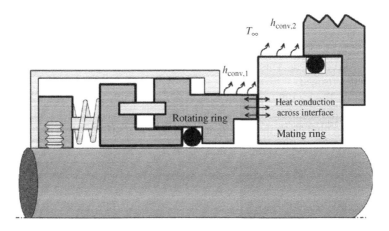

Figure 16.25   Heat transfer in mechanical seal.

Table 16.3 Typical values for convective heat transfer coefficient based on 10.16 cm (4 in.) seal OD running at 3600 rpm

Fluid	Convective heat transfer coefficient, h_{conv} (Btu/h ft °F, W/m² K)	Allowable temperature rise, ΔT (°F, °C)
Water	4800 (27 256)	15 (8.3)
Propane	1300 (7382)	5 (2.8)
Lube oil	500 (2839)	30 (16.6)
Air	15 (85)	—

Source: Buck, 1999. Reproduced with permission of John Wiley & Sons.

primary rings. The frictional heat generated at the interface between the rings is conducted into the mating and primary rings according to the material properties of the rings, often described by a so-called heat partitioning factor. As shown in Figure 16.25, heat conducted into the rings is also lost from the rings to the flush fluid by convection.

To determine the face temperature, convection heat transfer should be coupled to analysis of the flow in a chamber, which is beyond the scope of this chapter. In reality, the heat convective coefficient on the rotating ring is not the same as that of the mating ring. However, many simple analyses often assume a single, average heat transfer coefficient for both rings. In what follows, we first present a relatively simple procedure for estimating temperature rise across the seal face based on an approach proposed by Buck (1989, 1999). Next, we present a method for estimating the interface temperature as well as the heat partitioning factor that takes into account the ring shape designs.

Based on field experience and results of experiments with a seal of OD = 10.16 cm (4 in.) diameter operating at 3600 rpm, Buck gives representative convective heat transfer coefficients, h_{conv}, for several applications in Table 16.3. Also shown is the allowable temperature rise for the flush fluid, based on which the minimum rate of flush can be estimated. This information is meant to provide basic guidelines, and the manufacturer data sheets should be consulted for more appropriate minimum flush rates.

Assuming that all the heat is carried away by the flush supplied to the seal at the rate of $\dot{m}$, heat loss and allowable temperature rise can be described by the following equation:

$$Q = \dot{m}\, C_p\, \Delta T \tag{16.38}$$

where C_p is the specific heat. This equation is useful in estimating the flush rate $\dot{m}$. However, Q must first be determined.

To determine Q, Buck (1999) proposes treating the primary ring or the mating ring as a simple rectangular fin, with heat flux generated at the interface entering the fin and convected into the surrounding fluid along the fin length. Assuming that the end tip of the fin is insulated, the following expression for the heat transfer can be derived:

$$Q = \sqrt{h_{conv}pkA}\ \tanh(mL)\,(T - T_\infty) \tag{16.39}$$

where k is the thermal conductivity, p the seal circumference, A the cross-sectional area of the rectangular fin, and $mL = \sqrt{(h_{conv}p/kA)}\, L$. Note that T varies along the length of the fin, being highest at the face exposed to the heat flux. The mL product can be put in terms of the seal width, w, as follows: $mL = (L/w)\sqrt{h_{conv}w/k}$.

Buck applies the above idealization to both the primary and mating rings in contact and suggests the following expression for the total heat transfer from both rings exposed to a fluid medium with the convective heat transfer coefficient h_{conv}:

$$Q = \pi D \sqrt{h_{conv}} w(\sqrt{k_p} + \sqrt{k_m})(T - T_\infty) \tag{16.40}$$

where k_p and k_m are the thermal conductivities of the primary and mating rings, respectively. The seal width, w, can be taken as the difference between the OD and ID of the nose primary ring divided by 2. In the above equation, the temperature difference represents the *average temperature rise* at the seal face. Note that at the interface, $T = T_{face}$ and heat generation $Q = E_p$ as given in Equation (16.35). Therefore, the average face temperature can be estimated using the following equation:

$$T_{face} = \frac{E_p}{\pi D_m \sqrt{h_{conv}} w(\sqrt{k_p} + \sqrt{k_m})} + T_\infty \tag{16.41}$$

where D_m is the mean diameter, w is the face width, and the fluid temperature in the chamber around the seal rings is represented by T_∞. Note that while L does not explicity appear in Equation (16.41), the relationship of $(L/w)\sqrt{h_{conv}w/k} \geq 2$ must be satisfied for the seal rings. The procedure is illustrated in Example 16.3.

Example 16.3 An inside mechanical seal using a carbon graphite primary ring and tungsten carbide mating ring is in a system with $P_1 = 500$ psi water service pressure at 100 °F with the shaft rotating at $N = 3600$ rpm. The primary ring has the following specifications: $D_p = 2.6$ in., $d_p = 2.0$ in., and shaft shoulder diameter $B = 2.2$ in. The seal head spring pressure is 28 psi.

(a) Determine the balance ratio and seal face pressure, assuming linear pressure profile.
(b) Determine the PV factor.
(c) Determine the power loss.
(d) Estimate the opening force.
(e) Recommend a flush rate and inlet supply temperature.
(f) Estimate the average seal face temperature and comment if there is danger of boiling across the face.

(a) Using Equation (16.27), the balance ratio is

$$b = \frac{(D_p^2 - B^2)}{(D_p^2 - d_p^2)} = \frac{(2.6^2 - 2.2^2)}{(2.6^2 - 2.0^2)} = 0.70$$

Using a typical pressure gradient factor $k = 0.5$, the seal face pressure is obtained from Equation (16.32):

$$P_f = \Delta P(b - k) + P_{sp} = (500 - 0)(0.70 - 0.5) + 28 = 128 \, \text{psi}$$

Note that the face pressure is roughly four times smaller than the water service pressure.

(b) To determine the PV factor, we first compute the mean velocity, V_m:

$$\omega = \frac{2\pi N}{60} = \frac{2\pi(3600)}{60} = 377 \ 1/s$$

$$D_m = \frac{D_p + d_p}{2} = \frac{2.6 + 2.0}{2} = 2.3 \ \text{in.}$$

$$V_m = D_m\omega/2 = 2.3(377)/2 = 433.5 \ \text{in./s}$$

Using Equation (16.33), the PV factor is

$$PV = P_f V_m = (128)(433.5) = 5.54 \times 10^4 \ \text{psi .in./s} = 2.77 \times 10^5 \ \text{psi .ft/min}$$

(c) To determine the power loss, the face area must be calculated:

$$A_f = \pi D_m \left(\frac{D_p - d_p}{2}\right) = \pi(2.3) \left(\frac{2.6 - 2.0}{2}\right) = 2.168$$

From Table 16.2, the friction coefficient is estimated to be 0.07. The power loss is

$$E_p = f P_f A_f V_m = 0.07(2.168)(2.82 \times 10^5) = 4.28 \times 10^4 \ \text{ft lbf/min} = 1.30 \ \text{hp}$$

Note that if the mating ring were made of silicon carbide, then $f = 0.02$ and the power loss would reduce to only 0.37 hp.

(d) The opening force can be estimated using the face pressure:

$$F_o = k A_f \Delta P = (0.5)(2.168)(500) = 542 \ \text{lbf}$$

(e) To determine the flush rate, using Table 16.3, the recommended temperature rise should be below 15 °F for applications involving water. The flush flow rate is

$$\dot{m} = \frac{E_p}{C_p . \Delta T} = \frac{(1.274 \ \text{hp})(2544.4 \ \text{Btu/(hp h)})}{1 \ \text{Btu/lbm °F} (15 \ \text{°F})} = 220.51 \ \text{lbm/h} = 0.444 \ \text{gpm}$$

Therefore, roughly 0.5 gpm is recommended for this application. This means that for water temperature in the seal chamber to remain around 100 °F in the vicinity of the seal rings, the water should be injected into the seal at 92.5 °F and exit at 107.5 °F, to satisfy the prescribed 15 °F temperature rise.

(f) The average temperature rise is given by Equation (16.41). However, we must first check the following conditions. For a rotating ring,

$$\frac{L}{w} \sqrt{\frac{h_{conv} w}{k}} = \frac{1}{0.3} \sqrt{\frac{4800(0.3/12)}{8}} = 12.91 > 2.0$$

For a stationary ring,

$$\frac{L}{w}\sqrt{\frac{h_{conv}w}{k}} = \frac{2}{0.3}\sqrt{\frac{4800(0.3/12)}{45}} = 10.89 > 2.0$$

$$T_{face} = \frac{E_p}{\pi D_m\sqrt{h_{conv}w}(\sqrt{k_p} + \sqrt{k_m})} + T_\infty$$

$$= \frac{3300\ Btu/h}{\pi(2.3/12)\sqrt{(4800)(0.3/12)}(\sqrt{8} + \sqrt{45})} + 100 = 153\,^\circ F$$

where average thermal conductivity values from Table 16.3 were used. Note that this average temperature is below the boiling temperature of water and there should not be a concern of flashing across the face.

Extension of Fin Analysis

The concept of fin analysis can be further extended to include various seal-ring geometry effects as well as determining heat partitioning between the seal rings (Luan and Khonsari, 2006). The efficiency, η, of a fin is defined as the ratio of actual heat transferred and the heat which would be transferred if the entire fin area were at base temperature. Assuming that one end of the fin is at base temperature and the other end is insulated, the efficiency is

$$\eta = \frac{\tanh(mL)}{mL} \tag{16.42}$$

where $m = \sqrt{h_{conv}P/kA_c}$, which for a cylindrical seal-ring shape with $A_c = \pi(r_o^2 - r_i^2)$ becomes

$$mL = \left(\frac{L}{W}\right)\sqrt{\frac{2hr_oW}{k(r_o + r_i)}} \tag{16.43}$$

Parameter W represents the difference of the inner and outer radius of the cylindrical fin. Note that the efficiency of a seal fin varies with length, width, thermal conductivity, and the convective heat transfer coefficient. In applications that involve low values of mL, the fin efficiency is high.

An alternative definition of fin efficiency is

$$\eta = \frac{Q}{h_{conv}A_h(T_{face} - T_\infty)} \tag{16.44}$$

where Q is the heat flux at the base of the seal fin, and A_h is the heat convection area. This equation is valid for each of the rings, or fins. The amount of heat conducted into each of these rings can be calculated by determining an approximate heat partitioning factor. Let η_1 and η_2 represent the efficiency of the rotating ring and stationary ring respectively:

$$Q_1 = \eta_1 h_{\text{conv1}} A_{\text{h1}} (T_{\text{face}} - T_\infty)$$ (16.45)

$$Q_2 = \eta_2 h_{\text{conv2}} A_{\text{h2}} (T_{\text{face}} - T_\infty)$$ (16.46)

The partitioning factor can be estimated using Equations (16.45) and (16.46):

$$\gamma = \frac{Q_1}{Q_2} = \frac{\eta_1 h_{\text{conv1}} A_{\text{h1}}}{\eta_2 h_{\text{conv2}} A_{\text{h2}}}$$ (16.47)

The total frictional heat is

$$E_p = PVA_f f = Q_1 + Q_2$$ (16.48)

where P is pressure, V represents the rotation speed, A_f is the friction area, and f is the friction coefficient. Using the definition of the partitioning factor γ, Equation (16.48) can be written as follows:

$$E_p = Q_2 + \gamma Q_2$$ (16.49)

where

$$Q_2 = \frac{E_p}{1+\gamma}, \qquad Q_1 = \frac{\gamma E_p}{1+\gamma}$$ (16.50)

Using the concept of fin efficiency, the total frictional heat is given by

$$E_p = Q_1 + Q_2 = (\eta_1 h_{\text{conv1}} A_{\text{h1}} + \eta_2 h_{\text{conv2}} A_{\text{h2}})(T_{\text{face}} - T_\infty)$$ (16.51)

Solving Equation (16.51) for the average base temperature T_b yields

$$T_{\text{face}} = \frac{E_p}{\eta_1 h_{\text{conv1}} A_{\text{h1}} + \eta_2 h_{\text{conv2}} A_{\text{h2}}} + T_\infty$$ (16.52)

To determine the surface temperature of the rings the heat transferred by convection must be considered. Heat loss from the primary and the mating ring can be evaluated as follows:

$$Q_1 = h_{\text{conv1}} A_{\text{h1}} (T_{\text{s1}} - T_\infty)$$ (16.53)

$$Q_2 = h_{\text{conv2}} A_{\text{h2}} (T_{\text{s2}} - T_\infty)$$ (16.54)

where T_{s1} and T_{s2} are the average temperature at the outer surface of the primary and mating ring, respectively. The heat transferred by the rings becomes

$$Q_1 = \frac{\gamma E_p}{1+\gamma} = h_{\text{conv1}} A_{\text{h1}} (T_{\text{s1}} - T_\infty)$$ (16.55)

$$Q_2 = \frac{E_p}{1+\gamma} = h_{\text{conv2}} A_{\text{h2}} (T_{\text{s2}} - T_\infty)$$ (16.56)

Solving for the temperatures yields

$$T_{s1} = \frac{\gamma E_p}{h_{conv1} A_{h1}(1+\gamma)} + T_\infty \qquad (16.57)$$

$$T_{s2} = \frac{E_p}{h_{conv2} A_{h2}(1+\gamma)} + T_\infty \qquad (16.58)$$

Basic Seal Ring Shapes

While seal rings come in many possible shapes, four different shapes can characterize the wide range of ring configurations used in industry. Figures 16.26–16.29 show the heat transfer efficiencies for shapes 1–4. Note that the cases considered in shapes 2–4 all have $l = W = W'$, where $W = r_o - r_i$, is the radial width of the seal face where heat flux is applied to, and length L represents the axial length of the wetted outer surface of the rotating or stationary ring. Shape 2 is primarily for modeling the mating ring since its radius is normally greater than that of the rotating ring. Shape 1, and shapes 3 and 4 with typical 'wear nose,' pertain to the rotating ring.

The following examples illustrate the procedure for utilizing the fin efficiency concept.

Example 16.4 Estimate the face temperature and the heat partitioning factor for the seal specifications given in Example 16.3.

For the rotating ring with $r_i = 1.0$ in. and $r_o = 1.3$ in., shape 1 can be applied.

$$\frac{L_1}{W} = 3.33$$

$$mL_1 = \left(\frac{L_1}{W}\right)\sqrt{\frac{2h_{conv} r_o W}{k_1(r_o + r_i)}} = 13.71$$

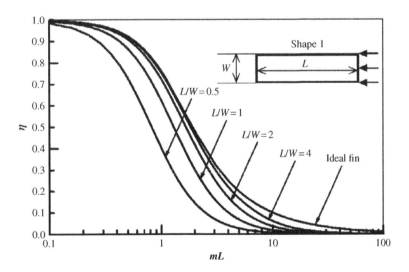

Figure 16.26 Fin efficiency for shape 1.

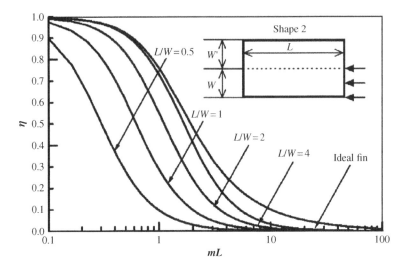

Figure 16.27   Fin efficiency for shape 2.

Using the values of mL_1 and L_1/W, the heat transfer efficiency can be calculated using the heat transfer efficiency for shape 1 (Figure 16.26):

$$\eta_1|_{L/W=2} \cong 0.02 \text{ and } \eta_1|_{L/W=4} \cong 0.036$$
$$\eta_1|_{L/W=3.33} \cong 0.031$$

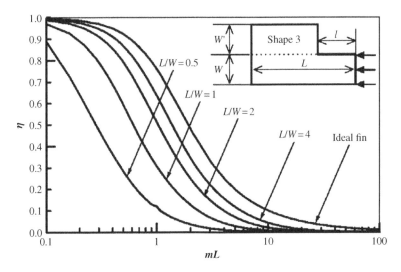

Figure 16.28   Fin efficiency for shape 3.

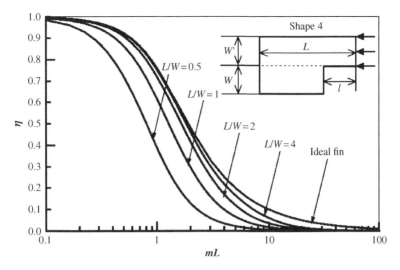

Figure 16.29   Fin efficiency for shape 4.

Next, consider the stationary ring which can be represented by shape 1. Similarly,

$$\frac{L_2}{W} = 4.0$$

$$mL_2 = \left(\frac{L_2}{W}\right)\sqrt{\frac{2h_2 r_o W}{k_2(r_o + r_i)}} = 6.94$$

Using the values of mL_2 and L_2/W, the heat transfer efficiency can be calculated from the figure of heat transfer efficiency of shape 1 as

$$\eta_2 \cong 0.11$$

The heat convection areas for the rotating and stationary rings are

$$A_{h1} = 8.17 \, \text{in.}^2$$
$$A_{h2} = 9.80 \, \text{in.}^2$$

The heat partitioning factor is

$$\gamma = \frac{\eta_1 h_1 A_{h1}}{\eta_2 h_2 A_{h2}} = 0.235$$

From Example 16.3, the total heat generation is $E_p = 3300$ BTU/h. Using Equation (16.52), the average temperature at the contact face between the rings can be estimated as follows:

$$T_{face} = \frac{E_p}{\eta_1 h_1 A_{h1} + \eta_2 h_2 A_{h2}} + T_\infty = 174 \, {}^{\circ}\text{F}$$

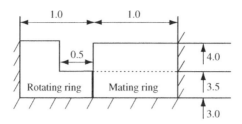

Figure 16.30   Example 16.5.

Example 16.5   Estimate the heat partition factor for a seal using a rotating and stationary ring pair with dimensions shown in Figure 16.30. The process fluid is water at 300 K and the effective heat convection coefficients for the two rings are: $h_1 = h_2 = 1.0 \times 10^4$ W/m² K. The rotating ring is made of carbon graphite with heat conductivity $k_1 = 8.6$ W/m K and the stationary ring is silicon carbide with heat conductivity $k_2 = 138.5$ W/m K. Determine:

(a) Heat partitioning factor
(b) Average rotating and mating surface temperature.

In this example, shape 2 should be applied to the mating ring and shape 3 to the rotating one.
For the rotating ring with $r_i = 3.0$ cm and $r_o = 3.5$ cm,

$$\frac{L}{W} = \frac{L}{r_o - r_i} = 2$$

Using Equation (16.43),

$$mL = \left(\frac{L}{W}\right)\sqrt{\frac{2h_1 r_o W}{k(r_o + r_i)}} = 5.0$$

Referring to Figure 16.28, the fin efficiency is $\eta_1 \cong 0.072$.
Next consider the mating ring which can be represented by shape 2. Similarly,

$$\frac{L}{W} = \frac{L}{r_o - r_i} = 2$$

Note that the width W refers to the rubbing or contact face of the two rings.

$$mL = \left(\frac{L}{W}\right)\sqrt{\frac{2h_2 r_o W}{k(r_o + r_i)}} = 1.25$$

Referring to Figure 16.27, the fin efficiency is $\eta_2 \cong 0.43$.
The heat convection areas for the rotating and stationary rings are

$$A_{h1} = 35.33 \, \text{cm}^2 \quad \text{and} \quad A_{h2} = 36.90 \, \text{cm}^2$$

Using Equation (16.47), the heat partition factor is

$$\gamma = \frac{q_1}{q_2} = \frac{\eta_1 h_1 A_{h1}}{\eta_2 h_2 A_{h2}} = 0.1603$$

$$E_p = QA_f = 1.18 \times 10^5 \times \pi((4 \times 10^2)^2 - (3 \times 10^2)^2) = 259.50\,\text{W}$$

Using Equations (16.57) and (16.58), the average temperatures at the outer surface of the rotating ring and rotating rings are:

$$T_{s1} = \frac{\gamma E_p}{h_1 A_{h1}(1+\gamma)} + T_\infty = \frac{0.164 \times 259.50}{1 \times 10^4 \times 25.13 \times 10^4 \times (1+0.164)} + 300 = 301\,\text{K} = 28\,°\text{C}$$

$$T_{s2} = \frac{E_p}{h_2 A_{h2}(1+\gamma)} + T_\infty = \frac{259.50}{1 \times 10^4 \times 37.70 \times 10^4 \times (1+0.164)} + 300 = 306\,\text{K} = 33\,°\text{C}$$

Using Equation (16.52), the average interface temperature is estimated as follows:

$$T_{face} = \frac{E_p}{\eta_1 h_{conv1} A_{h1} + \eta_2 h_{conv2} A_{h2}} + T_\infty = 314\,\text{K} = 41\,°\text{C}$$

Problems

16.1 Consider the flow on an incompressible flow in a visco seal. The fluid is assumed incompressible and Newtonian and the flow is steady and viscous forces dominate such that the flow is laminar. The Navier-Stokes equation subject to normal thin film flow assumption reduced to the following:

$$\frac{\partial^2 u}{\partial z^2} = \frac{1}{\mu}\frac{\partial P}{\partial \xi}$$

$$\frac{\partial^2 v}{\partial z^2} = \frac{1}{\mu}\frac{\partial P}{\partial \eta}$$

where $\xi = x\cos\alpha + y\sin\alpha$ and $\eta = y\cos\alpha - x\sin\alpha$.

Derive the appropriate expression for the flow velocity profile in the groove and along the land of a visco seal.

16.2 Determine the laminar and turbulent seal coefficient for a range of critical Reynolds numbers ranging from 20 to 10^4 for a visco seal with the following specifications:

$$\alpha = 3.86; \quad t = 0.2586; \quad \beta = 7.31; \quad \gamma = 0.5; \quad D = 1.243\,\text{in.}; \quad c = 0.00029\,\text{in.};$$

and $Re_c = 400$

16.3 Estimate the heat partitioning coefficient and average temperature at the outer surface of cylindrically shaped mating and rotating rings without a wear nose. The process fluid is water at 300 K and the rotational speed is 1800 rpm. The heat generation rate is assumed to be $q = 1.18 \times 10^5\,\text{W/m}^2$ and the heat convection coefficients for the two rings are

$h_1 = h_2 = 1.0 \times 10^4$ W/m^2 K. The thermal conductivity of the rotating and mating rings are $k_1 = 8.6$ W/m K and $k_2 = 138.5$ W/m K, respectively. The pressure acting on the seal ring is $P = 1.79 \times 10^5$ N/m^2 and $f = 0.1$.

16.4 Seal rings of a mechanical seal take on a tapered angle β with outer and inner pressures P_o and P_i (assume atmospheric) as shown in Figure 16.31. The film thickness, $h = h(r)$, can be described as

$$h(r) = (h_i - r_i\beta) + \beta r$$

(a) Starting from the axis-symmetric Reynolds of the form

$$\frac{d}{dr}\left(rh^3\frac{dP}{dr}\right) = 0,$$

show that the pressure distribution can be described by the following expression:

$$P(r) = C_1\left(-\frac{1}{h_i - r_i\beta}\ln\left(\frac{h}{r}\right) + \frac{1}{h} + \frac{h_i - r_i\beta}{2}\frac{1}{h^2}\right) + C_2$$

where C_1 and C_2 are given by

$$C_1 = \frac{P_o}{\left(\frac{1}{h_i - r_i\beta}\ln\frac{r_o h_i}{r_i h_o} + \left(\frac{1}{h_o} - \frac{1}{h_i}\right) + \frac{h_i - r_i\beta}{2}\left(\frac{1}{h_o^2} - \frac{1}{h_i^2}\right)\right)}$$

$$C_2 = -C_1\left(-\frac{1}{h_i - r_i\beta}\ln\frac{h_i}{r_i} + \frac{1}{h_i} + \frac{h_i - r_i\beta}{2}\frac{1}{h_i^2}\right)$$

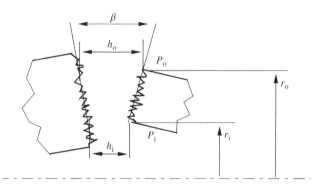

Figure 16.31   Problem 16.4: Schematic of seal rings with a tapered angle.

(b) Show that for a narrow seal ring,

$$P(r) = \frac{P_\mathrm{o}}{\dfrac{1}{h_\mathrm{o}^2} - \dfrac{1}{h_\mathrm{i}^2}} \left(\frac{1}{h^2} - \frac{1}{h_\mathrm{i}^2} \right)$$

(c) Derive an expression for the opening force.

References

Barber, J.R. 1969. 'Thermoelastic Instabilities in the Sliding of Conforming Solids,' *Proceedings of the Royal Society of London, Series A*, Vol. 312, pp. 381–394.

Buchter, H.H. 1979. *Industrial Sealing Technology*, John Wiley & Sons, Inc., New York.

Buck, G. 1999. 'Estimating Heat Generation, Face Temperature and Flush Rate for Mechanical Seals,' *Proceedings of Pump Users Expo '99*, pp. 167–172.

Buck, G. 2001. 'Materials for Seal Faces,' *Pumps and Systems*, pp. 22–26.

Buck, G. and Gabriel, R. 2007a. 'Circulation Systems for Single and Multiple Seal Arrangements,' *Pumps and Systems*, May, pp. 43–53.

Buck, G. and Gabriel, R. 2007b. 'Circulation Systems for Single and Multiple Seal Arrangements – Part Two,' *Pumps and Systems*, June, pp. 72–80.

Buck, G. and Gabriel, R. 2007c. 'Circulation Systems for Single and Multiple Seal Arrangements – Part Three,' *Pumps and Systems*, July, pp. 36–53.

Buck, G.S. 1989. 'Heat Transfer in Mechanical Seals,' *Proceedings, 6th Int. Pump User Symposium*, College Station, Tex., pp. 9–15.

Burton, R.A., Nerlikar, V., and Kilaparti, R. 1973. 'Thermoelastic Instability in a Seal-Like Configuration,' *Wear*, Vol. 24, pp. 177–188.

Chupp, R.E., Prior, R.J., Loewenthal, R.G., and Menendez, R.P. 1997. 'Advanced Seal Development for Large Industrial Gas Turbines,' Paper No. AIAA 97-2731, presented at 33rd AIAA/ASME/SAE/ASEE Joint Propulsion Conference & Exhibition, Seattle, Wa, 6–9, July.

Dinc, S., Demiroglu, M., Turnquist, N., Mortzheim, J. Goetze, G., Maupin, J. Hopkins, J., Wolfe, C., and Florin, M. 2002. 'Fundamental Design Issues of Brush Seals for Industrial Applications,' *ASME Journal of Turbomachinery*, Vol. 294, pp. 293–300.

Egli, A. 1935. 'The Leakage of Steam through Labyrinth Seals,' *Transactions ASME*, Vol. 57, pp. 115–122.

Etsion, I. 2005. 'State of the Art in Laser Surface Texturing,' *ASME Journal of Tribology*, Vol. 127, pp. 248–253.

Etsion, I. and Kligerman Y. 1999. 'Analytical and Experimental Investigation of Laser-Textured Mechanical Seal Faces,' *Tribology Transactions*, Vol. 42, pp. 511–516.

Etsion, I., Halperin, G., Brizmer, V., and Kligerman, Y. 2004. 'Experimental Investigation of Laser Surface Textured Parallel Thrust Bearings,' *Tribology Letters*, Vol. 17, pp. 295–300.

Fesanghary, M. and Khonsari, M.M. 2011. 'On the Shape Optimization of Self-Adaptive Grooves,' *STLE Tribology Transactions*, Vol. 54, pp. 256–264.

Fesanghary, M. and Khonsari, M.M. 2012. 'Topological and Shape Optimization of Thrust Bearings for Enhanced Load Carrying Capacity,' *Tribology International*, Vol. 53, pp. 12–2.

Fesanghary, M. Khonsari, M.M. 2013. 'On the Optimum Groove Shapes for Load-Carrying Capacity Enhancement in Parallel Flat Surface Bearings: Theory and Experiment,' *Tribology International*, Vol. 67, pp. 254–262.

Hendricks, R.C., Wilson, J., Wu. T., and Flower, F. 1998. 'Two-Way Brush Seals Catch a Wave,' *Mechanical Engineering, ASME*, Nov.

Horve, L.A. 1992. 'Understanding the Sealing Mechanism of the Radial Lip Seal for Rotating Shafts,' *Proceedings, 13th BHRG International Conference on Fluid Sealing*, B. S. Nau (ed.), Kluwer, Dordrecht, The Netherlands, pp. 5–19.

Jagger, E.T. 1957. 'Rotary Shaft Seals: the Sealing Mechanism of Synthetic Rubber Seals Running at Atmospheric Pressure,' *Proceedings of the Institution of Mechanical Engineers*, Vol. 171, pp. 597–616.

Jang, J.Y. and Khonsari, M.M. 2000. 'Thermoelastic Instability with Consideration of Surface Roughness and Hydrodynamic Lubrication,' *ASME Journal of Tribology*, Vol. 22, pp. 725–732.

Jang, J.Y. and Khonsari, M.M. 2003. 'A Generalized Thermoelastic Instability Analysis,' *Proceedings of the Royal Society of London, Series A*, Vol. 459, pp. 309–329.

Jang, J.Y. and Khonsari, M.M. 2004a. 'Thermoelastic Instability of Two-Conductor Friction System Including Surface Roughness,' *ASME Journal of Applied Mechanics*, Vol. 71, pp. 57–68.

Jang, J.Y. and Khonsari, M.M. 2004b. 'On the Growth Rate of Thermoelastic Instability,' *ASME Journal of Tribology*, Vol. 126, pp. 50–55.

Jang, J.Y. and Khonsari, M.M. 2013. 'Thermoelastic Instability in Mechanical Systems with Provision for Surface Roughness,' in *Encyclopedia of Elasticity*, Y.W. Chung and Q.J. Wang (eds.), Springer Science, New York.

Johnson, R.L. and Schoenherr, K. 1980. 'Seal Wear,' *Wear Control Handbook*, M.B. Peterson and W.O. Winer (eds). ASME Publication, New York.

Khonsari, M.M. and Somanchi, A.K. 2005. 'Mechanical Seal Having a Double-Tier Mating Ring,' US Patent number US6942219 B2.

Khonsari, M.M. and Somanchi, A.K. 2007. 'Mechanical Seal Having a Single-Piece, Perforated Mating Ring,' US Patent number US7252291 B2.

Khonsari, M.M. and Xiao, N. 2017. 'Mechanical Seal with Textured Sidewall,' US Patent 9568106B2

Khonsari, M.M. and Xiao, N. 2017. 'Cooled Seal,' US Patent 9,599,228B2

Kovalchenko, A., Ajayi, O., Erdemir, A., Fenske, G., and Etsion, I. 2005. 'The Effect of Laser Surface Texturing on Transitions in Lubrication Regimes During Unidirectional Sliding Contact,' *Tribology International*, Vol. 38, pp. 219–225.

Lebeck, A.O. 1980. 'The Effect of Ring Defection and Heat Transfer on the Thermoelastic Instability of Rotating Face Seals,' *Wear*, Vol. 59, pp. 121–133.

Lebeck, A.O. 1991. *Principles and Design of Mechanical Face Seals*. John Wiley & Sons, Inc., New York.

Luan, Z. and Khonsari, M.M. 2006. 'Numerical Simulations of the Flow Field around the Rings of Mechanical Seals,' *ASME Journal of Tribology*, Vol. 128, pp. 559–565.

Luan, Z. and Khonsari, M.M. 2007a. 'Computational Fluid Dynamic Analysis of Turbulent Flow within a Mechanical-Face Seal Chamber,' *ASME Journal of Tribology*, Vol. 129, pp. 120–128.

Luan, Z. and Khonsari, M.M. 2007b. 'Heat Transfer Analysis in Mechanical Seals Using Fin Theory,' *IMechE Journal of Engineering Tribology*, Vol. 221, pp. 717–725.

Ludwig, L.P., Strom, T.N., and Allen G.P. 1966. 'Gas Ingestion and Sealing Capacity of Helical Groove Fluid Film Seal (Viscoseal) Using Sodium and Water as Sealed Fluids,' NASA Lewis Research Center, TN D-3348, Cleveland, Ohio.

Müller, H.K. and Nau, B.S. 1998. *Fluid Sealing Technology*. Marcel Dekker, New York.

McGrew, J.M. and McHugh, J.D. 1964. 'Analysis and Test of the Screw Seal in Laminar and Turbulent Operation,' *ASME Journal of Basic Engineering*, Paper 64-LUBS-7, ASME.

Nau, B.S. 1985. 'Rotary Mechanical Seals in Process Duties: an Assessment of the State of the Art,' *Proceedings of the Institution of Mechanical Engineers*. Vol. 99 (A1), p. 17.

Netzel, J. 1999. 'Panel Discussion: Asset Management and Methods to Improve Equipment Reliability in Pumps and Other Turbomachinery,' presented at the STLE/ASME Tribology Conference in Florida.

Netzel, J.P. 1980. 'Observation of Thermoelastic Instability in Mechanical Face Seals,' *Wear*, Vol. 59, pp. 135–148.

Peng, Z-C., Khonsari, M.M., and Pascovici, M.D. 2003. 'On the Thermoelastic Instability of a Thin-Film-Lubricated Sliding Contact: A Closed-Form Solution,' *Proceedings of the. Institution of Mechanical Engineers. Part J: Journal of Engineering Tribology*, Vol. 217, pp. 197–204.

Qiu, Y. and Khonsari, M.M. 2009. 'On the Prediction of Cavitation in Dimples Using a Mass-Conservative Algorithm,' *ASME Journal of Tribology*, Vol. 131, 041702-1:1–11.

Qiu, Y. and Khonsari, M.M. 2011a. 'Performance Analysis of Full-Film Textured Surfaces with Consideration of Roughness Effects,' *ASME Journal of Tribology*, Vol. 133, 021704:1–10.

Qiu, Y. and Khonsari, M.M. 2011b. 'Experimental Investigation of Tribological Performance of Laser Textured Stainless Steel Rings,' *Tribology International*, Vol. 44, pp. 635–644.

Qiu, Y. and Khonsari, M.M. 2011c. 'Investigation of Tribological Behavior of Annular Rings with Grooves,' *Tribology International*, Vol. 44, pp. 1610–1619.

Qiu, Y. and Khonsari, M.M. 2012. 'Thermohydrodynamic Analysis of Spiral Groove Mechanical Face Seals for Liquid Applications,' *ASME Journal of Tribology*, Vol. 134, 021703:1–11.

Sabini, E. 1999. 'Panel Discussion: Asset Management and Methods to Improve Equipment Reliability in Pumps and Other Turbomachinery,' presented at the STLE/ASME Tribology Conference in Florida.

Salant, R.F. 1996. 'Elastohydrodynamic Model of the Rotary Lip Seal,' *ASME Journal of Tribology*, Vol. 118, pp. 292–296.

Salant, R.F. 1999. 'Theory of Lubrication of Elastomeric Rotary Shaft Seals,' *IMechE Journal of Engineering Tribology*, Vol. 213, pp. 189–201.

Salant, R.F. and Rocke, A.H. 2004. 'Hydrodynamic Analysis of the Flow in a Rotary Lip Seal Using Flow Factors,' *ASME Journal of Tribology*, Vol. 126, pp. 156–161.

Schoenherr, K. 1965. 'Design Terminology Mechanical End Face Seals,' John Crane Mechanical Maintenance Training Center, Morton Grove, Ill.

Shen, C. and Khonsari, M.M. 2013. 'Effect of Dimple's Internal Structure on Hydrodynamic Lubrication,' *Tribology Letters*, Vol. 52, pp. 415–430.

Shen, C. and Khonsari, M.M. 2016. 'Texture Shape Optimization for Seal-Like Parallel Surfaces: Theory and Experiment,' *STLE Tribology Transactions*, Vol. 59, pp. 698–706.

Smith, M.I. and Fuller, D.D. 1956. 'Journal-Bearings Operations at Superlaminar Speeds,' *Transactions of ASME*, Vol. 78, pp. 469–474.

Stair, W.F. 1965. 'Analysis of Visco Seal. Part I – The Concentric Laminar Case,' NASA CR-285, Washington, DC.

Stair, W.F. 1983. 'Dynamic Seals,' *CRC Handbook*, Vol. II, E.R. Booser (ed.), pp. 581–622.

Stair, W.F. and Hale, R.H. 1966. 'Analysis of the Visco Seal. Part II – The Concentric Turbulent Case,' Contract report prepared for NASA, the University of Tennessee, Department of Mechanical Engineering, NsG-S87.

Wang, X. and Kato, K. 2003. 'Improving the Anti-seizure Ability of SiC Seal in Water with RIE Texturing,' *Tribology Letters*, Vol. 14, pp. 275–280.

Wang, X., Kato, K., Adachi, K., and Aizawa, K. 2001. 'The Effect of Laser Texturing of Sic Surface on the Critical Load for the Transition of Water Lubrication Mode from Hydrodynamic to Mixed.' *Tribology International*, Vol. 34, pp. 703–711.

Wheelock, E.A. 1981. 'High Pressure Radial Lip Seals for Rotary and Reciprocating Application,' *Lubrication Engineering*, Vol. 37, p. 332.

Winkler, A, Otschik, J., and Schicktanz, R. 2011. 'Mechanical Seal Assembly with Integrated Heat Transfer Unit,' US Patent Office, US2011/0169225 A1.

Xiao, N. and Khonsari, M.M. 2012. 'Thermal Performance of Mechanical Seals with Textured Side Wall,' *Tribology International*, Vol. 45, pp. 1–7.

Xiao, N. and Khonsari, M.M. 2013. 'A Review of Mechanical Seals Heat Transfer Augmentation Techniques,' *Journal of Patents on Mechanical Engineering*, Vol. 6, pp. 87–96.

Xiao, N. and Khonsari, M.M. 2014. 'Thermal Performance of Mechanical Face Seal with Built-In Heat Pipe,' *Institution of Mechanical Engineers Journal of Engineering Tribology*, Vol. 228, pp. 498–510.

Xiao, N. and Khonsari, M.M. 2015a. 'Improving Bearings Thermal and Tribological Performance with Built-In Heat Pipe,' *Tribology Letters*, Vol. 57, pp. 31–42.

Xiao, N. and Khonsari, M.M. 2015b. 'Recent Development in Heat Transfer Augmentation Design for Reducing Surface Temperature in a Lubrication System,' *Recent Patents on Mechanical Engineering*, Vol. 8, pp. 148–153.

17

Condition Monitoring and Failure Analysis

Proper use of regular monitoring and failure analysis goes far in providing improved designs, improved reliability, better operating practices, and more reliable lubrication. This chapter covers various means for monitoring bearing performance, identifying bearing failures, and correction methodologies.

17.1 Installation Analysis

Bearing troubles often result from improper maintenance; contamination involving dirt, water, and corrosive atmospheres; or excessive load and vibration. Troubles associated with improper installation are not discussed in this chapter since their assessments call for system inspections that include many miscellaneous issues such as problems with structures and foundations.

Poor maintenance itself frequently causes bearing failures. Troubles stemming from poor maintenance normally do not show up at a single position or on a single machine, but in a rash of difficulties throughout the area. Inaccessibility of drain plugs and other fixtures may encourage poor lubrication practices. Poor storage facilities for lubricants may result in their contamination with dirt and water. Plant-wide use of inexpensive uninhibited oils—in place of high-quality lubricants for applications that require specific properties—may result in sludging, acidity, and wear in heavy-duty machinery.

Ambient conditions about the installation also merit careful consideration. High temperature raises concerns with oil oxidation and sludging, reducing the oil life by one-half for each $10\,°C$ ($18\,°F$) increase in temperature (see Chapter 2). Water and its condensation in very humid environments often promote corrosion, break down vital additives, and cause oil sludging. Where water cannot be avoided, a suitable oil with rust-inhibiting capability should be selected and means provided in the oil reservoir to remove the water with a centrifuge, by evaporation, or by filtration.

Corrosive gases and mists in the atmosphere containing sulfur, chlorine and other chemicals are also detrimental. They should either be eliminated with proper ventilation or appropriate oil and bearing materials selected that can resist the contaminated atmosphere.

Applied Tribology: Bearing Design and Lubrication, Third Edition. Michael M. Khonsari and E. Richard Booser.
© 2017 John Wiley & Sons Ltd. Published 2017 by John Wiley & Sons Ltd.
Companion Website: www.wiley.com/go/Khonsari/Applied_Tribology_Bearing_Design_and_Lubrication_3rd_Edition

Dirt should be avoided wherever possible. Solid contaminants are harmful both in causing wear in the bearing itself as well as in thickening of the lubricant and interfering with its flow through filters, orifices, pumps, wicks, and porous metal bearings. If the presence of dirt cannot be avoided, precautions to minimize its effect should be taken with seals in bearing housings and in the lubricating-oil system. A last resort in dealing with a dirty environment is to supply clean air under slight positive pressure to bearing-housing chambers and oil reservoirs so that air circulation inside the system forces dirt away rather than have dirty exterior air coming in.

17.2 On-Line Monitoring

Continuous monitoring of machinery behavior is essential to maintain reliable performance. Common techniques involve use of sensors, instruments, and recorders for bearing and oil temperatures, and signal processing for vibration and noise (Figure 17.1). In the discussion that follows, guidelines are also presented pertaining to periodic analysis for the deterioration and contamination of lubricants.

Temperature

The traditional and still common measure for impending bearing trouble is temperature rise. Bearing temperature is commonly monitored with a resistance temperature detector (RTD) with its output fed to a central control panel in large industrial systems such as the turbine generators in electric power stations, steel rolling mills, or paper mills. On the other hand, thermocouples are predominately used for monitoring bearing temperature in many electric motors and simpler machinery (Figure 17.2).

These sensors are mounted in the expected high-temperature zone of bearings and in oil drains to monitor any erratic behavior. Drain temperature is also a good indicator of the

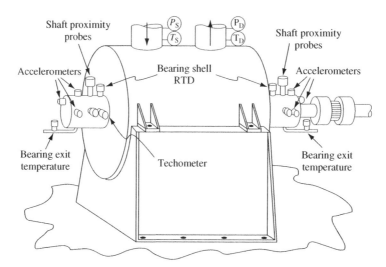

Figure 17.1 Turbomachine instrumentation for monitoring vibration and temperature. *Source*: Marscher, 1997. Reproduced with permission of Taylor & Francis.

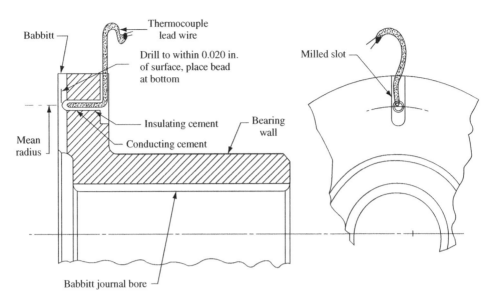

Babbitt

Thermocouple
lead wire

Drill to within 0.020 in.
of surface, place bead
at bottom

Milled slot

Mean
radius

Insulating cement

Conducting cement

Bearing
wall

Babbitt journal bore

Figure 17.2   Thermocouple installation in thrust face of electric motor bearing for fast response to temperature changes. *Source*: Adapted from Elwell and Booser, 1969.

effective (mean) bearing temperature and closely approximates the runner surface temperature. For a hydrodynamic journal bearing, the maximum temperature location is in the vicinity of the minimum oil-film thickness zone: commonly on the axial centerline about 20–30 degrees of rotation beyond the load line (see Chapter 8). With simple gravity loading in the bearing from the weight of a large rotor, an RTD would be desirably placed some 25 degrees beyond the bottom centerline of the bearing. If an alignment question were involved, two sensors might be installed – one at each end of the bearing. Difference in temperature axially from one end of the bearing to the other then provides a measure of bearing misalignment with its journal.

With a ball or roller bearing, a thermocouple is commonly spring-loaded to contact the outside diameter surface directly at the load line. With minor loss in response time, but with better continuity of support for the outer bearing ring, the thermocouple well might be ended within the bearing housing just short of the bearing housing bore.

With a circulating oil system, oil drain temperature rise also gives a direct indication of any change either in frictional power loss in the bearings, or a change in the rate of oil flow itself. A warning signal would commonly be activated at 10 °C above the normal operating range for both oil drain and bearing temperature sensors. Machine shutdown might be initiated for an abnormal rise of 20 °C.

Vibration

Rotating machinery problems can also be triggered by increased vibration, hence the necessity of monitoring its occurrence. Figure 17.1 illustrates an installation of appropriate vibration and associated transducers, plus bearing oil drain temperature monitors.

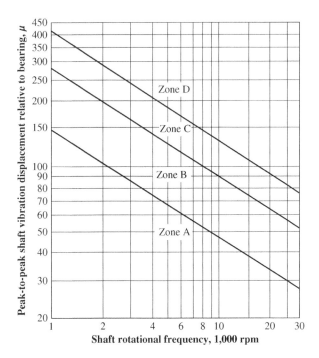

Figure 17.3 ISO 7919 recommended maximum relative shaft displacements for coupled industrial machines. *Source*: Adapted from Bloch, 2006.

Standards for maximum peak-to-peak shaft displacement relative to stationary bearings are given in Figure 17.3. For a 3600 rpm turbine generator, for instance, Zone A (for new machine operation) recommends normal baseline vibration under 80 microns (0.0032 in. or 3.2 mils). Zone B with up to 150 micron vibration (< 6 mils), is normally acceptable for long-term operation. Zone C (< 9 mils) would call for remedial action, and machine damage would be expected in Zone D.

Tighter limits are generally called for in experienced-based practice in the United States (Bloch, 2006). For all speeds up to 5000 rpm, the limit for new equipment (as in Zone A) is 2 mils. Up to 3.5 mils is acceptable for long-term operation (equivalent to Zone B), and marginal operation extends to 5.5 mils. Above 5000 rpm these limits drop with increasing speed until they bottom out above 20 000 rpm at about one-quarter the 0–5000 rpm limits.

ISO 7919 recommends an initial alarm setting for a 3600 rpm unit of 4.7 mils (3.2 baseline limit for Zone A plus 25% of the 6 mil Zone B limit). The trip setting would use the 9.0 mil limit on Zone C. Alternatively, Bloch suggests setting the alarm 50% above the normal level for a new machine operating at low vibration levels (e.g. 1.5 mils peak-to-peak). This is far below the 4.7 mils recommended by ISO. If that level is exceeded, the change might indicate a growing problem and should be investigated.

Troubleshooting with vibration monitors generally involves analyses of accelerometer output by a fast Fourier transform (FFT) as to the vibration level over the range from several Hertz to beyond the pass frequency of the blades, vanes or other distinctive rotor features of the machine. Barring an inherently faulty design, most vibration problems can be corrected

Table 17.1 Common vibration excitations in rotating machinery

Frequency	Source
0.05–0.40X	Diffuser stall
0.42–0.49X	Rotor dynamic instability
0.50X	Rub or looseness
0.60–0.95X	Impeller stall
1X	Imbalance
1X + 2X	Alignment
Number of vanes X	Vane passing
Number of blades X	Blade passing

Note: Resonance is a possible factor in final vibration level. X is rotational frequency. Imbalance, for instance, in a 3600 rpm rotor ($X = 60$ Hz) would commonly excite vibration at a frequency of $1X = (1)(60) = 60$ Hz.
Source: Marscher, 1997. Reproduced with permission of Taylor & Francis.

by rotor rebalancing, by alignment of the coupling for warmed-up and normal operating conditions, and by observations of response in fluid machines over their pressure and flow range (Marscher, 1997).

Supplementary excitation of a rotating machine and its structure by experimental modal analysis (EMA) is also useful. EMA involves striking the unit with an instrumented impact hammer to observe natural frequencies of a rotating machine, its rotor system, and attached piping and system components. Results provide information on resonance frequencies and whether all system and supporting structures have adequate separation between exciting and natural frequencies. Table 17.1 gives typical excitation frequencies and their likely sources.

Distinctive 'shaft whip' instability is a troublesome self-excited vibration common in oil-film cylindrical-bore journal bearings as rotor speeds approach and exceed twice the first vibrational critical frequency of the rotor. This rotor speed involves the journal floating into a large orbit in synchronism with the approximately half-speed of the bearing oil film which then loses much of its centering force. Corrective measures include increasing unit bearing loading, often above about 200 psi on the projected bearing area, a lemon-shaped bearing bore, or ideally a tilt-pad journal bearing. Inadequate oil feed to a plain cylindrical journal bearing may also reduce oil-film stiffness so as to introduce erratic vibration. Chapter 8 gives analytical tools for oil-film vibration response.

17.3 Oil Analysis

Regular analyses of oil taken from sampling points representative of concentration of contaminates provide an excellent means for avoiding unforeseen lubrication problems and machinery failures. Oil analysis also provides a useful record of any loss in oil quality: increase in oil oxidation, additive depletion, and contamination by water or solvents. In addition to oil quality, periodic sampling enables operating personnel to make proper assessment as to the remaining useful service life of lube oils and on possible wear and corrosion problems throughout the system (Lockwood and Dalley, 1992).

Table 17.2 Recommended oil sampling frequencies

	Operating hours
Diesel engines – off highway	150
Transmissions, differentials, final drives	300
Hydraulics – mobile equipment	200
Gas turbines – industrial	500
Steam turbines	500
Air/gas compressors	500
Chillers	500
Gearboxes – high speed/duty	300
Gearboxes – low speed/duty	1000
Bearings – journal and rolling element	500
Aviation reciprocating engines	25–50
Aviation gas turbines	100
Aviation gearboxes	100–200
Aviation hydraulics	100–200

Source: Fitch, 1997. Reproduced with permission of Taylor & Francis.

Sampling intervals in operating hours as recommended for different classes of machines are given in Table 17.2. These intervals are commonly adjusted for environmental conditions, site contamination, and for the need for cleanliness in a particular machine. For example, sampling intervals should be short for rotating machinery in a coal-burning electric power house with a heavy concentration of fly-ash dust in its atmosphere. With a clean atmosphere in a natural gas-fired power house in continuous operation, and without repeated start–stop cycles that could result in added journal bearing wear, sampling time of 500 operating hours for the steam turbine (Table 17.2) might reasonably be increased by a factor of 3–4.

Sampling points for the oil in hydraulic and circulating lube systems are located on return lines from motors, valves, cylinders, compressors and other machine components before reaching filters. To be suitably representative of the concentration of particulates, water, and other contaminants, samples should be withdrawn into a properly clean bottle from a turbulent zone such as a piping elbow. Prior to and during sampling, machines should be run at normal speeds, temperatures, and load cycles.

Figure 17.4 shows routine sampling points in a main return oil line for general trend analysis. Follow-up sampling from individual zones provides for diagnosing conditions in particular machinery components. Splash, oil-ring, and flood lubricated machine elements are best sampled from drain plugs after flushing or using an off-line sampler.

Oil properties to be monitored typically vary somewhat with machinery applications as given in Table 17.3. The suggested targets and warning limits should routinely involve comparison with new oil properties. A common rule of thumb warning limit is a drop in oxidation inhibitor concentration to 50% or less of its fresh oil value. At this point, however, buildup of oil-soluble contaminants in severe service may place the oil beyond its half-life. For milder conditions in steam turbine service, on the other hand, many years of useful oil life may still be available.

While ASTM test procedures provide the basis for periodic backup checks, much quicker automated instrumental techniques are now in common use. Infrared (IR) spectral analyses are

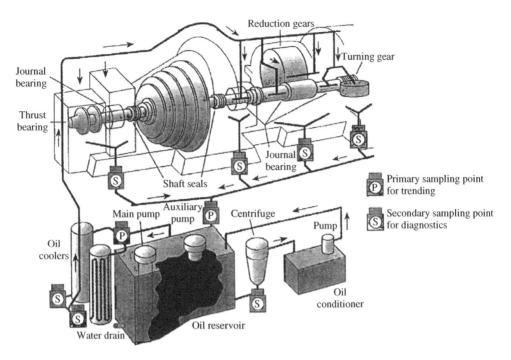

Figure 17.4 Primary oil sampling locations for large circulating system. *Source*: Fitch, 1997. Reproduced with permission of Taylor & Francis.

particularly useful in evaluating oil condition. Intensity of light absorption by a used oil sample at distinct frequencies within the IR spectrum gives information on (1) the general lubricant type such as paraffinic or naphthenic mineral oil, or a synthetic; (2) remaining additive concentration; (3) buildup of oxidation products; and (4) presence of free water, solvents, and other contaminants (see Table 17.4). The FT-IR computerized monitoring instrument uses a Fourier transform calculation to extract IR frequency/intensity points (or area under spectrum peaks) for comparison with new oil. The entire spectrum for a used oil sample can be collected and analyzed in less than a minute.

Similar automated units are available for measuring viscosity, total acidity number (TAN), and the opposite alkalinity base number (TBN) for diesel engine oils. Emission spectrographic wear monitoring of oil samples also provides a supplementary indication as to the remaining concentration of any metallo-organic antiwear and corrosion additives.

17.4 Wear Monitoring

Diagnosis of oil contaminants provides an important indication of the health of oil-wetted machine components. A useful technique for determining if wear has reached alarming levels is *emission spectrographic analysis*. In one type of unit, a film of the oil sample is carried on a graphite wheel to a high voltage arc in a narrow gap with a sharp graphite rod electrode. Each metallic element in the oil is excited by the arc to emit its own characteristic spectrum of light. Intensity of these light bands is then converted to a report on element concentrations

Table 17.3 Suggested warning limits for industrial oils

Property	ASTM test method	Steam and Gas Turbines D4378	Hydraulic and Circulating	Diesel engines	Gears	Gas and air compressors
Viscosity change at 40 °C, (% max.)	D445	20	20	30	20	20
Total acid number (mg KOH/g above initial)	D664, D974	0.2 steam, 0.5 gas				
Total base number (% of new oil)	D2896, D4739			50		
Particulates (electronic counter)	ISO 4406	Consult supplier				
Water content (max. %)	D1744, D4928	0.2	0.2			
Oxidation inhibitor (% of new oil)	Infrared	50	50			50
EP/antiwear additive (% of new oil)	Infrared				50	
Rust inhibitor	D665	Fail test	Fail test			
Soot (%)	Infrared			2-3		

by a computer. The following are possible sources of elements in such an analysis (Lukas and Anderson, 1997):

Element	Possible sources
Wear and contaminants	
Aluminum	Pistons, pivots, bearings
Copper	Bronze bearings and seals
Iron	Rotor and seal surfaces, ball and roller bearings, gears, rust
Lead	Bearings
Silicon	Sand, dust and dirt
Tin	Babbitt and bronze bearings
Oil soluble additives	
Calcium	Detergent, rust inhibitor
Phosphorus	Antiwear, antioxidant
Silicon	Antifoam
Zinc	Antiwear

Warning concentrations commonly represent a 25–50% change from the new oil level: a rise representing a warning of wear or contamination, a decrease indicating the loss of oil additive content.

Table 17.4   FT-IR Infrared absorption bands for oil analysis

Material	Absorption (cm^{-1})	Interferences
Water	3300–3500, 1600	Glycol, oxidants
Gasoline	750	
Fuel	500–1000, 1600	Unsaturation in lubricants
Oxidized lubricant	1700–1750, 1100–1200, 1630, 1270	VI improvers, alkenes
Glycol coolant	880, 1040,1080	Oxidation, antioxidant
Antiwear additive	950–1050	Aromatics
Sulfate detergent	1100–1200	Oxidation
Pour-point depressant	1732	Oxidation
Dispersant: succinimide	1707	Oxidation
Dispersant: Mannich base	1650	Oxidation
VI improver: polymethacrylate	1732	Oxidation
VI improver: styrene-isoprene	700	Fuel
Soot	Baseline near 200, compared with background	

Source: Lukas and Anderson, 1997. Reproduced with permission of Taylor & Francis.

Emission spectroscopy provides analysis only of soluble metallic elements and fine suspended metallic particles below about 10–15 microns in size (well below the visible range). Other monitors such as in-line magnetic plugs for collecting wear particles, magnetic screening of particles by ferrography (Lockwood and Dalley, 1992), and ultrasonic methods also find use for detailed microscopic and metallographic examination. These instruments often provide follow-up to an emission spectrograph for obtaining the identity of the machine elements and mechanisms in which wear is being encountered.

17.5 Ball and Roller Bearing Failure Analysis

Initial selection of ball and roller bearings is often based on a fatigue life load rating to withstand the repeated contact stress of the rolling elements with their raceways. This type of failure involves subsurface nucleated fatigue in the zone of maximum shear stress as illustrated in Figure 17.5. Nevertheless, field experience shows that 90–95% of bearings removed from service have suffered one of the other failures modes reviewed in Table 17.5 (Derner and Pfaffenberger, 1984). Surface asperities and debris dents may combine with local corrosion and chemical effects to create microspalls, which then progress to pits and cracks and ultimately to bearing failure (ASME, 2003). Wear resulting from dried grease and contamination has been observed to be the common cause of failure in industrial electric motors.

Preliminary check in its installation of a questionable noisy bearing should include wear measurement of radial and axial play made by rocking the shaft in its bearing support, noting roughness during slow hand rotation, and careful visual examination to determine any underlying problem. Noise and vibration analysis are also useful starting points in an analysis (see Chapter 14). Once the bearing is dismounted, examination of shaft and housing mounting surfaces, raceways, rolling elements, and cages may reveal any problems with corrosion, fatigue, dried grease, and accumulation of contaminants.

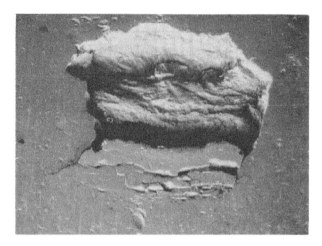

Figure 17.5 Subsurface nucleated spall on roller bearing inner ring. *Source*: Derner and Pfaffenberger, 1984. Reproduced with permission of Taylor & Francis.

Table 17.5 Common failure modes for ball and roller bearings

Failure mode	Cause	Result
Brinneling	Dents in race or balls caused by impact or improper loading	Rough running, surface origin fatigue
Cracked rings	Improper fits, abuse, cracked mounting	Catastrophic failure
Electric discharge	Flow of electric current through bearing	Rough running, fluting
Fretting	Small amplitude motion causes indenting in race or wear of mounting surfaces	Rough running, loss of bearing retention on shaft or in housing
Galvanic action	Electrolytic (salt water in oil), metal higher in galvanic series (copper, aluminum)	Pitting, leading to surface origin fatigue
Grease hardening, increased viscosity oil	Overheating, oil bleed loss, excess bearing cavity ventilation, high ambient temperature	Excessive friction, wear, cage failure, surface origin failure
Lock-up	Loss of clearance due to shaft heating	Steady state: preload results in inner ring turning on shaft and severe shaft damage. Intermittent: premature fatigue
Lubricant decomposition	Decomposition products	Surface pitting, varnish deposits
Peeling, frosting	Low viscosity lubricant, insufficient EHL film	Surface origin fatigue, increased clearance, rough operation
Race deformation	Overload and/or overheating	Increased clearance, cage failure, jam

(*continued*)

Table 17.5 (*Continued*)

Failure mode	Cause	Result
Rib wear and denting	Heavy impact, excessive thrust, abrasive contaminants, loss of or sparse lubricant	Loss of roller control where rib guided, cage failure, jam
Roller end wear	Skewing of cylindrical rollers under bearing misalignment, poor lubrication, contaminants	Slow deterioration leading to roller jamming, increased clearance
Rolling element banding	Excessive thrust load or overheating	Increased internal clearance, rough running on restart
Rolling element scoring	Hard particles trapped in retainer pockets causing deep cuts or scratches	Surface origin fatigue
Rusting	Humid or salt environment, with insufficient protection by lubricant	Rough running, surface origin fatigue
Smearing	Material transfer during skidding, loss of traction in load zone, very light load, rapid acceleration. Smearing of roller ends and guide flange	Rough running, surface origin fatigue. Overheating, loss of guidance, increased axial float
Wear of cages	Misalignment, contaminants, inadequate lubrication	Cage breakage, cage drop, jam
Wear of rolling elements	Loss of or sparse lubricant, abrasive contamination	Bearing jam by debris, excessive internal clearance, retainer breakage

Source: Derner and Pfaffenberger, 1984. Reproduced with permission of Taylor & Francis.

Once the bearing is removed and dismantled, after drilling out the rivets in a ball bearing cage or cutting in two the outer bearing ring with an abrasive cutting disk, inspection should follow with a magnifying glass or preferably with an adjustable binocular microscope starting with about 5× magnification. Damage photos and diagrams available from bearing manufacturers will commonly point to the basic problem involved (see also Jendzurski, 1997).

Ball path width and its angular location in the raceways of common single-row, deep-groove ball bearings will often allow an approximation of the load (see Figure 17.6). If contamination is involved, laboratory spectrographic analysis may point to the source. Dry, hard grease would indicate inadequate relubrication or elevated temperature. While metallurgical defects are rarely a factor in current engineering use of rolling bearings, a check for hardness and inclusions in the steel should be made for demanding applications involving questionable fatigue life.

Since evidence of the mechanism of damage and failure is usually maintained in the hard steel surfaces involved, the source of operating problems with ball and roller bearings can be identified visually in 80–90% of the cases with the aid of a magnifying glass or low power binocular microscope. If a bearing has suffered especially severe damage to the extent that the initiating event is obliterated, then examination of companion bearings operating in similar service will often give clues as to the root cause of the trouble. Aid from the equipment

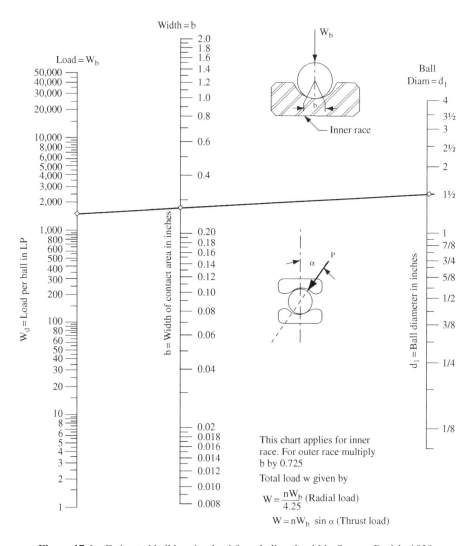

Figure 17.6 Estimated ball bearing load from ball path width. *Source*: Barish, 1939.

manufacturer and the bearing supplier is also helpful in pinpointing the cause of troublesome field problems.

17.6 Oil-Film Bearing Failure Analysis

The mechanisms by which oil-film bearings encounter trouble can frequently be identified by careful visual and microscopic surface examination. When supplemented by chemical analyses, plus hardness and metallographic checks, these examinations usually lead to the source of the trouble and to remedies such as given in Table 17.6. Fatigue, wear, contamination, corrosion, erosion and electrical damage represent common problems.

Table 17.6 Oil-film bearing troubleshooting

Cause	Remedy
I. Wear, scoring, Galling	
1. Dirt	Change oil, new filter, remove contaminant source, provide flushing chamfers at ends of oil grooves, softer bearing material to embed dirt or harder journal to reduce its effect
2. Boundary-film condition	Higher oil viscosity, antiwear or extreme-pressure oil, more compatible bearing material, bearing design for greater oil-film thickness
3. Insufficient lubrication	Add oil; increase inlet pressure; check pump, filter, cooler, piping
4. Load too high	Increase viscosity, antiwear or extreme-pressure oil, larger bearing, stronger bearing material, harder shaft
5. Too low oil viscosity	Higher viscosity, cool oil, increase oil flow, modify bearing design for higher load capacity
6. Misalignment	Align bearing with journal
7. Poor material combination	Select compatible bearing-journal materials, increase shaft hardness, hardened sleeve, additive oil
8. Rough journal	Polish or grind journal
9. Improper grooving	Remove grooves from bearing load zone
10. Vibration	Balance rotor; increase viscosity; use pivoted pad, elliptical, three-lobe, or pressure-groove bearing
11. Excessive journal deflection	Increase journal support, reduce deflecting load
12. Journal eccentricity	Replace journal
13. Electrical pitting	Remove current source, insulate bearing, ground journal
II. Corrosion	
1. Corrosive contamination	Remove source, seal bearing, use corrosion-resistant lubricant
2. Moisture	Remove source, rust-inhibited oil
3. Corrosive lubricant	Change oil periodically to avoid buildup of oxidation products, avoid unnecessary extreme-pressure oils, use tin babbitt or aluminum bearings
III. Overheating	
1. Insufficient oil film	Increase viscosity, increase bearing clearance, more oil-distributing grooves with larger chamfers, higher supply pressure, increase oil orifice size
2. Improper oil viscosity	Select for load, speed, and temperature
3. Heat from surroundings	Externally cool oil system, water cooling, insulate bearing, better ventilation
4. High load	Reduce load and belt tension, use antiwear or extreme-pressure oil, stronger bearing material, design for higher load
5. High speed	Decrease viscosity; increase clearance and oil flow; change to pivoted-pad bearing with leading-edge feed groove, or to elliptical or overshot groove design

(*continued*)

Table 17.6   (*Continued*)

Cause	Remedy
IV. Fatigue	
1. Overload	Larger bearing, stronger bearing material
2. Rotor vibration	Balance rotor; pivoted-pad, elliptical, three-lobe, or pressure-groove bearing; rotor support damping
3. Local high stress concentration	Check for corrosion, edge loading, electrical pits, rough surface finish. Use higher-viscosity oil
4. High bearing temperature	Design for higher oil flow, higher-temperature bearing material
5. Cavitation erosion	Relocate oil holes and grooving, fatigue-resistant material, higher viscosity, lower temperature
V. Seizure	
1. Too little clearance	Increase clearance, account for temperature and thermal expansion effects
2. Insufficient lubrication	Larger oil supply, higher viscosity, extreme-pressure lubricant, modify feed grooves and orifices
3. Too high load	Increase bearing area, compounded oil, higher viscosity
4. Overheating	External cooling, ample oil flow
5. Unsuitable bearing material	Compatible bearing material with satisfactory strength and thermal properties

Fatigue

Fatigue problems predominate in machines with reciprocating bearing loads such as automobile engines, and those units that experience rotor imbalance such as turbomachines. Oscillating oil-film pressure exceeding about half of the yield stress eventually initiates fatigue cracks in the maximum shear stress zone just below the surface of the bearing metal.

With babbitt bearings, these fatigue cracks progress down into the bearing metal at right angles to the bearing surface. The cracks then progress parallel to, but just above, the bond line of babbitt with its backing material until they meet with other radial cracks. In this manner, loose pieces of babbitt lining are created as shown in Figure 17.7. These wear fragments may be transferred to other parts of the same bearing and are often carried by the oil to other bearings in the system. There they often generate local stress concentrations, hot spots, and promote further fatigue. A fatigue load limit in oil-film pressure of about 5–10 MPa (750–1500 psi) with thick babbitt can be raised several fold by using only a thin babbitt layer to transfer the maximum shear stress down into a stronger backing bearing material (see Chapters 4 and 8).

With higher-strength bronzes, aluminum, and silver bearing materials, fatigue usually progresses as a more localized surface pitting. These pits are generally much smaller than the sections loosened by fatigue from a babbitt-lined bearing.

Wear

Wear in the form of removal of surface material occurs in a bearing whose surface is not adequately separated from its journal by a lubricant film. With marginal lubrication, dramatic

Figure 17.7   Fatigue failure of babbitt bearing. *Source*: Adapted from Burgess, 1953.

reduction in wear rate may result by utilizing sulfur, phosphate, and zinc antiwear or extreme pressure additives capable of coating copper, steel and other reactive surfaces. Rate of wear commonly drops to a negligible level with any factors that increase the thickness of the lubricant film above the range of 3–10 times the combined surface roughness profile of the bearing and its mating surface (see Chapter 3).

A number of other factors that significantly contribute to wear include the following. *Contamination* by debris carried with oil is a common source of abrasive action and mechanical wear (Khonsari and Booser, 2006). This debris is often introduced during operation of a new machine as manufacturing residue such as weld spatter, grinding abrasives, and foundry sand. These particulates can commonly be picked out from their embedment in a bearing surface for analysis (see Figure 17.8).

Particles smaller than a babbitt lining surface are commonly heated by rubbing of the passing journal, and then embed in the softened babbitt with a deleterious effect limited to only minor polishing of the journal surface. Larger hard particles may bridge from the bearing backing to the journal, score the shaft, locally melt the babbitt as they are dragged through it to form a wormlike trough, and the melting and flow of the bearing material may then cause overall thermal breakdown, loss of internal bearing clearance and seizure (Pascovici and Khonsari, 2001).

Figure 17.8   Small contaminant hard particles commonly embedded harmlessly in soft babbitt surface. *Source*: Adapted from Burgess, 1953.

Wire-wool and related *self-propagating* wear may be encountered if abnormal hardening is induced in either debris or bearing surfaces. High-chromium steel journals undergo this machining-type damage as any steel chip or weld spatter entering the bearing clearance cuts an initial circumferential wire-wool-like sliver from the journal. This sliver is heated by the cutting action, work hardens, and also forms hard oxide and carbide inclusions by reacting with oil and its dissolved air. The hardened sliver and secondary slivers thus generated accumulate as a hard scab on the babbitt bearing surface to machine a much larger groove in the journal (Figure 17.9).

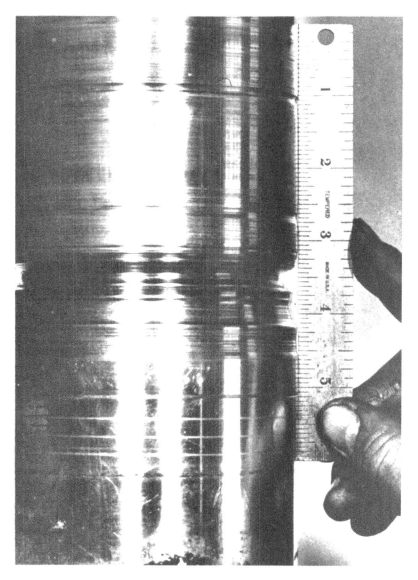

Figure 17.9   Wire-wool damage to 12% chromium steel turbine journal. *Source*: Booser, 1967. Reproduced with permission of ASLE transactions.

To avoid this self-propagating wear problem with high-chromium steel rotors in high-temperature and corrosion-resistant service, protective overlays and low-alloy steel sleeves are employed. General introduction of 3% chromium steel rotors by British steam turbine builders in the 1950s–1960s resulted in a rash of similar problems (Dawson and Fidler, 1962). With chromium content lowered to 1.5–2.5%, improved oil system cleanliness and careful filtration have minimized difficulties.

Chlorine extreme-pressure gear oil additives have induced similar damage in marine equipment (Wolfe *et al.*, 1971). The chlorine appears to promote corrosion and oxide hardening of any low-alloy steel slivers. Change to phosphate gear oil additives alleviated the problem.

Hardening of tin babbitt bearing surfaces has also led to self-propagating wear (Bryce and Roehner, 1961). With salt water contamination in oil during marine exposure, galvanic action has generated tin-oxide inclusions within tin babbitt to give a file-hard surface. As a result, the bearing becomes incapable of embedding dirt particles or accommodating misalignment.

Corrosion

Corrosion of a bearing results from attack by reactive chemicals in the oil stream. These may be either corrosive materials picked up by the oil from its surroundings, by reactive chemicals and additives in the oil, or by oil oxidation products.

The most common sources of corrosive attack are oxidation products in the oil itself: organic acids and peroxides that corrode lead, copper, cadmium and zinc bearing materials. Corrosion of lead in lead babbitt, copper–lead and leaded bronze bearings has been the most troublesome (see Figure 17.10). These lead-containing materials are in wide use and may undergo relatively rapid reaction with oil oxidation products, especially in the presence of water. Tin babbitt, aluminum and silver are not normally affected by these oxidation products.

Fortunately, lead corrosion problems have been almost eliminated in many applications by including up to 10% tin alloying in lead babbitt, using indium coatings over the lead, incorporating improved oil oxidation resistance, and practicing periodic oil changes. Such measures have brought lead babbitt into wide use in many types of demanding applications.

Unfortunately, oil additives may themselves contribute to corrosion. Sulfur compounds provide an essentially mild corrosive action to generate a protective low-shear-strength surface layer on steel and copper alloy surfaces, so as to prevent welding and destructive contact with the journal material. Since this corrosive action continues slowly even under normal operating

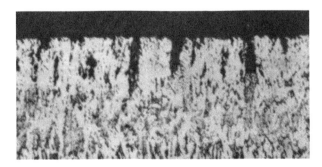

Figure 17.10   Lead corroded from copper–lead bearing by oxidized oil. *Source*: Adapted from Burgess, 1953.

conditions – and is often aggravated by the presence of water contamination – extreme-pressure type oils should be employed only where necessary. Possible corrosion of copper conductors has usually discouraged their use in turbine oils for power-generating equipment. These additives give no reaction and no advantage with tin babbitt and are to be avoided because of corrosion of silver bearings used for their high fatigue strength in some diesel engines.

Erosion

Erosion damage involves formation of pits or voids in the bearing surface by fluid action. A cure usually involves change in oil groove pattern and other geometry features to avoid sudden changes in the direction of high-velocity oil streams to reduce abrasion by the impinging action of any debris in the oil.

Cavitation erosion, especially with the oscillating load on the bearings of gear sets, sometimes results in fatigue pitting from the rapid formation and collapse of high-pressure gas bubbles. Use of a thin layer – 25–50 microns (0.001–0.002 in.) – of babbitt on a steel or other high-fatigue-strength backing metals often cures this problem.

Electrical Damage

Pitting of bearings and journals from passage of electric sparks can be encountered in motors and generators, and from static charge buildup in turbo-machinery. With either a.c. or d.c. voltage differences of 0.2–1.0 or more between a shaft and a bearing housing, sparks are likely to generate a frosted appearance from microscopic pits on the journal surface or somewhat larger melted craters in the softer bearing material (Figure 17.11).

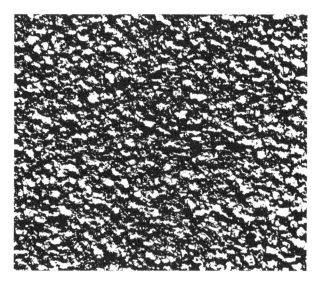

Figure 17.11   Pitting craters of approximately 0.025–0.075 micron (0.001–0.003 in.) diameter in thrust bearing babbitt made by electrostatic spark discharges from steam turbine rotor. *Source*: Adapted from General Electric, 1951.

With thicker bearing oil films (than in ball bearings, for example) and with more limited voltage spikes in wound-rotor variable frequency drives, voltages up into the 10–30 range may be tolerated before encountering spark damage. These drives apply pulsed voltage waves to give an average off-ground potential to the rotor winding with a corresponding capacitance discharge to the stator frame (Sharke, 2000).

Reaction with wet steam droplets raises electrostatic charge on a steam turbine rotor at a rate of some 12 000 V/s. This charge is typically dissipated in less than 0.1 ms as a low-energy spark discharge when the rotor voltage reaches 30 000–100 000 (Gruber and Hansen, 1959). Since these sparks involve currents only in the milliampere range, troublesome bearing operation is often encountered only following months of satisfactory operation. The resulting craters of the order of 0.001–0.003 in. in diameter in the minimum film thickness area of a bearing surface can commonly be identified at 10–30 magnification by the melted appearance at the bottom of each electrical pit. Surges of high current such as that from an attached welding machine, on the other hand, create large craters.

Other sources for electrical discharge in bearings are: (1) magnetic dissymmetry in a machine giving flux variation with angular position of the shaft, (2) shaft magnetization, or homopolar effect, involving unbalanced ampere-turns encircling the shaft to cause its axial magnetization, and (3) electrostatic potentials developed by a charged lubricant, a charged belt, or impinging particles (Brothwell, 2006; Boyd and Kaufman, 1959).

As a first step to prevent electrical sparking damage, consideration should be given to eliminating the source of the bearing currents. If this requires impractical redesign or system changes, consideration should be given to insulating bearings and all metallic connections such as oil lines to eliminate electrical conducting paths between the bearing and its support and surroundings. For electrostatic and related buildup of rotor voltage, grounding the shaft with a separate brush or metallic strap will often prevent the rotor from reaching a sparking potential (Brothwell, 2006). Occasionally increased oil-film thickness (as with increased viscosity and lower unit bearing loading), finer oil filtration, oil drying, or a change in oil characteristics or bearing geometry will discourage spark discharges.

References

ASME. 2003. *Life Rating for Modern Rolling Bearings, ASME Tribology Division Technical Committee*, Trib-Vol. 14, ASME, New York.

Barish, T. 1939. 'Ball Bearing Troubles,' *Product Engineering*, Vol. 10, pp. 88–93.

Bloch, H.P. 2006. 'Vibration Limits: is there a Consensus?,' *Hydrocarbon Processing*, April, pp. 9–10.

Booser, E.R. 1967. 'Influence of Lubricant Additives on Wire-Wool Type Bearing Failures with 12 per cent Chromium Steel Journals,' *ASLE Transactions*, Vol. 23(8), pp. 325–328.

Boyd, J. and Kaufman, H.F. 1959. 'The Causes and Control of Electrical Currents in Bearings,' *Lubrication Engineering*, Jan., pp. 28–35.

Brothwell, I. 2006. 'Identifying, Monitoring and Eliminating Rotating Equipment Shaft Voltages,' *Hydrocarbon Processing*, Vol. 85(5), pp. 47–60.

Bryce, J.B. and Roehner, T.G. 1961. 'Corrosion of Tin Base Babbitt Bearings in Marine Steam Turbines,' *Transactions of the Institute of Marine Engineers*, Vol. 73(11), p. 373.

Burgess, P.B. 1953. 'Mechanisms of Sleeve Bearing Failures,' *Lubrication Engineering*, Vol. 9, pp. 309–312.

Dawson, P.H. and Fidler, F. 1962. 'The Behaviour of Chromium Steel in Large High-Speed Bearings,' *A.E.I. Engineering*, Vol. 2(2), pp. 54–62.

Derner, W.J. and Pfaffenberger, E.E. 1984. 'Rolling Element Bearings,' in *Handbook of Lubrication*, Vol. II. CRC Press, Boca Raton, Fla, pp. 495–537.

Elwell, R.C. and Booser, E.R. 1969. 'Design Guidelines for Flat-Pad Thrust Bearings,' *Machine Design*, Vol. 41(20), pp. 41–46.

Fitch, J.C. 1997. 'Elements of an Oil Analysis Program,' in *Tribology Data Handbook*, CRC Press, Boca Raton, Fla, pp. 875–888.

Gruber, J.M. and Hansen, E.F. 1959. 'Electrostatic Shaft Voltage on Steam-Turbine Rotors,' *Journal of Engineering for Power*, ASME, Jan., pp. 97–110.

Jendzurski, T. 1997. 'Ball and Roller Bearing Trouble Shooting,' in *Tribology Data Handbook*, CRC Press, Boca Raton, Fla, pp. 956–985.

Khonsari, M.M. and Booser, E.R. 2006. 'Effect of Contaminants on the Performance of Hydrodynamic Bearings,' *Proceedings* IMechE, *Journal of Engineering Tribology*, Vol. 220, Part J, pp. 419–428.

Lockwood, F.E. and Dalley, R. 1992. 'Lubricant Analysis,' *ASM Handbook, ASM International, Materials Park, Ohio*, Vol. 18, pp. 299–312.

Lukas, M. and Anderson, D.P. 1997. 'Laboratory Used Oil Analysis Methods,' in *Tribology Data Handbook*, CRC Press, Boca Raton, Fla, pp. 897–911.

Marscher, W.D. 1997. 'Rotating Machinery Vibration and Condition Monitoring,' in *Tribology Data Handbook*, CRC Press, Boca Raton, Fla, pp. 944–955.

Pascovici, M.D. and Khonsari, M.M. 2001. 'Scuffing Failure of Hydrodynamic Bearings due to an Abrasive Contaminant Partially Penetrated in the Bearing Over-layer,' *ASME Journal of Tribology*, Vol. 123(2), pp. 430–433.

Sharke, P. 2000. 'Current Jam,' *Mechanical Engineering*, ASME, May, pp. 64–67.

Wolfe, G.F., Cohen, M., and Gibbs, G.H. 1971. 'The Occurrence and Prevention of Machining-Type Bearing Failures in Marine Steam Turbines,' *Transactions of the Institute of Marine Engineers*, presented at 12–13 Nov. 1970 Annual Meeting, Society of Naval Architects and Marine Engineers.

Appendix A

Unit Conversion Factors

Following are conversion factors for various units of measure, primarily to international SI units. (See NIST Special Publication 811, *Guide for the Use of the International System of Units*, US Government Printing Office, Washington, DC, 1991.)

Factors are given as a number equal to or greater than 1 and less than 10. This number is followed by the letter E (for exponent), plus or minus two digits which indicate the power of 10 by which the number must be multiplied. For instance:

$$2.54 \, E + 02 \quad \text{is} \quad 2.54 \times 10^2$$

Conversion from one non-SI unit to another non-SI unit can be carried out in two steps, as illustrated:

$$1 \, \text{in.} = 2.54 \, E{-}02 \, \text{m}$$
$$1 \, \text{ft.} = 3.048 \, E{-}01 \, \text{m}$$

Combining:

$$1 \, \text{ft} = \frac{3.048 \, E{-}01}{2.54 \, E{-}02 \, \text{in.}} = 12 \, \text{in.}$$

Quantity	British in.-lb-s or traditional unit	Multiply by this conversion factor	To obtain this SI unit
Acceleration	in./s^2	2.54 E + 02	m/s^2 (meter/second^2)
	ft./s^2	3.048 E–01	m/s^2
Gravitational free fall, g	32.1740 ft./s^2 (386.1 in./s^2)		9.80665 m/s^2
Area	in.^{2}	6.4516 E–04	m^2 (meter)

Applied Tribology: Bearing Design and Lubrication, Third Edition. Michael M. Khonsari and E. Richard Booser.
© 2017 John Wiley & Sons Ltd. Published 2017 by John Wiley & Sons Ltd.
Companion Website: www.wiley.com/go/Khonsari/Applied_Tribology_Bearing_Design_and_Lubrication_3rd_Edition

Quantity	British in.-lb-s or traditional unit	Multiply by this conversion factor	To obtain this SI unit
Bending moment, torque	$lb_f \cdot in.$	1.129848 E–01	$N \cdot m$ (newton-meter)
	$dyne \cdot cm$	1.00 E–07	$N \cdot m$
Density	$lb_m/in.^3$	2.7679905 E + 04	kg/m^3 (kilogram/m^3)
	g/cm^3	1.00 E + 03	kg/m^3
Energy	Btu (mean)	1.05587 E + 03	$N \cdot m$ (joule)
	calorie (mean)	4.19002 E + 00	$N \cdot m$
	erg	1.00 E–07	$N \cdot m$
	$in. \cdot lb_f$	1.11298 E–01	$N \cdot m$
	watt second	1.00 E + 00	$N \cdot m$
Force	dyne	1.00 E–10	N (newton)
	kg_f	9.80665 E + 00	N
	lb_f	4.4482216152605 E + 00	N
Length	angstrom	1.00 E–10	m (meter)
	foot	3.048 E–01	m
	in.	2.54 E–02	m
	in.	25.4	mm
	micron (micrometer)	1.00 E–06	m
	mil (0.001 in.)	2.54 E–05	m
	mil (0.001 in.)	2.54 E + 01	micron
Mass	lb_m	4.5359237 E–01	kg (kilogram)
Power	Btu/s	1.054350264488 E + 03	W (watt)
	calorie/s	4.184 E + 00	W
	$ft \cdot lb_f/s$	1.3558179 E + 00	W
	horsepower (550 $ft \cdot lb_f/s$)	7.4569987 E + 02	W
Pressure or stress	$lb_f/in.^2$	6.8947572 E + 03	N/m^2 (pascal, p_a) ($1.00 \; N/m^2 = 1.00 \; p_a$)
	$dyne/cm^2$	1.00 E–01	N/m^2
Specific heat capacity	Btu/lb·°F	4.1868 E + 03	$J/(kg \cdot K)$
Speed or velocity	in./s	2.54 E–02	m/s (meter/second)
Temperature	°C (Celsius)	(°C + 273.15).	K (Kelvin)
	°F (Fahrenheit)	[(5/9)(°F + 459.67)]	K
	°F (Fahrenheit)	[(5/9)(°F – 32)]	°C (Celsius)
	°R (Rankine)	5/9	K
Thermal conductivity	Btu/hr · ft ·° F	1.730735 E + 00	$W/(m \cdot K)$
Viscosity	(See Appendix B)		
Volume	fluid ounce (US)	2.957353 E–05	m^3 (meter)^3
	gallon (UK liquid)	4.546087 E–03	m^3
	gallon (US dry)	4.40488377086 E–03	m^3
	gallon (US liquid)	3.785411784 E–03	m^3
	$in.^3$	1.6387064 E–05	m^3
	liter	1.00 E–03	m^3

Appendix B

Viscosity Conversions

Unfortunately, the many units for expressing viscosity, the internal friction or stiffness of a fluid, are quite confusing. The fundamental SI unit is the *Pascal-second, Pa · s* (or the equivalent *Newton − seconds/meter^2*), while the reyns (pound force × seconds/in.^{2}) is used in engineering calculations employing British inch-pound-second units.

With most laboratory instruments, oil density provides the driving force for flow. Oil efflux time is then proportional to *kinematic viscosity*, which is absolute viscosity divided by density. The *centistoke* is the most common kinematic viscosity unit used in reporting oil properties. The significance of viscosity and its units is reviewed further in Chapter 2, which also discusses the sometimes dramatic influence of temperature, pressure and shear rates.

British in.-lb-s or traditional unit	Multiply by this conversion factor	To obtain this SI or another unit
Dynamic or absolute viscosity (units: force · time/area)		
dyne · s/cm^2 (poise, P)	1.00 E–01	Pa · s (also N · s/m^2)
cP (mPa · s)	1.00 E–03	Pa · s
lb$_f$ · s/in.^{2} (reyns)	6.895 E + 03	Pa · s
lb$_{(mass)}$/(in · s)	1.786 E + 01	Pa · s
P	1.000 E + 02	cP
cP	1.450 E–07	lb$_f$ · s/in.^{2} (reyns)
Pa.s	1.450 E–04	lb$_f$ · s/in.^{2} (reyns)
Kinematic viscosity (absolute viscosity/density)		
cm^2/s (Stokes, St)	1.000 E–04	m^2/s (preferred unit)
mm^2/s (centistokes, cSt)	1.000 E–06	m^2/s
in.^{2}/s	6.452 E–04	m^2/s
cSt	1.550 E–03	in.^{2}/s
Kinematic to absolute viscosity conversions		
cSt	1.450 E–07 · density (g/cm^3)	lb$_f$ · s/in.^{2} (reyns)
cSt	density	cP

Applied Tribology: Bearing Design and Lubrication, Third Edition. Michael M. Khonsari and E. Richard Booser.
© 2017 John Wiley & Sons Ltd. Published 2017 by John Wiley & Sons Ltd.
Companion Website: www.wiley.com/go/Khonsari/Applied_Tribology_Bearing_Design_and_Lubrication_3rd_Edition

British in.-lb-s or traditional unit	Multiply by this conversion factor	To obtain this SI or another unit
Absolute to kinematic viscosity conversions		
Pa · s	1.000 E–03/density (g/cm^3)	m^2/s
P (poise)	1/density	St (Stoke)
P	1.000 E + 02/density	cSt
cP	1/density	cSt

Many viscosity data and specifications for lubricants are still supplied only in Saybolt Universal Seconds (SUS): the efflux time in seconds for a test oil to drain down through a specified capillary. At high viscosities (above about 215 SUS) these SUS values are directly proportional to kinematic viscosity. At lower viscosities, entrance and exit flow losses give higher efflux times. ASTM method D2161 gives detailed tables and equations for conversions between SUS and kinematic viscosity values in centistokes. The following give approximate conversions for several ranges (from *Texaco Lubrication*, Vol. 52(3), 1996).

$$cSt = 0.224\ SUS - 185/SUS \qquad \text{For range from 34 to 115 SUS}$$
$$cSt = 0.223\ SUS - 1.55 \qquad \text{For range from 115 to 215 SUS}$$
$$cSt = 0.2158\ SUS \qquad \text{Above 215 SUS}$$

Index

Abbott–Firestone curve, surface texture, 77
abrasive wear
    coefficient, 115–116
    rate, 115–116
absolute viscosity
    classifications, 30–31
    conversion factors, 29
    defined, 29–30
    index, 33
    pressure relations, 34–36
    temperature relations, 31–34
additives breakdown, 633
adjusted rating life, 486–490
air bearing materials, 145–149
air-oil systems, 553
aluminum, oil-film bearing materials, 142–143
American Bearing Manufacturers Association
        (ABMA), 471–472, 477, 486
American Gear Manufacturers Association
        (AGMA), 30
American Society of Mechanical Engineers
        (ASME), 3, 71, 79
American Society for Testing and Materials
        (ASTM), 30
amplitude density function, 77
amplitude parameters, 67
angular-contact ball bearings, 467, 486, 491, 532,
        534
angular misalignment, rolling element bearings,
        476, 499, 526
antiwear and extreme pressure agents, lubricant
        fortification, 25
Archard's adhesive wear equation, 104–108
arrow of time, 120

asperity
    plasticity index, 88–90
    single, elastic deformation of, 98
atmospheric pressure, pressure–viscosity
        coefficients, 35–37
average roughness (R_a), surface texture
        parameters, 66–67
axial clearance, rolling element bearings, 499
axial groove, journal bearing lubricant supply,
        287–288

babbitt melting, 632
babbitts, oil-film bearing materials, 139–140
balance ratio, 596–597, 605
balanced versus unbalanced seals, 595–596
ball bearings, types of, 463–467
ball contact, curvature and, rolling element
        bearings, 496
ball and roller bearing failure analysis, 627–630
ball-raceway contacts, surface stresses and
        deformation, 500–503
bearing area curve, 77
bearing frequencies, vibration, 517
bearing jamming, 629
bearing materials
    dry and semilubricated, 144
        carbon-graphite, 145
        plastics, 144–145
        rubber, 145
        wood, 145
    high-temperature, 138
    oil-film
        aluminum, 142–143
        babbitts, 139–141

Applied Tribology: Bearing Design and Lubrication, Third Edition. Michael M. Khonsari and E. Richard Booser.
© 2017 John Wiley & Sons Ltd. Published 2017 by John Wiley & Sons Ltd.
Companion Website: www.wiley.com/go/Khonsari/Applied_Tribology_Bearing_Design_and_Lubrication_3rd_Edition

bearing materials (*Continued*)
   cast iron and steel, 143
   copper alloys, 141–142
   silver, 143
   zinc, 143
  rolling materials, 152
  selection factors
   compatibility, 136
   corrosion resistance, 138
   embedability and conformability,
    136
   strength, 138
   thermal properties, 138–139
bearing principles
  classifications, 12
  comparisons, 14
  economics, 17
  environmental requirements, 17
  mechanical requirements, 14–16
  modernization of applications, 18–19
bearing ratio, 77
bearing stiffness, journal bearings, 300–304
bellows mechanism, 592
Bingham solid, lubricating greases, 45
blade passing frequency, 623
boundary conditions
  journal bearings
   full-Sommerfeld, 262–263
   generally, 261
   half-Sommerfeld, 265
   Sommerfeld number defined, 264–265
   Swift–Stieber (Reynolds), 270–273
  Reynolds equation, 269
boundary dimensions, rolling element bearing
   selection, 472
boundary lubrication, principles, 11–12
brinneling, 628
brush seals, 579–580
buffer fluids, 587–590, 594
burnishing process, 115
bushing shaft seals, 575

cage
  considerations, 514–516
  noise of, 520–522
  speed of, 522–524
cage failure, 628
cage jamming failure, 628–629
capacitance discharge, 637
capillary tube viscometer, 175–176

capillary-compensated hydrostatic bearings,
   382–384
carbon-graphite, dry and semilubricated bearing
   materials, 145
Cartesian coordinates, 162–163, 166–167
  conservation of energy, viscous flow, 167
  conservation of mass, viscous flow, 162
  conservation of momentum, viscous flow, 163
  squeeze-film bearings, 196
cast iron, oil-film bearing materials, 143
cavitation erosion, 632, 636
cavitation phenomena, journal bearings, 269
centralized lubricant supply systems, 60
chamfer dimensions
  internal clearance, 474–476
  precision classifications, 476–477
circular step thrust bearings
  flow rate requirement, 378
  friction torque, 379
  load capacity, 378
  optimization, 380
  power losses, 379
  pressure distribution and load, 376–378
  stiffness, 376
  thermal effects, 380–382
circulating oil systems
  lubricant supply, 56
  rolling element bearings, 554
circumferential groove, journal bearing lubricant
   supply, 258
clearance seals, 575–581
closed-form solutions
  gas bearings, 401
  limiting cases and, Reynolds equation,
   401–403
closing force, 593, 597
codes, rolling element bearing selection, 213
compatibility, bearing materials selection, 136
compensated hydrostatic bearings
  capillary-compensated, 382
  design procedure for, 386
  orifice-compensated, 384–385
compressibility, density and, lubricants, 39–40
computer hard disk, 148–149
condition monitoring, 619–637
cone-and-plate viscometer, 178
conformability, bearing materials selection,
   136–138
conservation laws
  Reynolds equation, mass, 193

viscous flow
  energy, 167
  mass, 162
  momentum, 163
contact seals, 589–592
contact between surfaces, 79–84. *See also* surface
    texture and wear interaction
contacting methods, surface texture measurement,
    72–73
contamination, 619–620, 623, 626, 629, 630–631,
    633, 635
conversion tables
  units, 639
  viscosity, 641
copper alloys, oil-film bearing materials, 141–142
corrosion, 619, 623, 625–629, 632, 635–636
corrosion, hydrostatic bearings, 375
corrosion resistance, bearing materials selection,
    138
corrosive
  contamination, 631
  wear, 117
cost, bearing principles, 12
Couette flow, viscous flow, conservation of
    energy, 169
cracked rings, 628
curvature, ball contact and, rolling element
    bearings, 496–498
cutoff lengths, surface texture measurement, 73
cylindrical coordinates
  conservation of energy, viscous flow, 167
  conservation of mass, viscous flow, 162
  conservation of momentum, viscous flow,
    163
  squeeze-film bearings, 331
cylindrical roller bearings, 467
cylindrical viscometer, 176–177

debris denting, 627, 629
deep-groove single-row ball bearings, 463
deformation, surface stresses and, rolling element
    bearings, 7
delamination theory of wear, 111–114
  wear coefficient interpretation of, 113–114
density, compressibility and, lubricants, 39
density wedge, Reynolds equation, 195
detergents, lubricant fortification, 25
diametral (radial) internal clearance, rolling
    element bearings, 474–476
diffuser stall, 623

disk-oiled bearings, partially starved oil-film
    bearings, 445–446
dispersants, lubricant fortification, 25
double seal, 592, 594
drop-feed oiling, semilubricated bearings, 451
dry and semilubricated bearings
  materials
    carbon-graphite, 433
    plastics, 433–436
    porous metal bearings, 436–438
    rubber, 434
    wood, 434
  minimal oil supply
    drop-feed oiling, 451
    mist, air-oil, and grease feed, 453
    wick oiling, 451–453
  overview, 144–145
  partially starved bearing analysis, 446
  partially starved oil-film bearings, 439–440
    disk-oiled bearings, 445–446
    oil-ring bearings, 440
    ring-speed and oil delivery, 442
  temperature, 437–439
dry sliding, principles, 7–8
dynamic seals, 573–574

eccentricity ratio, minimum film thickness and,
    journal bearing design, 277–278
economics, bearing principles, 17
elastic deformation, of single asperity, contact
    between surfaces, 79–80
elasticity, rolling element bearings, 518
elastohydrodynamic lubrication (EHL)
  principles, 10
  rolling element bearings, 6
elastohydrodynamic pressure–viscosity
    coefficients, calculation of, 36–7
electrical damage, 636–637
elevated temperature, 629
emission spectrographic wear monitoring, 625
energy equation, nondimensional, viscous flow,
    180–182
energy, conservation laws, viscous flow, 161
entropy, 119–122
  dissipative systems, 120
  equation, 120
  flow, 121
environmental requirements, bearing principles,
    17
erosion, 636

experimental modal analysis (EMA), 623
externally pressurized feed, principles, 9
extreme pressure agents, lubricant fortification, 25

failure analysis, 627–637
failure analysis, rolling element bearings, 619
falling body viscometer, 178
fatigue, 97, 632
   analysis, 110–111
   bearing principles, 18
   life, 138
   strength, 138
   surface, 117–119
   theory of adhesive wear, 109–110
ferrofluid seals, 580–581
ferrography, 627
filling-notch ball bearings, 467
film thickness profile, journal bearings, 278
finite thrust gas bearings
   rectangular, 404
   sector-pad, 404
   tilting-pad design, 407
flat-land thrust bearings, 238–239
   description, 238
   performance of, 238–239
flaws, defined, surface texture, 66
floating bushing seal leakage, 575–577, 579
floating bushing seals, 575–577
floating ring seals, 575
flooded condition, starved condition compared,
      journal bearing design, 313
fluid-film bearings
   gas bearings, 395–427
   history, 213
   hydrostatic bearings, 373–393
   journal bearings, 255–324
   Reynolds equation, 189–218
   squeeze-film bearings, 329–370
   thrust bearings, 221–252
   viscous flow, 161–188
fluid-film lubrication, principles, 8–10
flush plan, 593, 601–603
foam inhibitors, lubricant fortification, 25
foil gas bearing, 423–430
   analysis, 425–427
   bump foil, 426
   limiting load and speed analysis, 427–429
   top foil, 426
fracture wear, 119
frequencies, bearings, vibration, 517
fretting, 628

friction, 98–100
   bearing materials selection, 4–5
   history, 4
   power loss and
      bearing principles, 9
      piston rings, 360
   rolling element bearings, 6
friction modifiers, lubricant fortification, 25
friction moments, rolling element bearings, 531
friction torque, circular step thrust bearings, 379
FT-IR, 627
full-arc plain journal bearings, with infinitely long
      approximation (ILA), 260
full-Sommerfeld boundary condition
   half-Sommerfeld condition compared, 265
   journal bearings, 261

galling, 123
gas bearings, 395–430
   applications of, 395–396
   closed-form solutions, 403
   equation of state, 395
   finite thrust bearings
      rectangular, 404
      sector-pad, 404
      tilting-pad design, 408
   journal bearings
      finite, 277–278
      tilting-pad, 255, 258
   Reynolds equation, 396–399
   viscous properties, 395–396
gas ingestion in visco-seals, 585
gaseous cavitation, journal bearings, 269
geometric characterization, of surface texture,
      65–67
geometric squeeze, Reynolds equation, 194
gravity, order-of-magnitude analysis, viscous
      flow, 189–190
   greasehardening, 628
   lubrication, 455–456
   mechanical properties, 50
   oils in, 49
   rolling element bearings maintenance, 51, 619
   thickeners, 49–50
Greenwood–Williamson model (GW), 85–86

half-Sommerfeld boundary condition
   full-Sommerfeld condition compared, 260–264
   journal bearing, 267
hyaluronic acid, 343
hybrid parameters, surface texture, 71

hydraulic lift, hydrostatic bearings, 390
hydrodynamic lubrication, principles, 10
hydrostatic bearings, 373–393
 applications, 373–374
 circular step thrust bearings, 376
  flow rate requirement, 378
  friction torque, 379
  load capacity, 378
  optimization, 380
  power losses, 379
  pressure distribution and load, 376–377
  stiffness, 378
  thermal effects, 380
 compensated
  capillary-compensated, 382
  configurations, generalization to, 387
  design procedure for, 386
  hydraulic lift, 390
  orifice-compensated, 384
  overview, 391
 types and configurations, 375–376
hydrostatic oil lifts, 374–375
hydrostatic thrust bearings, description, 376

impeller stall, 623
infinitely long approximation (ILA)
 full-arc plain journal bearings, 260
 Reynolds equation, 260
infinitely long journal bearings, gas bearings,
  405
infinitely long tapered-step and slider bearings,
  gas bearings, 403
infinitely short approximation (ISA), 204–206
infrared (IR) spectral analyses, 624
inside mechanical seals, 594, 605
installation analysis, 619–620
instrument ball bearings, 467
instrumentation, 620
interfacial seals, 573–574
internal clearance, chamfer dimensions, 474–476
International Organization for Standardization
  (ISO), viscosity, 30–31
interstitial seals, 573–574
 ISO, 7919, 622

Jackson–Green model (JG), 86–87
journal bearings, 255–324
 attitude angle for configurations, 255
 boundary conditions
  full-Sommerfeld, 262
  generally, 261

half-Sommerfeld, 265
 Sommerfeld number defined, 264–265
 Swift–Stieber (Reynolds), 270–273
cavitation phenomena, 269
design guides
 bearing load and dimensions, 313–314
 eccentricity and minimum film thickness,
  314–315
 effective temperature, 306–307
 flooded and starved conditions compared,
  312
 maximum temperature, 307–308
 misalignment and shaft deflection, 320–322
 operating clearance, 315–316
 turbulent and parasitic loss effects, 312
film thickness profile, 256
finite bearing design and analysis, 277–279
full and half short solutions, 275
full-arc plain (with ILA), 260
gas bearings
 finite, 277–286
 infinitely long, 260–261
 tilting-pad, 258
infinitely short approximation (ISA), 273–274
lubricant supply methods, 287
 axial groove, 287–288
 circumferential groove, 289
 flow considerations, 289
 supply hole, 287
performance parameters, 279
principles, 300–304
stiffness vibration, and oil-whirl instability,
  300–304
types of, 258

KE model, 82
kinematic viscosity, defined, 29
Kogut–Etsion model (KE), 86
kurtosis, 69

labyrinth shaft seals, 578–579
laminar, 60, 169, 173, 176–177, 189, 198, 201,
  250–251, 278, 304, 312, 385–386, 453,
  577–578, 581–584, 587, 602
lantern ring packing seals, 591
lay, 66
leakage of packing seals, 590–592
life limit
 bearing principles, 12
 maintenance, bearing principles, 15
 oil life, 15

life limit (*Continued*)
  rolling element bearings
    adjusted rating life, 485–490
    grease life, 556–559
    load-life relations, 478–481
    wear, 535–537
lift-off speed, 313
limiting cases, closed-form solutions and,
        Reynolds equation, 203–206
line contact, rolling element bearings, 540–550
lip seals, 589–590
load
  bearing principles, 15
  circular step thrust bearings, 376
  rolling element bearings, 15
load-life relations, rolling element bearing
        selection, 478
local expansion, Reynolds equation, 194–195
lubricant friction, rolling element bearings, 529
lubricants, 23–61
  base stock categories, 24
  density and compressibility, 39–40
  grease
    mechanical properties, 50–52
    oils in, 49
    thickeners, 49–50
  history, 23
  mineral oils, 23–25
  oil life, 46–48
  self-contained units, 54–60
  solid, 52
  supply methods
    centralized systems, 60
    circulating oil systems, 55–60
    journal bearings, 255. *See also* journal
        bearings
    partially starved oil-film bearings, 439
    rolling element bearings, 16
    synthetic oils, 25
  thermal properties, 40–42
  viscosity
    classifications, 29
    defined, 29
    elastohydrodynamic pressure–viscosity
        coefficients, 36–37
    pressure and, 34–36
    shear rate relations (non-Newtonian), 42–45
    temperature and, 31–34
    viscoelastic effect, 45
lubricated wear, 125

lubricating grease, viscosity-shear rate relations,
        45
lubrication, bearing principles, 16
lubrication regime, surface roughness, 96–97

maintenance, bearing principles, 17
Mannich base, 624
mass conservation law
  Reynolds equation, 190
  viscous flow, 161
mating (stationary) rings, 592–594, 600,
        603–606, 608–609, 612, 614
mean peak spacing (S_m), surface texture
        parameters, 69
measurement, of surface texture, 72–75. *See also*
        surface texture and wear interaction
mechanical properties, greases, 50–52
mechanical requirements, bearing principles, 14
micro-contact of surface asperities, 80–84
  elasto-fully plastic deformation of, 82
  KE model, 82
  spherical asperity, 80
  ZMC model, 82
mild wear coefficient, 123
mineral oils, lubricants, 23–25
miniature ball bearings, 467
minimum film thickness, 314–315
misalignment limit, 526
mist systems, 453–455
moderate wear, 123
moisture contamination, 631
momentum, conservation laws, viscous flow, 163
multiple (segmented) floating ring seals, 575

nanotribology and surface effects, 6–7
Navier–Stokes equations, Reynolds equation, 190
needle roller bearings, 468–469
noise, rolling element bearings, 520
noncontacting methods, surface texture
        measurement, 75
nondimensional energy equation, viscous flow,
        180–182
nondimensional flow equation, viscous flow,
        179–180
nondimensional Reynolds equation, 201–202
nonplanar squeeze-film bearings
  expressions for geometries, 338–343
  finite surfaces, 343–348
  sphere approaching plate, 337–338
nonpusher seal, 593

normal squeeze, Reynolds equation, 194
numerical method, Reynolds equation, 211–213

oil analysis, 623–625
oil delivery, bearings, 442–445
oil film bearing failure analysis, 630–637
oil-film bearing materials
    aluminum, 142–143
    babbitts, 139–141
    cast iron and steel, 143
    copper alloys, 141–142
    silver, 143
    zinc, 143
oil oxidation, 619, 623, 635
oil sampling frequency, 624
oil sampling points, 625
oil sludging, 619
oil-whirl instability, journal bearings, 300–304
on-line monitoring, 620–623
opening force, 593, 596, 605–606, 615
operating temperature, rolling element bearings,
    537–539
optimization of seal head coefficient, 582–583
order-of-magnitude analysis (viscous flow)
    gravity contribution, 183
    inertia and viscous terms compared, 182–183
    pressure contribution, 184
    pressure and viscous forces compared,
        184–186
    orifice-compensated hydrostatic bearings,
        384
O-rings, 487, 593
outside mechanical seals, 594
overload, 628, 632
oxidation inhibitors, lubricant fortification, 24

packing seals, 590–591
packing seals automatic-type, 591–592
packing seals compression, 590
packing seals jam-packed type, 590
parallel disk viscometer, 176
parameters, surface texture, 67–71
parasitic power loss
    journal bearing design, 312
    thrust bearings, 223
partially starved oil-film bearings, 439–442
    analysis, 438–439
    disk-oiled bearings, 445–446
    oil-ring bearings, 440–442
    ring-speed and oil delivery, 442–445

particle embedding, 635
peak sharpness (kurtosis), surface texture
    parameters, 70–71
peak-to-peak shaft displacement, 622
peak-to-valley height, surface texture parameters,
    68–69
performance parameters, Reynolds equation,
    202–203
Petroff's formula, viscous flow, 173–175
physical stretch, Reynolds equation, 193
physical wedge, Reynolds equation, 193–194
piston rings
    description, 360–366
    friction force and power loss, 366–369
pivoted-pad thrust bearings
    description, 232–236
    performance of, 232
planar squeeze-film bearings
    elliptical disks, 333–334
    finite surfaces, 343–348
    generalization for, 334–337
    two parallel circular disks, 332–333
plastics, dry and semilubricated bearings,
    144–145
point contact, rolling element bearings, 541–544
polymer viscosity index improvers, shear rate
    relations (non-Newtonian), 42–45
pour-point depressant, 627
pour-point depressants lubricant fortification,
    25
power loss, 583, 598–599, 621
    friction and, 366
        bearing principles, 14
        piston rings, 360–366
    parasitic, thrust bearings, 223
precision machinery, heavily loaded, hydrostatic
    bearings, 373
pressure
    order-of-magnitude analysis, viscous flow,
        182
    viscosity and, 184
pressure gradient factor, 597, 605
pressure–velocity (PV) factor, 598, 603
pressure–viscosity coefficient, calculation of,
    36–37
primary (rotor) rings, 604

radial contact seal, 589–592
radial (diametral) internal clearance, rolling
    element bearings, 474–476

radially loaded rolling bearings
    load distribution, 508
    thrust loaded combined, load distribution,
        510–512
Rayleigh step bearing, Reynolds equation,
        206–211
remaining additive concentration, 625
resistance temperature detector (RTD), 206
Reynolds equation, 189–218
    application (Rayleigh step bearing), 206–211
    assumptions, 189–190
    derivations
        boundary conditions, 191
        conservation of mass, 191
        cylindrical coordinates, 197
        general equation, 195–196
        Navier–Stokes equations, 190–191
        standard equation, 196
    gas bearings, 396–403
    limiting cases and closed-form solutions,
        203–206
    nondimensionalization, 202–202
    numerical method, 211–213
    performance parameters, 202–203
    squeeze-film bearings, 197
    surface roughness, 199–201
    Swift–Stieber (Reynolds) boundary condition,
        journal bearings, 270
    turbulent flows, 197–199
rib wear, 629
ring-speed, oil delivery and, partially starved
        oil-film bearings, 442–445
roller bearings, types of, 467–468
roller-raceway line contacts, surface stresses and
        deformation, 504–506
rolling element bearing selection
    adjusted rating life, 486–491
    ball bearings, types of, 463–467
    boundary dimensions, 472
    chamfer dimensions, 472–474
    internal clearance, 474–476
    load-life relations
        combined radial and thrust load, 481–484
        minimum load, 485–486
        varying load, 484–485
    nomenclature, 469
    precision classifications, 476–477
    roller bearings, types of, 467–468
    shaft and housing fits, 477–478

    static load capacity, 491–492
    thrust bearings, types of, 468–469
rolling element bearings
    cage considerations, 514–516
    elasticity, 518
    failure analysis, 619
    friction, 529–531
    friction moments, 531–535
    history, 463
    internal geometry
        angular misalignment, 499–500
        axial clearance, 499
        curvature and ball contact, 496–498
        point and line contact, 496
        radial (diametral) internal clearance,
            498–499
    load distribution
        combined radial and thrust loads, 510–512
        radially loaded, 508–510
        thrust loaded, 510
    load limit, 524
    lubrication
        EHL lubrication, 540
        grease life, 556–559
        grease selection and application, 555–556
        greasing and regreasing, 559–561
        oil application, 553–554
    misalignment limit, 526
    noise, 420
    oil change intervals
        oil viscosity selection, 550–553
        solid lubricants, 561
    operating temperature, 537–539
    speed of cage and rolling elements, 513–514
    speed limit, 522–524
    subsurface stresses, 506–508
    surface stresses and deformation
        ball-raceway contacts, 500–503
        roller-raceway line contacts, 504–506
    temperature limit, 525
    vibration, 516–517
    wear, 535–537
rolling friction, rolling element bearings,
        529
root-mean-square roughness (R_q), surface texture
        parameters, 67–69
rotational viscometers, 176–179
rotor vibration, journal bearings, 300–304
rough surfaces, 87–88

roughness
    defined, 65. *See also* surface texture and wear
        interaction
rubber, dry and semilubricated bearing materials,
    145
rust inhibitor, 626
rust inhibitors, lubricant fortification, 24

screw pump, 587–588
scuffing failure, 123, 374
seal chamber, 601, 606
seal classifications, 573–574
seal face flatness, 600
seal face materials, 595, 600
seal face surface finish, 600, 632
seal face temperature, 603–613
seal failure, 573, 598
seal flushing system, 600–603
seal gland, 599
seal heat generation, 598–599
seal leakage, 579
seal materials and coefficient of friction,
    600
seal power loss, 598–599, 605–606, 621
seal ring shapes, 609–613
seals API flush plans, 602
seals and shields, ball bearings, 467
sector-pad finite thrust gas bearings, 404–407
seizure, 123, 632–633
self-contained lubricant supply, 537
self-excited vibration, 523
self-propagating wear, 634–635
sensors, 620–621
shaft deflection, misalignment and, journal
    bearing design, 321
shaft and housing fits, rolling element bearing
    selection, 477
shaft magnetization, 637
shaft whip instability, 523
shear rate relations (non-Newtonian), viscosity,
    42
signal processing, 620
silver, oil-film bearing materials, 143
single asperity, elastic deformation of, contact
    between surfaces, 79–80
skewness, 69–70
skid/skidless instruments, surface texture
    measurement, 72–73
sleeve bearings, *see* journal bearings

sliding friction, rolling element bearings, 529
Society of Automotive Engineers (SAE),
    viscosity, 30–31
soft elastohydrodynamic lubrication (EHL),
    principles, 11
solid lubricants
    description, 52
    rolling element bearings, 561
Sommerfeld number, defined, 264–265
soot, 626–627
space requirement, bearing principles, 16
spacing and shape parameters, surface texture,
    69
speed
    bearing principles, 15
    rolling element bearings, grease life, 556–559
speed limit, rolling element bearings, 522–524
spiral groove, 289
spring-mounted thrust bearings
    description, 237–238
    performance of, 238
squeeze film, Reynolds equation, 197
squeeze-film action, principles, 329
squeeze-film bearings, 329–370
    equations governing, 330–331
    finite surfaces
        nonplanar, 348–354
        planar, 343–348
    nonplanar
        expressions for geometries, 338–343
        sphere approaching plate, 337–338
    overview, 329
    piston rings
        description, 360
        friction force and power loss, 366–369
    planar
        elliptical disks, 333–334
        generalization for, 334–337
        two parallel circular disks, 332–333
stadium mover/converter, hydrostatic bearings,
    375
starved condition, flooded condition compared,
    journal bearing design, 313
static load capacity, rolling element bearing
    selection, 491
static seals, 573
statistical descriptions, surface texture, 77
steady film, Reynolds equation, 197
steel, oil-film bearing materials, 139

step thrust bearings
    description, 236
    performance of, 236
strength, bearing materials selection, 138
stuffing box, 590–591
stylus sensitivity, 74–75
subsurface stresses, rolling element bearings,
        506–508
sulfate detergent, 627
supply hole, journal bearing lubricant supply, 287
surface drag loss, thrust bearings, 247
surface effects, history, 6–7
surface fatigue wear, 117–119
surface pitting, 628, 632
surface stresses, deformation and, rolling element
        bearings, 500–502
surface texture and wear interaction
    contact between surfaces, 79–80
        asperity plastic deformation, 80–82
        elastic deformation of single asperity, 80–81
    geometric characterization, 65–67
    lubrication regime, 96–97
    measurement
        contacting methods, 72
        noncontacting methods, 75
    parameters
        amplitude, 67
        hybrid, 71–72
        spacing and shape, 69–70
    Reynolds equation, 396–399
        statistical descriptions, 77
        symbols, 78
Swift–Stieber (Reynolds) boundary condition,
        journal bearings, 270
symbols, surface texture, 78–79
synthetic oils, lubricants, 25

tandem mechanical seals, 592, 595
tapered roller bearings, 468
tapered-land thrust bearings
    description, 226
    performance of, 227
tapered-step and slider bearings, infinitely long,
        gas bearings, 403
temperature
    circular step thrust bearings, 376–382
    dry and semilubricated bearing, 306–307
    flash temperature in sliding contacts, 93
    gas bearings, viscous properties, 396
    high-temperature bearing materials, 149–151

hydrostatic bearings, 380–382
    monitoring, 620–621
    Reynolds equation, thermohydrodynamic
        analyses, 213
    rolling element bearings, 523–524
    tapered-land bearings, 222
    viscosity and, 31–34
thermal breakdown, 633
thermally induced seizure, 316
thermocouple, 621
thermodynamics of wear, 119–123
thermohydrodynamic analyses, Reynolds
        equation, 213
thickeners, greases, 49–50
through-flow loss, thrust bearings, 246
thrust bearings, 221–251
    design factors, 224–225
    flat-land thrust bearings, 238–239
    parasitic power loss, 246–249
    performance analysis, 225–226
    pivoted-pad thrust bearings, 232–236
    principles, 221
    spring-mounted thrust bearings, 237–238
    step thrust bearings, 236–237
    tapered-land thrust bearings, 226–228
    turbulence, 249–251
    types of, 221–223. See also finite thrust gas
        bearings; fluid-film bearings
thrust loaded rolling bearings
    load distribution, 508
    radially loaded combined, load distribution,
        510
tilting-pad journal bearings, design of, 418–423
tilting-pad thrust gas bearings, design of, 407
total acidity number (TAN), 625
total base number (TBN), 626
tribology
    defined, 3
    history of, 3–4
    principles of, 7–12. See also bearing principles
tribometer tests, 121
troubleshooting, 622, 631
turbulent flows
    journal bearing design, 307–308
    Reynolds equation, 197
    thrust bearings, 251
turbulent seal flow coefficient, 197–198

unbalanced mechanical seals, 592, 595–596
unit conversion factors, 639

vane passing, 623
vapor cavitation, journal bearings, 269–270
vibration, 619–623
    journal bearings, 300–304
    rolling element bearings, 517
vibration monitoring, 621, 623
visco-seal friction resistance coefficient, 583–584
visco seals, 581–588
viscoelastic effect, description, 45
viscometers, 175–179
viscosity
    classifications, 30–31
    defined, 29
    elastohydrodynamic pressure–viscosity
        coefficients, 36–37
    pressure and, 34–36
    shear rate relations (non-Newtonian), 42
    temperature and, 31–4
    viscoelastic effect, 45
viscosity conversion table, 641–642
viscosity index improvers
    lubricant fortification, 25
    shear rate relations (non-Newtonian), 42
viscosity index, use of, 33–34
viscosity–temperature coefficient, calculation of,
        34
viscous flow, 161–188
    conservation laws, 161
        energy, 167
        mass, 162
        momentum, 163–164
    energy equation, nondimensionalization of,
        180–182
    flow equation, nondimensionalization of,
        179–180
    order-of-magnitude analysis, 182
        gravity contribution, 183
        inertia and viscous terms compared,
            182–183

pressure contribution, 184
pressure and viscous forces compared,
        184–186
Petroff's formula, 173–175
viscometers, 175

warning limits, 624–625
water contamination, 635
water content, 626
waviness, 66
wear
    abrasive, 114
    adhesive, 100–104
    classification of, 123–124
    coefficient, 16
    delamination theory of, 111–114
    entropy flow, 121–122
    equation, 104–108
    fracture wear, 119
    fragment of, 112
    lubricated, 125
    maps, 123–124
    monitoring, 625–627
    progression of, 125–126
    reduction, 127–128
    rolling element bearings, 535–537
    thermodynamics of, 119–123
    types, 123
whirl instability, gas bearings, finite journal
        bearings, 417–418
wick oiling, semilubricated bearing,
        451
wire wool wear, 634
wood, dry and semilubricated bearing materials,
        144

zinc, oil-film bearing materials, 143
ZMC model, 82

Printed and bound by CPI Group (UK) Ltd, Croydon, CR0 4YY